KEYS TO THE NEMATODE PARASITES OF VERTEBRATES

KEYS TO THE NEMATODE PARASITES OF VERTEBRATES

Archival Volume

edited by

ROY C. ANDERSON
ALAIN G. CHABAUD
SHEILA WILLMOTT

www.cabi.org

CABI is a trading name of CAB International

CABI
Nosworthy Way
Wallingford
Oxfordshire OX10 8DE
UK

Tel: +44 (0)1491 832111
Fax: +44 (0)1491 833508
E-mail: cabi@cabi.org
Website: www.cabi.org

CABI
745 Atlantic Avenue
8th Floor
Boston, MA 02111
USA

Tel: +1 617 682 9015

A catalogue record for this book is available from the British Library, London, UK.

ISBN-13: 978 1 84593 572 6

First printed in 2009.
Transferred to Print on Demand in 2018.

Typeset by MTC, Manila, The Philippines
Printed and bound by CPI Group (UK) Ltd, Croydon, CR0 4YY

Contents

FOREWORD

We are extremely pleased that the CIH Keys to the Nematode Parasites of Vertebrates are being reprinted. CABI's decision to re-issue the keys to the parasitic nematodes, members of an extremely large animal phylum, is in response to a persistent demand and need for means to identify parasitic nematodes as well as a resurgent interest in the origins and evolution of parasitism, demonstrated by the publication of numerous recent articles on this topic.

The keys were first published between 1974 and 1983, with individual volumes appearing as they were completed. They were not published in a systematic order and therefore it is frequently necessary to consult several volumes to obtain information on a single order, a feature now seen to detract from their utility. In addition, phylogenetically unrelated taxa were included in the same volume (such as Volume 9 on the Rhabditoidea, Dioctophymatoidea, Trichinelloidea and Muspiceoidea).

The present re-issue by default follows the original classification of the nematodes adopted by Anderson, Chabaud and Willmott. Since the original publication of the keys, some 300 additional genera have been described and these new genera are included in a supplement by Lynda Gibbons, which largely follows the original classification but presents the new genera in a phylogenetically ordered fashion.

In the intervening 35 years since the first publication of the keys, novel methods for establishing the relationships of living organisms based on DNA sequence data have been applied to nematodes by workers such as Blaxter, de Ley and Mitreva. The question arises as to whether it is still pertinent to reproduce a key based on methodologies that might now be considered to be relatively old. We believe that it is indeed worthwhile, particularly with regard to recent advances in molecular phylogeny.

The speculative, non-cladistic phylogeny proposed by Chabaud in Fig. 1.68 of the first volume of the keys suggested relationships between 24 families, superfamilies and orders of the Secernentea. Four distinct groups were identified: a large basal group containing the cosmocercoids from which the Ascaridida and Spirurida evolved, with the Oxyurida representing a divergent lineage. Separately, the Strongylida as well as two small families, the Rhabdiasidae and Strongyloididae, arose from common ancestors, the free-living rhabditids. In recent molecular studies, the Chromadorida are considered to be basal to previously monophyletic Secernentea and the free-living rhabditids have been subdivided. However, among the parasitic taxa, the groupings have not changed greatly. Among the clades identified by Blaxter and his colleagues based on molecular data, branches 2–6 of Chabaud (Strongylida) correspond to Clade 5 identified by molecular methods, branches 24–25 (Rhabdiasidae and Strongyloididae) to Clade 4 and branches 8–23 (Cosmocercoidea, Spirurida, Ascaridida) to Clade 3. It is therefore not necessary to abandon the earlier classification used in the keys, which was based on morphological and biological features.

However two discrepancies do exist between the two approaches. Whereas the Oxyurida form part of Clade 3 in the work of Blaxter and his colleagues, they diverge earlier in the phylogeny proposed by Chabaud, a position that takes account of the haploid males and differences in the third-stage larvae. A second feature of the phylogeny presented in the keys is the origin of the Spirurida from the Seuratoidea, rather than a separate origin from the free-living rhabditids, which gave rise to the Ascaridida in the opinion of Chitwood and Chitwood. This begs the question as to whether the Seuratoidea is a natural group and whether it represents a step in intermediate hosts becoming an essential part of the life cycles of nematodes. Apparent discrepancies arising from different methods of analysis are therefore important in identifying key questions in the evolutionary biology of nematodes and highlight the significance of taking into account phylogenies based on morphological and biological characteristics as well as those based on molecular methods. For this reason alone, the re-issuing of publications containing significant insights into nematode evolution is of value.

The publication of the keys to the nematodes was followed by a comprehensive text by Roy C. Anderson, now in its second edition, of the life cycles of nematodes (*Nematode Parasites of Vertebrates: Their Development and Transmission*. CAB International, 2000). This invaluable overview of nematode life cycles follows the taxonomic arrangement adopted in the keys and represents a significant advancement in consolidating the diverse literature on the complexity of nematode life cycles, providing a source of fundamental data for analysing the evolution of these cycles.

We hope that, in addition to meeting the need for a means of identification of animal-parasitic nematodes, the re-issuing of the keys will also facilitate investigations into both the evolution of nematodes and parasitism within the phylum by helping to link recent molecular insights with those provided to date in numerous morphological studies. We also hope that the keys will help to establish links between workers specializing in animal-parasitic nematodes and those focusing on free-living nematodes.

Odile Bain
Parasitologie comparée et Modèles expérimentaux,
Muséum National d'Histoire Naturelle, Paris, France
Ian Beveridge
Faculty of Veterinary Science,
University of Melbourne, Australia
Marie-Claude Durette-Desset
Département de systématique et d'evolution,
Muséum National d'Histoire Naturelle, Paris, France
Alain Chabaud
Parasitologie comparée et Modèles expérimentaux,
Muséum National d'Histoire Naturelle, Paris, France

GENERAL INTRODUCTION

In 1955, the decision was taken by the Commonwealth Governments to set up a taxonomic and identification service in the Commonwealth Bureau (now Institute) of Helminthology, comparable in every way to those already existing in the Commonwealth Mycological Institute and the Commonwealth Institute of Entomology. The Commonwealth Institute of Helminthology was also charged with two additional responsibilities, firstly to revise and bring up to date the then standard work on vertebrate nematode taxonomy, i.e. Yorke & Maplestone's "Nematode Parasites of Vertebrates" and secondly to build up reference collections of helminths which would be available to those working in countries without centres of taxonomic research or large museum collections but where there was a need for the identification of helminth parasites, at least to generic level.

In the event, neither of these two special responsibilities proved to be practicable, not the least of the reasons for this being that the quantity of material sent for identification absorbed all the time available and more than kept pace with the modest increases in staff which were possible. Additionally, specimens only too often were received in a condition which made it impossible to use them for reference collections, even if identifiable.

Consequently it became apparent that there was a real and increasing need for up-to-date keys to the taxa of parasitic nematodes of vertebrates. Many of those now existing are, to a greater or lesser extent, out of date and are not easy to obtain. Progress in systematics inevitably renders the preparation and use of dichotomous keys more and more difficult. In the early stages of the development of a classification the choice of a single, easily observed, character could be somewhat arbitrary, for the main object was to facilitate the easy and accurate recognition of species; this philosophy motivated the classic work of Yorke & Maplestone and even more recent treatises. However, the aim of modern systematics is not only the easy and accurate recognition of species but also the identification of their lines of evolution and their affinities in order ultimately to arrive at a deeper understanding of the group being studied. This more modern approach is not, as some maintain, of theoretical interest only but has practical value since our concept of systematic relationships can, and often does, have a decisive influence on our approach to the immunology, physiology and chemotherapy of a group.

Modern research on systematics of nematodes is based on a more and more delicately defined morphology and on increasingly sophisticated morphological studies, as well as utilizing data on the palaeontology and biogeography of the hosts and information on ecology, immunology and biochemistry. It is hardly surprising, therefore, that the use of a single feature to characterize a genus or family, however judiciously that feature is chosen, is generally quite inadequate to reflect the real complexity of nematode relationships. Only by means of associations of characters (any one of which could be inconstant) can the taxonomy of genera and higher taxa be defined and a more natural classification achieved. This does not mean that certain special characters cannot be emphasized in keys but only that, considered in isolation, they may be misleading.

The use, in keys, of generic definitions which are too narrow leads to the proliferation of genera simply because some authors believe it is necessary to create a new genus each time a newly discovered species does not fit comfortably into a dichotomous key. Eventually, a generic revision which ought to be relatively simple becomes terribly difficult because related species are scattered among a variety of essentially similar genera, often widely separated in a key and under varied and unexpected names. The conflict between "lumpers" and "splitters" is still with us and only a classification which is as nearly natural as possible can resolve this. For old synonymies, recognized by all authors, and for most references earlier than 1960 the reader should consult the classical works.

In the preparation of these keys, some taxa based on type species of dubious status have been deliberately omitted. These will be listed in the introduction to each key. There is an unfortunate tendency among inexperienced workers to elevate poorly known species to generic and even higher rank in order to accommodate them in a key and this should be resisted because it can, and does, cause great confusion. Even when a new species, apparently related, has been discovered and adequately described, the taxonomic problem raised by the questionable genus cannot be solved because the type species which is, in fact, the only important point in the definition of the genus, is then definitively fixed. At the same time, the discovery of a new species cannot easily be used to solve the taxonomic problem presented by the poorly described genus. Ultimately, of course, the simple need to identify a species is itself frustrated and our understanding of phylogenetic relationships is put in jeopardy.

The aim of the present series is two-fold. Firstly we hope to provide a working tool for those in the field, the laboratory, the medical, veterinary or biological department who need to know "what is this worm" and secondly to help to establish a classification which is not simply a series of pigeon-holes into which nematodes can be separated but which gives a real insight into their relationships and is, therefore, of value to the specialist.

Roy C. Anderson
Department of Zoology, College of Biological Sciences,
Guelph, Ontario, Canada
Alain G. Chabaud
Laboratoire de Zoologie,
43 Rue Cuvier, 75231 Paris, Cedex 05, France
Sheila Willmott
Commonwealth Institute of Helminthology,
103 St. Peter's Street, St. Albans, Herts AL1 3EW, England

1 GLOSSARY AND KEYS TO SUBCLASSES

GLOSSARY

by SHEILA WILLMOTT

ala (pl. alae)	thin, cuticular projection or fin, running longitudinally, usually lateral or sublateral, frequently paired *cervical alae* confined to the anterior end *caudal alae* on the posterior end of male
amphidelphic	having two opposed sets of female reproductive organs, one extending anteriorly and the other posteriorly to the vulva
amphids	pair of glandular sensory organs situated laterally in the cephalic region and opening through the cuticle—see Fig. (viii)
annulations	deep, transverse grooves occurring at regular intervals in the cuticle giving the body a segmented appearance—see Fig. (i)
annules	the intervals or rings of cuticle, between the annulations—see Fig. (i)
area rugosa	ornamentation of the cuticle, sometimes present on the ventral surface in front of the cloaca on the coiled part of the posterior extremity of the male—see Fig. (ii)

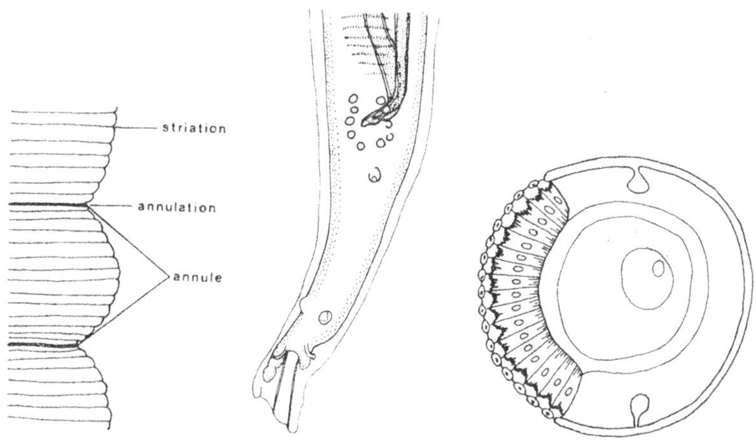

(i) (ii) (iii)

bacillary band	a modification of the hypodermis, consisting of longitudinal rows of columnar cells that have pore-like openings to the surface of the cuticle, occurring in Trichuroidea (=hypodermal glands of some authors)—see Fig. (iii)
bacillary layer	a non-vibratile form of ciliary lining of the intestine—see Fig. (iv)
ballonets	cuticular inflations in the cephalic region assuming a swollen band-shape immediately posterior to the lips (=head bulb of some authors)

bosses	small, round or oval, blister-like inflations of the cuticle
buccal cavity	part of digestive canal between oral opening and oesophagus; the stoma
buccal capsule	anterior enlarged portion of the buccal cavity with heavily sclerotized walls
bulb	posterior part of muscular oesophagus, generally swollen, containing a valvular apparatus—see Fig. (xi)
bursa copulatrix (=copulatory bursa)	modified caudal ala or alae found in males of some nematodes, may be circular or oval, often divided into two symmetrical or asymmetrical lateral lobes, separated by a dorsal lobe, and supported by rays or papillae—see Fig. (v)
caecum	a blind diverticulum or pouch from the intestine—see Fig. (vi)

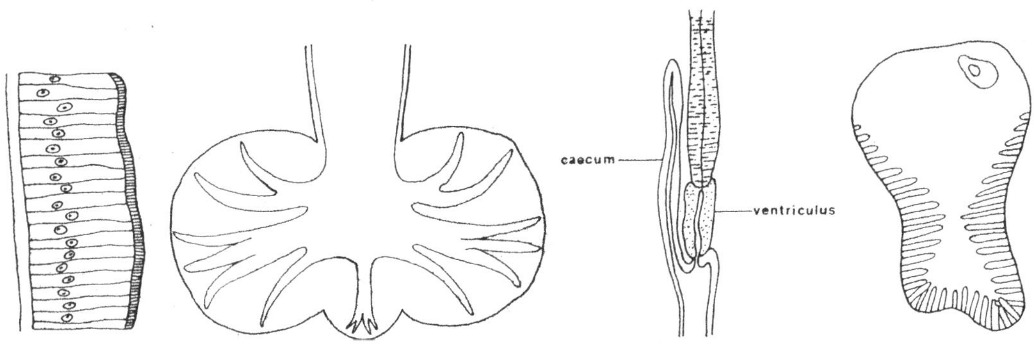

(iv) (v) (vi) (vii)

cardia	oesophago-intestinal valve
cephalic papillae	see under "head papillae"
cervical	applied to structures connected with "neck" region
cheilorhabdion	wall of cheilostom
cheilostom	first portion of the stoma of a rhabditid-like nematode
circumoesophageal	encircling the oesophagus
circumoral	encircling the mouth
circomyarian	a type of muscle cell in which the muscle fibre completely surrounds the sarcoplasm
claviform	club-shaped
cloaca	the common chamber into which the intestinal and genital canals open
coelomyarian	body musculature consisting of cells of which the internal non-contractile part is well developed and protrudes into the pseudocoelom; the muscle fibres extend to the sides of the muscle cells for varying distances—see Fig. (vii)
copulatory bursa	see bursa copulatrix
cord (chord)	dorsal, ventral or lateral longitudinal thickenings of the hypodermis
cordon	longitudinal, cuticular cord-like thickening extending posteriorly from the mouth and may be straight, recurved or form loops, present mainly in the spiruroid family Acuariidae

corona radiata	the border of the labial region divided into a series of leaf-like structures, found in certain strongyles. There may be two circles of leaf-like structures termed the internal and external corona (=internal and external leaf crowns of some authors)
corpus	anterior end of the oesophagus, often separated from the posterior bulb by the isthmus—see Fig. (xi)
deirids	a pair of sensory organs found laterally in the cervical region and usually protruding above the surface of the cuticle (=cervical papillae of some authors)
didelphic	with two sets of female reproductive organs
diorchic	with two testes
diverticulum	a tube or sac, blind at distal end, branching off from a canal or cavity
epaulets	specialized, ribbon-shaped, paired bands of cuticle at anterior end (= cordons of most authors)

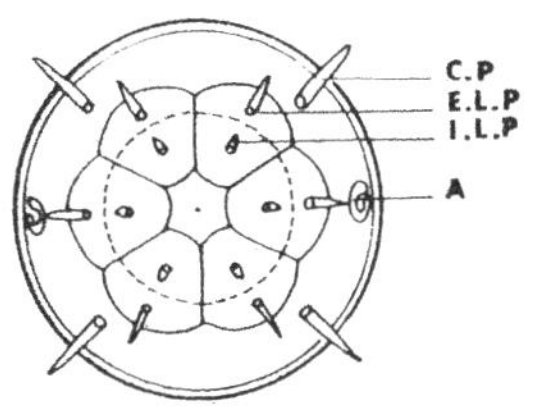

(viii)

genital papillae	tactile papillae or setae in the anal region of the male which may be preanal, postanal or caudal in position (see also pedunculated papillae, rosettes, plectanes)
gubernaculum	an accessory male copulatory piece which is formed by sclerotization of the dorsal wall of the spicule pouch, variable in shape but generally with incurved margins
head papillae	tactile sensory organs usually located on the lips or labial region, including two circles of six labial papillae and one circle of four cephalic papillae—see Fig. (viii) *cephalic papillae*—outer circle of four head papillae (=latero-ventral and latero-dorsal of Chitwood & Wehr) *externo-labial papillae*—median circle of six head papillae (=dorso-dorsal, ventro-ventral and ventro-lateral of Chitwood & Wehr) *interno-labial papillae*—inner circle of six head papillae
interlabia	cuticular outgrowth (neoformations) originating at the base of the lips or pseudolabia and extending between them, occurring in some ascarids and spirurids
isthmus	middle part of a muscular oesophagus, often constricted—see Fig. (xi)
labia	lobes or lips (primitively six) which surround the mouth
monodelphic	with one set of female reproductive organs
monorchic	with one testis
oesophago-intestinal valve	situated at opening between the oesophagus and intestine (=cardia of some authors)

opisthodelphic	uteri parallel directed posteriad
ovejector (ovijector)	part of the female genital system between the end of the uterus and the vulva, modified to aid in the expulsion of the eggs
oxyuroid (or bulboid) oesophagus	cylindrical anteriorly and terminating in a basal bulb—see Fig. (ix)

(ix)

pedunculated papillae	modified, stalked, genital papillae in anal region of male
pharynx	narrow, posterior part of the buccal cavity with thick sclerotized walls
phasmids	pair of glandular sensory organs situated laterally in the caudal region and opening to the surface by a slit or pore
plaques	cuticular "warts"
platymyarian	body musculature consisting of cells not protruding individually into pseudocoelom—see Fig. (x)
plectanes	cross striated cuticular plates functioning as supports for the genital papillae in some males
polydelphic	with more than two sets of female reproductive organs
preanal sucker	ventral, pre-cloacal structure, sucker-like in form
prodelphic	uteri parallel directed anteriad
protostom	second part of the stoma of a rhabditid-like nematode
pseudobulb	muscular swelling of the oesophagus without valvular arrangement
pseudolabia	cuticular outgrowths (neoformations) arising around the oral opening and which, during development, overlie and then replace the primitive lips. The anterior extremity of numerous Spirurida has two lateral pseudolabia (=probolae of some authors)
rays	genital papillae and their accompanying muscles embedded in the bursa of some males

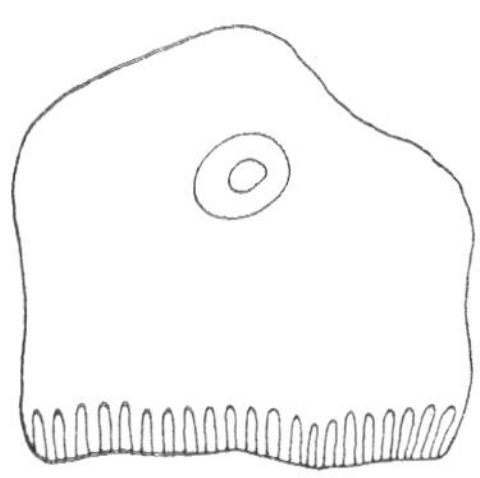

(x)

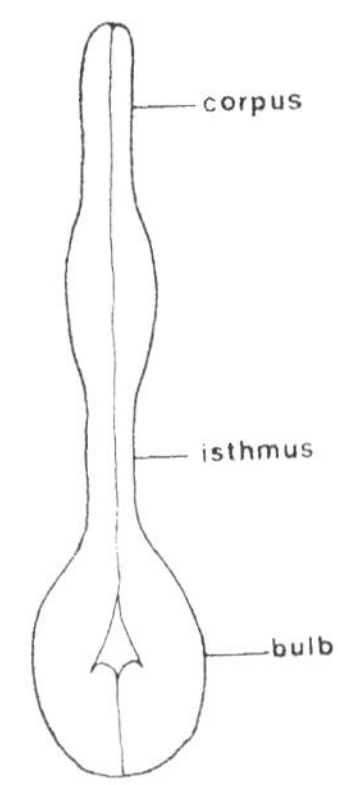

(xi)

rhabditoid (rhabditiform) oesophagus — oesophagus consisting of two swellings, one forming the end of the corpus, the other posteriorly forming the bulb. The region between the corpus and the bulb is known as the isthmus—see Fig. (xi)

ridges — raised cuticular areas which run the length of the body, very pronounced in some trichostrongyloids—see Fig. (xii)

rosettes — punctuation patterns of cuticle surrounding genital papillae

spicule — sclerotized, accessory copulatory organ of male nematodes, usually paired but sometimes single

stichocyte — glandular oesophageal cell not incorporated into oesophageal tissue—see Fig. (xiv)

stichosome — collection of stichocytes arranged in a longitudinal row—see Fig. (xiv)

striations — fine transverse grooves occurring at regular intervals—see Fig. (i)

strongyloid (filariform) oesophagus — slender, cylindrical, without bulb—see Fig (xiii)

supplements — in male, ventromedian papillae anterior to cloaca, as well as an adanal pair of papillae in some nematodes

synlophe — the ensemble of enlarged longitudinal or oblique cuticular ridges which serve to hold the nematodes in place on the gut wall, found in numerous trichostrongylids

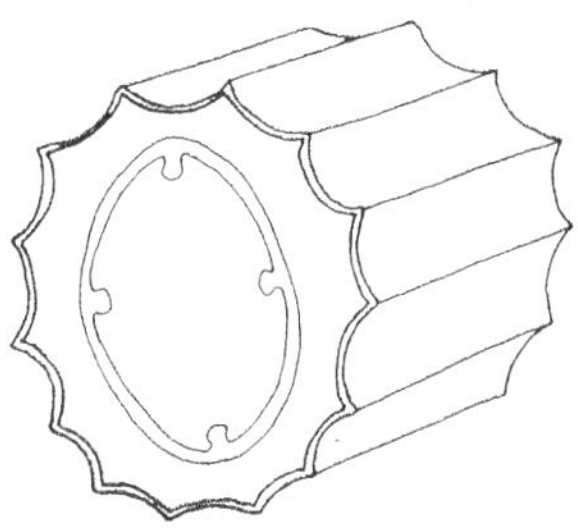

(xii)

(xiii)

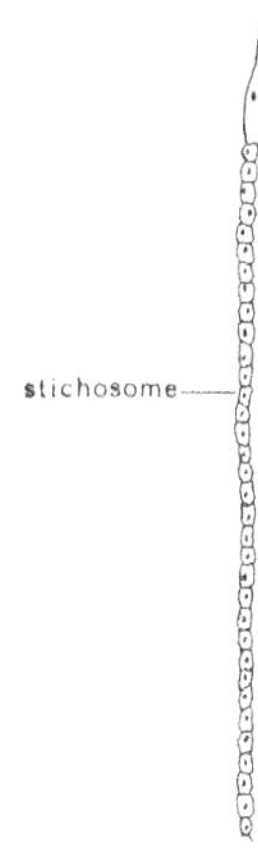

(xiv)

telamon — slightly sclerotized, immovable formation of complicated shape in the ventral and lateral cloacal walls that also aids in directing the spicules during copulation

telostom — posterior part of the stoma of a rhabditid-like nematode

trichuroid oesophagus — narrow muscular tube with associated glandular stichosome made up of stichocytes—see Fig. (xiv)

trophosome — reserve organ formed by the transformation of the digestive canal; often appears to be syncytial

ventriculus — glandular modification of the distal portion of the oesophagus of some nematodes, may have a solid appendage of varying length extending posteriorly dorsal to the intestine (ventricular appendix)

2*

CLASS NEMATODA

KEYS TO SUBCLASSES, ORDERS AND SUPERFAMILIES

by

ALAIN G. CHABAUD*

KEY TO SUBCLASSES

1-(2) Caudal papillae absent or few in number (Fig. 1.1).
Excretory system without lateral canals and terminal duct not lined with cuticle. Phasmids absent. Oesophagus cylindrical or with oesophageal glands free in pseudocoelom

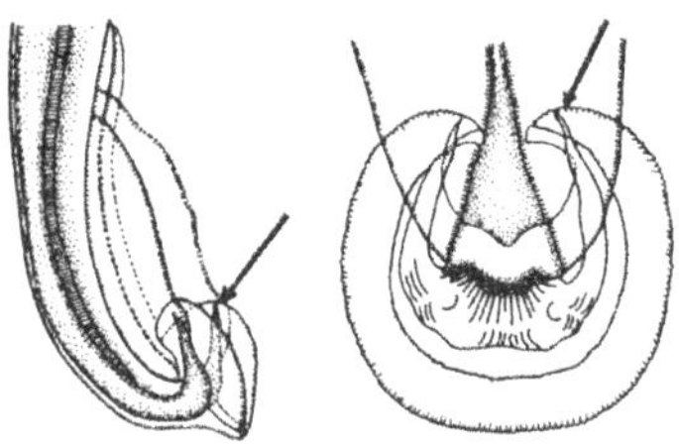

1.1

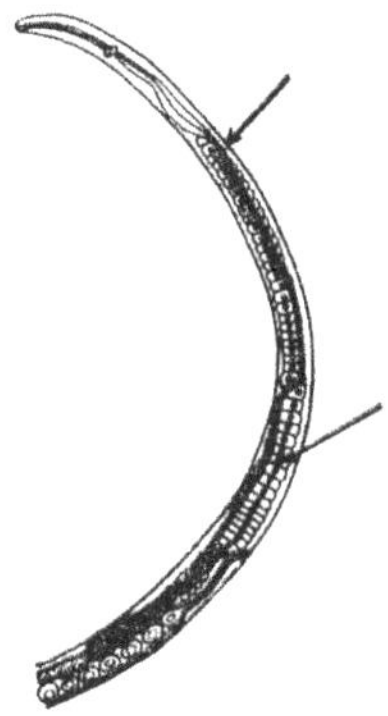

1.2

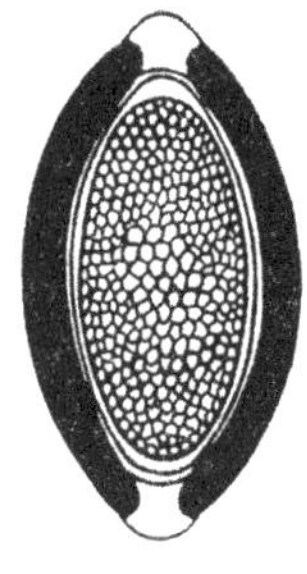

1.3

and forming stichosome (Fig. 1.2) or trophosome (Fig. 1.6). Eggs usually unsegmented with plug at either pole (Fig. 1.3) or hatching *in utero*. First-larval stage often with stylet and usually infective to final host

ADENOPHOREA

2-(1) Caudal papillae almost always numerous (the basic number 21) (Fig. 1.4).
Excretory system with lateral canals and terminal canal lined with cuticle. Phasmids present. Oesophagus never in form of stichosome. Eggs without polar plugs, rarely operculate at one pole. The beginning of the third larval stage infective to the final host.

SECERNENTEA

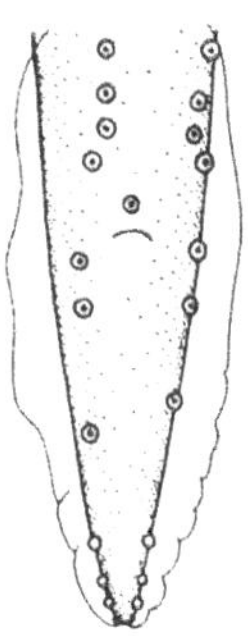

1.4

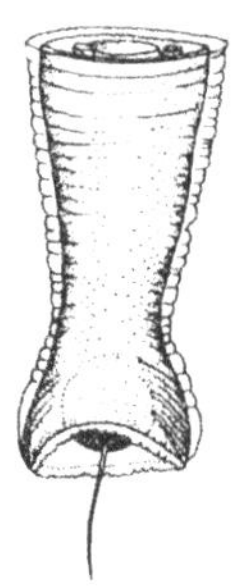

1.5

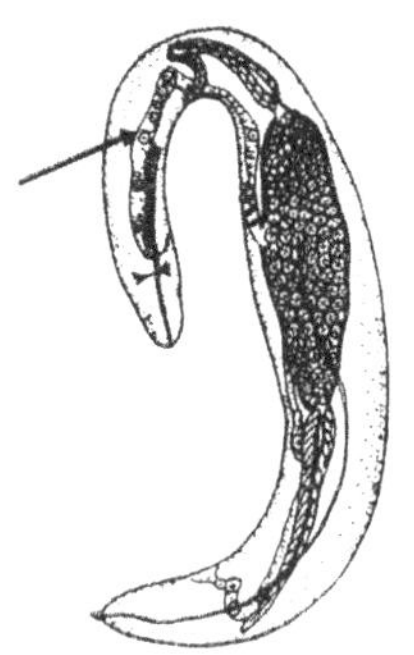

1.6

* Museum National d'Histoire Naturelle, Paris.

I. SUBCLASS ADENOPHOREA

ORDER ENOPLIDA

Key to Superfamilies

1-(2) Oesophagus cylindrical and well developed; stichosome or trophosome absent. Male tail modified to form ventral sucker (Fig. 1.5). Monodelphic.

Dioctophymatoidea

2-(1) Stichosome (Fig. 1.2) or trophosome (Fig. 1.6) present.

3-(4) Intestine and rectum well developed. Sexes separate. Monodelphic

Trichuroidea

4-(3) Alimentary tract reduced. Trophosome present (Fig. 1.6). Amphidelphic. Male unknown.

Muspiceoidea

II. SUBCLASS SECERNENTEA (Fig. 1.68)

Key to Orders

1-(2) Ovejector weakly developed. Vagina transverse and without well developed sphincters (Fig. 1.7).
Parasitic female parthenogenetic (except in *Parastrongyloides*) with cylindrical and frequently long oesophagus. Parasitic generation alternating with free-living generation of males and females which have rhabditoid oesophagi. Parasites of lungs of amphibians or small intestine of other vertebrates.

RHABDITIDA

2-(1) Ovejector complex with well developed sphincters. No free-living sexual generations.

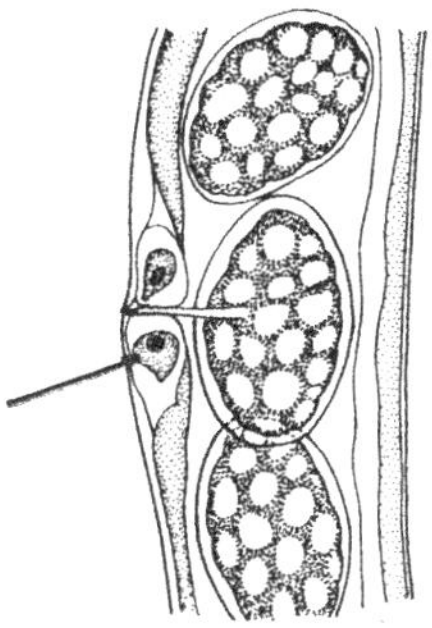

1.7

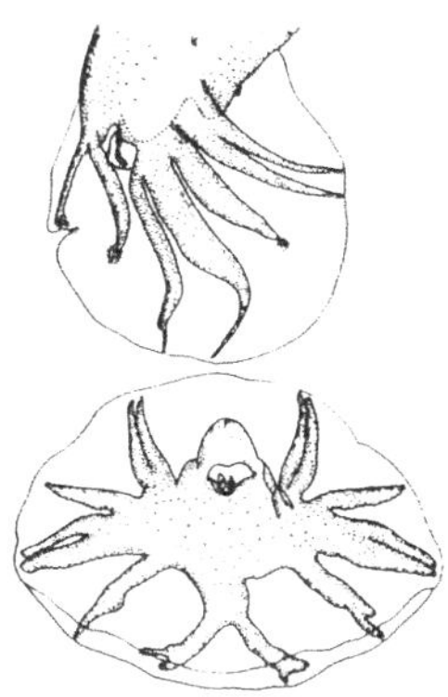

1.8

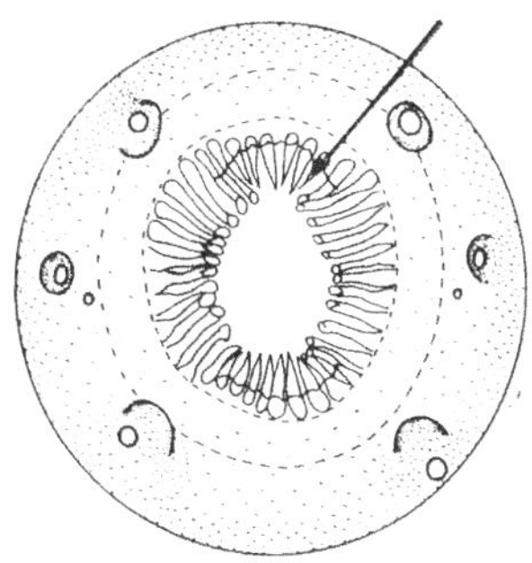

1.9

3-(4) Male with caudal bursa (Fig. 1.8).
Anterior extremity without lips, with 6 lips, or with *corona radiata* (Fig. 1.9). Buccal capsule variable in size and form, surrounded by oesophageal tissue. Oesophagus of larvae rhabditoid. Oesophagus of adult claviform. Excretory system with H-shaped tubular arrangement and 2 subventral glands (Fig.1.10). Spicules similar in size and morphology. Stages preinfective to final host (first, second, beginning of third), free-living or parasitic in invertebrates (e.g. annelids, molluscs). Parasites mainly of amphibians, reptiles, birds and mammals, extremely rare in fishes.

STRONGYLIDA

1.10

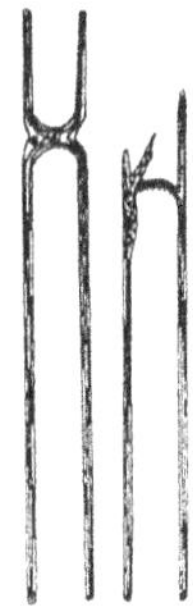

1.11

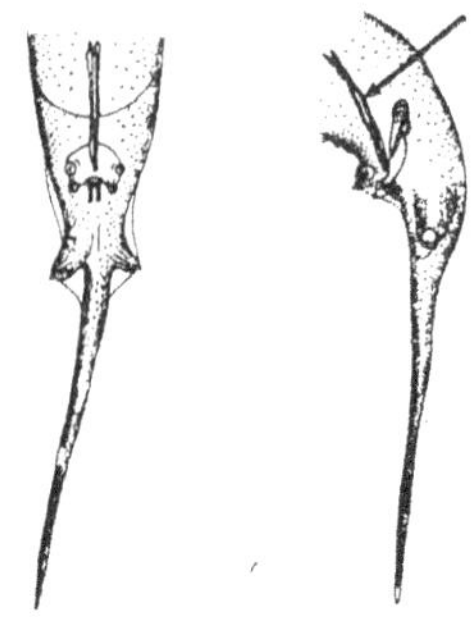

1.12

4-(3) Male without caudal bursa. A single ventral excretory gland present (Fig. 1.11).

5-(6) Male with reduced number of caudal papillae (Fig. 1.12). Generally only one spicule. Parasites of colon or rectum.
Body short and stout (Fig. 1.13). Oesophagus with bulb. Pre-anal sucker absent except in *Hoplodontophorus*). Female with large embryonated eggs often flattened on one side (Fig. 1.14). Monoxenous with two moults in egg.

OXYURIDA

1.13

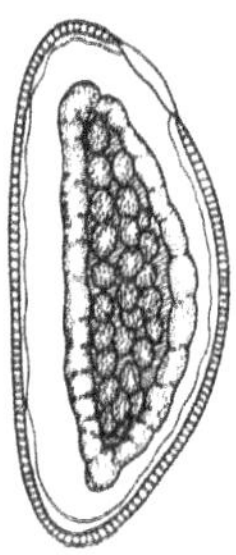

1.14

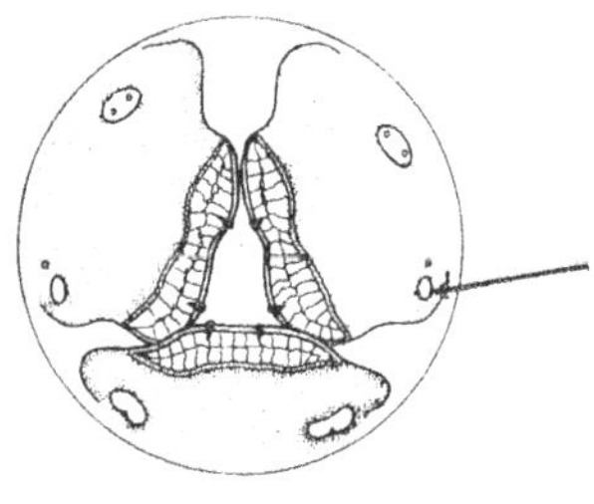

1.15

6-(5) Nematodes lacking most of the above characters.

7-(8) Anterior extremity triradiate (except in some Seuratoidea).
Lateral, external labial papillae present (Fig. 1.15). With 2 to 3 pairs of caudal papillae dorso-lateral in position (Fig. 1.16).

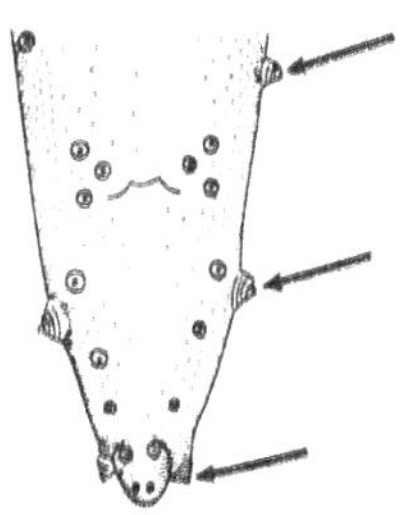

1.16

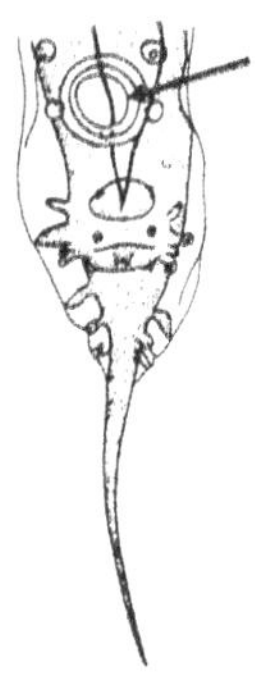

1.17

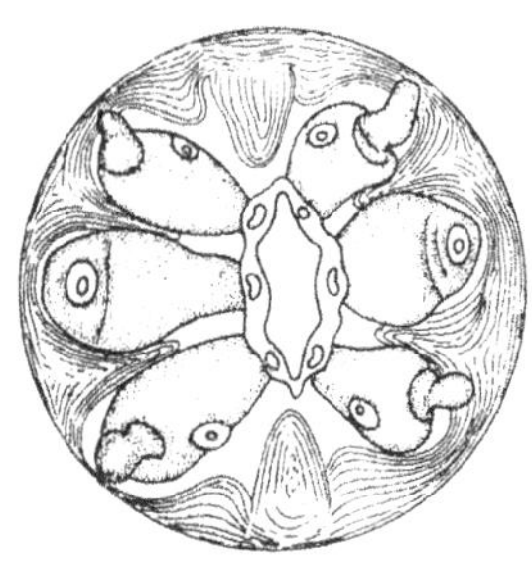

1,18

Oesophagus variable in form but not divided into short muscular and long glandular parts. Pre-anal sucker in male present (Fig. 1.17), or absent. Usually found in intestine of final host. Except in Subuluroidea, development of larval stages preinfective for final host does not take place entirely in an intermediate host.

ASCARIDIDA

8-(7) Anterior extremity bilaterally symmetrical. Lateral, external labial papillae absent (Fig. 1.18). Caudal papillae always ventral or ventrolateral in position (Fig. 1.4). Oesophagus divided, with anterior muscular part shorter than posterior glandular part (Fig. 1.19), division sometimes indistinct. Pre-anal sucker not present. Parasites of anterior part of gut (oesophagus, stomach, rarely duodenum) or tissues and tissue spaces. Larval stages preinfective for final host develop entirely in intermediate host.

SPIRURIDA

ORDER RHABDITIDA

One superfamily

Rhabditoidea

ORDER STRONGYLIDA

Key to Superfamilies

1-(2) Anterior extremity with 2 lateral jaws (Fig. 1.20).
Lips and *corona radiata* absent. Intestinal parasites of snakes and lizards.

Diaphanocephaloidea

2-(1) Anterior extremity without lateral jaws.
3-(6) Buccal capsule large, strongly cuticularized.
4-(5) Buccal capsule sub-globular, never hexagonal in transverse section. Lips and *corona radiata* absent. Oral opening unarmed or with teeth and cutting plates (Fig. 1.21).
Parasites of intestine of mammals.

Ancylostomatoidea

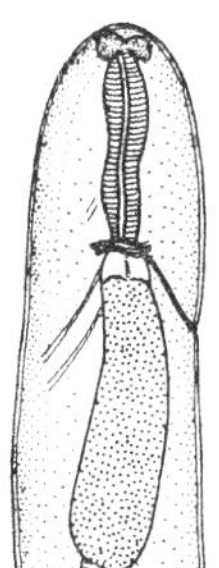

1.19

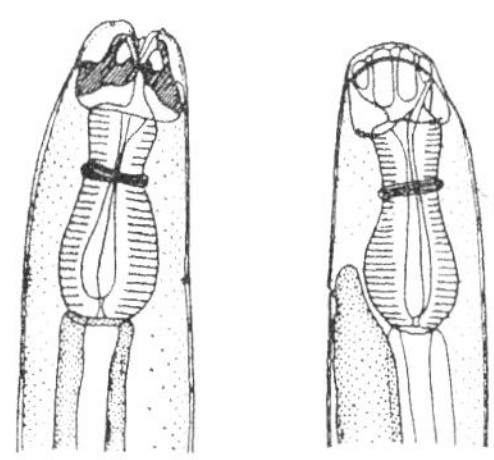

1.20

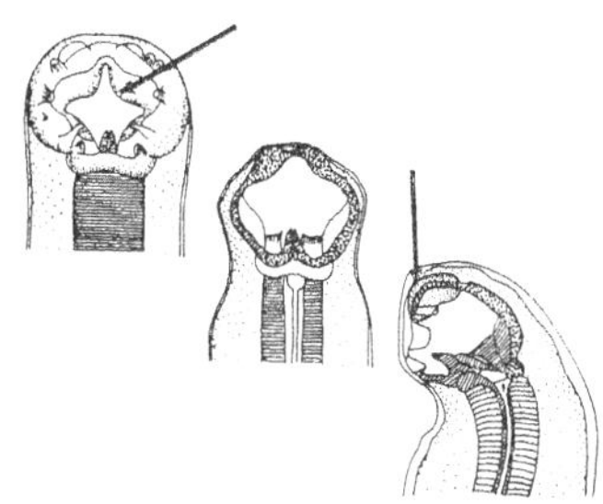

1.21

5-(4) Buccal capsule varied in form. Oral opening sometimes hexagonal (Fig. 1.22), sometimes surrounded by 6 lips (Fig. 1.23) or *corona radiata* (Fig. 1.9), and without teeth and cutting plates.
Parasites of intestine, more rarely of respiratory and urinary system of birds and mammals

Strongyloidea

6-(3) Buccal capsule absent or small and weakly cuticularized.
7-(8) Parasites of intestine (except Dictyocaulidae). Bursa well developed, at least lateral lobes. Monoxenous.

Cephalic vesicle often present (Fig. 1.24). Longitudinal cuticular ridges frequently present (Fig. 1.25). First-stage larva often with valved oesophageal bulb and conical tail ending in simple point (Fig. 1.26).

Trichostrongyloidea

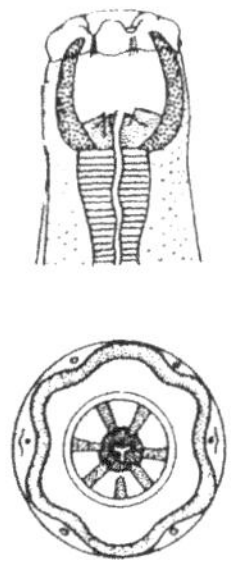

1.22

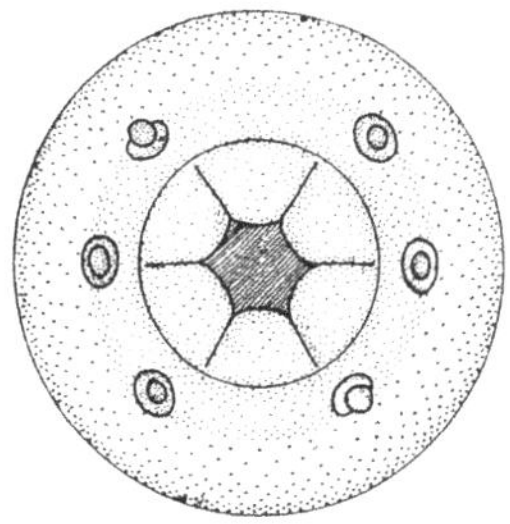

1.23

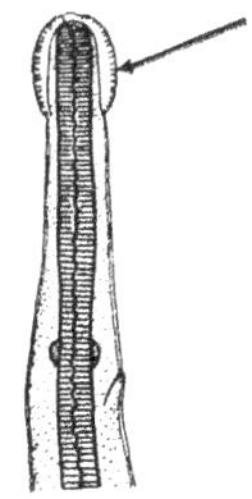

1.24

8-(7) Parasites mainly of respiratory system, less commonly of other parts of body of mammals. Bursa often reduced (Fig. 1.27). Heteroxenous, mainly in gastropods, sometimes in oligochaetes and, rarely, in vertebrates.
Cephalic vesicle and longitudinal cuticular ridges absent. Outer layer of cuticle frequently fluid-filled forming teguminal sheath. First-stage larva without valved oesophageal bulb and usually with short undulating tail (Fig. 1.28).

Metastrongyloidea

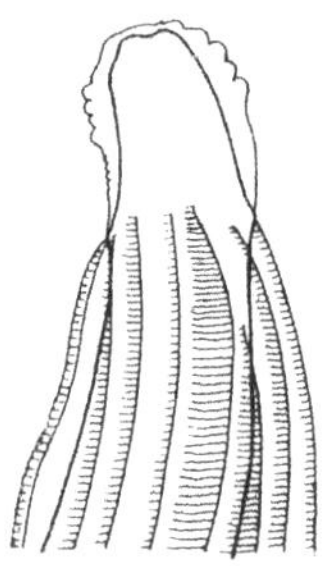

1.25

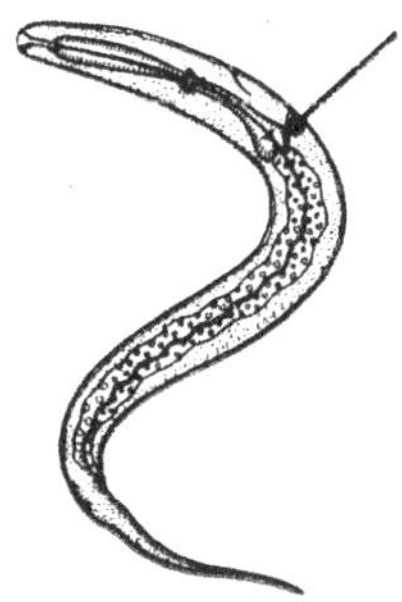

1.26

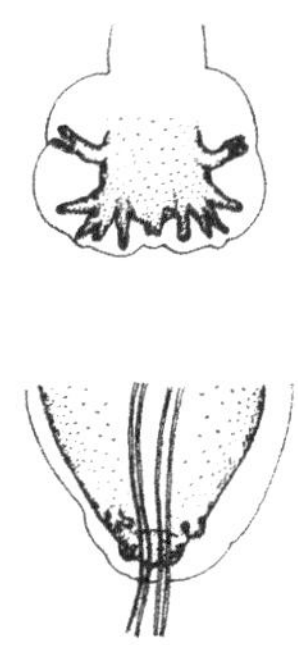

1.27

ORDER OXYURIDA

One superfamily **Oxyuroidea**

ORDER ASCARIDIDA

Key to Superfamilies

1-(8) Buccal cavity absent or weakly developed, without complex cuticularized posterior part. Larval stages preinfective (first, second and beginning of third) to the final host not occurring entirely in intermediate host.

2-(5) Lips present or absent, when present variable in number and form. Platymyarian (Fig. 1.29). Eggs hatching *in utero* or eggs with delicate shells deposited by female.
First moult generally outside egg. Usually small worms, less than 1 cm. long.

3-(4) Oesophagus composed of a cylindrical corpus, anteriorly differentiated distinct pharyngeal part, a subspherical or elongated isthmus and a valved bulb possessing uninucleate gland cells (Fig. 1.30). **Cosmocercoidea**

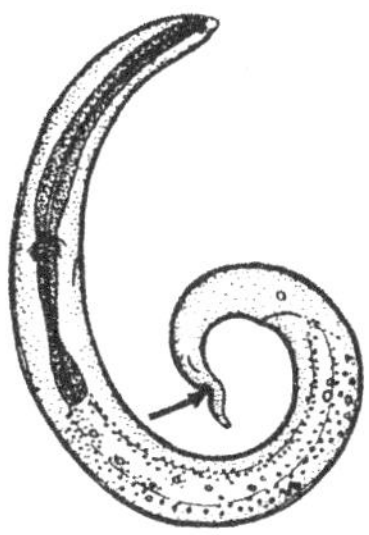

1.28

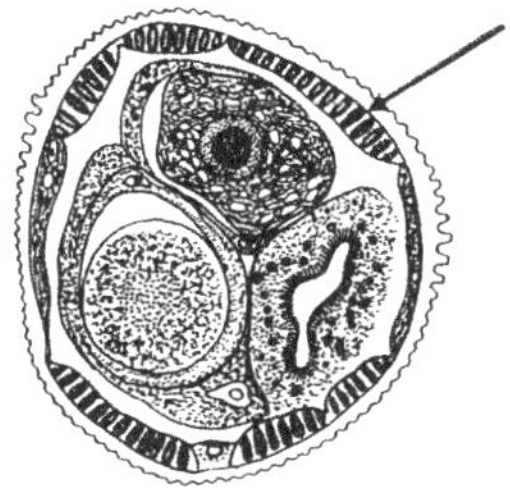

1.29

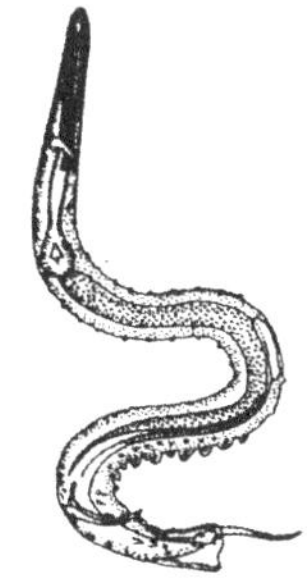

1.30

4-(3) Oesophagus short, simple and cylindrical, or short and divided into two parts having or not having same diameter (Fig. 1.31).
Pharyngeal part of oesophagus present or absent. Lips absent, or markedly reduced and variable. Oral opening sometimes elongated dorsoventrally (Fig. 1.32).
Seuratoidea

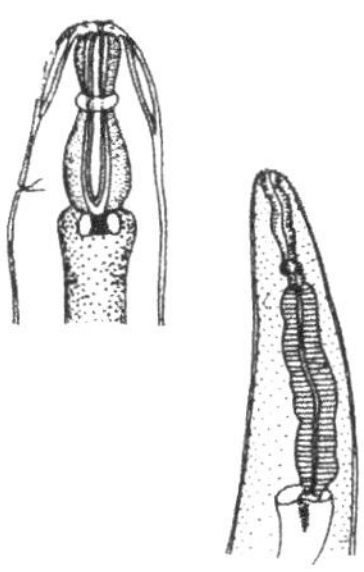

1.31

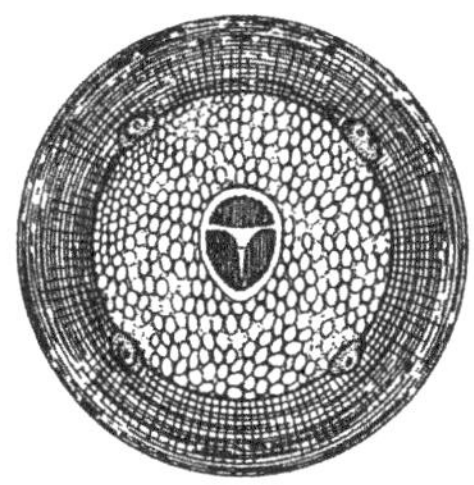

1.32

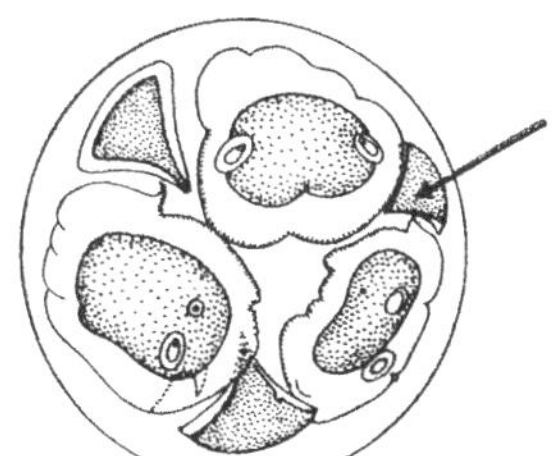

1.33

5-(2) With 3 well defined lips usually of great size, sometimes separated by interlabia (Fig. 1.33). Coelomyarian (Fig. 1.34). Eggs thick-shelled and not embryonated when deposited (Fig. 1.35).
First moult inside eggs, generally large worms, more than 1 cm. long.

6-(7) Pre-anal sucker present, surrounded by cuticularized ring (Fig. 1.36).
Oesophagus with claviform corpus, short isthmus, and valved bulb with binucleated subventral oesophageal glands (Fig. 1.37) or oesophagus cylindrical. Caeca absent.
Heterakoidea

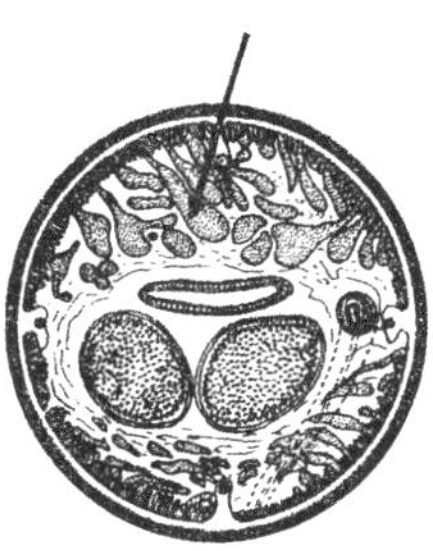

1.34

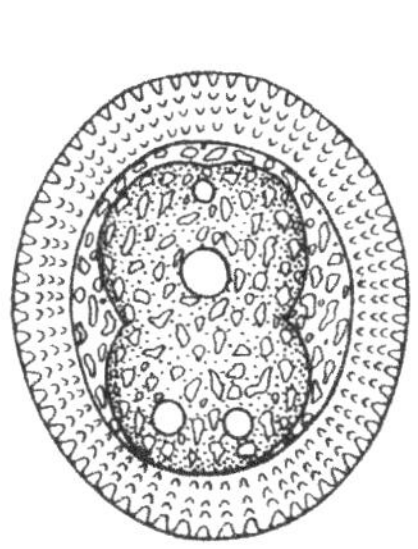

1.35

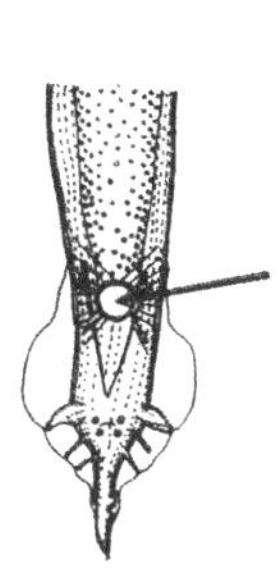

1.36

7-(6) Pre-anal sucker absent.
Oesophagus simple and cylindrical or terminated by swelling without valves (Fig. 1.38), containing uninucleate gland cells. Caeca present (Fig. 1.39) or absent.
Ascaridoidea

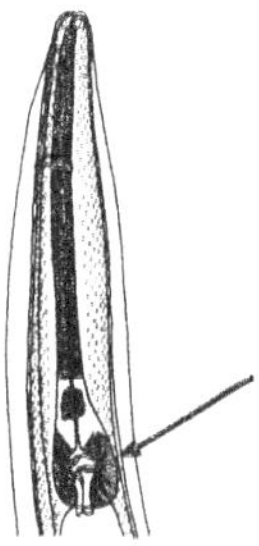

1.37

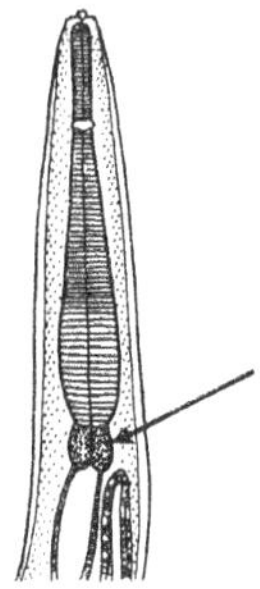

1.38

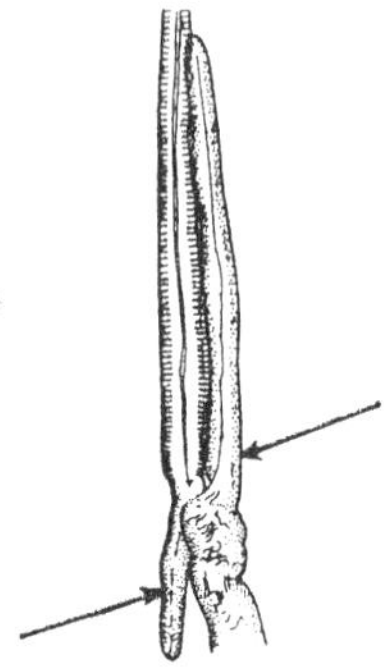

1.39

8-(1) Buccal cavity well cuticularized with an anterior piece circular, hexagonal or triangular in transverse section, and posterior part with complex cuticularized apparatus (Fig. 1.40). Pre-anal sucker present in male, generally without clearly defined ring (Fig. 1.41). Oral opening hexagonal or triangular, without salient lips. Eggs with thick shells embryonated when deposited. Larval stages preinfective to final host (first, second, beginning of third) occurring entirely in intermediate host.

Subuluroidea

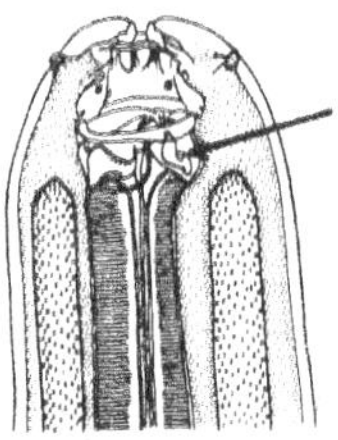

1.40

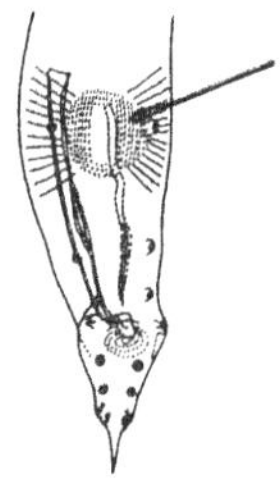

1.41

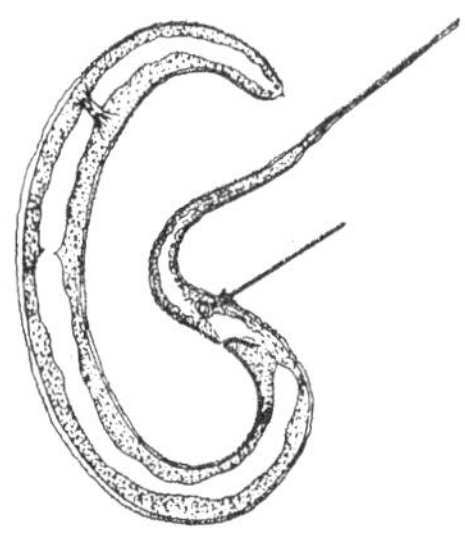

1.42

ORDER SPIRURIDA

Key to Suborders

1-(2) Larvae without cephalic hooks, tail generally long and pointed, usually with conspicuous phasmids containing broad cavities and prominent pores (Fig. 1.42). Oesophageal glands uninucleate (except in *Philonema*).
Parasites of gut of cold-blooded vertebrates or other organs of all classes of vertebrates. Intermediate hosts copepods.

CAMALLANINA

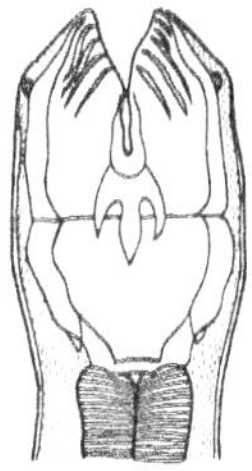

1.43

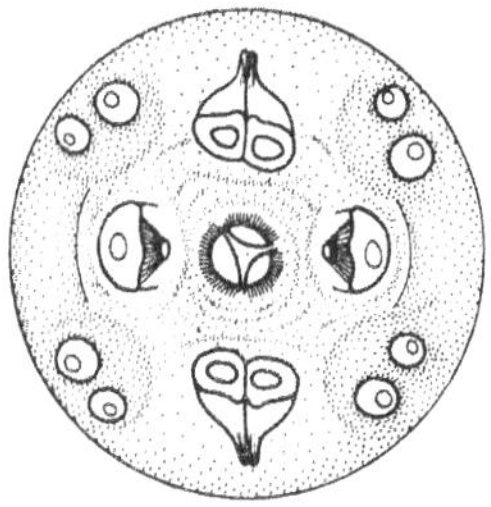

1.44

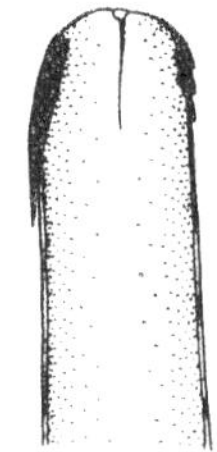

1.45

2-(1) Larvae usually with cephalic hooks or spines (Fig. 1.45) and inconspicuous pore-like phasmids. Oesophageal glands multinucleate.
Parasites of gut or tissues of all classes of vertebrates. Intermediate hosts invertebrates other than copepods (except Gnathostomatoidea).

SPIRURINA

SUBORDER CAMALLANINA

Key to Superfamilies

1-(2) Buccal cavity well developed (Fig. 1.43). Internal labial papillae tiny.
Parasites of gut.

Camallanoidea

2-(1) Buccal cavity weakly developed. Internal labial papillae prominent (Fig. 1.44).
Not usually parasites of digestive tract.

Dracunculoidea

SUBORDER SPIRURINA

Key to Superfamilies

1-(5) Buccal cavity little cuticularized (except in Desmidocercidae of air sacs of piscivorous birds).

2-(14) With two massive lateral pseudolabia (Fig. 1.46).
Parasites of alimentary tract.

3-(4) Pseudolabia trilobed (Fig. 1.47). Cephalic and outer labial papillae separate.
Cuticle on inner face of each pseudolabium thick, generally folded into rounded tooth-like formations which fit into corresponding folds on adjacent pseudolabium (Fig. 1.47). Anterior extremity sometimes swollen into bulb (Fig. 1.48). Intermediate hosts at least in part, a copepod or mollusc.

Gnathostomatoidea

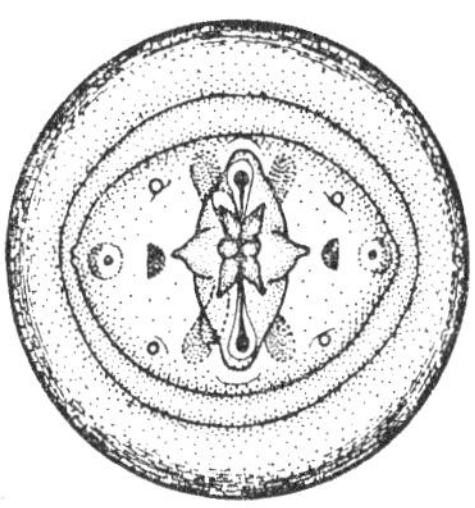

1.46

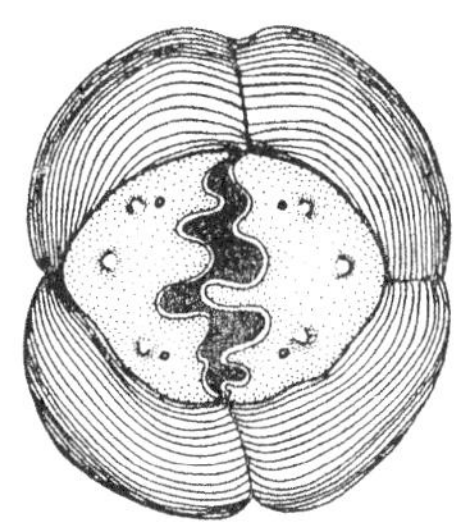

1.47

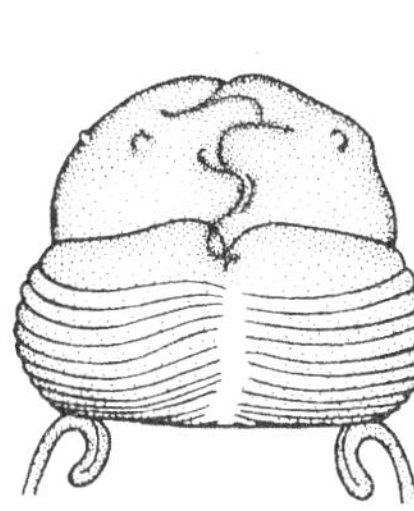

1.48

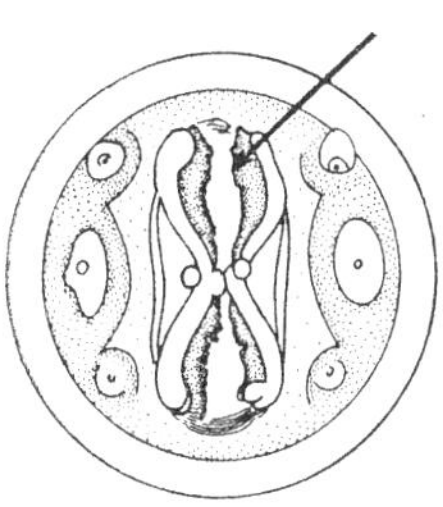

1.49

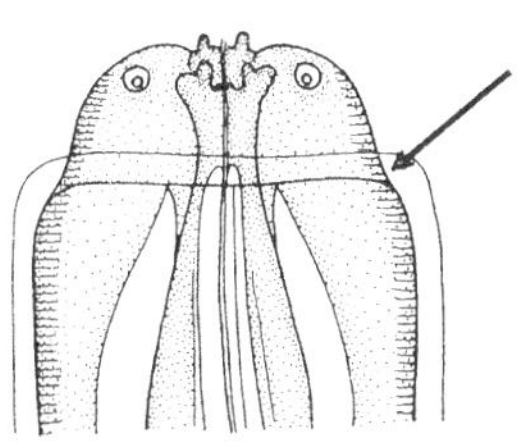

1.50

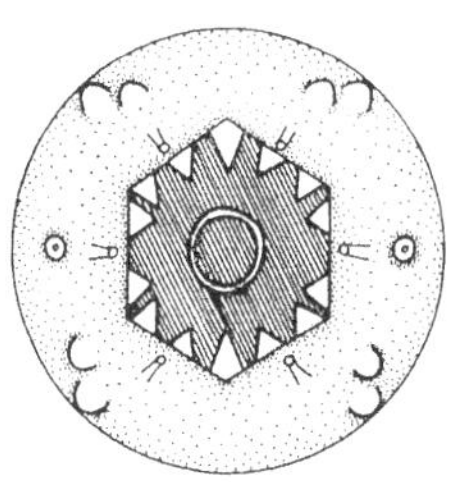

1.51

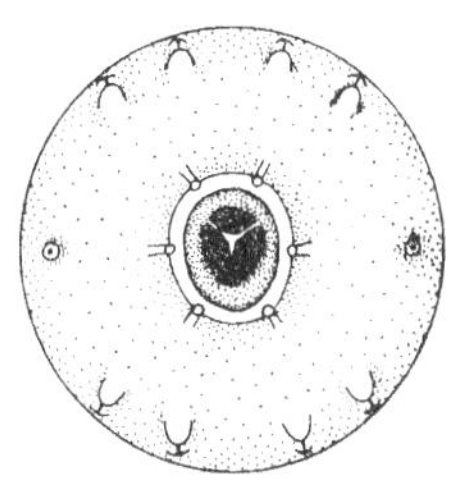

1.52

4-(3) Pseudolabia not lobed (Fig. 1.46). Cephalic and outer labial papillae fused.
Pseudolabia with variable number of teeth on their free borders (Fig. 1.49). Body cuticle immediately behind pseudolabia often expanded to form collarette (Fig. 1.50). Intermediate hosts insects.

Physalopteroidea

5-(1) Buccal cavity well cuticularized (except in Pneumospirurinae of lungs of mammals).

6-(11) Oral opening hexagonal (Fig. 1.51) or oval (Fig. 1.52).
Pseudolabia absent. Internal labial papillae well developed.

7-(8) Body ornamented along its length with two longitudinal rows of spines (Fig. 1.53).

Rictularioidea

8-(7) Body without two longitudinal rows of spines.

9-(10) Nematodes with most of following characters:
Buccal cavity variable, sometimes long and cylindrical (Fig. 1.54). Caudal alae absent. Caudal papillae (Fig. 1.55) not arranged as in typical spirurid. Vulva markedly anterior or posterior in position.
Parasites of orbit of birds and mammals, lungs of mammals or intestine of fishes.

Thelazioidea

1.53

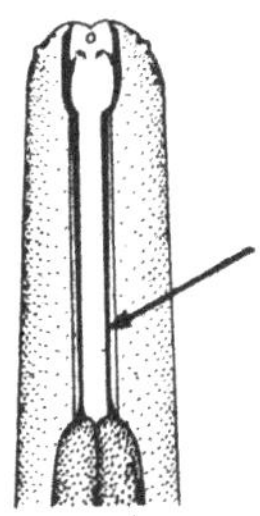

1.54

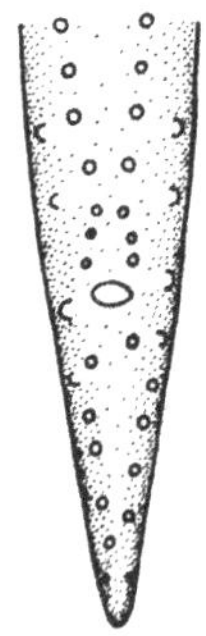

1.55

10-(9) Nematodes with most of the following characters: Buccal cavity well cuticularized, never long and cylindrical (Fig. 1.56). Caudal alae present. Caudal papillae arranged as in typical spirurid (Fig. 1.57). Vulva rarely markedly anterior or posterior in position.

Spiruroidea

11-(6) Oral opening with well-defined bilateral symmetry (at least in male) (Fig. 1.58). Pseudolabia present (Fig. 1.58). Internal labial papillae small or absent.

12-(13) Pseudolabia not covering entire cephalic surface, median lips still visible (Fig. 1.58).
4 outer labial papillae often visible on lips. Buccal cavity rarely elongated and cylindrical. Cephalic cuticular ornamentation present (Fig. 1.59) or absent.

Habronematoidea

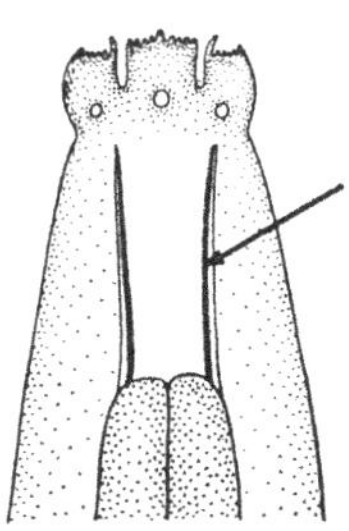

1.56

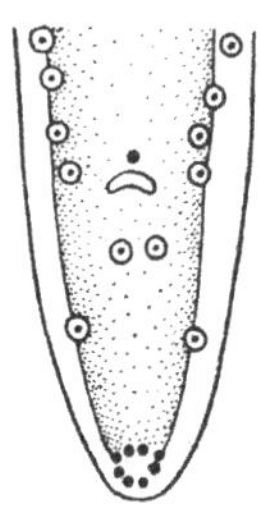

1.57

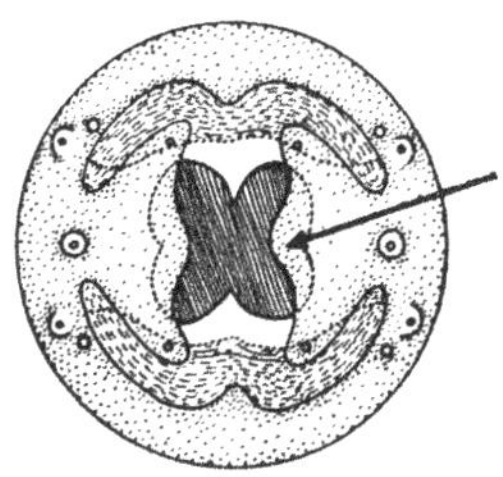

1.58

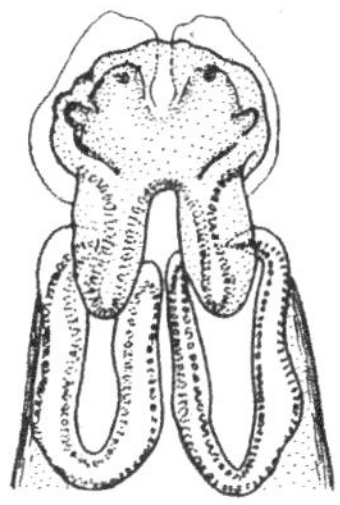

1.59

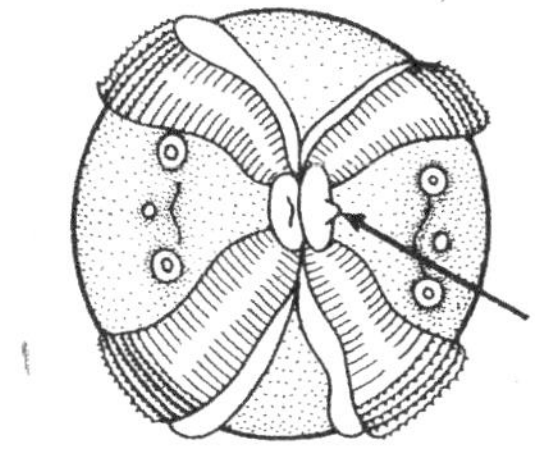

1.60

1.61

13-(12) Pseudolabia large, involving entire cephalic surface (Fig. 1.60).
Outer labial papillae fused with cephalic papillae. Buccal cavity generally elongated and cylindrical. Cuticular cephalic ornamentation present (Fig. 1.61). Parasites of gizzard of birds, sometimes of insectivores.

Acuarioidea

14-(2) Pseudolabia absent.
Parasites of tissues and tissue spaces of all classes of vertebrates except fish.

15-(16) Egg usually thin shelled, with poorly differentiated larva (microfilaria) (Fig. 1.62) (except Filariidae).

1.62

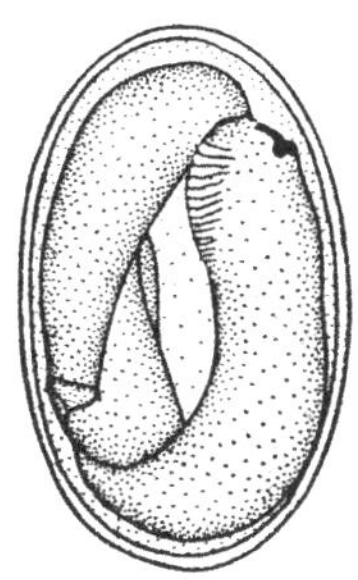

1.63

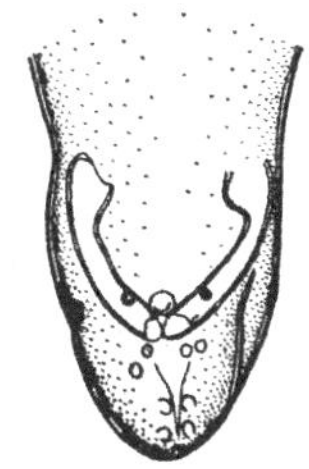

1.64

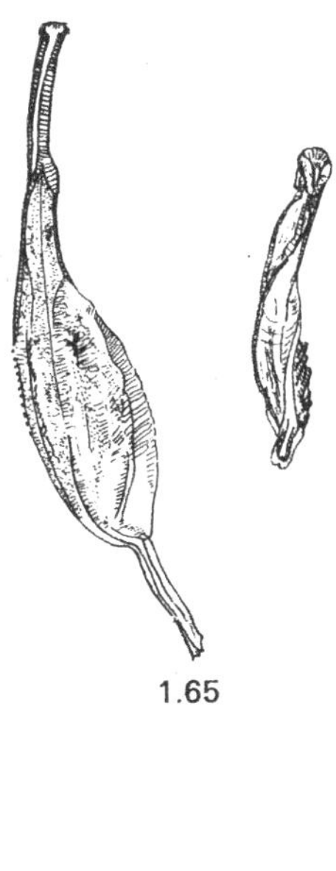

1.65

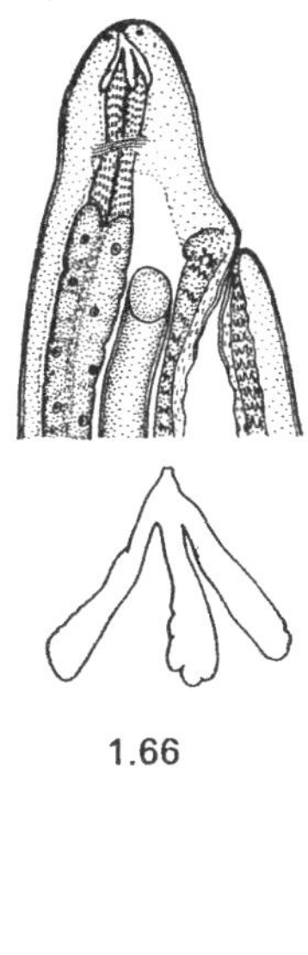

1.66

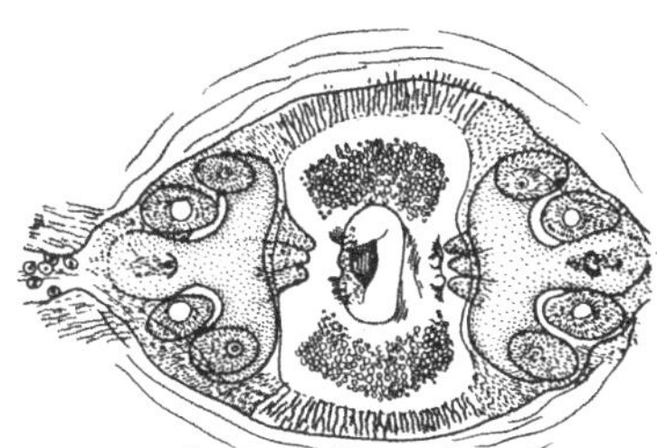

1.67

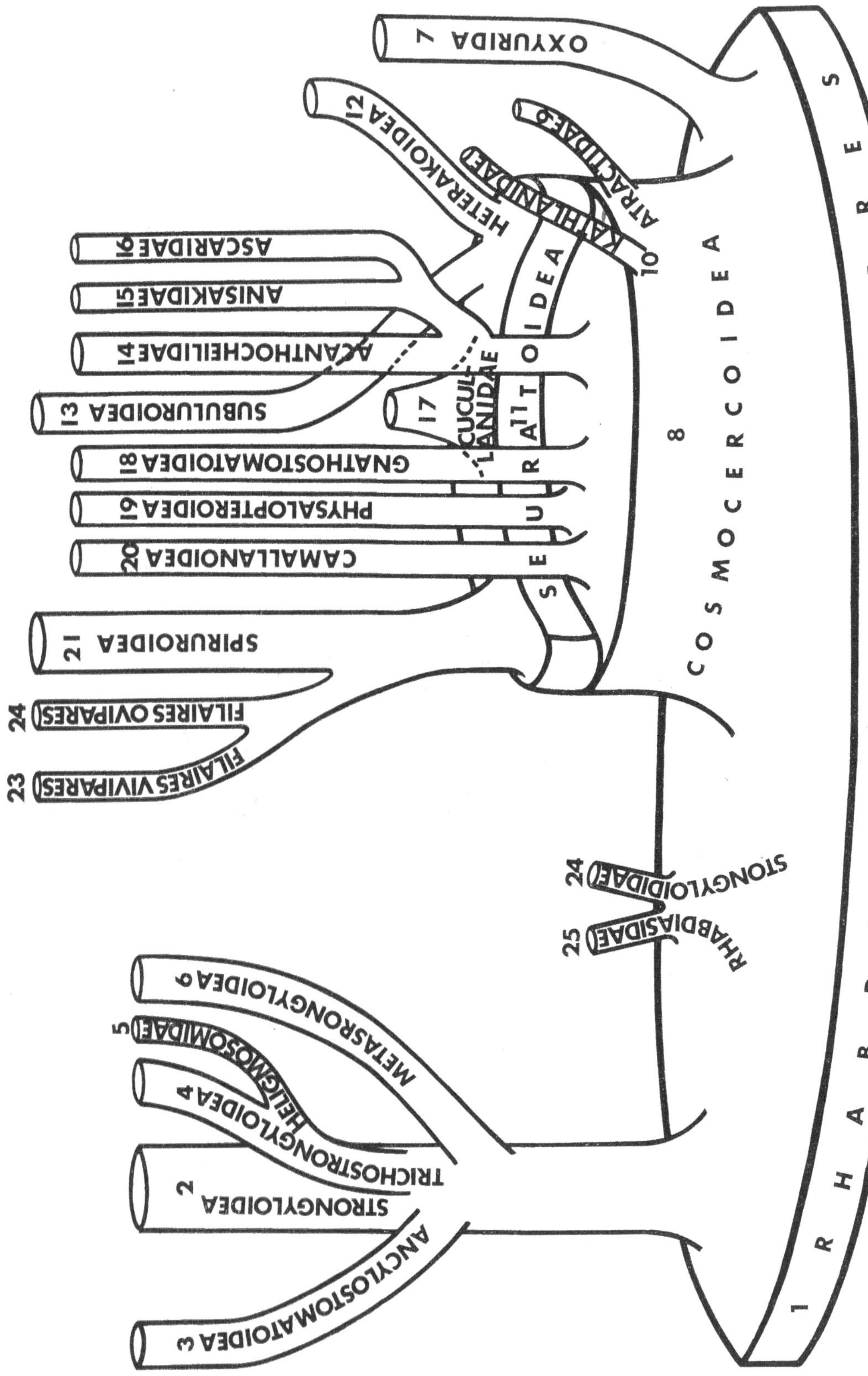

FIG. 1.68. Phylogenetic tree of the subclass Secernentea indicating the possible relationships between the major systematic groups adopted in the classification used herein. See Corrigenda on opposite page.

Spicules variable in length but always dissimilar in structure. Transmitted by haematophagous arthropod vectors.
Parasites of tissues and tissue spaces of all classes of vertebrates except fishes.

Filarioidea

16-(15) Egg thick-shelled, with fully differentiated first-stage larva (Fig. 1.63).

17-(18) Spicules short, arcuate, similar in size and structure (except Desmidocercidae) (Fig. 1.64).
Cephalic extremity usually without well developed cuticular structures (except Tetracheilonematidae).
Parasites of air sacs and subcutaneous tissues (especially of head and neck region) of birds.

Aproctoidea

18-(17) Spicules usually long, markedly dissimilar in size and structure, generally complex, alate (Fig. 1.65).
Usually with well developed cephalic cuticular structures in form of tridents (Fig. 1.66), circumoral elevations and epaulets (Fig. 1.67). First-stage larva with delicate cephalic and caudal spines. Intermediate hosts coprophagous or herbivorous insects.
Parasites of air sacs of reptiles and birds.

Diplotriaenoidea

Fig. 1.68 CORRIGENDA

for METASRONGYLOIDEA **6** *read* METASTRONGYLOIDEA **6**

for STONGYLOIDIDAE **24** *read* STRONGYLOIDIDAE **24**

for FILAIRES OVIPARES **24** *read* FILAIRES OVIPARES **22**

2 ENOPLIDA

DIOCTOPHYMATOIDEA, TRICHINELLOIDEA AND MUSPICEOIDEA

by ROY C. ANDERSON[1] and ODILE BAIN[2]

DIOCTOPHYMATOIDEA

The superfamily includes only four genera of adenophorean nematodes associated with the kidney of carnivores (*Dioctophyme*) and the alimentary tract of carnivorous birds (*Hystrichis, Eustrongylides*) and mammals (*Soboliphyme*). The group includes only a few species but they seem to have wide geographical distributions. For example, *Dioctophyme renale* is known world wide and *Soboliphyme* is found in North America and western Asia.

The worms are generally large and frequently associated with cyst-like cavities in the host. *D. renale* occurs normally in a cyst produced from the right kidney of the host. *Hystrichis* and *Eustrongylides* spp. live in cavities in the proventriculus.

Most members of the superfamily are known to produce resistant unembryonated eggs which develop in water into small, stylet-bearing first-stage larvae. *D. renale, Hystrichis* spp. and *Eustrongylides* spp. develop in aquatic oligochaetes which serve as intermediate hosts (Karmanova, 1956, 1960, 1965; Mace & Anderson, 1975). Paratenic hosts, generally frogs and fishes, ensure transmission to the carnivorous final host.

Mirandonema Kreis, 1945, reported from the intestine of *Nasua rufa* from Brazil, is regarded as a synonym of *Dioctophyme. D. renale* has been reported in *Nasua* spp. on several occasions in South America (Grosso *et al.*, 1944; Lutz, 1924; Proença, 1935) and the description of *M. intestinalis* Kreis fails to mention characters which would distinguish this species from *D. renale* which is common in South America.

DIOCTOPHYMATOIDEA
Key to families

1-(2) Muscular cephalic sucker present (Fig. 9.31).
Soboliphymatidae

2-(1) Muscular cephalic sucker absent.
Dioctophymatidae

Family *SOBOLIPHYMATIDAE* Petrov, 1930

One genus.
Parasites of intestine of carnivores, especially Mustelidae in North America and western Asia.
Soboliphyme Petrov, 1930

Family *DIOCTOPHYMATIDAE* (Railliet, 1915)
Key to subfamilies

1-(2) Vulva in oesophageal region.
Extremely large nematodes.
Parasites of kidney (abnormally of peritoneal cavity) of carnivores, especially Mustelidae and Canidae.
Dioctophymatinae

2-(1) Vulva in posterior part of body, terminal or subterminal and near anus (Fig. 9.32, 9.33).
Moderately sized nematodes.
Parasites of proventriculus of fish-eating birds.
Eustrongylinae

[1] Department of Zoology, University of Guelph, Guelph, Ontario, Canada.
[2] Museum National d'Histoire Naturelle, Paris, France.

9.31

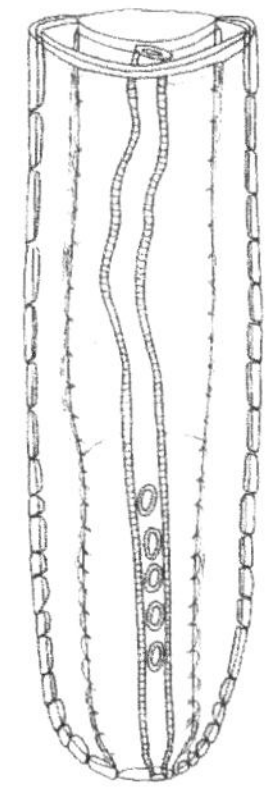

9.32

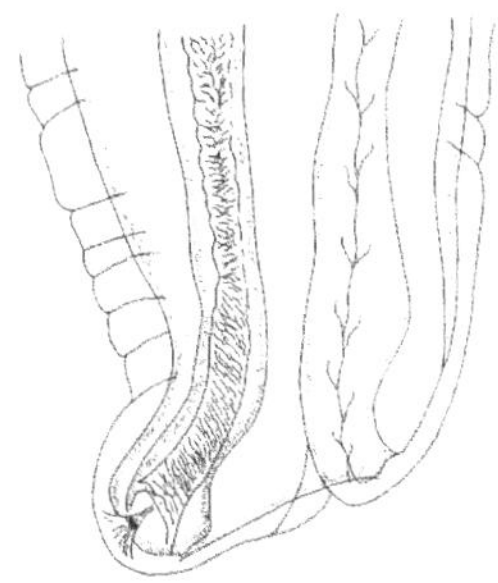

9.33

9.34

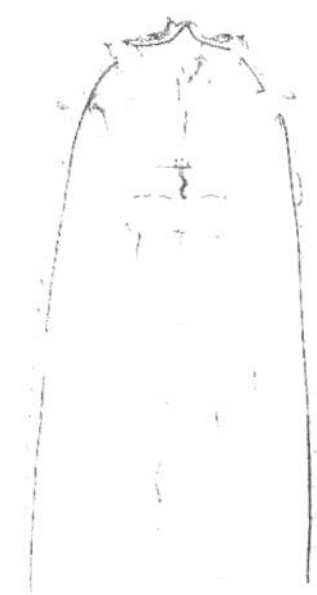

9.35

FIG. 9.31. *Soboliphyme baturini* Petrov, 1930, female, lateral view (after Petrov, 1930).

FIG. 9.32. *Eustrongylides tubifex* (Nitzsch, 1819), posterior end female, ventral view (after Fastzkie & Crites, 1963).

FIG. 9.33. *Hystrichis coronatus* Molin, 1861, caudal end female (after Karmanova, 1955).

FIG. 9.34. *Hystrichis coronatus* Molin, 1861, anterior extremity (after Karmanova, 1955).

FIG. 9.35. *Eustrongylides tubifex* (Nitzsch, 1819) anterior extremity male, lateral view (after Fastzkie & Crites, 1963).

Subfamily **Dioctophymatinae** (Castellani & Chalmers, 1910)

One genus (*D. renale*, the giant kidney-worm).

Dioctophyme Collet-Meygret, 1802
(= *Mirandonema* Kreis, 1945)

Subfamily **Eustrongylinae** Chitwood & Chitwood, 1937

1-(2) Cephalic region generally dilated, with rows of posteriorly directed spines (Fig. 9.34).
Hystrichis Dujardin, 1845

1-(2) Cephalic region not dilated, without rows of spines (Fig. 9.35).
Eustrongylides Jägerskiöld, 1909

REFERENCES

(Dioctophymatoidea)

GROSSO, A. M., PRIETO & STROBINO, E. 1944. Dioctophimosis en dos especies de nuestra fauna autoctona. *Gac. vet.*, **6**, 2–12.

KARMANOVA, E. M. 1956. [An interpretation of the biological cycle of the nematode *Hystrichis tricolor* Dujardin, 1845, a parasite of domestic and wild ducks.] *Dokl. Akad. Nauk, SSSR*, **111**, 245–247. [In Russian.]

———. 1960. [The life-cycle of the nematode *Dioctophyme renale* (Goeze, 1782) parasitic in the kidneys of carnivorous animals and man.] *Dokl. Akad. Nauk. SSSR*, **132**, 1219–1221. [In Russian.]

———. 1965. [Intermediate hosts of *Eustrongylides excisus*, a parasite of aquatic birds.] *Trudy gel'mint. Lab.*, **15**, 86–87. [In Russian.]

KREIS, H. A. 1945. Beiträge zur Kenntnis parasitischer Nematoden. XIII. Parasitische Nematoden aus den Tropen. *Rev. Suisse Zool.*, **52**, 551–596.

LUTZ, A. 1924. Sur le *Dioctophyme renalis*. *C.r. Séanc. Soc. Biol.*, **90**, 696.

MACE, T. F. A. & ANDERSON, R. C. 1975. Development of the giant kidney worm, *Dioctophyme renale* (Goeze, 1782) (Nematoda: Dioctophymatoidea). *Can. J. Zool.*, **53**, 1552–1568.

PROENÇA, M. C. 1935. Sobre dois casos de *Dioctophyme renale* em cães no Rio de Janeiro. *Bolm. Vet. Exército, Brasil*, **2**, 50–51.

TRICHINELLOIDEA

Roman's (1965) classification is retained since it seems to reflect the evolutionary lines of the superfamily as revealed by the most recent work. Three families and five subfamilies are recognized herein.

Capillaria petruschewskii (Schulman, 1948) is probably the most primitive species in the superfamily since its glandular oesophagus consists of three stichosomes (Kutzer & Otte, 1966; Moravec, 1980). The parasite is found in the liver of Cyprinidae, Percidae and other fish raising the possibility that the trichinelloids are derived from ancestors that were initially tissue parasites of freshwater fish.

Trichuridae

This family includes the more primitive forms. We recognize three subfamilies.

Capillariinae

The genus *Capillaria* has a large number of synonyms (Butterworth & Beverley-Burton, 1980) resulting from attempts to divide the genus taxonomically. *Paratrichosoma crocodilus* Ashford & Muller, 1978 of the skin of crocodiles was originally placed in the Trichosomoididae. It has, however, a long cloaca, a spicule and a cirrus and falls into the synonymy of *Capillaria* (see Spratt, 1982). Species are widespread throughout the vertebrate groups and they are found in many different organ systems (e.g., urogenital tract, viscera, respiratory system). Species are extremely diverse morphologically. For example, some primitive forms have three stichosomes (one dorsal and two lateral, each with 40 stichocytes). In other species there are two recognizable stichosomes although they may be fused (see *C. catastomi* Pearse, 1924 redescribed by Bell & Beverley-Burton, 1981); in these species the two kinds of stichocytes alternate regularly. In the most highly specialized species there is a single homogeneous row of stichocytes (approximately 40) although the alternation of two types of stichocytes may appear in larvae (Wright, 1961). In addition to differences in the stichosome, the number of bacillary bands varies from four (two lateral and two median as also found in free-living Dorylaimidae) to one. Also, some species have post-oesophageal coelomocytes and other species lack them. The cloaca is short or long. These various features noted in species of *Capillaria* may eventually permit a proper revision of the genus. Clearly, the characters used in the past to divide the genus (e.g., spicule present or absent, cirrus with or without spines or tubercles) have not been able to provide acceptable genera or subgenera

Members of the Capillariinae are fundamentally oviparous. *C. philippinensis* Chitwood *et al.*, 1968 behaves abnormally in man; in addition to oviparous females, there are viviparous females (Chitwood *et al.*, 1972). This feature may be aberrant since man is not the normal host of the parasite.

Trichurinae

The subfamily contains the unique genus *Trichuris* which has evolved a single row of stichocytes (40 to 200), one bacillary band ventral to the oesophagus and a cloaca with thick muscles in the pre-spicular portion. The whipworms have adapted themselves to the large intestine of mammals.

Trichosomoidinae

Members of this subfamily probably departed early from the Capillariinae line. They retain a primitive oesophagus in which the 60 to 150 stichocytes are not in a perfect row and traces of two and perhaps three stichosomes are detectable. In members of the subfamily the male has a short cloaca and lacks a spicule. They occur in the mucosa of mammals and lay lenticular, embryonated eggs.

Anatrichosoma is the more primitive genus since the male is of normal size. Members of the genus occur in marsupials, Tupaidae, primates and Muridae. In *Trichosomoides* the male is greatly reduced and occurs usually in the uterus of the female worm. The species occur entirely in Muridae.

Trichinellidae

The family contains only *Trichinella* of the intestine of mammals. The genus has some striking similarities to the Trichuridae, i.e., two lateral bacillary bands (Kim & Ledbetter, 1980), and males have a cirrus and a vestigial spicule (Wu, 1955). Nevertheless, the family is justified because of a number of special features of *Trichinella* such as the position of the vulva near the middle of the oesophagus, viviparity, encapsulation of larvae in muscles of the host, and direct nutrition of the oocytes in contact with the host's intestine (Herbaut *et al.*, 1979).

Cystoopsidae

The family is characterized by the presence of two stichosomes, an atrophied anus and the anterior position of the vulva. It is divided into two subfamilies. The Cystoopsinae have uninuclear stichocytes and occur in freshwater fish. The Dioctowittinae, originally placed in the Dioctophymatoidea (see Chabaud & Le Van Hoa, 1960), have multinucleate stichocytes (Bain & Ghadirian, 1967) and are parasites of reptiles.

The genera *Trichuroides* Ricci, 1949 and *Sclerotrichum* Rudolphi, 1819 are of uncertain status and not included in the keys.

TRICHINELLOIDEA

Key to families

1-(4) Intestine normal, tubular in form (Fig. 9.36).
Anus present.
Vulva not near nerve ring (Fig. 9.37).
Egg without membranous envelope or polar filaments (Fig. 9.38).
One to three stichosomes present (Fig. 9.39).

2-(3) Vulva near end of oesophagus.
Female oviparous (with one rare exception).

Trichuridae

3-(2) Vulva near mid-region of oesophagus (Fig. 9.40).
Female viviparous.

Trichinellidae

4-(1) Intestine modified to form sac or trophosome (Fig. 9.41).
Anus absent.
Vulva near nerve ring.
Eggs with membranous envelope (Fig. 9.42) or with polar filaments (Fig. 9.43).
Two stichosomes present.

Cystoopsidae

Family *TRICHURIDAE* (Ransom, 1911) Railliet, 1915

Key to subfamilies

1-(4) Stichocytes regularly aligned throughout length of oesophagus, similar in form.
Eggs generally not embryonated.

2-(3) Cloaca with thin muscular wall anterior and posterior to point of entry of spicule (Fig. 9.44).
20 to 60 stichocytes present in 1–3 rows.
Male caudal end straight or curved ventrally.
Body usually filiform but, exceptionally, post-oesophageal region markedly expanded.

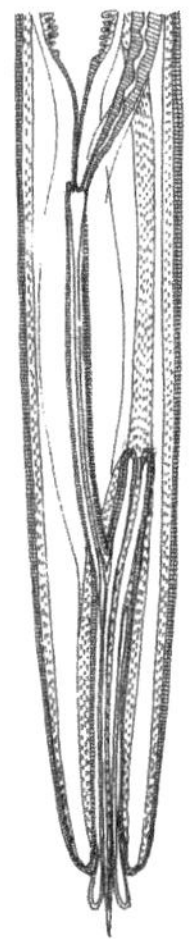

9.36

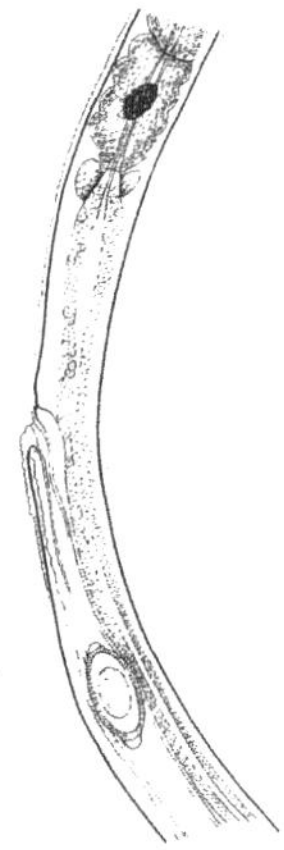

9.37

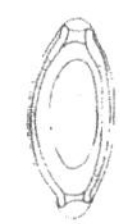

9.38

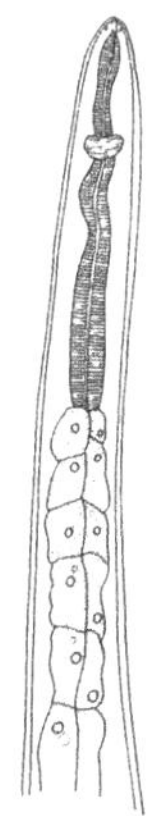

9.39

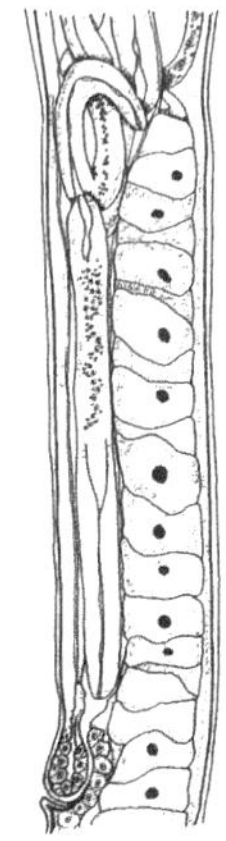

9.40

9.41

FIG. 9.36. *Trichuris suis* (Schrank, 1788) (= *T. crenatus* Rudolphi, 1809) caudal extremity male, ventral view (after Rauther, 1918).

FIG. 9.37. *Capillaria putorii* (Rudolphi, 1819), vulvar region (after Butterworth, 1976, M.Sc. thesis, University of Guelph).

FIG. 9.38. *Capillaria putorii* (Rudolphi, 1819), egg (after Butterworth, 1976, M.Sc. thesis, University of Guelph).

FIG. 9.39. *Capillaria petruschewskii* (Shulman, 1848), anterior extremity (after Moravec, 1980).

FIG. 9.40. *Trichinella spiralis* (Owen, 1835), vulva region (after Wu, 1955).

FIG. 9.41. *Cystoopsis acipenseris* Wagner, 1867, male, lateral view (after Janicki & Rasin, 1930).

Oesophageal region generally shorter than post-oesophageal region.
Coelomocytes usually present behind oesophagus (Fig. 9.37) or absent.
Cirrus usually present, often barely cuticularized.
Parasites of skin, viscera, spleen, respiratory and excretory system of vertebrates.

Capillariinae

3-(2) Cloaca with thick muscular wall anterior to point of entry of spicule; wall thinner posterior to point of entry (Fig. 9.36).
Single broad bacillary band present, ventral to oesophagus (Fig. 9.45).
Stichocytes in single row (? 40–200).
Caudal end of male curved dorsally.
Post-oesophageal region of body markedly thickened.
Oesophageal region longer than post-oesophageal region.
Coelomocytes absent behind oesophagus.
Spicule present, well developed.
Cirrus usually with spines or tubercles.
Parasites of large intestine of mammals.

Trichurinae

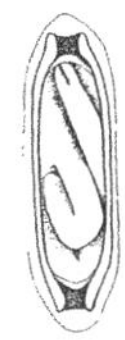

9.42

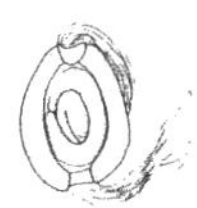

9.43

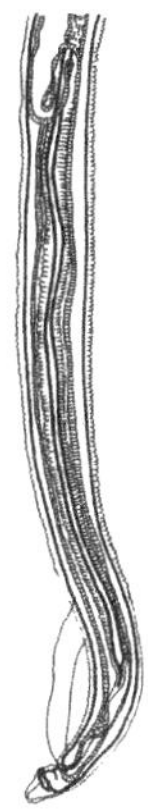

9.44

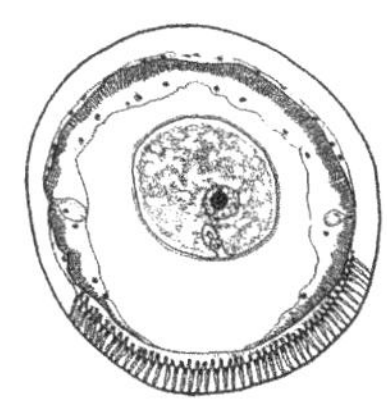

9.45

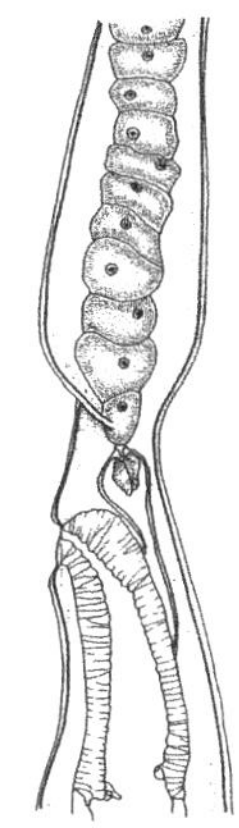

9.46

FIG. 9.42. *Cystoopsis acipenseris* Wagner, 1867, mature egg (after Janicki & Rasin, 1930).

FIG. 9.43. *Dioctowittus denisoniae* Jones, 1978, egg (after Jones, 1978).

FIG. 9.44. *Capillaria putorii* (Rudolphi, 1819) caudal end male, lateral view (after Butterworth, 1976, M.Sc. thesis, University of Guelph).

FIG. 9.45. *Trichuris suis* (Schrank, 1788) (= *T. crenatus* Rudolphi, 1809), cross-section through oesophageal region (after Rauther, 1918).

FIG. 9.46. *Anatrichosoma cynomolgi* Smith & Chitwood, 1954, vulvar region (after Conrad & Wong, 1973).

4-(1) Stichocytes irregularly aligned in posterior region of oesophagus (Fig. 9.46), elongate in anterior region and short in posterior region.
Eggs embryonated when laid.
60 to 150 stichocytes present.
Post-oesophageal coelomocytes present or absent.
Cloaca short (maximum 300 μm) (Fig. 9.47).
Cuticular ornamentation frequently present in form of cephalic vesicles and bosses.
Parasites of various tissues (mucosa, paracloacal glands, cornea) of marsupials, Tupaidae, Primates, Muridae.

Trichosomoidinae

Subfamily **Capillariinae** Railliet, 1915

One genus.

Capillaria Zeder, 1800
(= *Trichosoma* Rudolphi, 1819; = *Trichosomum* Creplin, 1829; = *Liniscus* Dujardin, 1845; = *Calodium* Dujardin, 1845; = *Eucoleus* Dujardin, 1845; = *Thominx* Dujardin, 1845; = *Hepaticola* Hall, 1916; = *Capillostrongyloides* Freitas & Lent, 1935; = *Skrjabinocapillaria* Skarbilovitsch, 1946; = *Aonchotheca* Lopez-Neyra, 1947; = *Gessyella* Freitas, 1959; = *Pterothominx* Freitas, 1959; = *Pseudocapillaria* Freitas, 1959; = *Ritaklossia* Freitas, 1959; = *Pearsonema* Freitas & Mendonça, 1960; = *Orthothominx* Freitas & Silva, 1960; = *Paracapillaria* Mendonça, 1963; = *Schulmanela* Ivashkin, 1964; = *Armocapillaria* Gagarin & Nazarova, 1966; = *Paratrichosoma* Ashford & Muller, 1978).

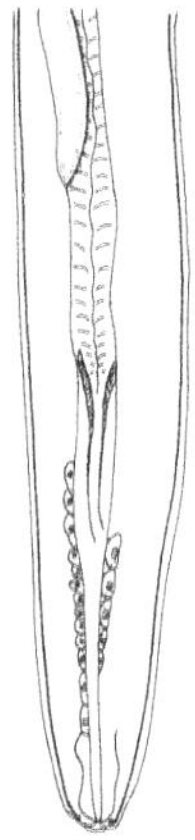

9.47

9.48

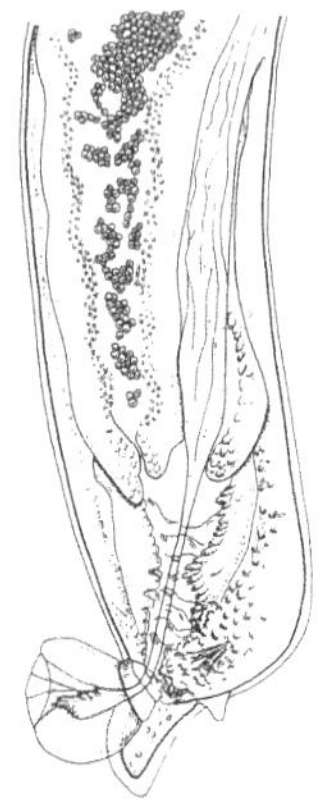

9.49

9.50

FIG. 9.47. *Anatrichosoma gerbillis* (Bernard, 1969) Spratt (= *Skrjabinocapillaria rodentium*), caudal end male, ventral view (after Spratt, in press).

FIG. 9.48. *Trichosomoides nasalis* Biocca & Aurizi, 1961, female, lateral view (after Biocca & Aurizi, 1961).

FIG. 9.49. *Trichinella spiralis* (Owen, 1835), caudal end male, lateral view (after Wu, 1955).

FIG. 9.50. *Cystoopsis acipenseri* Wagner, 1867, female (after Janicki & Rasin, 1930).

Subfamily **Trichurinae** Ransom, 1911

One genus (whipworm).

Trichuris Roederer, 1761
(= *Trichocephalos* Goeze, 1782; = *Trichocephalus* Schrank, 1788; = *Mastigodes* Zeder, 1800; *Buckleyuris* Sarwar, 1959; = *Salamia* Sarwar, 1959; = *Rudolphia* Sarwar, 1959).

Subfamily **Trichosomoidinae** Hall, 1916

Key to genera

1-(2) Male as long as, or only slightly shorter than, female.
Male living independent of female.
Spicule and cirrus absent or vestigial (Fig. 9.47).
Parasites of mucosa (stomach, buccal, nasal), paracloacal glands and cornea of marsupials, tupaids, primates and murids.
Anatrichosoma Smith & Chitwood, 1954

2-(1) Male much smaller than female.
Male lodged in uterus of female worm (Fig. 9.48).
Spicule and cirrus absent.
Parasites of nasal and urinary mucosa of murids.
Trichosomoides Railliet, 1895
(= *Trichodes* Linstow, 1874)

Family *TRICHINELLIDAE* Ward, 1907

One genus (Fig. 9.49).
Parasites of intestinal mucosa of mammals.

Trichinella Railliet, 1895
(= *Trichina* Owen, 1835)

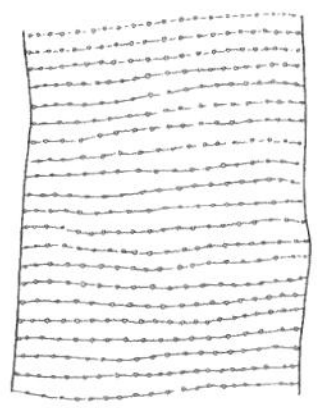

9.51

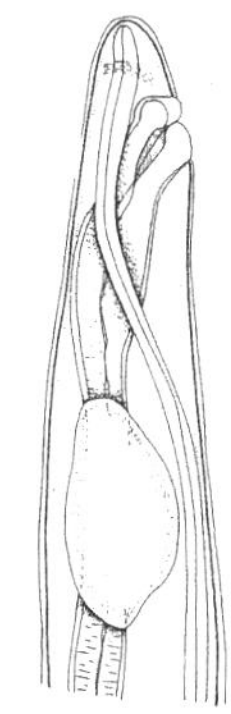

9.52

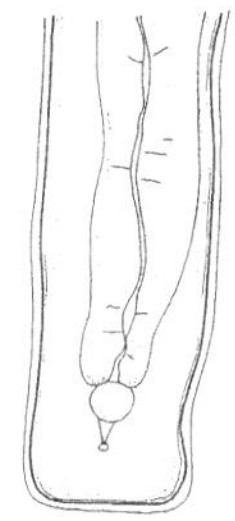

9.53

FIG. 9.51. *Cystoopsis acipenseris* Wagner, 1867, cuticle (after Janicki & Rasin, 1930).

FIG. 9.52. *Dioctowittus chabaudi* Bain & Ghadirian, 1967, anterior extremity female, lateral view (after Bain & Ghadirian, 1967).

FIG. 9.53. *Dioctowittus chabaudi* Bain & Ghadirian, 1967, caudal end male, ventral view (after Bain & Ghadirian, 1967).

Family *CYSTOOPSIDAE* Skrjabin, 1923

Key to subfamilies

1-(2) Female with body divided into thin anterior region and posterior region expanded to form vesicle (Fig. 9.50).
Stichocytes unicellular, not voluminous.
Intestine dilated into sac.
Cuticle of anterior region of female with circular rows of small tubercles (Fig. 9.51).
Genital tubes with numerous convolutions.
Spicule and, possibly, cirrus present.
Eggs in membranous oval envelope. **Cystoopsinae**

2-(1) Female with cylindrical undifferentiated body.
Stichocytes voluminous and syncitial (Fig. 9.52).
Intestine transformed into trophosome.
Cuticle of female without tubercles.
Genital tubes not markedly convoluted.
Spicule and cirrus absent (Fig. 9.53).
Eggs lenticular with tufts of polar filaments. **Dioctowittinae**

Subfamily **Cystoopsinae** (Skrjabin, 1923)

One genus.
Parasites of conjunctiva of ganoid fish (*Acipenser*).
Cystoopsis Wagner, 1867

Subfamily **Dioctowittinae** Chabaud & Le Van Hoa, 1960

One genus.
Parasites of body cavity of reptiles.
Dioctowittus Chabaud & Le Van Hoa, 1960

Acknowledgement

We are especially grateful for the advice of Dr. David Spratt, CSIRO, Australia, on the systematics of the Capillariinae and for the opportunity to read material he has in press.

REFERENCES

(Trichinelloidea)

ASHFORD, R. W. & MULLER, R. 1978. *Paratrichosoma crocodilus* n.gen. n.sp. (Nematoda: Trichosomoididae) from the skin of the New Guinea crocodile. *J. Helminth.*, **52,** 215–220.

BAIN, O. & GHADIRIAN, E. 1967. Description d'une nouvelle espèce *Dioctowittus* (Nématode) et note sur la position systématique du genre. *Annls Parasit. hum. comp.*, **42,** 643–650.

BELL, D. A. & BEVERLEY-BURTON, M. 1981. The taxonomy of *Capillaria* spp. (Nematoda: Trichuroidea) in North American freshwater fishes. *Syst. Parasitol.*, **2,** 157–169.

CHABAUD, A. G. & LE VAN HOA. 1960. Nématodes parasites de reptiles. Aphasmidiens. *Explòr. Parc Nat. Upemba. Mission G.F. de Witte.* Fasc. (65), 59–64.

CHITWOOD, M. B., VELASQUEZ, C. & SALAZA, M. G. 1968. *Capillaria philippinensis* sp.n. (Nematoda: Trichinellidae) from the intestine of man in the Philippines. *J. Parasit.*, **54,** 368–371.

CROSS, J. H., BANZON, T., CLARKE, M. D., BASACA SERVILLA, V., WATTEN, R. H. & DIZON, J. J. 1972. Studies on the experimental transmission of *Capillaria philippinensis* in monkeys. *Trans. R. Soc. trop. Med. Hyg.*, **66,** 819–827.

FREITAS, J. F., TEIXEIRA DE. 1959. Esbôco de novo arranjo sistemático os nematódeos capilarineos (Trichuroidea). *Atas Soc. Biol. Rio de J.*, **3,** 4–7.

——— & MENDONÇA, J. M. 1960. Novo nematodéo parasito de *Procyon cancrivorus* Cuv.: *Pearsonema pearsoni* gen. nov. sp. nov. (Trichuroidea, Capillariidae). *Atas Soc. Biol. Rio de J.*, **4,** 63–66.

——— & JORGE DA SILVA, A. A. 1960. Alguns nematódeos parasitos de *Gallus gallus* dom. (L.) no estado de Bahia. *Mem. Inst. Oswaldo Cruz*, **58,** 189–207.

GAGARIN, V. G. & NAZAROVA, N. S. 1966. [*Armocapillaria moschiferi* n.g., n.sp. (Nematoda: Capillaridae) from *Moschus moschiferus*.] *Mater. Nauch. Konf. Vsesoyuz. Obshch. Gel'mint.*, Part 3, 66–71. [In Russian.]

HERBAUT, C., SLOMIANNY, C., VERNES, A. & BIGUET, J. 1979. Spécialisation du reticulum endoplasmique au contact de l'intestin postérieur, chez les ovocytes de *Trichinella spiralis* (Nématode, Trichuroide). *Annls Parasit. hum. comp.*, **54,** 237–242.

IVASHKIN, V. M. 1964. [On the position of the species *Hepaticola petruschewksii* Schulman, 1948, in the system Trichocephalata.] *Mater. Nauch. Konf. Vsesoyuz, Obshch. Gel'mint.*, Part 1, 135–137. [In Russian.]

KIM, C. W. & LEDBETTER, M. C. 1980. Surface morphology of *Trichinella spiralis* by scanning electron microscopy. *J. Parasit.*, **66,** 75–81.

KUTZER, E. & OTTE, E. 1966. *Capillaria petruschewskii* (Schulman, 1948): Morphologie, Biologie und pathogene Bedeutung. *Z. Parasitkde*, **28,** 16–30.

MENDONÇA, J. M. 1963. Sôbre dois Capilariineos parasitos de peixes (Nematoda, Trichuroidea). *Mem. Inst. Oswaldo Cruz*, **61,** 321–327.

MORAVEC, F. 1980. Revision of nematodes of the genus *Capillaria* from European freshwater fishes. *Folia Parasitol. (Praha)*, **27,** 309–324.

ROMAN, E. *in* GRASSE, P. P. 1965. Traité de Zoologie. *Némathelminthes*, **4**(2), 731. Paris: Masson et Cie.

SPRATT, D. M. 1982. *Anatrichosoma* Smith and Chitwood and *Trichosomoides* Railliet (Nematoda: Trichinelloidea) from Australian native mammals, with note on *Skrjabinocapillaria* Skarbilovitch. *Aust. J. Zool.* (in press).

SPRATT, D. M. 1982. Redescription of trichurid nematodes from Australian vertebrates. *Aust. J. Zool.* (in press).

WRIGHT, K. A. 1961. Observation on the life cycle of *Capillaria hepatica* (Bancroft, 1893) with a description of the adult. *Can. J. Zool.*, **38,** 167–182.

WU, L. Y. 1955. Studies on *Trichinella spiralis*. 1. Male and female reproductive systems. *J. Parasit.*, **41,** 40–47.

MUSPICEOIDEA

The superfamily includes a number of highly specialized, viviparous, adenophorean genera found in the tissues of birds (*Robertdollfusa*), marsupials (*Durikainema*), mice (*Muspicea*) and bats (*Riouxgolvania, Pennisia, Lukonema*). It has been suggested that the group belongs to the Dorylaimina *sensu* Chitwood, and has affinities with the Trichinelloidea and Mermithoidea.

Following Bain & Chabaud (1979) the genera are separated into the families Robertdollfusidae and Muspiceidae. The former consists of *Robertdollfusa*, found in the eye of corvines or the brain of falconids, and *Durikainema* found in blood vessels of kangaroos; the latter consists of *Muspicea* of the tissues of *Mus* and three genera from the wings of bats.

Genera of the Muspiceidae can be separated on the basis of several characters (e.g., "phasmids", larva, vulva) but we prefer to use the structure of the oesophageal glands as the primary key character (Bain & Chabaud, 1979).

The means of reproduction and transmission are apparently diverse. There are separate sexes or protandry. The vulva is atrophied and the larvae escape directly by rupture of the body-wall in the head region or the vulva is functional and the larvae pass through the vulva and then migrate between two layers of body cuticle and escape by the head region. Transmission is suspected to be by cannibalism, cutaneous penetration or during grooming and lactation.

MUSPICEOIDEA
Key to families

1-(2) Alimentary tract obscure (Fig. 9.54).
Slender worms 5–6 mm long.
Body cavity occupied by uterine pouch extending most of the length of body.
Vulva near middle of body.
Larva in uterine pouch, 800–950 µm long; gut undifferentiated.
Parasites of birds and of marsupials.

Robertdollfusidae

2-(1) Alimentary tract clearly defined, with ventrolateral oesophageal glands; one dorsal oesophageal gland modified as excretory organ and trophosome.
Stout worms 3–4 mm long.
Vulva functional near middle of body or atrophied.
Larva in uterine pouch 300–400 µm long; gut present with oesophagus ending in large dorsal gland and trophosome.
Parasites of tissues of mouse and of wings of bats.

Muspiceidae

Family *ROBERTDOLLFUSIDAE* Chabaud & Campana, 1950

1-(2) Head without cuticular cephalic inflation, papillae or amphids, musculature atrophied (Fig. 9.54).
Male unknown.
Parasites of Corvidae and Falconidae.

Robertdollfusa Chabaud & Campana, 1950
(= *Encephalonema* Parukhin & Oshmarin, 1960)

2-(1) Head with cuticular cephalic inflation, elongated papillae and amphids, musculature well developed (Fig. 9.55).
Male with single spicule and long attenuated tail.
Parasites of Macropodidae.

Durikainema Spratt & Speare, 1982

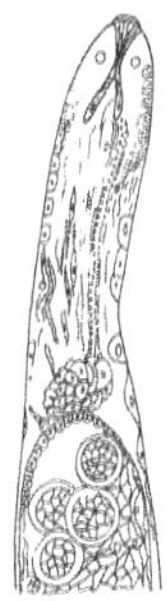

9.54

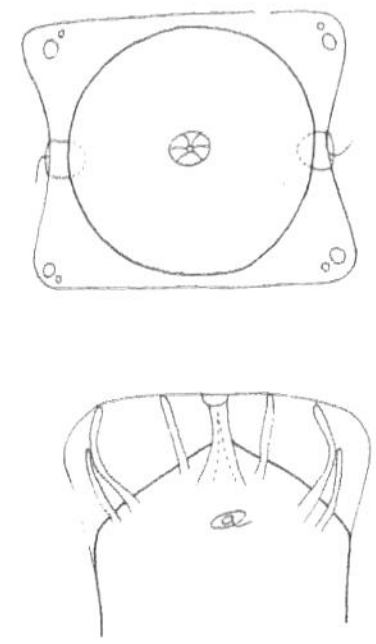

9.55

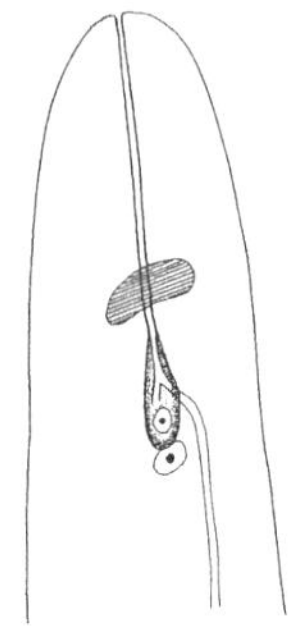

9.56

FIG. 9.54. *Robertdollfusa paradoxa* Chabaud & Campana, 1950, anterior extremity female, lateral view (after Chabaud & Campana, 1950).

FIG. 9.55. *Durikainema macropi* Spratt, 1982, cephalic extremity, *en face* and lateral views (after Spratt, 1982).

FIG. 9.56. *Riouxgolvania*, anterior extremity (diagrammatic) (after Bain & Chabaud, 1979).

Family *MUSPICEIDAE*

Key to genera

1-(2) Ventral primary oesophageal glands absent (Fig. 9.56).
"Phasmids" small.
Vulva functional.
Larva with buccal armature.
Protandrous hermaphrodites.
Larvae leave parent through body wall.
Infection of host probably by skin penetration.
Parasites of bats in Europe and Australia.
Riouxgolvania Bain & Chabaud, 1968

2-(1) Ventral primary oesophageal glands present.

3-(4) Two separate ventral primary oesophageal glands present (Fig. 9.57).
"Phasmids" highly modified to form external absorptive organs.
Vulva present.
Dioecious.
Larva unknown.
Parasites of bats in Africa.
Pennisia Bain & Chabaud, 1979

4-(3) Four separate or fused ventral primary oesophageal glands present.

5-(6) Separate ventral primary oesophageal glands present (Fig. 9.58).
"Phasmids" small.
Vulva atrophied.
Larva without buccal armature.
Protandrous hermaphrodites.
Larva expelled through anterior end of worm.
Infection probably *per os*.
Parasites of mice (*Mus*) and other rodents.
Muspicea Sambon, 1925
(= *Fascionema* Nasarova & Fedorov, 1979)

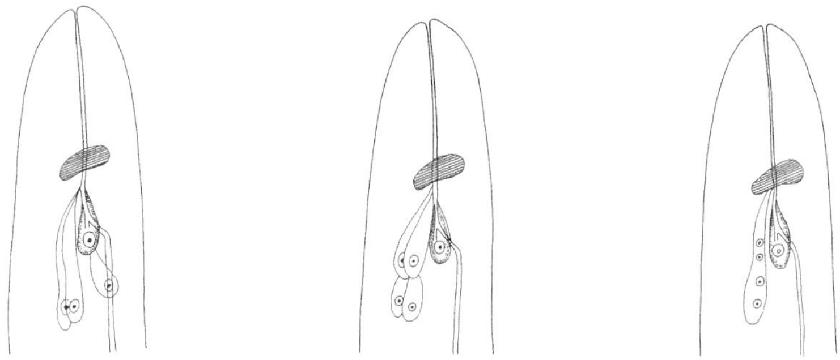

9.57 9.58 9.59

FIG. 9.57. *Pennisia*, anterior extremity (diagrammatic) (after Bain and Chabaud, 1979).

FIG. 9.58. *Muspicea*, anterior extremity (diagrammatic) (after Bain & Chabaud, 1979).

FIG. 9.59. *Lukonema*, anterior extremity (diagrammatic) (after Bain & Chabaud, 1979).

6-(5) Fused ventral primary oesophageal glands present (Fig. 9.59) forming single multinucleate body.
"Phasmids" highly modified to form internal absorptive organs.
Vulva functional.
Larvae with buccal armature.
Protandrous hermaphrodites.
Larvae leave parent through body wall.
Infection of host probably through skin.
Parasites of bats in South America.

Lukonema Chabaud & Bain, 1974

REFERENCES
(Muspiceoidea)

BAIN, O. & CHABAUD, A. G. 1968. Description de *Riouxgolvania rhinolophi* n.g. n.sp., nématode parasite de *Rhinolophe*, montrant les affinités entre Muspiceoidea et Mermithoidea. *Annls Parasit. hum. comp.*, **43**, 45–50.

——— ———. 1979. Sur les Muspiceidae (Nematoda: Dorylaimina). *Annls Parasit. hum. comp.*, **54**, 207–225.

BRUMPT, E. 1930. *Muspicea borreli* Sambon, 1925 et cancers des souris. *Annls Parasit. hum. comp.*, **8**, 309–343.

CHABAUD, A. G. & BAIN, O. 1974. Données nouvelles sur la biologie des nématodes Muspiceidae fournies par l'étude d'un parasite de chiroptères: *Lukonema lukoschusi* n.gen., n.sp. *Annls Parasit. hum. comp.*, **48**, 819–843.

——— & CAMPANA, Y. 1950. Nouveau parasite remarquable par l'atrophie de ses organes: *Robertdollfusa paradoxa* (Nematoda, *incertae sedis*). *Annls Parasit. hum. comp.*, **25**, 325–334.

——— & LE VAN HOA. 1962. Nématodes parasites de reptiles, Aphasmidiens. *Explor. Parc Nat. Upemba. Mission G. F. de Witte.* Fasc. (65), 59–64.

NASAROVA, N. S. & FEDOROV, K. P. 1979. *In:* Ryzhikov, K. M. [Key to the helminths of rodents of the USSR. Nematodes and acanthocephalans.] Moscow: Izdatel'stvo "Nauka". [In Russian.]

SAMBON, L. W. 1925. Researches on the epidemiology of cancer made in Iceland and in Italy (July–October 1924). *J. trop. Med. Hyg.*, **28**, 39–71.

SPRATT, D. M. & SPEARE, R. 1982. *Durikainema macropodi* gen. et sp. nov. (Muspiceoidea: Robertdollfusidae) a remarkable nematode parasite of kangaroos (Macropodidae). *Annls Parasit. hum. comp.* (in press).

3 RHADITIDA

RHABDITOIDEA

by ROY C. ANDERSON[1] and ODILE BAIN[2]

RHABDITOIDEA

The Rhabditoidea includes species which are free-living saprophagous forms rarely associated with vertebrates and species which are true parasites. The Rhabditidae and the Cephalobidae are mainly of the former group. The Cylindrocorporidae includes both free-living and parasitic species. The Angiostomatidae includes species parasitic in amphibians or terrestrial gastropods. The Rhabdiasidae and Strongyloididae are exclusively parasitic in vertebrates and both families are heterogonic.

Rhabditidae

The family is divided into the Rhabditinae and the Diploscapterinae.

Diploscapterinae

The subfamily is mainly associated with plant material but *Diploscapter coronata* (Cobb, 1893) Cobb, 1913 has been reported in the stomach of man when there was an absence or a deficiency of HCl (Yokagawa, 1936; Chandler, 1938).

Rhabditinae

A number of species of this subfamily have been associated with vertebrates, including man. In most instances the precise relationship is obscure but often it is obviously a question of the contamination of faeces or urine passed from the host. Some of these reports have resulted in the description of species subsequently shown to be synonyms of well-known free-living forms, e.g., *Rhabditis tenuicaudata, R. genitalis, R. hominis, R. usei, R. schachtiella, R. donbass, Rhabditella multipara*. Some free-living rhabditids readily invade decomposing carcasses. There are rare reports of *Rhabditis terricola* and *R. lucianii* in salamanders (Chitwood, 1933).

Pelodera strongyloides is commonly associated with lesions in soiled skin of livestock and companion animals (Chitwood, 1932, 1933; Brizard & Euzeby, 1949; Levine *et al.*, 1950; Rhode *et al.*, 1953; Grano & Olmsted, 1953; Hayes, 1953, Clark & Belding, 1956; Slusarski *et al.*, 1956; Jones, 1963; Willers, 1970; Bergeland *et al.*, 1976). The dauerlarva of this species is also commonly found in the orbits of rodents in the temperate world (Rausch, 1952; Stammer, 1956; Dollfus *et al.*,

[1] Department of Zoology, University of Guelph, Guelph, Ontario, Canada.
[2] Museum National d'Histoire Naturelle, Paris, France.

1961; Osche, 1963; Poinar, 1965; Kinsella, 1967; Cross & Santana, 1974; Cliff *et al.*, 1978; Cliff & Anderson, 1980).

Some rhabditines have been reported in laboratory saline used for diluting biological specimens.

It is important that these free-living genera be recognized when they appear in faeces, urine or elsewhere, to avoid confusion with the developmental stages of such heterogonic genera as *Strongyloides* and *Rhabdias*.

The key includes all the genera and subgenera which have been associated with vertebrates. The species usually involved are given below with their synonyms. The classification and synonyms follow Osche (1952), Dougherty (1955), Baker (1962) and Goodey (1963).

(1) *Rhabditis (Rhabditis) axei* (Cobbold, 1884) Dougherty, 1955.
= *Pelodera axei* Cobbold, 1884; = *Rhabditis usei* Watanabe, 1927; = *Rhabditis gracilis* Shingareva, Demidova & Kudriavstev, 1928; = *Rhabditis tenuicaudata* Menzel & Stefański in Stefański, 1917; = *Rhabditella multipara* Li, 1951.
(2) *Rhabditis (Rhabditis) terricola* Dujardin, 1845 (Osche, 1952).
(3) *Rhabditis (Rhabditis) lucianii* Maupas, 1919 (Dougherty, 1955).
(4) *Rhabditis (Pellioditis) pellio* (Schneider, 1866) Bütschli, 1873.
= *Pelodera pellio* Schneider, 1866; = *Leptodera pellio* (Schneider, 1866) Ward, 1903; = *Rhabditis (Choriorhabditis) pellio* (Schneider, 1866) Bütschli, 1873 (Osche, 1952); = *Rhabditis genitalis* Scheiber, 1880.
(5) *Rhabditoides inermis* (Schneider, 1866) Dougherty, 1955.
= *Leptodera inermis* Schneider, 1866; = *Rhabditis (Telorhabditis) inermis* (Schneider, 1866) Oerley, 1880 (Osche, 1952); = *Rhabditis hominis* Kobayashi, 1920; = *Rhabditis faecalis* Watanabe, 1922; = *Rhabditis schachtiella* Skrjabin & Shults, 1926.
(6) *Pelodera (Pelodera) strongyloides* (Schneider, 1860) Schneider, 1866.
= *Pelodytes strongyloides* Schneider, 1860; = *Rhabditis strongyloides* (Schneider, 1860) Oerley, 1880; = *Rhabditis (Rhabditis) strongyloides* (Schneider, 1860) Oerley, 1880 (Osche, 1952); = *Leptodera strongyloides* (Schneider, 1860) Schneider, 1866.
(7) *Pelodera (Pelodera) teres* Schneider, 1866 (Dougherty, 1955).
= *Rhabditis teres* (Schneider, 1866) Bütschli, 1873 (Osche, 1952); = *Rhabditis (Rhabditis) teres* (Schneider, 1866) Bütschli, 1873; = *R. donbass* Skrjabin, Shults & Serbinov, 1926.
(8) *Caenorhabditis clavopapillata* (Kreis & Faust, 1933) Dougherty, 1955.
= *Rhabditis clavopapillata* Kreis & Faust, 1933; = *Rhabditis (Caenorhabditis) clavopapillata* Kreis & Faust, 1933 (Osche, 1952).

One species, *Leptodera niellyi* Blanchard, 1885, found in pustules in man, is a *species inquirenda* since only larvae were found; it may, however, belong to one of the species listed above.

Cephalobidae

Most cephalobids are free-living saprophagous nematodes associated with decaying plant material and only rarely with vertebrates.

Micronema deletrix Anderson & Bemrick, 1965 was originally found in a nasal tumour in a horse. It has been reported several times since in various parts of the body of the horse, including the central nervous system, and fatal cases of cerebrospinal nematodiasis involving this worm have been reported in man (Johnson & Johnson, 1966; Stone *et al.*, 1970; Ferris *et al.*, 1972; Rubin & Woodward, 1974; Hoogstraten & Young, 1975; Powers & Benz, 1977; Shadduck *et al.*, 1979). The worm is highly pathogenic especially if the central nervous system is invaded.

The vinegar eel, *Turbatrix aceti* (Müller, 1783) Peters, 1927, is a rare contaminant of the human genital tract. Infections are related apparently to the practice of using contaminated vinegar as part of a douche (Stiles & Frankland, 1902).

Tricephalobus gingivalis (Stefański, 1954) Dougherty, 1955 (= *Rhabditis gingivalis*) found in the gums of a horse in Poland is of doubtful status. Males are unknown and Anderson & Bemrick (1965) suggest it may be a species of *Micronema*. There is considerable doubt if it belongs in *Tricephalobus* which was regarded as a synonym of *Trilabiatus* by Goodey (1963). *T. gingivalis* is regarded herein as a *species inquirenda* and neither *Tricephalobus* nor *Trilabiatus* is included in the following key.

Cylindrocorporidae

This family includes species which are free-living and saprophagous and a few species found in the intestine of vertebrates. The latter fall into the genus *Longibucca* Chitwood, 1933 and include the generitype *L. vivipara* Chitwood, 1933 from the stomach of the snake *Pseudoboa cloelia*, *L. eptesica* Elsea, 1953 from the stomach of the bat *Myotis lucifugus* and *L. lasiura* McIntosh & Chitwood, 1933 from the small intestine of the bat *Lasiurus borealis*.

Angiostomatidae

The family contains the genus *Angiostoma* which includes *A. limacis* Dujardin, the generitype, from terrestrial gastropods, and *A. plethodontis* Chitwood, 1933 from the intestine of the salamander, *Plethodon cinereus*. Thus, this genus, like *Cosmocercoides* Wilkie, 1930, is found in both amphibians and gastropods (Anderson, 1960; Baker, 1978).

Rhabdiasidae

Members of the family are heterogonic parasites found in the lungs of amphibians and reptiles. The parasitic form is hermaphroditic and markedly different in form to the free-living generation (Baker, 1978, 1979). According to Baker (1980) the family consists of *Pneumonema, Acanthorhabdias, Entomelas* and *Rhabdias*.

Strongyloididae

The Strongyloididae consists of heterogonic forms generally with an hermaphroditic parasitic stage (in *Strongyloides* and *Leipernema*) markedly different from the free-living stages (Little, 1966a, b). In *Parastrongyloides*, however, the parasitic form is dioecious. Species of *Strongyloides* are extremely common and widespread parasites of vertebrates.

RHABDITOIDEA

Key to families

1-(4) Oesophagus with prominent terminal valved bulb (Fig. 9.1).

2-(3) Buccal cavity cylindrical with clearly defined unjointed rhabdions (Fig. 9.12).
Male with caudal alae forming bursa.
Mainly soil-dwelling saprophagous forms. Rarely accidental parasites of vertebrates.
Rhabditidae

3-(2) Buccal cavity with jointed rhabdions (Fig. 9.2).
Male without caudal alae.
Mainly soil-dwelling saprophagous forms. Rarely accidental parasites of vertebrates.
Cephalobidae

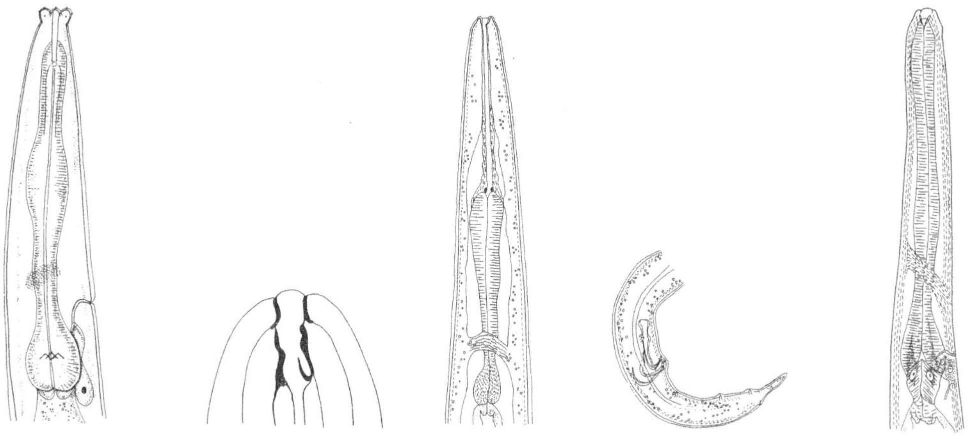

9.1 9.2 9.3 9.4 9.5

FIG. 9.1. *Pelodera strongyloides* (Schneider, 1860) anterior end female, lateral view (after Cliff & Anderson, 1980).

FIG. 9.2. *Turbatrix aceti* (Müller, 1783), cephalic extremity, lateral view (after Goodey, 1963).

FIG. 9.3. *Longibucca lasiura* McIntosh & Chitwood, 1934, anterior end (after McIntosh & Chitwood, 1934).

FIG. 9.4. *Longibucca lasiura* McIntosh & Chitwood, 1934, caudal extremity male, lateral view (after McIntosh & Chitwood, 1934).

FIG. 9.5. *Angiostoma plethodontis* Chitwood, 1933, anterior end, lateral view (after Chitwood, 1933).

4-(1) Oesophagus in parasitic form without terminal valved bulb.

5-(6) Buccal cavity long, with unjointed rhabdions (Fig. 9.3).
Oesophagus clavate.
Male without caudal alae (Fig. 9.4).
Vulva posterior in position.
Mainly free-living saprophagous forms with some species parasitic in gut of amphibians, reptiles and mammals.

Cylindrocorporidae

6-(5) Buccal cavity in parasitic form much reduced or capsuliform with thick walls.

7-(8) Parasitic forms dioecious.
Oesophagus with terminal swelling (pseudobulb) (Fig. 9.5).
Buccal cavity capsuliform (Fig. 9.6).
Male with leptoderan bursa, monorchic (Fig. 9.7).
Vulva near middle of body.
Didelphic, uteri opposed.
Parasites of intestine of salamanders (*Plethodon*) and terrestrial gastropods.

Angiostomatidae

8-(7) Parasitic form hermaphroditic (except *Parastrongyloides*).
Oesophagus long and slender or short and clavate, without terminal swelling (pseudobulb).
Heterogonic. Free-living generation rhabditid-like and saprophagous.

9-(10) Parasitic form stout, generally with markedly inflated cuticle (Fig. 9.8).
Oesophagus short and clavate (Fig. 9.9).
Vulva near middle of body.
Parasites of lungs of amphibian and reptiles.

Rhabdiasidae

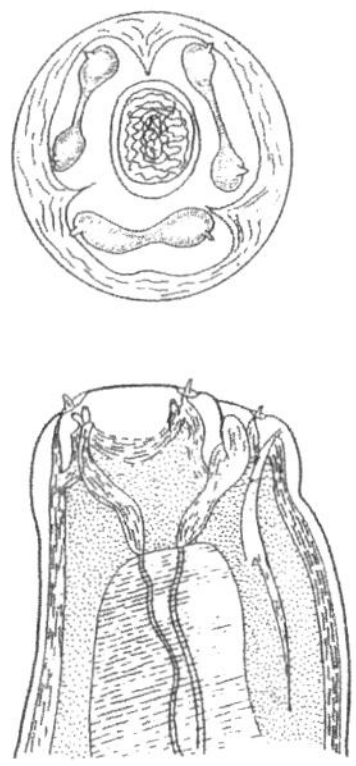

9.6

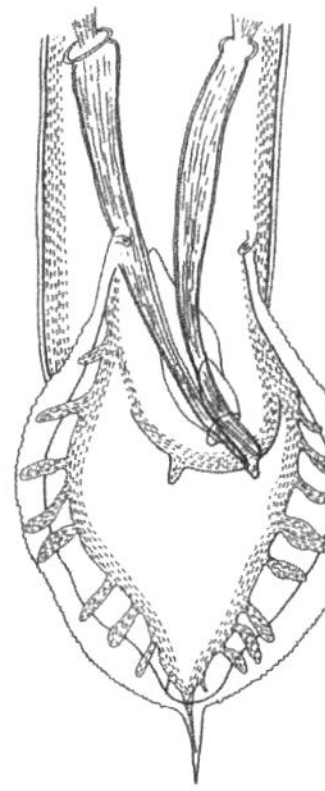

9.7

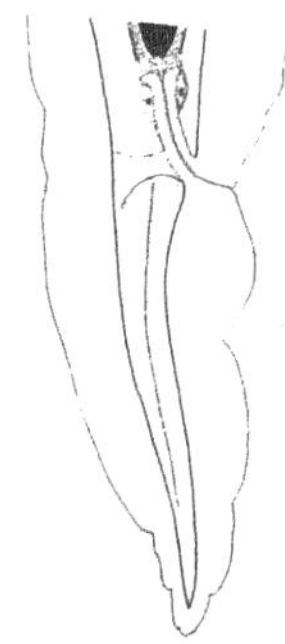

9.8

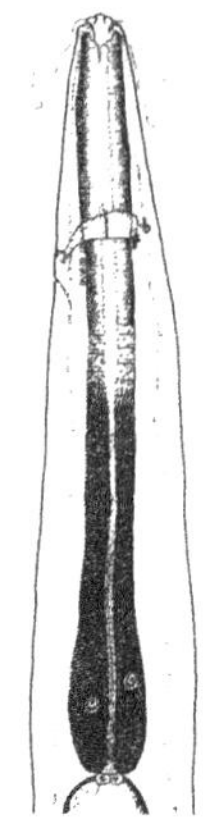

9.9

FIG. 9.6. *Angiostoma plethodontis* Chitwood, 1933, cephalic extremity, *en face* and lateral views (after Chitwood, 1933).

FIG. 9.7. *Angiostoma plethodontis* Chitwood, 1933, caudal end male, ventral view (after Chitwood, 1933).

FIG. 9.8. *Rhabdias americanus* Baker, 1978, posterior end parasitic form, lateral view (after Baker, 1978).

FIG. 9.9. *Rhabdias americanus* Baker, 1978, anterior end parasitic form, lateral view (after Baker, 1978).

10-(9) Parasitic form long and slender with uninflated cuticle (Fig. 9.10).
Oesophagus long and slender.
Vulva in posterior quarter of body.
Parasites of gut of vertebrates other than fish.

Strongyloididae

Family *RHABDITIDAE* Oerley, 1880

Key to subfamilies

1-(2) Three to six lips present.
Glottoid apparatus present, usually with teeth or tubercles.

Rhabditinae

2-(1) Dorsal and ventral lips replaced by paired hooks (Fig. 9.11).
Glottoid apparatus absent.

Diploscapterinae

Subfamily **Diploscapterinae** Chitwood & Chitwood, 1937

One genus.
Rare inhabitant of stomach devoid of or deficient in HCl.

Diploscapter Cobb, 1913

Subfamily **Rhabditinae** (Oerley, 1880) Micoletzky, 1922

Key to genera

1-(4) Metarhabdions each with 3 or 5 tubercles (Fig. 9.12).
Lips closed.
Oesophageal collar present.
Spicules separate.

Rhabditis Dujardin, 1845
(= *Pelodytes* Schneider, 1859 not Bonaparte, 1841;
= *Cuticularia* van der Linde, 1938)

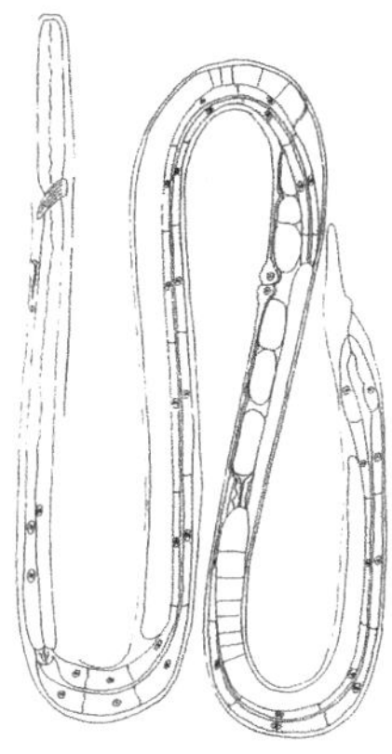

9.10

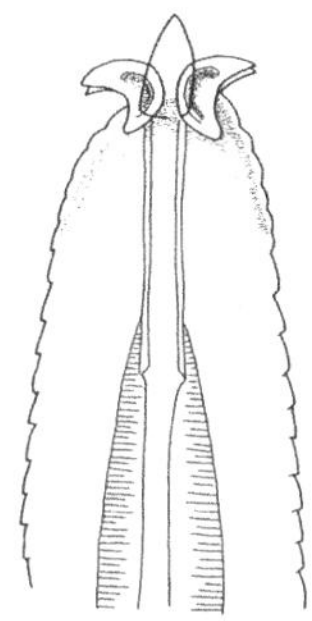

9.11

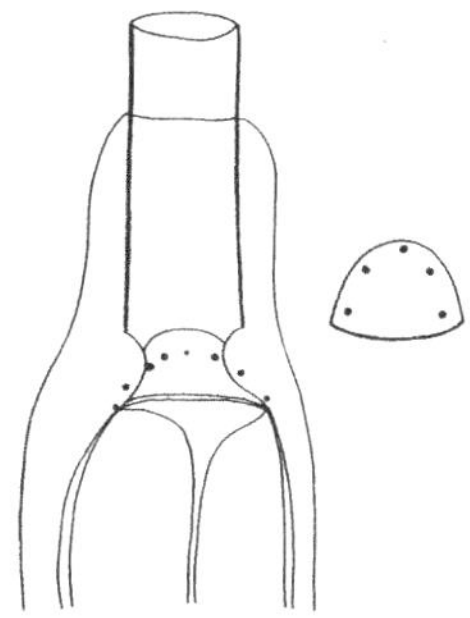

9.12

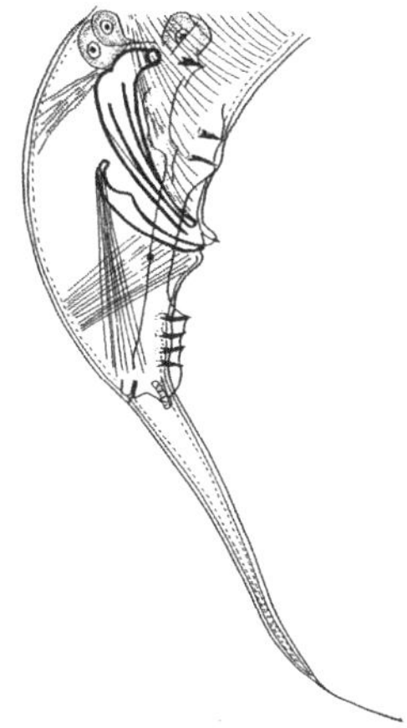

9.13

FIG. 9.10. *Strongyloides stercoralis* (Bavay, 1876), parasitic form, lateral view (after Little, 1966).

FIG. 9.11. *Diploscapter coronata* Cobb, 1893, cephalic extremity (after Chandler, 1938).

FIG. 9.12. *Rhabditis maxima* Völk, buccal region (after Osche, 1952).

FIG. 9.13. *Rhabditis (Rhabditella) axei* (Cobbold, 1884), caudal extremity male, lateral view (after Steiner, 1943).

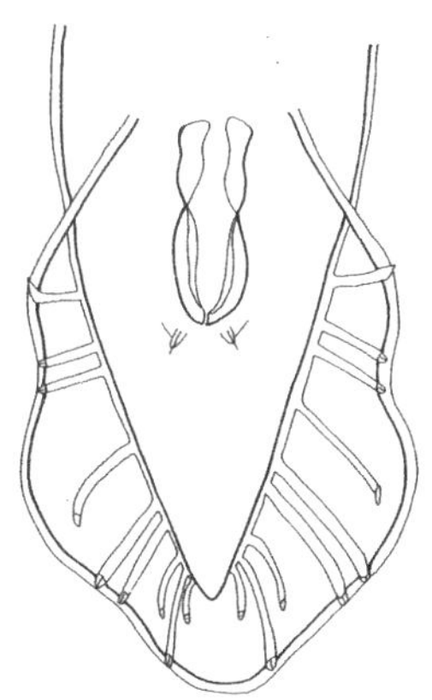

9.14

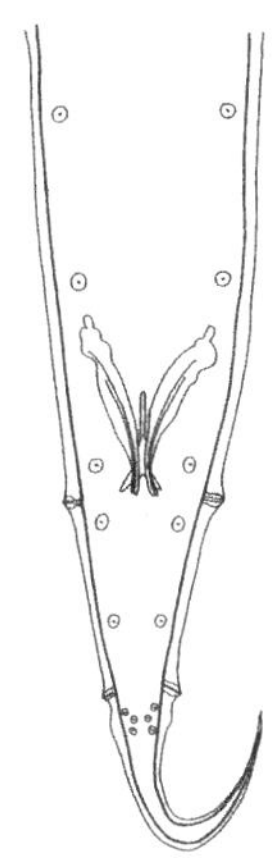

9.15

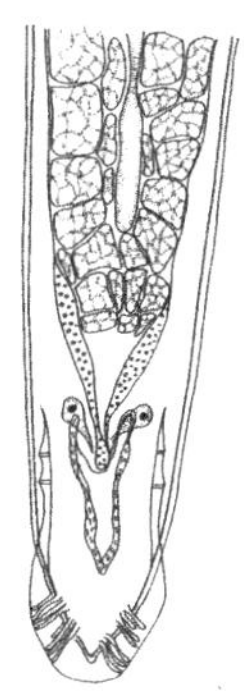

9.16

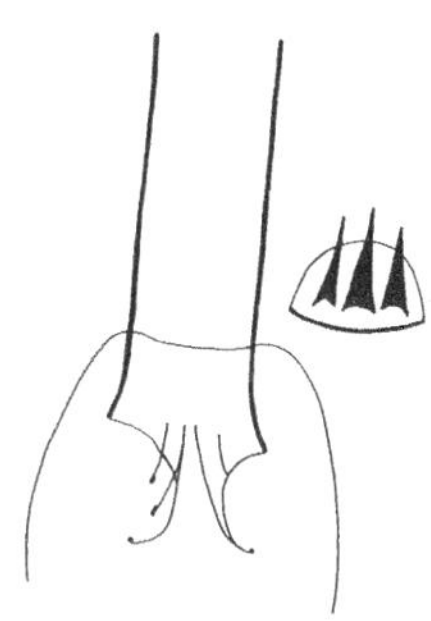

9.17

FIG. 9.14. *Rhabditis (Pellioditis) friderici* Hirschmann 1952, caudal extremity male, ventral view (after Hirschmann, 1952).

FIG. 9.15. *Rhabditoides longispina* (Reiter, 1928), caudal extremity male, ventral view (after Goodey, 1929).

FIG. 9.16. *Pelodera strongyloides* (Schneider, 1860), caudal extremity male, ventral view (after Cliff & Anderson, 1980).

FIG. 9.17. *Pelodera strongyloides* (Schneider, 1860), buccal region (Osche, 1952).

2-(3) Male tail leptoderan, bursa reduced, tail very long (Fig. 9.13).
Female tail long and conical.
Occasional contaminant of faeces of dogs and monkeys, less commonly of urine of man.
subgenus *Rhabditella* (Cobb, 1929)

3-(2) Male tail peloderan, bursa large (Fig. 9.14).
Female tail conical or dome-shaped with spine.
Rare contaminant of human vagina.
subgenus *Pellioditis* (Dougherty, 1953)

4-(1) Metarhabdions each with 2 to 3 teeth or spines.

5-(6) Male tail leptoderan, bursa much reduced (Fig. 9.15).
Female tail long and conical to dagger-shaped.
Spicules separate, gubernaculum forked.
Lips closed.
Oesophageal collar present.
Occasional contaminant of faeces of man and animals (dogs, rabbits, rats).
Rhabditoides T. Goodey, 1926
(= *Telorhabditis* Osche, 1952)

6-(5) Male tail peloderan (Fig. 9.16).

7-(8) Metarhabdions each with 3 teeth (Fig. 9.17).
Large plump worms.
Lips large and distinct, open.
Oesophageal collar present.
Female tail conical or dome-shaped, with short spine.
Spicules usually fused distally.
Caudal papillae not radially arranged.
Rare inhabitant of soiled skin of animals (e.g. dogs, cattle and horses), causing dermatitis.
Rather common inhabitant (in dauerlarval stage) of orbits of wild ground-dwelling rodents.
Pelodera Schneider, 1866
(= *Cruznema* Artigas, 1927 = *Epimedides* Gurierrez, 1949)

8-(7) Metarhabdions each with 2 teeth (Fig. 9.18).
Lips closed, not distinct.
Oesophageal collar present.
Spicules separate (Fig. 9.19).
Tail conical, rather long.
Three pairs of preanal papillae usually present.
Occasional inhabitants of soiled perianal and abdominal hair of dogs and monkeys.
Caenorhabditis (Osche, 1952) Dougherty, 1953

Family *CEPHALOBIDAE* (Filipjev, 1934) Chitwood & Chitwood, 1934

Key to genera

1-(2) Oesophagus with median bulb (Fig. 9.20).
Vulva in caudal region of body.
Monodelphic, prodelphic, uterus with one egg (Fig. 9.21).
Female tail elongate and ending in blunt tip.
Male unknown.
Rare pathogen of tissues, especially oral and nasal, and the central nervous system of animals and man.
Micronema Körner, 1954

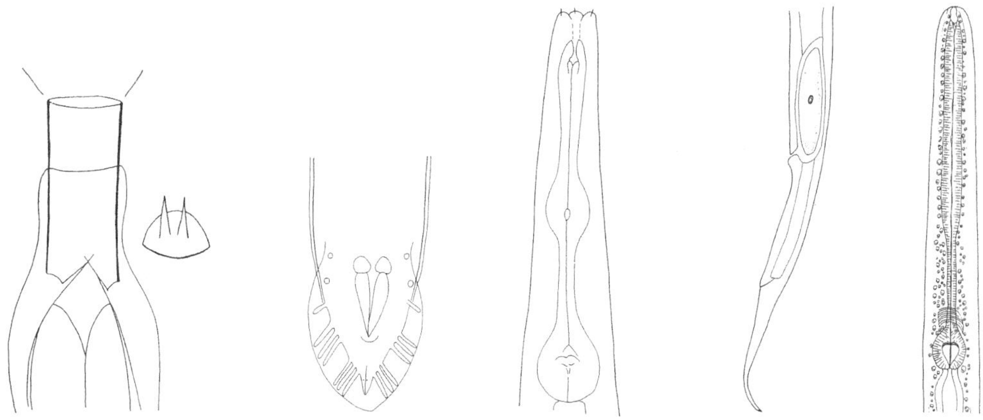

9.18 9.19 9.20 9.21 9.22

FIG. 9.18. *Caenorhabditis dolichura* (Schneider, 1866), buccal region (after Osche, 1952).
FIG. 9.19. *Caenorhabditis dolichura* (Schneider, 1866), caudal extremity male, ventral view (after Völk, 1950).
FIG. 9.20. *Micronema minutum* Körner, 1954, anterior end female (after Körner, 1954).
FIG. 9.21. *Micronema minutum* Körner, 1954, caudal extremity female, lateral view (after Körner, 1954).
FIG. 9.22. *Turbatrix aceti* (Müller, 1783) anterior end (after de Mann, 1910).

2-(1) Oesophagus without median bulb (Fig. 9.22).
Vulva near middle of body.
Monodelphic, prodelphic, uterus with four eggs.
Post-vulvar uterine sac present.
Tail attenuated to delicate point in both sexes (Fig. 9.23).
Spicules short, gubernaculum present.
Rare contaminant of urogenital tract of man.

Turbatrix Peters, 1927

Family *CYLINDRICORPORIDAE* Goodey, 1939

One genus.
Parasites of gut of amphibians, reptiles and bats.

Longibucca Chitwood, 1953

Family *ANGIOSTOMATIDAE* Blanchard, 1895

One genus.
Parasites of intestine of salamanders (*Plethodon*).

Angiostoma Dujardin, 1845
(= *Limaconema* Mengert, 1953)

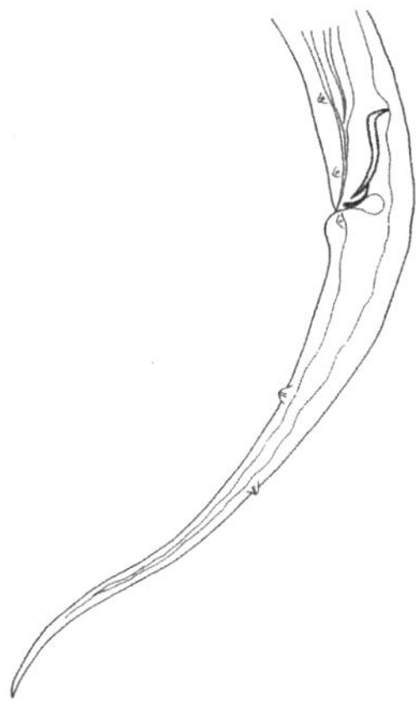

9.23

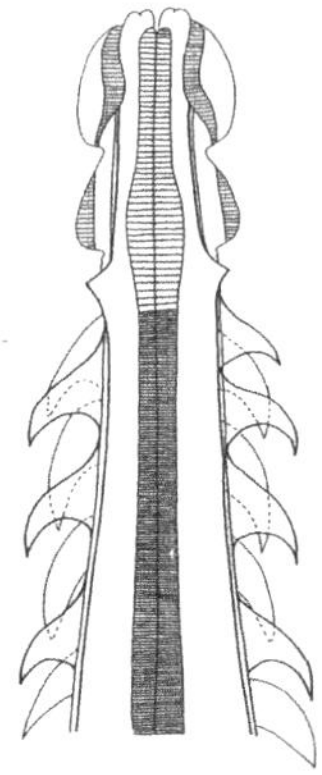

9.24

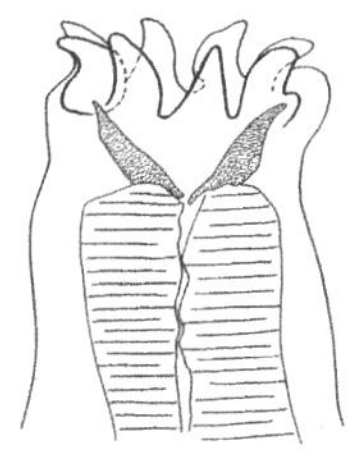

9.25

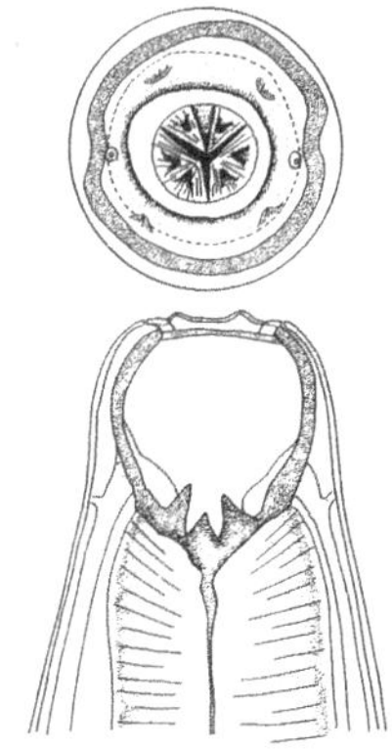

9.26

FIG. 9.23. *Turbatrix aceti* (Müller, 1783) caudal extremity male, lateral view (after de Mann, 1910).

FIG. 9.24. *Pneumonema tiliquae* Johnston, 1916, anterior end, ventral view (after Yorke & Maplestone, 1926).

FIG. 9.25. *Acanthorhabdias acanthorhabdias* Pereira, 1927, cephalic extremity (after Pereira, 1927).

FIG. 9.26. *Entomelas entomelas* (Dujardin, 1845) cephalic extremity of sub-gravid adult, *en face* and lateral view (after Baker, 1980).

Family *RHABDIASIDAE* Railliet, 1916

Key to genera

1-(2) Anterior half of body with two rows of large spines diminishing in size posteriorly (Fig. 9.24).
Cuticle behind cephalic extremity expanded to form cervical alae.
Parasites of lungs of reptiles (*Tiliqua*) in Australia.
Pneumonema Johnston, 1916

2-(1) Cuticular spines on body absent.
Cuticle of body smooth or with irregular folds.

3-(4) Lips modified to form prominent spine-like projections (Fig. 9.25).
Buccal cavity broad with thick walls.
Parasites of lungs of reptiles (*Liophis*) in South America.
Acanthorhabdias Pereira, 1927

4-(3) Lips not modified to form spine-like projections.

5-(6) Buccal cavity subspherical with thick dense walls of uniform thickness (Fig. 9.26).
Cephalic extremity usually broader than body.
Onchia usually present on posterior wall of buccal cavity.
Parasites of lungs, body and pericardial cavities of reptiles.
Entomelas Travassos, 1930
(= *Angiostoma* Dujardin, 1845 in part; = *Hexadontophorus* Kreis, 1939; = *Kurilonema* Szczerbak & Sharpilo, 1969; = *Paraentomelas* Sharpilo, 1976)

6-(5) Buccal cavity reduced with thin wall devoid of onchia (Fig. 9.27).
Outer layers of cuticle usually inflated.
Cephalic extremity without lips or with six lips or pairs of pseudolabia.
Onchia absent from wall of buccal cavity.
Parasites of lungs and body cavity of amphibians and reptiles.

Rhabdias Stiles & Hassall, 1905
(= *Ophiorhabdias* Yamaguti, 1943; = *Shorttia* Singh & Ratnamala, 1977)

Family *STRONGYLOIDIDAE* Chitwood & McIntosh, 1934

1-(2) Dioecious parasites.
Male with short, arcuate spicules and gubernaculum (Fig. 9.28).
Male tail with one pair of postanal papillae and one preanal median papilla.
Heterogonic.
Parasites of intestines of insectivores and marsupials.

Parastrongyloides Morgan, 1928

2-(1) Parasitic form hermaphroditic.

3-(4) Buccal cavity well developed, with three pairs of teeth projecting from end of oesophagus (Fig. 9.29).
Parasites of edentates (*Manis*).

Leipernema Singh, 1976

4-(3) Buccal cavity reduced, without teeth (Fig. 9.30).
Parasites of amphibians, reptiles, birds and mammals.

Strongyloides Grassi, 1879

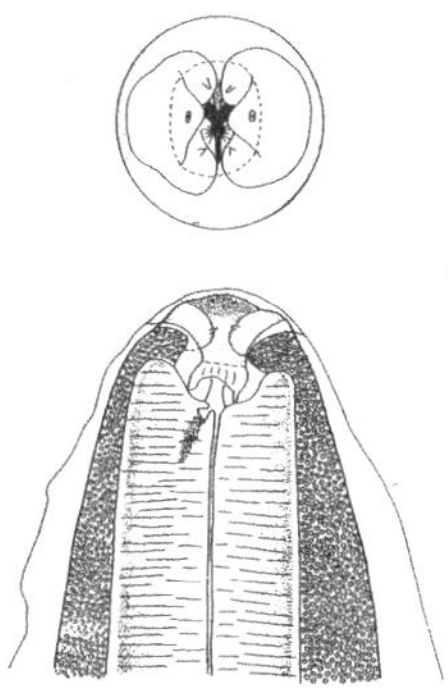

9.27

9.28

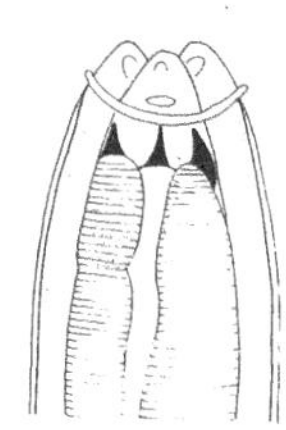

9.29

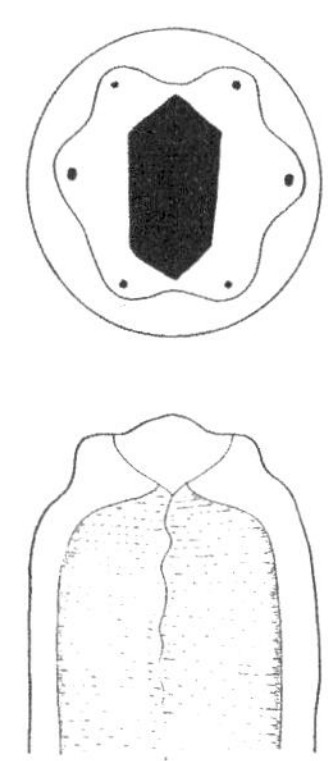

9.30

FIG. 9.27. *Rhabdias americanus* Baker, 1978, cephalic extremity of parasitic form, *en face* and lateral views (after Baker, 1978).

FIG. 9.28. *Parastrongyloides trichosuri* Mackerras, 1959, caudal extremity male, lateral view (after Mackerras, 1959).

FIG. 9.29. *Leipernema leiperi* Singh, 1976, cephalic extremity parasitic form, lateral view (after Singh, 1976).

FIG. 9.30. *Strongyloides stercoralis* (Bavay, 1876) cephalic extremity of parasitic form, *en face* and lateral view (after Little, 1966).

REFERENCES

(Rhabditoidea)

ALSTAD, A. D., BERG, I. E. & SAMUEL, G. 1979. Disseminated *Micronema deletrix* infection in the horse. *J. Am. Vet. med. Ass.*, **174,** 264–266.

ANDERSON, R. C. 1960. On the development and transmission of *Cosmocercoides dukae* of terrestrial molluscs in Ontario. *Can. J. Zool.*, **38,** 801–825.

ANDERSON, R. V. & BEMRICK, W. J. 1965. *Micronema deletrix* n.sp., a saprophagous nematode inhabiting a nasal tumour of a horse. *Proc. helminth. Soc. Wash.*, **32,** 74–75.

ARTIGAS, P. & PACHECO, G. 1933. *Strongyloides myopotami* n.sp. (Nematoda) *C. r. Séanc. Soc. Biol.*, **112,** 406–407.

ARTIGAS, P. T., ARAUJO, P. & GRAEIRO, A. 1973. Redescrição de *Acanthorhabdias acanthorhabdias* Pereira, 1927 (Nematoda: Rhabditoidea). *Arq. Inst. Biol. São Paulo,* **40,** 33–37.

BAKER, A. D. 1962. *Checklists of the nematode superfamilies Dorylaimoidea, Rhabditoidea, Tylenchoidea, and Aphelenchoidea.* Leiden: E. J. Brill.

BAKER, M. R. 1978. Transmission of *Cosmocercoides dukae* (Nematoda: Cosmocercoidea) to amphibians. *J. Parasit.*, **64,** 765–766.

———. 1978. Morphology and taxonomy of *Rhabdias* spp. (Nematoda: Rhabdiasidae) from reptiles and amphibians of southern Ontario. *Can. J. Zool.*, **56,** 2127–2141.

———. 1979. The free-living and parasitic development of *Rhabdias* spp. (Nematoda: Rhabdiasidae) in amphibians. *Can. J. Zool.*, **57,** 161–178.

———. 1979. Seasonal population changes in *Rhabdias ranae* Walton, 1929 (Nematoda: Rhabdiasidae) in *Rana sylvatica* of Ontario. *Can. J. Zool.*, **57,** 179–183.

———. 1980. Revision of *Entomelas* Travassos, 1930 (Nematoda: Rhabdiasidae) with a review of genera in the family. *Syst. Parasitol.*, **1,** 83–90.

BALLANTYNE, R. J. & PEARSON, J. C. 1963. The taxonomic position of the nematode *Pneumonema tiliquae. Aust. J. Sci.*, **25,** 498.

BARŮS, V. 1968. Parasitic nematodes of birds of the family Icteridae (Passeriformes) in Cuba. *Folia Parasit., Praha,* **15,** 131–146.

BERGELAND, M. E., TODD, K. S. & OHLENDORF, L. R. 1976. Dermatitis in sheep caused by *Pelodera strongyloides. Proc. helminth. Soc. Wash.*, **43,** 230–231.

BRENES, R. R. & BRAVO HOLLIS, M. 1959. Hemintos de la República de Costa Rica VIII. Nematoda 2. Algunos nematodes de *Bufo marinus marinus* (L), y algunas consideraciones sobre los generos *Oxysomatium* y *Aplectana. Revta Biol. trop.*, **7,** 35–55.

BRIZARD, A. & EUZÉBY, J. 1949. Un cas de dermatose vermineuse du chien. *Revue Méd. vét.*, **100,** 82–86.

BUCHWALDER, R. 1960. Zum koprologischen Nachweis einer *Strongyloides-infektion. Mh. VetMed.*, **15,** 149–150.

CHABAUD, A. G., BRYGOO, E. R. & PETTER, A. J. 1961. Description et caractères biologiques de deux nouveaux *Rhabdias* malgaches. *Annls Parasit. hum. comp.*, **36,** 752–763.

CHANDLER, A. C. 1938. *Diploscapter coronata* as a facultative parasite of man, with a general review of vertebrate parasitism by rhabditoid worms. *J. Parasit.*, **30,** 44–45.

CHITWOOD, B. G. 1932. The association of *Rhabditis strongyloides* with dermatitis in dogs. *N. Am. Vet.*, **13,** 35–40.

———. 1933. On some nematodes of the superfamily Rhabditoidea and their status as parasites of reptiles and amphibians. *J. Wash. Acad. Sci.*, **23,** 508–520.

CHU, T. 1936. A review of the status of the reptilian nematodes of the genus *Rhabdias* with a redescription of *Rhabdias fuscovenosa* var. *catanensis* (Rizzo, 1902) new rank. *J. Parasit.*, **22,** 130–139.

CLARK, D. T. & BELDING, S. A. 1956. Rhabditic dermatis in a dog in Michigan. *Mich. St. Univ. Vet.*, **17,** 48–53.

CLIFF, G. M., ANDERSON, R. C. & MALLORY, F. F. 1978. Dauerlarvae of *Pelodera strongyloides* (Schneider, 1866) (Nematoda: Rhabditidae) in the conjunctival sac of lemmings. *Can. J. Zool.*, **56,** 2117–2121.

——— & ANDERSON, R. C. 1980. Development of *Pelodera strongyloides* (Schneider, 1860) Schneider, 1866 (Nematoda: Rhabditidae) in culture. *J. Helminth.*, **54,** 135–146.

CROSS, J. H. & SANTANA, F. J. 1974. *Pelodera strongyloides* (Schneider) in the orbit of Taiwan rodents. *Chin. J. Microbiol.*, **7,** 137–138.

CRUSZ, H. & SANMUGASUNDERAM, V. 1974. Parasites of the relict fauna of Ceylon III. Nematodes from a rhacophorid frog and reptiles of the hill country. *Annls Parasit. hum. comp.*, **48,** 767–795.

DOLLFUS, R. M., DESPORTES, C., CHABAUD, A. G. & CAMPANA-ROUGET, Y. 1961. Station expérimentale de parasitologie de Richelieu (Indre-et-Loire). Contributions à la faune parasitaire régionale. Chapitre II. Liste des parasites par ordre systématique. E. Nématodes. *Annls Parasit. hum. comp.*, **36,** 303–311.

DOUGHERTY, E. C. 1955. The genera and species of the subfamily Rhabditinae Micoletzky, 1922 (Nematoda): A nomenclatural analysis–including an addendum on the composition of the family Rhabditidae Örley, 1880. *J. Helminth.*, **29,** 105–152.

ELSEA, J. R. 1953. Observations on the morphology and biology of *Longibucca eptesica* n.sp. (Nematoda: Cylindrocorporidae) parasites in the bat. *Proc. helm. Soc. Wash.*, **20,** 65–76.

FERNANDES, B. M. M. & DE SOUZA, S. V. 1974. Redescrição de *Acanthorhabdias acanthorhabdias* Pereira, 1927. *Mems Inst. Oswaldo Cruz,* **72,** 291–292.

FERRIS, D. H., LEVINE, N. D. & BEAMER, P. D. 1972. *Micronema deletrix* in equine brain. *Am. J. vet. Res.*, **33,** 33–38.

FLEURY, G. C. 1973. Discovery of adult specimens of *Strongyloides stercoralis* and *Rhabditis* sp. in human faeces. *Revta Inst. Adolfo Lutz,* **33,** 35–59.

GOLDSMID, J. M. 1967. *Rhabditis (Rhabditella) axei* in the urine of an African in Rhodesia. *J. Helminth.*, **41,** 305–308.

GOODEY, J. B. 1963. *Soil and Freshwater Nematodes.* Methuen & Co. Ltd. London. John Wiley & Sons Inc. N.Y.

GRANO, E. & OLMSTED, R. C. 1953. Dermatitis caused by *Rhabditis* larvae. *N. Am. Vet.*, **34,** 873–874.

GREEVE, J. H. 1969. *Strongyloides elephantis* sp.n. from an Indian elephant, *Elephas indicus. J. Parasit.*, **55,** 498–499.

GUPTA, S. P. 1960. Nematode parasites of vertebrates of East Pakistan. VI. *Amplicaecum cacopi, Thelandros* sp. *Rhabdias ranae* and *Oswaldocruzia melanosticti* sp. nov. from frogs. *Can. J. Zool.*, **38,** 745–750.

HARA, M., NAKAZATO, H., ARAKI, T. & IWATA, S. 1974. [A case of urinary tract infection by a nematode larva (*Rhabditis* sp.?)] *Jap. J. Parasit.*, **23** (2, Suppl.): 48 [in Japanese].

HARTWICH, G. 1972. Über *Rhabdias bufonis* (Schrank, 1788) und die Abtrennung von *Rhabdias dossei* nov. spec. (Nematoda, Rhabdiasidae). *Mitt. zool. Mus. Berl.*, **48,** 401–414.

———. 1975. Schlauchwürmer, Nemathelminthes; Rund- oder Fadenwürmer, Nematoda; Parasitische Rundwürmer von Wirbeltieren. I. Rhabditida und Ascaridida. Die Tierwelt Deutschlands, 62. Teil. Jena, East Germany: VEB Gustav Fischer Verlag, 256 pp.

HAYES, I. E. 1953. Nematode dermatitis in a dog caused by *Rhabditis strongyloides. N. Am. Vet.*, **34,** 571.

HEYDON, G. M. & GREEN, A. K. 1931. Some worm infestations of man in Australia. *Med. J. Aust.*, **i,** 619–628.

HOOGSTRATEN, J. & YOUNG, W. G. 1975. Meningoencephalomyelitis due to the saprophagous nematode, *Micronema deletrix. J. Can. Sci. Neurol.*, **2,** 121–126.

JOHNSON, K. H. & JOHNSON, D. W. 1966. Granulomas associated with *Micronema deletrix* in the maxillae of a horse. *J. Am. vet. med. Ass.*, **149,** 155–159.

JONES, J. R. 1963. The cause of dermatitis in dogs. *Can. J. Zool.*, **45,** 136–139.

KINSELLA, J. M. 1967. Helminthes of microtines in western Montana. *Can. J. Zool.*, **43,** 269–274.

KIPNIS, R. M. & TODD, K. S. 1977. *Pelodera strongyloides* in the urine of a cat. *Feline Practice,* **7** (16), 18–19.

KLOSS, G. R. 1971. Alguns *Rhabdias* (Nematoda) de *Bufo* no Brasil. *Papéis Dep. Zool. S. Paulo,* **24,** 1–52.

KREIS, H. A. 1932. Studies on the genus *Strongyloides* (Nematodes). *Am. J. Hyg.*, **16,** 450–491.

KUROCHKIN, Y. V. & GUSKOV, E. P. 1963. [A new nematode species from the eye of *Natrix natrix*] [In: Helminths of man, animals and plants and their control. Papers on helminthology presented to Academician K. I. Skrjabin on his 85th birthday] Moscow: Izdatelstvo Akad. Nauk. SSSR, pp. 183–185. [In Russian.]

LAZAREVSKAYA, S. L. 1965. [The evolution-morphological analysis of the structure of Rhabditidae]. *Helminthologia,* **6,** 281–298. [In Russian.]

LEVINE, N. D., MILLER, L. J., MORRILL, C. C. & MANSFIELD, M. E. 1950. Nematode dermatitis in cattle associated with *Rhabditis. J. Am. vet. med. Ass.*, **116,** 294–296.

———, BIRCH, C. L., DOLOWY, W. C. & MCKINNEY, R. E. 1963. *Rhabditis axei,* a pseudoparasitic nematode of the dog. *J. Am. vet. med. ass.*, **142,** 1404–1406.

LITTLE, M. D. 1966a. Seven new species of *Strongyloides* (Nematoda) from Louisiana. *J. Parasit.*, **52,** 85–97.

LITTLE, M. D. 1966b. Comparative morphology of six species of *Strongyloides* (Nematoda) and redefinition of the genus. *J. Parasit.*, **52**, 69–84.

LU, S. C. 1934. On *Rhabdias*, a genus of parasitic Nematoda of Nanking. *Sinensia (Nanking)*, **5**, 164–172.

MACKERRAS, M. J. 1959. *Strongyloides* and *Parastrongyloides* (Nematoda: Rhabdiasoidea) in Australian marsupials. *Aust. J. Zool.*, **7**, 87–104.

MCINTOSH, A. & CHITWOOD, B. G. 1934. A new nematode, *Longibucca lasiura* n.sp. (Rhabditoidea, Cylindrogasteridae) from a bat. *Parasitology*, **26**, 138–140.

MORGAN, D. O. 1928. *Parastrongyloides winchesi* gen. et sp. nov. A remarkable new nematode parasite of the mole and the shrew. *J. Helminth.*, **6**, 79–86.

NADTOCHII, E. V. & RASSAKAZOVA, T. T. 1971. [Nematodes in shrews in some far-eastern territories.] *Trudy Severo-Vostochnogo Kompleksnoga Instituta Dal'nevost, Tsentr. Akademii Nauk SSSR*, **42**, 93–99. [In Russian.]

OLIVEIRA RODRIGUES, H. DE. 1968. Sôbre uma nova espécie do gênero *Strongyloides* Grassi, 1879 (Nematoda, Rhabdiasoidea). *Atas Soc. Biol. Rio de J.*, **12**, 31–32.

OSCHE, G. 1952. Systematik und Phylogenie der Gattung *Rhabditis* (Nematoda). *Zool. Jb. Abt. Syst.*, **81**, 190–280.

———. 1963. Morphological, biological and ecological considerations in the phylogeny of parasitic nematodes. In: *The lower Metazoa, Comparative Morphology and Phylogeny*. Dougherty, E. C., Brown, N. C., Hanson, E. D. & Hartman, W. D. (Editors). Berkeley, USA: University of California Press, pp. 283–302.

PEREIRA, C. 1927. Fauna helminthologica dos ophideos brasileiros. *Bolm Biol., S. Paulo*, **10**, 177–185.

———. 1928. Fauna helminthologica dos ophideos brasileiros (2°). *Bolm Biol., S. Paulo*, **11**, 13–22.

———. 1929. *Strongyloides ophidae* n.sp. *Bolm Biol., S. Paulo*, **15**, 16–17.

POINAR, G. O. 1965. Life history of *Pelodera strongyloides* (Schneider) in the orbits of murid rodents in Great Britain. *Proc. helminth. Soc. Wash.*, **32**, 148–151.

POWERS, R. D. & BENZ, G. W. 1977. *Micronema deletrix* in the central nervous system of a horse. *J. Am. vet. med. Ass.*, **170**, 175–177.

QUENTIN, J. C. 1969. *Parastrongyloides chrysochloris* n.sp. anguillule parasite d'un insectivore et d'un rongeur d'Afrique équatoriale. *Annls Parasit. hum. comp.*, **44**, 157–163.

RAO, P. N. & SINGH, S. N. 1968. A note on a new species of *Strongyloides* from the toad (Nematoda, Strongyloididae). *Papéis Dep. Zool., S. Paulo*, **21**, 111–113.

RAUSCH, R. 1952. Studies of the helminth fauna of Alaska. XI. Helminth parasites of microtine rodents. Taxonomic considerations. *J. Parasit.*, **38**, 418–442.

RHODE, E. A., JASPER, D. E., BAKER, N. F. & DOUGLAS, J. R. 1953. The occurrence of *Rhabditis* dermatitis in cattle. *N. Am. Vet.*, **34**, 634–637.

RUBIN, H. L. & WOODARD, J. C. 1974. Equine infection with *Micronema deletrix. J. Am. vet. med. Ass.*, **165**, 156–258.

SAKAMOTO, T. & YAMASHITA, J. 1970. Studies on strongyloidiasis of the peacock. II. *Strongyloides pavonis* n.sp. (Nematoda: Strongyloididae) from the green peafowl *Pavo muticus* Linnaeus. *Jap. J. vet. Res.*, **18**, 163–171.

SCHMIDT, G. D. & KUNTZ, R. E. 1972. Nematode parasites of Oceanica. XVIII. *Caenorhabditis avicola* sp.n. (Rhabditidae) found in a bird from Taiwan. *Proc. helminth. Soc. Wash.*, **39**, 189–191.

SHARPILO, V. P. & VAKKER, V. G. 1972. [A new species of the genus *Entomelas* (Nematoda, Rhabdiasidae) a lung parasite of *Ophisaurus apodus*.] *Vestnik Zoologii*, **6**, 86–88. [In Russian.]

———. 1973. [First finding of members of the genus *Strongyloides* Grassi, 1879 (Nematoda Strongyloididae) in reptiles of Europe, Transcaucasia and Middle Asia.] *Dopovidi Akademiyi Nauk Ukrayins'koyi RSR*, B, No. 11, 1047–1050. [In Ukranian.]

———. 1976. [Parasitic worms of the reptilian fauna of the USSR, systematics, chorology, biology.] Kiev, USSR; "Naukova Dumka", 287 pp. [In Russian.]

SHINOHARA, T. 1960. [Studies on *Rhabditis* (Nematoda, Rhabditidae) I. *Rhabditis* spp. obtained from human faeces. II. *Rhabditis* spp. found in the alimentary organs of *Fruticicola (Acusta) sieboldiana* Pfeiffer and *Limax (Limacus) flavus* Linné.] *J. Kurume med. Ass.*, **23**, 2777–2819. (In Japanese.]

SCHLOTTHAUER, C. F. & ZOLLMAN, P. E. 1955. The occurrence of *Rhabditis strongyloides* in association with dermatitis in a dog. *J. Am. vet. med. Ass.*, **127**, 510–511.

SCHWARTZMAN, R. M. 1964. Rhabditic dermatitis in the dog. *J. Am. vet. med. Ass.*, **145,** 25–28.

SHADDUCK, J. A., UBELAKER, J. & VAN QU TELFORD. 1979. *Micronema deletrix* meningoencephalitis in an adult man. *Am. J. clin. Path.*, **72,** 640–643.

SINGH, S. N. 1976. On a new nematode *Leipernema leiperi* n.g., n.sp. (Strongyloididae), parasitic in the pangolin *Manis pentadactyla* from Hyderabad, India. *J. Helminth.*, **50,** 267–274.

——— & RATNAMALA, R. 1977. On a new genus and new species of rhabdiasoid nematode *Shorttia shortii* n.g., n.sp. infesting lungs of amphibians. *Indian J. Helminth.* (1975, publ. 1977), **27,** 132–138.

SLUSARSKI, W., MARKIEWICZ, K. & STEFAŃSKI, W. 1956. On a case of verminous dermatitis caused by the invasion of *Rhabditis strongyloides* Schneider 1866 in a dog. *Acta parasit. pol.*, **4,** 163–174.

STAMMER, H. J. 1956. Die Parasiten deutscher Kleinsäuger. *Zool. Anz.* (Suppl.), **19,** 362–390.

STEFÁNSKI, W. 1954. *Rhabditis gingivalis* sp.n. parasite trouvé dans un granulome de la gencive chez un cheval. *Acta parasit. pol.*, **1,** 329–336.

STILES, C. W. & FRANKLAND, W. A. 1902. A case of vinegar eel infection in the human bladder. *Bur. Animal Industry Bull.*, No. 35, Washington, D.C., 35 pp.

STONE, W. M., STEWART, T. B. & PECKHAM, J. C. 1970. *Micronema deletrix* Anderson and Bemrick, 1965 in the central nervous system of a pony. *J. Parasit.*, **56,** 986–987.

SUDHAUS, W. 1976. Nomenklatorische Bemerkungen über Arten und Gattungen der Unterfamilie Rhabditinae *sensu lato* (Rhabditidae, Nematoda). *Nematologica*, **22,** 49–61.

UOTANI, K. 1960. [Studies on *Rhabditis* sp. isolated from human urine. 3. Experimental infection to animals.] *Jap. J. Parasit.*, **9,** 328–335. [In Japanese.]

VARJÚ, L. 1963. *Strongyloides*—studien. I. Der gegenwärtige Stand unserer Kenntnisse im Spiegel der Literatur. *Acta vet. hung.*, **13,** 267–280.

VICENTE, J. J. & SANTOS, E. DOS. 1976. Fauna helmintológica de *Leptodactylus ocellatus* (L., 1758) de Volta Redonda, Estado de Rio de Janeiro. *Atas Soc. Biol. Rio de J.*, **18,** 27–42.

WILLERS, W. B. 1970. *Pelodera strongyloides* in association with canine dermatitis in Wisconsin. *J. Am. vet. med. Ass.*, **156,** 319–320.

WILLIAMS, R. W. 1960. Observations on the life history of *Rhabdias sphaerocephala* Goodey, 1924 from *Bufo marinus* L., in the Bermuda Islands. *J. Helminth.*, **34,** 93–98.

YAMAGUTI, S. 1943. *Rhabdias (Ophiorhabdias) horigutti* n. subg. n.sp. (Nematoda) from the lung of a Japanese snake *Natrix tigrina. Annotnes zool. jap.*, **22,** 8–10.

YOKOGAWA, S. 1936. [A human case of accidental parasitism of *Diploscapter coronata* (Cobb, 1893) Cobb, 1913.] *Zool. Mag., Tokyo,* **48,** 507–512. [In Japanese.]

YUEN, P. H. 1965. Some studies on the taxonomy and development of some rhabdisoid and cosmocercoid nematodes from Malayan amphibians. *Zool. Anz.*, **174,** 275–298.

4 STRONGYLIDA

DIAPHANOCEPHALOIDEA AND ANCYLOSTOMATOIDEA

by J. RALPH LICHTENFELS*

DIAPHANOCEPHALOIDEA

The Diaphanocephaloidea are hookworm-like parasites of the digestive tracts of snakes and rarely of lizards. They are characterized by a large, complex buccal capsule and a dorsoventrally elongated oral opening giving the buccal capsule a bivalved appearance (Figs. 8.44 to 8.46, 8.48). The oral opening is surrounded by a cuticular membrane that in *Diaphanocephalus* and a few species of *Kalicephalus* bears numerous tiny elements of a corona radiata. A well-developed dorsal gutter extends through the buccal capsule wall and communicates with a peri-oral groove that surrounds the oral opening and may extend to ventral and basal portions of the buccal capsule (Figs. 8.44 to 8.46, 8.48, 8.49). The buccal capsule is composed of four pairs of sclerotized plates. A single lateral plate forms each side of the buccal capsule. Dorsally and ventrally a pair of small anterior plates and a large posterior plate form the buccal capsule wall. The anterior edges of the lateral plates are curled medially to form a ridge that delimits the peri-oral groove anteriorly. A posterior ridge and two membranous cuticular processes also help to form the peri-oral groove in *Diaphanocephalus*.

More than 50 species of Diaphanocephaloidea have been described. However, the revisions of Freitas & Lent (1938) and Schad (1962) reduced the number of recognized species to about 33 and the number of genera to two.

Diaphanocephalus was established by Diesing (1851) for a single species from a Brazilian lizard. Molin (1861) redescribed the species and provided a new generic diagnosis. Rudolphi & Henry (1909) combined *Diaphanocephalus* and *Kalicephalus*. However, since Ortlepp's (1923) redefinition of *Kalicephalus* and *Diaphanocephalus* both genera have been recognized by most helminthologists. Freitas & Lent (1938) reviewed the genus *Diaphanocephalus* and added a second species to the genus. *Diaphanocephalus* spp. have been reported only from South American lizards of the genus *Tupinambis* but two lots of *Diaphanocephalus*, which were collected in Brazil from mammals (*Proechimys* and *Didelphis*), have been found in the U.S. National Parasite Collection.

Schad (1962) arranged 22 of 23 species and numerous subspecies of *Kalicephalus* into six "groups". The 23rd species (*K. willeyi*) and species described since 1962 can also be placed in Schad's six "groups". In this treatment emendations of Schad's group names are used for three subgenera, older generic names are used for two subgenera, and a sixth subgenus is named in Schad's honour. The subgeneric designations are made primarily to provide a greater utility for the key to genera and subgenera.

Schad's (1962) definitions of amphidelphic and prodelphic are adopted herein. Regardless of whether the ovejectors are convergent or divergent, the nematode is prodelphic if the posterior uterus extends anteriorly past the anterior ovejector and amphidelphic if it does not. Although earlier workers had agreed that the disposition of the uteri was a useful character for determining species, Schad (1962) and Ghadirian (1968) found species with both prodelphic and amphidelphic uteri. The present study found a close relationship between vulva position and disposition of the uteri. Therefore, the prodelphic subgenus (*Schadius*) includes only species that are prodelphic and have the vulva closer to the anus than to mid-body in mature specimens. The type species is *K. (Schadius) schadi* Ogden, 1966.

* U.S. Department of Agriculture, Science and Education Administration, Agricultural Research, Animal Parasitology Institute, Beltsville, Maryland, U.S.A. 20705.
Original drawings by Robert B. Ewing.

Most workers have not recognized the genera *Occipitodontus* and *Kalicephaloides*. We agree with Schad (1962) and others that the characteristics used to erect *Occipitodontus* and *Kalicephaloides* are of less than generic level. We believe, however, that both names should be recognized at the subgeneric level. *Occipitodontus* was erected by Ortlepp (1923) for his new species (*K. fimbriatus*). The subgenus (*Occipitodontus*) includes species which Schad (1962) designated as the Fimbriatus group. The type species is *K. (Occipitodontus) fimbriatus*. *Kalicephaloides* was proposed by Yeh (1956) for *K. minutus* (Baylis & Daubney, 1922) a species with unequal spicules. The subgenus (*Kalicephaloides*) includes species which Schad (1962) designated as the Bungari group. The type species is *K. (Kalicephaloides) bungari* (MacCallum, 1918) (= *K. minutus*). Redefinitions of (*Occipitodontus*), (*Kalicephaloides*) and the other subgenera can be derived from the key to genera and subgenera.

DIAPHANOCEPHALOIDEA (Travassos, 1920 fam.)

One family *DIAPHANOCEPHALIDAE* Travassos, 1920

Key to genera and subgenera

1–(2) Peri-oral groove delimited by both anterior and posterior ridges (Figs. 8.44–8.46).
Two membranous cuticular processes at base of buccal cavity.
Spicules without alae.
Copulatory bursa bell-shaped (Fig. 8.47).
Thick muscular dorsal bulge anterior to copulatory bursa (Fig. 8.47).
Parasites of small intestine of South American lizards (*Tupinambis*).

Diaphanocephalus Diesing, 1851
(= *Strongylus* Mueller, 1780 of Rudolphi, 1819, in part;
= *Sclerostomum* Diesing, 1851 of Dujardin, 1845, in part;
= *Kalicephalus* Molin, 1861 of Railliet & Henry, 1909, in part)

2–(1) Peri-oral groove delimited only by anterior ridge (Fig. 8.48).
Buccal cavity without membranous cuticular processes at base.
Spicules alate.
Copulatory bursa lobate (Figs. 8.51, 8.52).
Without muscular prebursal bulge.
Parasites of snakes and, rarely, lizards; cosmopolitan.

Kalicephalus Molin, 1861
(= *Strongylus* Mueller, 1780 of Rudolphi, 1819;
= *Sclerostoma* Rudolphi, 1809 of Dujardin, 1845;
= *Diaphanocephalus* Diesing, 1851, in part;
= *Ankylostoma* Lutz, 1885 of Blanchard, 1886, in part;
= *Dochmius* Dujardin, 1845 of Stossich, 1895, in part;
= *Uncinaria* Froelich, 1789 of *Allesandrini*, 1905, in part;
= *Camallanus* Railliet & Henry, 1915 of MacCallum, 1918;
= *Occipitodontus* Ortlepp, 1923;
= *Kalicephaloides* Yeh, 1956)

3–(4) Vulva closer to anus than to mid-body in mature specimens.
Female prodelphic (i.e. posterior uterus extending anteriorly past anterior ovejector).
Spicules equal.
Parasites of snakes and, rarely, lizards; cosmopolitan.

Kalicephalus (*Schadius*) n. subg.

4–(3) Vulva closer to mid-body than to anus in mature specimens.
Female usually amphidelphic.
Spicules equal or unequal.

5–(6) Spicules unequal.
Parasites of intestine of Oriental snakes.

Kalicephalus (*Kalicephaloides*) (Yeh, 1956, genus)

6–(5) Spicules equal.

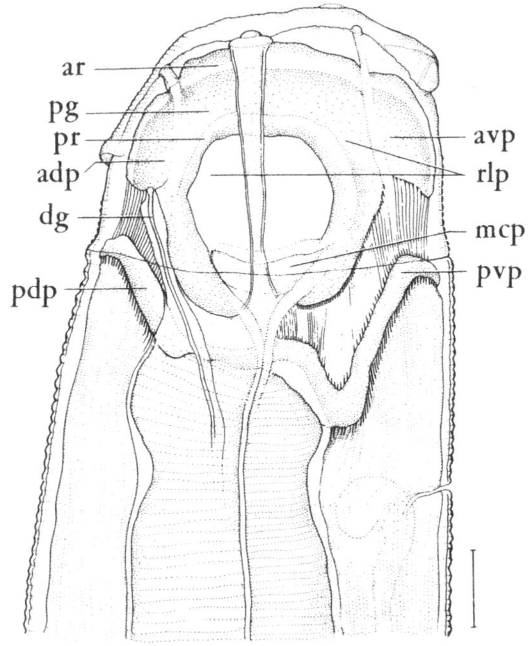

8.44

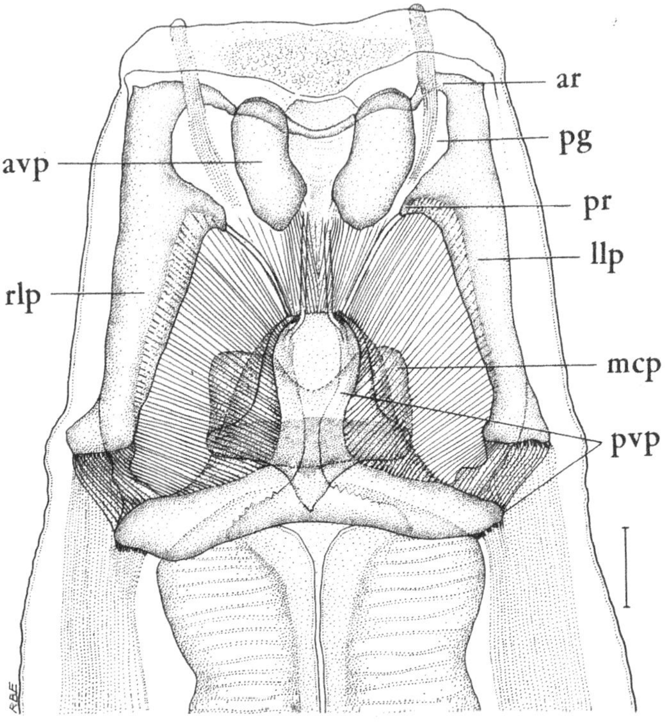

8.45

FIG. 8.44. *Diaphanocephalus galeatus* (Rudolphi, 1819), cephalic extremity, lateral view. (Original.) (Scale bar 50 µm.)

FIG. 8.45. *D. galeatus*, cephalic extremity, dorsal view. (Original.) (Scale bar 50 µm.)

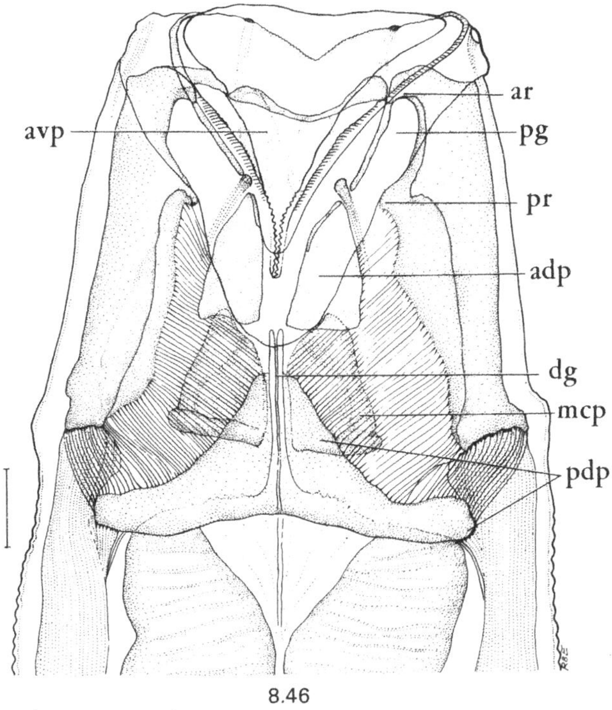

8.46

FIG. 8.46. *D. galeatus*, cephalic extremity, ventral view. (Original.) (Scale bar 50 µm.) (adp = anterior dorsal plate, ar = anterior ridge of peri-oral groove, avp = anterior ventral plate, dg = dorsal gutter, llp = left lateral plate, pdp = posterior dorsal plate, pg = peri-oral groove, pr = posterior ridge of peri-oral groove, pvp = posterior ventral plate, rlp = right lateral plate, mcp = membranous cuticular processes.)

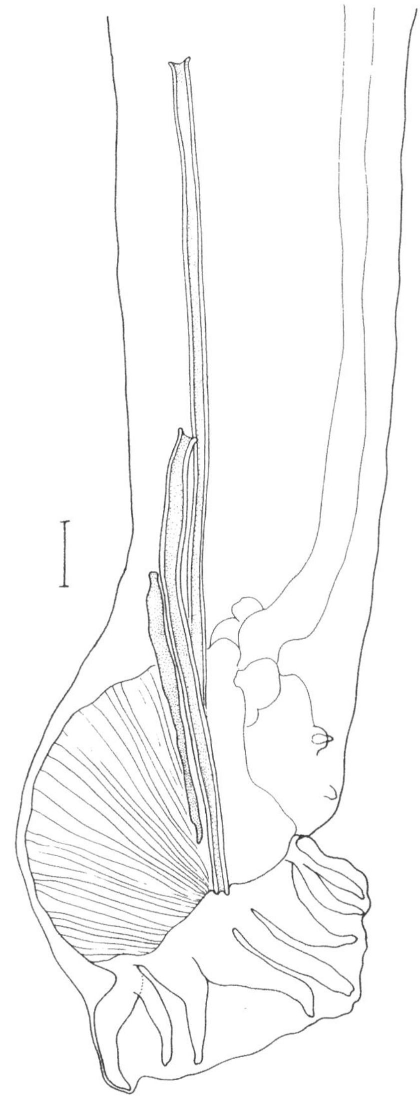

8.47

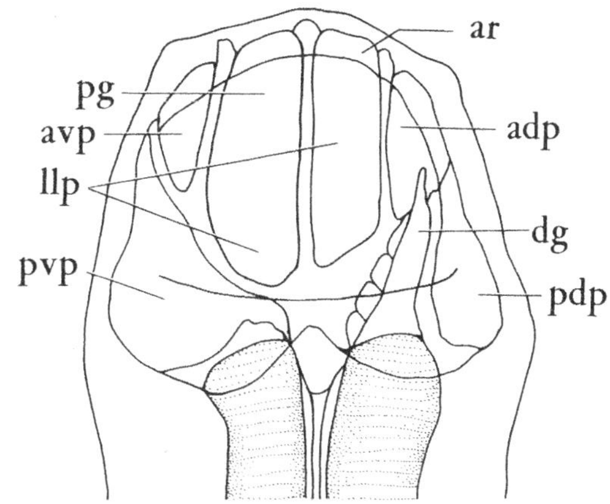

8.48

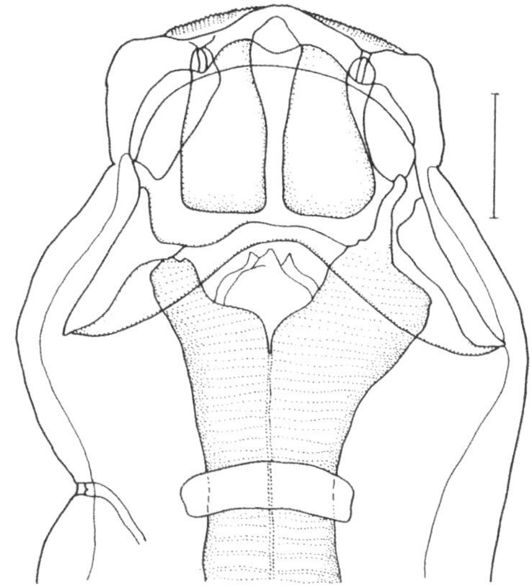

8.49

FIG. 8.47. *D. galeatus*, male posterior extremity, lateral view. (Original.) (Scale bar 50 μm.)

FIG. 8.48. *Kalicephalus* (generalized), cephalic extremity, lateral view. (After Schad, 1962.) (adp = anterior dorsal plate, ar = anterior ridge of peri-oral groove, avp = anterior ventral plate, dg = dorsal gutter, llp = left lateral plate, pdp = posterior dorsal plate, pg = peri-oral groove, pvp = posterior ventral plate.)

FIG. 8.49. *Kalicephalus* (*Occipitodontus*) *fimbriatus* (Ortlepp, 1923) cephalic extremity, lateral view. (After Schad, 1962.) (Scale bar 100 μm.)

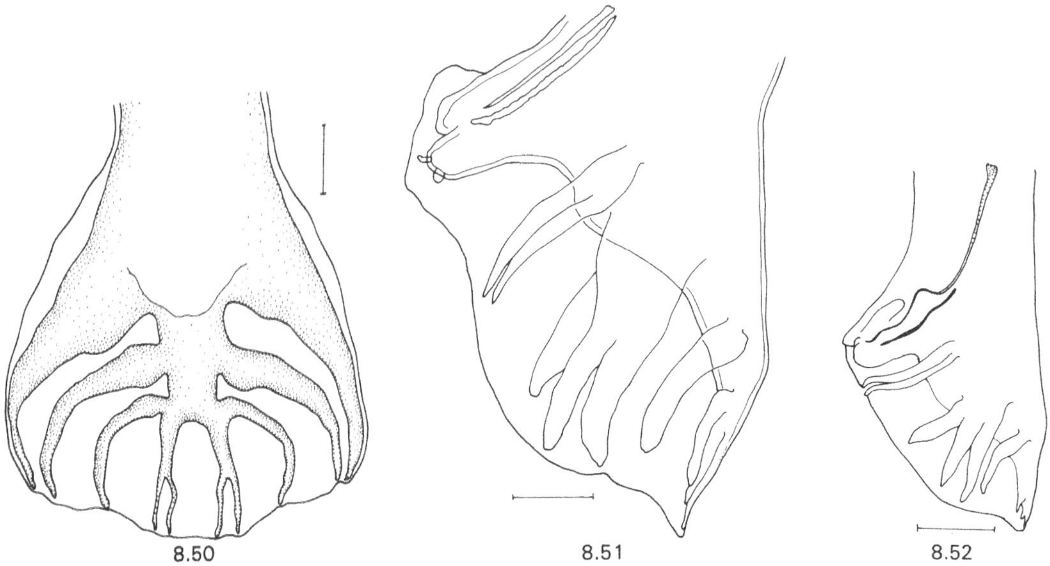

8.50 8.51 8.52

FIG. 8.50. *Kalicephalus (Rectiphiloides) rectiphilus* Harwood, 1932, dorsal ray of copulatory bursa, dorsal view. (Original.) (Scale bar 50 μm.)

FIG. 8.51. *Kalicephalus (Inermiformis) inermis* Molin, 1861, copulatory bursa, lateral view. (After Schad, 1962.) (Scale bar 100 μm.)

FIG. 8.52. *Kalicephalus (Variabiliformis) variabilis* Schad, 1962, copulatory bursa, lateral view. (After Schad, 1962.) (Scale bar 100 μm.)

7–(8) Dorsal bursal ray with primary branches closer to externodorsals than to secondary branches (Fig. 8.50).
Parasitic in rectum of Oriental and New World snakes.
Kalicephalus (*Rectiphiloides*) Schad, 1962, emend.

8–(7) Dorsal bursal ray with all branches closer together than to externodorsals.
Parasites of digestive tract other than rectum.

9–(10) Buccal capsule with prominent projections of posterior dorsal and ventral plates directed posteriorly (Fig. 8.49).
Rudimentary corona radiata present (Fig. 8.49).
Parasites of stomach and intestine of Oriental and Australasian snakes.
Kalicephalus (*Occipitodontus*) (Ortlepp, 1923, genus)

10–(9) Buccal capsule without prominent projections of posterior plates.
Corona radiata absent.

11–(12) Spicules long (rarely under 0.5 mm).
Lateral rays of copulatory bursa parallel for most of length (Fig. 8.51).
Parasites of oesophagus, stomach and intestine of New World snakes.
Kalicephalus (*Inermiformis*) Schad, 1962, emend.

12–(11) Spicules short (rarely over 0.5 mm).
Lateral rays of copulatory bursa divergent (Fig. 8.52).
Parasites of oesophagus, stomach and intestine of snakes of warmer regions of Europe, Asia, Africa and Australia.
Kalicephalus (*Variabiliformis*) Schad, 1962, emend.

REFERENCES

BARUS, V. & COY OTERO, A. 1978. Nematodes parasitizing Cuban snakes (Ophidia). *Věst. čsl. Spol. zool.*, **42**, 85–100.

COOPER, J. E. 1971. Disease in East African snakes associated with *Kalicephalus* worms (Nematoda: Diaphanocephalidae). *Vet. Rec.*, **89**, 385–388.

CRUSZ, H. & SANMUGASUNDERAM, V. [1974]. Parasites of the relict fauna of Ceylon. III. Nematodes from a rhacophorid frog and reptiles of the hill country. *Annls Parasit. hum. comp.*, **48**, 767–795.

FERNANDES, M. P. M & ARTIGAS, P. T. 1975. *Kalicephalus subulatus* Molin, 1861 (Nematoda, Diaphano cephalidae). Confirmação desta espécie; informacoes sobre sua dispersão geográfica e enumeraçãdo de serpentes parasitadas. [English summary.] *Mems Inst. Butantan,* **39**, 103–121.

FREITAS, J. F. T. & LENT, H. 1938. Pesquisas helminthologicas realisadas no Estado do Pará. V. Genero *Diaphanocephalus* Diesing, 1851 (Nematoda: Strongyloidea). *Mems Inst. Oswaldo Cruz,* **33**, 423–432.

GHADIRIAN, E. 1968. Nématodes parasites d'ophidiens malgaches. *Mém. Mus. natn. Hist. nat., Paris, Serie A, Zool.*, **54**, 1–54.

JONES, H. I. 1979. Gastrointestinal nematodes, including three new species, from Australian and Papua New Guinean Pythons. *Proc. helminth. Soc. Wash.*, **46**, 1–14.

LE-VAN-HOA & PHAM-NGOC-KHUE. 1968. Les kalicéphales, parasites des vipères, *Agkistrodon rhodostoma* B. du Sud-Vietnam. *Bull. Soc. Path. exot.*, **61**, 661–672.

NGUYEN-VAN-AI. [1965]. Aperçu sur la faune helminthologique au Sud-Viêtnam. *Rap. a. Fonct. tech. Inst. Pasteur, Viet-Nam* (1963), 52–56.

OGDEN, C. G. 1966. On some parasitic nematodes from reptiles, mainly from Ceylon. *J. Helminth.*, **40**, 81–90.

ORTLEPP, R. J. 1923. Observations on the nematode genera *Kalicephalus, Diaphanocephalus,* and *Occipitodontus* g.n., and on the larval development of *Kalicephalus philodryadus* sp. n. *J. Helminth.*, **1**, 165–189.

OSHMARIN, P. G. & DEMSHIN, N. I. 1972. [The helminths of domestic and some wild animals in Vietnam.] Trudy Biologo-Pochyennogo Instituta Dal'nevostochnyi Nauchnyi Tsentr AN SSSR (Issledovaniya po faune, sistematike biokhimii gel'mintov Dal'nego Vostoka), **11**, 5–115. [In Russian.]

POPOVA, T. I. 1960. [Strongyloids of animals and man. Cloacinidae, Stephanuridae, Diaphanocephalidae.] [In Russian.] Osnovy Nematodologii, Vol. 9. Akad. Nauk SSSR, Moscow.

SCHAD, G. A. 1962. Studies on the genus *Kalicephalus* (Nematoda: Diaphanocephalidae) II. A taxonomic revision of the genus *Kalicephalus* Molin, 1861. *Can. J. Zool.*, **40**, 1035–1165.

———. 1964. Studies on the genus *Kalicephalus* (Nematoda: Diaphanocephalidae). New records of taxonomic and ecological interest. *Can. J. Zool.*, **42**, 1143–1145.

UBELAKER, J. E. & DAILEY, M. D. 1966. Observations on *Kalicephalus subulatus* Molin, 1861 from Nicaragua and Mexico. *Revta Biol. trop.*, **14**, 47–53.

WIDMER, E. A. 1967. Helminth parasites of the prairie rattlesnake, *Crotalus viridis* Rafinesque, 1818, in Weld County, Colorado. *J. Parasitol.*, **53**, 362–363.

YEH, L.-S. 1956. On *Kalicephalus hongkongensis* n. sp. from *Elaphe moellendorffi* and the erection of a new genus, *Kalicephaloides*. *J. Helminth.*, **30**, 203–210.

ANCYLOSTOMATOIDEA

The Ancylostomatoidea or hookworms are small to medium-sized, thick-bodied, bursate nematodes that are blood-sucking parasites of the small intestine of mammals. The large, globular, dorsally orientated buccal capsules of the Ancylostomatoidea are devoid of both lips and corona radiata. In classification schemes based on characteristics of the buccal capsule, the Ancylostomatoidea have sometimes been grouped with the Strongyloidea because both superfamilies have large globular buccal capsules (Yorke & Maplestone, 1926; Chitwood, 1969). Both Dougherty (1951) and Schulz (1952), however, emphasized the presence in the Ancylostomatoidea of an ovejector of the trichostrongyloid type and concluded that the Ancylostomatoidea are more closely related to the Trichostrongyloidea than to the Strongyloidea. The Ancylostomatoidea also resemble the Trichostrongyloidea in characteristics of the life-cycle, copulatory bursa and female tail, and in that both superfamilies are parasitic in the small intestine rather than the large, where most of the Strongyloidea are found. Chabaud (1965, 1974) described the Ancylostomatoidea as follows: "Buccal capsule sub-globular, never hexagonal in transverse section. Lips and corona radiata absent. Oral opening unarmed or with teeth and cutting plates." This broad definition of the superfamily (closely following Dougherty, 1951) accommodates several genera without cutting plates or with reduced cutting plates in the oral opening, and it is accepted herein as the definition of the Ancylostomatoidea.

There have been two widely different theories on the classification of the genera of the Ancylostomatoidea—that of Looss (1905, 1911) and that of Lane (1917). Looss (1911) established two subfamilies: (i) the Ancylostomatinae, with a dorsal gutter, oblong buccal capsule with sutures on the ventral wall and one dorsal ray, and (ii) the Bunostominae, with a dorsal tooth (or cone), round buccal capsule without ventral sutures and two dorsal rays. Lane (1917) proposed a new classification based on characteristics of the teeth in the oral opening. He recognized two different subfamilies: (i) the Ancylostominae, with ventral teeth, and (ii) the Necatorinae, with ventral cutting plates in the oral opening. Lane's system has been almost universally accepted although most authors (Chabaud, 1965; Chitwood, 1969) have recognized *Uncinaria* rather than *Necator* as the type genus for Lane's subfamily with ventral cutting plates (Uncinariinae thus replacing Necatorinae). However, after a comparison of numerous characters of the genera of the Ancylostomatoidea, Rep (1963a, 1963b, 1964) demonstrated that Looss's system was far more successful than Lane's in grouping together genera determined by Rep to share numerous characters. Rep (1963a), however, recognized as Ancylostomatidae only nine genera with well-defined cutting plates or teeth in the oral opening.

The present study confirms Rep's conclusion that Looss's system (greatly modified herein) is far superior to the generally accepted system of Lane (1917) but includes 18 genera (including one new genus) and six subgenera in the family Ancylostomatidae. Three genera (*Agriostomum* Railliet, 1902, *Cyclodontostomum* Adams, 1933, and *Hypodontus* Mönnig, 1929) included by many authors in the Ancylostomatoidea were previously moved (Lichtenfels, 1980) to the Strongyloidea. *Globocephaloides* Yorke & Maplestone, 1926, included by some previous workers in the Ancylostomatoidea, is placed in the Amidostomatidae following Beveridge (1979). The new classification proposed herein easily accommodates genera without cutting plates or teeth

TABLE I. Key characters and character states for the phylogenetic analysis of subfamilies, tribes, and genera of the Ancylostomatoidea by taxonomic level.

CHARACTERS	STATES	
	Generalized	Specialized
I. **Subfamilies**		
Duct of dorsal oesophageal gland	Dorsal cone	Dorsal gutter
Gubernaculum	Present	Absent
Dorsal ray	Shallow bifurcation	Deep bifurcation
Vulva position	Posterior	Anterior
Female tail	Unspined	Spined
II. **Tribes**		
Anterior extremity, orientation	Anteriorly	Dorsally
Buccal capsule anterior edge	Unarmed	Armed
Armature of oral opening	Cutting plates	Teeth
Buccal capsule	One piece	Composed of articulated plates
Development of armature of oral opening	Well-developed	Poorly developed or rudimentary
Buccal collar	Present	Absent
Oesophagus as percentage of body length	10%	5%
III. **Genera**		
Vulva position	Near midbody	Toward extremity
Buccal capsule shape	Subglobular	Other
Buccal capsule size	Fills cephalic space	Other
Cuticle	Finely striated	Serrated
Plates forming buccal capsule	8 to 10	Reduced number
Deirids, location	Oesophageal	Level of buccal capsule
Dorsal cone, size	Large	Reduced
Spicule length	Less than one-third body length	More than one-half body length
Spicule thickness	Thin	Thick
Dorsal lobe, size	Similar to size of laterals	Other
Bursa symmetry	Symmetrical	Asymmetrical
Ovejector, proportions of sections	Vestibula longer than sphincters	Sphincters longest section
Subdorsal supports for dorsal cone	Absent	Present
Externodorsal rays, origin	Anterior to bifurcation of dorsal ray	At or posterior to bifurcation of dorsal ray
Intestinal caecum	Absent	Present

(*Globocephalus* and *Acheilostoma*) or with small cutting plates (*Tetragomphius* and *Bathmostomum*). Specimens of all 18 genera and six subgenera considered valid in this work were studied.

Table I lists the characters used in the key presented herein. Each character and character state is identified as either generalized (ancestral) or specialized (derived). This listing should aid the reader in understanding the following brief discussion of the nematode groups and their proposed phylogenetic relationships reflected in the keys and in Fig. 8.1.

ANCYLOSTOMATIDAE

Eighteen genera separated into two subfamilies are placed in this family.

Ancylostomatinae

This subfamily includes those genera of the Ancylostomatoidea with a dorsal gutter, a gubernaculum, a dorsal bursal ray with a shallow median fissure (and, therefore, a single main stem for most of its length), a terminal spine on the female tail and the vulva in the posterior half of the body. The eight genera with six subgenera are separated into four tribes.

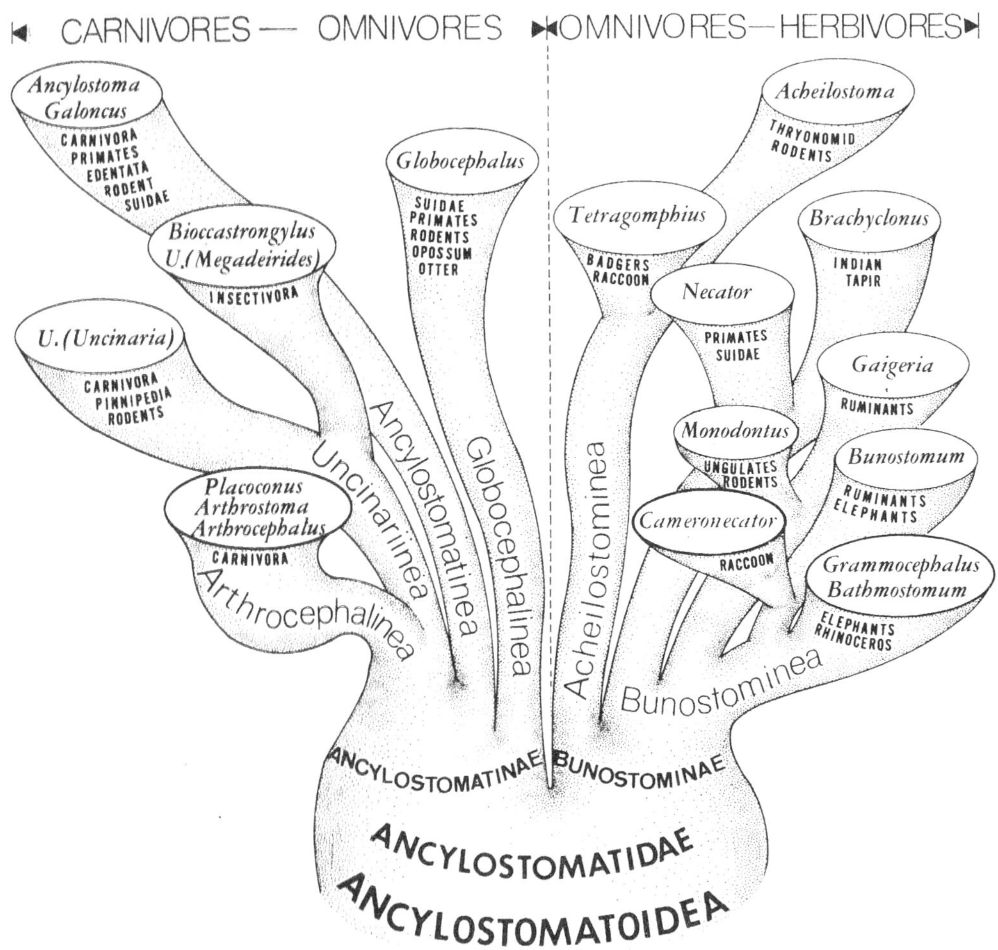

1

FIG. 8.1 Dendrogram of possible evolutionary relationships of the Ancylostomatoidea. (Original.)

Globocephalinea: The genus *Globocephalus* lacks teeth or cutting plates in the oral opening, the anterior extremity is only slightly inclined dorsally, and the vulva is near mid-body. *Characostomum* and *Raillietostrongylus* are regarded as synonyms of *Globocephalus*. *Characostomum* included species from monkeys and a rodent and was previously separated from *Globocephalus* because of a lack of subventral lancets. *Raillietostrongylus* included a single species from pigs and was distinguished by its bicuspid subventral lancets. However, considerable variation in the degree of development of the subventral lancets in *Globocephalus* spp. from pigs has been described by Ortlepp (1964); and Rep (1963b) and others have pointed out the variability in number and morphology of ventral lancets in *Bunostomum*. Accordingly, the presence or absence or degree of development of subventral lancets are not considered to be generic characters in this group of nematodes.

Various authors (Skrjabin *et al.*, 1952; Chabaud, 1965; Durette-Desset *et al.*, 1976) have grouped *Globocephaloides* Yorke & Maplestone, 1926 with *Globocephalus*. However, recent studies (Inglis, 1968; Beveridge, 1979) have correctly placed *Globocephaloides* in the Amidostomatidae.

Ancylostomatinea: This tribe includes two genera and four subgenera with teeth in the oral opening and without the buccal capsule divided into articulating plates. The genus *Ancylostoma* is divided into four subgenera (cf. Biocca & LeRoux, 1958) on the basis of number of teeth in the oral opening, arrangement of lateral bursal rays and relative size of bursal lobes. Two of the four subgenera are also characterized by distinctive geographical distributions.

The genus *Galoncus* has been recognized by most previous workers. However, Rep (1963a) argued that *Galoncus* was probably a synonym of *Ancylostoma*. While studying specimens of *Galoncus* from a tiger and a black leopard several differences were observed that distinguish *Galoncus* from *Ancylostoma*. In contrast to the latter genus, *Galoncus* has a small cup-shaped buccal capsule that fills only about half of the cephalic space, a thin cuticle with serrated annules and a female tail that narrows immediately behind the anus.

Arthrocephalinea: This tribe includes three genera with buccal capsules made up of articulating plates. The monotypic *Arthrocephalus* is the only member of the Ancylostomatinae that has a dorsal cone rather than a dorsal gutter. Otherwise the three genera are quite similar, differing most markedly in the number of plates that form the buccal capsule and in their geographical distribution.

Uncinariinea: This tribe includes two genera and two subgenera with cutting plates in the oral opening and without division of the buccal capsule into articulating plates. *Uncinaria* is separated into two subgenera following Chabaud, Bain & Houin (1966). The subgenus *Megadeirides* is recognized for two species with exceptionally large deirids—one from an insectivore from Madagascar and one from a lemur from Borneo. *Bioccastrongylus* also has large deirids and is from an insectivore from Madagascar, but it differs from *Uncinaria* in characteristics of the buccal capsule and oesophagus.

Bunostominae

This subfamily includes ten genera with a tooth-like dorsal cone supporting the duct of the dorsal oesophageal gland. They lack a gubernaculum and a terminal spine on the female tail. They usually have a dorsal bursal ray with two long stems (except *Brachyclonus indicus* Railliet & Henry, 1910 and *Monodontus giraffae* Yorke & Maplestone, 1926). The vulva is in the anterior half of the body (except *Tetragomphius*). The Bunostominae can be divided into two tribes on the basis of morphological characteristics.

Acheilostominea: The two genera of this tribe have the anterior end only slightly inclined dorsally, with a buccal collar and without (or with very small) cutting plates in the oral opening. *Acheilostoma* is parasitic in large African thryonomyid rodents, and *Tetragomphius* is found in Asian badgers and a raccoon.

A study of the types of *Acheilostoma simpsoni* Leiper, 1911 and *A. paranecator* Travassos & Horta, 1915 has shown that the latter species does not belong in the genus *Acheilostoma*. Leiper (1911) regarded the most distinctive feature of his genus to be the complete absence of teeth or cutting plates in the oral opening. Both African species from rodents (*A. simpsoni* and *A. moucheti* Railliet, 1918) fit this description. As indicated in the original description, however, the South American species (*A. paranecator*) from an ass (*Equus asinus*) has cutting plates in the mouth and also differs from the two African species in having much shorter spicules, a thicker and smoother cuticle and an oesophagus-to-body-length ratio of 1:10 rather than 1:20. Accordingly, *A. paranecator* cannot be accommodated in *Acheilostoma*. The type specimens of *A. paranecator* are in poor condition and the species must be considered a *species inquirenda* until additional specimens become available.

Bunostominea: The eight genera of this tribe have well-developed cutting plates in a dorsally directed buccal capsule. They lack a buccal collar. *Necator* and *Brachyclonus* are characterized by the presence of subdorsal plates flanking the dorsal cone and elongate vestibula in the ovejector. *Necator* is parasitic in primates, pigs, rhinoceroses and a pangolin, and *Brachyclonus* occurs in the Indian tapir. The remaining six genera all lack subdorsal plates and have short vestibula in the ovejectors.

Two genera with exceptionally large dorsal bursal lobes (*Gaigeria* and *Monodontus*) are both parasites of ruminants of Africa. *Gaigeria* is also found in ruminants of India, and *Monodontus* in ruminants and rodents of North and South America and in a peccary in South America. The generic concepts of *Monodontus* (=*Monodontella*) and *Gaigeria* follow Chitwood & Jordan (1965).

Grammocephalus and *Bathmostomum*, which are almost exclusively parasites of elephants, share the specialized characteristics of short thick spicules, a posterior origin of the externodorsal rays of the bursa and greatly enlarged sac-like sphincters in the ovejectors.

A new genus, *Cameronecator*, is proposed for *Necator urichi* Cameron, 1936 from *Procyon cancrivorus* from Trinidad and Brazil. The new genus is named in honour of the late Dr. Thomas W. M. Cameron for his many outstanding contributions to helminthology.

Cameronecator n. g.

Generic Diagnosis: Ancylostomatoidea, Ancylostomatoidae, Bunostominae, Bunostominea. Small hookworms with exceptionally large deirids (Fig. 8.38). Buccal capsule fills cephalic space, deeper than wide. Dorsal cone high on wall of buccal capsule, without subdorsal plates. Subventral lancets present. Oesophagus claviform. Excretory pore, deirids, and nerve ring near middle of oesophagus. Vulva near midbody. Ovejectors with very short vestibula, well-developed sphincters and very short infundibula (Fig. 8.40). Spicules filiform, gubernaculum absent. Rays of bursa symmetrical, prebursal papillae present. Dorsal lobe of bursa much smaller than laterals, externodorsal rays longer than dorsal ray (Fig. 8.39).

Cameronecator is similar to *Bunostomum* in lacking subdorsal plates and morphology of the ovejector. These same characteristics separate it from *Necator*. Differences between the new genus and other genera of the Bunostominea are indicated in the key to genera.

Evolution and Hosts

The origin of the Ancylostomatoidea is obscure. They are exclusively parasites of mammals and, with the exception of a single species in a South American opossum, are parasites of placental mammals.* Previous authors (Dougherty, 1951; Schulz, 1952; Durette-Desset *et al.*, 1976) believed

* *Hypodontus* Mönnig, 1929 and *Globocephaloides* Yorke & Maplestone, 1926, previously included in the Ancylostomatoidea, have been placed elsewhere. Lichtenfels (1980) placed *Hypodontus* in the Strongyloidea, and Beveridge (1979) placed *Globocephaloides* in the Amidostomatidae.

the Ancylostomatoidea probably shared a common ancestor with the Amidostomatidae which are trichostrongyloid parasites of reptiles, birds and Australian marsupials, with a single species in a South American placental mammal. However, current studies of Australian Amidostomatidae may provide new interpretations of the relationships of strongylid superfamilies (M. C. Durette-Desset and A. C. Chabaud, Museum of Natural History, Paris, personal communication).

Host relationships (Fig. 8.1) and parasite morphology provide limited insight into evolution within the Ancylostomatoidea. A relationship was observed, however, between host diet and the two subfamilies of the Ancylostomatoidea. Carnivorous hosts are parasitized only by the Ancylostomatinae, herbivorous hosts by the Bunostominae and omnivorous hosts by both subfamilies (Fig. 8.1). Furthermore, the primitive *Globocephalus* and *Tetragomphius* are among the genera found only in omnivorous hosts. According to Kowalski (1971) ancestors of present-day carnivorous and herbivorous placental mammals evolved from more primitive omnivorous stock in the late Cretaceous. The observed association between host diet and nematode subfamilies supports the classification proposed herein and the hypothesis that the Ancylostomatoidea evolved with early omnivorous mammals in the Cretaceous and blossomed with the mammals in the Eocene.

A hypothetical ancestral form for the Ancylostomatoidea might be expected to have a dorsal cone (as found in all species of the Bunostominae, one species of the Ancylostomatinae and in most of the Amidostomatidae) and to have the anterior end inclined slightly dorsally or not at all. If this hypothetical ancestor is accepted as a starting point, the most similar genus may be *Acheilostoma* in large African rodents. *Acheilostoma* appears to be the most primitive genus of the Bunostominae because of its slightly inclined oral opening without teeth or cutting plates and its cuticular buccal collar. The thryonomyid rodent hosts of *Acheilostoma* are also primitive, dating to the Oligocene (Kowalski, 1971).

The evolution of the Bunostominae appears to have resulted in one tribe (Acheilostominea) with generalized morphology in omnivorous badgers and raccoons and in herbivorous rodents, and another (Bunostominea) with more specialized morphology in omnivorous primates, pigs, rodents and raccoons, and herbivorous ruminants, elephants, rhinoceroses and tapirs (Fig. 8.1).

Globocephalus, with its generalized morphology, may be the most primitive member of the Ancylostomatinae. *Globocephalus*, however, appears to represent a side branch in the family tree of the Ancylostomatinae (Fig. 8.1). With seven or eight species in pigs, one in a large African rodent, one in a South American opossum, one in otters and two in primates, *Globocephalus* may have been isolated from the rest of the Ancylostomatinae in the ancestors of pigs and, only later, have become established in its other hosts. If *Globocephalus* had evolved with the non-porcine hosts more differences betwen the species would be expected to have evolved.

Ancylostoma occurs in a wide range of hosts. However, only the Carnivora appear to harbour significant numbers of species, and more differences among the species would be expected if *Ancylostoma* had evolved in its non-carnivore hosts. Rather, *Ancylostoma* has probably only secondarily colonized lemurs, primates, edentates and the aardvark.

Arthrostoma Cameron, 1927 is probably the oldest of the three genera of the specialized tribe Arthrocephalinea, because of its relatively generalized morphology (Key to Ancylostomatinae and Table I) and large number of species in a broad range of hosts. *Arthrostoma* has seven species in Felidae, Mustelidae, Canidae and Viverridae of Asia and some Pacific Islands, including both suborders of the Carnivora. *Arthrocephalus* Ortlepp, 1925 has a single species in Viverridae of Asia and Africa. *Placoconus* Webster, 1956 has two species in Procyonidae, Mustelidae or Ursidae, all members of the suborder Arctoidea, in North and South America.

Among the Uncinariinea, *Bioccastrongylus* in insectivores of Madagascar and *Uncinaria* (*Megadeirides*) in insectivores of Madagascar and lemurs of Borneo have been isolated from *Uncinaria* (*Uncinaria*) long enough to have evolved significant differences. The morphology of the nematodes (Key to Ancylostomatinae and Table I) is interpreted to indicate that *Uncinaria* (*Uncinaria*) is more similar to the ancestral stock of the tribe Uncinariinea.

ANCYLOSTOMATOIDEA

One Family *ANCYLOSTOMATIDAE* (Looss, 1905)

Key to subfamilies

1-(2) Duct of dorsal oesophageal gland usually in dorsal gutter (Figs. 8.2, 8.5 to 8.8, 8.16, 8.20, 8.22) on inner surface of buccal capsule (except *Arthrocephalus*).
Gubernaculum present.
Dorsal ray of copulatory bursa usually with two short stems (Fig. 8.17).
Vulva posterior to mid-body.
Female tail with terminal spine (Fig. 8.13).

Ancylostomatinae

2-(1) Duct of dorsal oesophageal gland in toothlike dorsal cone (Figs. 8.24, 8.25, 8.26, 8.29, 8.31, 8.33, 8.35, 8.42).
Gubernaculum absent.
Dorsal ray of bursa usually with two long stems (Figs. 8.28, 8.32, 8.34, 8.36, 8.39, 8.43) (except *Brachyclonus*) (Fig. 8.30).
Vulva usually in anterior half of body (except *Tetragomphius*).
Female tail blunt or pointed, but without terminal spine.

Bunostominae

Subfamily **Ancylostomatinae** Looss, 1905

Key to tribes, genera and subgenera

1-(2) Anterior extremity directed anterodorsally.
Anterior edge of buccal capsule without teeth or cutting plates (Fig. 8.2).
Vulva slightly posterior to mid-body.
Parasites of Suidae (*Sus, Potamochoerus*), Primates (Cercopithecidae), Mustelidae (*Lutra*), opossum (*Metachirops*), rodents (*Cricetomys*) and Cervidae (*Sika*).

GLOBOCEPHALINEA (Travassos & Vogelsang, 1932, subfam.)
One genus *Globocephalus* Molin, 1861
(= *Cystocephalus* Railliet, 1895; = *Characostomum* Railliet, 1902; = *Crassisoma* Alessandrini, 1909; = *Raillietostrongylus* Lane, 1923)

2-(1) Anterior extremity bent dorsally.
Oral opening with teeth (Figs. 8.3 to 8.8) or cutting plates (Figs. 8.19, 8.20, 8.22).
Vulva usually at junction of middle and posterior thirds of body.

3-(12) Oral opening armed with teeth (Figs. 8.3 to 8.8).

ANCYLOSTOMATINAE (Looss, 1905 subfam.)

4-(5) Buccal capsule small and cup-shaped, occupying about half of cephalic space (Figs. 8.3, 8.4).
Cuticle thin and serrated (Figs. 8.3, 8.4).
Parasites of submucosal cysts of intestine of Felidae.

Galoncus Railliet, 1918
(= *Strongylus tubaeformis* of Linstow, 1789)

5-(4) Buccal capsule large and globular, occupying entire cephalic space (Figs. 8.5 to 8.8).
Cuticle thick and unserrated.
Parasites of intestinal lumen of Carnivora, Primates, Edentata, Rodentia and Suidae.

Ancylostoma (Dubini, 1843)
(= *Agchylostoma* Dubini, 1843; = *Dochmius* Dujardin, 1845, in part; = *Anchylostomum* Diesing, 1851; = *Diploodon* Molin, 1861; = *Ankylostomum* Bugnion, 1881; = *Ankylostoma* Lutz, 1885; = *Ceylancylostoma* Lane, 1916)

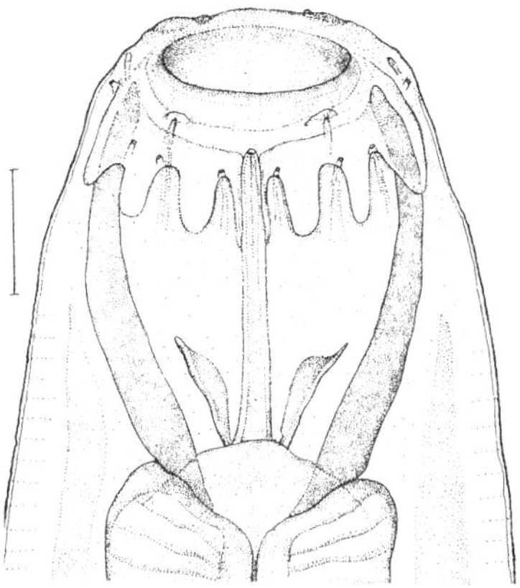

8.2

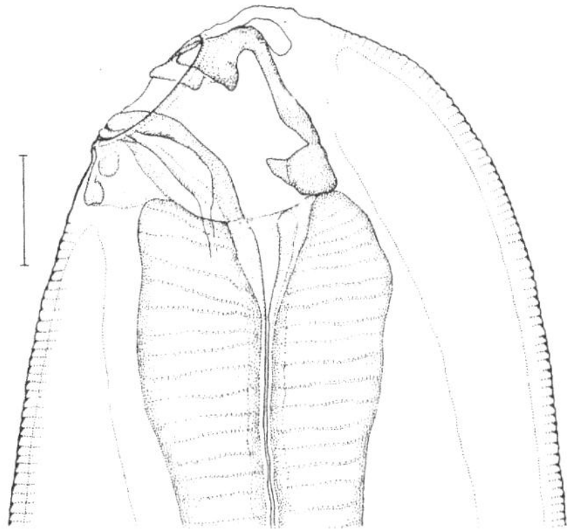

8.3

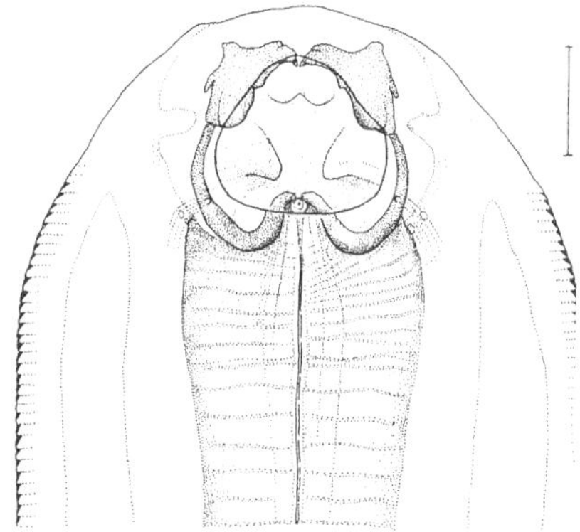

8.4

FIG. 8.2. *Globocephalus longemucronatus* Molin, 1861, cephalic extremity, dorsal view. (Original.) (Scale bar 50 μm.)

FIG. 8.3 *Galoncus perniciosus* (Linstow, 1885), cephalic extremity, lateral view. (Original.) (Scale bar 50 μm.)

FIG. 8.4. *Galoncus perniciosus* (Linstow, 1885), cephalic extremity, dorsal view. (Original.) (Scale bar 50 μm.)

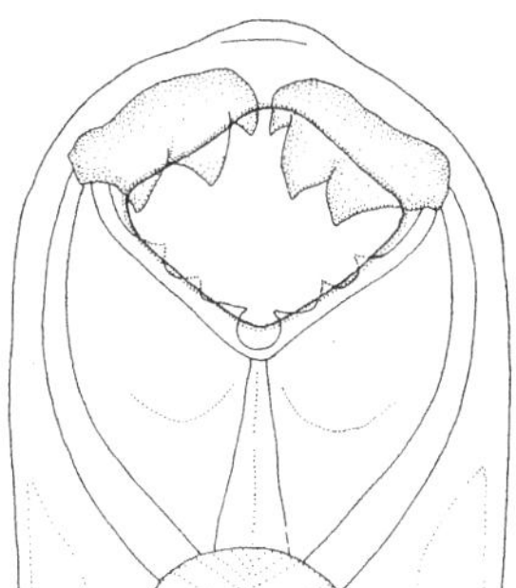

8.5

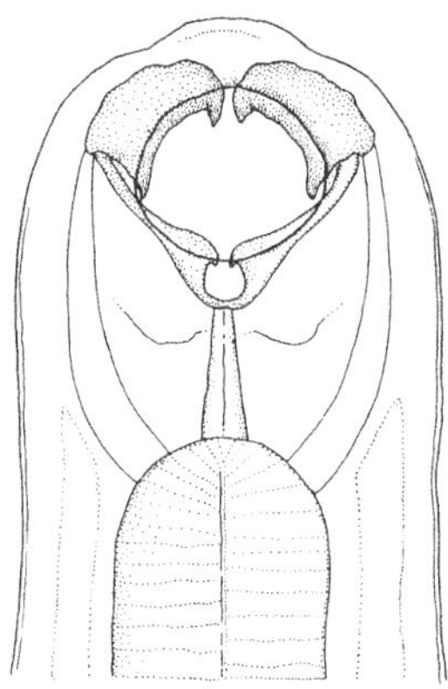

8.6

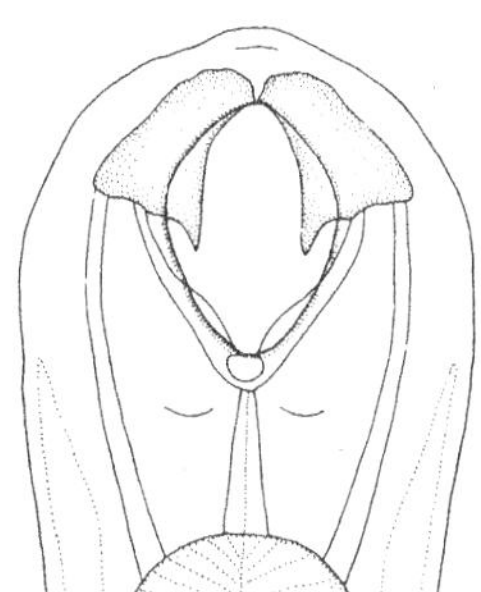

8.7

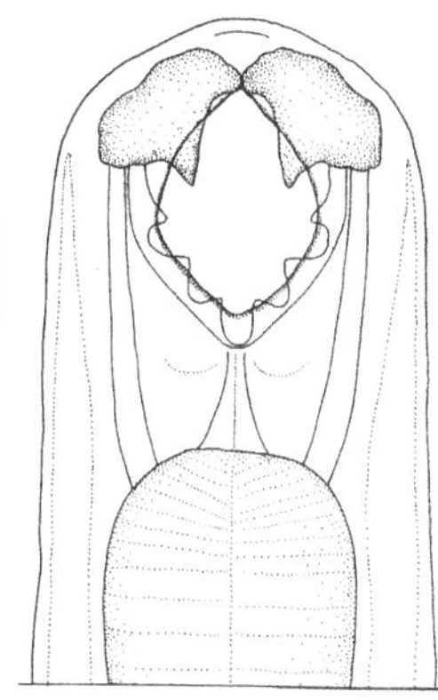

8.8

FIG. 8.5. *Ancylostoma* (*Ancylostoma*) Lane, 1916, cephalic extremity, dorsal view. (After Biocca & LeRoux, 1958.)

FIG. 8.6. *Ancylostoma* (*Ceylancylostoma*) Lane, 1916, cephalic extremity, dorsal view. (After Biocca & LeRoux, 1958.)

FIG. 8.7. *Ancylostoma* (*Afrancylostoma*) Biocca & LeRoux, 1958, cephalic extremity, dorsal view. (After Biocca & LeRoux, 1958.)

FIG. 8.8. *Ancylostoma* (*Amerancylostoma*) Biocca & LeRoux, 1958, cephalic extremity, dorsal view. (After Biocca & LeRoux, 1958.)

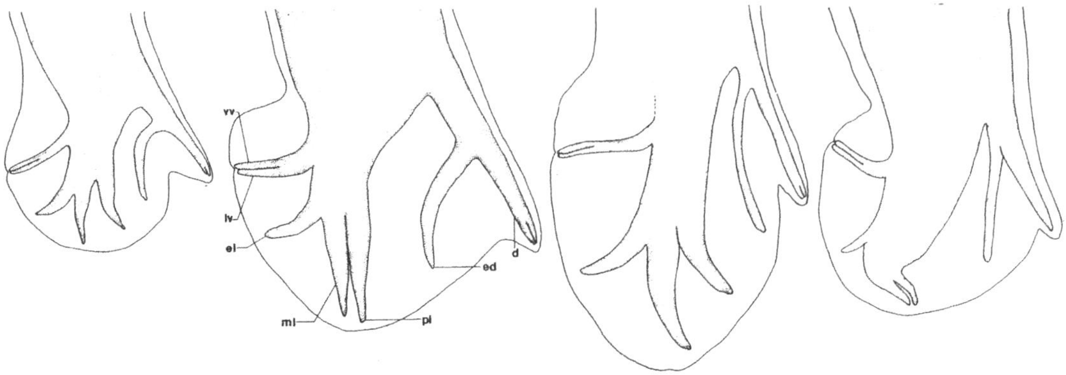

8.9 8.10 8.11 8.12

FIG. 8.9. *Ancylostoma* (*Ancylostoma*) copulatory bursa, lateral view. (After Biocca & LeRoux, 1958.) (vv = ventroventral, lv = lateroventral, el = externolateral, ml = mediolateral, pl = posterolateral, ed = externodorsal, d = dorsal.)

FIG. 8.10. *Ancylostoma* (*Ceylancylostoma*) copulatory bursa, lateral view. (After Biocca & LeRoux, 1958.)

FIG. 8.11. *Ancylostoma* (*Afrancylostoma*) copulatory bursa, lateral view. (After Biocca & LeRoux, 1958.)

FIG. 8.12. *Ancylostoma* (*Amerancylostoma*) copulatory bursa, lateral view. (After Biocca & LeRoux, 1958.)

6-(7) Oral opening with three pairs of ventrolateral teeth, most ventral pair sometimes rudimentary.
One or more pairs of dorsolateral teeth (Fig. 8.5).
Bursa with short lateral lobe and divergent lateral rays (Fig. 8.9).
Parasites of Carnivora (Canidae, Felidae, Ursidae, Procyonidae and Mustelidae), Primates (man, apes, monkeys, lemurs), Edentata, Suidae(*Sus*) and the aardvark; cosmopolitan.
Ancylostoma (*Ancylostoma*) Lane, 1916

7-(6) Oral opening with two pairs of ventrolateral teeth, more ventral pair sometimes rudimentary.

8-(9) Two well-developed pairs of ventrolateral teeth present (Fig. 8.6).
Externolateral ray of bursa divergent from mediolateral and posterolateral rays which are parallel (Fig. 8.10).
Externodorsal ray usually originating near middle of dorsal ray.
Parasites of Carnivora (Felidae, Viverridae, Canidae, Ursidae and Mustelidae), Primates (man, *Nycticebus*), Rodentia (Muridae, Sciuridae); Indo-Malayan region.
Ancylostoma (*Ceylancylostoma*) Lane, 1916

9-(8) More ventral pair of lateroventral oral teeth rudimentary (Figs. 8.7, 8.8).
Externodorsal ray originating near base of dorsal ray (Figs. 8.11, 8.12).

10-(11) Only one pair of dorsolateral teeth present (Fig. 8.7).
Lateral lobe of bursa long with divergent lateral rays (Fig. 8.11).
Parasites of Carnivora (Felidae, Canidae, Ursidae, Procyonidae, Viverridae and Mustelidae), Edentata (*Dasypus*) and Rodentia (*Mus*); cosmopolitan.
Ancylostoma (*Afrancylostoma*) Biocca & LeRoux, 1958

11-(10) More than one pair of dorsolateral teeth present (Fig. 8.8).
Lateral lobe of bursa short with short externolateral ray directed ventrally and with parallel

mediolateral and posterolateral rays longer and directed dorsally (Fig. 8.12). Parasites of Felidae in South America.

Ancylostoma (*Amerancylostoma*) Biocca & LeRoux, 1958

12-(3) Oral opening with cutting plates (Figs. 8.19, 8.20, 8.22).

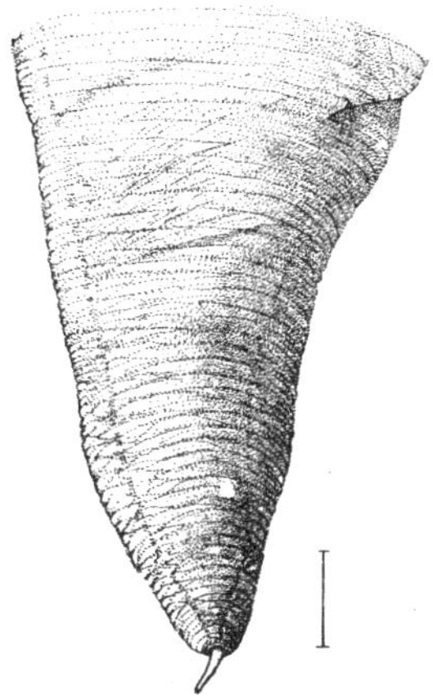

8.13

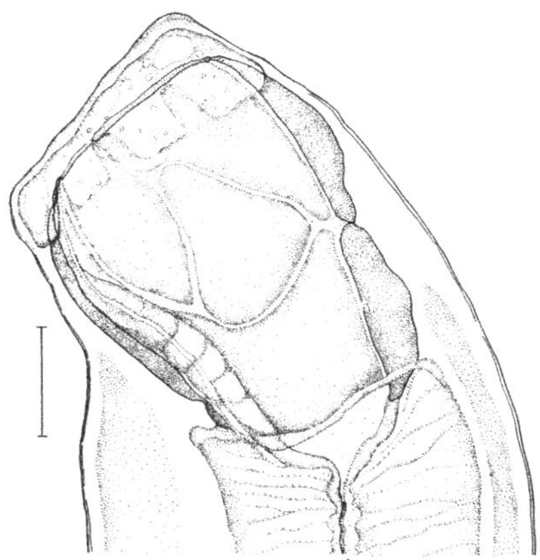

8.14

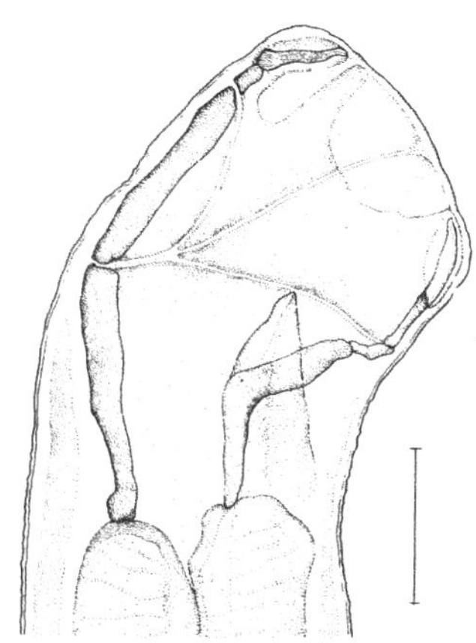

8.15

FIG. 8.13. *Ancylostoma* (*A.*) *caninum* (Ercolani, 1859), female tail, lateral view. (Original.) (Scale bar 50 μm.)

FIG. 8.14. *Arthrostoma vampira* Schmidt & Kuntz, 1968, cephalic extremity, lateral view. (Original.) (Scale bar 50 μm.)

FIG. 8.15. *Arthrocephalus gambiensis* Ortlepp, 1925, cephalic extremity, lateral view. (Original.) (Scale bar 50 μm.)

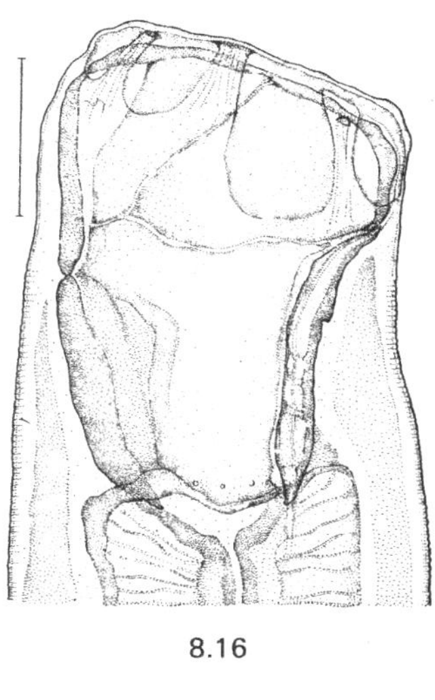

8.16

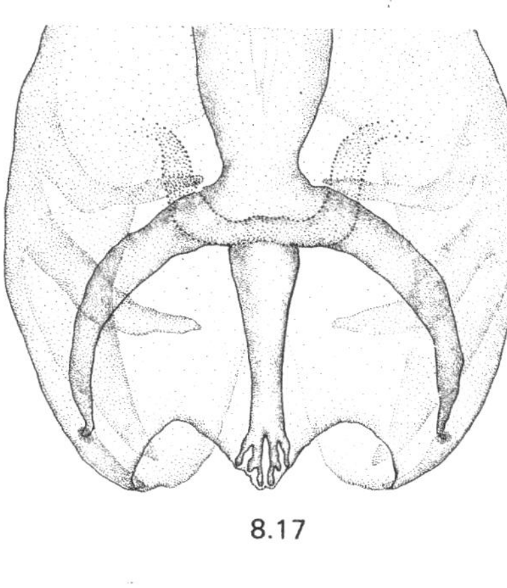

8.17

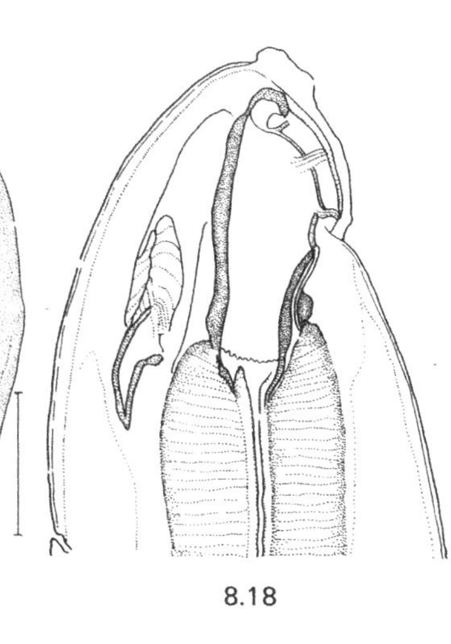

8.18

FIG. 8.16. *Placoconus lotoris* (Schwartz, 1925), cephalic extremity, lateral view. (Original.) (Scale bar 50 μm.)

FIG. 8.17. *Placoconus lotoris*, dorsal bursal rays. (Original.) (Scale bar 50 μm.)

FIG. 8.18. *Bioccastrongylus bioccai* Chabaud, Petter & Brygoo, 1961, cephalic extremity, lateral view. (After Chabaud, Petter & Brygoo, 1961).

13-(18) Buccal capsule formed by five to ten articulating plates (Figs. 8.14, 8.15, 8.16).
Cutting plates in oral opening, poorly developed.
ARTHROCEPHALINEA (Schmidt & Kuntz, 1968 subfam.)

14-(15) Buccal capsule consisting of eight or ten articulating plates (Fig. 8.14).
Parasites of Felidae, Viverridae, Mustelidae or Canidae in Asia or Pacific Oceanica.
Arthrostoma Cameron, 1927
(= *Arthrocephaloides* Yamaguti, 1961; = *Uncinaria* Frœlich, 1789, in part; = *Necator* Stiles, 1903, in part)

15-(14) Buccal capsule consisting of five or six articulating plates (Figs. 8.15, 8.16).

16-(17) Dorsal cone present (Fig. 8.15).
Buccal capsule consisting of six articulating plates.
Parasites of Viverridae in Asia and Africa.
Arthrocephalus Ortlepp, 1925

17-(16) Dorsal gutter present.
Buccal capsule consisting of five articulating plates (Fig. 8.16).
Parasites of Procyonidae, Mustelidae, and Ursidae in North and South America.
Placoconus Webster, 1956
(= *Arthrocephalus* Ortlepp, 1925, in part)

18-(13) Walls of buccal capsule continuous, not composed of separate plates (Figs. 8.3, 8.4, 8.18, 8.21).
Cutting plates in oral opening well developed.
UNCINARIINEA (Stiles, 1903 subfam.)

19-(20) Buccal capsule funnel-shaped, occupying about half of cephalic space (Fig. 8.18).
Oral opening with only two small ventrolateral cutting plates (Fig. 8.19).
Oesophagus only about twice as long as thick.
Deirids at level of buccal capsule.
Vulva near mid-body.
Parasites of insectivores (*Setifer*) of Madagascar.
Bioccastrongylus Chabaud & Petter, 1961

20-(19) Buccal capsule globular, occupying entire cephalic space (Figs. 8.20, 8.21).
Oral opening armed with two ventrolateral and two dorsolateral cutting plates (Fig. 8.20).
Oesophagus about four times as long as thick.
Deirids at level of oesophagus.
Vulva at junction of middle and posterior thirds of body.
Uncinaria Froelich, 1789
(= *Dochmoides* Cameron, 1924; = *Dochmius* Dujardin, 1845, in part)

21-(22) Deirids exceptionally large (Fig. 8.22).
Parasites of insectivores (*Tenrec*) of Madagascar and (*Tupaia*) of Borneo.
Uncinaria (*Megadeirides*) Chabaud, Bain & Houin, 1966

22-(21) Deirids not exceptionally large.
Parasites of Carnivora (Mustelidae, Canidae, Ursidae, Procyonidae and Felidae), Pinnipedia (Otariidae) and Rodentia (Muridae); occasionally in domestic pigs and man; cosmopolitan.
Uncinaria (*Uncinaria*) Froelich, 1789

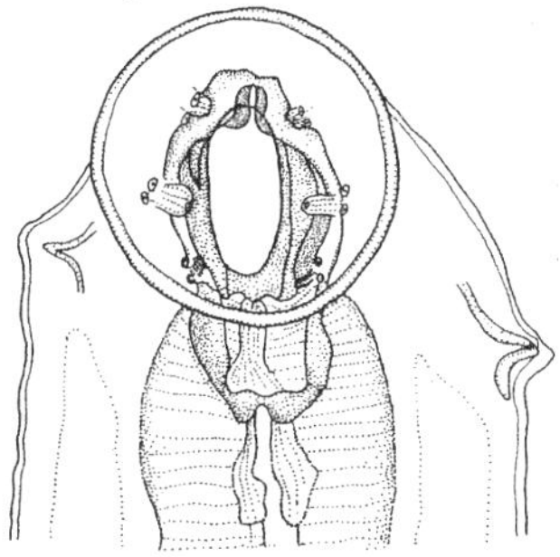

8.19

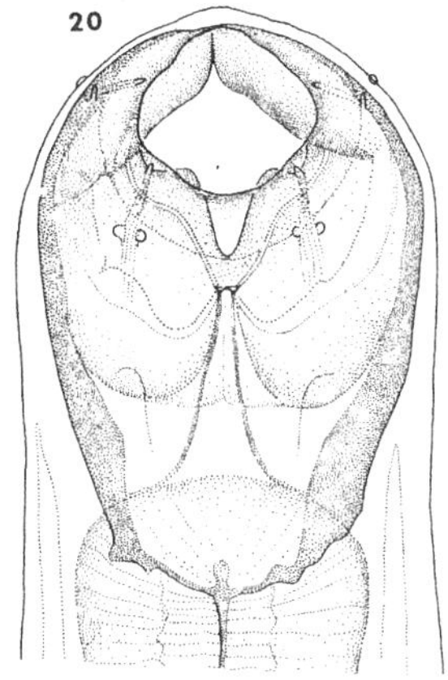

8.20

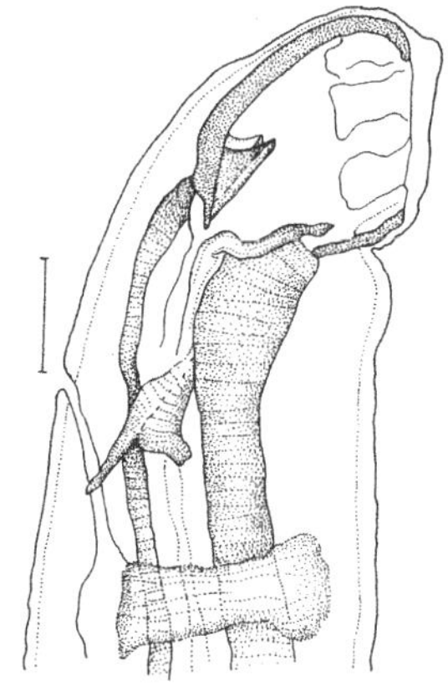

8.21

FIG. 8.19. *Bioccastrongylus bioccai* Chabaud, Petter & Brygoo, 1961, cephalic extremity, dorsal view. (After Chabaud, Petter & Brygoo, 1961.)

FIG. 8.20. *Uncinaria (U.) stenocephala* (Railliet, 1884), cephalic extremity, dorsal view. (Original.) (Scale bar 50 µm.)

FIG. 8.21. *Uncinaria (Megadeirides) bauchoti* Chabaud, Brygoo & Tcheprakoff, 1964, cephalic extremity, lateral view. (After Chabaud, Brygoo & Tcheprakoff, 1964.) (Scale bar 50 µm.)

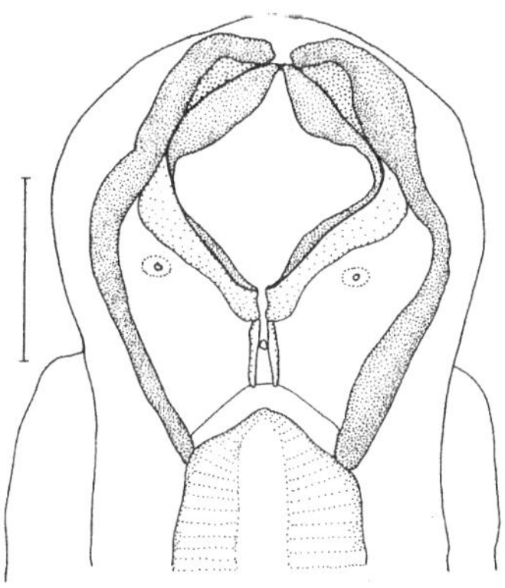

8.22

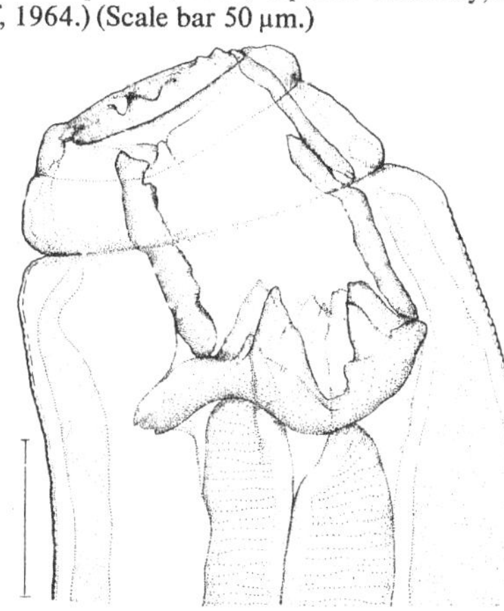

8.23

FIG. 8.22. *Uncinaria (Megadeirides) bauchoti* Chabaud, Brygoo & Tcheprakoff, 1964, cephalic extremity, dorsal view. (After Chabaud, Brygoo & Tcheprakoff, 1964.) (Scale bar 50 µm.)

FIG. 8.23. *Tetragomphius arctonycis* Jansen, 1968, cephalic extremity, lateral view. (Original.) (Scale bar 50 µm.)

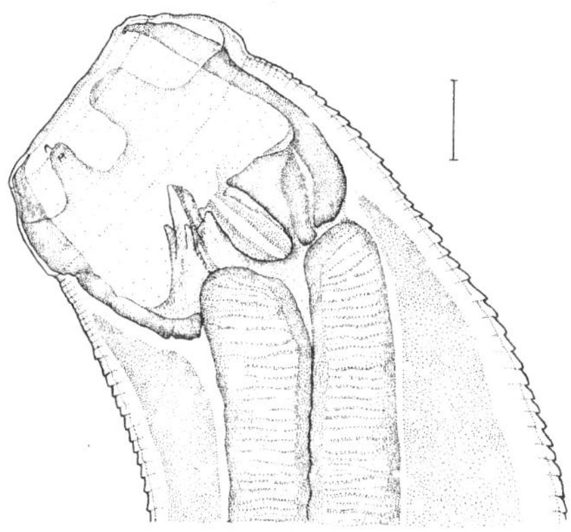

8.24

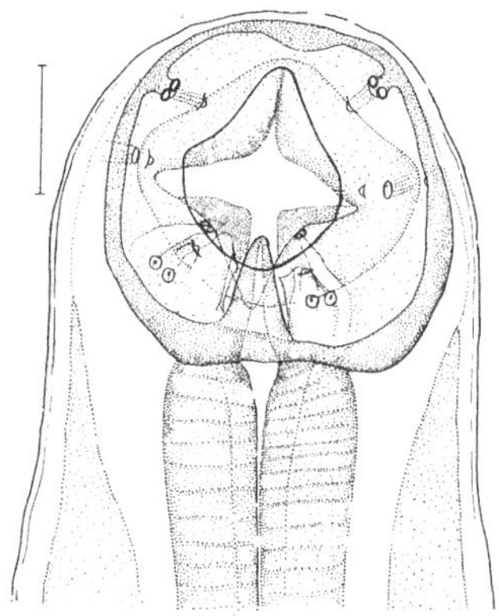

8.25

FIG. 8.24. *Acheilostoma simpsoni* Leiper, 1911, cephalic extremity, lateral view. (Original.) (Scale bar 50 µm.)

FIG. 8.25. *Necator americanus* (Stiles, 1902), cephalic extremity, dorsal view. (Original.) (Scale bar 50 µm.)

Subfamily **Bunostominae** (Railliet & Henry, 1909, Tribe) Looss, 1911

Key to tribes and genera

1-(4) Anterior extremity anterodorsally directed, dorsal inclination very slight.
Oral opening without cutting plates (Fig. 8.24) or with small inconspicuous ventral cutting plates (Fig. 8.23).
Buccal collar present.
Oesophagus about 5% of body length.
ACHEILOSTOMINEA n. tribe

2-(3) Buccal capsule cup-shaped (Fig. 8.23).
Dorsal cone small.
Vulva in posterior third of body.
Spicules more than half length of body.
Parasites of Asian badgers (*Arctonyx* and *Meles*) and a raccoon (*Procyon*) in the Calcutta Zoo.
Tetragomphius Baylis & Daubney, 1923

3-(2) Buccal capsule subglobular (Fig. 8.24).
Dorsal cone large.
Vulva at mid-body.
Spicules less than half length of body.
Parasites of "large African rodents" including (*Thryonomys*).
Acheilostoma Leiper, 1911

4-(1) Dorsal inclination of anterior extremity pronounced (Figs. 8.26, 8.29, 8.33, 8.35, 8.38).
Oral opening usually with well-developed cutting plates (Figs. 8.25, 8.26, 8.29, 8.31).
Buccal collar absent.
Oesophagus usually about 10% of body length.
BUNOSTOMINEA Railliet & Henry, 1909

5-(8) Dorsal cone supported by two subdorsal plates (Figs. 8.25, 8.26, 8.29).
Ovejectors with elongate vestibula (Fig. 8.27).

6-(7) Dorsal ray of bursa much shorter than lateral rays (Fig. 8.28).
Median fissure of lateral ray deep.
Subdorsal plates usually bifid (Figs. 8.25, 8.26).
Parasites of Primates (man, apes), Suidae (*Sus, Phacochoerus, Potamochoerus*); rarely of rhinoceroses (*Diceros*, *Rhinoceros*), pangolins (*Manis*) and dogs (*Canis*).
Necator Stiles, 1903

7-(6) Dorsal ray of bursa almost as long as lateral rays (Fig. 8.30).
Median fissure of dorsal ray shallow.
Subdorsal plates each with single point (Fig. 8.29).
Parasites of Indian tapir (*Tapirus*).
Brachyclonus Railliet & Henry, 1910

8-(5) Dorsal cone usually not supported by subdorsal plates (Figs. 8.31, 8.33, 8.35, 8.38).
Ovejectors with short vestibula (Figs. 8.37, 8.40).

9-(12) Dorsal bursal lobe larger than lateral lobes (Figs. 8.32, 8.34).

10-(11) Rays of bursa symmetrical (Fig. 8.32).
Parasites of ruminants (Bovidae, Antilocapridae) of Africa and India.
Gaigeria Railliet & Henry, 1910

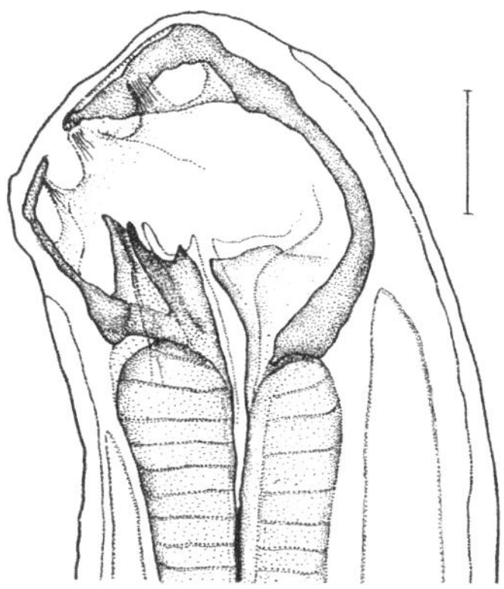

8.26

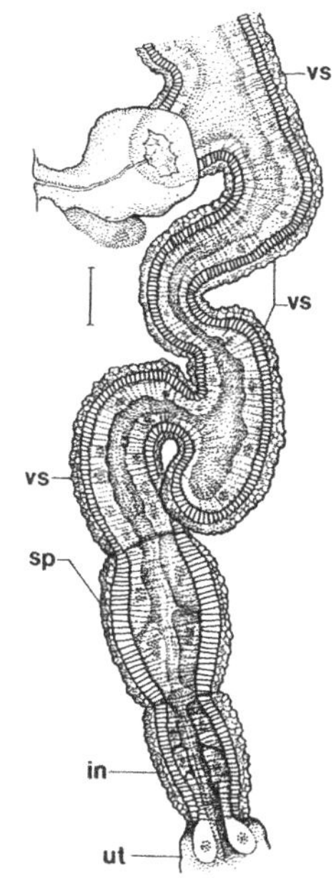

8.27

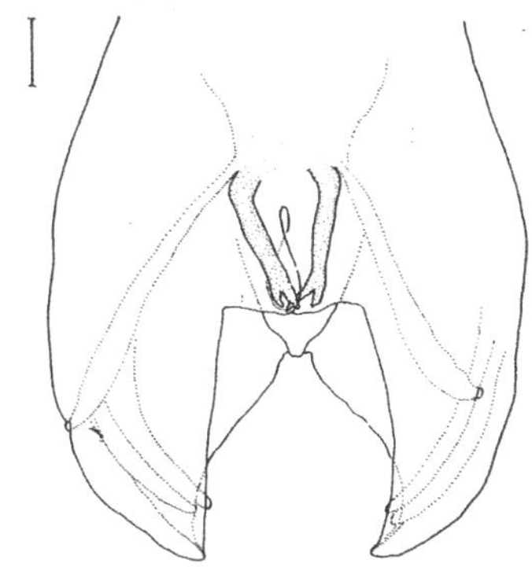

8.28

FIG. 8.26. *Necator americanus*, cephalic extremity, lateral view. (Original.) (Scale bar 50 μm.)

FIG. 8.27. *Necator americanus*, ovejector, lateral view, showing elongate vestibule (vs), sphincter (sp), infundibulum (in), and uterus (ut). (Original.) (Scale bar 50 μm.)

FIG. 8.28. *Necator americanus*, dorsal bursal rays. (Original.) (Scale bar 50 μm.)

11-(10) Rays of bursa asymmetrical (Fig. 8.34).
Parasites of rodents (*Sigmodon, Mesomys, Dasyprocta, Capromys*), ruminants of North and South America (*Odocoileus*) and Africa (Giraffidae) and peccaries (*Tayassu*) and tapirs (*Tapirus*) of South America.

Monodontus Molin, 1861
(= *Eumonodontus* Railliet & Henry, 1910;
= *Monodontella* Yorke & Maplestone, 1926)

12-(9) Dorsal bursal lobe smaller than lateral lobes (Figs. 8.36, 8.39, 8.43).

13-(16) Buccal capsule globular with large dorsal cone (Figs. 8.35, 8.38).
Spicules filiform.
Externodorsal rays arising anterior to branching of dorsal ray (Figs. 8.36, 8.39).

14-(15) Buccal capsule with subventral lancets extending from bottom (Fig. 8.35).
Deirids normally developed.
Vulva at junction of anterior and middle thirds of body.
Dorsal ray usually asymmetrical (Fig. 8.36).
Parasites of ruminants (Bovidae, Antilocapridae, Cervidae) and elephants (*Loxodonta, Elephas*); cosmopolitan.

Bunostomum Railliet, 1902
(= *Bustomum* Lane, 1917;
= *Monodontus* Molin, 1861, in part)

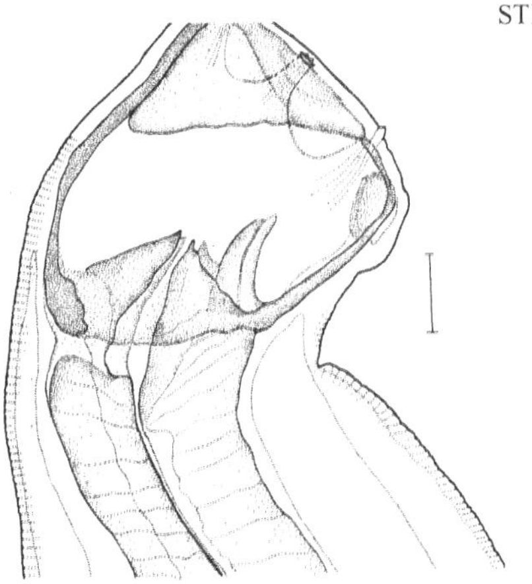

8.29

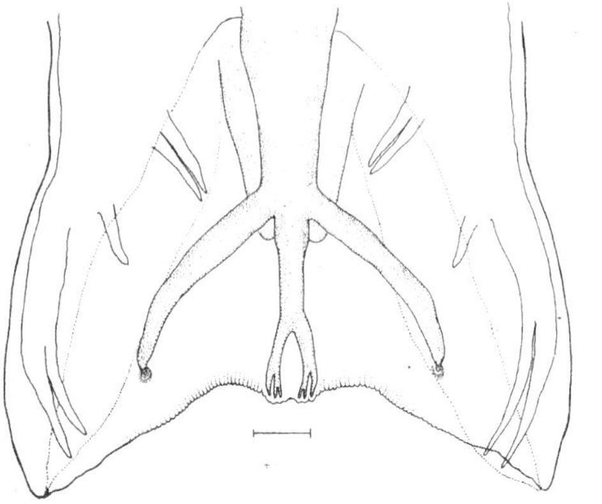

8.30

FIG. 8.29. *Brachyclonus indicus* Railliet & Henry, 1910, cephalic extremity, lateral view. (Original.) (Scale bar 50 μm.)

FIG. 8.30. *Brachyclonus indicus*, dorsal bursal rays. (Original.) (Scale bar 50 μm.)

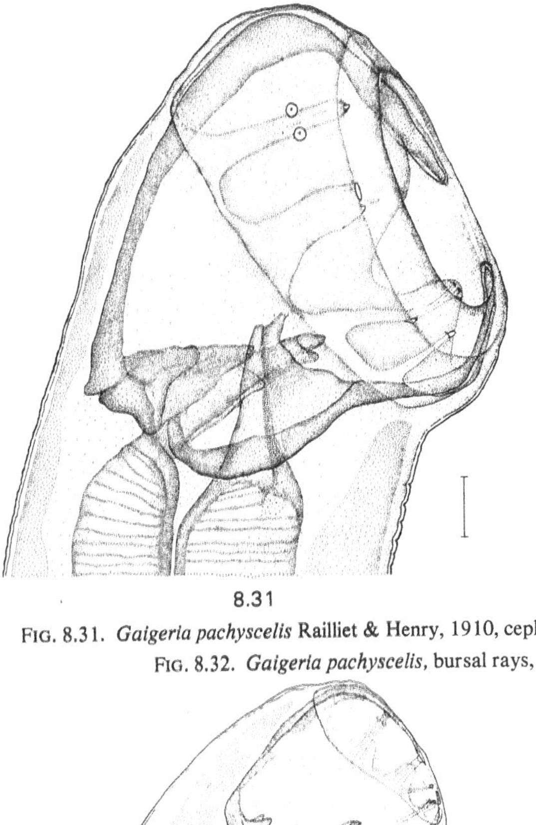

8.31

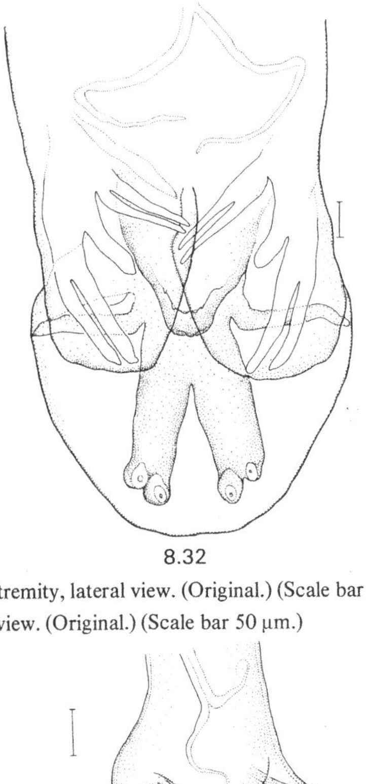

8.32

FIG. 8.31. *Gaigeria pachyscelis* Railliet & Henry, 1910, cephalic extremity, lateral view. (Original.) (Scale bar 50 μm.)

FIG. 8.32. *Gaigeria pachyscelis*, bursal rays, dorsal view. (Original.) (Scale bar 50 μm.)

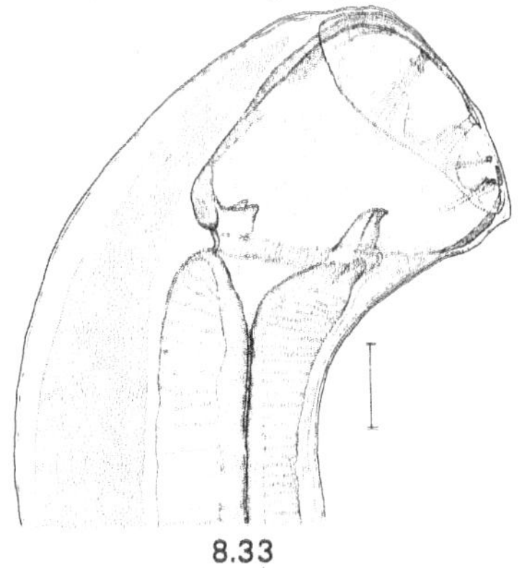

8.33

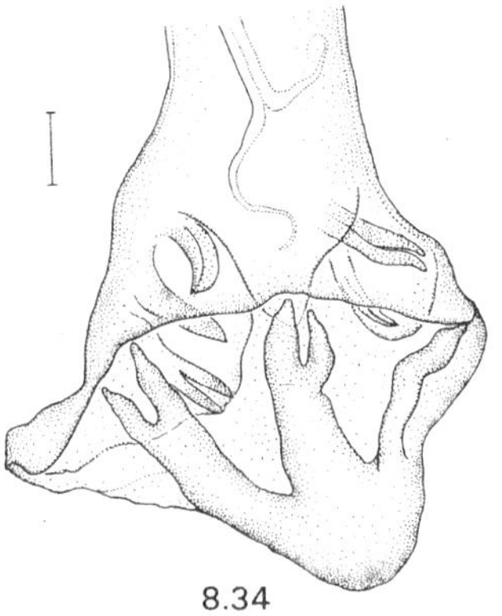

8.34

FIG. 8.33. *Monodontus louisianensis* Chitwood & Jordan, 1965, cephalic extremity, lateral view. (Original.) (Scale bar 50 μm.)

FIG. 8.34. *Monodontus louisianensis*, bursal rays, ventral view. (After Chitwood & Jordan, 1965.) (Scale bar 50 μm.)

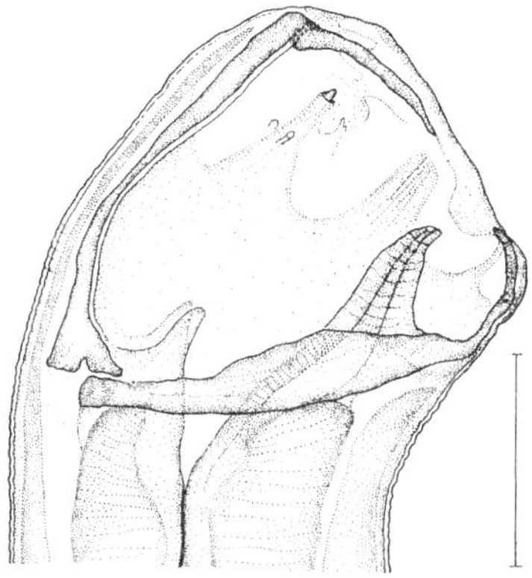

8.35

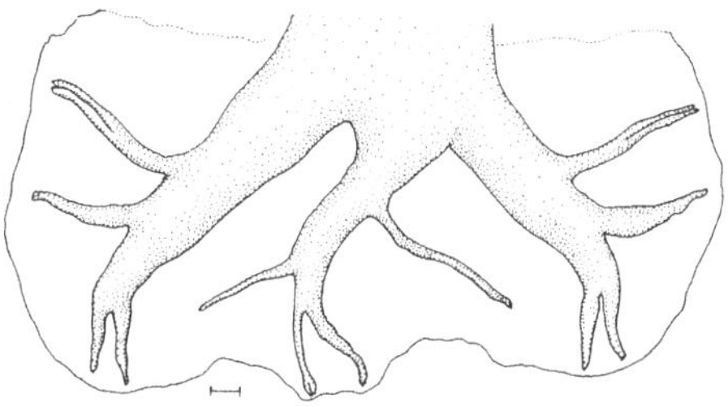

8.36

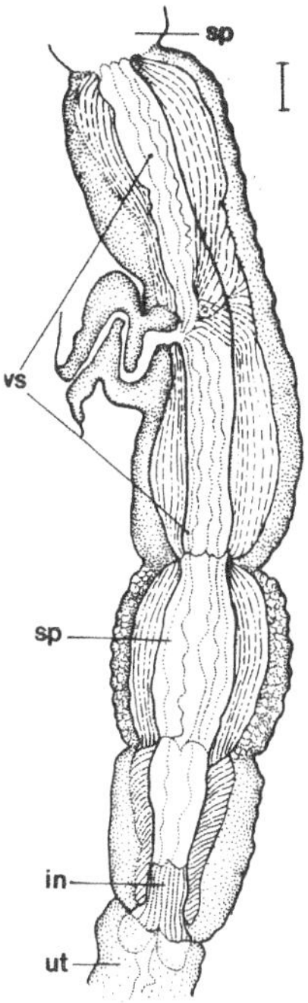

8.37

FIG. 8.35. *Bunostomum trigonocephalum* (Rudolphi, 1808), cephalic extremity, lateral view. (Original.) (Scale bar 50 μm.)

FIG. 8.36. *Bunostomum trigonocephalum,* dorsal bursal rays. (After Ransom, 1911.) (Scale bar 50 μm.)

FIG. 8.37. *Bunostomum trigonocephalum*, ovejector, lateral view, showing vestibula (vs), sphincters (sp), infundibula (in), and uteri (ut). (Original.) Scale bar 50 μm.)

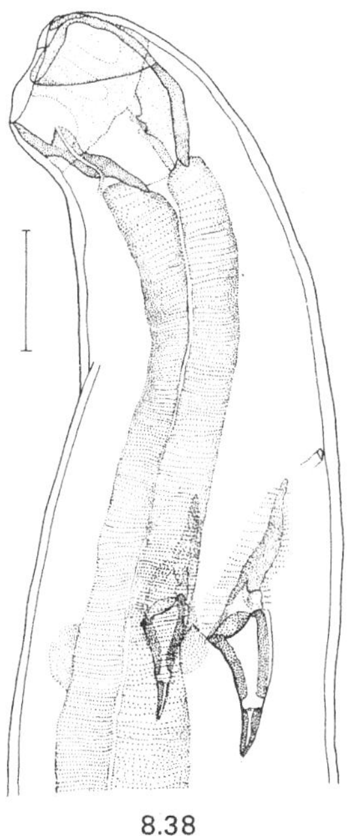

8.38

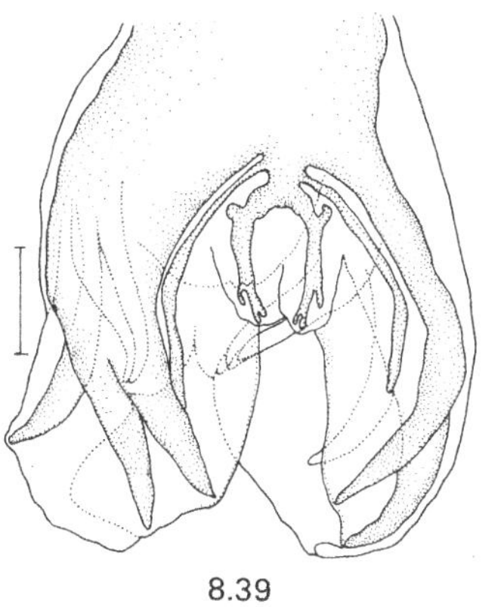

8.39

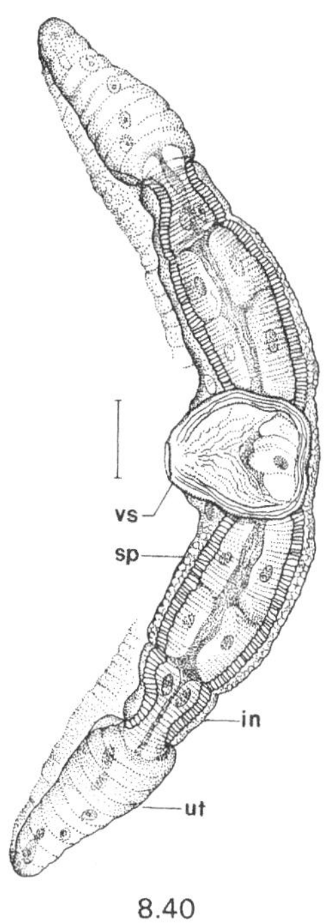

8.40

FIG. 8.38. *Cameronecator urichi* (Cameron, 1936) n. g., n. comb., cephalic extremity, lateral view. (After Freitas, 1951.) (Scale bar 50 μm.)

FIG. 8.39. *Cameronecator urichi,* dorsal bursal rays. (After Freitas, 1951.) (Scale bar 50 μm.)

FIG. 8.40. *Cameronecator urichi,* ovejector, lateral view, showing vestibula (vs), sphincters (sp), infundibulum (in), and uterus (ut). (Original.) (Scale bar 50 μm.)

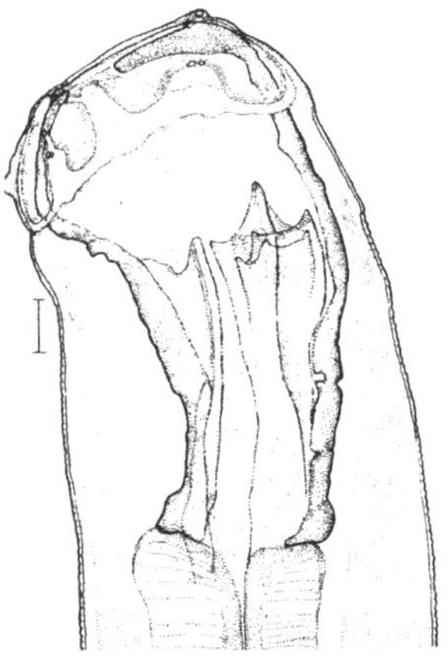

8.41

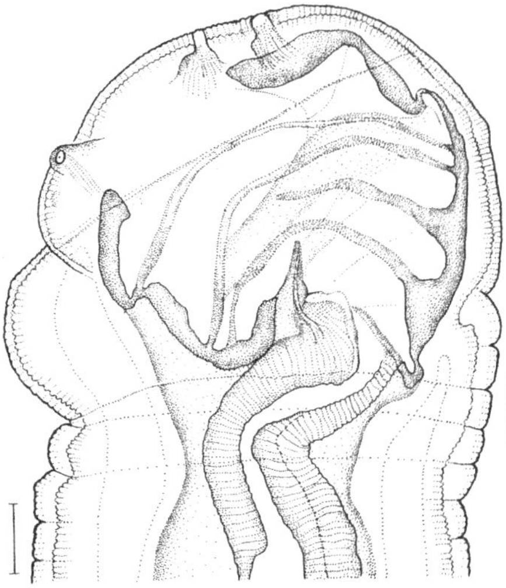

8.42

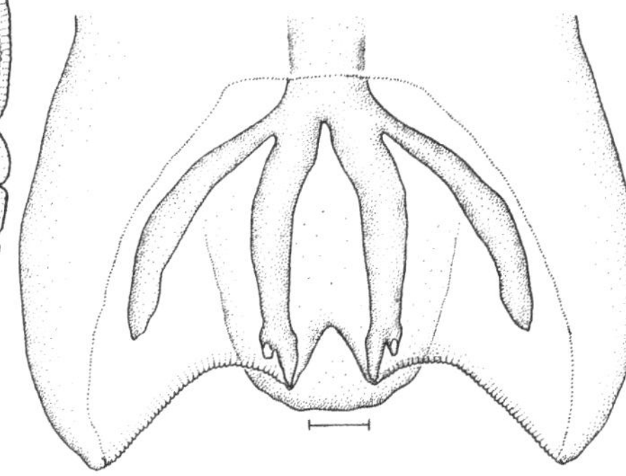

8.43

FIG. 8.41. *Grammocephalus clathratus* (Baird, 1868), cephalic extremity, lateral view. (Original.) (Scale bar 50 µm.)

FIG. 8.42. *Bathmostomum sangeri* (Cobbold, 1879), cephalic extremity, lateral view. (Original.) (Scale bar 50 µm.)

FIG. 8.43. *Bathmostomum sangeri*, dorsal bursal rays. (Original.) (Scale bar 50 µm.)

15-(14) Buccal capsule globular with subventral lancets on ventral wall (Fig. 8.38).
Deirids exceptionally large (Fig. 8.38).
Vulva near mid-body.
Dorsal ray symmetrical (Fig. 8.39).
Parasites of crab-eating raccoon (*Procyon cancrivorus*); South America, West Indies.
Cameronecator n. g.

16-(13) Buccal capsule long and funnel shaped or globular with network of internal ridges and small dorsal cone (Figs. 8.41, 8.42).
Externodorsal rays arising posterior to branching of dorsal ray or nearly so (Fig. 8.43).

17-(18) Buccal capsule elongate and funnel-shaped (Fig. 8.41).
Intestinal caecum directed anteriorly from oesophageal-intestinal junction.
Parasites of stomach or bile duct of elephants (*Loxodonta, Elephas*) and rhinoceroses (*Diceros*).
Grammocephalus Railliet & Henry, 1910

18-(17) Buccal capsule globular, with small dorsal cone and shelf-like ridges on inner surface of posterior walls (Fig. 8.42).
Intestinal caecum absent.
Parasites of Indian elephant (*Elephas*).
Bathmostomum Railliet & Henry, 1909

REFERENCES

BALASINGAM, E. 1964. Studies on the life cycle and developmental morphology of *Placoconus lotoris* (Schwartz, 1925) Webster, 1956 (Ancylostomidae: Nematoda). *Can. J. Zool.*, **42,** 869–902.

BEVERIDGE, I. 1979. A review of the Globocephaloidinae Inglis (Nematoda: Amidostomatidae) from macropodid marsupials. *Aust. J. Zool.*, **27,** 151–175.

BIOCCA, E. & LEROUX, P. L. 1958. Subdivisione del genere *Ancylostoma* (Dubini, 1843) in quattro sottogeneri. *Atti Accad. naz. Lincei Rc.*, Ser. 8, **23,** 470–477.

BOTTO, C. & MANE-GARZON, F. 1975. Sobre una nueva subespecie del genero *Uncinaria* (Nematoda, Strongyloidea) de *Otaria flavescens* Shaw y la especiacion en el genero *Uncinaria* en pinipedos Americanos. *Revta Biol. Uruguay*, **3,** 127–141.

CHABAUD, A. G. 1965. Ordre des Strongylida. *In:* Grassé, P. P. (Editor), *Traité de Zoologie*, Vol. 4, fasc. 3, pp. 869–933. Paris: Masson et Cie.

——— 1974. Keys to subclasses, orders and superfamilies. *In:* Anderson, R. C., Chabaud, A. G. & Willmott, S., Editors, *CIH Keys to the Nematode Parasites of Vertebrates.* No. 1, pp. 6–17. Farnham Royal, Bucks, England: Commonwealth Agricultural Bureaux.

———, BAIN, O. & HOUIN, R. 1966. Nématodes de Potamocheres malgaches. *Annls Parasit. hum. comp.*, **41,** 599–606.

———, BRYGOO, E. R. & TCHEPRAKOFF, R. 1964. Nématodes parasites d'insectivores malgaches. *Bull. Mus. natn. Hist. nat., Paris*, 2e sér., **36,** No. 2, 245–261.

——— & DURETTE-DESSET, M. C. 1975. *Uncinaria* (*Megadeirides*) *olseni* n. sp. Nématode à caractères archaiques parasite d'un *Tupaia* à Borneo. *Annls Parasit. hum. comp.*, **50,** 789–793.

———, PETTER, A. J. & BRYGOO, E. R. 1961. Trois nématodes parasites de hérissons malgaches. *Bull. Soc. zool., Fr.*, **86,** 38–51.

CHITWOOD, M. B. 1969. *The systematics and biology of some parasitic nematodes.* In: Florkin, M. & Scheer, B. T. (Editors). *Chemical Zoology.* Vol. III. *Echinodermata, Nematoda and Acanthocephala.* New York & London: Academic Press, Inc., pp. 223–244.

——— & JORDAN, H. E. 1965. *Monodontus louisianensis* sp. n. (Nematoda: Ancylostomatidae) a hookworm from the white-tailed deer, *Odocoileus virginianus* (Zimmermann), and a key to the species of *Monodontus. J. Parasit.*, **51,** 942–944.

DAENGSVANG, S., TINGPALOPONG, M. & LICHTENFELS, J. R. 1975. A record of *Tetragomphius arctonycis* Jansen, 1968 (Nematoda: Ancylostomatidae) from the hog-badger, *Arctonyx collaris,* of Trang Province, Southern Thailand. *S. East Asian J. trop. Med. publ. Hlth*, **6,** 287–289.

DOUGHERTY, E. C. 1951. Evolution of zooparasitic groups in the phylum Nematoda, with special reference to host-distribution. *J. Parasit.*, **37,** 353–378.

DURETTE-DESSET, M. C., DENKE, M. A. & MURUA, R. 1976. Présence chez un rongeur du Chili d'un nématode Inglamidinae (sub. fam. nov.) appartenant aux Amidostomatidae, famille connue des Mammifères d'Australie. *Annls Parasit. hum. comp.*, **51,** 453–460.

FREITAS, J. F. T. 1951. Alguns "Strongyloidea" parasitos de "*Procyon cancrivorus*" (Cuv.) (Nematoda). *Revta brasil. Biol.*, **11,** 189–202.

INGLIS, W. G. 1968. The geographical and evolutionary relationships of Australian trichostrongyloid parasites and their hosts. *J. Linn. Soc. (Zool.)*, **47,** 327–347.

JANSEN, J., Jr. 1968a. *Tetragomphius arctonycis* n. sp. from the pancreatic ducts of a Mustelid, *Arctonyx collaris. J. Helminth.*, **42,** 53–56.

———. 1968b. On *Arthrostoma spatulatum* n. sp. from the small intestine of *Pardofelis marmorata marmorata. J. Helminth.*, **42,** 305–308.

KALKAN, A. & HANSEN, M. F. 1966. *Ancylostoma taxideae* sp. n. from the American badger, *Taxidea taxus taxus. J. Parasit.*, **52,** 291–294.

KHALIL, L. F. 1973. *Characostomum howelli* n. sp. (Nematoda: Strongylidae) from the giant African rat *Cricetomys gambianus* and other helminths from Tanzania. *J. Helminth.*, **47,** 283–287.

KHANUM, A. 1967. A new nematode parasite *Ancylostoma sindhensis* (Ancylostomidae: Strongyloidea) from intestine of the common striped squirrel, *Funambulus pennanti argentescens* Wroughton, from Lower Sind. *Pakist. J. Sci.*, **19,** 105–108.

KOWALSKI, K. 1971. *Mammals. An outline of theriology.* (Translated English version available as TT71-54045 from N.T.I.S., U.S. Dept. Commerce, Springfield, VA 22151, USA).

LANE, C. 1917. *Bunostomum kashinathi* and the Ancylostomidae. *Indian J. med. Res.*, **4,** 414–439.

LICHENFELS, J. R. 1980. Keys to genera of the Superfamily Strongyloidea. *In:* Anderson, R. C., Chabaud, A. G. & Willmott, S. (Editors), *CIH Keys to the Nematode Parasites of Vertebrates.* No. 7, Farnham Royal, Bucks, England. Commonwealth Agricultural Bureaux.

LOOSS, A. 1905. The anatomy and life history of *Agchylostoma duodenale* Dub. A monograph. *Rec. Egypt. Govt Sch. Med.*, **3,** 1–158.

———. 1911. The anatomy and life history of *Agchylostoma duodenale* Dub. A monograph. Pt. 2: The development in the free state. *Rec. Egypt. Govt Sch. Med.*, **4,** 159–613.

OHBAYASHI, M., SUZUKI, Y. & ARAKI, J. 1974. *Tetragomphius melis* n. sp. (Nematoda: Ancylostomidae) from the Japanese badger, *Meles meles anakuma* Temminck. *Jap. J. vet. Res.*, **22,** 43–46.

OLSEN, O. W. 1968. *Uncinaria rauschi* (Strongyloidea: Nematoda), a new species of hookworms from Alaskan bears. *Can. J. Zool.*, **46,** 1113–1117.

ORTLEPP, R. J. 1964. Observations on helminths parasitic in warthogs and bushpigs. *Onderstepoort J. vet. Res.*, **31,** 11–38.

REP, B. H. 1963a. On the polyxenia of the Ancylostomidae and the validity of the characters used for their differentiation. *Trop. geogr. Med.*, **15,** 173–218.

———. 1963b. On the polyxenia of Ancylostomidae and the validity of the characters used for their differentiation (II). *Trop. geogr. Med.*, **15,** 271–316.

———. 1964. A new system for the diagnosis of Ancylostomidae, especially the human hookworm species. *Trop. geogr. Med.*, **16,** 354–369.

———, VETTER, J. C. M. & EIJSKER, M. 1968. Cross breeding experiments in *Ancylostoma braziliense* De Faria, 1910, and *A. ceylanicum* Looss, 1911. *Trop. geogr. Med.*, **20,** 367–378.

SCHMIDT, G. D. & KUNTZ, R. E. 1968. Nematode parasites of Oceanica. III. *Arthrostoma vampira* sp. n., with a reconsideration of those hookworms having articulated buccal capsules. *J. Parasit.*, **54,** 372–375.

SCHULZ, R. S. 1952. Phylogeny of Strongylates. *In:* Skrjabin, K. I. (Editor), *Key to Parasitic Nematodes, Vol. III. Strongylata.* Moscow: Akad. Nauk. SSSR, pp. 13–20. (Translated English version available as OTS60-51062 from N.T.I.S., U.S. Dept. Commerce, Springfield, VA 22151, USA.)

SETASUBAN, P. 1976. Light microscopy and scanning electron microscopy of *Bathmostomum sangeri* Cobbold, 1879, of elephants. *S. East Asian J. trop. Med. publ Hlth*, **7,** 390–394.

SINGH, N. & PANDE, B. P. 1966. Notes on nematode parasites of Indian mongoose, *Herpestes mungo. Annls Parasit. hum. comp.*, **41,** 467–485.

SKRJABIN, K. I., SHIKHOBALOVA, N. P., SCHULZ, R. S., POPOVA, T. I., BOEV, S. N. & DELYAMURE, S. L. 1952. Key to parasitic Nematodes. III. Strongylata. Moscow: Akad. Nauk, SSSR. 883 pp. (Translated English version available as OTS60-51062 from N.T.I.S., U.S. Dept. Commerce, Springfield, VA 22151, USA.)

THATCHER, V. E. 1971. Some hookworms of the genus *Ancylostoma* from Colombia and Panama. *Proc. helminth. Soc. Wash.*, **38,** 109–116.

UNI, S., ISEKI, M. & TAKADA, S. 1977. Scanning electron microscopy of the sense organs and the copulatory bursa of *Ancylostoma duodenale* and *Ancylostoma caninum. Jap. J. Parasit.*, **26,** 157–167.

WEBSTER, G. A. 1956. *Placoconus:* A new genus for *Arthrocephalus lotoris* (Schwartz, 1925) Chandler, 1942. *Can. J. Zool.*, **34,** 99–103.

YOSHIDA, Y. & ARIZONO, N. 1976. *Arthrostoma miyazakiense* (Nagayosi 1955) comb. n., a parasite of the raccoon-like dog, *Nyctereutes procyonoides,* with a key to the genus *Arthrostoma* (Nematoda: Ancylostomatidae). *J. Parasit.*, **62,** 766–770.

———, KONDO, K., KURIMOTO, H., FUKUTOME, S. & SHIRASAKA, S. 1974. Comparative studies on *Ancylostoma braziliense* and *Ancylostoma ceylanicum.* III. Life history in the definitive host. *J. Parasit.*, **60,** 636–641.

———, KONDO, K., OKADA, S., OKAMOTO, K., KURIMOTO, H., ODA, K. & SHIMADA, Y. 1974. Morphology and life history of *Ancylostoma kusimaense* Nagayosi, 1955. *Jap. J. Parasit.*, **23,** 187–200.

5 STRONGYLIDA

STRONGYLOIDEA

by J. RALPH LICHTENFELS*

INTRODUCTION

Members of the Strongyloidea are almost exclusively parasites of mammals, and the superfamily includes many important parasites of domestic animals. It is an ancient group that has been associated with its hosts for more than 75 million years. Excepting the Syngamidae, the Strongyloidea are very host-specific. Authors of earlier classifications of the Strongyloidea relied heavily on the morphology of buccal capsules. Separation of families and subfamilies was difficult and arbitrary and there was little indication of the relation between parasite and host evolution below the superfamily level (Chabaud, 1965; Yamaguti, 1961; Popova, 1955, 1958). Early in the present work it became obvious that additional characters would be needed for a classification that would reflect evolutionary patterns and have greater predictive value. Accordingly, representative specimens of all available genera were studied. Of the 78 genera (87 including genera and subgenera) of the Strongyloidea considered valid in this work, only *Archeostrongylus* Biocca & Ferretti, 1958, *Rhabditostomum* Chabaud & Krishnasamy, 1976, *Caballonema* Abuladze, 1937, *Lemurostrongylus* Chabaud, Brygoo & Petter, 1961 and *Macropicola* Mawson, 1978 were not studied. Other genera of uncertain classification or doubtful status that are not classified herein include: *Buccostrongylus* Johnston & Mawson, 1939, *Paraoesophagostomum* Scheidegger & Kreis, 1934, *Skrjabinodentus* Tshoijo, 1957 and *Tridentoinfundibulum* Tshoijo, 1957.

In addition to morphology, other data used in developing the new classification described herein were host evolution, zoogeography and site of parasitism. Some may find the new system radical, but it should be at least as easy to use for identifications as previous systems and, if it correctly reflects evolutionary relationships, it will have significantly improved predictive value.

KEY MORPHOLOGICAL CHARACTERS

Two types of ovejectors were found within the Strongyloidea. Also, the type of ovejector is correlated with the morphology of the dorsal ray and the host group. This new combination of characters greatly facilitates the separation of natural groups of the Strongyloidea in the following keys.

Ovejectors: Type I ovejectors are usually Y-shaped, but may be opened to a straight line arrangement in the few genera with an anteriorly placed vulva (Fig. 7.1). Type II ovejectors are usually J-shaped with the posterior branch turned anteriorly (Fig. 7.2).

* U.S. Department of Agriculture, Science and Education Administration, Agricultural Research, Animal Parasitology Institute, Beltsville, Maryland 20705.

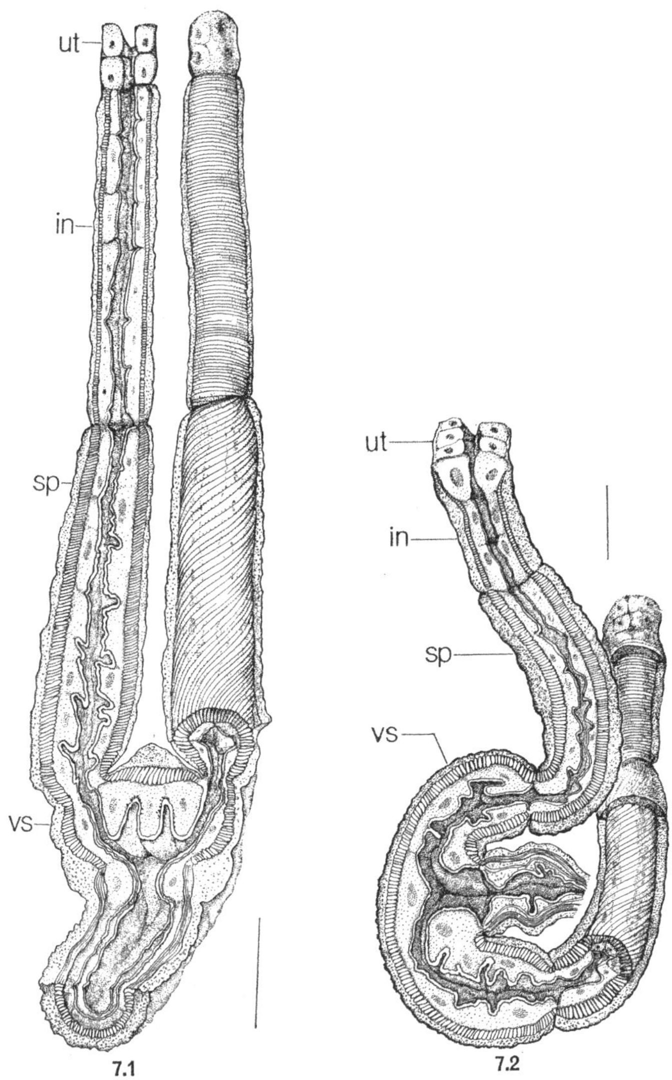

FIG. 7.1. Ovejector of *Cyathostomum labiatum* (Looss, 1902) showing uteri (ut) and vestibula (vs), sphincters (sp) and infundibula (in) of Y-shaped (Type I) ovejector. (Original.) (Scale bar 100 μm.)

FIG. 7.2. Ovejector of *Oesophagostomum* (*O.*) *dentatum* (Rudolphi, 1803) showing uteri (ut) and vestibula (vs), sphincters (sp) and infundibula (in) of kidney-shaped (Type II) ovejector. (Original.) (Scale bar 100 μm.)

Both types of ovejectors consist of three parts: (i) a thick-walled vestibule communicating with the vagina, (ii) paired thick-walled sphincters and (iii) thinner-walled infundibula which connect with the uteri.

All three parts of the ovejector have a cuticular lining, a cellular middle layer and an outer muscular layer. All muscular parts of the ovejector are covered by a fluffy coat about as thick as the muscular layer which may consist of the cell bodies of the muscles.

Type I and Type II ovejectors also differ as follows: (i) the vestibule is small and round or slightly Y-shaped in the Type I ovejectors (Figs. 7.1, 7.3–7.11) but is larger and often kidney-shaped in the J-shaped (Type II) ovejectors (Figs. 7.2, 7.12–7.20); (ii) the sphincters are relatively larger than the vestibule in the Y-shaped ovejectors but smaller than the vestibule in the J-shaped ovejectors; and (iii) the infundibula are usually similar in size with only slightly thinner walls than the sphincters in the Y-shaped ovejectors, but are usually significantly smaller with much thinner walls than the sphincters in the J-shaped ovejectors. The two types of ovejectors are usually recognized easily by their orientation, but variations do occur. A few Type II ovejectors are almost Y-shaped (*Phacochoerostrongylus* (Fig. 7.19), *Rhabditostomum* and *Woodwardostrongylus*) but they can be recognized as Type II by the elongated vestibule and greatly reduced infundibula. The completely opposed Type I ovejectors of *Strongylus* (Fig. 7.9) are believed, following Popova (1955), to result from the forward position of the vulva and the thick body size in which the Y-shaped orientation has become almost T-shaped. The Type I ovejectors of *Theileriana* (Fig. 7.11) and *Quilonia* (Fig. 7.10) may be similar to the ancestral stock that gave rise to the Type II ovejectors of the Chabertiidae, especially as found in *Trachypharynx* (Fig. 7.12).

The results of the trend among the genera with J-shaped ovejectors towards increasing relative size and thickness of the kidney-shaped vestibule and reduction of the size and thickness of the sphincters and, especially, the infundibula can be seen in the ovejectors of *Chabertia* (Fig. 7.16), *Oesophagostomum* (Fig. 7.2) and *Kuntzistrongylus* n. g. (Fig. 7.15).

Dorsal ray: Two kinds of dorsal ray were observed. Type I has three rami on each side of the median fissure and Type II has two rami on each side of the median fissure. With few exceptions, Strongyloidea with a Type I ovejector have a Type I dorsal ray (Figs. 7.21–7.32)

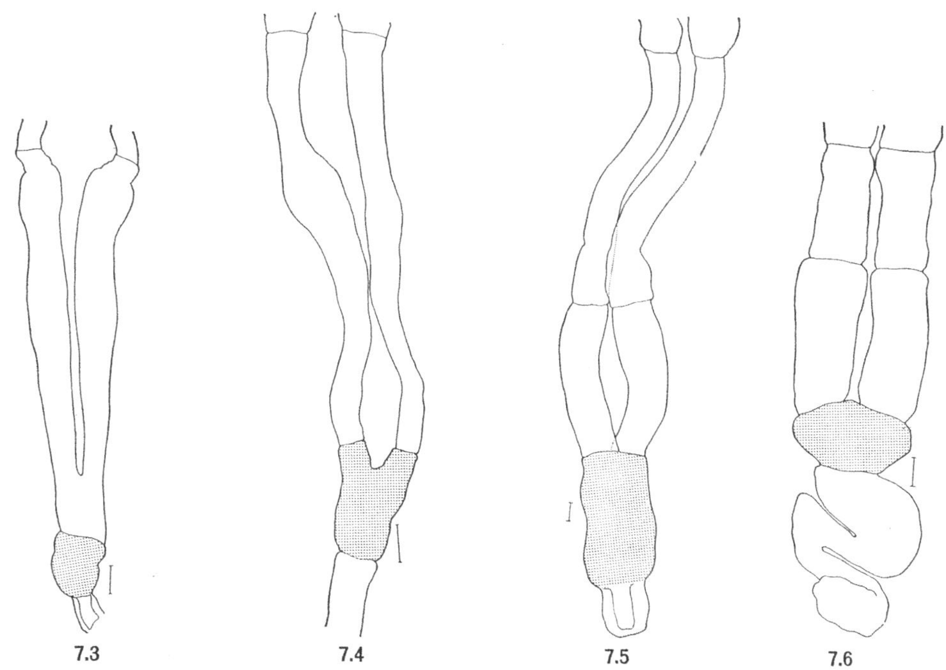

FIG. 7.3. Ovejector of *Sauricola sauricola* Chapin, 1924. (Original.) (Vestibula stippled; scale bar 50 μm.)

FIG. 7.4. Ovejector of *Choniangium epistomum* (Piana & Stazzi, 1900). (Original.) (Vestibula stippled; scale bar 50 μm.)

FIG. 7.5. Ovejector of *Murshidia murshidia* Lane, 1914. (Original.) (Vestibula stippled; scale bar 50 μm.)

FIG. 7.6. Ovejector of *Triodontophorus serratus* (Looss, 1900). (Original.) (Vestibula stippled; scale bar 50 μm.)

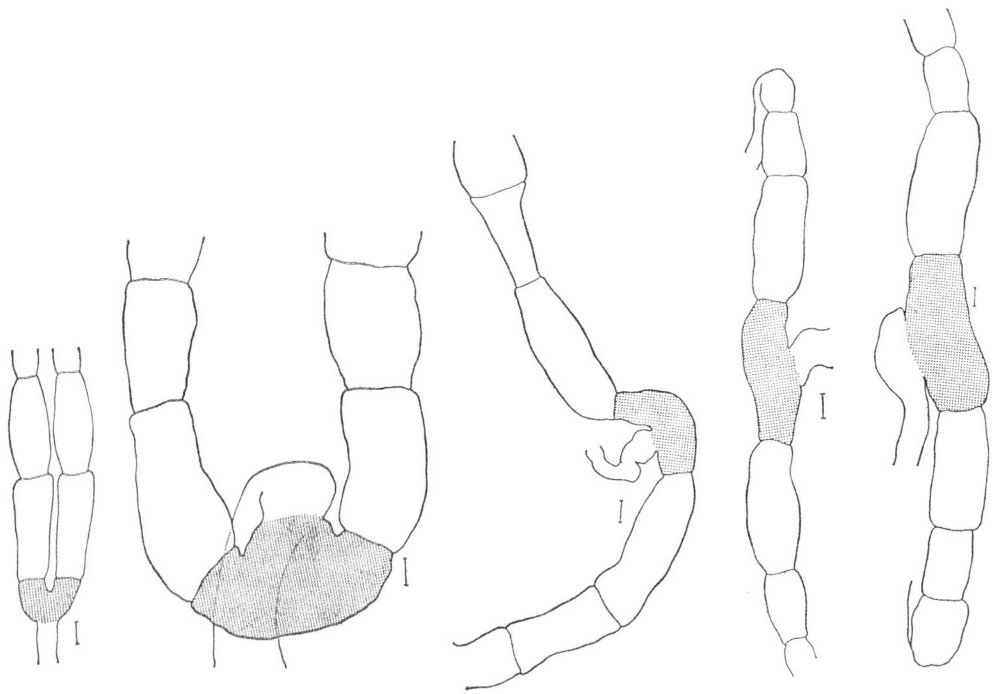

7.7 7.8 7.9 7.10 7.11

FIG. 7.7. Ovejector of *Phascolostrongylus turleyi* Canavan, 1931. (Original.) (Vestibula stippled; scale bar 50 μm.)

FIG. 7.8. Ovejector of *Paramacropostrongylus typicus* Johnston & Mawson, 1940. (Original.) (Vestibula stippled; scale bar 50 μm.)

FIG. 7.9. Ovejector of *Strongylus vulgaris* (Looss, 1900). (Original.) (Vestibula stippled; scale bar 50 μm.)

FIG. 7.10. Ovejector of *Quilonia* sp. (Original.) (Vestibula stippled; scale bar 50 μm.)

FIG. 7.11. Ovejector of *Theileriana brachylaima* (Linstow, 1901). (Original.) (Vestibula stippled; scale bar 50 μm.)

and those with a Type II ovejector also have a Type II dorsal ray (Figs. 7.37–7.52). A major exception to this correlation is found in the four genera of the Australian subfamily Phascolostrongylinae and the two Australian genera in the Strongylinae, *Macropicola* and *Hypodontus*. All Australian Strongyloidea have a Type II dorsal ray; but in those with a Type I ovejector the externodorsal rays arise from the dorsal ray (Fig. 7.36) and in those with a Type II ovejector the externodorsal rays arise separately from the dorsal ray (Figs. 7.45–7.48) (except *Corollostrongylus*, Fig. 7.52).

There is somewhat more variability in the type of dorsal ray among related genera and species than is found in the type of ovejector. Chabaud (1957) and Witenberg (1925) illustrated the variation in the degree of splitting of the proximal branches of the dorsal ray among the species of *Murshidia* (Fig. 7.25). The species with an evolutionary trend toward a longer dorsal lobe (subgenus *Pteridopharynx*) showed a reduction in splitting of the proximal branches. The species of *Kiluluma* show a variability in splitting of the proximal branches of the dorsal ray (Fig. 7.30) which is similar to that found among species of *Murshidia*. The dorsal ray of *Chapiniella variabilis* (Chapin, 1924) of tortoises also has three main branches each of which usually bifurcates (Fig. 7.26). *Chapiniella* and *Kiluluma* have in common several other characters considered primitive including thin Y-shaped ovejectors, broad cuticular annules and few

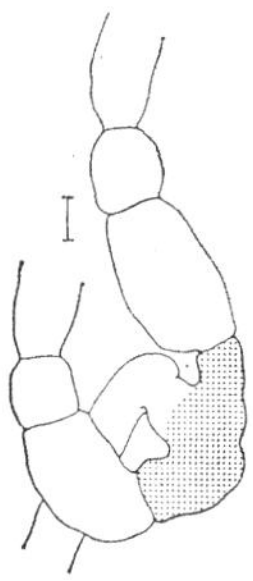

7.12

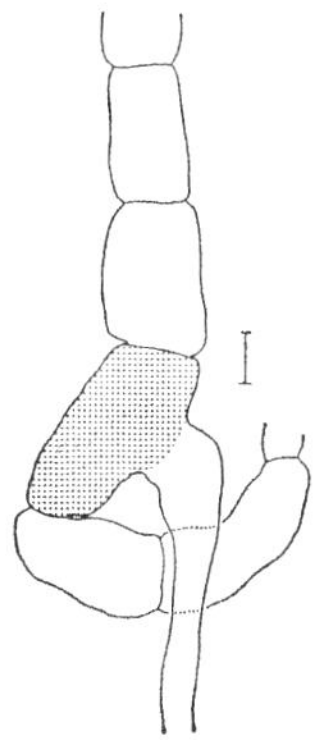

7.13

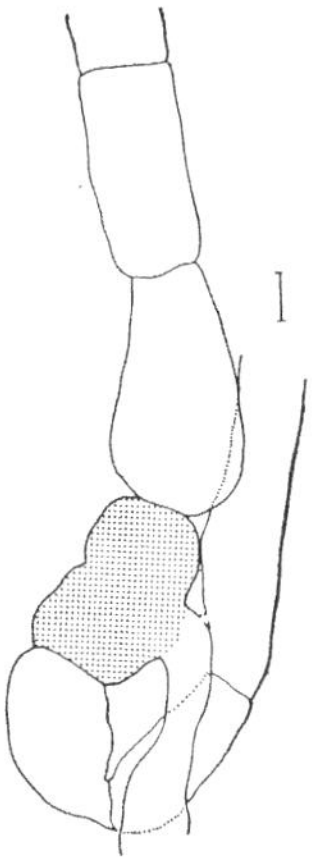

7.14

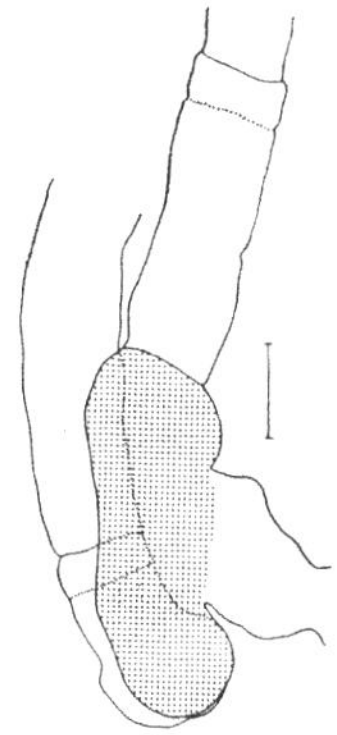

7.15

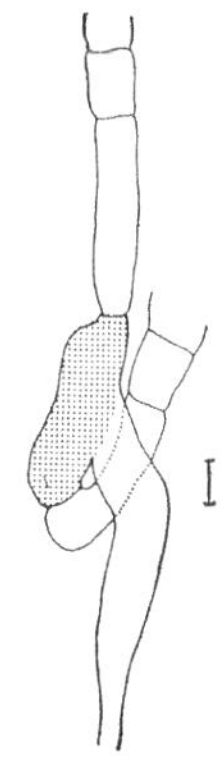

7.16

FIG. 7.12. Ovejector of *Trachypharynx natalensis* Ortlepp, 1962. (Original.) (Vestibula stippled; scale bar 50 μm.)

FIG. 7.13. Ovejector of *Ransomus rodentorum* Hall, 1916. (Original.) (Vestibula stippled; scale bar 50 μm.)

FIG. 7.14. Ovejector of *Castorstrongylus castoris* Chapin, 1925. (Original.) (Vestibula stippled; scale bar 50 μm.)

FIG. 7.15. Ovejector of *Kuntzistrongylus selfi* (Schmidt & Kuntz, 1975). (Original.) (Vestibula stippled; scale bar 50 μm.)

FIG. 7.16. Ovejector of *Chabertia ovina* (Fabricius, 1788). (Original.) (Vestibula stippled; scale bar 50 μm.)

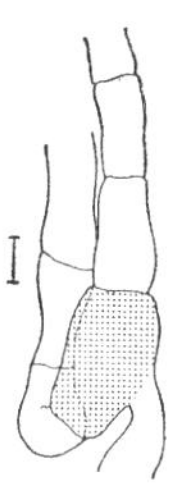

7.17

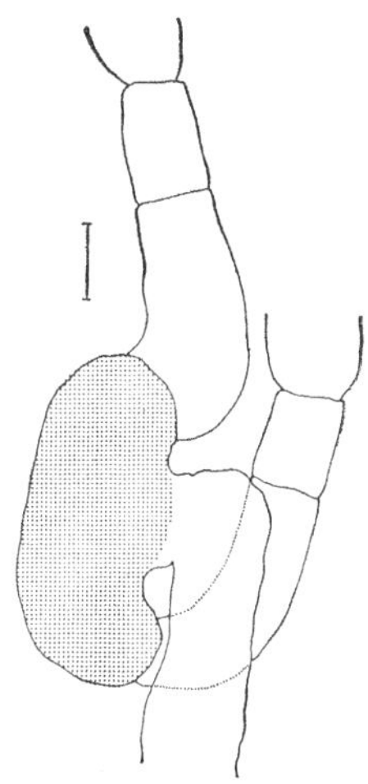

7.18

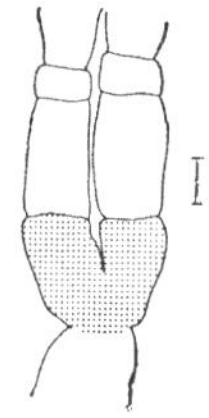

7.19

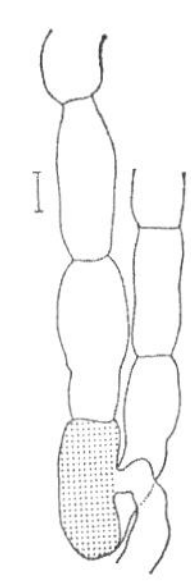

7.20

FIG. 7.17. Ovejector of *Rugopharynx* sp. (Original.) (Vestibula stippled; scale bar 50 μm.)

FIG. 7.18. Ovejector of *Cloacina* sp. (Original.) (Vestibula stippled; scale bar 50 μm.)

FIG. 7.19. Ovejector of *Phacochoerostrongylus hylochoeri* (Van den Berghe, 1943). (Original.) (Vestibula stippled; scale bar 50 μm.)

FIG. 7.20. Ovejector of *Corollostrongylus hypsiprymnodontis* Beveridge, 1978. (Original.) (Vestibula stippled; scale bar 50 μm.)

elements in the corona radiata. These primitive genera may be similar to the ancestral stock from which evolved the more specialized genera with thicker Y-shaped ovejectors with differentiated portions, narrow cuticular striae and, usually, dorsal rays with six rami. Among genera with a Type I ovejector, only *Sauricola* (Fig. 7.34) in tortoises, *Theileriana* (Fig. 7.35) in *Hyrax* and six genera in Australian marsupials have dorsal rays with four rather than six rami. The dorsal ray of *Decrusia* (Fig. 7.34) does not fit well in either the Type I or Type II groups.

Host groups: Most of the Strongyloidea are parasitic in mammals but a few genera are parasitic in birds and turtles. The evolution of the Strongyloidea is believed to be closely related to the evolution of mammals (Lichtenfels, 1979). If the 78 genera of the Strongyloidea are superimposed on a dendrogram of mammalian evolution (Lichtenfels, 1979), one finds a striking correlation between nematode morphological type and host groups. Nematodes with Type I ovejectors and dorsal rays are mostly parasitic in Perissodactyla, Proboscidea and Hyracoidea (23 genera in horses, elephants, hyraxes, rhinoceroses and tapirs) and in vombatid and macropodid marsupials (six genera), with a few species of only two genera parasitic in Artiodactyla. Two genera with Type I ovejectors (one of which has a Type I dorsal ray) are parasites of turtles (the only Strongyloidea in reptiles) and three Type I genera are parasites of ostriches and rheas.

Strongyloidea with Type II ovejectors and dorsal rays are parasites of Artiodactyla (10 genera in deer, cattle, sheep and pigs), macropodid marsupials (18 genera), rodents (six genera) and primates (four genera).

In addition to the gut parasites, *Stephanurus* is parasitic in the kidney tissue of swine and five genera are parasitic in the respiratory system of mammals (two genera in rodents, cats, bovids and elephants) and birds (three genera). The Strongyloidea parasitic in the respiratory system are apparently more site-specific than host-specific.

A dendrogram illustrating the possible evolution of the Strongyloidea (Fig. 7.53) was constructed from the above morphological and host data. Groups of nematodes are presumed to have separated when their hosts separated. With the use of the new key characters described above, some genera are placed in new groups under old group names. Detailed comparisons with previous systems are not given (see Lichtenfels, 1979), but some changes are noted under the following discussion of each family and subfamily, and genera can easily be located in the dendrogram (Fig. 7.53). Because the subfamilies are all believed to have been isolated in their host groups (for the Syngamidae in their tissue sites) before the end of the Cretaceous about 65 million years ago, consideration was given to elevating the subfamilies to family level. This elevation was avoided because of the similarity among genera of the Strongylinae and Cyathostominae of horses, among Strongylinae and Cyathostominae of marsupials and among genera of the Chabertiinae and Oesophagostominae. Separation of these subfamilies is on morphological evidence alone, and similarities in buccal capsules might be due to convergence within the families.

STRONGYLIDAE

Most Strongyloidea with Type I ovejectors and Type I dorsal rays are placed in this family. The six Australian genera with Type I ovejectors and Type II dorsal rays are also placed in the Strongylidae. Characteristics of the buccal capsule are used to separate three subfamilies.

Strongylinae

Strongylidae with large globular or subglobular buccal capsules are placed here. The 11 genera are parasites of equids, elephants, ostriches or Australian marsupials. These are believed to be ancient parasites dating perhaps to the Permian 205 million years ago when birds and mammals are believed to have evolved from reptilian stock. This hypothesis is supported by the survival of few species of the 11 genera. Part of the evidence against such

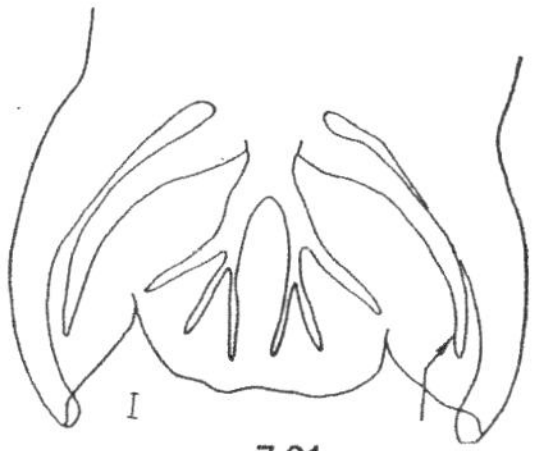

7.21

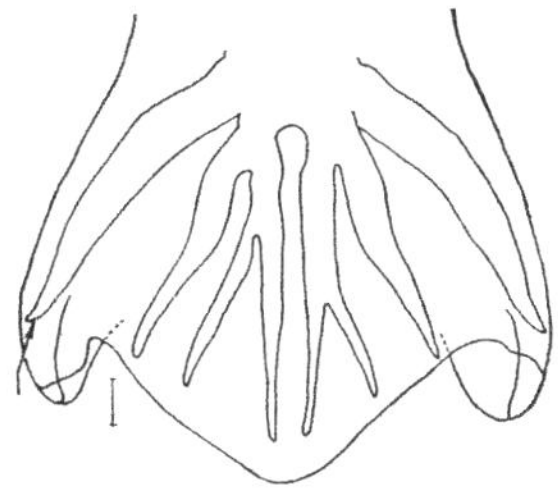

7.22

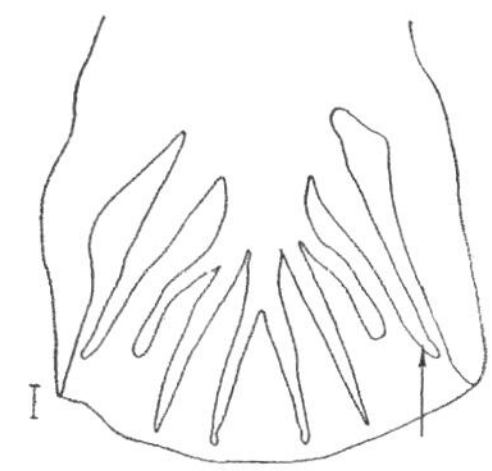

7.23

FIG. 7.21. Dorsal bursal rays of *Strongylus equinus* Mueller, 1780. (Original.) (Arrow indicates externodorsal ray; scale bar 50 μm.)

FIG. 7.22. Dorsal bursal rays of *Triodontophorus serratus* (Looss, 1900). (Original.) (Arrow indicates externodorsal ray; scale bar 50 μm.)

FIG. 7.23. Dorsal bursal rays of *Choniangium epistomum* (Piana & Stazzi, 1900). (Original.) (Arrow indicates externodorsal ray; scale bar 50 μm.)

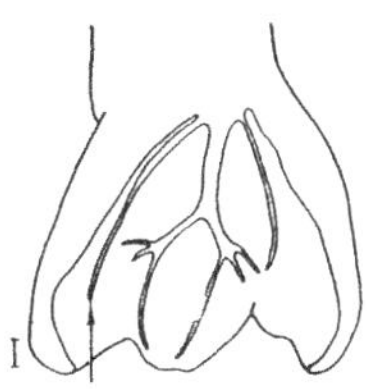

7.24

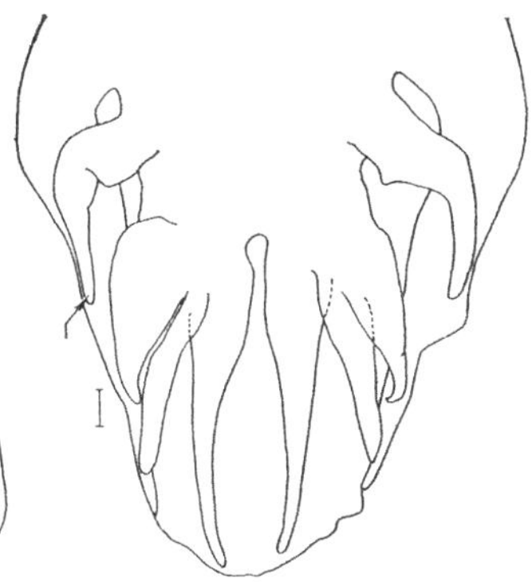

7.25

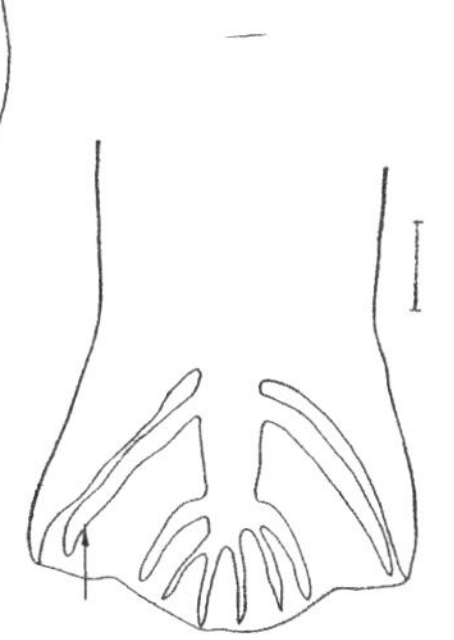

7.26

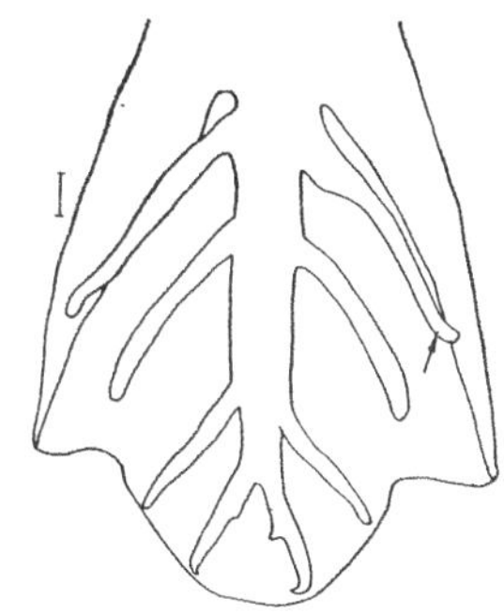

7.27

FIG. 7.24. Dorsal bursal rays of *Codiostomum struthionis* (Horst, 1885). (After Skrjabin *et al.*, 1952.) (Arrow indicates externodorsal ray; scale bar 50 μm.)

FIG. 7.25. Dorsal bursal rays of *Murshidia murshidia* Lane, 1914. (Original.) (Arrow indicates externodorsal ray; scale bar 50 μm.)

FIG. 7.26. Dorsal bursal rays of *Chapiniella variabilis* (Chapin, 1925). (Original.) (Arrow indicates externodorsal ray; scale bar 50 μm.)

FIG. 7.27. Dorsal bursal rays of *Eucyathostomum longesubulatum* Molin, 1861. (Original.) (Arrow indicates externodorsal ray; scale bar 50 μm.)

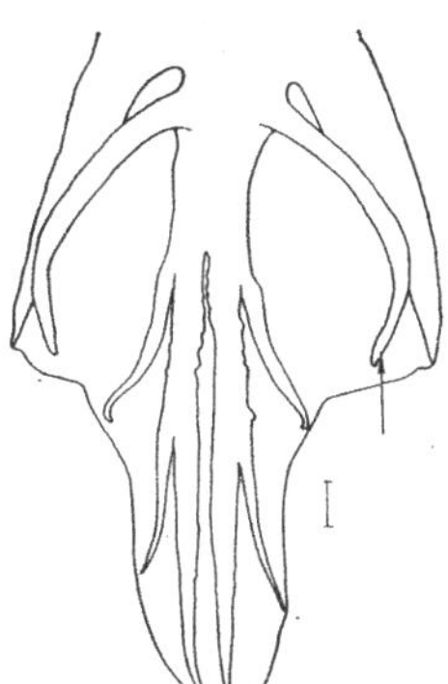

7.28

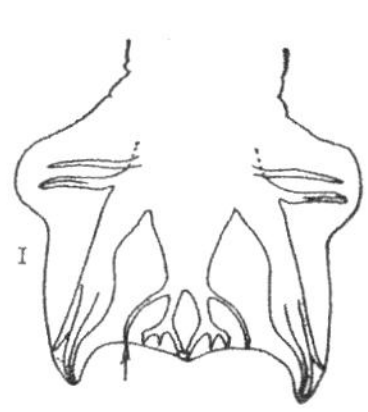

7.29

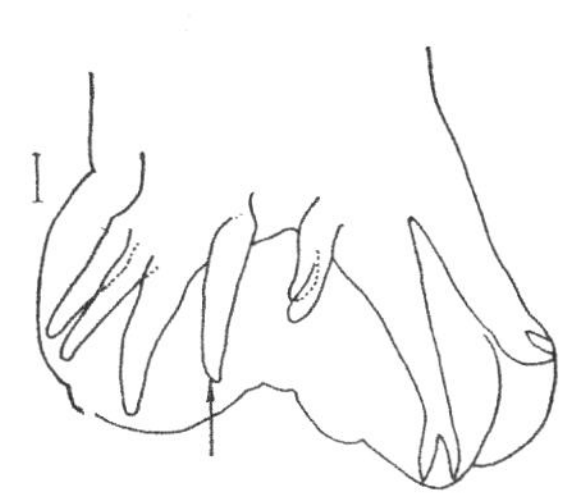

7.30

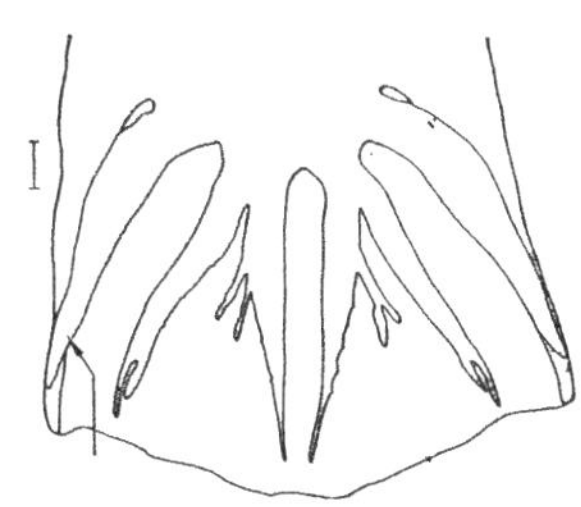

7.31

FIG. 7.28. Dorsal bursal rays of *Cyathostomum coronatum* Looss, 1900. (Original.) (Arrow indicates externodorsal ray; scale bar 50 μm.)

FIG. 7.29. Dorsal bursal rays of *Khalilia sameera* (Khalil, 1922). (After Ogden, 1966b.) (Arrow indicates externodorsal ray; scale bar 50 μm.)

FIG. 7.30. Dorsal bursal rays of *Kiluluma longipene* (Molin, 1861). (Original.) (Arrow indicates externodorsal ray; scale bar 50 μm.)

FIG. 7.31. Dorsal bursal rays of *Cylindropharynx intermedia* Theiler, 1923. (Original.) (Arrow indicates externodorsal ray; scale bar 50 μm.)

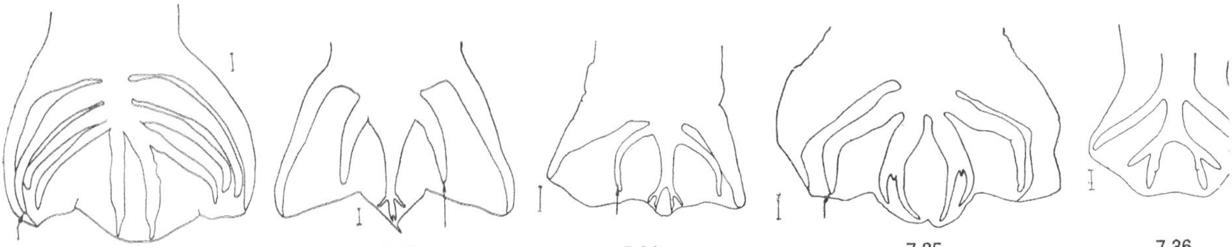

7.32 7.33 7.34 7.35 7.36

FIG. 7.32. Dorsal bursal rays of *Poteriostomum imparidentatum* Quiel, 1919. (Original.) (Arrow indicates externodorsal ray; scale bar 50 μm.)

FIG. 7.33. Dorsal bursal rays of *Decrusia additicta* Railliet, Henry & Bauche, 1914. (Original.) (Arrow indicates externodorsal ray; scale bar 50 μm.)

FIG. 7.34. Dorsal bursal rays of *Sauricola sauricola* Chapin, 1924. (Original.) (Arrow indicates externodorsal ray; scale bar 50 μm.)

FIG. 7.35. Dorsal bursal rays of *Theileriana brachylaima* (Linstow, 1901). (Original.) (Arrow indicates externodorsal ray; scale bar 50 μm.)

FIG. 7.36. Dorsal bursal rays of *Macropicola ocydromi* Mawson, 1978. (After Mawson, 1978a.) (Arrow indicates externodorsal ray; scale bar 50 μm.)

an ancient origin of the Strongylinae considered by Dougherty (1951) has been overcome by the recent discovery of representatives of the Strongylinae in Australian marsupials. The Australian species *Macropicola ocydromi*, discovered by Mawson, 1978, and *Hypodontus macropi* Mönnig, 1929 are almost unique in the subfamily in lacking a corona radiata (Fig. 7.55). *Macropicola* is placed in the Strongylinae because of similarity in the three oesophageal teeth, anteriorly directed oral opening, preanal vulva and the Y-shaped ovejectors and dorsal ray (Fig. 7.36) identical to those of other Australian Strongylidae. For the same reasons *Macropicola* is placed in the Strongylinae; *Hypodontus* Mönnig, 1929, parasitic in macropodid marsupials, is also placed in the Strongylinae rather than the Ancylostomatoidea. *Hypodontus* has been redescribed by Beveridge (1978c), who recognizes a single species *H. macropi* Mönnig, 1929. The ovejectors of *Hypodontus* are the typical Y-shaped strongyline type, the vulva is preanal, the dorsal ray is similar to those of other Australian Strongylidae and there are three similar conical teeth in the oesophageal funnel (Fig. 7.56). The oral opening is inclined anteroventrally rather than anterodorsally as in the Ancylostomatoidea. The evidence listed supports the conclusion that buccal capsule similarities between *Hypodontus* and genera of the Ancylostomatoidea are due to convergence (Inglis, 1968; Beveridge, 1978c) and further, that *Hypodontus* is directly related to genera of the Strongylinae.

Bidentostomum Tshoijo, 1957 was studied during a recent visit to Moscow at the Central Helminthological Museum of the All Union Institute of Helminthology. Figures were not made, but G. M. Dvoinos will redescribe the genus soon (personal communication).

Cyathostominae

Because a dichotomy in the nomenclature of this subfamily has existed for many years between those accepting *Cyathostomum* Molin, 1861 and those instead using *Trichonema* Cobbold, 1874 as the basis for the subfamily, an explanation is presented here for the use of *Cyathostomum*. The numerous references cited in the following excerpt from Lichtenfels (1975) can be found in that paper.

> "The genus *Cyathostomum* was described for the small to medium-sized strongylids of horses by Molin (1861) who lumped them all in one species, *C. tetracanthum* (Mehlis,

1831). Cobbold (1874) later described *Trichonema arcuata* for a larval form of *C. tetracanthum*—in reality a group of species. Looss (1900) described 12 new species in the genus *Cyathostomum*, selecting *C. tetracanthum*, the most common species, as type of the genus. Much controversy has developed over which genus, *Cyathostomum* or *Trichonema*, is valid. Both Cobbold (1875) and Looss (1902) abandoned their genera in favor of others. American workers (Foster, 1936, 1937; McIntosh, 1951) eventually accepted the validity of *Cyathostomum* based on the International Code of Zoological Nomenclature which clearly recognizes genera as valid even when the name differs in only one letter from another genus name. In the case of *Cyathostomum*, the existence of another nematode genus, *Cyathostoma* Blanchard, 1849, had convinced many workers that *Cyathostomum* was a homonym. The arguments for accepting *Cyathostomum* were summarized by McIntosh (1951), long after he had settled the question for himself and his co-workers and had submitted the case to the International Commission of Zoological Nomenclature in 1932. The Commission ruled (Hemming, ed., 1943) that *Cyathostomum* was not a homonym of *Cyathostoma*.

Unfortunately, many workers, especially Skrjabin and his students and co-workers, have followed the arguments of LeRoux (1924) that *Cyathostomum* is a *nomen nudum* and have recognized *Trichonema* as the name for this genus. According to the International Commission's Opinion, Looss' (1900) restriction of *C. tetracanthum* to the species most commonly found in horses and asses in Egypt could only be questioned if there was reasonable doubt as to whether this species was among those studied by Mehlis (1831). We now know that *C. tetracanthum* does occur in Europe so no reasonable doubt should exist that *Cyathostomum* is valid."

Strongylidae with small cylindrical buccal capsules and Type I ovejectors are grouped in the Cyathostominae. In most genera the dorsal rays are also Type I. Of 17 genera, many with numerous species, only a few species occur in hosts other than Perissodactyla. Two species of *Murshidia* occur in the warthog and a few species of *Eucyathostomum* occur in Artiodactyla. Two genera (*Sauricola* and *Chapiniella*) with a total of five species have been described from tortoises (*Testudo*) in South America and Asia. The species from tortoises have been placed by most previous workers in a separate subfamily (Sauricolinae) characterized by the occurrence of outpocketings of the intestine. These outpocketings could not be found in type specimens of the type species of *Chapiniella* and *Sauricola* however. Although the genera in tortoise may have been separated from other Cyathostominae since the Permian, it is considered preferable not to separate subfamilies or other groups solely on the basis of host. A single genus *Theileriana* with three species parasitic in hyraxes has a Type II dorsal ray (Fig. 7.35) and what appear to be intermediate ovejectors (Fig. 7.11). Because the ovejectors of *Theileriana* resemble those of *Quilonia* (Fig. 7.10) of elephants and rhinoceroses more than the similar ovejectors of *Trachypharynx* (Fig. 7.12), and because of host relationships, *Theileriana* is placed in the Cyathostominae.

Phascolostrongylinae n. subf.

This group of four genera differs from other Strongylidae with small cylindrical buccal capsules by having a Type II dorsal ray and by parasitizing vombatid and macropodid Australian marsupials. They differ from Australian Chabertiidae with small cylindrical or funnel-shaped buccal capsules (Cloacininae) by occurring mostly in the intestine rather than the stomach; they parasitize vombatid marsupials and have dorsal gutters, externodorsal rays that arise from the dorsal ray and Type I rather than Type II ovejectors. This new subfamily is erected for *Phascolostrongylus*, *Oesophagostomoides*, *Macropostrongyloides* and *Paramacropostrongylus*.

These four genera have been redescribed recently by Beveridge (1978b) and Beveridge & Mawson (1978). Some variation from the subfamily characteristics are found in two species (*Macropostrongyloides dissimilis* and *Paramacropostrongylus typicus*) that occur in the stomach of their hosts and have T-shaped rather than Y-shaped ovejectors. Because of the structure of the ovejectors of *P. typicus* (Fig. 7.8), it is considered to be Type I; thus, this species fits the subfamily description. *M. dissimilis* was not examined in the present study.

CHABERTIIDAE

This family includes nematodes with Type II ovejectors and dorsal rays. They are parasitic in artiodactyls, rodents, primates and macropodids. Characteristics of the buccal capsule and other cephalic and cervical characteristics are used to separate three subfamilies.

Chabertiinae

Chabertiidae with large thick-walled globular or subglobular buccal capsules are placed in this subfamily. The seven genera are parasitic in the intestines of ruminants, primates, rodents or macropodid marsupials and include only eight species in the seven genera. Some previous workers (Yamaguti, 1961; Popova, 1958) have placed *Ternidens* Railliet & Henry, 1909 or *Chabertia* Railliet & Henry, 1909, or both, in the Oesophagostominae because of the presence of a cervical groove at the excretory pore. The presence or absence of a cervical groove is a variable character within the Oesophagostominae. A more pronounced cervical groove is present in *Okapistrongylus* Berghe, 1937 than in *Chabertia*. This classification agrees with that of Chabaud (1965) and Popova (1955) in placing *Chabertia* with the forms with a large buccal capsule and with Diaouré (1964) and Chabaud (1965) in placing *Ternidens* with them. At present no characteristics except those of the buccal capsule are available for separating genera with large thick-walled globular buccal capsules from those with small cylindrical buccal capsules. An additional character that may aid in distinguishing genera with large globular or subglobular buccal capsules is the presence of a sclerotized buccal collar around the anterior end of the buccal capsule (Figs. 7.54–7.63, 7.88, 7.96). This structure is present in greatly reduced form in smaller strongyloids but it is much hypertrophied in those with the enlarged buccal capsules.

An error was discovered in the description of *Ransomus* Hall, 1916 that has been perpetuated by subsequent workers. The mouth of *Ransomus* is directed anterodorsally rather than anteroventrally as stated by Hall (1916). Hall did not observe the excretory pore. The types of *Ransomus* were examined and the excretory pore and the dorsal gutter leave no doubt of the anterodorsal orientation of the mouth (Fig. 7.92).

Cyclodontostomum Adams, 1933, parasitic in rats, and *Agriostomum* Railliet, 1902, parasitic in ruminants, are believed to be members of the Chabertiinae rather than the Ancylostomatoidea where they were previously placed. Rep (1963) pointed out that the two genera have more in common with the Strongyloidea than with the Ancylostomatoidea. Both genera have short buccal capsules with lancet-like teeth in the oesophageal funnel, rather than in the buccal capsule, and the dorsal ray, position of the vulva, and shape and structure of the ovejector are as found in the Chabertiinae. Some species of *Agriostomum* have a distinct cervical groove as found in some Chabertiidae. The hook-like teeth of *Cyclodontostomum* (Fig. 7.96) and *Agriostomum* (Fig. 7.95) are believed to be homologous to the hook-like elements of the internal corona radiata of *Castorstrongylus* (Fig. 7.93), *Corollostrongylus* (Fig. 7.94), *Trachypharynx* (Fig. 7.111) and other members of the Chabertiidae. Although both genera had previously been reported from the small intestine of their hosts, recent workers have found them to occur in the large intestine.

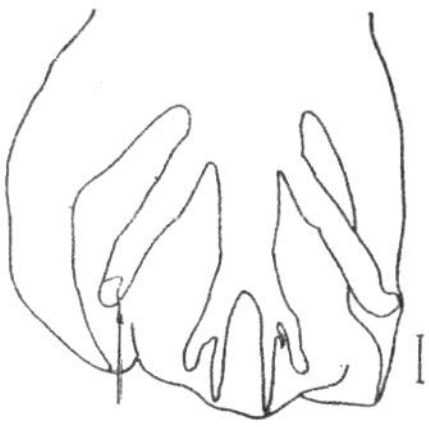

7.37

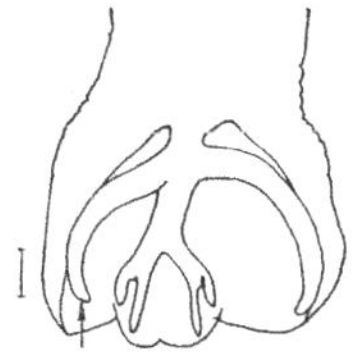

7.38

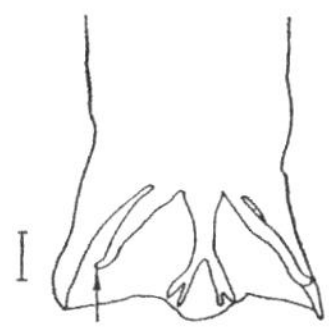

7.39

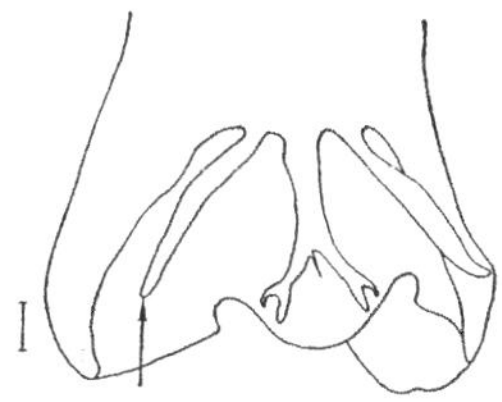

7.40

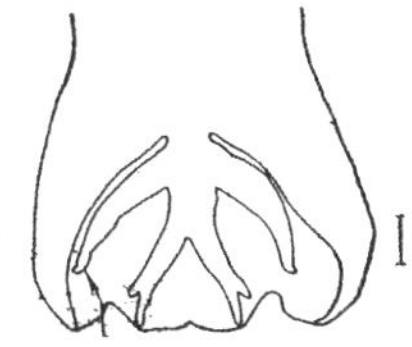

7.41

FIG. 7.37. Dorsal bursal rays of *Chabertia ovina* (Fabricius, 1788). (Original.) (Arrow indicates externodorsal ray; scale bar 50 μm.)

FIG. 7.38. Dorsal bursal rays of *Ternidens deminutus* (Railliet & Henry, 1905). (Original.) (Arrow indicates externodorsal ray; scale bar 50 μm.)

FIG. 7.39. Dorsal bursal rays of *Ransomus rodentorum* Hall, 1916. (Original.) (Arrow indicates externodorsal ray; scale bar 50 μm.)

FIG. 7.40. Dorsal bursal rays of *Castorstrongylus castoris* Chapin, 1925. (Original.) (Arrow indicates externodorsal ray; scale bar 50 μm.)

FIG. 7.41. Dorsal bursal rays of *Oesophagostomum (O.) dentatum* (Rudolphi, 1803). (Original.) (Arrow indicates externodorsal ray; scale bar 50 μm.)

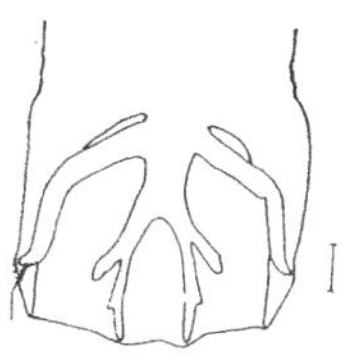

7.42

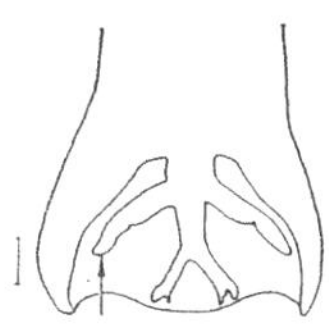

7.43

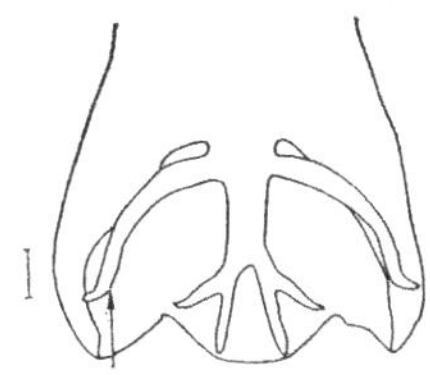

7.44

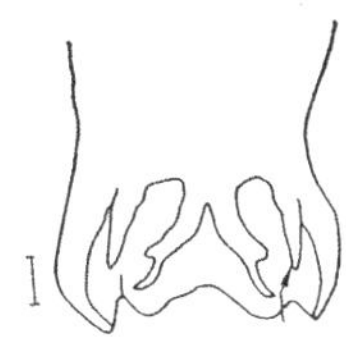

7.45

FIG. 7.42. Dorsal bursal rays of *Trachypharynx natalensis* Ortlepp, 1962. (Original.) (Arrow indicates externodorsal ray; scale bar 50 μm.)

FIG. 7.43. Dorsal bursal rays of *Kuntzistrongylus selfi* (Schmidt & Kuntz, 1975). (Original.) (Arrow indicates externodorsal ray; scale bar 50 μm.)

FIG. 7.44. Dorsal bursal rays of *Bourgelatia diducta* Railliet, Henry & Bauche, 1919. (Original.) (Arrow indicates externodorsal ray; scale bar 50 μm.)

FIG. 7.45. Dorsal bursal rays of *Cloacina* sp. (Original.) (Arrow indicates externodorsal ray; scale bar 50 μm.)

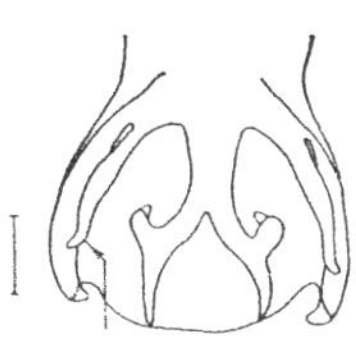

7.46

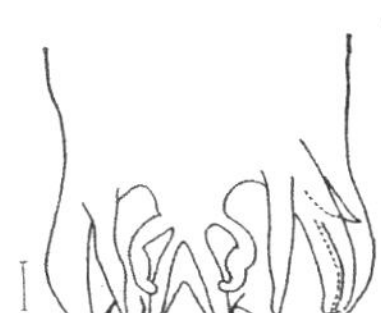

7.47

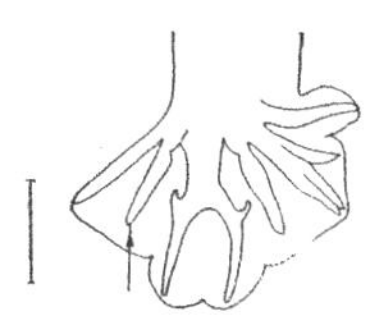

7.48

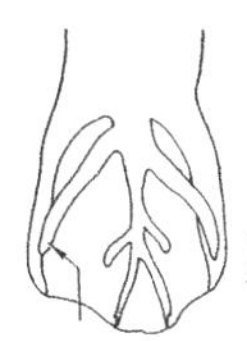

7.49

FIG. 7.46. Dorsal bursal rays of *Rugopharynx* sp. (Original.) (Arrow indicates externodorsal ray; scale bar 50 μm.)

FIG. 7.47. Dorsal bursal rays of *Zoniolaimus setifera* Cobb, 1898. (Original.) (Arrow indicates externodorsal ray; scale bar 50 μm.)

FIG. 7.48. Dorsal bursal rays of *Macropostrongylus macropostrongylus* Yorke & Maplestone, 1926. (After Mawson, 1977b.) (Arrow indicates externodorsal ray; scale bar 50 μm.)

FIG. 7.49. Dorsal bursal rays of *Phacochoerostrongylus pricei* Schwartz, 1928. (Original.) (Arrow indicates externodorsal ray; scale bar 50 μm.)

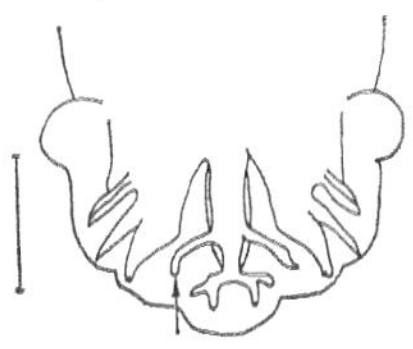

7.50

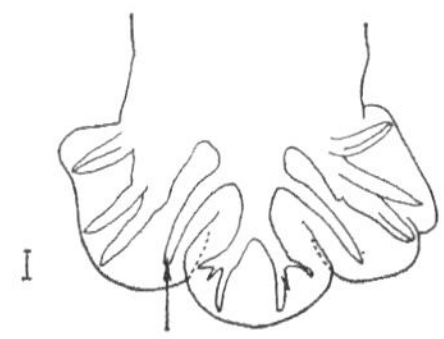

7.51

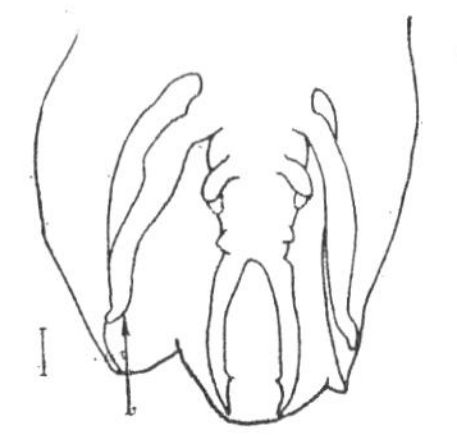

7.52

FIG. 7.50. Dorsal bursal rays of *Rhabditostomum traguli* (Maplestone, 1932). (After Chabaud & Krishnasamy, 1976.) (Arrow indicates externodorsal ray; scale bar 50 μm.)

FIG. 7.51. Dorsal bursal rays of *Daubneyia roubaudi* (Daubney, 1926). (After Troncy, Graber & Thal, 1973.) (Arrow indicates externodorsal ray; scale bar 50 μm.)

FIG. 7.52. Dorsal bursal rays of *Corollostrongylus hypsiprymnodontis* Beveridge, 1978. (Original.) (Arrow indicates externodorsal ray; scale bar 50 μm.)

Hook-like elements of the internal corona radiata similar to those of *Agriostomum* were found in *Schulzinema miroljubowi* Krastin, 1937 from the large intestine of *Cervus nippon*. Specimens were studied by the author in the Central Helminthological Museum of the All-Union Institute of Helminthology in Moscow. *Schulzinema* is very similar to *Agriostomum* except that in *Schulzinema* the hooks are smaller, the oesophageal teeth are larger, there is an external corona radiata and the oral opening is directed anteriorly. A short dorsal gutter and a perioral ring were found to extend less than half of the depth of the buccal cavity. The lining of the buccal capsule has vertical ridges. Drawings were not prepared.

Oesophagostominae

This subfamily includes all Chabertiidae with small cylindrical buccal capsules (except those parasitic in Australian marsupials). Oesophagostominae are parasites of ruminants, pigs, primates or rodents. The genera can be grouped on the basis of the presence or absence of a cervical groove at the excretory pore.

The genera with a cervical groove (*Oesophagostomum*, *Rhabditostomum*, *Bourgelatioides* and *Daubneyia*) have been studied recently by Chabaud & Durette-Desset (1973), Troncy, Graber & Thal (1973) and Chabaud & Krishnasamy (1976). The key presented herein relied heavily on the key to *Oesophagostomum* and its subgenera presented by Chabaud & Durette-Desset (1973).

Five genera of the Oesophagostominae lack a cervical groove and are parasites of pigs, lemurs or rodents. Although recent workers (Troncy, Graber & Thal, 1973) have considered *Phacochoerostrongylus* Schwartz, 1928 a synonym of *Bourgelatia* Railliet, Henry & Bauche, 1919, both genera are recognized herein. *Bourgelatia*, with a single species in domestic swine, can be separated from *Phacochoerostrongylus*, with two species in warthogs, by differences in the buccal capsule, ovejector, vagina length and dorsal ray. The genus *Trachypharynx* Leiper, 1911 includes two species parasitic in large African rats (*Thryonomys* and *Cricetomys*) and a single species from a bird (*Cercomela*) reported by Vuylsteke (1953). The latter was probably an accidental infection. Among the Chabertiinae the ovejectors (Fig. 7.12) and buccal capsule (Fig. 7.111) of *Trachypharynx* are most similar to those found in some Strongylidae; these features indicate that *Trachypharynx* may be the most primitive member of the subfamily. Among the hosts of the Chabertiinae, the large thryonomyid rodent hosts of *Trachypharynx* are also primitive, dating to the Oligocene (Kowalski, 1971).

Lemurostrongylus residuus Chabaud, Brygoo & Petter, 1961, from a lemur, is placed in the Oesophagostominae because of its bursa with externodorsal rays arising from the dorsal ray and its Type II ovejector. These characteristics place *Lemurostrongylus* with the Chabertiidae of other primates rather than with those of the Australian marsupials where it was originally placed.

A new genus, *Kuntzistrongylus*, is proposed for *Oesophagostomum selfi* Schmidt & Kuntz, 1975 from *Rattus coxinga* from Taiwan. The new genus is named in honour of Dr. Robert E. Kuntz whose vast collections include the only known specimens of this and many other helminth parasites.

Kuntzistrongylus n. g.

Generic Diagnosis: Strongyloidea, Chabertiidae, Oesophagostominae. Medium sized, stout worms without cervical groove. Head with cushion-like cephalic expansion (Fig. 7.112). Oral opening round with internal and external corona radiata. Internal corona inserted near base of buccal capsule. Cephalic papillae short and conical. Buccal capsule wider anteriorly. Extra chitinous supports absent. Dorsal gutter absent. Oesophagus short, thick, enlarged posteriorly. Cervical papillae near middle of oesophageal swelling. Bursa short with thick rays. Ventral rays parallel, extending to open ventral margin of bursa. Ventrolateral ray shorter than other laterals. Externodorsal ray arising from stem of dorsal ray (Fig. 7.43). Prebursal papillae present. Genital cone well developed with two large dorsal papillae and a single ventral papilla. Spicules with wide alae. Gubernaculum absent. Vulva well anterior to anus. Vagina short, ovejector with kidney-shaped vestibule, muscular sphincter and exceptionally short infundibulum (Fig. 7.15).

Cloacininae

This subfamily of the Chabertiidae, consisting of 18 genera found only in the stomachs of macropodid marsupials, is characterized by small, cylindrical, funnel-shaped or ring-shaped buccal capsules, externodorsal bursal rays arising separately from the dorsal ray, and a relatively long oesophagus (one tenth to one quarter of the body length). The 18 genera and 97 species (Mawson, personal communication) are organized into four tribes. The key presented herein, although differing somewhat in characters used and in placement of some genera, relied heavily upon recent revisions of several genera by Mrs. Patricia Mawson Thomas (Mawson, 1974, 1975, 1977a, 1977b, 1978b) and upon an unpublished key to Australian Strongyloidea graciously loaned by Mrs. Thomas.*

The genus *Woodwardostrongylus* Wahid, 1964 was returned to the Cloacininae from the Amidostomatidae on the advice of A. G. Chabaud, Marie-Claude Durette-Desset and Ian Beveridge (personal communication) who recently studied fresh specimens in Australia. The ovejector, bursa and cephalic papillae are clearly strongyloid rather than trichostrongyloid. Although the ovejector is T-shaped (nearly Y-shaped) it is clearly Type II because of the apparent lack of infundibula and presence of sphincters that have significantly thinner walls than the vestibule.

Considerable work on the Australian Strongyloidea is currently under way and the systematics of this group of nematodes is far from settled. *Buccostrongylus* Johnston & Mawson, 1939 was regarded by Kung (1948) as a synonym of *Zoniolaimus* but Mawson (personal communication) believes it is a good genus with at least two good species. *Buccostrongylus* is not included herein. According to Mawson (1977a) a new genus will be described for *Cyclostrongylus clelandi* Johnston & Mawson, 1939.

* Without the unselfish efforts of Mrs. Thomas and Dr. Ian Beveridge, who both supplied specimens and prepublication copies of manuscripts, the coverage of the Australian Strongyloidea would have been far less complete.

DELETROCEPHALIDAE

Deletrocephalus and *Paradeletrocephalus* occur in the large intestine of the South American ratite bird, *Rhea americana*. Chitwood & Chitwood (1938) described a primitive cephalic papillae pattern related to that found in the syngamids and placed *Deletrocephalus* in that group with which they share the characteristics of a cup-shaped hexagonal mouth and numerous teeth at the base of the buccal capsule. Skrjabin *et al.* (1952) and Chabaud (1965) also placed the two genera in the subfamily Deletrocephalinae in the Syngamidae. Freitas & Lent (1947) and Yamaguti (1961) placed *Deletrocephalus* and *Paradeletrocephalus* in the Strongylidae. The ovejectors of *Paradeletrocephalus* are Y-shaped and relatively undifferentiated, resembling closely the primitive ovejectors of *Sauricola* (Fig. 7.3) and *Choniangium* (Fig. 7.4). The ovejectors of *Deletrocephalus* are also Y-shaped, but are shorter and more heavily muscled. Dorsal rays of the two genera are quite variable, but that of *Deletrocephalus cesarpintoi* Vaz, 1936 is similar to that of *Chapiniella* (Fig. 7.26) of turtles. Because of the presence of a dorsal gutter and a perioral ring, Y-shaped ovejectors, vulva near anus, and parasitism of the intestine of their host, *Deletrocephalus* and *Paradeletrocephalus* are separated from Syngamidae and placed in a separate family intermediate between the Syngamidae and the Strongylidae. The Deletrocephalidae appears to be an ancient primitive group that may have been isolated in ratite birds for more than 200 million years (Fig. 7.53).

SYNGAMIDAE

Evolutionary parasitologists agree that the syngamids represent a primitive group that may have evolved with the Reptilia although it is now limited to birds and mammals (Dougherty, 1951; Skrjabin *et al.*, 1952). The general lack of fusion of the cephalic papillae of the outer circle (Chitwood & Chitwood, 1938), the hexagonal mouth and buccal capsule and the forward position of the vulva (in all but *Stephanurus*) provide evidence for the primitiveness of the syngamids. Most workers have included *Deletrocephalus* and *Paradeletrocephalus* in this family, but herein they are separated as a family intermediate between the Syngamidae and the Strongylidae. Dougherty (1951) and Chabaud (1965) included *Acheilostoma* Leiper, 1911 in the Syngamidae, but we place it in the Ancylostomatoidea. *Archeostrongylus* is placed in the subfamily Archeostrongylinae established by Chabaud (1958) for this syngamid genus. Following most workers *Stephanurus* is placed in a separate subfamily.

Syngaminae

The key to genera of the Syngaminae presented herein was influenced greatly by the work of Baruš & Tenora (1972a) who evaluated systematic characters of the groups and recognized six genera. Five of the six genera recognized by Baruš & Tenora are also given generic rank, but *Hovorkonema* Turemuratov, 1963 is regarded as a subgenus of *Cyathostoma* Blanchard, 1849.

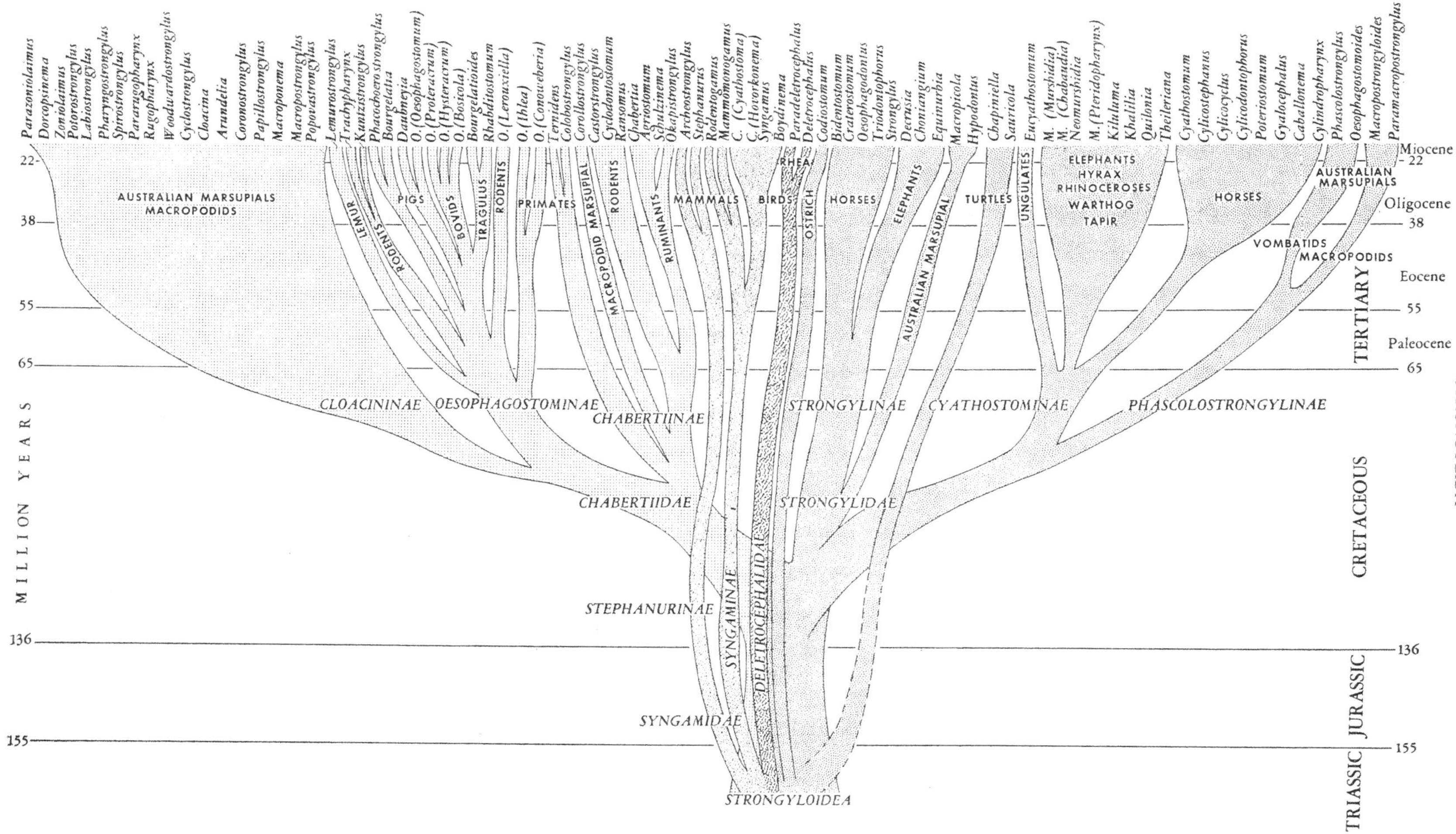

7.53

FIG. 7.53. Dendrogram of possible evolutionary relationships of 87 genera and subgenera of the Strongyloidea (Nematoda). (Original.)

STRONGYLOIDEA

Key to families

1-(4) Oral opening round or oval, usually with well-developed corona radiata or lips.
Cephalic papillae including 4 submedians.
Buccal capsule not hexagonal in cross section.
Parasites of gastro-intestinal tract.

2-(3) Paired ovejectors of female reproductive tract Y-shaped, usually with 2 anterior branches parallel (Figs. 7.1, 7.3–7.8) or rarely T-shaped with anterior and posterior branches completely opposed (Figs. 7.9–7.11).
Dorsal ray of male copulatory bursa usually with 3 rami on each side of median fissure (Figs. 7.21–7.36) (except Phascolostrongylinae).
Parasites of Perissodactyla, Australian marsupials, ostriches, tortoises or, rarely, of Artiodactyla.
Strongylidae

3-(2) Paired ovejectors of female reproductive tract opposed at origin but with posterior muscular arm turned anteriorly (Figs. 7.2, 7.12–7.20).
Dorsal ray of male copulatory bursa usually with 2 rami on each side of median fissure (Figs. 7.37–7.52).
Parasites of Artiodactyla, macropodid marsupials, rodents or primates.
Chabertiidae

4-(1) Oral opening hexagonal; corona radiata or lips rudimentary or absent.
Cephalic papillae of outer circle 10 in number, submedians paired but not fused.
Buccal capsule usually cup-shaped, hexagonal in cross section, with teeth at base (except *Archeostrongylus*).
Parasites of respiratory system of birds or mammals, urinary system of swine or intestine of *Rhea* or *Hystrix*.

5-(6) Unsclerotized dorsal gutter and prominent perioral groove present.
Copulatory bursa with long, thin tapering rami.
Vulva near anus.
Parasites of intestine of South American ratites (*Rhea*).
Deletrocephalidae

6-(5) Dorsal gutter and perioral groove absent.
Copulatory bursa reduced with short, thick, blunt rami.
Vulva near midbody (except *Stephanurus*).
Parasites of respiratory system of birds or mammals, urinary system of swine or intestine of *Hystrix*.
Syngamidae

Family *STRONGYLIDAE* Baird, 1853

Key to subfamilies

1-(2) Buccal capsule globular or subglobular (Figs. 7.54–7.63).
Parasites of equids, elephants, Australian marsupials or ostriches.
Strongylinae

2-(1) Buccal capsule cylindrical.

3-(4) Single corona radiata with few elements or with few tooth-like structures at anterior edge of buccal capsule (Figs. 7.65–7.68).
Dorsal ray with 2 rami on each side of median fissure.
Parasites of intestine of vombatid or macropodid marsupials.

Phascolostrongylinae

4-(3) Single or double corona radiata usually present with numerous elements (Figs. 7.70–7.76) or rarely with few fleshy lips or corona radiata elements originating from base of buccal capsule (Figs. 7.77–7.82).
Dorsal ray usually with 3 rami on each side of median fissure (Figs. 7.25–7.32).
Parasites of Perissodactyla; rarely of Artiodactyla or tortoises.

Cyathostominae

Subfamily **Strongylinae** Railliet, 1885

Key to genera

1-(2) Buccal capsule more than twice as deep as wide (Fig. 7.54).
Oral opening directed somewhat dorsally.
Single corona radiata surrounding oral opening.
Parasites of Indian elephant (*Elephas indicus*).

Choniangium Henry & Bauche, 1914
(= *Asifia* Lane, 1914)

2-(1) Buccal capsule less than twice as deep as wide.
Oral opening directed anteriorly or nearly so.

3-(6) Corona radiata absent.
Oesophageal funnel with 3 large teeth.
Dorsal ray with 4 distal rami (Fig. 7.36).
Dorsal gutter branches in midlength.
Parasites of Australian marsupials.

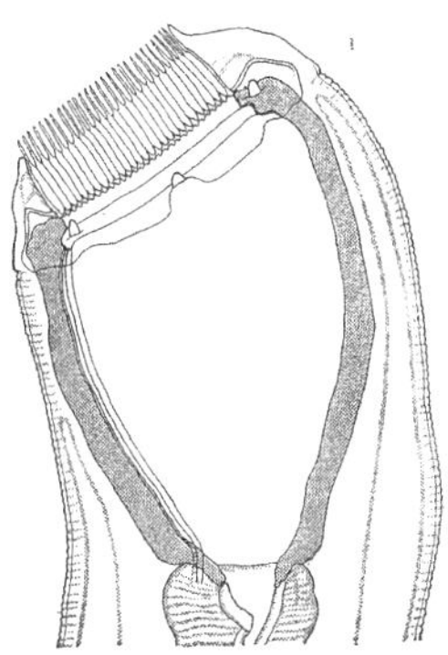

7.54

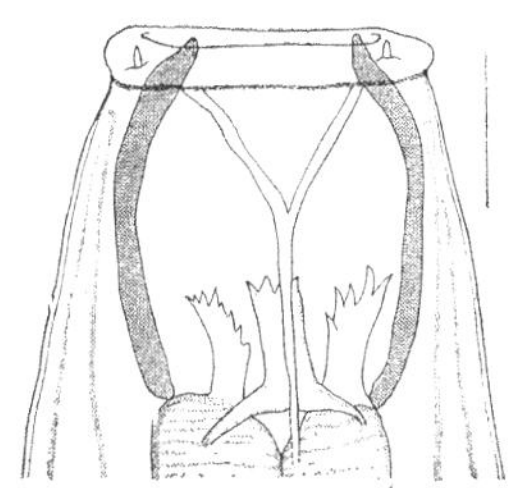

7.55

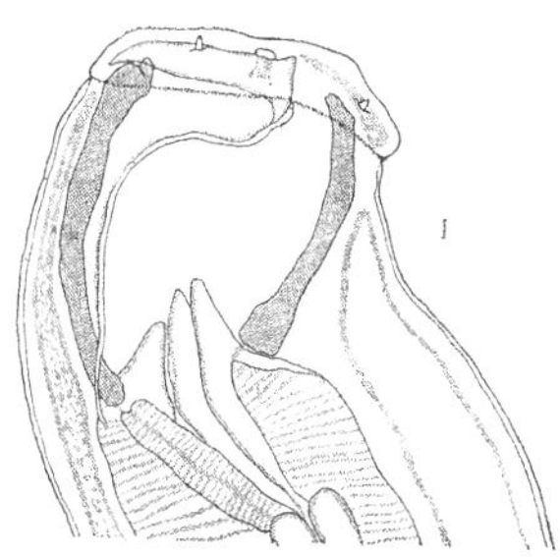

7.56

FIG. 7.54. *Choniangium epistomum* (Piana & Stazzi, 1900), cephalic extremity, lateral view. (Original.) (Scale bar 10 μm.)

FIG. 7.55. *Macropicola ocydromi* Mawson, 1978, cephalic extremity, dorsal view. (After Mawson, 1978a.) (Scale bar 100 μm.)

FIG. 7.56. *Hypodontus macropi* Mönnig, 1929, cephalic extremity, lateral view. (Original.) (Scale bar 10 μm.)

4-(5) Oral opening directed anteriorly.
Oesophageal teeth spiny (Fig. 7.55).
Macropicola Mawson, 1978

5-(4) Oral opening directed anteroventrally.
Oesophageal teeth conical (Fig. 7.56).
Hypodontus Mönnig, 1929

6-(3) Corona radiata present.
Buccal capsule variable.
Dorsal ray with 6 distal rami (Figs. 7.21–7.24).
Dorsal gutter extending without branching to perioral ring at base of internal corona radiata.
Parasites of Eutheria.

7-(8) Cervical and cephalic region swollen to almost twice thickness of rest of body.
Elements of external corona radiata of 2 lengths (Fig. 7.57).
Externodorsal ray triramous.
Parasites of Indian elephant.
Equinurbia Lane, 1914

8-(7) Cervical and cephalic region not swollen.
Elements of external corona radiata of uniform length.
Externodorsal rays monoramous.

9-(10) Buccal capsule cup-shaped with short external corona radiata and 2 large subventral teeth (Fig. 7.58).
Dorsal ray with very short branches (Fig. 7.33).
Gubernaculum absent.
Collar of oral opening reduced and depressed.
Parasites of Indian elephant.
Decrusia Lane, 1914

7.57 7.58 7.59

FIG. 7.57. *Equinurbia sipunculiformis* (Baird, 1859), cephalic extremity, lateral view. (Original.) (Scale bar 10 μm.)

FIG. 7.58. *Decrusia additicta* Railliet, Henry & Bauche, 1914, cephalic extremity, lateral view. (Original.) (Scale bar 10 μm.)

FIG. 7.59. *Codiostomum struthionis* (Horst, 1885), cephalic extremity, lateral view. (Original.) (Scale bar 10 μm.)

10-(9) Buccal capsule globular, subglobular or funnel-shaped with few large elements or long slender elements in external corona radiata.
Dorsal ray with long slender branches and median fissure (Figs. 7.21–7.24).
Gubernaculum present.

11-(12) Internal corona radiata elements longer than external (Fig. 7.59).
Gubernaculum small and leaf-like.
Parasites of ostriches (*Struthio*).

Codiostomum Railliet & Henry, 1911

12-(11) Internal corona radiata shorter than external (except *Oesophagodontus*).
Gubernaculum large with ventral groove.
Parasites of equids.

13-(16) External corona radiata consisting of few large, broad elements.

14-(15) Buccal capsule without teeth.
Dorsal gutter strongly supported, salient (Fig. 7.60).

Craterostomum Boulenger, 1920

15-(14) Buccal capsule with 3 long oesophageal teeth extending to edge of oral opening.
Dorsal gutter absent.

Bidentostomum Tshoijo, 1957

16-(13) External corona radiata with slender elements.

17-(18) Buccal capsule funnel-shaped, with thickened posterior ring (Fig. 7.61).
Dorsal gutter absent.
Submedian cephalic papillae bifid.

Oesophagodontus Railliet & Henry, 1902
(= *Pseudosclerostomum* Quiel, 1919)

18-(17) Buccal capsule globular or subglobular, without posterior ring.
Dorsal gutter present.
Submedian cephalic papillae not bifid.

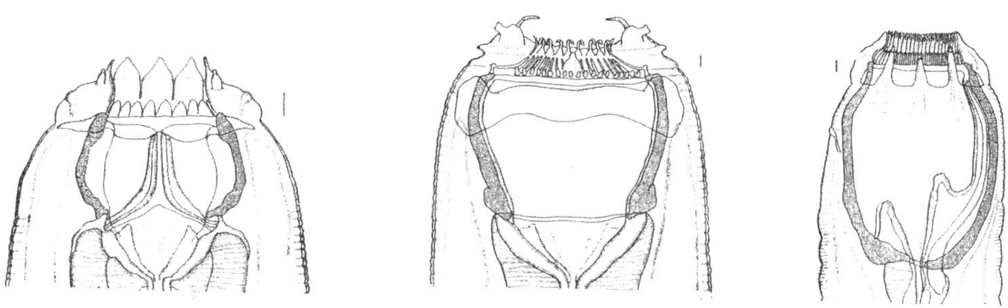

7.60 7.61 7.62

FIG. 7.60. *Craterostomum acuticaudatum* (Kotlan, 1919), cephalic extremity, dorsal view. (After Lichtenfels, 1975.) (Scale bar 10 μm.)

FIG. 7.61. *Oesophagodontus robustus* (Giles, 1892), cephalic extremity, lateral view. (After Lichtenfels, 1975.) (Scale bar 10 μm.)

FIG. 7.62. *Strongylus equinus* Mueller, 1780, cephalic extremity, lateral view. (After Lichtenfels, 1975.) (Scale bar 10 μm.)

19-(20) Buccal capsule deeper than wide.
Collar of oral opening high (Fig. 7.62).
Elements of corona radiata fine, needle-like.
Buccal teeth, if present, with rounded points.
Spicule tips straight.

Strongylus Mueller, 1780
(= *Sclerostoma* Rudolphi, 1809; = *Sclerostomum* Diesing, 1851;
= *Alfortia* (Railliet, 1923); = *Delafondia* (Railliet, 1923))

20-(19) Buccal capsule as wide as or wider than deep.
Collar of oral opening inflated or like flattened ring.
Elements of internal corona radiata plate-like.
Three teeth in buccal capsule, each with 3 main sharp points (Fig. 7.63).
Spicule tips pick-like (Fig. 7.64).

Triodontophorus Looss, 1902
(= *Triodontus* Looss, 1900)

Subfamily **Phascolostrongylinae** n. subf.

Key to tribes and genera

1-(4) Submedian papillae prominent.
External corona radiata elements large and triangular.
Oesophagus lined anteriorly with sclerotized oblong bodies and medially with comb-like structures.
Dorsal ray with proximal rami shorter than distal rami.

2-(3) Buccal capsule wider than deep with thick walls (Fig. 7.65).
External corona radiata with 12 elements.
Parasites of vombatid marsupials.

Phascolostrongylus Canavan, 1931,
emend. Beveridge, 1978

3-(2) Buccal capsule about as deep as wide with thin walls (Fig. 7.66).
External corona radiata with 8 elements.
Parasites of vombatid marsupials.

Oesophagostomoides Schwartz, 1928,
emend. Beveridge, 1978

4-(1) Submedian papillae minute.
External corona radiata elements small and tooth-like.
Oesophagus lining not prominently differentiated.
Dorsal ray with distal rami of equal length.

5-(6) Buccal capsule deeper than wide, with 4 large submedian teeth on anterior walls (Fig. 7.67).
Parasites of vombatid and macropodid marsupials.

Macropostrongyloides Yamaguti, 1961,
emend. Beveridge & Mawson, 1978

6-(5) Buccal capsule about as deep as wide, without 4 large teeth on anterior walls (Fig. 7.68).
Parasites of macropodid marsupials.

Paramacropostrongylus Johnston & Mawson, 1940,
emend. Beveridge & Mawson, 1978

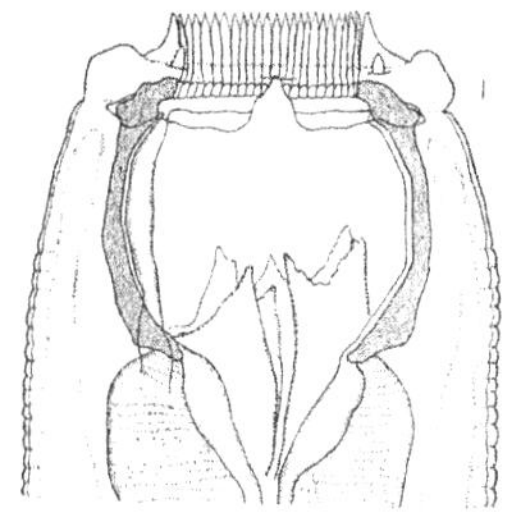

7.63

7.64

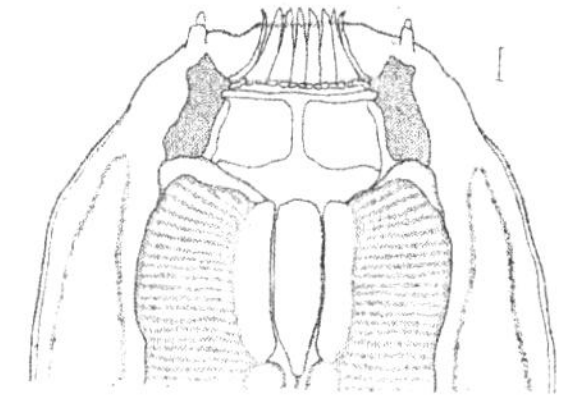

7.65

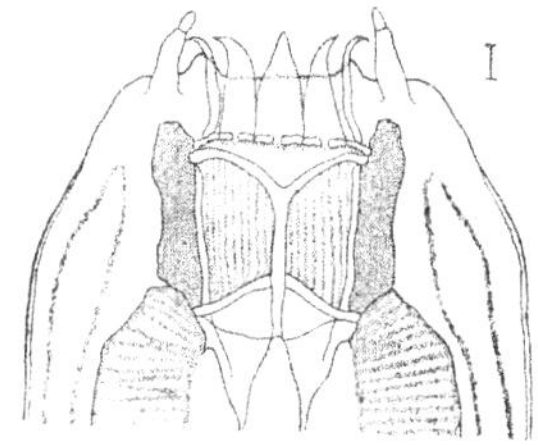

7.66

FIG. 7.63. *Triodontophorus serratus* (Looss, 1900), cephalic extremity, lateral view. (Original.) (Scale bar 10 μm.)
FIG. 7.64. Spicule tips of *Triodontophorus serratus*. (Original.) (Scale bar 10 μm.)
FIG. 7.65. *Phascolostrongylus turleyi* Canavan, 1931, cephalic extremity, dorsal view. (Original.) (Scale bar 10 μm.)
FIG. 7.66. *Oesophagostomoides giltneri* Schwartz, 1928, cephalic extremity, dorsal view. (Original.) (Scale bar 10 μm.)

Subfamily **Cyathostominae** Nicoll, 1927

Key to tribes, genera and subgenera

1-(16) Both internal and external coronae radiatae present.
Ovejectors well developed and parallel.
Spicules with pick-shaped tips (Fig. 7.64), without alae.
Gubernaculum massive, with longitudinal ventral groove and flanged transverse ventral notch (Fig. 7.69).
Parasites of equids.

CYATHOSTOMINEA

2-(3) Anterior end of oesophagus markedly dilated to surround 3 large sickle-shaped teeth (Fig. 7.70).

Gyalocephalus Looss, 1900

3-(2) Anterior end of oesophagus not greatly dilated or if dilated without sickle-shaped teeth.

4-(7) Cylindrical buccal capsule elongated, about twice or 3 times as deep as wide.

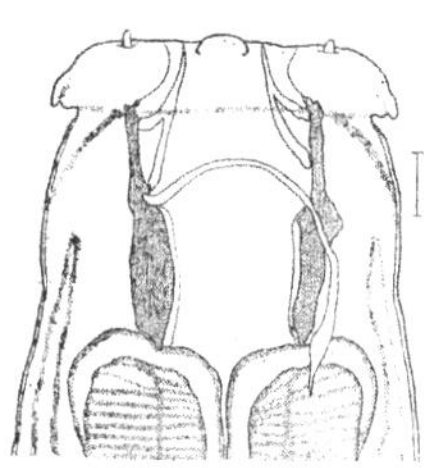

7.67

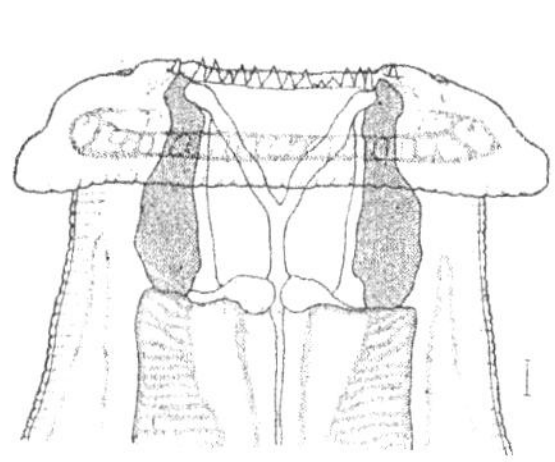

7.68

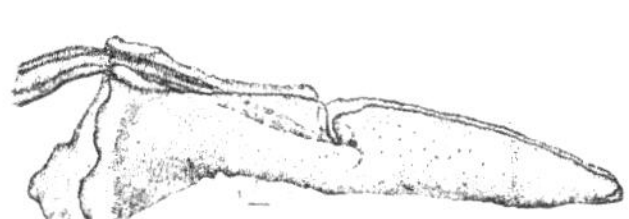

7.69

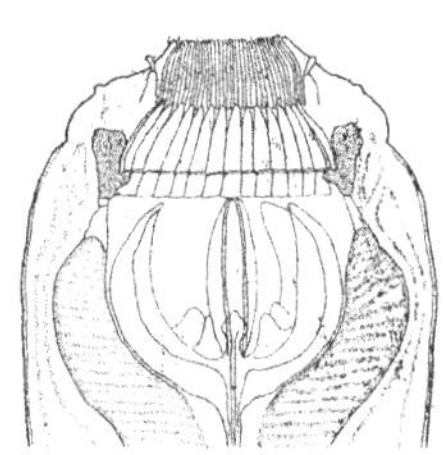

7.70

FIG. 7.67. *Macropostrongyloides lasiorhini* (Mawson, 1955), cephalic extremity, lateral view. (Original.) (Scale bar 10 μm.)
FIG. 7.68. *Paramacropostrongylus toraliformis* Beveridge & Mawson, 1978, cephalic extremity, dorsal view. (Original.) (Scale bar 10 μm.)
FIG. 7.69. Gubernaculum of *Cylicocyclus insigne* (Boulenger, 1917), lateral view. (Original.) (Scale bar 10 μm.)
FIG. 7.70. *Gyalocephalus capitatus* Looss, 1900, cephalic extremity, dorsal view. (After Lichtenfels, 1975.) (Scale bar 10 μm.)

5-(6) Dorsal gutter projecting well into buccal cavity.
Oesophageal funnel with 3 small teeth.
Internal corona radiata inconspicuous.

Caballonema Abuladze, 1937
(= *Sinostrongylus* Hsiung & Chao, 1949)

6-(5) Dorsal gutter not projecting into buccal cavity.
Oesophageal funnel without teeth.
Internal corona radiata with broad petals, salient.

Cylindropharynx Leiper, 1911

7-(4) Cylindrical buccal capsule not greatly elongated, not more than 1·5 times as deep as wide.

8-(11) Internal corona radiata elements as long as, or longer than, broader and usually fewer than elements of external corona radiata.

9-(10) Buccal capsule walls of uniform thickness or thicker anteriorly (Fig. 7.71).
Internal corona radiata inserted posterior to anterior edge of buccal capsule.
Dorsal ray of bursa bifurcated to level of proximal branches.
Female tail short with sharp tip.

Cylicodontophorus Ihle, 1922

10-(9) Buccal capsule walls thicker posteriorly than anteriorly (Fig. 7.72).
Internal corona radiata inserted at anterior edge of buccal capsule.
Dorsal ray of bursa bifurcated to most distal branches (Fig. 7.32).
Female tail long, tapering to blunt tip.

Poteriostomum Quiel, 1919
(= *Hexodontostomum* Ihle, 1920)

11-(8) Internal corona radiata elements shorter, usually narrower and more numerous than elements of external corona radiata.

12-(13) Buccal capsule with ring-like thickening posteriorly (Fig. 7.73).
Amphids usually prominent.

Cylicocyclus Ihle, 1922

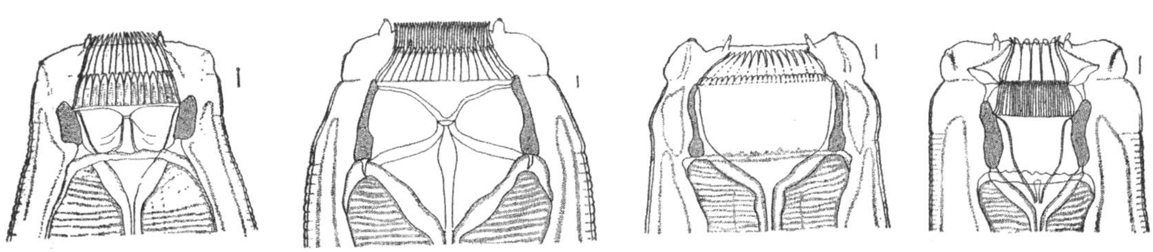

7.71 7.72 7.73 7.74

FIG. 7.71. *Cylicodontophorus bicoronatus* (Looss, 1900), cephalic extremity, dorsal view. (After Lichtenfels, 1975.) (Scale bar 10 μm.)

FIG. 7.72. *Poteriostomum imparidentatum* Quiel, 1919, cephalic extremity, dorsal view. (After Lichtenfels, 1975.) (Scale bar 10 μm.)

FIG. 7.73. *Cylicocyclus radiatus* (Looss, 1900), cephalic extremity, dorsal view. (After Lichtenfels, 1975.) (Scale bar 10 μm.)

FIG. 7.74. *Cyathostomum coronatum* Looss, 1900, cephalic extremity, dorsal view. (After Lichtenfels, 1975.) (Scale bar 10 μm.)

13-(12) Buccal capsule without ring-like thickening posteriorly.
Amphids not prominent.

14-(15) Internal corona radiata inserted deep on internal wall of buccal capsule, sometimes in uneven or sinuous line.
Extra chitinous supports for external corona radiata anterior to buccal capsule (Fig. 7.74).
Collar of oral opening high.
Buccal cavity uniformly cylindrical or slightly wider anteriorly.
Cyathostomum (Molin, 1861, *sensu stricto*) Lichtenfels, 1975
(= *Trichonema* Cobbold, 1874, in part;
= *Ershovinema* Tshoijo, 1957, in part)

15-(14) Internal corona radiata inserted at or near anterior edge of buccal capsule (Fig. 7.75).
Extra chitinous supports absent.
Collar of oral opening depressed.
Buccal cavity usually slightly narrower anteriorly.
Cylicostephanus (Ihle, 1922) Lichtenfels, 1975
(= *Petrovinema* Ershov, 1943; = *Schulzitrichonema* Ershov, 1943, in part;
= *Cylicotetrapedon* Ihle, 1925; = *Trichonema* Cobbold, 1874, in part;
= *Ershovinema* Tshoijo, 1957, in part)

16-(1) Corona radiata usually single.
Ovejectors usually parallel, long and thinly muscled (Fig. 7.3), rarely small and completely opposed (Figs. 7.10, 7.11).
Spicules alate, tips straight.
Gubernaculum small.

17-(24) Cuticle markedly inflated.
Elements of external corona radiata or cuticular lips usually few in number.

18-(19) Internal corona radiata present (Fig. 7.76).
Cuticular inflation confined to cervical region.
Cuticular annulation fine and shallow.
Parasites of ungulates in North and South America and Europe.
EUCYATHOSTOMINEA
Eucyathostomum Molin, 1861

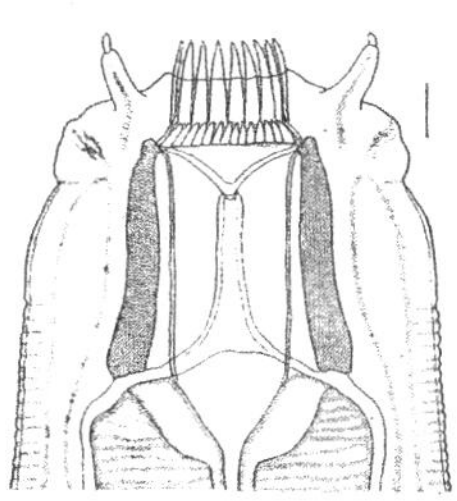

7.75

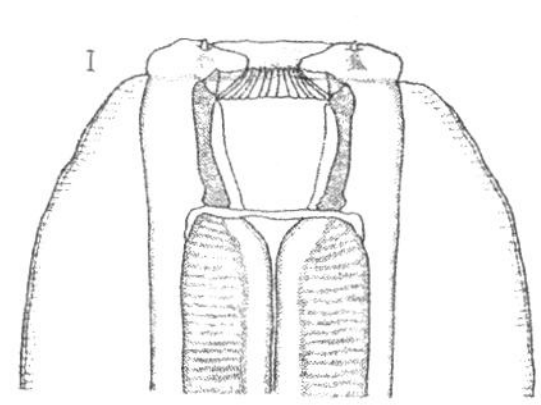

7.76

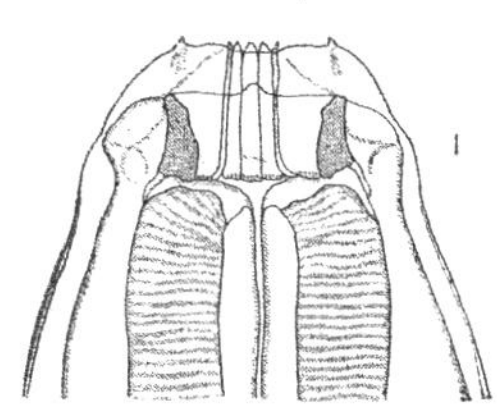

7.77

FIG. 7.75. *Cylicostephanus calicatus* (Looss, 1900), cephalic extremity, dorsal view. (After Lichtenfels, 1975.) (Scale bar 10 μm.)

FIG. 7.76. *Eucyathostomum longesubulatum* Molin, 1861, cephalic extremity, lateral view. (Original.) (Scale bar 10 μm.)

FIG. 7.77. *Kiluluma longipene* (Molin, 1861), cephalic extremity, lateral view. (Original.) (Scale bar 10 μm.)

19-(18) Internal corona radiata absent.
Cuticular inflation involving most of body.
Cuticular annulation broad and deep.

20-(23) Externodorsal ray originating from dorsal ray.
Corona radiata elements numerous (18 to 36).
Intestinal diverticula sometimes present.
Parasites of tortoises.
SAURICOLINEA

21-(22) Oesophagus short, thick (only twice as long as wide) and cylindrical.
Dorsal ray symmetrical with 2 bidigitate branches (Fig. 7.34).
Sauricola Chapin, 1924
(= *Echinopharynx* Thapar, 1925)

22-(21) Oesophagus slender (6 times as long as wide), enlarged slightly posteriorly.
Dorsal ray asymmetrical with 3 branches that are sometimes bidigitate (Fig. 7.26).
Chapiniella Yamaguti, 1961

23-(20) Externodorsal ray originating separately from dorsal ray (Fig. 7.30).
Corona radiata elements or cuticular lips few in number (6 to 8) (Fig. 7.77).
Parasites of rhinoceroses and South American tapir.
KILULUMINEA
One genus, *Kiluluma* Skrjabin, 1916

24-(17) Cuticle not inflated.
Elements of external corona radiata numerous.

25-(28) Ovejectors small, completely opposed (Figs. 7.10, 7.11).
QUILONIINEA

26-(27) Corona radiata elements and lining of buccal cavity attached to base of buccal capsule (Fig. 7.78).
Vulva near anus.
Vagina long and muscular.
Dorsal ray with 4 rami (2 each side) (Fig. 7.35).
Parasites of Hyracoidea.
Theileriana Mönnig, 1924

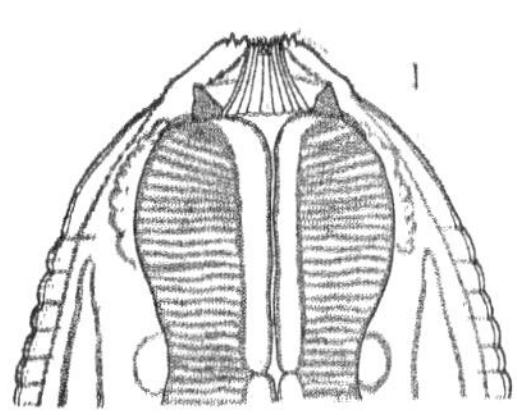
7.78

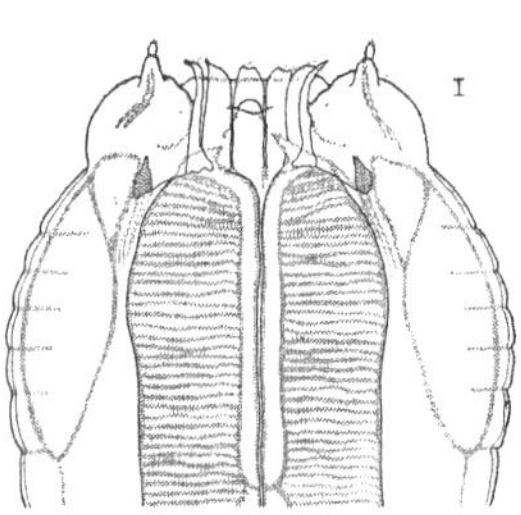
7.79

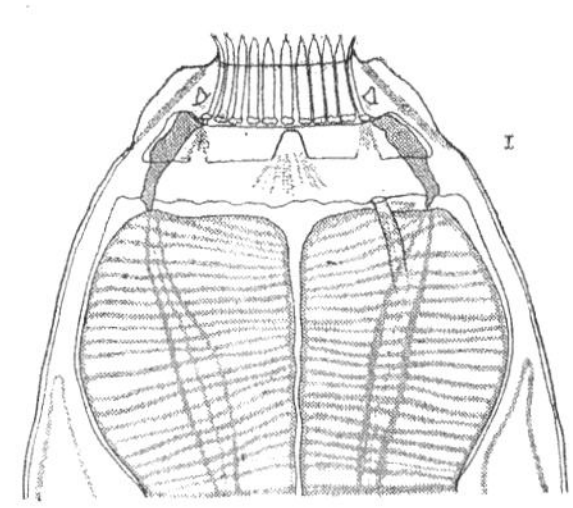
7.80

FIG. 7.78. *Theileriana brachylaima* (Linstow, 1901), cephalic extremity, lateral view. (Original.) (Scale bar 10 μm.)
FIG. 7.79. *Quilonia* sp. cephalic extremity, lateral view. (Original.) (Scale bar 10 μm.)
FIG. 7.80. *Khalilia sameera* (Khalil, 1922), cephalic extremity, lateral view. (Original.) (Scale bar 10 μm.)

27-(26) Buccal capsule walls not attached to lining of buccal cavity (Fig. 7.79).
Vulva in posterior third of body.
Vagina very short.
Dorsal ray usually with 6 rami (3 each side).
Parasites of elephants and rhinoceroses.

Quilonia Lane, 1914
(= *Evansia* Railliet, Henry & Joyeux, 1913;
= *Nematevansia* Ihle, 1919; = *Paraquilonia* Neveu-Lemaire, 1924)

28-(25) Ovejectors Y-shaped, parallel from origin.

MURSHIDIINEA

29-(30) Corona radiata inserted at anterior edge of buccal capsule.
Buccal capsule walls relatively thin. (Fig. 7.80).
Dorsal ray bifurcation deeper than proximal branches (Fig. 7.29).
Parasites of elephants and rhinoceroses.

Khalilia Neveu-Lemaire, 1924
(= *Amira* Lane, 1914; = *Amiroides* Strand, 1929)

30-(29) Corona radiata inserted on inside wall of buccal capsule.
Buccal capsule walls thick (Fig. 7.81).
Dorsal ray bifurcation usually extending to distal branches.

31-(32) Second corona radiata deep in buccal cavity (Fig. 7.82).
Female tail short.
Spicule tips bifid.
Parasite of South American tapir.

Neomurshidia Chabaud, 1957

32-(31) Corona radiata single (Fig. 7.81).
Female tail long.
Spicule tips usually simple.
Parasites of elephants, rhinoceroses and warthogs.

Murshidia Lane, 1914
(= *Buissonia* Neveu-Lemaire, 1924; = *Memphisia* Khalil, 1922;
= *Henryella* Neveu-Lemaire, 1924; = *Pterygopharynx* Witenberg, 1925)

33-(34) Elements of corona radiata usually fewer than 40 in number.
Buccal capsule round or oval in cross section.
Dorsal ray elongate.
Vulva usually very near anus.
Parasites of rhinoceroses and African elephants; one species in Indian elephant.

Murshidia (*Pteridopharynx*) (Lane, 1921) Chabaud, 1957

34-(33) Elements of corona radiata usually more than 40 in number.
Buccal capsule compressed laterally.
Dorsal ray short.
Vulva relatively distant from anus.

35-(36) Elements of corona radiata 80 or more in number.
Dorsal ray not greatly reduced.
Parasites of warthogs (*Hylochoerus* or *Phacochoerus*).

Murshidia (*Chabaudia*) Campana-Rouget, 1959

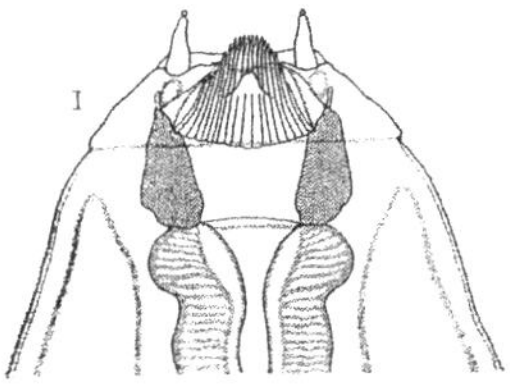

7.81

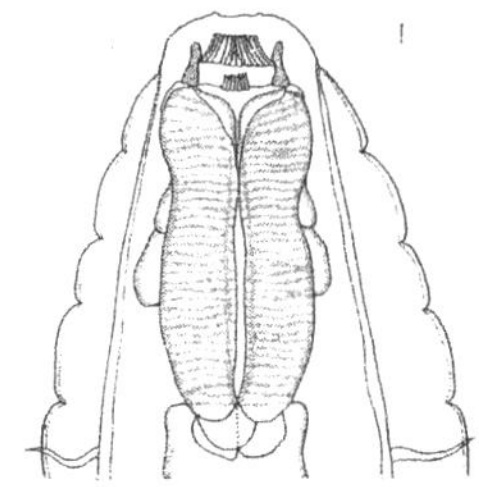

7.82

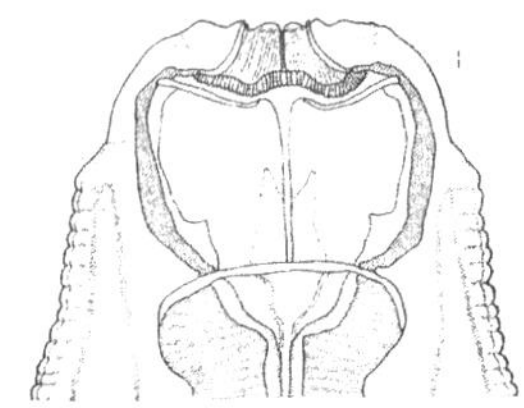

7.83

FIG. 7.81. *Murshidia murshidia* Lane, 1914, cephalic extremity, lateral view. (Original.) (Scale bar 10 μm.)

FIG. 7.82. *Neomurshidia monosticha* (Diesing, 1851), cephalic extremity, dorsal view. (After Travassos, 1930.) (Scale bar 10 μm.)

FIG. 7.83. *Deletrocephalus dimidiatus* Diesing, 1851, cephalic extremity, dorsal view. (Original.) (Scale bar 10 μm.)

36-(35) Elements of corona radiata fewer than 80 (except *M. falcifera* (Cobbold, 1882)).
Dorsal ray only slightly longer than, or similar in length to, laterals.
Parasites of African and Indian elephants.
Murshidia (*Murshidia*) (Lane, 1914) Chabaud, 1957

Family *DELETROCEPHALIDAE* Chitwood, 1969

Key to genera

1-(2) Elements of external corona radiata striated.
Internal corona radiata sinuous (Fig. 7.83).
Lining of buccal capsule without ridges.
Oesophagus almost cylindrical.
Parasites of *Rhea*. *Deletrocephalus* Diesing, 1851
(= *Quasistrongylus* Maplestone, 1932)

2-(1) External corona radiata absent.
Internal corona radiata not sinuous (Fig. 7.84).
Buccal capsule lining with vertical ridges.
Oesophagus flask-shaped.
Parasites of *Rhea*.
Paradeletrocephalus Freitas & Lent, 1947

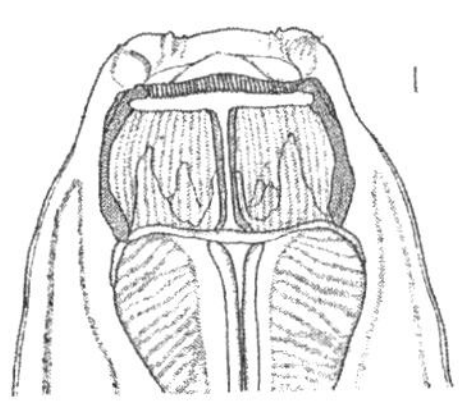

7.84

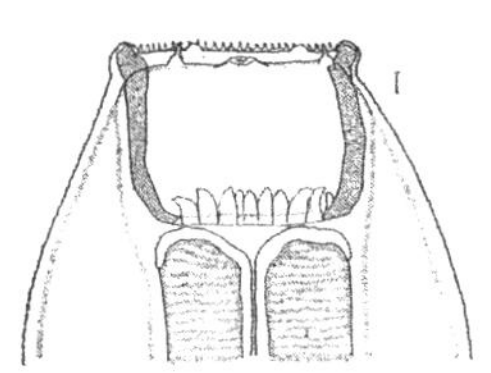

7.85

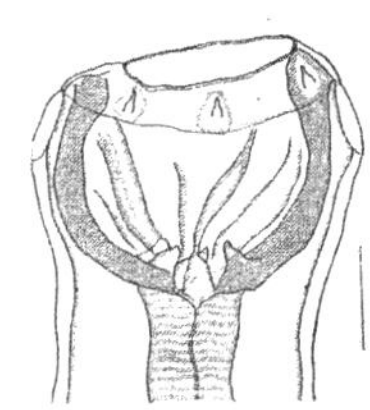

7.86

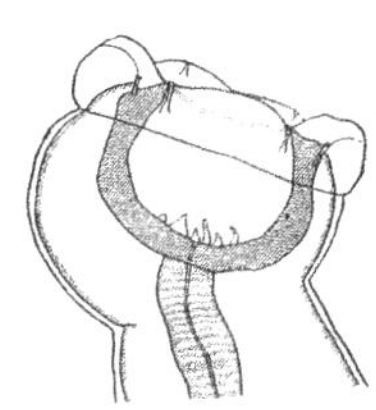

7.87

FIG. 7.84. *Paradeletrocephalus minor* (Molin, 1861), cephalic extremity, dorsal view. (Original.) (Scale bar 10 μm.)

FIG. 7.85. *Stephanurus dentatus* Diesing, 1839, cephalic extremity, lateral view. (Original.) (Scale bar 10 μm.)

FIG. 7.86. *Mammomonogamus laryngeus* (Railliet, 1899), cephalic extremity, lateral view. (After Baruš & Tenora, 1972.) (Scale bar 300 μm.)

FIG. 7.87. *Syngamus trachea* (Montagu, 1811), cephalic extremity, lateral view. (After Baruš & Tenora, 1972a.) (Scale bar 300 μm.)

Family *SYNGAMIDAE* Leiper, 1912

Key to subfamilies

1-(2) Rudimentary corona radiata present (Fig. 7.85).
Vulva near anus.
Bursa degenerate, subterminal.
Parasites of renal or perirenal tissue of domestic pig.
Stephanurinae Railliet, Henry & Bauche, 1919

2-(1) Corona radiata absent.
Vulva near midbody.
Bursa short but terminal.

3-(4) Buccal capsule cup-shaped with teeth at base.
Vulva anterior to midbody.
Parasites of respiratory system of birds or mammals.
Syngaminae Baylis & Daubney, 1926

4-(3) Buccal capsule cylindrical without teeth at base.
Vulva posterior to midbody.
Parasites of intestine of porcupine (*Hystrix*).
Archeostrongylinae Chabaud, 1958

Subfamily **Stephanurinae** Railliet, Henry & Bauche, 1919

One genus
Parasites of renal or perirenal tissue of domestic pig.
Stephanurus Diesing, 1839

Subfamily **Syngaminae** Baylis & Daubney, 1926

Key to genera and subgenera

1-(4) Buccal capsule with longitudinal ribs on inner wall (Fig. 7.86).
Eggs with or without opercula.
Parasites of mammals.

2-(3) Collar of oral opening absent.
Eggs without opercula.
Ends of bursal rays rounded.
Parasites of mammals in tropical regions.
Mammomonogamus Ryjikov, 1948

3-(2) Collar of oral opening present, well developed.
Eggs with opercula.
Ends of bursal rays pointed.
Parasites of rodents in east Palaearctic region.
Rodentogamus (Sadovskaja, 1950)

4-(1) Buccal capsule with smooth inner walls (Fig. 7.87).
Eggs with opercula.
Parasites of birds.

5-(8) Collar of oral opening present.
Dorsal ray as long as externodorsals.

6-(7) Distal ends of spicules not connected.
Spicules less than 0·1 mm long.
Syngamus Siebold, 1836
(= subgenus *Ornithogamus* Ryjikov, 1948)

7-(6) Distal ends of spicules connected.
Spicules more than 0·5 mm long.
Boydinema Baruš & Tenora, 1972

8-(5) Collar of oral opening absent.
Dorsal ray markedly longer than externodorsals.
Cyathostoma E. Blanchard, 1849
(= *Calcaronema* Hovorka & Macko, 1959)

9-(10) Dorsal ray extending beyond margin of bursa as thorn-like projection.
Spicules 0·08–0·40 mm long.
Cyathostoma (*Cyathostoma*) (Blanchard, 1849)

10-(9) Dorsal ray without central thorn-like projection, not extending beyond margin of bursa.
Spicules 0·45–0·80 mm long.
Cyathostoma (*Hovorkonema*) Turemuratov, 1963

Subfamily **Archeostrongylinae** Chabaud, 1958

One genus
Parasites of intestine of common or crested porcupine (*Hystrix*).
Archeostrongylus Biocca & Ferretti, 1958

Family *CHABERTIIDAE* (Popova, 1952 subfam.) n. fam.

Key to subfamilies

1-(2) Buccal capsule large, thick-walled, globular, or subglobular.
Parasites of intestine of ruminants, primates, rodents or macropodid marsupials.
Chabertiinae

2-(1) Buccal capsule small, relatively thin-walled, cylindrical, funnel-shaped or ring-shaped.

3-(4) Transverse cervical groove absent.
Externodorsal ray arising separately from dorsal ray.
Oesophagus usually more than one tenth of body length.
Parasites of stomach of macropodid marsupials.
Cloacininae

4-(3) Transverse cervical groove usually present at excretory pore (except tribe BOURGELATIINEA).
Externodorsal ray arising from common stem with dorsal ray.
Oesophagus usually only about one twentieth body length (except most of tribe BOURGELATIINEA).
Parasites of large intestine, small intestine or stomach of ruminants, primates, lemurs, pigs or rodents.
Oesophagostominae

Subfamily **Chabertiinae** Popova, 1952

Key to genera

1-(4) Head markedly narrower than cervical region.
Parasites of primates.

2-(3) Oral opening directed anteriorly.
Buccal capsule, excluding oesophageal funnel, much wider than deep (Fig. 7.88).
Teeth in oesophageal funnel small, lancet-shaped.
Parasites of *Colobus* spp.

Colobostrongylus Sandground, 1929

3-(2) Oral opening directed anterodorsally.
Buccal capsule as deep as wide (Fig. 7.91).
Three large tripartite oesophageal teeth project into buccal cavity.
Parasites of African and Asian apes, monkeys and man.

Ternidens Railliet & Henry, 1909

4-(1) Head as wide as or wider than cervical region.
Parasites of ruminants, rodents or Australian marsupials.

5-(8) Oral opening directed ventrally.
Elements of internal corona radiata very small, not hook-like.

6-(7) Head swollen to accommodate thick-walled buccal capsule twice as wide as deep (Fig. 7.89).
Three large oesophageal teeth extending into buccal capsule.
Parasites of okapi (*Okapia*).

Okapistrongylus Berghe, 1937

7-(6) Buccal capsule subglobular (Fig. 7.90).
Oesophageal funnel without teeth.
Parasites of ruminants.

Chabertia Railliet & Henry, 1909

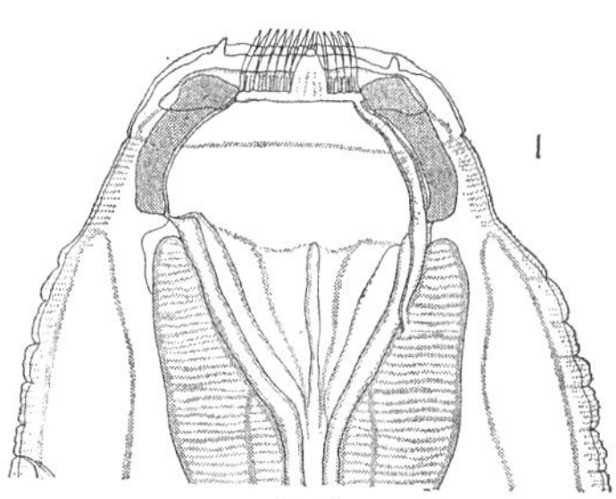

7.88

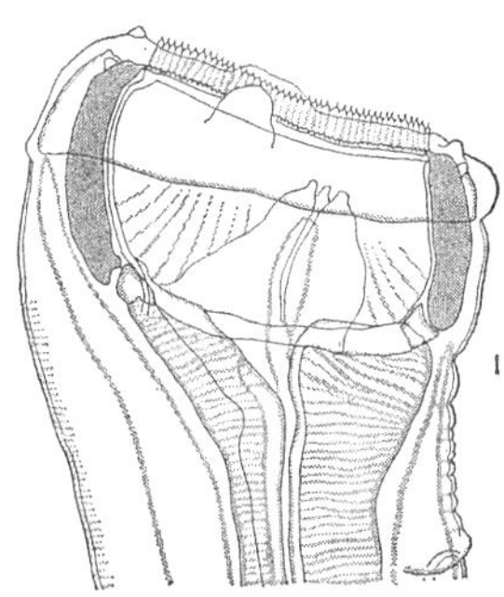

7.89

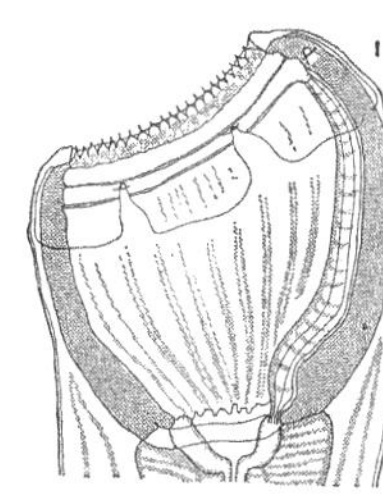

7.90

FIG. 7.88. *Colobostrongylus strongi* Sandground, 1929, cephalic extremity, lateral view. (Original.) (Scale bar 10 μm.)

FIG. 7.89. *Okapistrongylus epuluensis* Van den Berghe, 1937, cephalic extremity, lateral view. (Original.) (Scale bar 10 μm.)

FIG. 7.90. *Chabertia ovina* (Fabricius, 1788), cephalic extremity, lateral view. (Original.) (Scale bar 10 μm.)

7.91

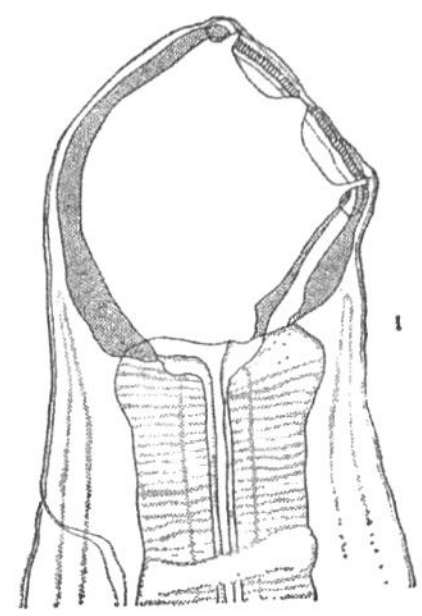

7.92

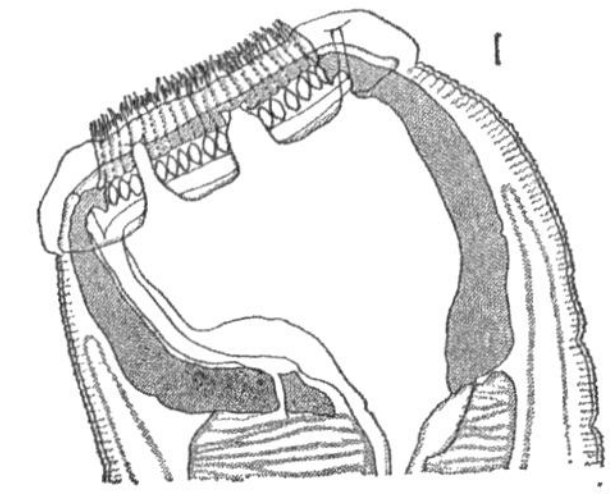

7.93

FIG. 7.91. *Ternidens deminutus* (Railliet & Henry, 1905), cephalic extremity, lateral view. (Original.) (Scale bar 10 μm.)

FIG. 7.92. *Ransomus rodentorum* Hall, 1916, cephalic extremity, lateral view. (Original.) (Scale bar 10 μm.)

FIG. 7.93. *Castorstrongylus castoris* Chapin, 1925, cephalic extremity, lateral view. (Original.) (Scale bar 10 μm.)

8-(5) Oral opening directed anterodorsally or anteriorly.
Internal corona radiata elements usually hook-like (Figs. 7.93–7.96) (but absent in *Ransomus*).

9-(14) Anterior edge of buccal capsule armed with large hook-like teeth some of which are paired or double (Fig. 7.95).

10-(13) Hook-like teeth paired in semilunar arrangement (Fig. 7.95).
Large lancets in large oesophageal funnel.
Parasites of ruminants.

11-(12) External corona radiata present.
Dorsal gutter extending half the depth of buccal cavity.
Oral opening directed anteriorly.
Parasites of Asian Cervidae.
Schulzinema Krastin, 1937

12-(11) External corona radiata absent.
Dorsal gutter absent.
Oral opening directed sharply anterodorsally.
Parasites of African and Asian Bovidae.
Agriostomum Railliet, 1902

13-(10) Hook-like teeth not in pairs; 16 to 24 in number (Fig. 7.96).
Small lancets in small oesophageal funnel.
Parasites of Malaysian rats (*Rattus*).
Cyclodontostomum Adams, 1933

14-(9) Anterior edge of buccal capsule either without hook-like teeth or internal corona radiata (Fig. 7.92), or internal corona radiata in form of numerous small hook-like elements (Figs. 7.93, 7.94).
Parasites of rodents or macropodid marsupials.

15-(18) Internal corona radiata in form of numerous, small hook-like elements.
External corona radiata elements long and slender.

16-(17) External corona radiata elements bifid (Fig. 7.93).
Gubernaculum absent.
Externodorsal rays arising separately from dorsal ray.
Female tail shorter than distance from vulva to anus.
Parasites of North American beaver (*Castor*).

Castorstrongylus Chapin, 1925

17-(16) Tips of external corona radiata elements not bifid (Fig. 7.94).
Gubernaculum present.
Externodorsal rays arising from dorsal ray.
Female tail more than twice as long as distance from vulva to anus.
Parasites of Australian marsupials (*Hypsiprymnodon*).

Corollostrongylus Beveridge, 1978

18-(15) Internal corona radiata absent.
External corona radiata in form of numerous small elements (Fig. 7.92).
Parasites of North American pocket gophers (*Thomomys*).

Ransomus Hall, 1916

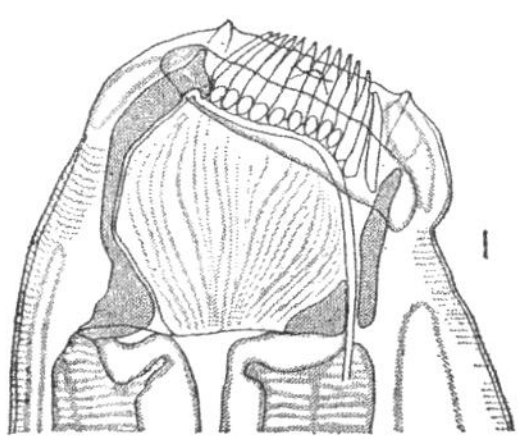
7.94

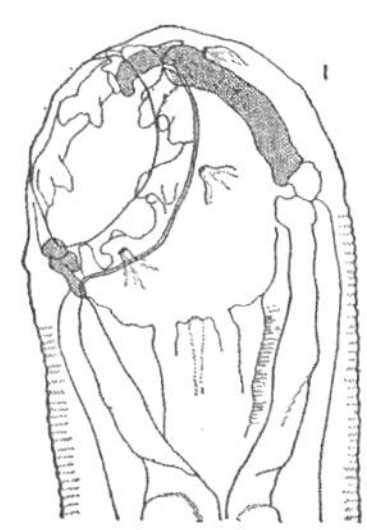
7.95

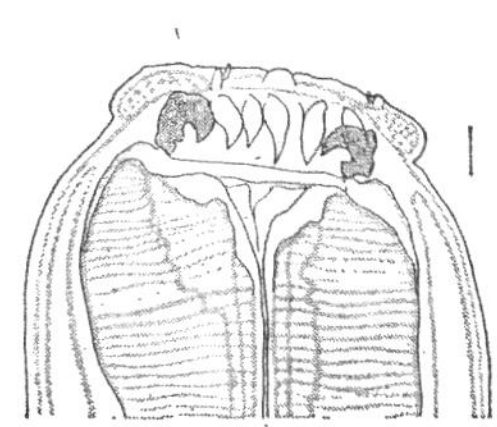
7.96

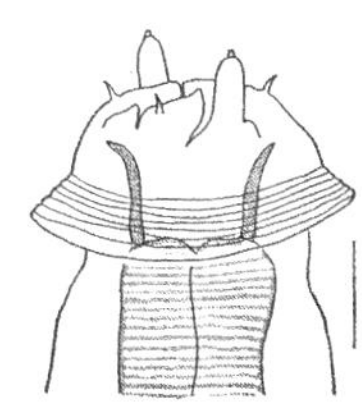
7.97

FIG. 7.94. *Corollostrongylus hypsiprymnodontis* Beveridge, 1978, cephalic extremity, lateral view. (After Beveridge, 1978a.) (Scale bar 10 μm.)

FIG. 7.95. *Agriostomum vryburgi* Railliet, 1902, cephalic extremity, lateral view. (Original.) (Scale bar 10 μm.)

FIG. 7.96. *Cyclodontostomum purvisi* Adams, 1933, cephalic extremity, lateral view. (Original.) (Scale bar 10 μm.)

FIG. 7.97. *Parazoniolaimus collaris* Johnston & Mawson, 1939, cephalic extremity, lateral view. (After Johnston & Mawson, 1939.) (Scale bar 10 μm.)

Subfamily **Cloacininae** Stossich, 1899

Keys to tribes and genera

1-(10) Oral opening surrounded by 6 to 8 fleshy lips (Fig. 7.97).

ZONIOLAIMINEA (Popova, 1952 fam.)

2-(5) Anterior end with postlabial collar (Fig. 7.97).

3-(4) Collar thin (cuticular), behind cephalic papillae.

Parazoniolaimus Johnston & Mawson, 1939

4-(3) Collar wide and thick, bearing amphids and cephalic papillae.

Dorcopsinema Mawson, 1977

5-(2) Collar absent.

6-(7) Oesophagus long and thin with terminal bulb.
Submedian papillae posterior to lips.
Zoniolaimus Cobb, 1898

7-(6) Oesophagus without bulb or, if bulb present, oesophagus relatively short.
Submedian papillae on lips.

8-(9) Oesophagus with elongate bulb with thick cuticular lining.
Cervical inflation present.
Potorostrongylus Johnston & Mawson, 1940

9-(8) Oesophagus usually without bulb; if bulb present narrower than preceding part of oesophagus.
Cervical inflation absent.
Labiostrongylus Yorke & Maplestone, 1926

10-(1) Oral opening without fleshy lips.

11-(22) Buccal capsule usually elongated and with striae (Fig. 7.98) (except *Spirostrongylus*) or, if short, very thick-walled.
Corona radiata, if present, external.
Submedian cephalic papillae usually setose.
PHARYNGOSTRONGYLINEA Popova, 1952

12-(17) Corona radiata present.

13-(14) Corona radiata with numerous pointed elements.
Cervical papillae in region of buccal capsule.
Pharyngostrongylus Yorke & Maplestone, 1926

14-(13) Corona radiata with rounded elements.
Cervical papillae in region of oesophageal bulb.

15-(16) Buccal capsule relatively short, without cross striations.
Worms spirally coiled.
Spirostrongylus Yorke & Maplestone, 1926
not Mönnig, 1926

16-(15) Buccal capsule elongated, with striated walls.
Worms straight.
Pararugopharynx Magzoub, 1964

17-(12) Corona radiata absent.

18-(21) Lateral alae absent.
Worms straight.

19-(20) Oral opening surrounded by labial and cephalic collars, latter with cephalic papillae (Fig. 7.98).
Rugopharynx Mönnig, 1926
(= *Spirostrongylus* Mönnig, 1926, *not* Yorke & Maplestone, 1926)

20-(19) Oral opening surrounded externally by dorsoventrally elongated ring of small sclerotized denticles.
Labial and cephalic collars absent.
Woodwardostrongylus Wahid, 1964
(= *Cristaceps* Mawson, 1971)

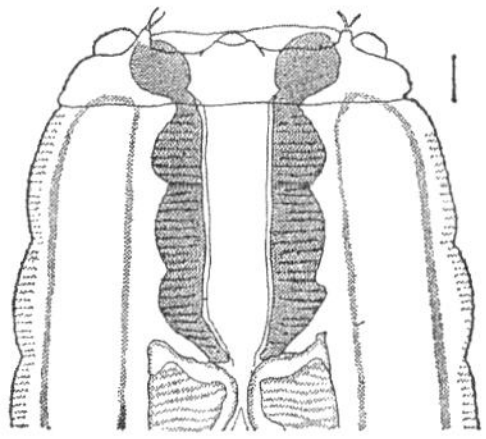

7.98

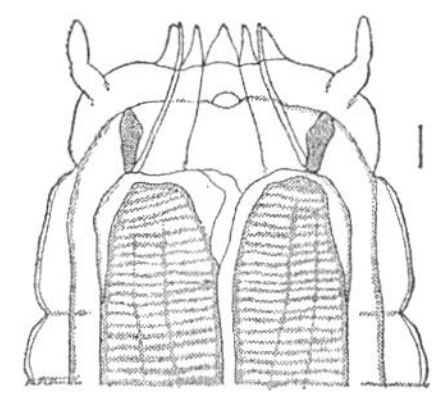

7.99

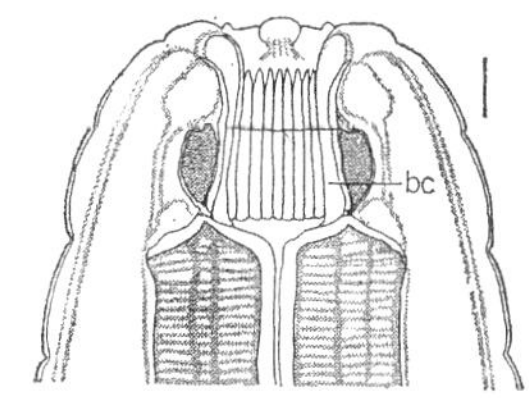

7.100

FIG. 7.98. *Rugopharynx* sp., cephalic extremity, lateral view. (Original.) (Scale bar 10 μm.)

FIG. 7.99. *Cloacina* sp., cephalic extremity, lateral view. (Original.) (Scale bar 10 μm.)

FIG. 7.100. *Coronostrongylus coronatus* Johnston & Mawson, 1939, cephalic extremity, lateral view. (Original.) (bc = buccal capsule.) (Scale bar 10 μm.)

21-(18) Lateral alae present.
Worms usually spirally coiled.
Cyclostrongylus Johnston & Mawson, 1939
(= *Oesophagonastes* Mawson, 1965)

22-(11) Buccal capsule not elongated, striated or thick-walled.

23-(26) Corona radiata present, occasionally of minute elements.
Submedian cephalic papillae fleshy, of 2 parts.
Oesophagus usually claviform or cylindrical.
CLOACININEA (Stossich, 1899, subf.)

24-(25) Corona radiata with 6 to 8 pointed elements arising at base of buccal capsule (Fig. 7.99).
Cloacina Linstow, 1898

25-(24) Corona radiata with 6 to 8 tiny elements arising near anterior end of buccal capsule.
Arundelia Mawson, 1977

26-(23) Corona radiata absent.
Oesophagus usually with posterior bulb.
Submedian papillae usually setose.
Buccal capsule thin-walled, sometimes unsclerotized or supported by ring-like thickenings or longitudinal ridges.
MACROPOSTRONGYLINEA

27-(28) Buccal capsule with large, thick, ring-like support near base (Fig. 7.100).
Cervical papillae and nerve ring posterior to middle of oesophagus.
Longitudinal ridges in buccal capsule numerous.
Coronostrongylus Johnston & Mawson, 1939

28-(27) Ring-like support for buccal capsule smaller or absent.
Cervical papillae and nerve ring anterior to middle of oesophagus.

29-(30) Prominent dorsal and ventral cuticular projections around oral opening (Fig. 7.101).
Papillostrongylus Johnston & Mawson, 1939

30-(29) Without dorsal and ventral cuticular projections around oral opening.

31-(32) Ventral lobes of bursa not joined.
Buccal capsule without longitudinal ridges.
Macroponema Mawson, 1978

32-(31) Ventral lobes of bursa joined.

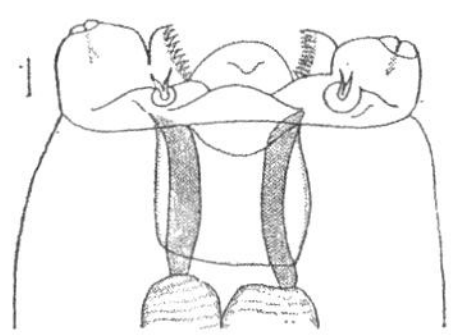

7.101

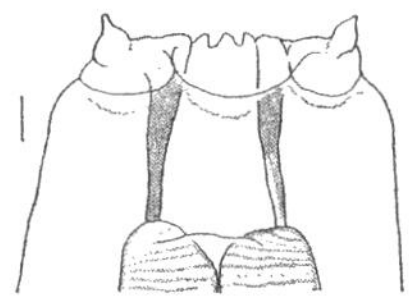

7.102

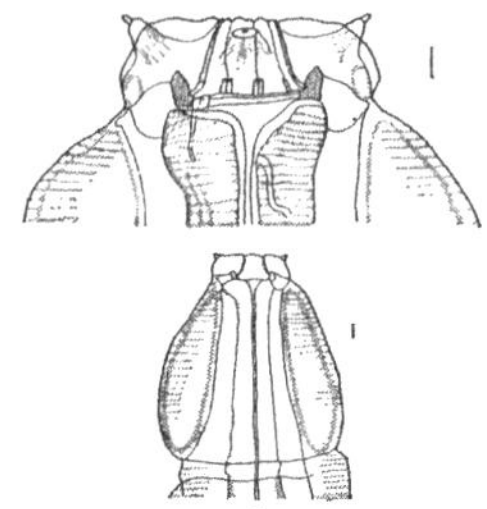

7.103

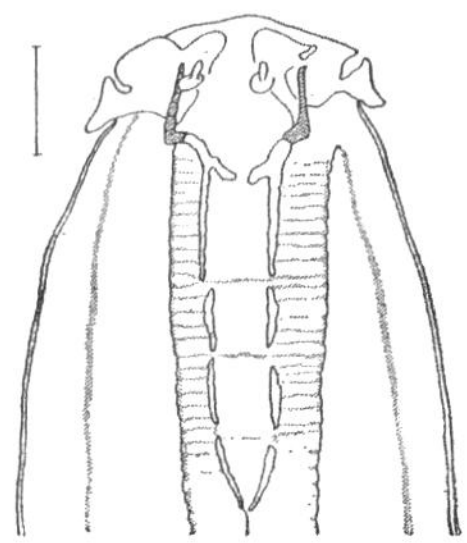

7.104

FIG. 7.101. *Papillostrongylus labiatus* Johnston & Mawson, 1939, cephalic extremity, dorsal view. (After Mawson, 1964.) (Scale bar 10 μm.)

FIG. 7.102. *Macropostrongylus macropostrongylus* Yorke & Maplestone, 1926, cephalic extremity, dorsoventral view. (After Mawson, 1977b.) (Scale bar 40 μm.)

FIG. 7.103. *Oesophagostomum* (*O.*) *dentatum* (Rudolphi, 1803); a. cephalic extremity, lateral view; b. anterior end, lateral view. (Original.) (Scale bar 10 μm.)

FIG. 7.104. *Rhabditostomum traguli* (Maplestone, 1932), anterior end, dorsoventral view. (After Chabaud & Krishnasamy, 1976.) (Scale bar 10 μm.)

33-(34) Buccal capsule with 8 internal longitudinal ridges.
Lateral cuticular projections present on cephalic collar (Fig. 7.102).
Buccal capsule usually thickest anteriorly or posteriorly.
Oesophagus nearly cylindrical with elongate posterior bulb.
Macropostrongylus Yorke & Maplestone, 1926, emend. Mawson, 1977
(= *Gelanostrongylus* Popova, 1952)

34-(33) Buccal capsule without internal ridges.
Cephalic collar without cuticular projections.
Buccal capsule thickest near midlength.
Oesophagus with more distinct isthmus and posterior expanded bulb.
Popovastrongylus Mawson, 1977

Subfamily **Oesophagostominae** Railliet, 1916

Key to tribes and genera

1-(20) Cervical inflation with transverse cervical groove at level of excretory pore (Fig. 7.103).

2-(5) Transverse cervical groove completely encircling body.
Parasites of mouse deer of Malaysia (*Tragulus*).
BOURGELATIOIDINEA

3-(4) Buccal capsule and oesophageal funnel consisting of 5 sections with funnel greatly elongated (Fig. 7.104).
Single corona radiata present with 6 lip-like internal elements.
Spicules with simple points.
Rhabditostomum Chabaud & Krishnasamy, 1976

4-(3) Oesophageal funnel not well developed, not elongated.
Internal and external coronae radiatae present (Fig. 7.105).
Spicules terminated by long coiled filaments (Fig. 7.106).
Bourgelatioides Chandler, 1931

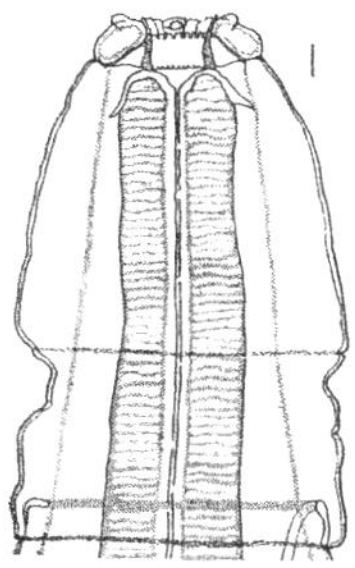

7.105

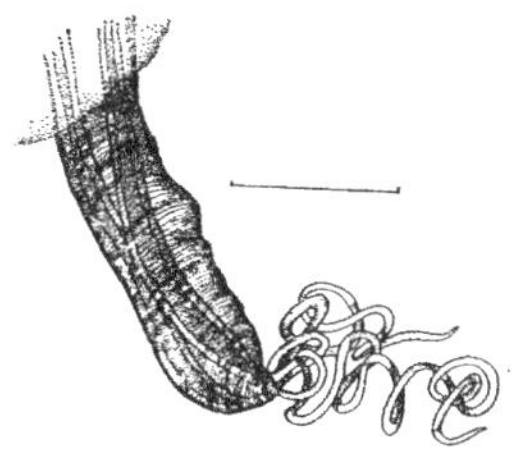

7.106

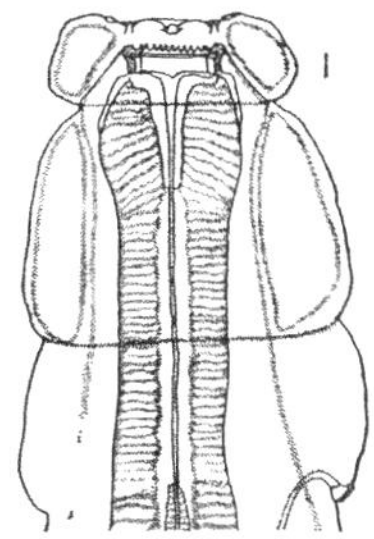

7.107

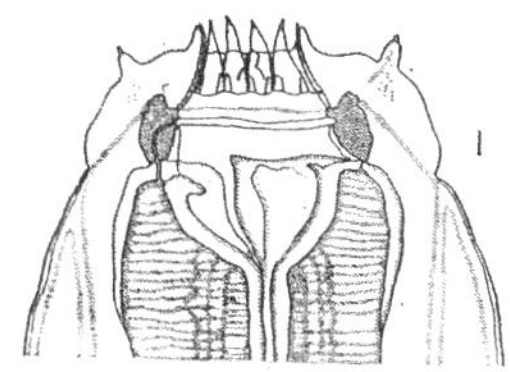

7.108

FIG. 7.105. *Bourgelatioides traguli* Chandler, 1931, anterior end, lateral view. (Original.) (Scale bar 10 μm.)

FIG. 7.106. Spicule tips of *Bourgelatioides traguli*. Original. (Scale bar 10 μm.)

FIG. 7.107. *Oesophagostomum* (*Bosicola*) *radiatum* (Rudolphi, 1803), anterior end, lateral view. (Original.) (Scale bar 10 μm.)

FIG. 7.108. *Oesophagostomum* (*Conoweberia*) *blanchardi* Railliet & Henry, 1912, cephalic extremity, lateral view. (Original.) (Scale bar 10 μm.)

5-(2) Transverse cervical groove confined to ventral side.

OESOPHAGOSTOMINEA

6-(7) External corona radiata with 6 to 8 broad elements.
Internal corona radiata present.
Parasites of African suids.

Daubneyia LeRoux, 1940

7-(6) External corona radiata with more than 8 elements (Fig. 7.103).
Internal corona radiata present.

Oesophagostomum Molin, 1861

8-(15) Oesophageal funnel without tubercles or teeth.

9-(14) Two coronae radiatae present.

10-(11) Oesophageal funnel well formed.
Buccal cavity shallow and usually uniformly cylindrical.
Parasites of Suidae.

Oesophagostomum (*Oesophagostomum*) Molin, 1861

11-(10) Oesophageal funnel simple or small.
Parasites of ruminants.

12-(13) Spicules long (more than 1·0 mm).
Vagina long.
Cervical papillae posterior to oesophageal expansion.
Parasites of ruminants.

Oesophagostomum (*Hysteracrum*) Railliet & Henry, 1913

13-(12) Spicules short (less than 1·0 mm).
Vagina short.
Cervical papillae anterior to oesophageal expansion.
Parasites of African or domestic Bovidae.

Oesophagostomum (*Proteracrum*) Railliet & Henry, 1913

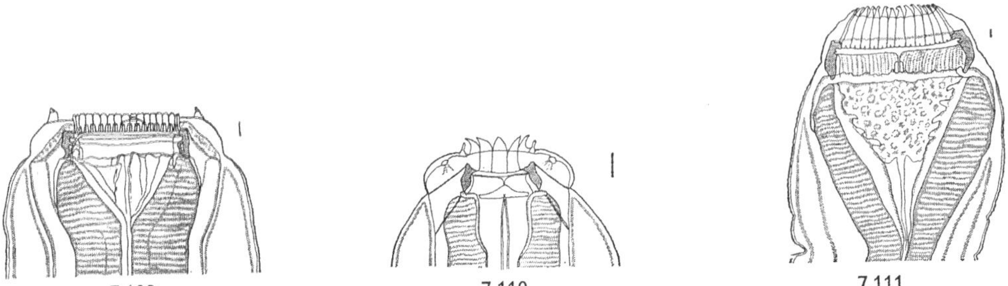

FIG. 7.109. *Oesophagostomum* (*Ihlea*) *stephanostomum* Stossich, 1904, cephalic extremity, lateral view. (Original.) (Scale bar 10 μm.)

FIG. 7.110. *Oesophagostomum* (*Lerouxiella*) *susannae* LeRoux, 1929, cephalic extremity, dorsal view. (After LeRoux, 1929.) (Scale bar 25 μm.)

FIG. 7.111. *Trachypharynx natalensis* Ortlepp, 1962, anterior end, dorsal view. (Original.) (Scale bar 10 μm.)

14-(9) One corona radiata present with numerous elements (38 to 45) (Fig. 7.107).
Oesophageal funnel well formed.
Cervical papillae anterior to oesophageal expansion.
Parasites of ruminants.
Oesophagostomum (*Bosicola*) Sandground, 1929

15-(8) Oesophageal funnel with tubercles or teeth.

16-(19) Buccal capsule without transverse processes internally.
Parasites of primates.

17-(18) External corona radiata with 10 to 15 elements.
Oesophageal funnel large with 3 concave walls, each with tooth (Fig. 7.108).
Cervical papillae anterior to oesophageal swelling.
Buccal capsule wider posteriorly.
Oesophagostomum (*Conoweberia*) Ihle, 1922

18-(17) External corona radiata with 30 to 40 elements.
Oesophageal funnel with 6 prominent plates (Fig. 7.109).
Cervical papillae posterior to oesophageal swelling.
Buccal capsule wider anteriorly.
Oesophagostomum (*Ihlea*) Travassos & Vogelsang, 1932

19-(16) Buccal capsule with transverse processes internally (Fig. 7.110).
External corona radiata with 10 to 12 elements.
Oesophageal funnel with 3 concave plates.
Cervical papillae at middle of oesophageal swelling.
Buccal capsule wider posteriorly.
Parasites of African rodents.
Oesophagostomum (*Lerouxiella*) Chabaud & Durette-Desset, 1973

20-(1) Transverse cervical groove absent.
Cervical inflation, if present, greatly reduced.
BOURGELATIINEA

21-(22) Anterior part of oesophagus greatly expanded (Fig. 7.111).
Parasites of large African cane rats (Thryonomyidae).
Trachypharynx Leiper, 1911

22-(21) Anterior part of oesophagus not greatly expanded.

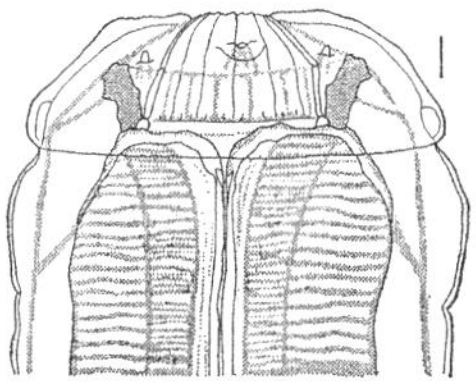
7.112

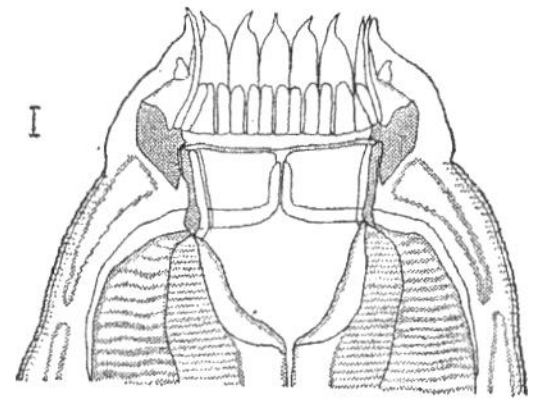
7.113

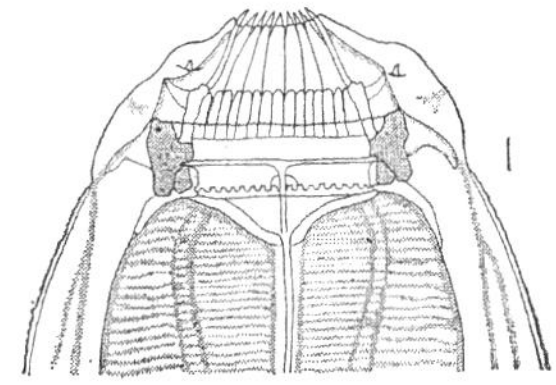
7.114

FIG. 7.112. *Kuntzistrongylus selfi* (Schmidt & Kuntz, 1975), cephalic extremity, lateral view. (Original.) (Scale bar 10 μm.)

FIG. 7.113. *Phacochoerostrongylus hylochoeri* (Van den Berghe, 1943), cephalic extremity, dorsal view. (Original.) (Scale bar 10 μm.)

FIG. 7.114. *Bourgelatia diducta* Railliet, Henry & Bauche, 1919, cephalic extremity, dorsal view. (Original.) (Scale bar 10 μm.)

23-(24) External corona radiata with 6 to 10 elements.
Cephalic papillae prominent.
Parasites of lemurs (*Hapalemur*).
Lemurostrongylus Chabaud, Brygoo & Petter, 1961

24-(23) External corona radiata with 19 or more elements.
Cephalic papillae not prominent.

25-(26) Extra chitinous supports absent (Fig. 7.112).
Elements of internal and external coronae radiatae equal in thickness.
Body length 24 to 33 times length of oesophagus.
Cuticle coarsely annulated.
Oesophagus cylindrical anteriorly, expanded posteriorly.
Dorsal gutter absent.
Parasites of Asian murid rodents (*Rattus*).
Kuntzistrongylus n.g.

26-(25) Large extra chitinous supports present (Figs. 7.113, 7.114).
Elements of internal corona radiata twice as thick as elements of external corona.
Body length 11 to 15 times that of oesophagus.
Cuticle finely annulated.
Dorsal gutter present.
Oesophagus expanded anteriorly and posteriorly.
Parasites of Suidae.

27-(28) Buccal capsule cylindrical; twice as wide as deep (Fig. 7.113).
Ovejectors Y-shaped; infundibula (Fig. 7.19) as in *Bourgelatia*.
Vagina long.
Dorsal ray long, bifurcated only half way to proximal branches (Fig. 7.49).
Parasites of warthogs (*Phacochoerus*).
Phacochoerostrongylus Schwartz, 1928

28-(27) Buccal capsule very short with ring-like walls; about 9 times as wide as deep (Fig. 7.114).
Ovejectors opposed.
Vagina short.
Dorsal ray short; bifurcated deeper than proximal branches (Fig. 7.44).
Parasites of domestic pig.
Bourgelatia Railliet, Henry & Bauche, 1919

REFERENCES

BARUŠ, V. & TENORA, F. 1972a. Notes on the systematics and taxonomy of the nematodes belonging to the family Syngamidae Leiper, 1912. *Acta Univ. Agric., Brno, Fac. Agron.*, **20**, 275–286.

——— & TENORA, F. 1972b. Notes on the evolution of nematodes belonging to the family Syngamidae Leiper, 1912. *Acta Univ. Agric., Brno, Fac. Agron.*, **20**, 287–295.

BEVERIDGE, I. 1978a. *Corollostrongylus hypsiprymnodontis* gen. et sp. n. (Nematoda: Strongylidae) from the rat-kangaroo, *Hypsiprymnodon moschatus* (Marsupialia). *J. Parasit.*, **64**, 657–660.

——— 1978b. A taxonomic revision of the genera *Phascolostrongylus* Canavan, and *Oesophagostomoides* Schwartz (Nematoda: Trichonematidae) from wombats. *Aust. J. Zool.*, **26**, 585–602.

——— 1978c. *Hypodontus macropi* Mönnig, 1929, a hookworm-like parasite of macropodid marsupials. (In Press).

——— & MAWSON, P. M. 1978. A taxonomic revision of the genera *Macropostrongyloides* Yamaguti and *Paramacropostrongylus* Johnston and Mawson (Nematoda: Trichonematidae) from Australian marsupials. *Aust. J. Zool.*, **26**, 763–787.

BIOCCA, E. & FERRETTI, G. 1958. *Archeostrongylus italicus* gen. nov. et sp. nov., nuovo nematode borsato parassita di *Hystrix cristata* in Italia Central. *Atti Accad. naz. Lincei Rc.*, Ser. 8, **23**, 467–472.

CHABAUD, A. 1957. Revue critique des nematodes du genre *Quilonia* Lane 1914 et du genre *Murshidia* Lane 1914. *Annls Parasit. hum. comp.*, **32**, 98–131.

———, 1958. Osservazioni sulla classificazione del sottordine Strongylina e creazione di una nuova sotto-famiglia: Archeostrongylinae. *Atti Accad. naz. Lincei Rc.*, **24**, 176–179.

——— 1965. Ordre des Strongylida. In: Grassé, P. P. *Traité de Zoologie*, Vol. 4, fasc. 3, pp. 869–933.

———, BRYGOO, E. R. & PETTER, A. J. 1961. Les nématodes parasites de lémuriens malgaches. IV. Description de deux nouveaux genres et observations sur *Protofilaria furcata* Chandler. *Bull. Mus. natn. Hist. nat., Paris*, 2e sér., **33** (5), 532–544.

——— & DURRETTE-DESSET, M. C. 1973. Description d'un nouveau nématode oesophagostome, parasite d'*Hyemoschus* au Gabon, et remarques sur le genre *Oesophagostomum*. *Bull. Mus. natn. Hist. nat., Paris*, 3e sér., No. 184, Zoologie 123, 1415–1424.

——— & KRISHNASAMY, M. 1976. Nématodes oesophagostomes parasites de *Tragulus javanicus*. *Bull. Mus. natn. Hist. nat., Paris*, 3e sér., No. 388, Zoologie 270, 721–727.

——— & ROUGEAUX, M. 1957. Remarques sur la dentition de *Syngamus trachea* (Montagu) et sur la place systématique des syngames. *Annls Parasit. hum. comp.*, **32**, 264–266.

——— & TCHEPRAKOFF, R. 1977. Sur *Chapiniella diazi* n.sp., strongylide parasite de *Testudo denticulata* au Vénézuela. *Bull. Mus. natn. Hist. nat., Paris*, 3e sér., No. 469, Zoologie 326, 765–769.

CHITWOOD, B. G. & CHITWOOD, M. B. 1938. *An Introduction to Nematology* (revised 1974). Baltimore, USA: University Park Press, 344 pp.

CHITWOOD, M. B. 1969. The systematics and biology of some parasitic nematodes. In: *Chemical Zoology.* Florkin, M. & Scheer, B. T. (Editors). Vol. III. *Echinodermata, Nematoda and Acanthocephala.* New York & London: Academic Press, Inc., pp. 223–244.

DIAOURE, A. 1964. Strongylides parasites de mammifères du Congo-Brazzaville (collection R. Rousselot). *Annls Parasit. hum. comp.*, **39**, 243–284.

DOUGHERTY, E. C. 1951. Evolution of zooparasitic groups in the phylum Nematoda, with special reference to host distribution. *J. Parasit.*, **37**, 353–378.

FREITAS, J. F. T. & LENT, H. 1947. Revisão da subfamília Deletrocephalinae Railliet, 1916 (Nematoda, Strongyloidea). *Rev. bras. Biol.*, **7**, 73–100.

HALL, M. C. 1916. Nematode parasites of mammals of the orders Rodentia, Lagomorpha, and Hyracoidea. *Proc. U.S. natn. Mus.*, **50**(2131), 1–258.

HOVORKA, J. & MACKO, J. 1959. *Calcaronema* gen. nov. a new genus of the subfamily Cyathostominae Nicoll, 1927 (Syngamidae Leiper, 1912) and the description of the new species *C. trifurcatum* sp. n. and *C. verrucosum* sp. n. *Helminthologia*, **1**, 103–112.

INGLIS, W. G. 1968. The geographical and evolutionary relationships of Australian trichostrongyloid parasites and their hosts. *J. Linn. Soc. (Zool.)*, **47**, 327–347.

KEAST, A. 1977. Historical biogeography of the marsupials. In: *The Biology of Marsupials*. Stonehouse, B. & Gilmore, D. (Editors). London: The Macmillan Press Ltd., pp. 69–95.

KOWALSKI, K. 1971. Mammals. An outline of theriology. [Translated English version. Available as TT71-54045, from N.T.I.S., Springfield, VA 22151, USA.]

KUNG, C. C. 1948. Some new nematodes from the Australian wallaby (*Macropus rufogrisea fruticus*) with a note on the synonymy of the genera *Zoniolaimus*, *Labiostrongylus* and *Buccostrongylus*. *J. Helminth.*, **22**, 93–108.

LICHTENFELS, J. R. 1975. Helminths of domestic equids. Illustrated keys to genera and species with emphasis on North American forms. *Proc. helminth. Soc. Wash.*, **42** (Special Issue), vi+92 pp.

——— 1979. A conventional approach to classification of the Strongyloidea, nematode parasites of mammals. *Amer. Zoologist.* (In Press).

MAGZOUB, A. 1964. Three nematode species (Strongyloidea: Trichoneminae) from Queensland wallabies. *Trans. R. Soc. S. Aust.*, **88**, 47–51.

MAWSON, P. M. 1964. Some Nematoda (Strongylina and Oxyurina) from kangaroos (*Macropus* spp.) from eastern Australia. *Parasitology*, **54**, 237–262.

——— 1965. Notes on some species of Nematoda from kangaroos and wallabies, including a new genus and three new species. *Parasitology*, **55**, 145–162.

——— 1972. Three new species of the genus *Cloacina* Linstow (Nematoda: Strongylata) from macropod marsupials. *Trans. R. Soc. S. Aust.*, **96**, 109–113.

——— 1974. The genus *Potorostrongylus* Johnston and Mawson (Nematoda: Trichonematidae) from macropod marsupials. *Trans. R. Soc. S. Aust.*, **98**, 135–137.

——— 1975. Two new species of the genus *Cloacina* (Nematoda: Strongylida) from the tammar, *Macropus eugenii*. *Trans. R. Soc. S. Aust.*, **99**, 39–42.

——— 1976. *Woodwardostrongylus obendorfi* new species (Nematoda: Amidostomatidae) from kangaroos. *Trans. R. Soc. S. Aust.*, **100**, 121–124.

——— 1977a. The genus *Cyclostrongylus* Johnston and Mawson (Nematoda: Trichonematidae). *Trans. R. Soc. S. Aust.*, **101**, 19–20.

——— 1977b. Revision of the genus *Macropostrongylus* and descriptions of three new genera: *Popovastrongylus*, *Dorcopsinema*, and *Arundelia* (Nematoda: Trichonematidae). *Trans. R. Soc. S. Aust.*, **101**, 51–62.

——— 1977c. *Cloacina cornuta* (Davey & Wood) and *C. caballeroi* sp. nov. (Nematoda: Cloacininae) from macropods from Papua. In: Excerta parasitológica en memoria del Doctor Eduardo Caballero y Caballero, Mexico: Pub. Espec. (4), Inst. Biol., Universidad Nacional Autónoma de Mexico, pp. 455–458.

——— 1978a. *Macropicola ocydromi* n. g., n. sp. (Nematoda: Strongylidae) from a Western Australian kangaroo. *Trans. R. Soc. S. Aust.*, **102**, 113–115.

——— 1978b. *Macroponema* (Nematoda: Trichonematidae) a new genus from macropod marsupials. *Int. J. Parasitol.*, **8**, 163–166.

OGDEN, C. G. 1966a. A new species of hookworm, *Agriostomum monnigi* (Nematoda), from the gemsbok, *Oryx gazella*. *Ann. Mag. nat. Hist.*, Year 1965, Ser. 13 **8**(93/94), 569–573.

——— 1966b. A revision of the genus *Khalilia* Neveu-Lemaire, 1924 (Nematoda: Strongyloidea). *Parasitology*, **56**, 471–480.

ORTLEPP, R. J. 1961. On *Trachypharynx natalensis* sp. nov. and some associated genera of nematodes. *Onderstepoort J. vet. Res.*, **29**, 159–163.

——— 1964. Observations on helminths parasitic in warthogs and bushpigs. *Onderstepoort J. vet. Res.*, **31**, 11–38.

POPOVA, T. I. 1955. Strongyloids of animals and man. Strongylidae. Vol. V. Essentials of Nematodology. Moscow: Akad. Nauk. SSSR. [Translated English version available as OTS64-11008, N.T.I.S., Springfield, VA 22151, USA.]

——— 1958. Strongyloids of animals and man. Trichonematidae. Vol. VII. Essentials of Nematodology. Moscow: Akad. Nauk. SSSR. [Translated English version available as TT65-50073 from N.T.I.S., Springfield, VA 22151, USA.]

——— 1960. [Strongyloids of animals and man. Cloacinidae, Stephanuridae, Diaphanocephalidae.] (In Russian.) Osnovy Nematodologii, Vol. IX. Moscow: Akad. Nauk. SSSR., 244 pp.

PURSGLOVE, S. R. 1976. *Eucyathostomum webbi* sp. n. (Strongyloidea: Cloacinidae) from white-tailed deer (*Odocoileus virginianus*). *J. Parasit.*, **62**, 574–578.

REP, B. H. 1963. On the polyxenia of the Ancylostomidae and the validity of the characters used for their differentiation. *Trop. geogr. Med.*, **15**, 173–218.

SCHMIDT, G. D. & KUNTZ, R. E. 1975. Nematode parasites of Oceanica. *Oesophagostomum* (*Conoweberia*) *selfi* sp. n., (Nematoda: Trichonematidae) from *Rattus coxinga coxinga* Swinhoe. *Proc. Okla. Acad. Sci.*, **55**, 163–165.

SKRJABIN, K. I., SHIKHOBALOVA, N. P., SCHULZ, R. S., POPOVA, T. I., BOEV, S. N. & DELYAMURE, S. L. 1952. Key to Parasitic Nematodes. III. Strongylata. Moscow: Akad. Nauk. SSSR. [Translated English version available as OTS60-51062 from N.T.I.S., Springfield, VA 22151, USA.]

TRONCY, P. M., GRABER, M. & THAL, J. 1973. Oesophagostominae des suidés sauvages d'Afrique centrale. *Bull. Mus. natn. Hist. Nat. Paris*, 3e sér., No. 184, Zoologie 123, 1425–1450.

——— & THAL, J. 1977. Description d'un nématode parasite du buffle africain: *Oesophagostomum* (*Proteracum*) *synceri* n. sp. *Bull. Mus. natn. Hist. Nat. Paris*, 3e sér., No. 428, Zoologie 298, 173–176.

TSHOIJO, U. 1959. [New helminths in equines.] *Trudy Gelmint. Lab.*, **9**, 346–361. [In Russian.]

TUREMURATOV, A. 1963. [Syngamids of pelecaniform birds of the Aral Sea and notes on the structure of this nematode family.] *Vest. karakalpaksk. Fil, AN Uzbek SSSR*, **13**, 45–50. [In Russian.]

VUYLSTEKE, C. 1953. Notes sur les nématodes parasites de l'éléphant d'Afrique. *Rev. Zool. Bot. afr.*, **48**, 213–239.

——— 1963. Nématodes parasites d'oiseaux. Exploration du Parc National de l'Upemba, Mission G. F. de Witte (1946–1949). Deuxième sér. Brussels, Fasc. **17** (1), 3–41.

WAHID, S. 1964. A preliminary revision of the genus *Pharyngostrongylus* Yorke and Maplestone, 1926. *J. Helminth*, **38**, 181–190.

YAMAGUTI, S. 1961. *Systema Helminthum.* Vol 3. *The nematode parasites of vertebrates.* New York: Interscience Publishers, Inc., 1261 pp.

6 STRONGYLIDA

TRICHOSTRONGYLOIDEA

by MARIE-CLAUDE DURETTE-DESSET*

Introduction

The superfamily Trichostrongyloidea (order Strongylida) consists of monoxenous parasites with reduced buccal capsules, which inhabit the stomach or small intestine of all classes of terrestrial vertebrates (dictyocaulids however, are lung parasites). They constitute the richest superfamily of parasitic nematodes in numbers of genera and species. The taxonomy of the Trichostrongyloidea is complex and difficult, not only because of the large number of species involved, but also because the worms are small and superficially similar in form.

The caudal bursa and synlophe have essential diagnostic characters and must be studied by means of certain techniques. The caudal bursa can be studied only after being fully extended in ventral view. This is accomplished by rolling the worm (in water, lactophenol, or a mixture of the two) between slide and coverslip while observing through the microscope. Fresh material or material fixed in 70% ethanol is most suitable; worms fixed in formalin are usually brittle and it is often impossible to extend the bursa fully.

The structure of the synlophe can be appreciated only by studying a transverse section generally taken through the midbody region. It is indispensable to recognize dorsal, ventral, right and left sides of the specimen. The section is, therefore, taken asymmetrically, i.e. with the anterior surface cut at a right angle, and the posterior surface at an oblique angle to the body axis. Since nematodes generally lie on their sides this procedure usually results in a section which is thicker ventrally than dorsally or vice versa. This asymmetry, as well as the fact that the posterior surface of the section is oblique to the body axis, permits recognition of dorsal, ventral, left and right sides of the worm. Specimens stored in 70% ethanol are ideal for such study; glycerine-alcohol gradually destroys the internal structure of the specimens and should be avoided.

Special terms used in Key

1. *Neodont Formation:* Cuticular oesophageal tooth projecting into oesophageal lumen and representing a neoformation (Fig. 10.1). This should not be confused with the opening of the dorsal oesophageal gland which may be on an elevation (dorsal oesophageal tooth) in the buccal cavity (Fig. 10.2).
2. *Synlophe* (see Willmott, 1974): Illustrations of transverse body sections given in the Key are oriented so that the anatomical dorsal, ventral, left and right sides of the worm correspond to the top, bottom, left and right sides respectively of the diagram.

   The apex of the ridges may be directed perpendicular to, or oblique to, the body surface. In the latter case, an axis of orientation exists which separates ridges pointing in opposite directions. This axis is:
   (a) *sagittal* if it passes through dorsal and ventral hypodermal cords (Fig. 10.3, A);
   (b) *frontal* if it passes through lateral hypodermal cords (Fig. 10.3, B);
   (c) *oblique* if it passes between median and lateral hypodermal cords (Fig. 10.3, C).

---

*Laboratoire de Zoologie (Vers) associé au CNRS, Muséum National d'Histoire Naturelle, Paris, France

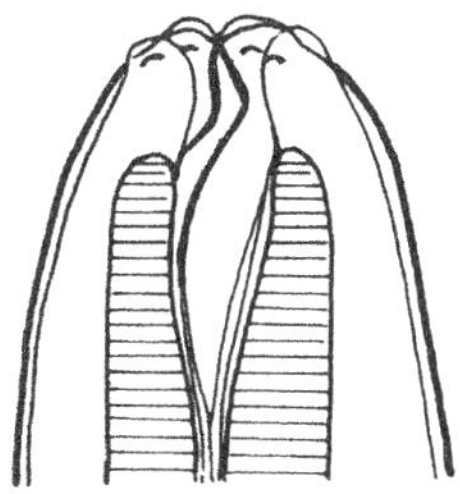

10.1

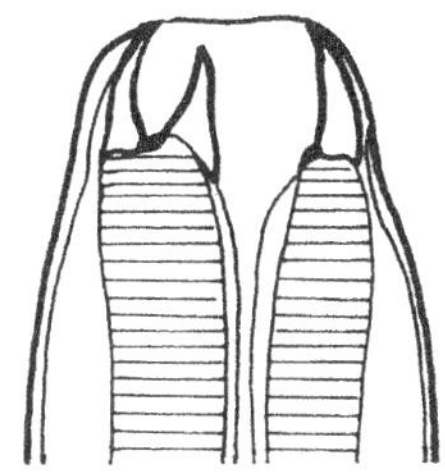

10.2

FIG. 10.1. Neodont formation, *Ashworthius perrilli* Chauhan, Pande & Singh, 1972, cephalic extremity, lateral view (after Chauhan, Pande & Singh, 1972).

FIG. 10.2. Dorsal oesophageal tooth, *Amidostomum fulicae* (Rudolphi, 1819), cephalic extremity, lateral view (after Wertheim & Durette-Desset, 1975).

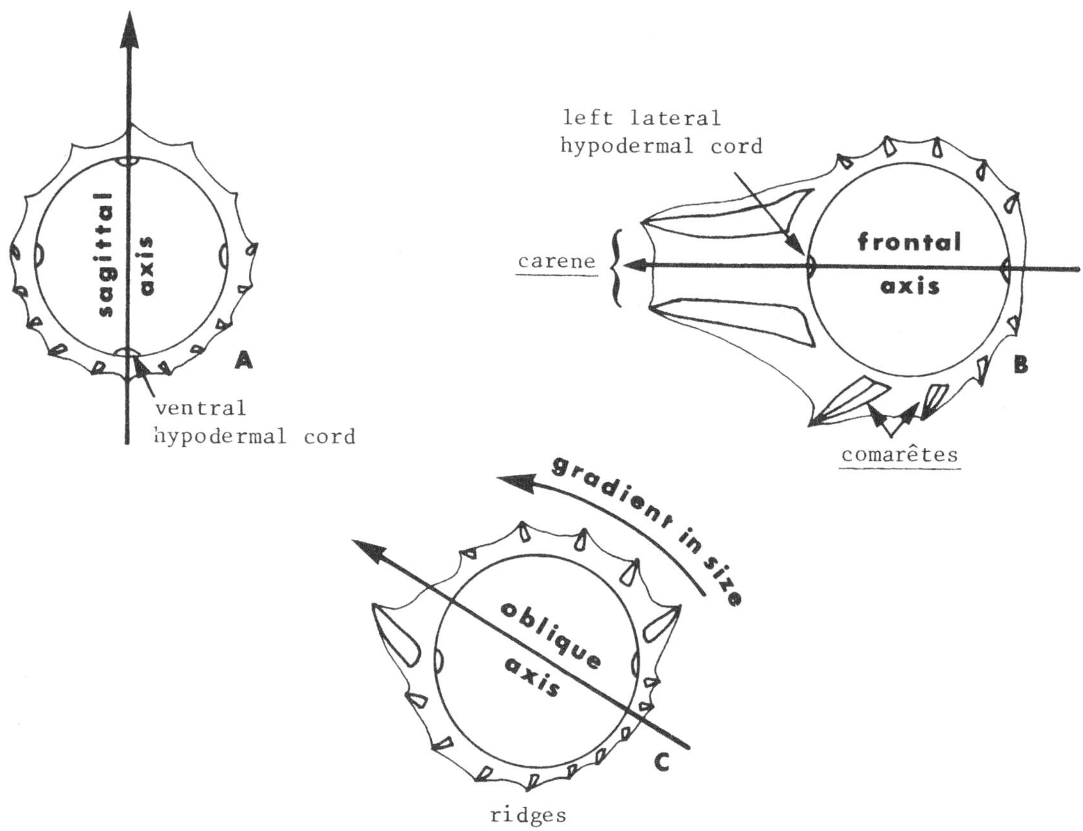

10.3

FIG. 10.3. Axes of orientation of cuticular ridges. A: sagittal, B: frontal, C: oblique.

In primitive species the axis of orientation is sagittal. During evolution, the axis may rotate, as much as 90°, to become frontal. In a cross section of the body there may be a gradient in size of the ridges. This gradient is always described in terms of diminishing size.

Comarêtes are highly developed ridges resulting from the fusion of several ridges.

A carene is a cuticular dilation on the left or dorsal left side of the body (Fig. 10.3B). Three types are recognized:

Type A: Carene supported by hypertrophied left lateral cuticular ridge; adjacent ridge on dorsal side also often hypertrophied.

Type B: Carene supported by two or more pairs of ridges larger than other ridges of synlophe.

Type C: Carene without ridges or with ridges equal in size to those on ventral left or dorsal right side of body.

3. *Caudal Bursa:* Nomenclature of the caudal bursa follows Chabaud *et al.* (1970) (Fig. 10.4).

Papilla Zero: generally unpaired, on anterior lip of genital cone.
Papilla 1: prebursal, at end of prebursal rays.
Papilla 2: ventroventral, at end of ray 2.
Papilla 3: lateroventral, at end of ray 3.
Papilla 4: externolateral, at end of ray 4.
Papilla 5: mediolateral, at end of ray 5.
Papilla 6: posterolateral, at end of ray 6.
Papilla 7: pair of papillae on posterior lip of genital cone.
Papilla 8: externodorsal, at end of ray 8.
Papillae 9 and 10: dorsal, at end of dorsal ray.

In some species, extradorsal rays (Inglis, 1968) exist between rays 8 and 9 (Fig. 10.55). Bursal rays may become grouped and the following types of bursal arrangements can exist (Fig. 10.5):

Type 2–1–2: Rays 2 and 3 and rays 5 and 6 grouped, ray 4 isolated (Fig. 10.5A).
Type 2–2–1: Rays 2 and 3 and rays 4 and 5 grouped, ray 6 isolated (Fig. 10.5C).
Type 1–3–1: Ray 2 isolated, rays 3 to 5 grouped, ray 6 isolated (Fig. 10.5B).
Type 3–2: Rays 2 to 4 and rays 5 and 6 grouped (Fig. 10.5D).
Type 2–3: Rays 2 and 3 and rays 4 to 6 grouped (Fig. 10.5E).
Bursae are illustrated in ventral view unless indicated otherwise.

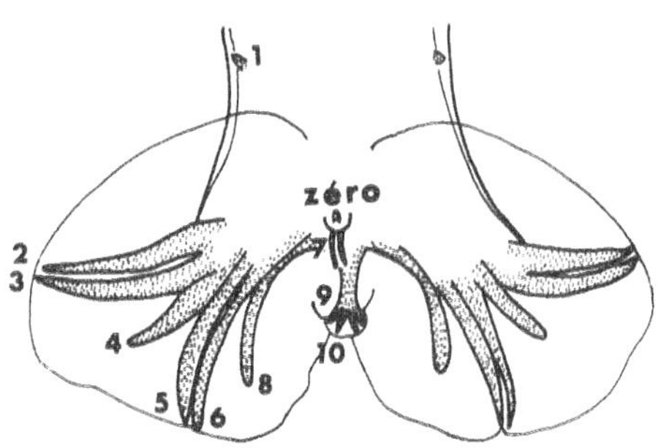

10.4

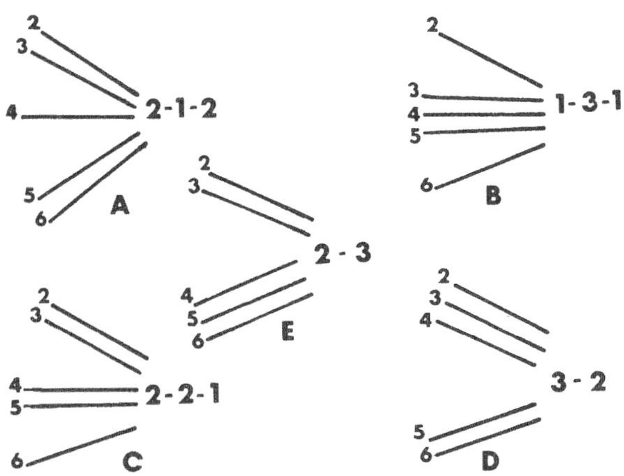

10.5

FIG. 10.4. Nomenclature of rays of caudal bursa (after Durette-Desset & Chabaud, 1981).

FIG. 10.5. Different types of caudal bursae.

4. *Ovejector:* The *vagina vera* is the unpaired ectodermal portion of the ovejector (Fig. 10.6). The vestibule or muscular vagina is the portion between the two sphincters (didelphic) or between the *vagina vera* and the sphincter (monodelphic). The infundibulum is situated between the sphincter and the uterus (Fig. 10.6).

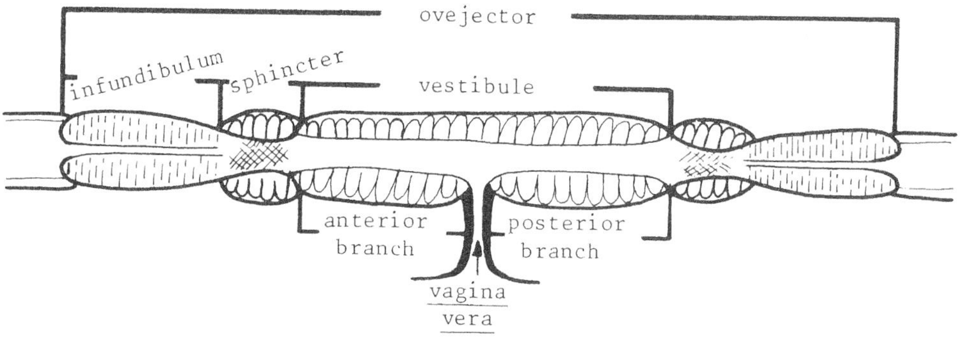

10.6

FIG. 10.6. Anatomy of ovejector.

Classification

Travassos (1937) published the first important monograph on the group. Skrjabin, Schikhobalova & Schulz (1954) devoted two volumes to the trichostrongyloids—one to didelphic forms, the other to the monodelphic forms.

Chabaud (1959) defined primitive and advanced states for the principal morphological characters and attempted to simplify the classification by relying on the degree of evolution of these characters. Mawson (1961, 1973) and Inglis (1968) made important contributions to the study of the Australian fauna. Durette-Desset (1971) proposed a classification of monodelphic forms which relied mainly on the synlophe. This system was extended to the entire superfamily by Durette-Desset & Chabaud (1977). After studying the Australian fauna, Durette-Desset & Chabaud (1981a) separated ancestral forms previously classified in the Amidostomatidae into different evolutionary lines and adopted the 14 families and 24 subfamilies which form the basis of the present Key (Fig. 10.266). Systematic problems raised by the Molineinae, Cooperiinae and Ostertagiinae were treated by Durette-Desset & Chabaud (1981b), and Durette-Desset (1982a and b); these publications contain most of the relevant references but subsequent changes are mentioned below.*

DICTYOCAULIDAE

Mertensinematinae

Parabatrachostrongylus Tantalean & Naupay, 1974 is considered a synonym of *Borrellostrongylus* Gutierrez, 1945 since its distinguishing characters (absence of a buccal capsule and a dorsal oesophageal tooth, and disposition of bursal rays) are identical to those of the latter genus.

MOLINEIDAE

Anoplostrongylinae

Freitas & Dobbin (1960) proposed *Didactyluris* for parasites of *Glossophaga soricina* (Chiroptera) in Brazil and illustrated the type species in 1962. The genus is close to *Bidigiticauda* in

*Gibbons & Khalil (1982) published a reclassification and Key to the genera of the Trichostrongylidae while the present article was in press.

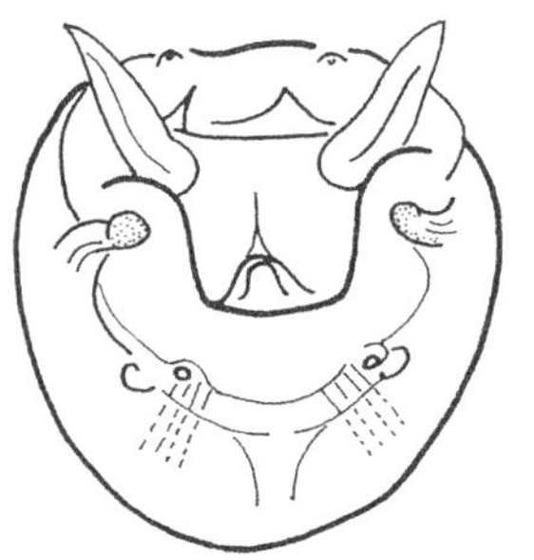

10.7

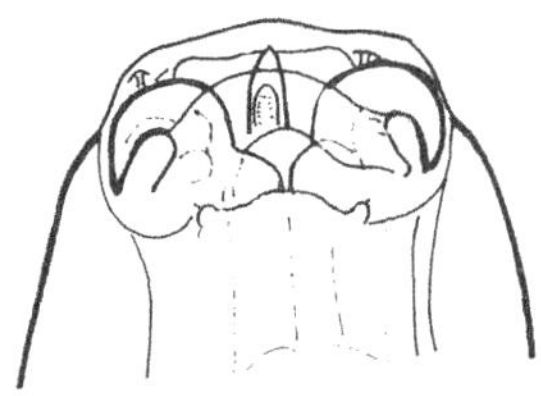

10.8

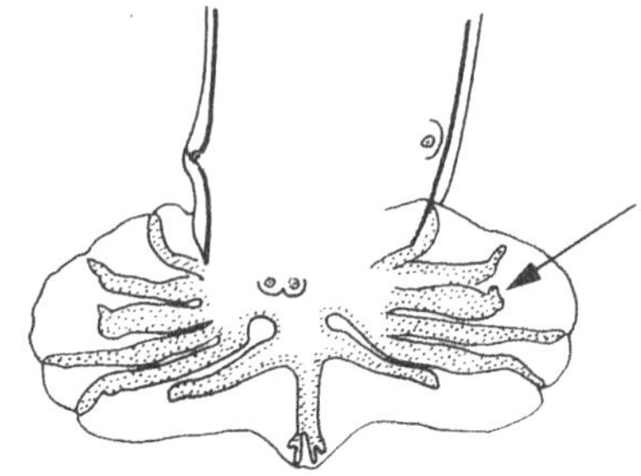

10.9

FIG. 10.7. *Strongylacantha glycirrhiza* (Van Beneden, 1873), cephalic extremity, apical view (after Desportes *in* Durette-Desset & Chabaud, 1975).

FIG. 10.8. *Strongylacantha glycirrhiza* (Van Beneden, 1873), cephalic extremity, ventral view (after Desportes *in* Durette-Desset & Chabaud, 1975).

FIG. 10.9. *Amidostomum acutum* (Lundahl, 1848), bursa (after Wertheim & Durette-Desset, 1975).

the form of the caudal bursa and the shape of the tail in the female and the two genera are considered synonyms.

Brevigraphidium Freitas & Mendonça, 1960 from Brazilian anteaters was described in a preliminary note without illustrations. This genus is probably close to *Paragraphidium* but is not sufficiently characterized to be included in the present Key.

Neohistiostrongylus Barus & Valle, 1967 is based on a poorly described species (*N. viguerasi* (Lopez-Neyra, 1946)) and cannot be classified.

Adolpholutzia Travassos, 1935 and *Dasypostrongylus* Travassos, 1935 are considered synonyms of *Moennigia* although their synlophes are unknown. The three genera occur in *Dasypus* in Brazil and in all three the bursa is similar and the posterior branch of the reproductive tract atrophied.

HELIGMOSOMIDAE

Nadtochyi (1976) proposed *Heligmoptera* to accommodate *H. myospalaxi* (Nadtochyi, 1970) of *Myospalax myospalax* (Spalacidae: Rodentia) in the USSR.

Genov & Janchev (1981) proposed *Dessetia* to accommodate *H. moldovensis* Andrejko, 1963 of *Spalax microphthalmus* and *S. leucodon* (Spalacidae: Rodentia) in Eastern Europe.

HELIGMONELLIDAE

Heligmonellinae

We propose the new genus *Sciuricola* for *Heligmonella moreli* and *H. dremomysi* from oriental sciurids. The new genus is distinguished from *Heligmonella* by the number of cuticular ridges, the relative length of rays 5 and 6 and the form of the dorsal lobe of the bursa.

Sciuricola n. gen.

Heligmonellidae. Heligmonellinae. Synlophe with 17–18 ridges. Bursa with ray 6 longer than ray 5 and with dorsal lobe well set off from rest of bursa. Parasites of oriental Sciuridae.

Type species: *Sciuricola moreli* (Gibbons, Durette-Desset & Daynes, 1977) n. comb. (= *Heligmonella dremomysi* Durette-Desset, 1974a; = *H. moreli* Gibbons, Durette-Desset & Daynes, 1977).

Other species: *Sciuricola dremomys* (Yen, 1973) n. comb. (= *Impalaia dremomys* Yen, 1973; = *Heligmonella dremomys* Gibbons, Durette-Desset & Daynes, 1977).

Pudicinae

Evandroia Travassos, 1937 is considered a synonym of *Pudica* Travassos & Darriba, 1929 since the synlophe of the type-species of these genera is identical.

Nippostrongylinae

In 1974 (Durette-Desset, 1974b) we redescribed *Longistriata carolinensis* Dikmans, 1935 and placed it in the genus *Boreostrongylus* Durette-Desset, 1971. However, we overlooked the fact that this species is the type of the subgenus *Longistriata (Carolinensis)* Travassos, 1937. Thus, *Carolinensis* must be elevated to the level of a genus and *Boreostrongylus* becomes its synonym.

Paraheligmonelloides Fukumoto, Kamiya & Suzuki, 1980 from Ethiopian lagomorphs is considered valid. It is close to *Carolinensis* and included in the Nippostrongylinae.

The caudal bursa of recently described species of *Tenorastrongylus* Durette-Desset, 1970 is almost identical to that of *H. murina* Baylis, 1928, type species of *Heligmonoides*. Although the synlophe of the latter has not been described, it is probably similar to that of *Tenorastrongylus*. *Tenorastrongylus* is, therefore, considered a synonym of *Heligmonoides*.

Gobindonema Sood & Parshad, 1974 of Indian Muridae belongs in the Nippostrongylinae but cannot be further classified.

Unclassified genera

Molinofilaria Vuylstěke, 1956 was reported from *Cercopithecus* in the Congo. It is similar to *Otostrongylus* De Bruyn, 1933 from bronchi and veins of pinnipeds. *Otostrongylus* is considered a member of the Crenostomatidae (Metastrongyloidea) (Anderson, 1978).

Trichostrongyloids described from elasmobranchs by Mawson (1954) and from teleosts by Nikolaeva (1969) and placed in *Ichthyostrongylus* Mawson, 1954 are insufficiently known to be classified. Although Mawson suggested that these are really marine parasites, further study is required and *Ichthyostrongylus* has not been included in the Key.

Shikhobalovia Ali & Deshpande, 1969 of rhinolophids in India cannot be classified since we were unable to obtain the description.

[*Neoheligmonoides* Sadovskaya, 1952 is mentioned by Skrjabin *et al.* (1952). The female is unknown and the caudal bursa undescribed. The genus may be close, or identical to, *Longistriata* (Schulz, 1926) and cannot be included in the Key.]

Only the female of *Heligmodentostoma* Skrjabin & Schulz, 1952 of *Microtus* in the USSR, is known and no illustrations exist. The genus cannot be classified.

Ierestrongylus Cameron, 1935 and *Mazanema* Cameron, 1935 of *Mazama* (Cervidae) in the Antilles are known only on the basis of poorly preserved males and are considered *genera inquirenda*. In both there is a strongly developed ventral ridge and the form of the caudal bursa suggests that, despite the host, they belong to the Pudicinae.

TRICHOSTRONGYLOIDEA

Key to families

1– (2) Conspicuous intrabuccal hooks present (Figs 10.7, 10.8).
Parasites of rhinolophoid bats. **Strongylacanthidae**

2– (1) Intrabuccal hooks absent.

3– (4) Ray 4 short and thick, with terminal portion oriented anteriorly (Fig. 10.9).
Female tail slender and without terminal spine (Fig. 10.10).
Cephalic extremity either with well developed buccal capsule and sensorial papillae located at the extremity of long peduncles (Fig. 10.11) or lacking well developed buccal capsule and with simple (Fig. 10.12) or denticular labial appendices (Fig. 10.13).
Parasites of aquatic birds.

Amidostomatidae

4– (3) Nematodes lacking most of these characters.
5–(16) Synlophe generally absent or if present bilaterally symmetrical (except *Sprattellus*).
Body usually not coiled.
Usually didelphic.

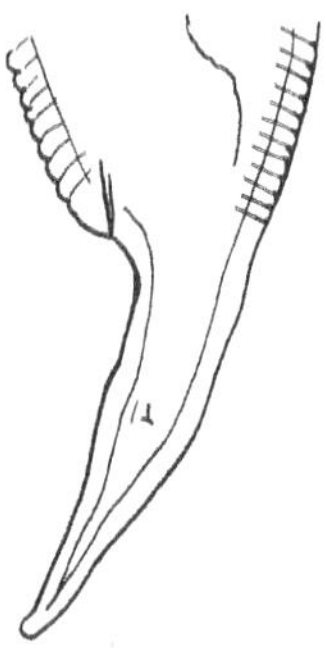

10.10

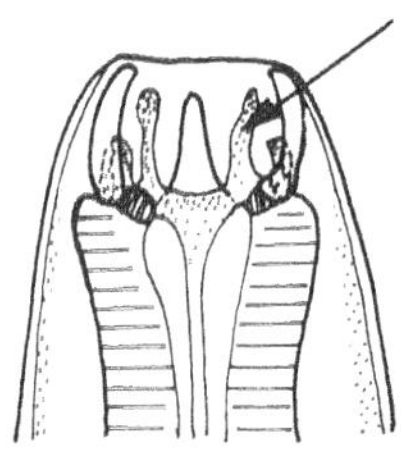

10.11

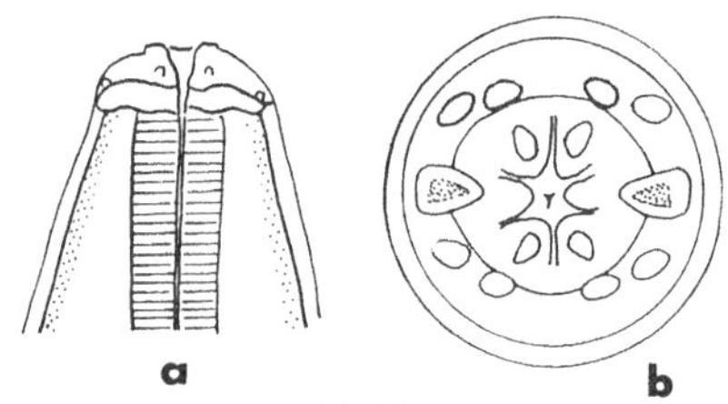

10.12

FIG. 10.10. *Epomidiostomum uncinatum* (Lundahl, 1848), tail end, female, lateral view (after Desportes, original).

FIG. 10.11. *Amidostomum fulicae* (Rudolphi, 1819), cephalic extremity, ventral view (after Wertheim & Durette-Desset, 1975).

FIG. 10.12. *Pseudamidostomum boulengeri* Maplestone, 1930, cephalic extremity, (a) lateral view, (b) apical view (after Ali, 1971).

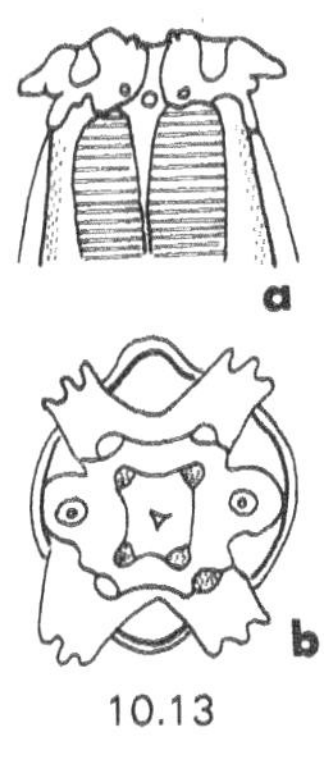

10.13

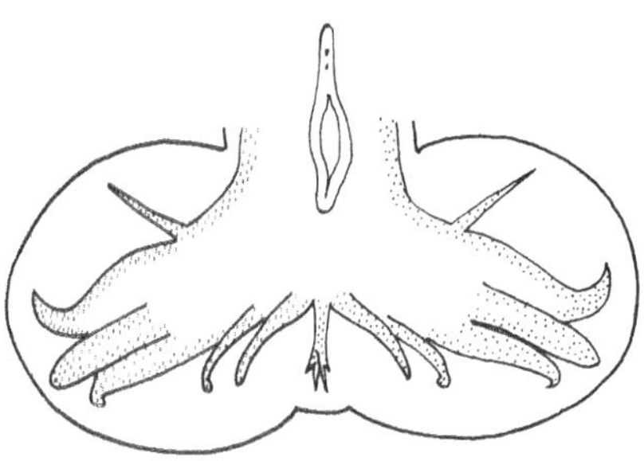

10.14

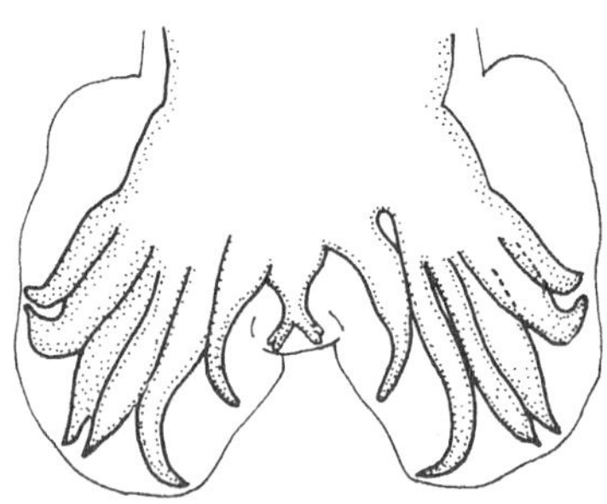

10.15

FIG. 10.13. *Epomidiostomum alii* Ali, 1971, cephalic extremity, (a) lateral view, (b) apical view (after Ali, 1971).

FIG. 10.14. *Trichostrongylus lerouxi* Biocca, Chabaud & Ghadirian, 1974, bursa (after Biocca, Chabaud & Ghadirian, 1974).

FIG. 10.15. *Teladorsagia kenyensis* (Gibbons & Khalil, 1980) n. comb., bursa (after Gibbons & Khalil, 1980).

6– (9) Nematodes with most of the following characters:
Cephalic vesicle absent (except *Dromaeostrongylus* and Cooperiinae.)
Synlophe absent or present but not elaborate.
Caudal spine of female absent.
Ray 2 well separated from ray 3 (except Graphidiinae, Ostertagiinae, and some Dromaeostrongylidae).
Ray 4 generally long, forming trident with rays 3 and 5 (Fig. 10.14) or grouped with ray 5 (Fig. 10.15).
Bursa usually with rays 2 and 3 or rays 5 and 6 hypertrophied (Figs. 10.16, 10.17).

7– (8) Dorsal ray very long (Fig. 10.18) or deeply divided (Fig. 10.19).
—if dorsal ray very long, synlophe present.
—if dorsal ray deeply divided, synlophe absent.
Parasites of ratite birds, neotropical birds and australasian marsupials.
Dromaeostrongylidae

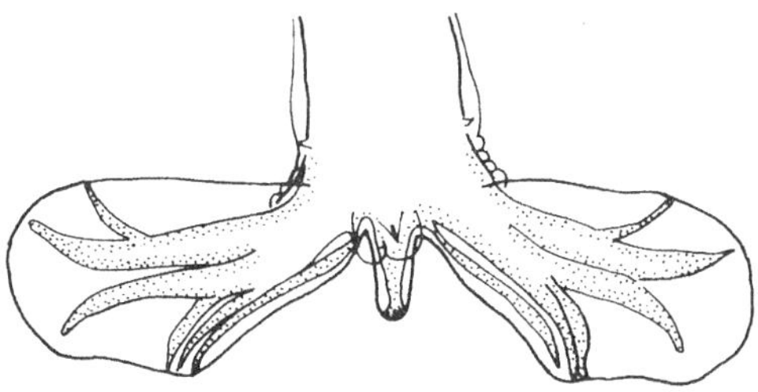

10.16

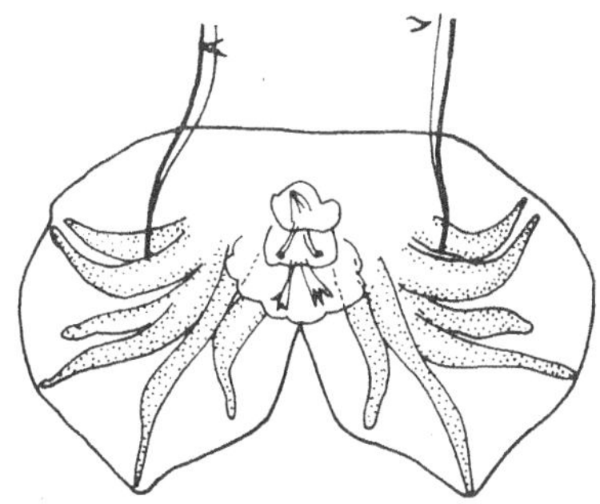

10.17

FIG. 10.16. *Ashwortius lerouxi* Diaouré, 1964, bursa (after Durette-Desset & Denké, 1978).
FIG. 10.17. *Spiculopteragia caballeroi* Chabaud, 1976, bursa (after Chabaud, 1976).

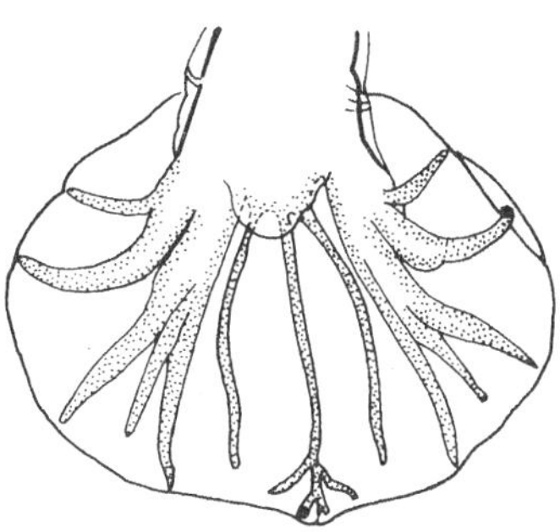

10.18

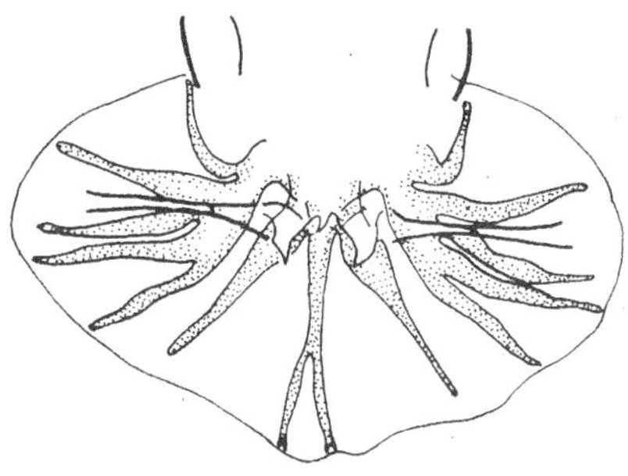

10.19

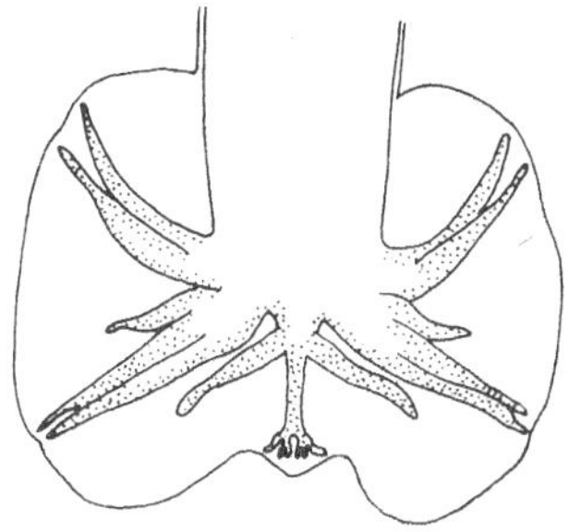

10.20

FIG. 10.18. *Dromaeostrongylus bicuspis* Lubimov, 1933, bursa (after Durette-Desset, 1980).
FIG. 10.19. *Filarinema flagrifer* Mönnig, 1929, bursa (after Cassone & Baccam, *Bull. Mus. nat. Hist. nat.*, in press).
FIG. 10.20. *Molineus felineus* Cameron, 1923, bursa (after Cameron, 1923).

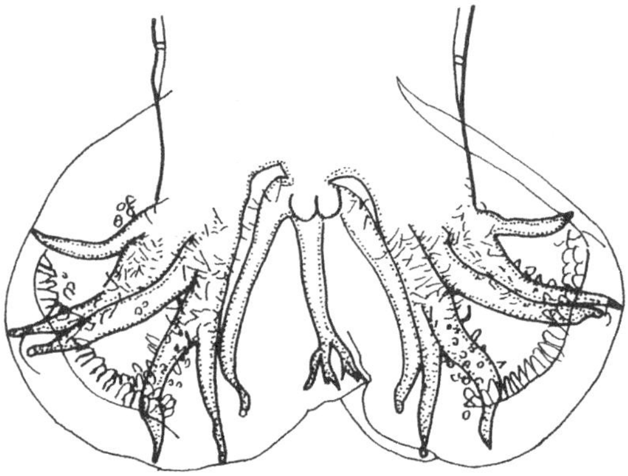

10.21

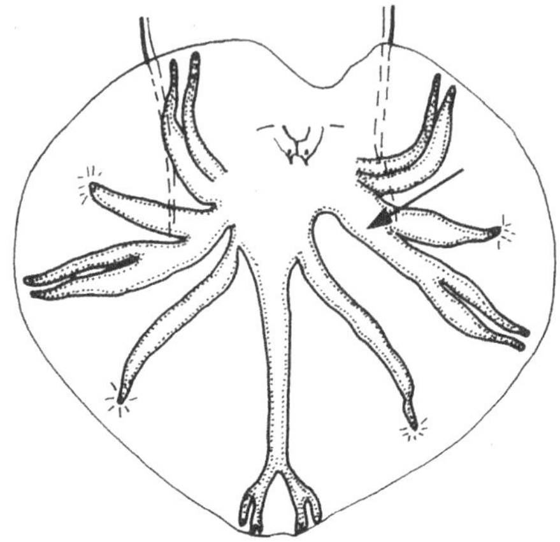

10.22

FIG. 10.21. *Molinostrongylus skrjabini* Skarbilovitsch, 1934, bursa (after Durette-Desset & Chabaud, 1975).
FIG. 10.22. *Amphibiophilus chabaudi* Puylaert, 1967, bursa (after Puylaert, 1967).

8– (7) Dorsal ray usually short and not deeply divided (Fig. 10.14).
—if dorsal ray very long, synlophe absent (except *Impalaia*, *Megacooperia* and some *Marshallagia* spp.)
—if dorsal ray deeply divided, synlophe present (except some *Marshallagia* spp.)
Parasites of eutherian mammals, mainly lagomorphs and ruminants.
Trichostrongylidae

9– (6) Nematodes with most of the following characters:
Cephalic vesicle present (except *Batrachostrongylus* and most Dictyocaulidae).
Synlophe present and usually elaborate (absent in Dictyocaulidae, *Schulzia, Poekilostrongylus* and *Filicapitis*).
Caudal spine of female present (except Amphibiophilidae and Dictyocaulidae).
Ray 2 close to ray 3.
Ray 4 generally short, sometimes widely separated from rays 2 and 3 and rays 5 and 6 (Fig. 10.20) sometimes grouped with ray 3 (Fig. 10.21), sometimes arising anterior to rays 5 and 6 (Fig. 10.22).
Bursa with rays 2 and 3 approximately of same size as rays 5 and 6 (Fig. 10.23).

10–(11) Dorsal ray long and ray 4 arising anterior to rays 5 and 6 (Fig. 10.22).
Buccal capsule present and ovejector with short *vagina vera*, or buccal capsule absent and ovejector with long *vagina vera*.
Parasites of amphibians and lagomorphs (Ochotonidae).
Amphibiophilidae

11–(10) Dorsal ray short or, if long, ray 4 arising at same level as rays 5 and 6.
Buccal capsule absent and ovejector with short *vagina vera*.

12–(13) Caudal bursa round (Figs. 10.23, 10.24).
Caudal spine of female absent.
Cephalic vesicle absent (except *Mertensinema*).
Synlophe absent.
Parasites of digestive tract of amphibians and reptiles and respiratory system of ungulates.
Dictyocaulidae

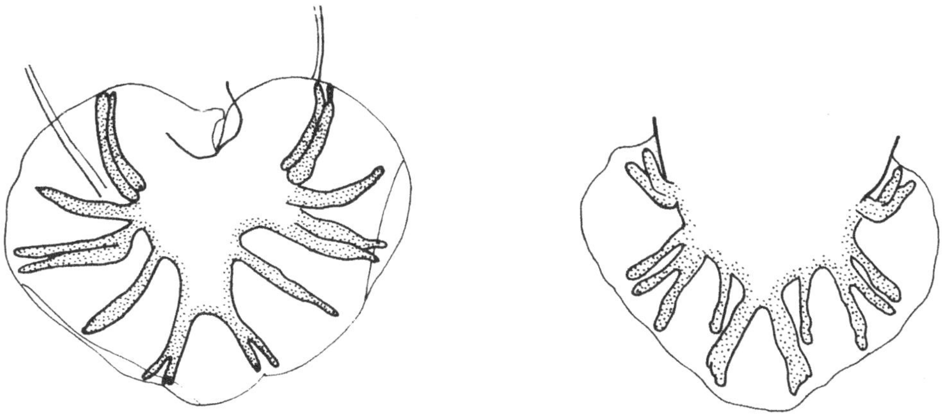

10.23 10.24

FIG. 10.23. *Mertensinema sepilokensis* Durette-Desset, 1980, bursa (after Durette-Desset, 1980).

FIG. 10.24. *Dictyocaulus arnfieldi* (Cobbold, 1884), bursa (after Skrjabin & Ershov, 1933).

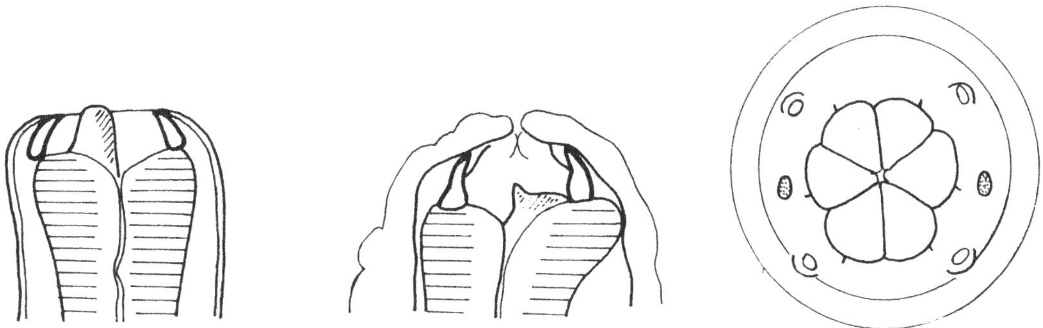

10.25 10.26 10.27

FIG. 10.25. *Tachynema spratti* Durette-Desset & Cassone, 1983, cephalic extremity, lateral view (after Durette-Desset & Cassone, 1983).

FIG. 10.26. *Nicollina queenslandensis* Durette-Desset & Cassone, 1983, cephalic extremity, lateral view (after Durette-Desset & Cassone, 1983).

FIG. 10.27. *Nicollina queenslandensis* Durette-Desset & Cassone, 1983, cephalic extremity, apical view (after Durette-Desset & Cassone, 1983).

13–(12) Caudal bursa rarely round.
Caudal spine usually present in female.
Cephalic vesicle present.
Synlophe present (except *Filicapitis*, *Schulzia* and *Poekilostrongylus*).

14–(15) Buccal ring absent.
Extradorsal rays absent.
Parasites of amphibians, reptiles, carnivores, Tenrecoidea, bats, Tupaiidae, primates, Xenarthra, Pholidota, Tubulidentata, archaic rodents, lagomorphs and ruminants.
Molineidae

15–(14) Buccal ring present (Fig. 10.25)
Extradorsal rays present (Fig. 10.55).
Parasites of monotremes and dasyuroid and peramelioid marsupials in Australia.
Mackerrastrongylidae

16– (5) Synlophe present (except Globocephaloidinae), not bilaterally symmetrical.
Body usually coiled in sinistral spiral.
Often monodelphic.

17–(20) Bursa of type 1–3–1 (Fig. 10.32) or 3–2 (Fig. 10.34).
Didelphic with ovejector divided unequally (Fig. 10.31) (except *Copemania*).
Buccal capsule well developed (Fig. 10.26) (except *Copemania*).
Dorsal oesophageal tooth present (except *Copemania*).
Oral opening surrounded by six lips (Fig. 10.27) or two jaw-like structures present (Figs. 10.28 to 10.30) (except *Copemania* and *Batrachonema*).
Parasites of amphibians, reptiles and Australian mammals.

18–(19) Caudal bursa of type 1–3–1 and dorsal lobe not hypertrophied (Fig. 10.33).
Oral opening surrounded by six lips or without lips (*Copemania*).
Parasites of amphibians (Malaysia), monotremes, rarely dasyuroid marsupials (*Copemania*).

Nicollinidae

19–(18) Caudal bursa of type 3–2 (Fig. 10.34) or dorsal lobe hypertrophied (Fig. 10.35).
Oral opening surrounded by six lips or formed by two jaw-like structures (Figs. 10.28 to 10.30).
Parasites of reptiles and Australian marsupials.

Herpetostrongylidae

20–(17) Bursa of type 2–1–2, 2–2–1 or 2–3.
Didelphic with ovejector divided equally or monodelphic.
Buccal capsule absent or very small (except *Inglamidum*, Fig. 10.37).
Dorsal oesophageal tooth usually absent (except *Inglamidum*).
Oral opening lacking lips and jaw-like structures (except *Inglamidum*, Fig. 10.36).
Parasites of birds and various mammals (not Australian mammals).

21–(22) Caudal bursa of type 2–2–1, 2–1–2 (Fig. 10.38) or irregular (Fig. 10.39).
Synlophe with three ventral-left ridges (Fig. 10.40) or ridges oriented from right to left on ventral and dorsal sides, but leaving free space on lateral sides (Figs. 10.41, 10.42).
Parasites of New World marsupials and caviomorph rodents.

Viannaiidae

22–(21) Nematodes lacking most of these characters.

23–(24) Caudal bursa of type 2–2–1 (Fig. 10.43).
Synlophe with frontal symmetry (Fig. 10.44).
Didelphic (except *Inglamidum*).
Parasites of American birds, bats and geomyoid rodents, more rarely of Old World birds and bats and neotropical cricetids (*Inglamidum*).

Ornithostrongylidae

24–(23) Nematodes lacking most of these characters.

25–(26) Axis of orientation of ridges subfrontal (Fig. 10.46).
Female tail with spine.
Caudal bursa generally of type 2–3 (Fig. 10.47).
Parasites of soricoid insectivores and oriental, holarctic, and neotropical rodents.

Heligmosomidae

26–(25) Axis of orientation of ridges almost always oblique (Figs. 10.48, 10.49).
Female tail lacking spine.
Caudal bursa generally of type 2–2–1 (Fig. 10.50).
Parasites of talpoid insectivores, lagomorphs and rodents.
Cosmopolitan.

Heligmonellidae

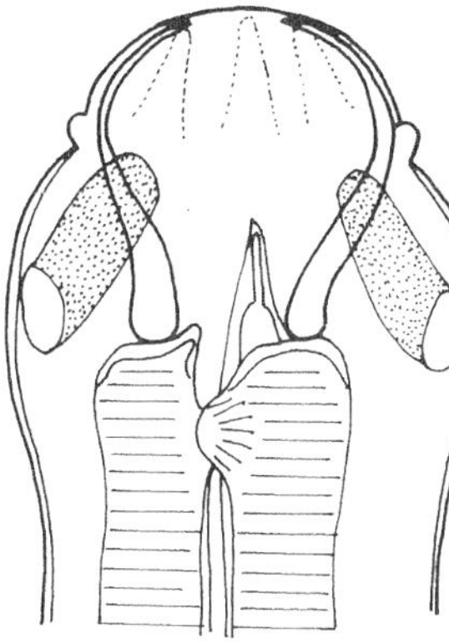

10.28

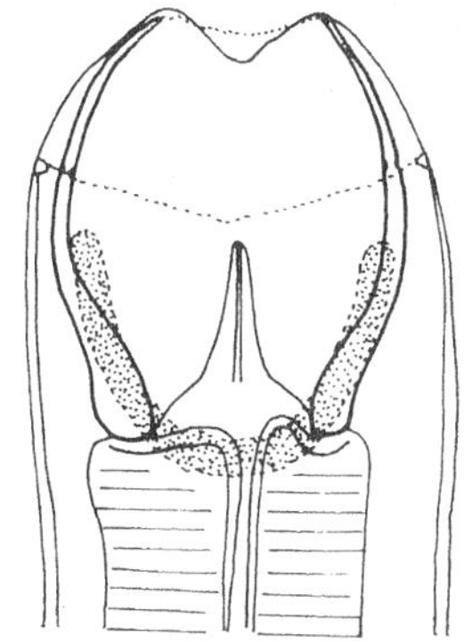

10.29

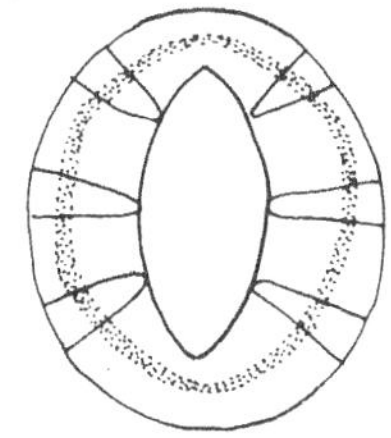

10.30

FIG. 10.28. *Globocephaloides macropodis* Yorke & Maplestone, 1926, cephalic extremity, lateral view (after Beveridge, 1979).

FIG. 10.29. *Globocephaloides macropodis* Yorke & Maplestone, 1926, cephalic extremity, dorsal view (after Beveridge, 1979).

FIG. 10.30. *Globocephaloides macropodis* Yorke & Maplestone, 1926, cephalic extremity, apical view (after Beveridge, 1979).

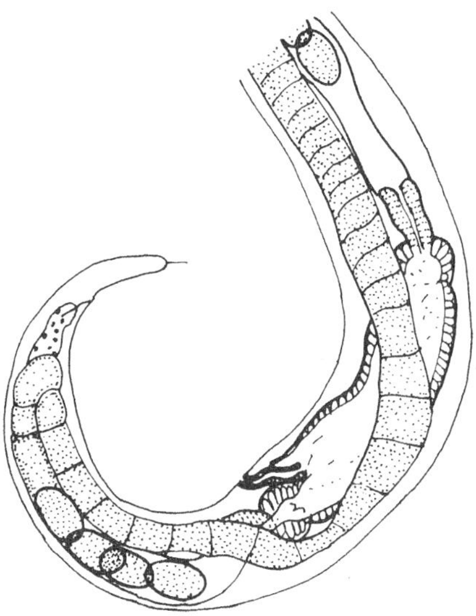

10.31

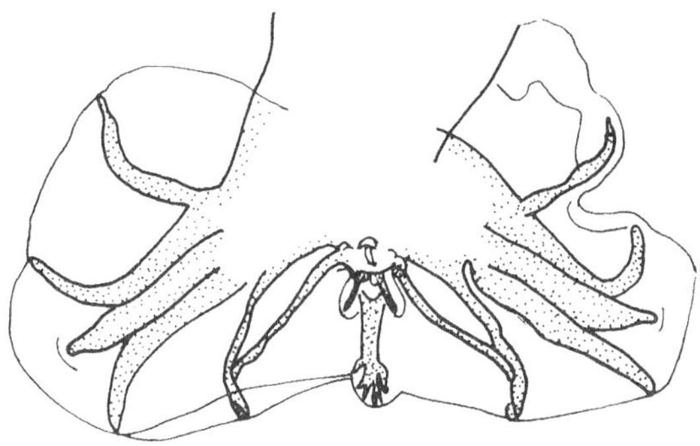

10.32

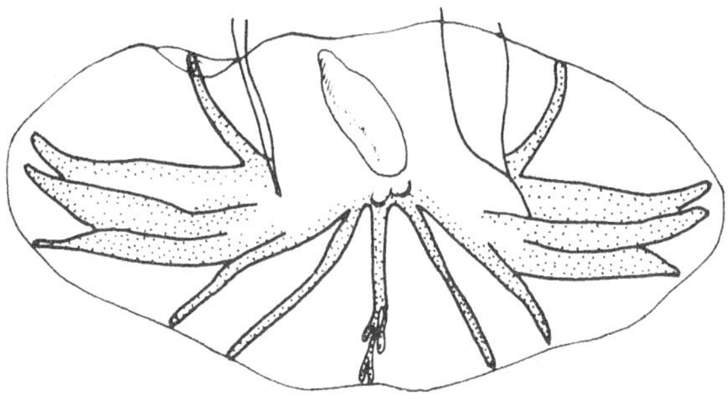

10.33

FIG. 10.31. *Nicollina queenslandensis* Durette-Desset & Cassone, 1983, ovejector, lateral view (after Durette-Desset & Cassone, 1983).

FIG. 10.32. *Copemania obendorfi* Durette-Desset & Beveridge, 1981, bursa (after Durette-Desset & Beveridge, 1981).

FIG. 10.33. *Nicollina echidnae* Durette-Desset & Cassone, 1983, bursa (after Durette-Desset & Cassone, 1983).

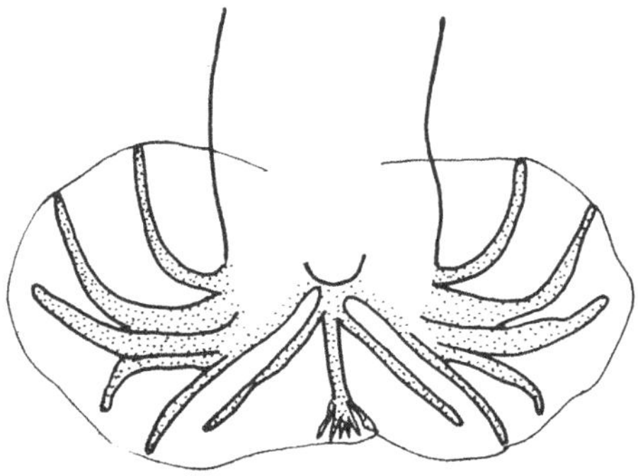

10.34

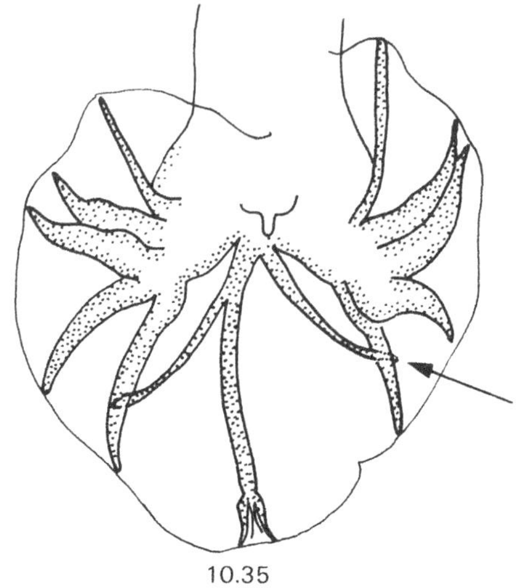

10.35

Fig. 10.34. *Woolleya antechini* Humphery-Smith & Durette-Desset, 1981, bursa (after Humphery-Smith & Durette-Desset, 1981).

Fig. 10.35. *Vaucherus vaucheri* Durette-Desset, 1980, bursa (after Durette-Desset, 1980).

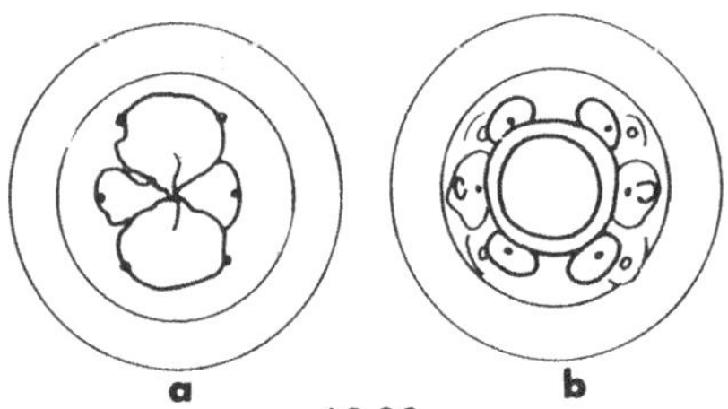

10.36

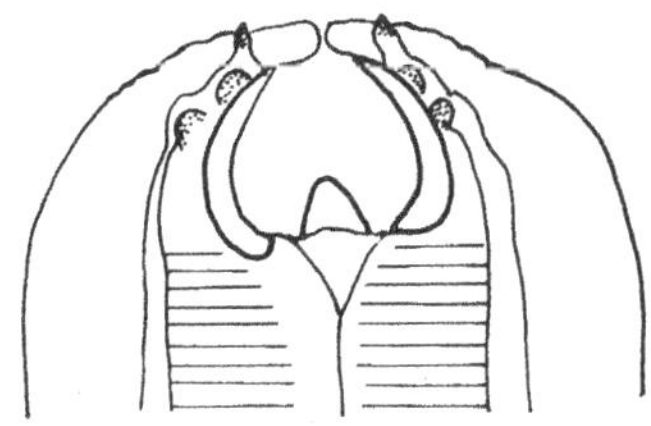

10.37

Fig. 10.36. *Inglamidum akodon* Durette-Desset, Denké & Murua, 1976, cephalic extremity, apical view, (a) superficial view, (b) deep view (after Durette-Desset, Denké & Murua, 1976).

Fig. 10.37. *Inglamidum akodon* Durette-Desset, Denké & Murua, 1976, cephalic extremity, dorsal view (after Durette-Desset, Denké & Murua, 1976).

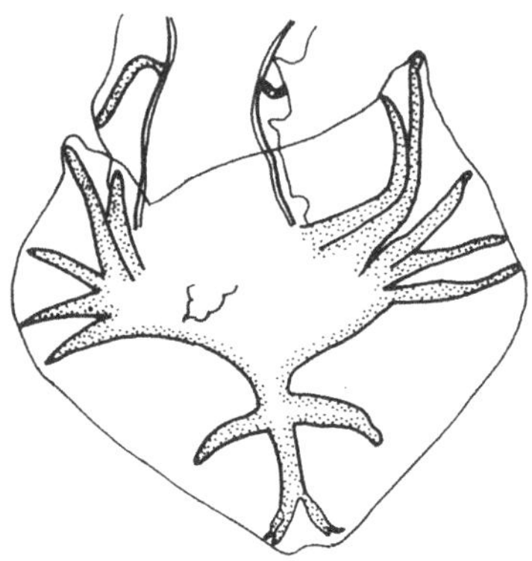

10.38

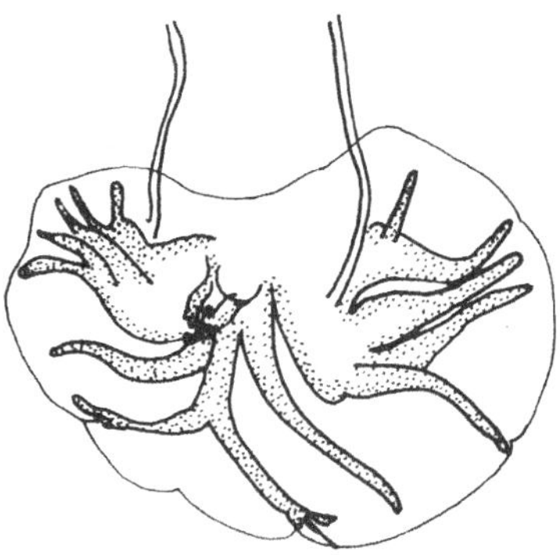

10.39

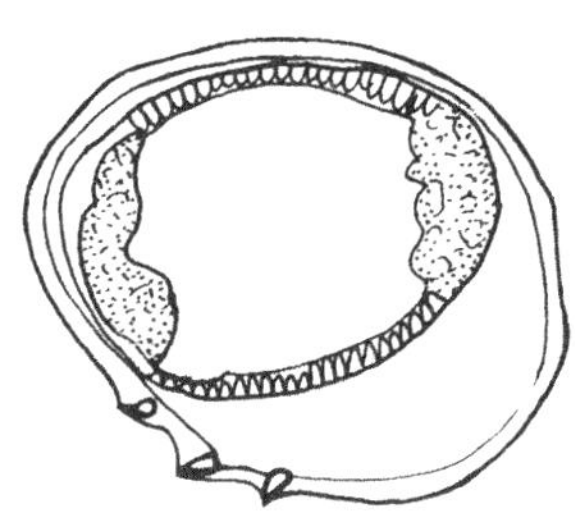

10.40

Fig. 10.38. *Viannaia viannai* Travassos, 1914, bursa (after Durette-Desset, 1968).

Fig. 10.39. *Viannella lenti* Durette-Desset, 1968, bursa (after Durette-Desset, 1968).

Fig. 10.40. *Viannaia monodelphisi* Durette-Desset, 1968, synlophe (after Durette-Desset, 1968).

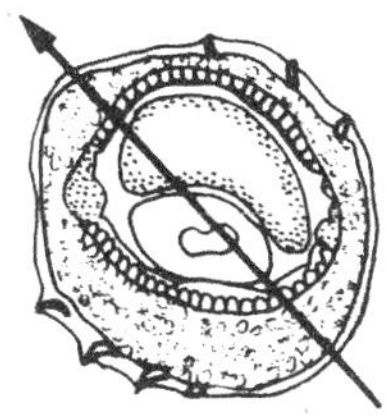

10.41

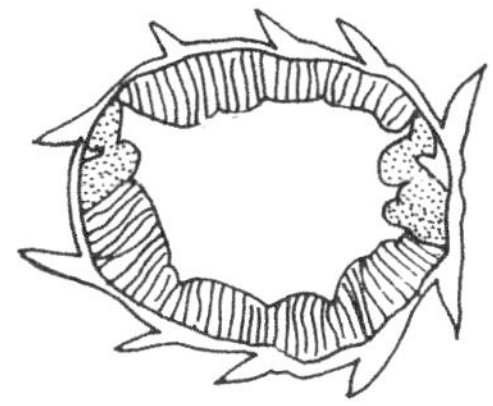

10.42

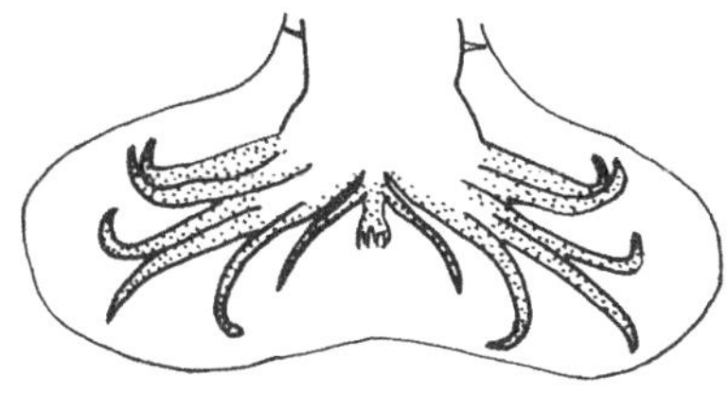

10.43

FIG. 10.41. *Viannella lenti* Durette-Desset, 1968, synlophe (after Durette-Desset, 1968).

FIG. 10.42. *Travassostrongylus travassosi* Durette-Desset, 1968, synlophe (after Durette-Desset, 1968).

FIG. 10.43. *Ornithostrongylus quadriradiatus* (Stevenson, 1904), bursa (after Travassos, 1921).

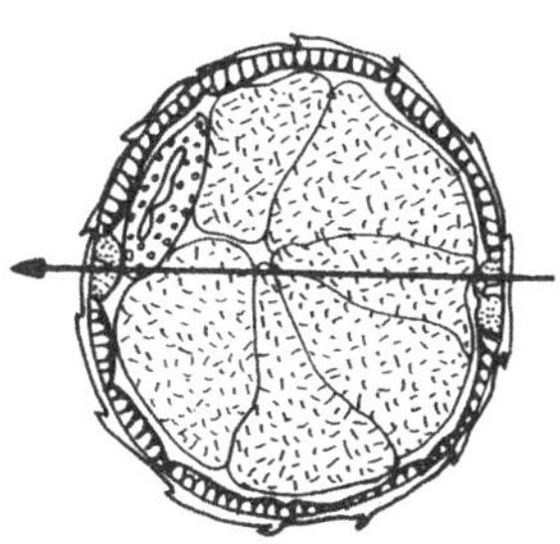

10.44

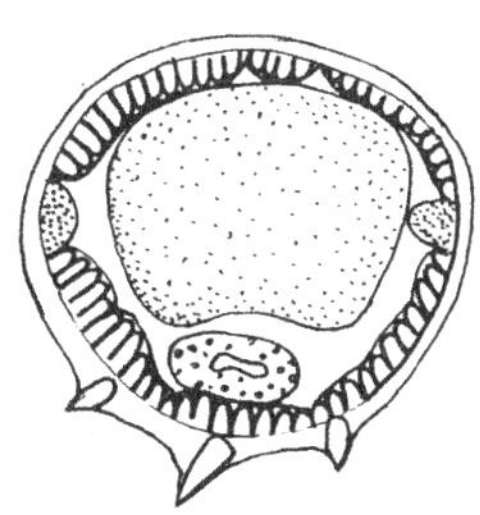

10.45

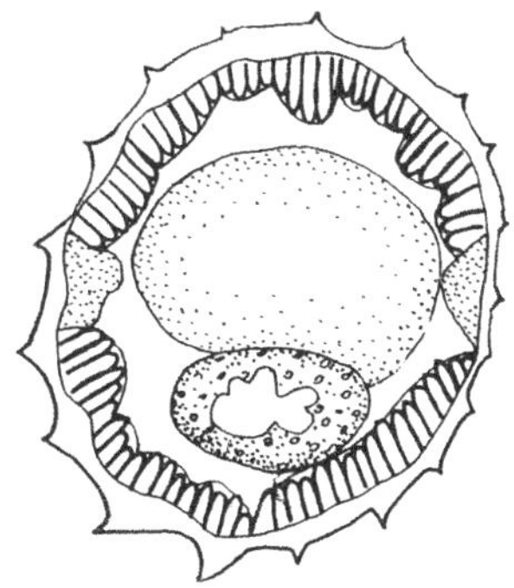

10.46

FIG. 10.44. *Ornithostrongylus fariai* Travassos, 1914, synlophe (after Durette-Desset & Pinto, 1977).

FIG. 10.45. *Suncinema murini* Durette-Desset, 1973, synlophe (after Durette-Desset, 1973).

FIG. 10.46. *Heligmosomoides wisconsinensis* Durette-Desset, 1967, synlophe (after Durette-Desset, 1967).

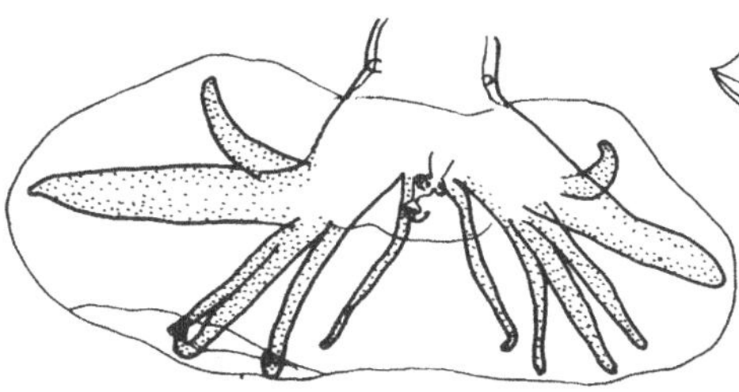

10.47

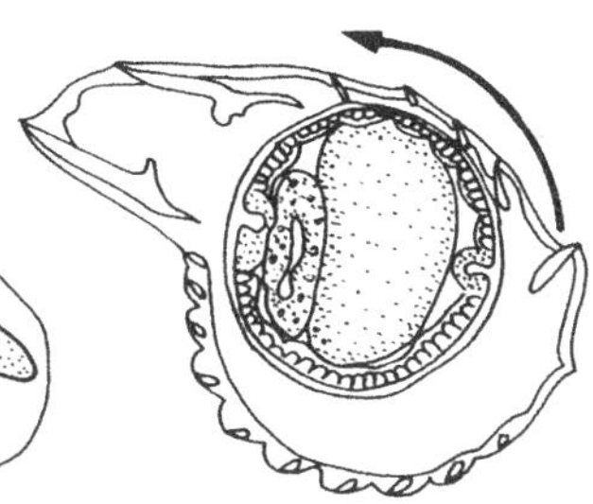

10.48

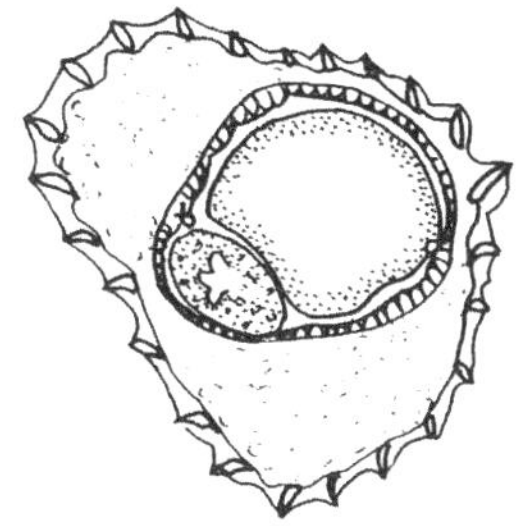

10.49

FIG. 10.47. *Heligmosomoides smirnovae* Durette-Desset, Rausch & Kayemba, 1980, bursa (after Durette-Desset, Rausch & Kayemba, 1980).

FIG. 10.48. *Paraheligmonina paratrifurcata* (Durette-Desset, 1970), synlophe (after Durette-Desset, 1970).

FIG. 10.49. *Hassalstrongylus hoineffae* (Durette-Desset, 1969), synlophe (after Durette-Desset, 1969).

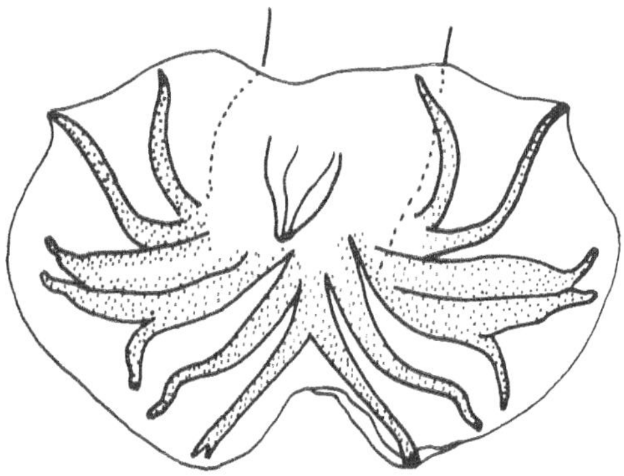

10.50

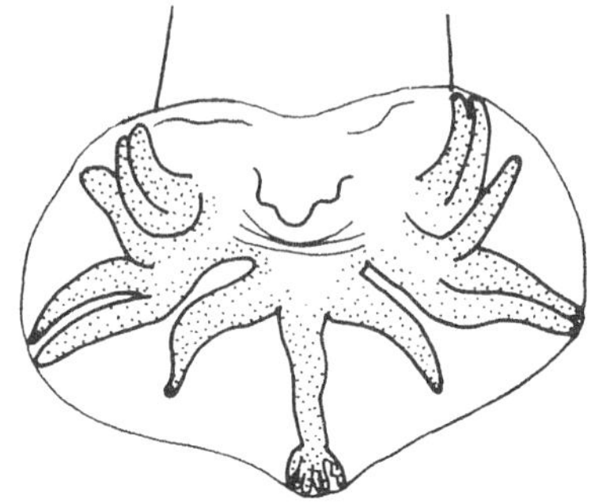

10.51

FIG. 10.50. *Fissicauda callosciuri* (Supperer & Kutzer, 1963), bursa (after Durette-Desset & Krishnasamy, 1976).
FIG. 10.51. *Strongylacantha glycirrhiza* (Van Beneden, 1873), bursa (after Desportes *in* Durette-Desset & Chabaud, 1975).

Family *STRONGYLACANTHIDAE*
(Yorke & Maplestone, 1926, subfam.) Chabaud, 1960

One genus (Figs. 10.7, 10.8, 10.51).
Parasites of Rhinolophoidea.

Strongylacantha Van Beneden, 1873

Family *AMIDOSTOMATIDAE*
(Travassos, 1919, subfam.) Baylis & Daubney, 1926

Key to subfamilies

1– (2) Buccal capsule long and well developed.
Cephalic sensorial papillae located at extremity of long peduncles (Fig. 10.11).
Amidostomatinae

2– (1) Buccal capsule short and poorly developed.
Lips with simple (Fig. 10.12) or denticular appendices (Fig. 10.13).
Epomidiostomatinae

Subfamily **Amidostomatinae**

One genus (Figs. 10.9, 10.11).
Parasites of aquatic birds.

Amidostomum Railliet & Henry, 1909

Subfamily **Epomidiostomatinae** Skrjabin & Schulz, 1937

Key to genera

1– (2) Cephalic extremity armed posteriorly with four protruding formations directed away from oral opening (Fig. 10.13).
Epomidiostomum Skrjabin, 1915

2– (1) Cephalic extremity without above formations (Fig. 10.12).
Pseudamidostomum Boulenger, 1926

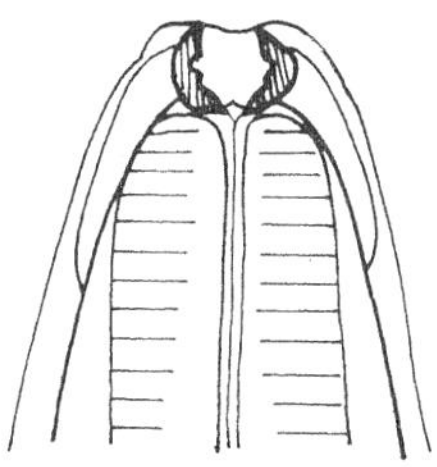

10.52

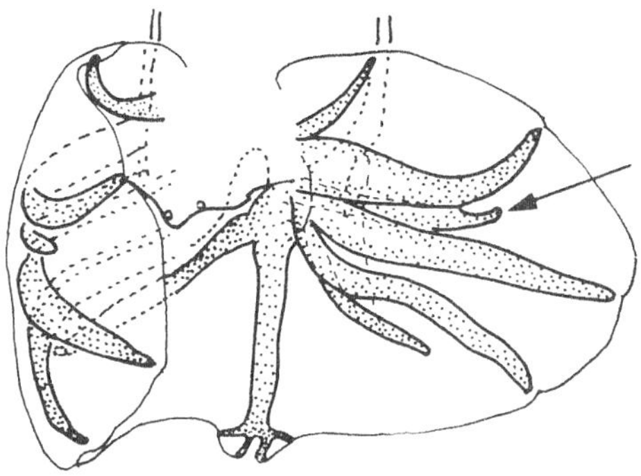

10.53

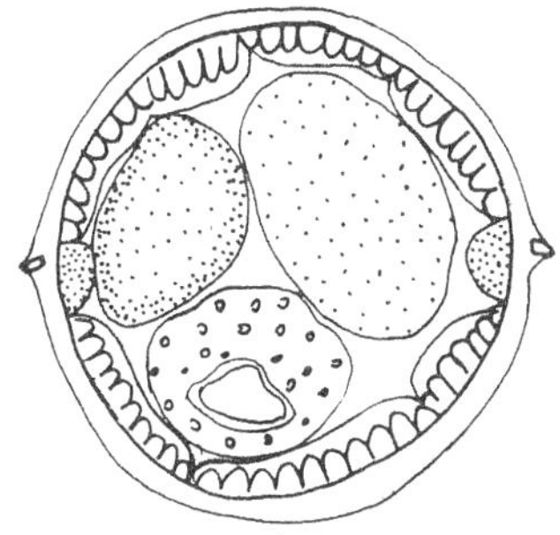

10.54

FIG. 10.52. *Paramidostomum pulchrum* Freitas & Mendonça, 1949, cephalic extremity (after Freitas & Mendonça, 1950).

FIG. 10.53. *Paramidostomum pulchrum* Freitas & Mendonça, 1949, bursa (after Freitas & Mendonça, 1950).

FIG. 10.54. *Peramelistrongylus skedastos* Mawson, 1960, synlophe (after Durette-Desset & Beveridge, 1981).

Family *DROMAEOSTRONGYLIDAE*
(Skrjabin & Schulz, 1937, tribe) Durette-Desset & Chabaud, 1981

Key to genera

1– (4) Dorsal ray divided at posterior extremity (Fig. 10.18).
Synlophe present.
Parasites of birds.

2– (3) Buccal capsule well developed (Fig. 10.52).
Cervical groove present.
Ray 4 shorter than ray 5 (Fig. 10.53).
Parasites of ventricle of birds (Anhimidae).
Paramidostomum Freitas & Mendonça, 1949

3– (2) Buccal capsule reduced.
Cervical groove absent.
Ray 4 longer than ray 5 (Fig. 10.18).
Parasites of intestine of australasian ratites (accidentally in the marsupial *Trichosurus*).
Dromaeostrongylus Lubimov, 1933

4– (1) Dorsal ray divided at or above posterior third (Fig. 10.19).
Synlophe absent or much reduced (Fig. 10.54).
Parasites of Australian marsupials.

5– (6) Extradorsal rays present (Fig. 10.55).
Synlophe reduced to two lateral ridges (Fig. 10.54).
Parasites of Dasyuroidea and Peramelioidea, rarely australasian rodents.
Peramelistrongylus Mawson, 1960

6– (5) Extradorsal rays absent.
Synlophe absent.
Parasites of Phalangeroidea.

7– (8) Dorsal ray divided at posterior third (Fig. 10.56).
Oesophagus with single tooth (Fig. 10.57).
Parasites of Phalangeridae.
Profilarinema Durette-Desset & Beveridge, 1981

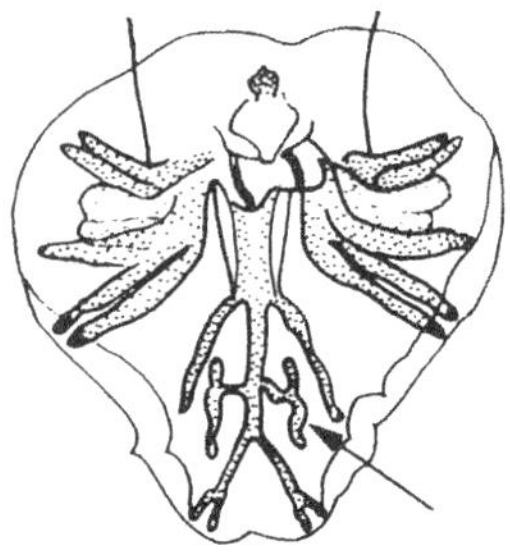

10.55

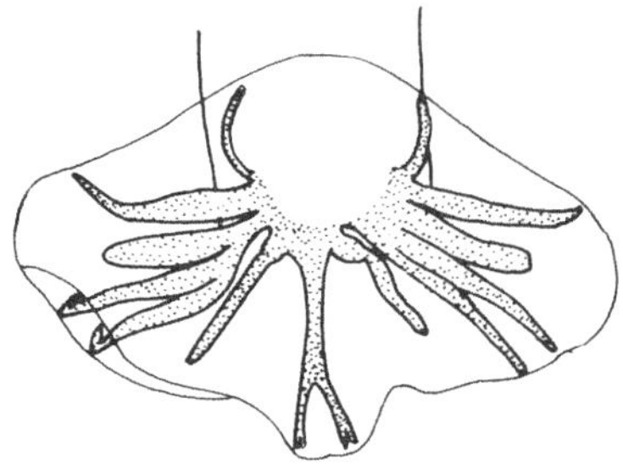

10.56

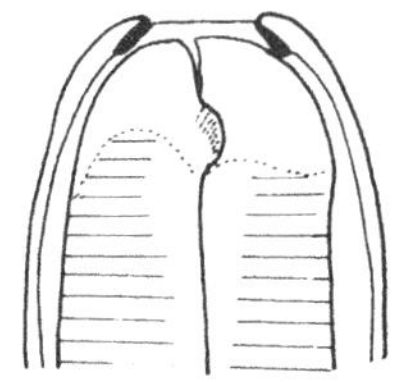

10.57

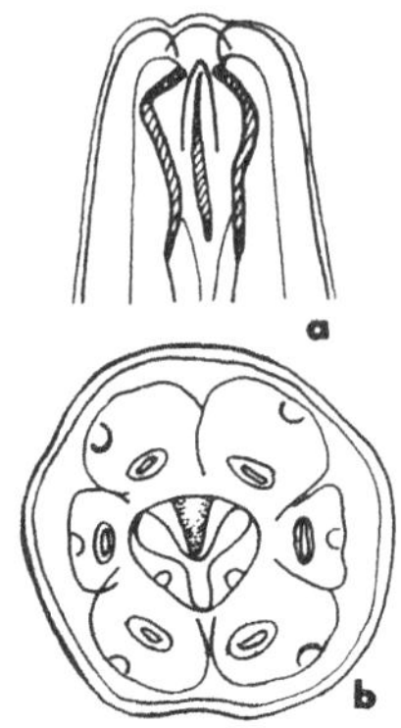

10.58

FIG. 10.55. *Peramelistrongylus skedastos* Mawson, 1960, bursa (after Durette-Desset & Beveridge, 1981).

FIG. 10.56. *Profilarinema hemsleyi* Durette-Desset & Beveridge, 1981, bursa (after Durette-Desset & Beveridge, 1981).

FIG. 10.57. *Profilarinema hemsleyi* Durette-Desset & Beveridge, 1981, cephalic extremity, lateral view (after Durette-Desset & Beveridge, 1981).

FIG. 10.58. *Filarinema flagrifer* Mönnig, 1929, cephalic extremity, (a) dorsal view, (b) apical view (after Cassone & Baccam, *Bull. Mus. nat. Hist. nat.*, in press).

8– (7) Dorsal ray divided above posterior third (Fig. 10.19).
Oesophagus with complex cuticular apparatus (Fig. 10.58).
Parasites of Macropodidae.

Filarinema Mönnig, 1929
(= *Asymmetricostrongylus* Nagaty, 1932)

Family *TRICHOSTRONGYLIDAE*
(Leiper, 1908 subfam.) Leiper, 1912

Key to subfamilies

1– (4) Distal ends of rays 2 and 3 curved towards each other; ray 2 markedly smaller than ray 3; ray 3 directed posteriorly then curving abruptly anteriorly (Fig. 10.59).

2– (3) Cephalic vesicle absent.
Parasites of ratites, rodents (*Thryonomys*, *Bathyergus*, *Atherurus*) and hyracoids in Africa, lemurs and rodents in Madagascar, lagomorphs and *Antilocapra* in North America.

Libyostrongylinae

3– (2) Cephalic vesicle present, usually differentiated into cephalic and cervical portions (Fig. 10.60).
Parasites of Bathyergidae and ruminants of Old World origin.

Cooperiinae

4– (1) Rays 2 and 3 not as above.

5– (8) Rays 2 and 3 approximately equal in length, close to each other and approximately parallel (Fig. 10.61).

6– (7) Buccal capsule reduced to ring (Fig. 10.62).
Dorsal ray long, thick at base and divided only at posterior extremity (Fig. 10.63).
Parasites of Tragulidae, Suidae, rarely ungulates and lagomorphs.

Graphidiinae

10.59

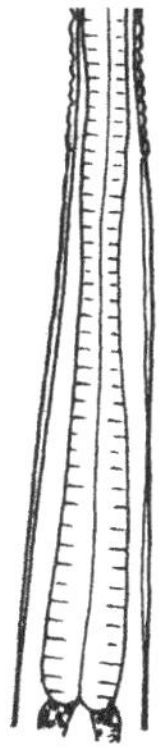
10.60

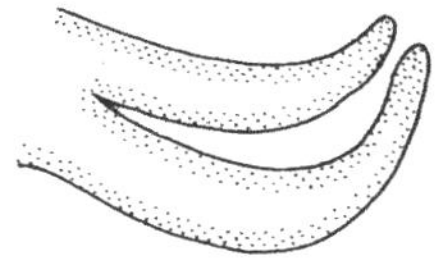
10.61

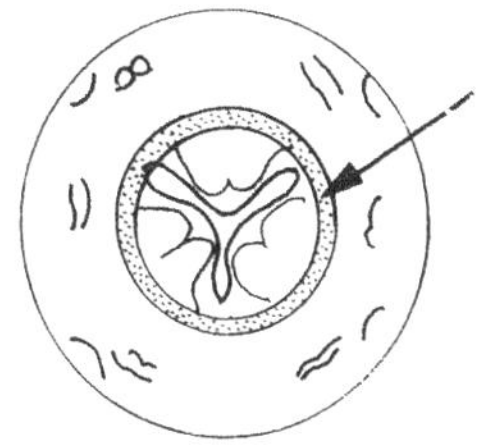
10.62

FIG. 10.59. Libyostrongylinae—Cooperiinae—Rays 2 and 3 of bursa (after Durette-Desset & Chabaud, 1977).

FIG. 10.60. *Cooperia hungi* Mönnig, 1931, cephalic extremity, lateral view (after Gibbons, 1981).

FIG. 10.61. Graphidiinae—Ostertagiinae—Rays 2 and 3 of bursa (after Durette-Desset & Chabaud, 1977).

FIG. 10.62. *Graphidium strigosum* (Dujardin, 1845), cephalic extremity, apical view (after Durette-Desset & Denké, 1978).

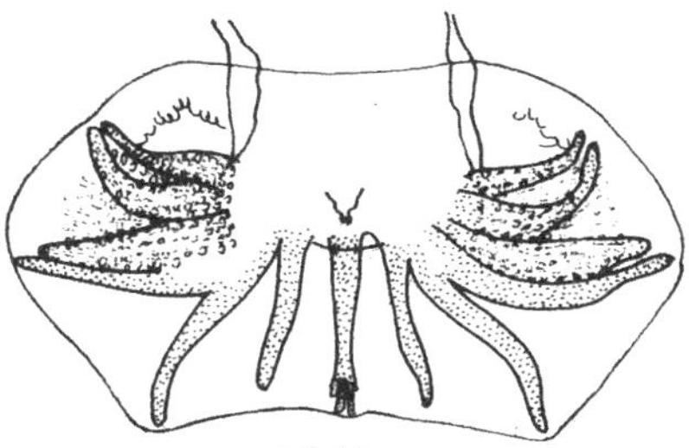
10.63

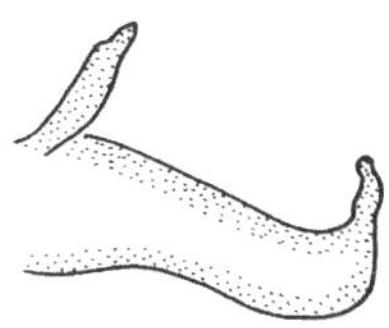
10.64

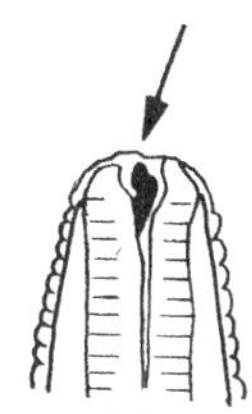
10.65

FIG. 10.63. *Hyostrongylus gabonensis* Durette-Desset & Chabaud, 1974, bursa (after Durette-Desset & Chabaud, 1974).

FIG. 10.64. Trichostrongylinae—Haemonchinae—Rays 2 and 3 of bursa (after Durette-Desset & Chabaud, 1977).

FIG. 10.65. *Haemonchus contortus* (Rudolphi, 1803), cephalic extremity, lateral view (after Gibbons, 1979).

7– (6) Buccal capsule absent.
Dorsal ray short (except certain *Marshallagia* spp.) and divided before posterior third (Figs. 10.15, 10.17).
Parasites of ruminants, mainly in Old World.

Ostertagiinae

8– (5) Ray 2 smaller than, and sharing common base with, ray 3; distal ends of rays 2 and 3 well separated (Fig. 10.64).

9–(10) Neodont formation absent.
Dorsal lobe not reduced, symmetrical (Fig. 10.14).
Parasites of caviomorphs, beavers, lagomorphs and ruminants; occasionally in various other mammals and man. Cosmopolitan.

Trichostrongylinae

10– (9) Neodont formation present (Fig. 10.65).
Dorsal lobe reduced and often asymmetrical (Fig. 10.66).
Parasites of ruminants (mainly in Old World), suiforms (cosmopolitan), lagomorphs (nearctic), *Nutria* (New World and Europe), and soricids (Japan); reported once in man.

Haemonchinae

Subfamily **Libyostrongylinae** Durette-Desset & Chabaud, 1977

Key to genera

1–(10) Dorsal ray well developed.
2– (7) Synlophe absent.
3– (6) Ray 4 without specialized sclerotized formations.
4– (5) Extremities of rays 3 and 4 closer together than those of rays 4 and 5 (Fig. 10.67). Parasites of ratites.
Libyostrongylus Lane, 1923
5– (4) Extremities of rays 3 and 4 further apart than those of rays 4 and 5 (Fig. 10.68). Parasites of archaic rodents (*Thryonomys*, *Bathyergus*, *Atherurus*), hyracoids, lagomorphs and gorillas in Africa.
Paralibyostrongylus Ortlepp, 1939
(= *Baylisiella* Skrjabin & Schulz, 1937)

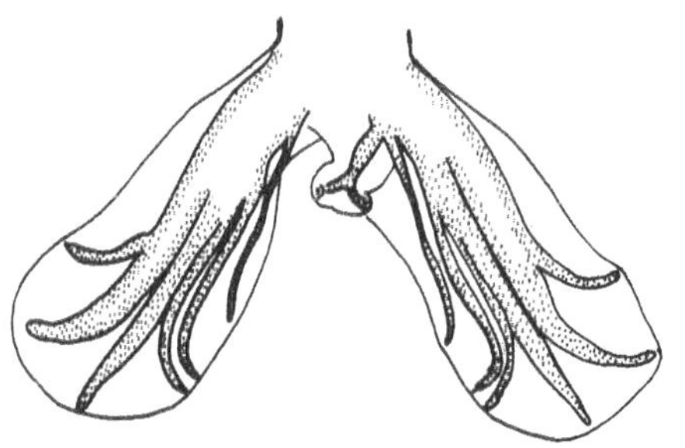

10.66

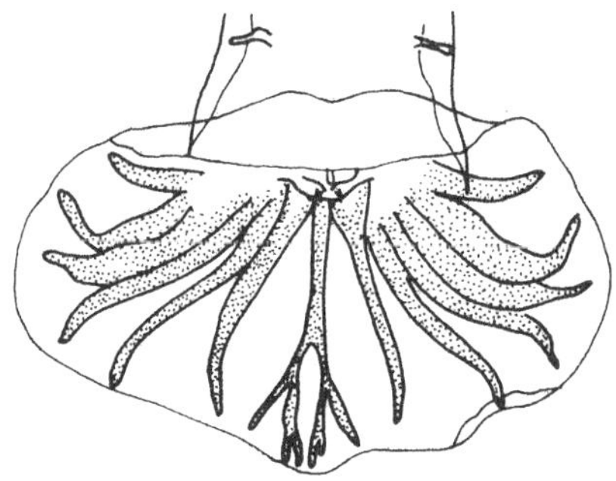

10.67

FIG. 10.66. *Haemonchus longistipes* Railliet & Henry, 1909, bursa (after Gibbons, 1979).
FIG. 10.67. *Libyostrongylus douglasii* (Cobbold, 1882), bursa (after Denké, 1968).

6– (3) Rays 4 with specialized sclerotized formations forming pincer when bursa closed (Fig. 10.69). Parasites of stomach of *Hypogeomys* (Cricetidae) in Madagascar.
Cnizostrongylus Chabaud, Durette-Desset & Houin, 1967
7– (2) Synlophe present.
8– (9) Ray 3 much more slender and longer than ray 4 (Fig. 10.70). Parasites of lemurs in Madagascar.
Pararhabdonema Kreis, 1945
9– (8) Rays 3 and 4 of same width and length (Fig. 10.71). Parasites of nearctic *Antilocapra* and *Ovis*.
Pseudostertagia (Orloff, 1933) Travassos, 1937
10– (1) Dorsal ray small (Fig. 10.72). Parasites of nearctic lagomorphs.
Obeliscoides Graybill, 1924
(= *Obeliscus* Graybill, 1923)

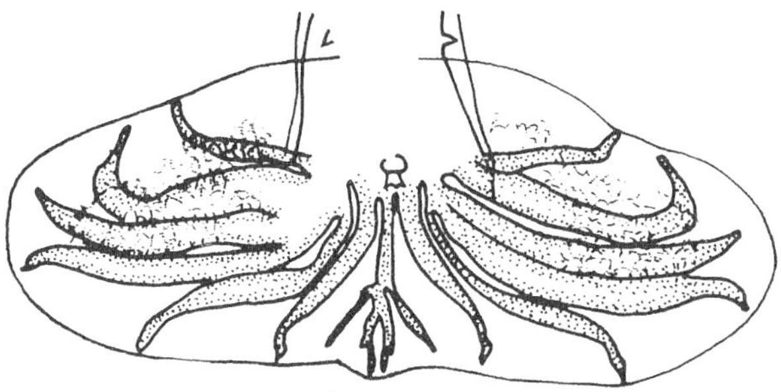

10.68

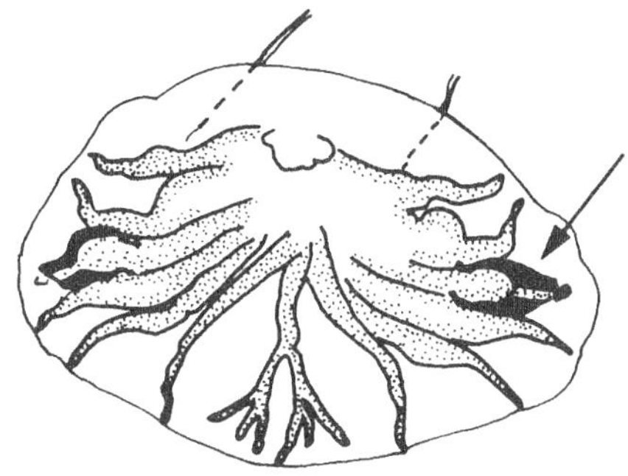

10.69

FIG. 10.68. *Paralibyostrongylus cassonei* Durette-Desset & Denké, 1968, bursa (after Durette-Desset & Denké, 1968).

FIG. 10.69. *Cnizostrongylus kleini* Chabaud, Durette-Desset & Houin, 1967, bursa (after Chabaud, Durette-Desset & Houin, 1967).

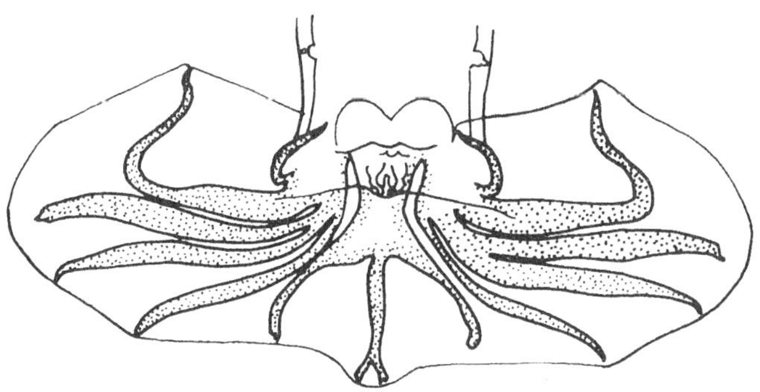

10.70

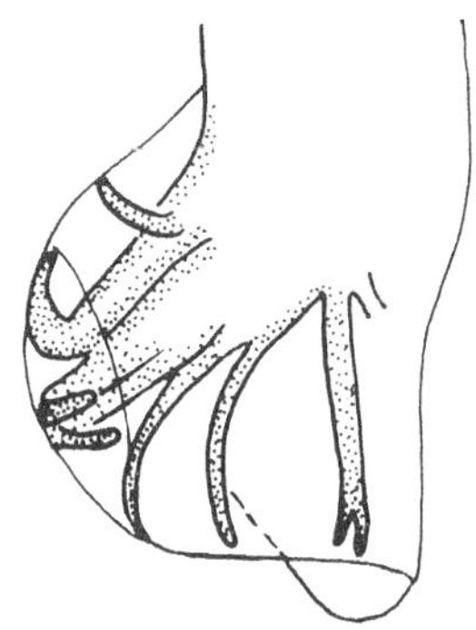

10.71

FIG. 10.70. *Pararhabdonema longistriata* Kreis, 1945, bursa (after Chabaud & Choquet, 1955).

FIG. 10.71. *Pseudostertagia bullosa* Ransom & Hall, 1912, bursa, lateral view (after Travassos, 1937).

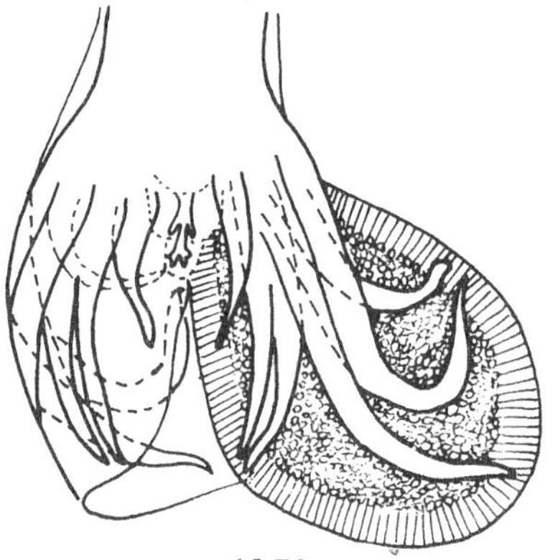

10.72

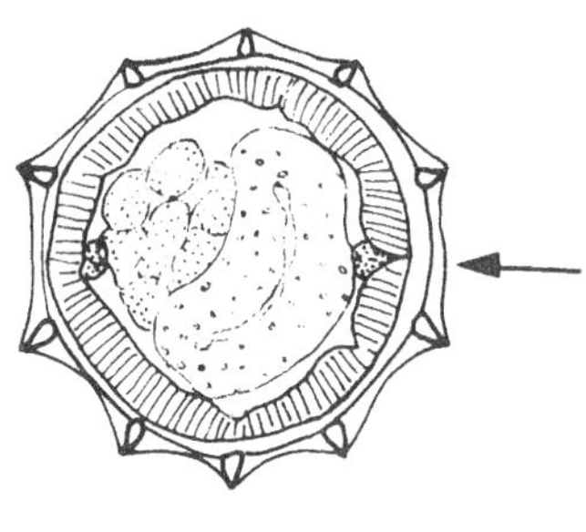

10.73

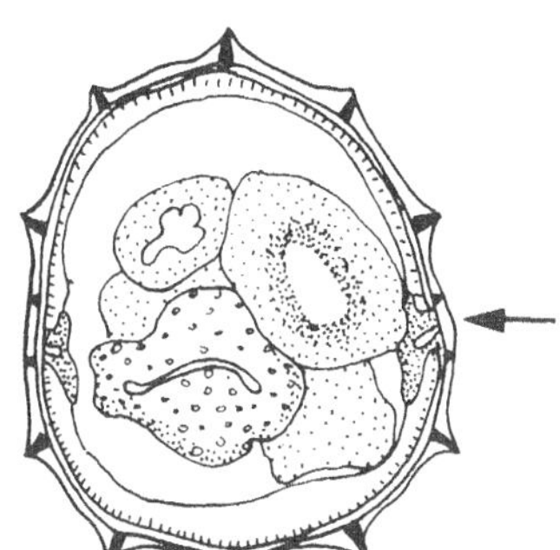

10.74

FIG. 10.72. *Obeliscoides cuniculi* (Graybill, 1923), bursa, dorsal view, (after Chandler, 1924).

FIG. 10.73. *Cooperia hepaticae* (Ortlepp, 1938), synlophe (after Gibbons, 1978).

FIG. 10.74. *Cooperia hungi* Mönnig, 1931, synlophe (after Gibbons, 1978).

Subfamily **Cooperiinae** (Skrjabin & Schulz, 1937, tribe)
Skrjabin & Schikhobalova, 1952

Key to genera

1–(10) Didelphic.
Gubernaculum absent.
2– (7) Ridges of synlophe equal in size (Fig. 10.73) or lateral ridges smaller than median ridges (Fig. 10.74).
3– (6) Dorsal ray long and divided in posterior third (Fig. 10.75).
4– (5) Cephalic vesicle double (Fig. 10.60).
Length of ovejector from sphincter to sphincter less than 500 μm.
Excretory pore in posterior half of oesophagus.
Parasites of ethiopian and oriental Tragulidae and *Muntiacus* (Cervidae).
Chabaudstrongylus Durette-Desset & Denké, 1978
5– (4) Cephalic vesicle simple.
Length of ovejector from sphincter to sphincter greater than 3.5 mm.
Excretory pore in anterior half of oesophagus.
Parasites of *Gazella* (Bovidae) of Africa.
Gazellostrongylus Yeh, 1956
6– (3) Dorsal ray long and divided above its posterior third (Fig. 10.76) or dorsal ray short (Fig. 10.77).
Parasites of Bovidae (mainly), Giraffidae, Camelidae. Cosmopolitan.
Cooperia Ransom, 1907
(= *Cooperioides* Daubney, 1933)
7– (2) Lateral ridges of synlophe more developed than median ridges (Fig. 10.78).
8– (9) Latero-dorsal ridges larger than latero-ventrals.
Lateral ridges directed dorsally (Fig. 10.78).
Ray 3 longer than ray 4 (Fig. 10.79).
Indentation on spicules present (Fig. 10.80).
Parasites of Old World Bovidae.
Paracooperia Travassos, 1935
(= *Schwartziella* Le Roux, 1935)

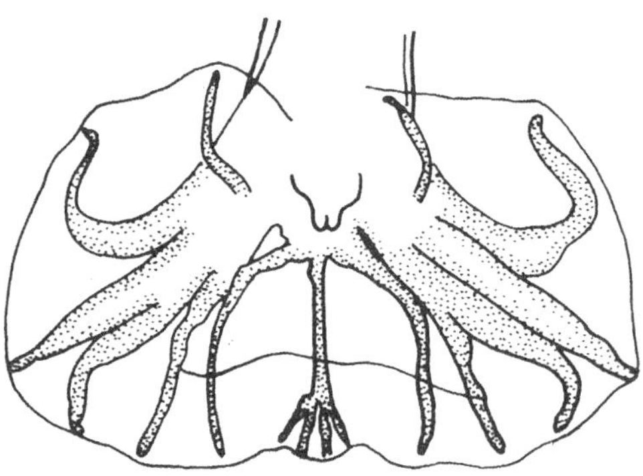

10.75

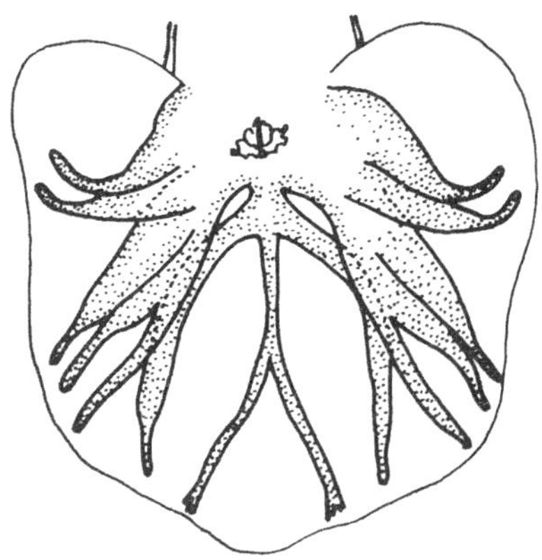

10.76

FIG. 10.75. *Chabaudstrongylus dubosti* (Durette-Desset & Chabaud, 1974), bursa (after Durette-Desset & Chabaud, 1974).

FIG. 10.76. *Cooperia kenyensis* (Daubney, 1933), bursa (after Daubney, 1933).

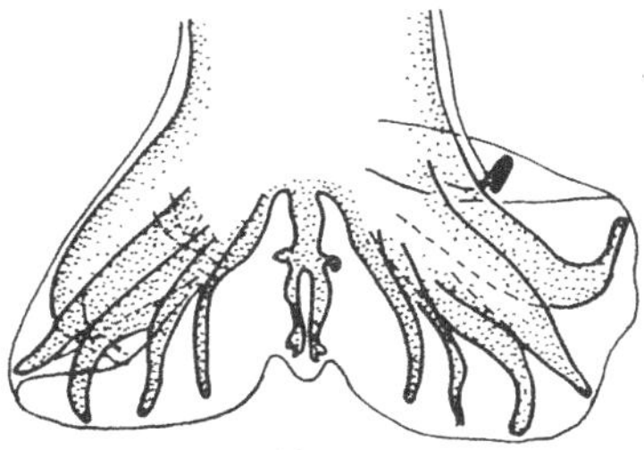

10.77

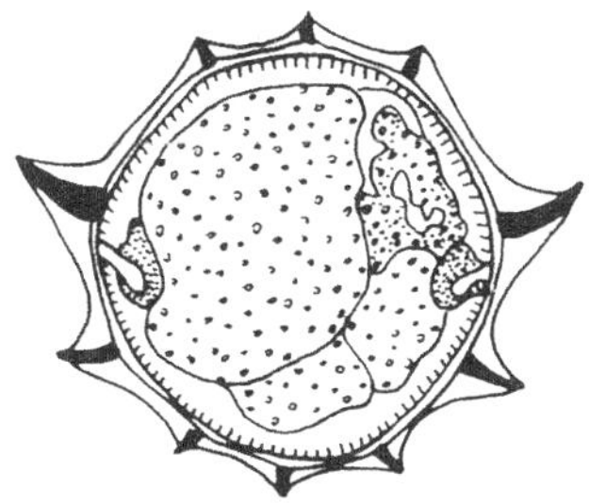

10.78

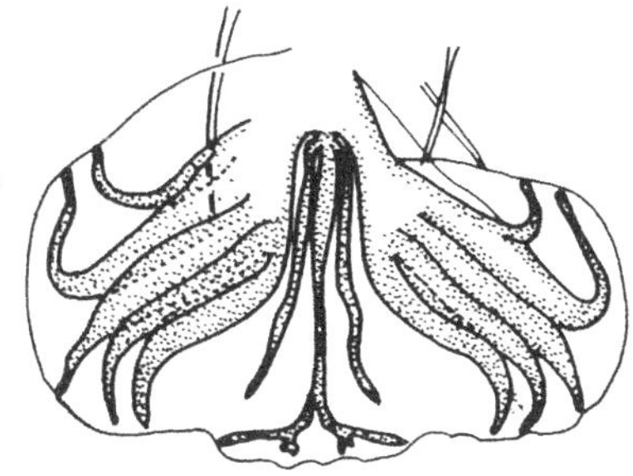

10.79

FIG. 10.77. *Cooperia yoshidai* Mönnig, 1939, bursa (after Gibbons, 1981).

FIG. 10.78. *Paracooperia serrata* (Mönnig, 1931), synlophe (after Gibbons, 1978).

FIG. 10.79. *Paracooperia nodulosa* (Schwartz, 1928), bursa (after Gibbons, 1978).

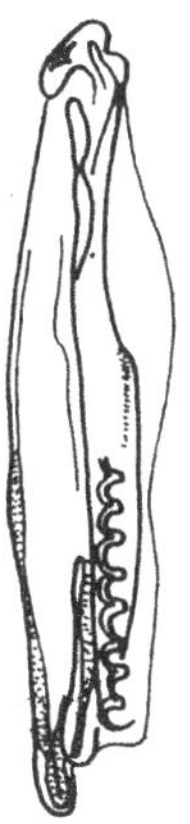

10.80

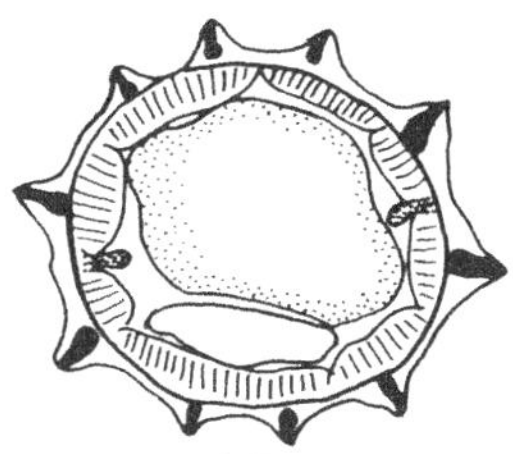

10.81

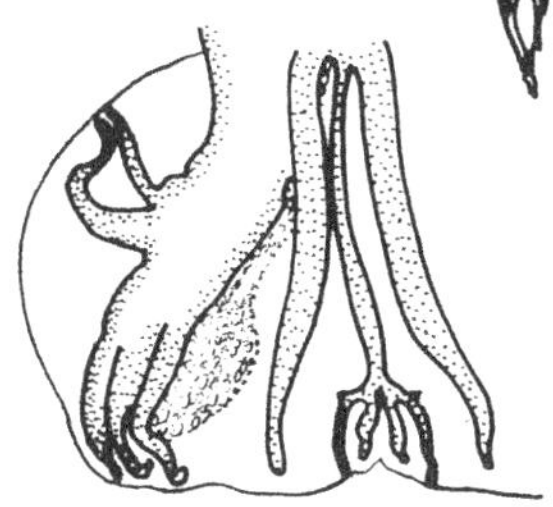

10.82

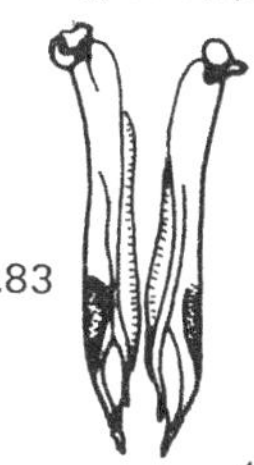

10.83

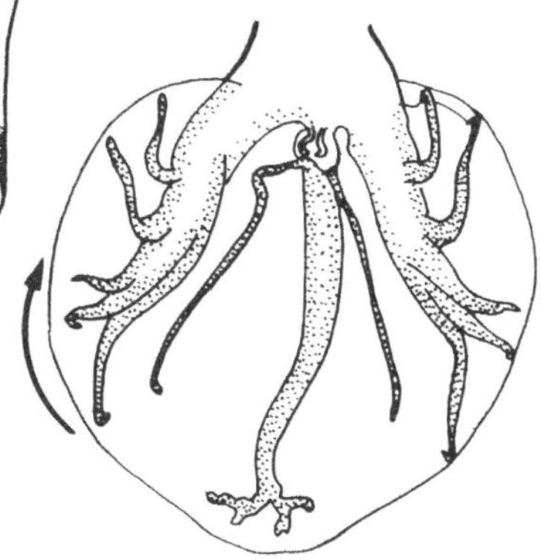

10.84

FIG. 10.80. *Paracooperia nodulosa* (Schwartz, 1928), spicule (after Gibbons, 1978).

FIG. 10.81. *Megacooperia woodfordi* Khalil & Gibbons, 1976, synlophe (after Khalil & Gibbons, 1976).

FIG. 10.82. *Megacooperia woodfordi* Khalil & Gibbons, 1976, bursa (after Khalil & Gibbons, 1976).

FIG. 10.83. *Megacooperia woodfordi* Khalil & Gibbons, 1976, spicules (after Khalil & Gibbons, 1976).

FIG. 10.84. *Impalaia tuberculata* Mönnig, 1923, bursa (after Gibbons, Durette-Desset & Daynes, 1977).

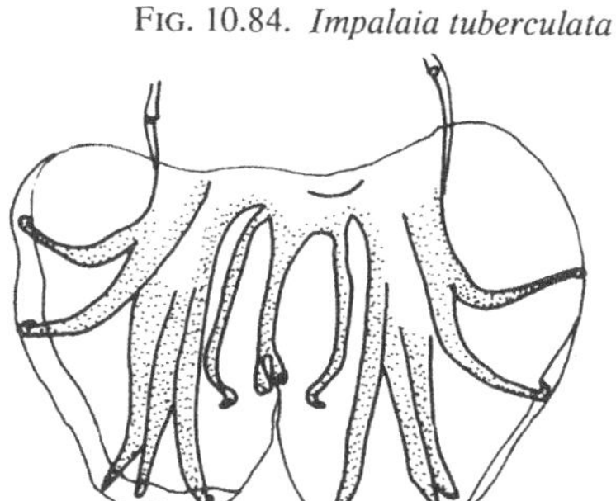

10.85

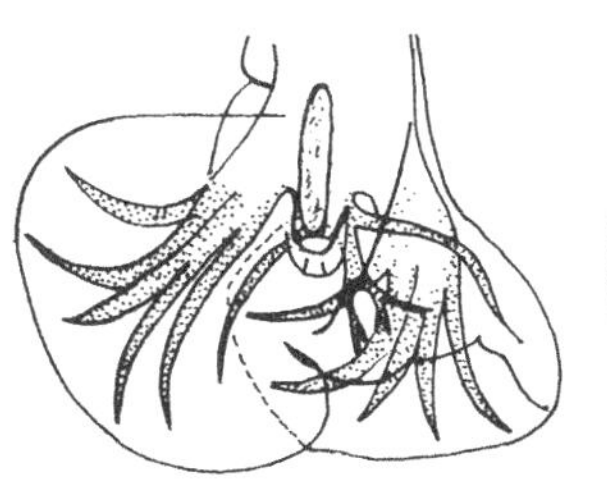

10.86

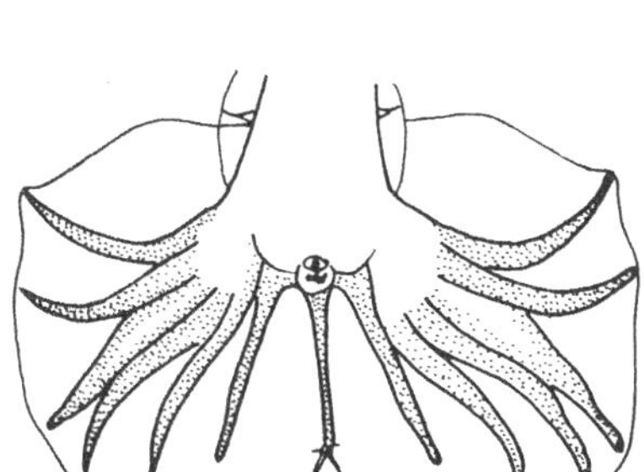

10.87

FIG. 10.85. *Ortleppstrongylus bathyergi* (Ortlepp, 1939), bursa (after Durette-Desset, 1970).

FIG. 10.86. *Minutostrongylus taurotragi* (Le Roux, 1936), bursa, dorsal view (after Le Roux, 1936).

FIG. 10.87. *Parostertagia heterospiculum* Schwartz & Alicata, 1933, bursa (after Schwartz & Alicata, 1933).

9– (8) Latero-dorsal and latero-ventral ridges equal in size.
Lateral ridges directed perpendicular to body (Fig. 10.81).
Ray 3 shorter than ray 4 (Fig. 10.82).
Indentation on spicules absent (Fig. 10.83).
Parasites of *Nesotragus* (Bovidae) in Kenya.
Megacooperia Khalil & Gibbons, 1976

10– (1) Monodelphic.
Gubernaculum present.

11–(12) Dorsal ray very long.
Ray 8 shorter than dorsal ray.
Rays 6 to 4 with gradient of decreasing size (Fig. 10.84).
Parasites of Bovidae, Camelidae, Giraffidae, rarely lagomorphs.
Ethiopian.
Impalaia Mönnig, 1923
(= *Anthostrongylus* Croveri, 1929)

12–(11) Dorsal ray short.
Ray 8 longer than dorsal ray.
Rays 6 to 4 without gradient in size (Fig. 10.85).

13–(14) Ray 4 shorter than ray 5 (Fig. 10.86).
Dorsal ray divided above posterior third.
Gubernaculum well developed.
Parasites of *Taurotragus* (Bovidae) in Africa.
Minutostrongylus Le Roux, 1936

14–(13) Ray 4 as long as ray 5 (Fig. 10.85).
Dorsal ray divided behind posterior third.
Gubernaculum poorly developed.
Parasites of *Bathyergus* (Bathyergidae) in Africa.
Ortleppstrongylus Durette-Desset, 1970

Subfamily **Graphidiinae** Travassos, 1937

Key to genera

1– (4) Cephalic vesicle absent.
Telamon absent.
Ends of rays 5 and 6 pointed in same direction (Fig. 10.87).

2– (3) Spicules long and slender.
Posterior branch of vestibule shorter than anterior (Fig. 10.88).
Parasites of palaearctic Leporidae.
Graphidium Railliet & Henry, 1909

3– (2) Spicules short and thick.
Branches of vestibule equal (Fig. 10.89).
Parasites of nearctic Suidae.
Parostertagia Schwartz & Alicata, 1933

4– (1) Cephalic vesicle present.
Telamon present.
Ends of rays 5 and 6 divergent (Fig. 10.63).
Parasites of ethiopian Leporidae and Tragulidae, neotropical Suidae, domesticated pigs, *Okapi* in Africa and *Antilope cervicapra* (Bovidae) in China.
Hyostrongylus Hall, 1921
(= *Bergheia* Drozdz, 1965)

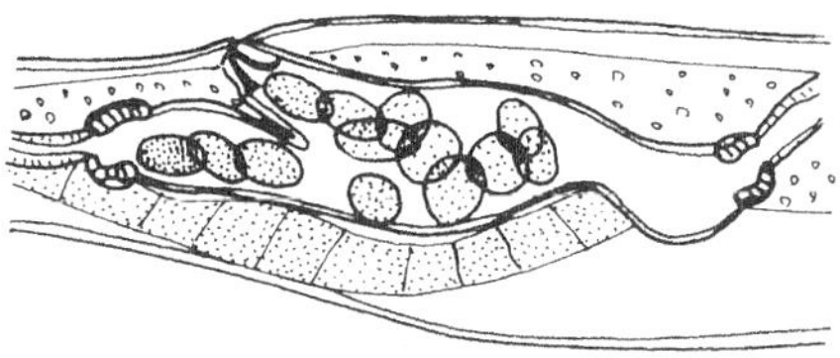

10.88

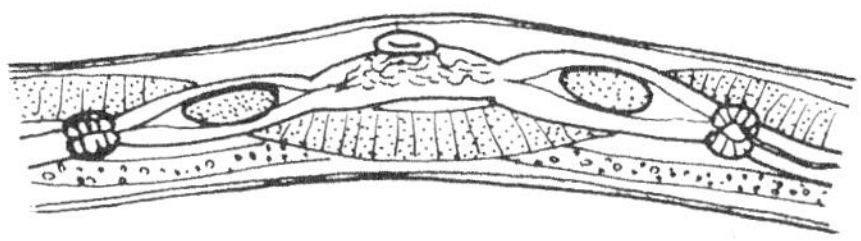

10.89

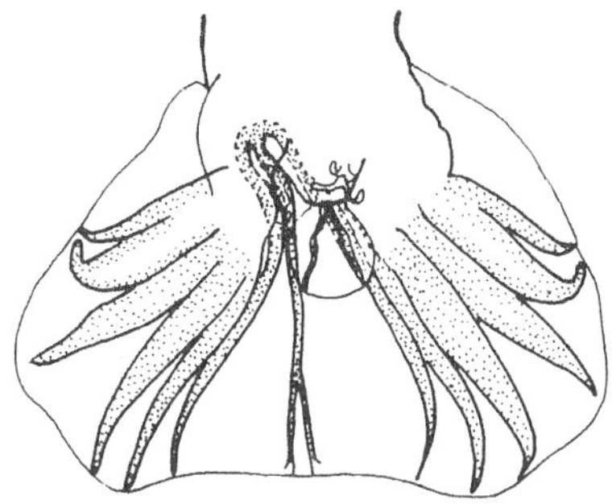

10.90

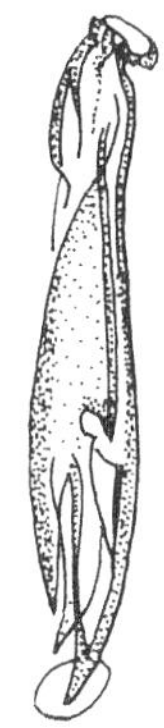

10.91

FIG. 10.88. *Graphidium strigosum* (Dujardin, 1845), ovejector, lateral view (after Durette-Desset & Denké, 1978).

FIG. 10.89. *Parostertagia heterospiculum* Schwartz & Alicata, 1933, ovejector, lateral view (after Schwartz & Alicata, 1933).

FIG. 10.90. *Marshallagia marshalli* (Ransom, 1907), bursa (after Ransom, 1907).

FIG. 10.91. *Marshallagia buriatica* (Konstantinova, 1934), spicules (after Konstantinova, 1934).

Subfamily **Ostertagiinae** (Skrjabin & Schulz, 1937, tribe) Lopez-Neyra, 1947

Key to genera

1– (6) Bursa of type 2–1–2 (Fig. 10.90).

2– (5) Ray 4 shorter than or as long as ray 5 (Figs. 10.90, 10.92).

3– (4) Rays 5 and 6 much longer than rays 2 and 3.
Dorsal ray slender, slightly shorter or of same length as ray 8 (Fig. 10.90).
Spicules thick (Fig. 10.91).
Parasites mainly of livestock, also camels, wild Bovidae and Cervidae.
Marshallagia (Orloff, 1933) Travassos, 1937
[= *Camelostrongylus* Orloff, 1933; = *Grosspiculagia* (Orloff, 1933) Sarwar, 1956; = *Orloffia* Drozdz, 1965; = *Ostertagiella* Andreeva, 1956]

4– (3) Rays 5 and 6 as long as or slightly longer than rays 2 and 3.
Dorsal ray thick, much shorter than ray 8 (Fig. 10.92).
Spicules slender (Fig. 10.93).
Parasites of wild Bovidae (Antilopinae, Hippotraginae), rarely sheep and goats.
Longistrongylus Le Roux, 1931
[= *Bigalkea* Mönnig, 1931; = *Bigalkenema* Ortlepp, 1963; = *Kobusinema* Ortlepp, 1963; = *Ostertagia (Costarcuata)* Schulz & Kadenazii, 1950; = *Pseudomarshallagia* (Roetti, 1941) Graber & Delavenay, 1978]

5– (2) Ray 4 longer than ray 5 (Fig. 10.94).
Parasites of holarctic Cervidae (Cervinae, Odocoileinae), cattle and goats.
Ostertagia Ransom, 1907
[= *Capreolagia* Schulz, Andreeva & Kadenazii, 1954; = *Gruhneria* Sarwar, 1956; = *Muflonagia* Schulz, Andreeva & Kadenazii, 1954; = *Ostertagiana* Dikov, 1963; = *Sjobergia* Sarwar, 1956; = *Skrjabinagia* (Kassimov, 1942) Altaev, 1952]

6– (1) Bursa of type 2–2–1 (Figs. 10.15, 10.17).

7– (8) Ray 4 shorter than ray 5 (Fig. 10.17).
Parasites of Cervidae (Muntiacinae, Cervinae and Odocoileinae), one species found in *Rupicapra* (Caprinae), oriental Tragulidae and livestock.
Spiculopteragia (Orloff, 1933) Travassos, 1937
[= *Altaevia* Sarwar, 1957; = *Apteragia* Jansen, 1958;
= *Mazamostrongylus* Cameron, 1935; = *Rinadia* Grigorian, 1951;
= *Sarwaria* Drozdz, 1965; = *Spiculopteragia* (*Petrowiagia*) Ruchljadew, 1961;
= *Spiculopteroides* Jansen, 1958]

8– (7) Ray 4 as long as or longer than ray 5 (Fig. 10.15).
Parasites of livestock, neotropical Camelidae, ethiopian Bovidae and Cervidae.
Teladorsagia Andreeva & Satubaldin, 1954
(= *Paramecistocirrus* Roetti, 1941; = *Stadelmania* Sarwar, 1956)

Subfamily **Trichostrongylinae** Leiper, 1912

Key to genera

1– (2) Synlophe absent.
Gubernaculum present (Fig. 10.14).
Parasites of birds, ruminants and lagomorphs throughout the world, and primitive rodents, especially in New World, rarely man.
Trichostrongylus Loos, 1905
(= *Buckleya* Sarwar, 1956; = *Cobboldostrongylus* Sarwar, 1956;
= *Gilesia* Sarwar, 1956; = *Probostrongylus* Sarwar, 1956)

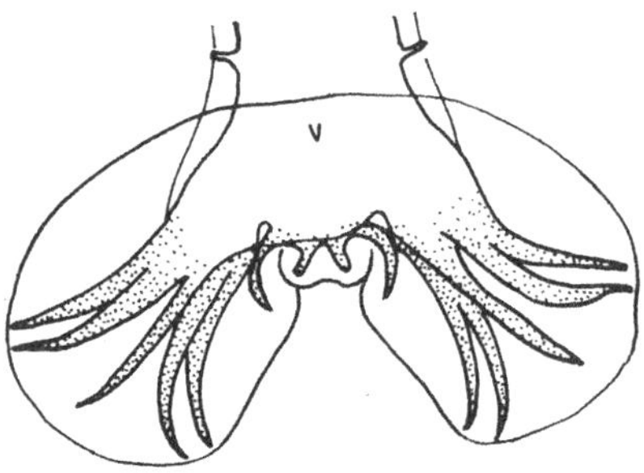
10.92

10.93

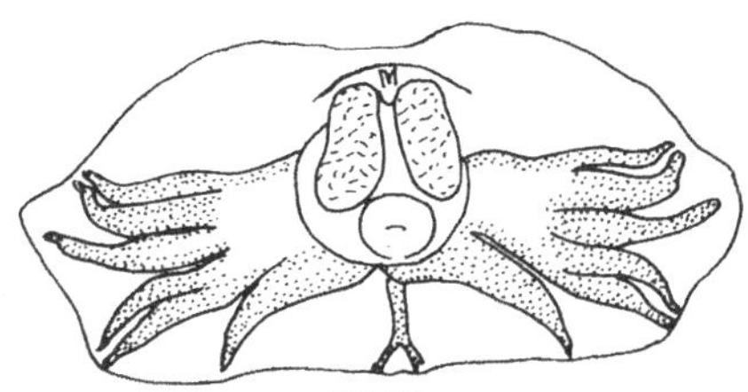
10.94

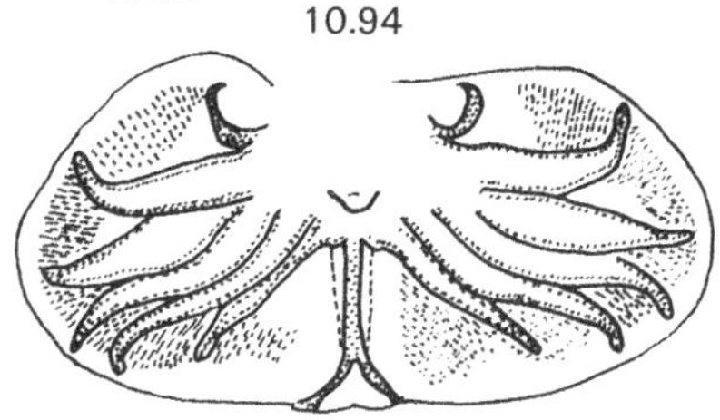
10.95

FIG. 10.92. *Longistrongylus meyeri* Le Roux, 1931, bursa (after Le Roux, 1931).
FIG. 10.93. *Longistrongylus meyeri* Le Roux, 1931, spicule (after Gibbons, 1977).
FIG. 10.94. *Ostertagia ostertagi* (Stiles, 1892), bursa (after Drozdz, 1965).
FIG. 10.95. *Graphidioides berlai* Travassos, 1943, bursa (after Travassos, 1943).

2– (1) Synlophe present.
Gubernaculum absent (Fig. 10.95).

3– (4) Spicules long (greater than 500 μm) and slender.
Parasites of stomach of caviomorph rodents and small intestine of neotropical primates.
Graphidioides Cameron, 1923

4– (3) Spicules short (less than 200 μm) and thick.
Parasites of stomach of Castoridae.
Travassosius Khalil, 1922

Subfamily **Haemonchinae** (Skrjabin & Schulz, 1937, tribe)
Skrjabin & Schulz, 1952

Key to genera

1–(12) *Corona radiata* absent.

2– (3) Buccal capsule present (Fig. 10.96).
Parasites of stomach of nearctic Leporidae.
Biogastranema Rohrbacher & Ehrenford, 1954

3– (2) Buccal capsule absent.

4– (5) Five intraoesophageal teeth present (Fig. 10.97).
Parasites of *Myopotamus coypus*.
Boehmiella Gebauer, 1932

5– (4) One intraoesophageal tooth present (Fig. 10.65).

6–(11) *Vagina vera* not hypertrophied (Fig. 10.98).

7–(10) Dorsal lobe of bursa not directed to left (Fig. 10.16).

8– (9) Synlophe present only in anterior half of body.
Parasites of Bovidae and Cervidae.
Ashworthius Le Roux, 1930

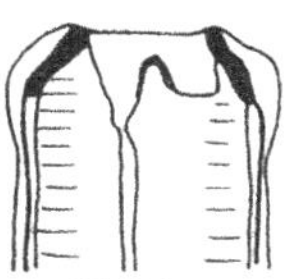

10.96

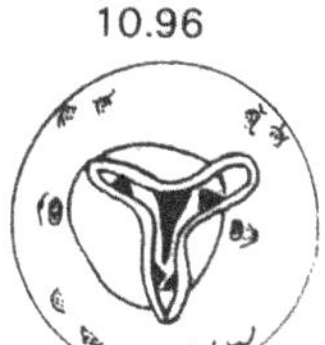

10.97

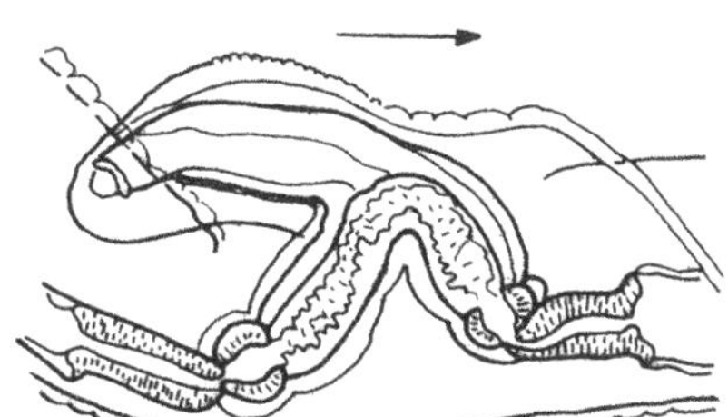

10.98

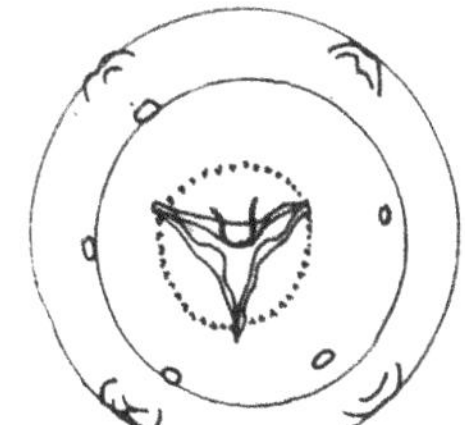

10.100

10.99

FIG. 10.96. *Biogastranema affinis* Rohrbacher & Ehrenford, 1954, cephalic extremity, lateral view (after Durette-Desset, 1978).

FIG. 10.97. *Boehmiella perichitinea* Gebauer, 1932, cephalic extremity, apical view (after Durette-Desset & Sutton, 1979).

FIG. 10.98. *Haemonchus similis* Travassos, 1914, ovejector, lateral view (after Gibbons, 1979).

FIG. 10.99. *Mecistocirrus digitatus* (Linstow, 1906), ovejector, lateral view (after Toschev, 1929).

FIG. 10.100. *Moguranema nipponicum* Yamaguti, 1941, apical view (after Durette-Desset, 1977).

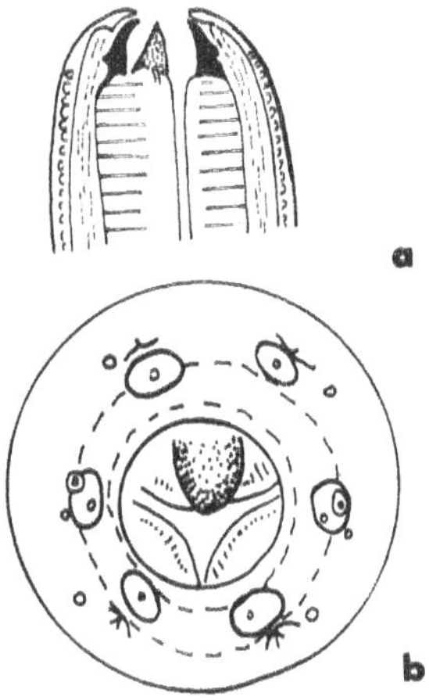

10.101

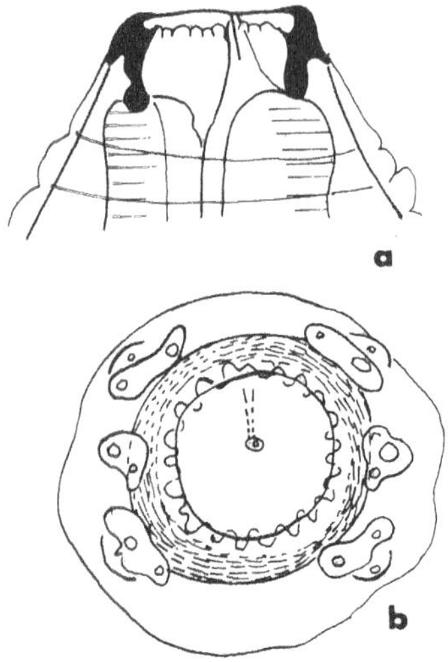

10.102

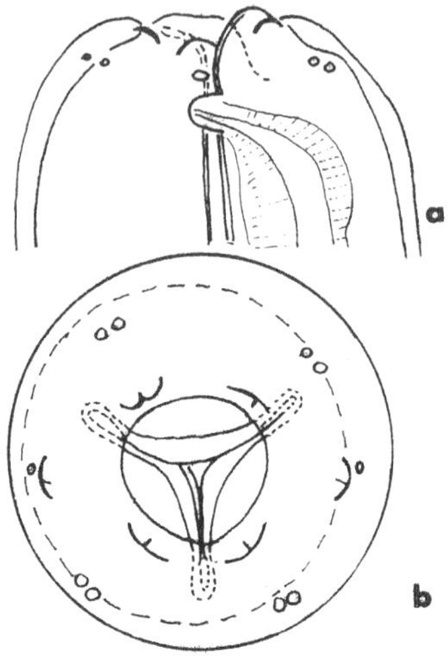

10.103

FIG. 10.101. *Amphibiophilus chabaudi* Puylaert, 1967, cephalic extremity, (a) lateral view (after Puylaert, 1967), (b) apical view (after Baker, 1981).

FIG. 10.102. *Graphidiella nepalensis* Durette-Desset & Tcheprakoff, 1983, cephalic extremity, (a) lateral view, (b) apical view (after Durette-Desset & Tcheprakoff, 1983).

FIG. 10.103. *Batrachostrongylus longispiculus* Yuen, 1963, cephalic extremity, (a) left lateral view, (b) apical view (after Baker, original).

9– (8) Synlophe present over entire body length.
Parasites of ethiopian suiforms (*Hippopotamus*).
Leiperiatus Sandground, 1930

10– (7) Dorsal lobe of bursa directed to left (Fig. 10.66).
Parasites of ethiopian Bovidae (mainly Hippotraginae and Antilopinae), Camelidae, Giraffidae and Cervidae (palaearctic Cervinae and oriental Muntiacinae).
Haemonchus Cobb, 1898

11– (6) *Vagina vera* hypertrophied (Fig. 10.99).
Parasites of cattle, goats, domestic pigs and man in Asia, Central America, Sumatra, USSR and India.
Mecistocirrus Railliet & Henry, 1912

12– (1) *Corona radiata* present (Fig. 10.100).
Parasites of Soricoidea (*Mogura*) in Japan.
Moguranema Yamaguti, 1941

Family *AMPHIBIOPHILIDAE* Durette-Desset & Chabaud, 1981

Key to genera

1– (4) Buccal capsule well developed (Fig. 10.101).
Ray 4 shorter than ray 5 (Fig. 10.22).
Vagina vera not extremely long.

2– (3) *Corona radiata* absent (Fig. 10.101).
Parasites of amphibians (Africa-Asia).
Amphibiophilus Skrjabin, 1916

3– (2) *Corona radiata* present (Fig. 10.102).
Parasites of holarctic lagomorphs (Ochotonidae).

Graphidiella Olsen, 1948

4– (1) Buccal capsule absent (Fig. 10.103).
Ray 4 as long as ray 5.
Vagina vera long (1.5 mm).
Parasites of amphibians (Malaysia).

Batrachostrongylus Yuen, 1963

Family *DICTYOCAULIDAE* (Skrjabin, 1933, subfam.) Skrjabin, 1941

Key to subfamilies

1– (2) Dorsal ray divided at midpoint (Fig. 10.23).
Parasites of digestive tract of amphibians and reptiles.

Mertensinematinae

2– (1) Dorsal ray divided at base (Fig. 10.24).
Parasites of respiratory system of ungulates.

Dictyocaulinae

Subfamily **Mertensinematinae** Sharpilo, 1976

Key to genera

1– (2) Cephalic vesicle present.
Spicules with three points.
Parasites of amphibians (USSR) and reptiles (Borneo).

Mertensinema Sharpilo, 1976

2– (1) Cephalic vesicle absent.
Spicules with two points.
Parasites of neotropical amphibians.

Borrellostrongylus Gutierrez, 1945
(= *Parabatrachostrongylus* Tantalean & Naupay, 1974)

Subfamily **Dictyocaulinae** Skrjabin, 1933

Key to genera

1– (2) Spicules thick, less than 1 mm long.
Buccal ring present (Figs. 10.104, 10.105).
Parasites of Bovidae, Cervidae, Equidae and Suidae. Cosmopolitan.

Dictyocaulus Railliet & Henry, 1907
(= *Micrurocaulus* Skrjabin, 1934)

2– (1) Spicules thin, more than 2 mm long.
Buccal ring absent.
Parasites of African Bovidae.

Bronchonema Mönnig, 1932

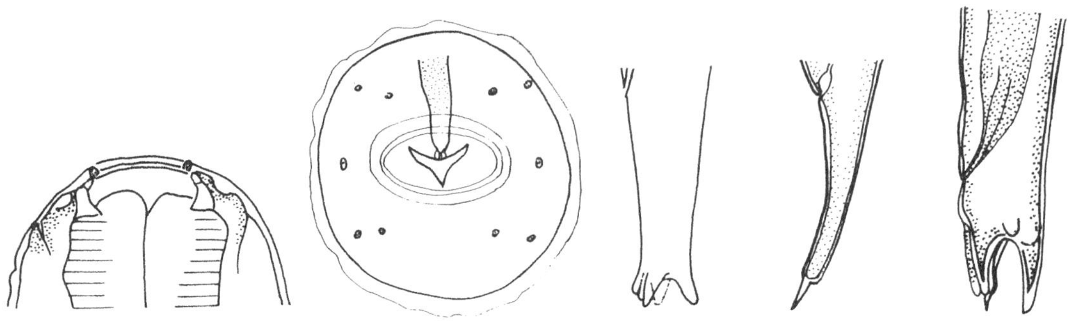

10.104 10.105 10.106 10.107 10.108

FIG. 10.104. *Dictyocaulus eckerti* Skrjabin, 1931, cephalic extremity, lateral view (original).

FIG. 10.105. *Dictyocaulus eckerti* Skrjabin, 1931, cephalic extremity, apical view (original).

FIG. 10.106. *Ollulanus skrjabini* Bundelev, 1950, female, tail end, lateral view (original).

FIG. 10.107. *Nochtia* (*Tenrecola*) *pilosus* (Chabaud, Brygoo & Tcheprakoff, 1964), female, tail end, lateral view (after Chabaud, Brygoo & Tcheprakoff, 1964).

FIG. 10.108. *Anoplostrongylus paradoxus* (Travassos, 1918), female, tail end, lateral view (after Barus & Valle, 1967).

Family *MOLINEIDAE*
(Skrjabin & Schulz, 1937, subfam.) Durette-Desset & Chabaud, 1977

Key to subfamilies

1– (2) Viviparous.
Tail of female without caudal spine but with five tubercles (Fig. 10.106).
Ollulaninae

2– (1) Oviparous.
Tail of female usually with one caudal spine (Fig. 10.107); tubercles often present (Fig. 10.108) (but if spine absent, tubercles also absent).

3– (6) Dorsal ray not divided at base.
Neodont formation absent.
Parasites of various mammals (not lagomorphs and ruminants).

4– (5) Synlophe, if present, with ridges not oriented from ventral to dorsal side (Figs. 10.109 to 10.111).
Parasites of amphibians, reptiles, carnivores, insectivores (Tenrecoidea, rarely Erinacoidea), Old World bats, Tupaiidae, primates, Pholidota, Tubulidentata and rodents (Sciuridae).
Molineinae

5– (4) Synlophe with ridges directed from ventral to dorsal side (Figs. 10.112, 10.113).
Parasites mainly of New World bats, Xenarthra, rarely neotropical Cricetidae, Old World bats and Tupaiidae.
Anoplostrongylinae

6– (3) Dorsal ray divided at base (Fig. 10.114) (except *Lamanema*).
Neodont formation present (Fig. 10.115) (except *Murielus*).
Parasites of lagomorphs and ruminants.
Nematodirinae

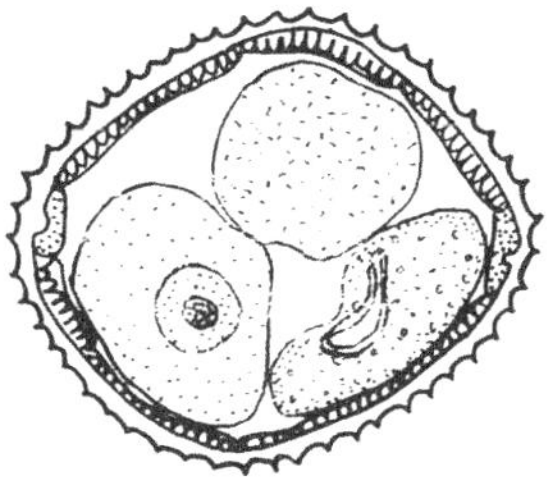

10.109

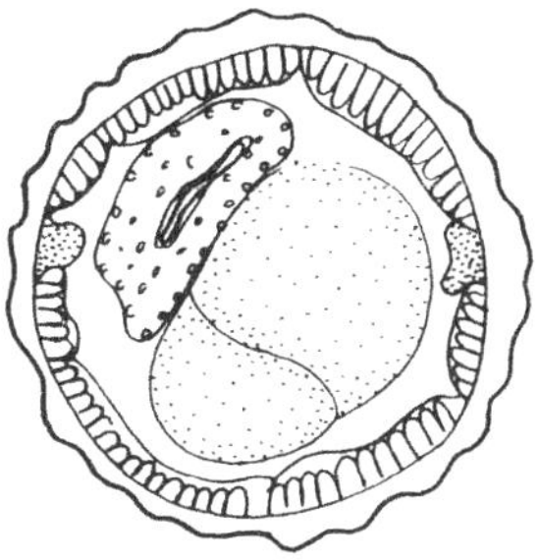

10.110

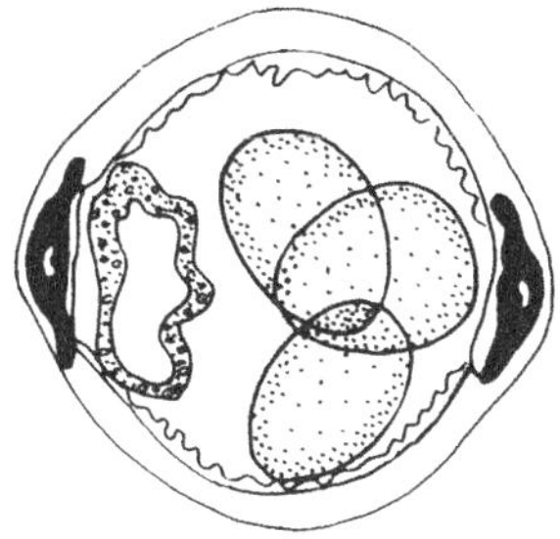

10.111

FIG. 10.109. *Oswaldocruzia johnstoni* Durette-Desset & Vaucher, 1979, synlophe (after Durette-Desset & Vaucher, 1979).

FIG. 10.110. *Brygoonema petteri* (Durette-Desset, 1973), synlophe (after Durette-Desset, 1973).

FIG. 10.111. *Molinostrongylus vespertilionis* Morozov & Spasski, 1961, synlophe (after Morozov & Spasski, 1961).

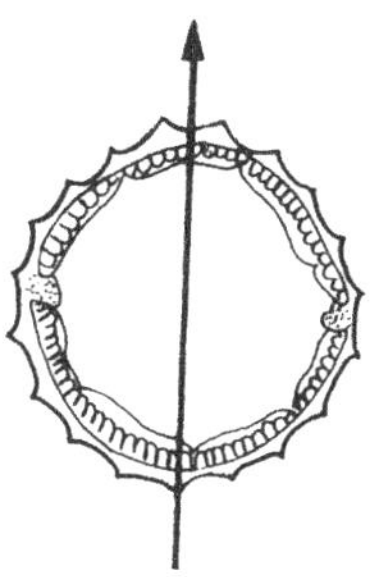

10.112

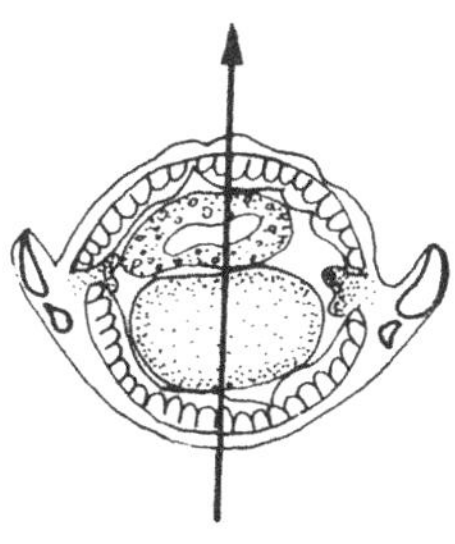

10.113

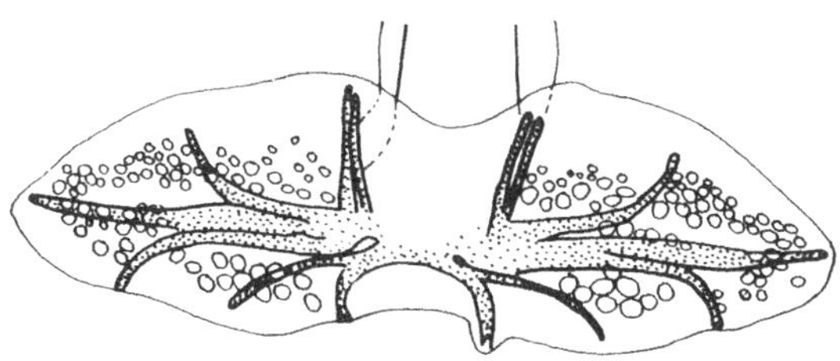

10.114

FIG. 10.112. *Anoplostrongylus paradoxus* (Travassos, 1918), synlophe (after Durette-Desset & Pinto, 1977).

FIG. 10.113. *Fontesia fontesi* Travassos, 1928, synlophe (after Durette-Desset, Chabaud & Cassone, 1977).

FIG. 10.114. *Rauschia triangularis* (Boughton, 1932), bursa (after Durette-Desset, 1979).

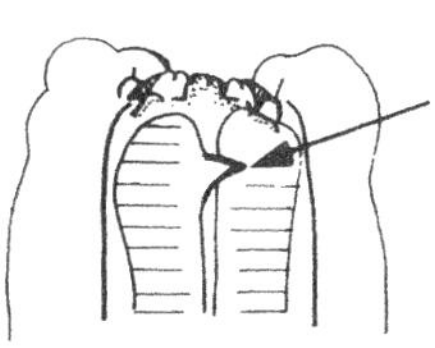

10.115

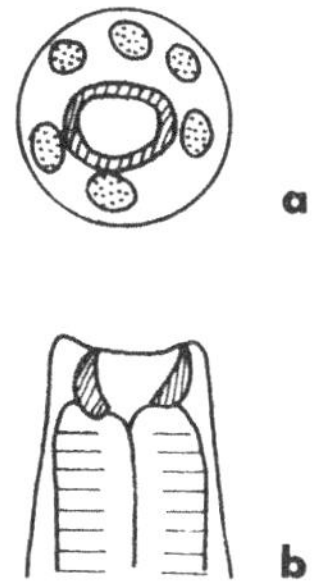

10.116

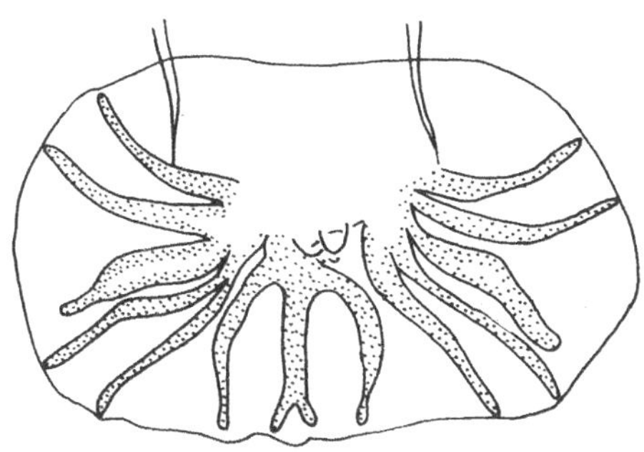

10.117

FIG. 10.115. *Nematodirus filicollis* (Rudolphi, 1802), neodont formation, lateral view (after Durette-Desset, 1979).

FIG. 10.116. *Ollulanus skrjabini* Bundelev, 1950, cephalic extremity, (a) apical view, (b) lateral view (original).

FIG. 10.117. *Ollulanus skrjabini* Bundelev, 1950, bursa (original).

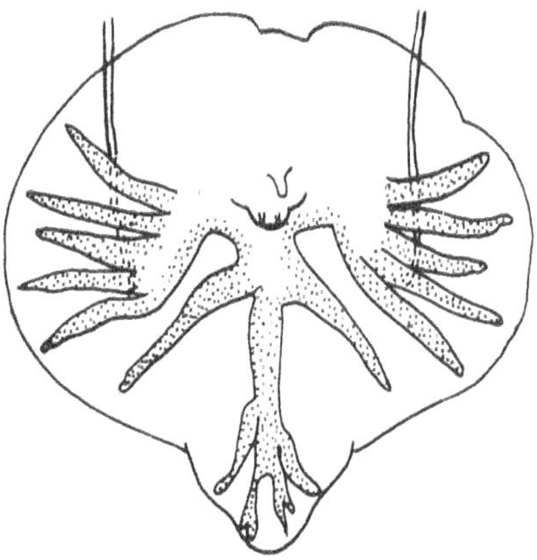

10.118

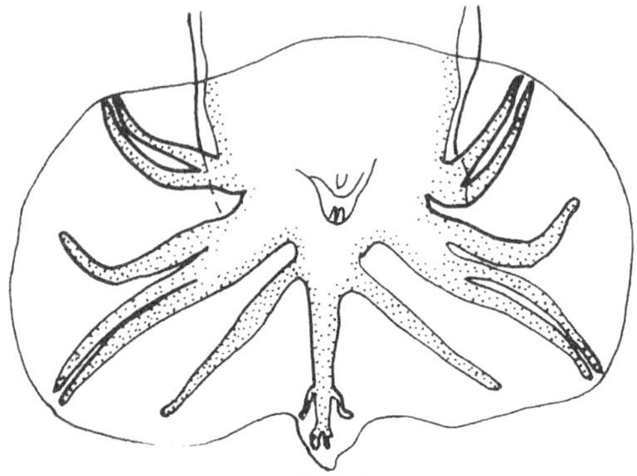

10.119

10.120

FIG. 10.118. *Schulzia* sp., bursa (after Baker, original).

FIG. 10.119. *Oswaldocruzia gracilipes* Durette-Desset & Vaucher, 1979, bursa (after Durette-Desset & Vaucher, 1979).

FIG. 10.120. *Trichoskrjabinia douglassi* Durette-Desset, 1978, female, tail end, lateral view (after Durette-Desset, 1978).

Subfamily **Ollulaninae** Hall, 1916

One genus (Figs. 10.106, 10.116, 10.117).
Mainly parasites of Felidae.

Ollulanus Leuckart, 1865

Subfamily **Molineinae** Skrjabin & Schulz, 1937

Key to genera

1– (2) Rays 2 to 6 shorter than dorsal ray.
Distal extremities of rays 2 to 6 nearly equidistant (Fig. 10.118).
Parasites of neotropical amphibians.

Schulzia Travassos, 1937

2– (1) Nematodes lacking above characters.

3–(10) Ray 4 arising anterior to rays 5 and 6.
Rays 2 and 3 parallel and rays 5 and 6 parallel (Figs. 10.119, 10.121).
Female tail with stout terminal point (Fig. 10.120).
Parasites of amphibians and reptiles.

4– (5) Synlophe absent.
Parasites of neotropical amphibians.

Poekilostrongylus Schmidt & Whittaker, 1975

5– (4) Synlophe present.
Parasites of amphibians and reptiles.

6– (9) Gubernaculum present.
Parasites of reptiles.

7– (8) Lateral rays slender and long.
Ray 4 well separated from ray 5 (Fig. 10.122).
Parasites of Chelonia.

Trichoskrjabinia Travassos, 1937

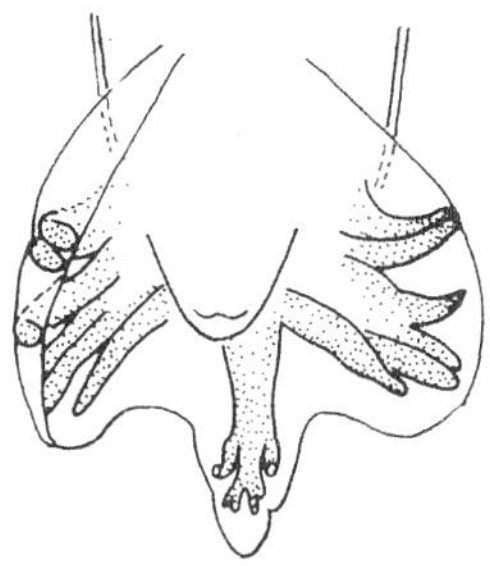

10.121

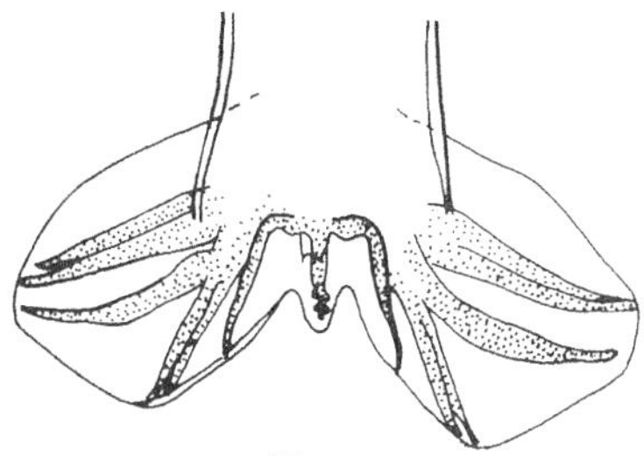

10.122

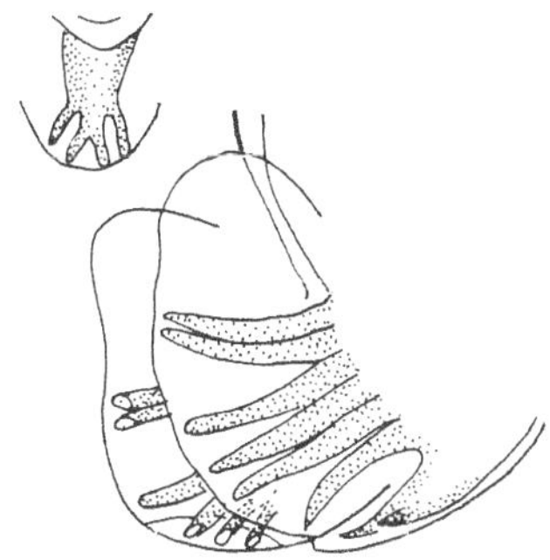

10.123

FIG. 10.121. *Poekilostrongylus puertoricensis* Schmidt & Wittaker, 1975, bursa (after Schmidt & Wittaker, 1975).
FIG. 10.122. *Trichoskrjabinia malayana* (Baylis, 1933), bursa (after Baylis, 1933).
FIG. 10.123. *Typhlopsia kratochvili* Barus & Otero, 1978, bursa, lateral view and dorsal ray (after Barus & Otero, 1978).

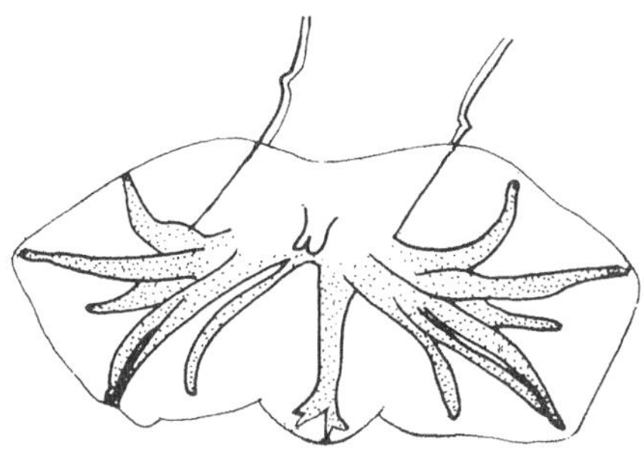

10.124

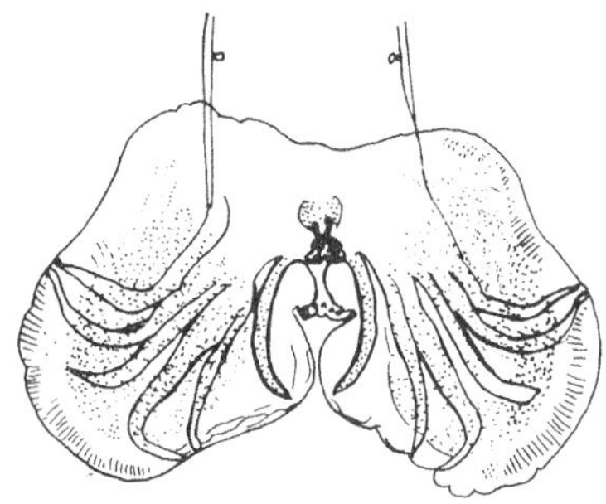

10.125

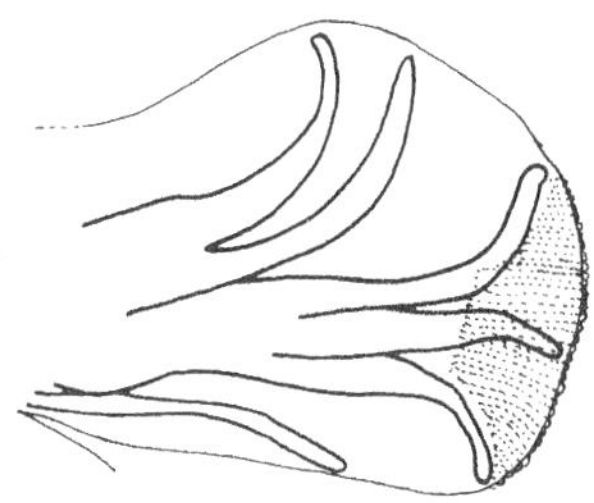

10.126

FIG. 10.124. *Brygoonema oesophagostomoides* (Chabaud, Brygoo & Tcheprakoff, 1964), bursa (after Chabaud, Brygoo & Tcheprakoff, 1964).
FIG. 10.125. *Pithecostrongylus teochii* (Quentin, 1970) n. comb., bursa (after Quentin, 1970).
FIG. 10.126. *Shattuckius shattucki* Sandground, 1938, lateral lobe of bursa (after Sandground, 1938).

8– (7) Lateral rays relatively thick and short.
Rays 4 and 5 close together (Fig. 10.123).
Parasites of Sauria.

Typhlopsia Barus & Otero, 1978

9– (6) Gubernaculum absent (Fig. 10.119).
Parasites of amphibians and rarely reptiles.

Oswaldocruzia Travassos, 1917
[= *Oswaldocruzia* (*Bialata*) Morishita, 1926]

10– (3) Ray 4 arising at same level as rays 5 and 6 (Figs. 10.20; 10.124 to 10.126).
Female tail with slender spine only (Fig. 10.107) or base of spine with tubercles (Figs. 10.127, 10.128).
Parasites of mammals.

10.127

10.128

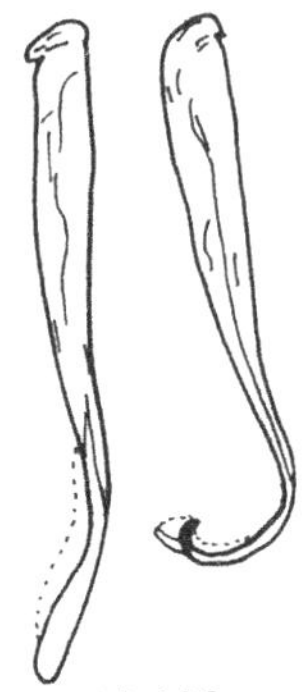

10.129

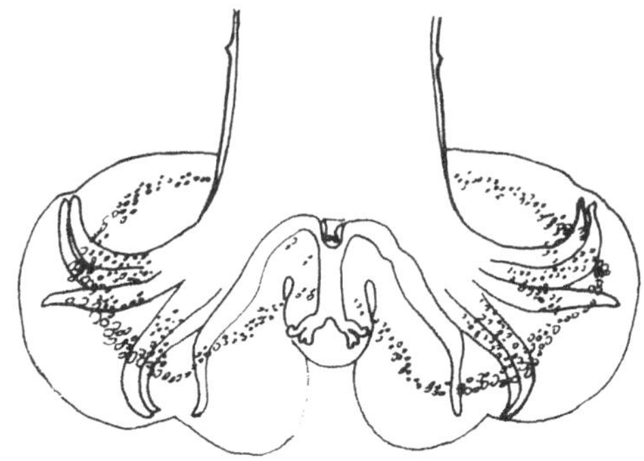

10.130

FIG. 10.127. *Molinostrongylus scotophili* Meszaros, 1973, female, tail end, lateral view (after Meszaros, 1973).

FIG. 10.128. *Pithecostrongylus serratus* (Ow-Yang, 1965) n. comb., female, tail end, lateral view (after Ow-Yang, 1965).

FIG. 10.129. *Angulocirrus orycteropi* Biocca & Le Roux, 1957, spicules (after Biocca & Le Roux, 1957).

FIG. 10.130. *Trichochenia meyeri* (Cameron & Myers, 1960), bursa (after Cameron & Myers, 1960).

11–(12) Spicules dissimilar (Fig. 10.129).
Parasites of Pholidota and Tubulidentata.
Angulocirrus Biocca & Le Roux, 1957

12–(11) Spicules similar.

13–(18) Rays 2 and 3 close together and parallel (Fig. 10.20).

14–(15) Terminal papillae of ray 4 equidistant from those of rays 3 and 5 (Fig. 10.20).
Parasites of fissipeds, neotropical primates, and rarely rodents.
Molineus Cameron, 1923
(= *Microstrongylus* Cameron, 1927;
= *Nematostrongylus* Cameron, 1928; = *Tenuostrongylus* Le Roux, 1933)

15–(14) Terminal papillae of ray 4 closer to that of ray 3 than to that of ray 5 (Fig. 10.130).

16–(17) *Corona radiata* absent.
Location intestinal.
Parasites of oriental and ethiopian Pholidota.
Trichochenia Kou, 1958
(= *Manistrongylus* Baer, 1959; = *Manistrongylus* Cameron & Myers, 1960;
= *Pholidostrongylus* Baer, 1959)

17–(16) *Corona radiata* present (Fig. 10.131).
Location most often peculiar, outside intestine.
Parasites of oriental and australasian rodents.
Hepatojarakus Yeh, 1955

18–(13) Rays 2 and 3 separated and not parallel (Fig. 10.124) (except *Brygoonema ogdeni*).

19–(32) Ovejector straight with branches of vestibule equal (Fig. 10.132).
Tail of female without tubercles (Fig. 10.107).
Parasites of insectivores (Tenrecoidea, Erinacoidea), lemurs and Old World monkeys and rodents.

20–(29) Spicules less than 200 µm long, usually deeply divided (Fig. 10.134).
Bursa with rounded dorsal lobe almost completely fused with lateral lobes.
Parasites of Tenrecoidea of Africa and Madagascar, ethiopian lemurs and monkeys, rarely rodents.

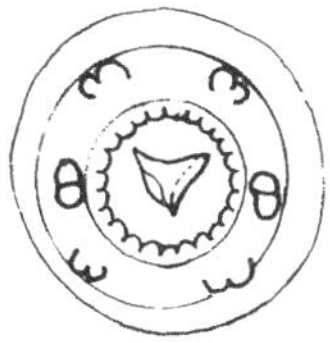
10.131

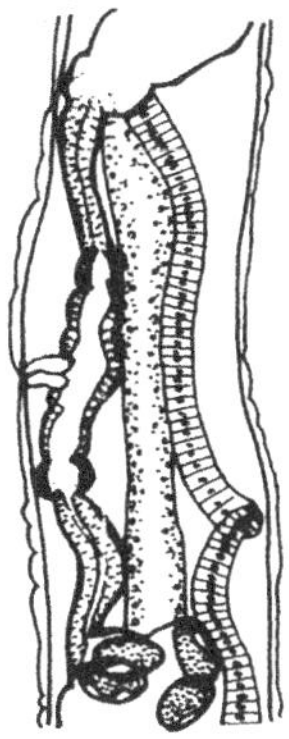
10.132

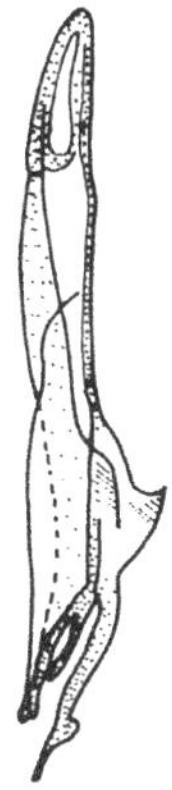
10.133

10.134

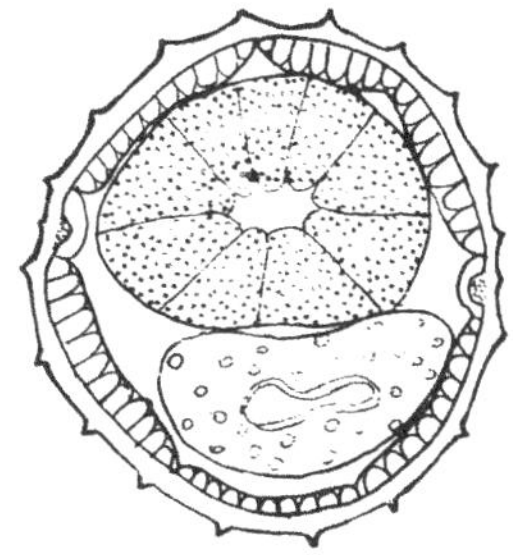
10.135

FIG. 10.131. *Hepatojarakus malayae* Yeh, 1955, cephalic extremity, apical view (after Durette-Desset & Chabaud, 1975).

FIG. 10.132. *Brygoonema petteri* (Durette-Desset, 1973), ovejector, lateral view (after Durette-Desset, 1973).

FIG. 10.133. *Pithecostrongylus* sp., spicule (original).

FIG. 10.134. *Brygoonema oesophagostomoides* (Chabaud, Brygoo & Tcheprakoff, 1964), spicule (after Chabaud, Brygoo & Tcheprakoff, 1964).

FIG. 10.135. *Brygoonema rogierae* (Durette-Desset, 1973), synlophe (after Durette-Desset, 1973).

21–(26) Rays 5 and 6 close together and parallel (Fig. 10.124).
22–(25) Lateral alae absent.
Spicules thick and deeply divided (Fig. 10.134).
Parasites of Tenrecoidea of Madagascar and Africa.
23–(24) Synlophe with all ridges perpendicular to body (Figs. 10.110, 10.135).
Parasites of Tenrecoidea of Madagascar.
Brygoonema Durette-Desset & Chabaud, 1981
24–(23) Synlophe with at least some ridges not oriented perpendicular to body (Fig. 10.136).
Parasites of ethiopian Tenrecoidea.
Hugotnema Durette-Desset & Chabaud, 1981
25–(22) Lateral alae present.
Spicules thin with filamentous points (Fig. 10.137).
Parasites of ethiopian Sciuridae.
Dollfusstrongylus Quentin, 1970
26–(21) Rays 5 and 6 not parallel and not close together (Figs. 10.125, 10.126).
27–(28) Ray 4 longer than ray 5.
Rays 2 to 6 divergent (Fig. 10.126).
Lateral alae absent.
Spicules with reduced terminal points (Fig. 10.138).
Parasites of *Solenodon* (Insectivora) of Santo Domingo.
Shattuckius Sandground, 1938

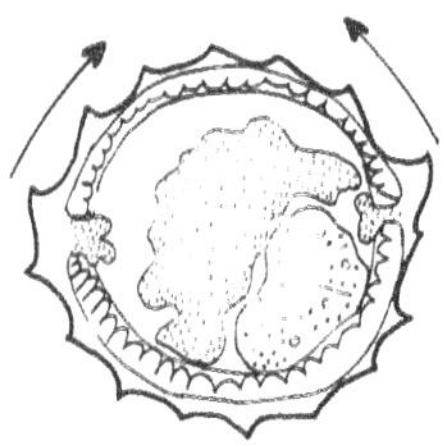

10.136

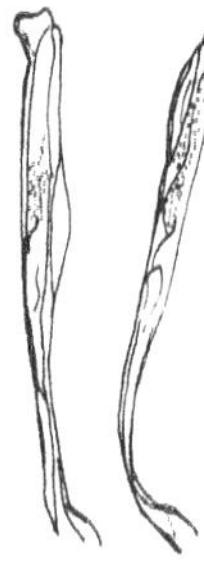

10.137

10.138

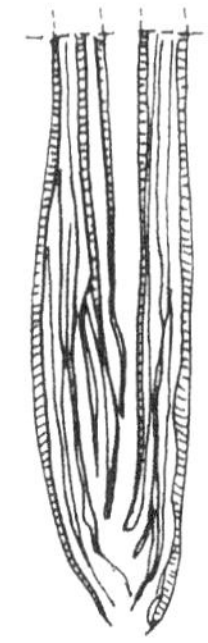

10.139

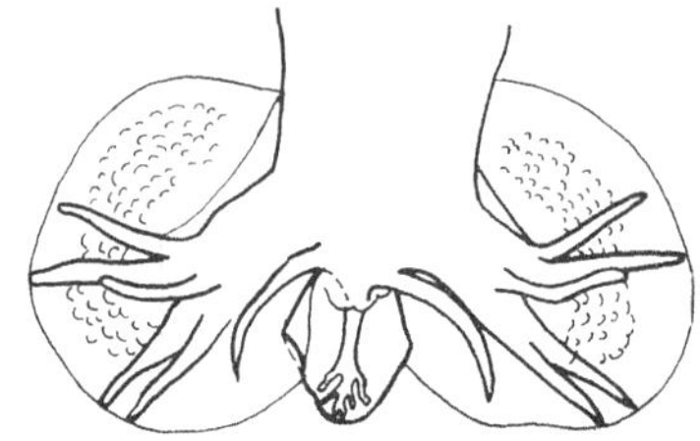

10.140

FIG. 10.136. *Hugotnema amberti* (Hugot, 1969), synlophe (after Hugot, 1969).

FIG. 10.137. *Dollfusstrongylus sciurei* Quentin, 1970, spicules (after Quentin, 1970).

FIG. 10.138. *Shattuckius shattucki* Sandground, 1938, spicule (after Sandground, 1938).

FIG. 10.139. *Nochtia (Nochtia) atelerixi* Quentin & Durette-Desset, 1983, ends of spicules (after Quentin & Durette-Desset, 1983).

FIG. 10.140. *Nochtia (Nochtia) nochti* Travassos & Vogelsang, 1929, bursa (after Travassos & Vogelsang, 1929).

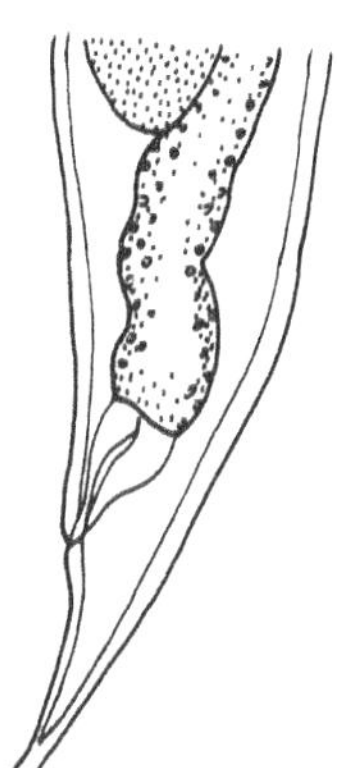

10.141

10.142

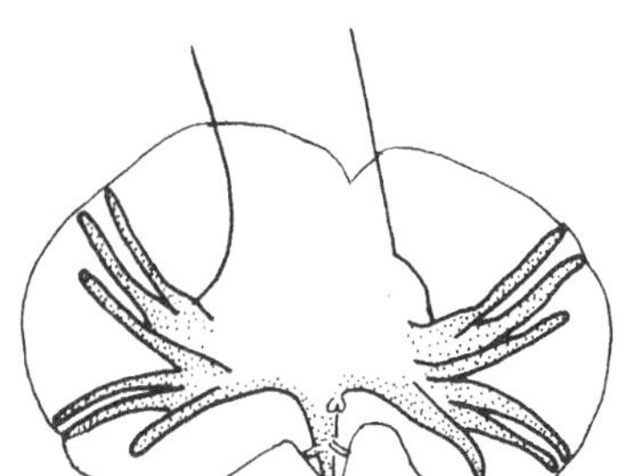

10.143

10.144

FIG. 10.141. *Nochtia (Nochtia) nochti* Travassos & Vogelsang, 1929, female, tail end, lateral view (after Travassos & Vogelsang, 1929).

FIG. 10.142. *Nochtia (Nochtia) nochti* Travassos & Vogelsang, 1929, spicules (after Travassos & Vogelsang, 1929).

FIG. 10.143. *Nochtia (Tenrecola) pilosus* (Chabaud, Brygoo & Tcheprakoff, 1964), bursa (after Chabaud, Brygoo & Tcheprakoff, 1964).

FIG. 10.144. *Nochtia (Tenrecola) lerouxi* (Chabaud, Brygoo & Tcheprakoff, 1964), end of spicule (after Chabaud, Brygoo & Tcheprakoff, 1964).

28–(27) Ray 4 shorter than ray 5 (Fig. 10.125).
Ends of rays 2 and 3 and rays 5 and 6 converging.
Lateral alae present.
Spicules with well-developed terminal points (Fig. 10.133).
Parasites of ethiopian lemurs and oriental insectivores, primates and rodents.
Pithecostrongylus Lubimov, 1930
(= *Hooperstrongylus* Lie Kian Joe & Ow Yang, 1963)

29–(20) Spicules more than 350 µm long, divided only at distal end (Fig. 10.139).
Bursa with triangular dorsal lobe well differentiated from lateral lobes (Fig. 10.140).
Parasites of Tenrecoidea of Madagascar, ethiopian Erinacoidea and oriental monkeys.
Nochtia Travassos & Vogelsang, 1929

30–(31) Tail of female without caudal spine (Fig. 10.141).
Ray 4 less than half as long as ray 3 (Fig. 10.140).
Spicules with one point (Fig. 10.142) or many points (Fig. 10.139).
Parasites of oriental primates and ethiopian Erinacoidea.
Nochtia (*Nochtia*) Travassos & Vogelsang, 1929

31–(30) Tail of female with caudal spine (Fig. 10.107).
Ray 4 more than half as long as ray 3 (Fig. 10.143).
Spicules with 2 points (Fig. 10.144).
Parasites of Tenrecoidea of Madagascar.
Nochtia (*Tenrecola*) Durette-Desset & Chabaud, 1981

32–(19) Ovejector kidney-shaped (Fig. 10.145) or with branches of vestibule unequal (Fig. 10.146).
Tail of female with 3 to 6 large tubercles (Fig. 10.127).
Parasites of bats and Tupaiidae.

33–(34) Branches of vestibule unequal (Fig. 10.146); ovejector not kidney-shaped.
Rays 3 and 4 close together throughout their length (Fig. 10.21).
Lateral alae well developed (Figs. 10.111, 10.147).
Parasites of Old World bats.
Molinostrongylus Skarbilovitch, 1934
(= *Chiroptostrongylus* Skarbilovitch *in* Orloff, 1933)

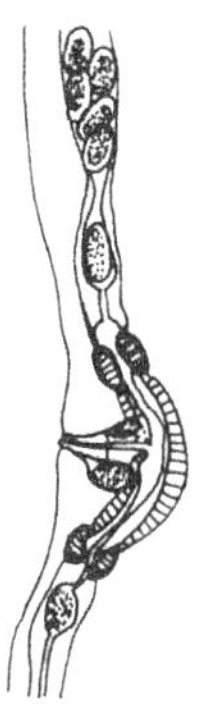
10.145

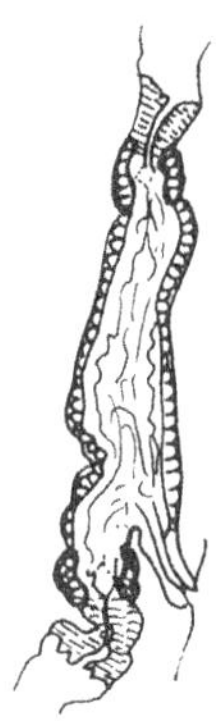
10.146

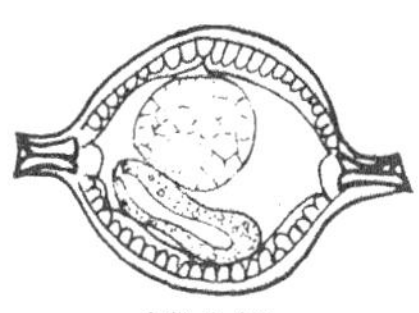
10.147

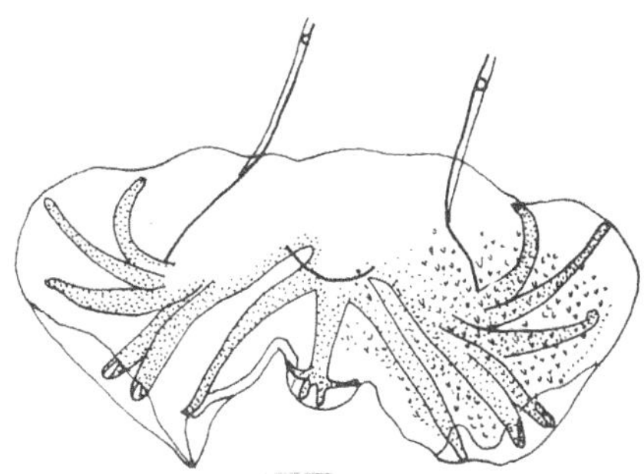
10.148

FIG. 10.145. *Nycteridostrongylus (Petiellus) petersi* Durette-Desset & Chabaud, 1975, ovejector, lateral view (after Durette-Desset & Chabaud, 1975).

FIG. 10.146. *Molinostrongylus aelleni* Durette-Desset & Chabaud, 1975, ovejector, lateral view (after Durette-Desset & Chabaud, 1975).

FIG. 10.147. *Molinostrongylus skrjabini* Skarbilovitch, 1934, synlophe (after Durette-Desset & Chabaud, 1975).

FIG. 10.148. *Tupaiostrongylus major* Durette-Desset, Palmieri, Purnomo & Cassone, 1981, bursa (after Durette-Desset, Palmieri, Purnomo & Cassone, 1981).

10.149

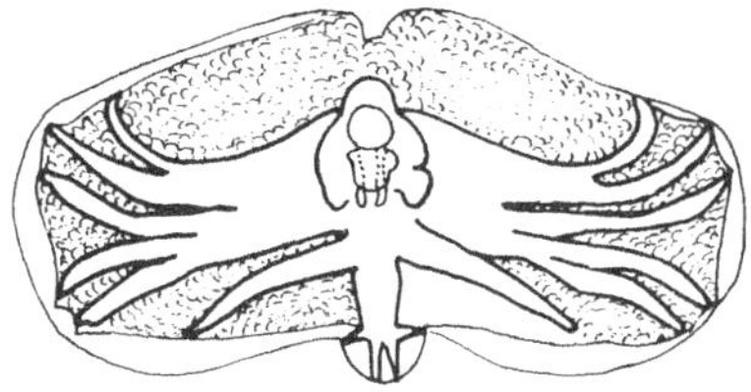

10.150

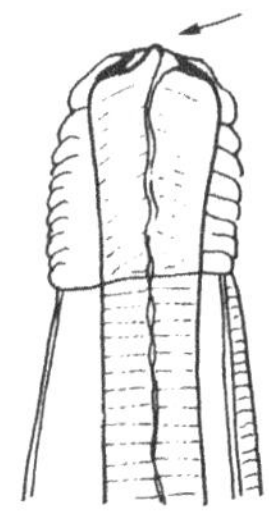

10.151

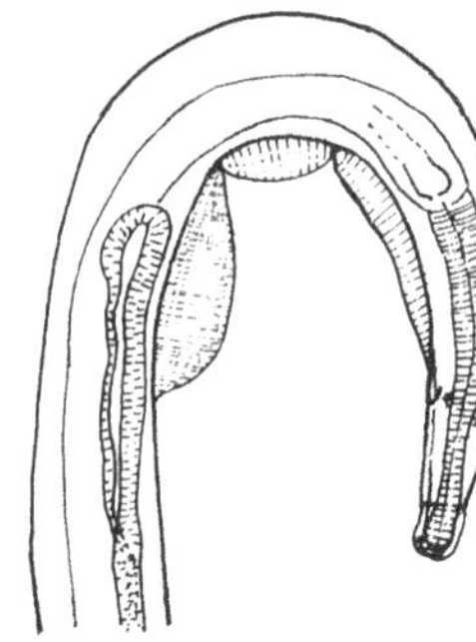

10.152

FIG. 10.149. *Nycteridostrongylus (P.) petersi* Durette-Desset & Chabaud, 1975, synlophe in the anterior part of the body (after Durette-Desset & Chabaud, 1975).

FIG. 10.150. *Nycteridostrongylus (P.) petersi* Durette-Desset & Chabaud, 1975, bursa (after Durette-Desset & Chabaud, 1975).

FIG. 10.151. *Nycteridostrongylus (N.) uncicollis* Baylis, 1930, cephalic extremity, lateral view (after Baylis, 1930).

FIG. 10.152. *Nycteridostrongylus (N.) uncicollis* Baylis, 1930, anterior extremity with cervical organ of fixation, lateral view (after Baylis, 1930).

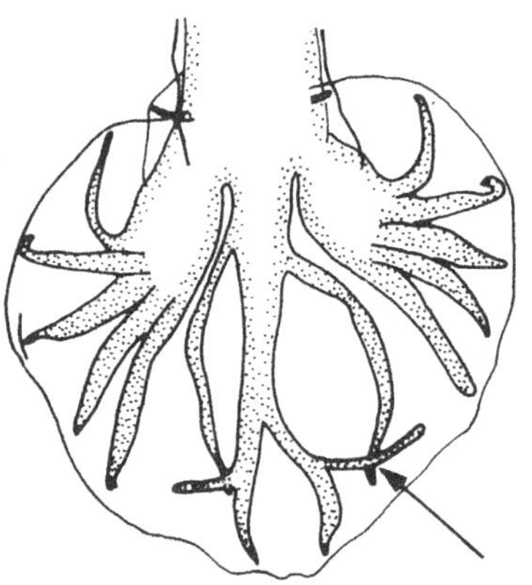

10.153

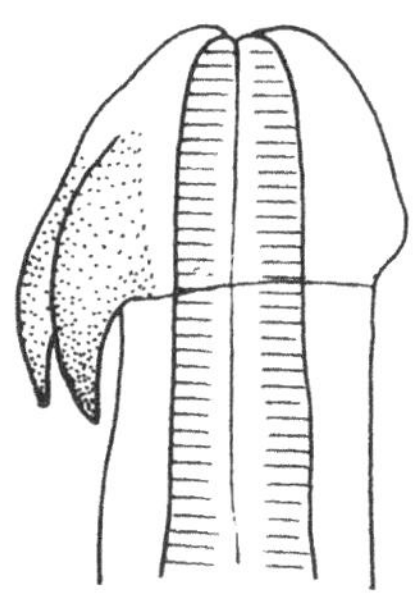

10.154

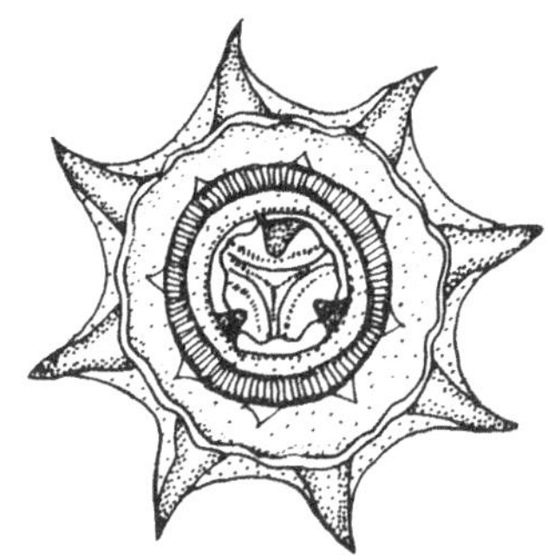

10.155

FIG. 10.153. *Bradypostrongylus inflatus* Molin, 1861, bursa (after Travassos, 1928).

FIG. 10.154. *Biacantha desmoda* Wolfgang, 1954, cephalic extremity, lateral view (after Wolfgang, 1954).

FIG. 10.155. *Histiostrongylus coronatus* Molin, 1861, cephalic extremity, apical view (after Vigueras, 1941).

34–(33) Branches of vestibule equal; ovejector kidney-shaped (Fig. 10.145).
Rays 3 and 4 not close together (Fig. 10.148).
Lateral alae reduced or absent.
Parasites of Tupaiidae and Old World bats.

35–(38) Synlophe with oblique cuticular ridges only on anterior portion of body (Fig. 10.149).
Ray 4 almost as long as ray 3 (Fig. 10.150).
Parasites of oriental and australasian bats and Tupaiidae.
Nycteridostrongylus Baylis, 1930

36–(37) Dorsal oesophageal tooth present (Fig. 10.151).
Cervical organ of fixation present (Fig. 10.152).
Parasites of oriental and australasian bats.
Nycteridostrongylus (*Nycteridostrongylus*) Baylis, 1930

37–(36) Oesophageal tooth absent.
Cervical fixation organ absent.
Parasites of Tupaiidae.
Nycteridostrongylus (*Petiellus*) Durette-Desset & Chabaud, 1975

38–(35) Synlophe with longitudinal cuticular ridges over entire length of body.
Ray 4 much shorter than ray 3 (Fig. 10.148).
Parasites of Tupaiidae.
Tupaiostrongylus Dunn, 1963

Subfamily **Anoplostrongylinae** Chandler, 1938

Key to genera

1– (2) Dorsal ray very long.
Extremity of ray 8 crossing ray 9 (Fig. 10.153).
Parasites of Bradypodidae and Myrmecophagidae.
Bradypostrongylus Price, 1928
(= *Pintoia* Travassos, 1928)

2– (1) Dorsal ray not very long.
Extremity of ray 8 not crossing ray 9.

3–(38) Head not armed or with 2 to 8 simple spines (Figs. 10.154, 10.155).
Cervical region smooth.
Female tail with simple spine or with spines and tubercles.
Parasites of Xenarthra and New World bats.

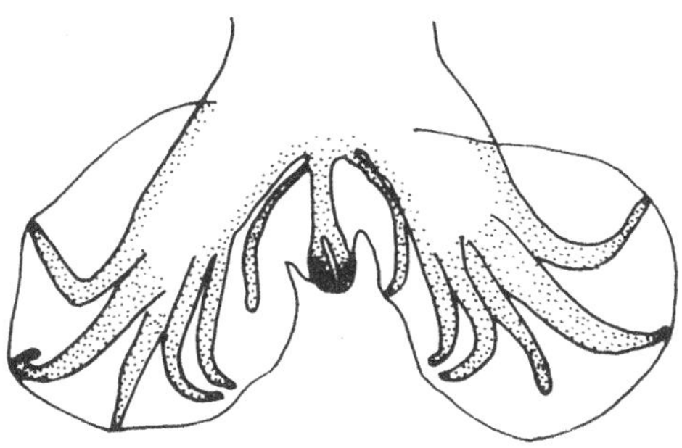

10.156

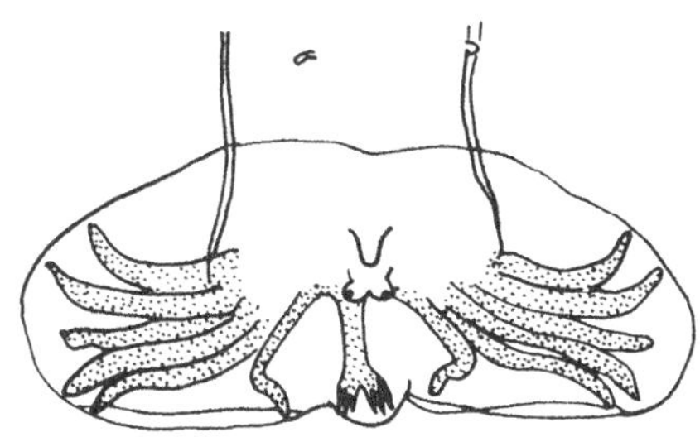

10.157

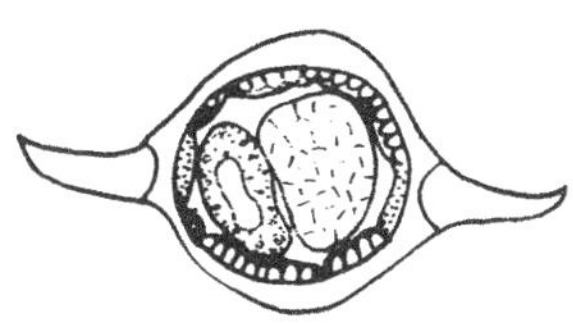

10.158

FIG. 10.156. *Moennigia michelae* Durette-Desset, Chabaud & Cassone, 1977, bursa (after Durette-Desset, Chabaud & Cassone, 1977).

FIG. 10.157. *Graphidiops costalimai* Lent & Freitas, 1938, bursa (after Durette-Desset, Chabaud & Cassone, 1977).

FIG. 10.158. *Delicata perronae* Durette-Desset, Chabaud & Cassone, 1977, synlophe (after Durette-Desset, Chabaud & Cassone, 1977).

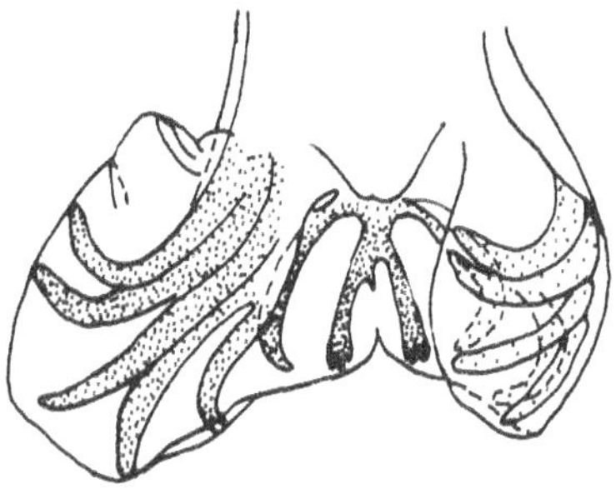

10.159

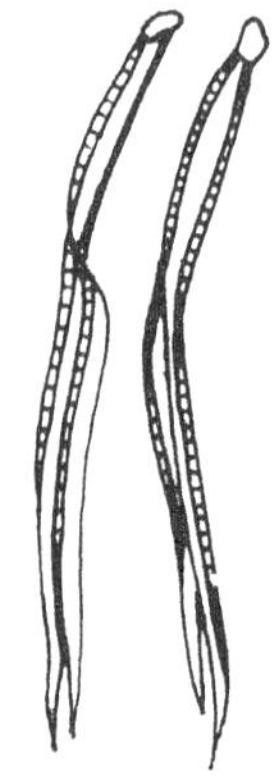

10.160

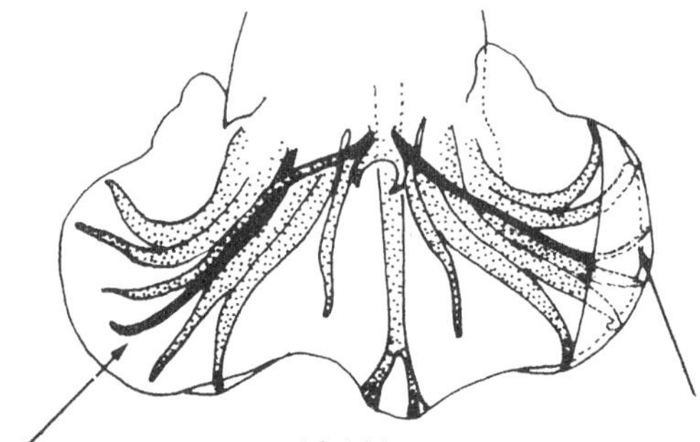

10.161

FIG. 10.159. *Caenostrongylus splendidus* Lent & Freitas, 1938, bursa (after Lent & Freitas, 1938).

FIG. 10.160. *Caenostrongylus splendidus* Lent & Freitas, 1938, spicules (after Lent & Freitas, 1938).

FIG. 10.161. *Paragraphidium pseudosexradiatum* Freitas & Mendonça, 1959, bursa (after Freitas & Mendonça, 1959).

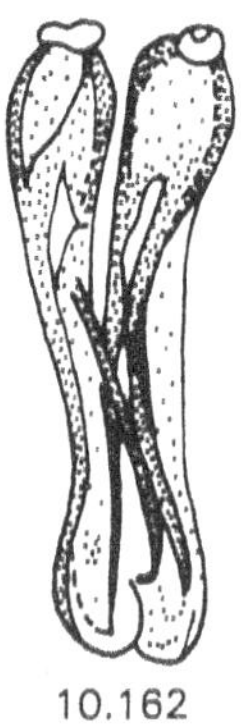

10.162

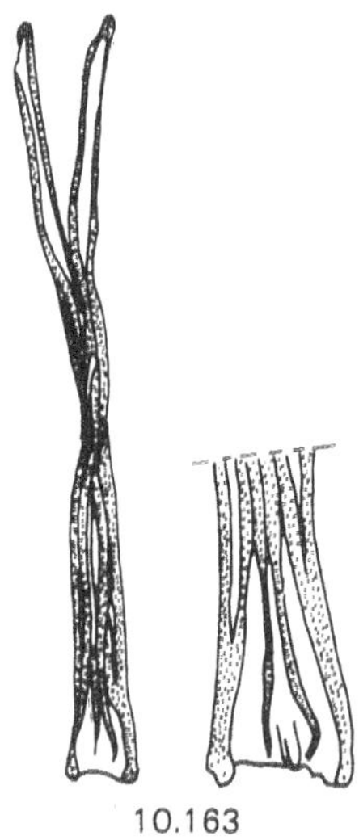

10.163

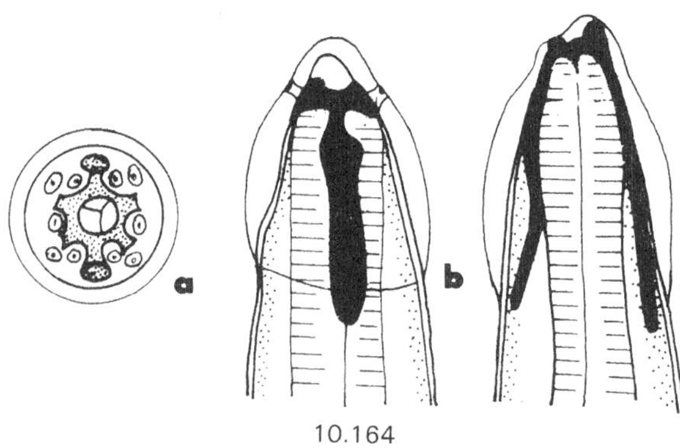

10.164

FIG. 10.162. *Maciela chagasi* Travassos, 1935, spicules (after Travassos, 1935).

FIG. 10.163. *Graphidiops costalimai* Lent & Freitas, 1938, spicules, distal extremities on right (after Durette-Desset, Chabaud & Cassone, 1977).

FIG. 10.164. *Fontesia fontesi* Travassos, 1928, cephalic extremity, (a) apical view, (b) median view, (c) lateral view (after Durette-Desset, Chabaud & Cassone, 1977).

4–(23) Female tail without tubercles.
Rays 2 and 3 usually larger than rays 5 and 6.
Extremities of rays 5 and 6 generally close together (Fig. 10.156).
Parasites of Xenarthra.

5–(22) Cephalic vesicle present.
Body not markedly slender anteriorly.

6–(19) Didelphic, both branches of female reproductive tract well developed.

7–(18) Ray 6 as long as or smaller than ray 5 (Figs. 10.156, 10.157).
Vulva far from anus.
Amphidelphic.

8–(17) Post-anal papillae (pair 7) small, papilliform.

9–(16) Cephalic extremity without cuticular ring, lacking cuticular plates.

10–(11) Tail of female rounded with caudal spine.
Lateral alae large (Fig. 10.158).
Parasites of intestine of Xenarthra.

Delicata Travassos, 1935

11–(10) Tail of female pointed, without terminal spine.
Lateral alae absent.
Parasites of stomach of Xenarthra.

12–(13) Dorsal ray divided in anterior half (Fig. 10.159).
Spicules short and slender (less than 200 μm) (Fig. 10.160).
Parasites of Myrmecophagidae.

Caenostrongylus Lent & Freitas, 1938

13–(12) Dorsal ray divided in posterior third (Fig. 10.157).
Spicules short and thick or long.

14–(15) Spicules short (less than 200 μm) and thick (Fig. 10.162).
Parasites of Dasypodidae.

Maciela Travassos, 1935

15–(14) Spicules long (more than 300 μm) and slender (Fig. 10.163).
Parasites of Myrmecophagidae and Bradypodidae.

Graphidiops Lent & Freitas, 1938

16– (9) Head with cuticular ring and 2 median plates (ventral and dorsal) (Fig. 10.164).
Parasites of Myrmecophagidae.

Fontesia Travassos, 1928

17– (8) Post-anal papillae (pair 7) as long as ray 4 (Fig. 10.161).
Parasites of Myrmecophagidae.

Paragraphidium Freitas & Mendonça, 1959

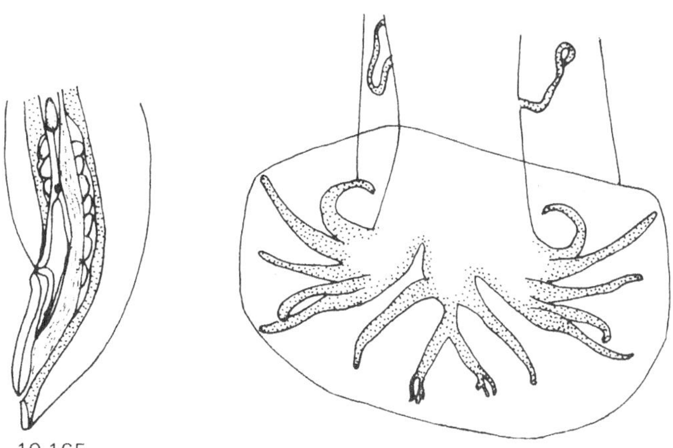

10.165 10.166

FIG. 10.165. *Trichohelix tuberculata* (Parona & Stossich, 1901), female, tail end, lateral view (after Ortlepp, 1922).
FIG. 10.166. *Trichohelix tuberculata* (Parona & Stossich, 1901), bursa (after Ortlepp, 1922).

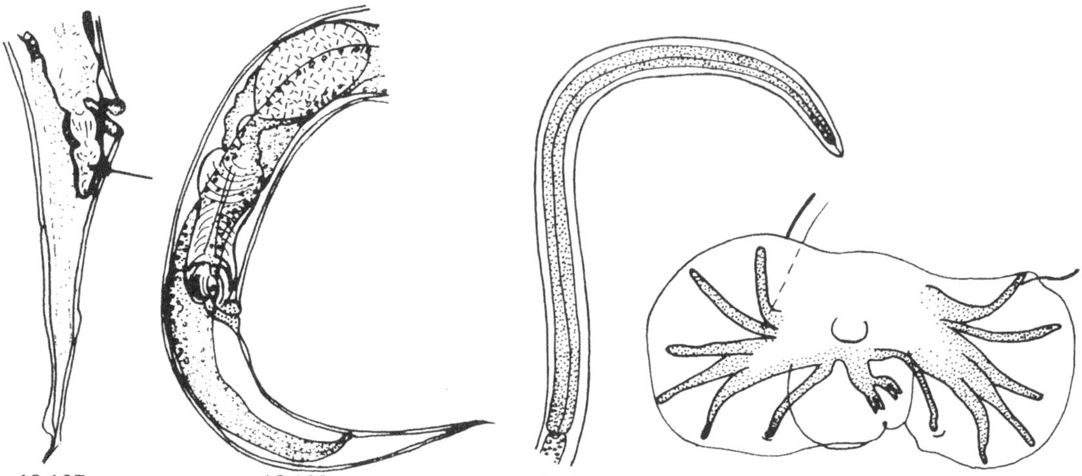

10.167 10.168 10.169 10.170

FIG. 10.167. *Moennigia lenteigneae* Durette-Desset, Chabaud & Cassone, 1977, female, posterior end, lateral view (after Durette-Desset, Chabaud & Cassone, 1977).

FIG. 10.168. *Trifurcata minuscula* Travassos, 1915, female, posterior end, lateral view (after Travassos, 1915).

FIG. 10.169. *Filicapitis longicollis* Travassos, 1949, anterior end (after Travassos, 1949).

FIG. 10.170. *Anoplostrongylus* sp., bursa (original).

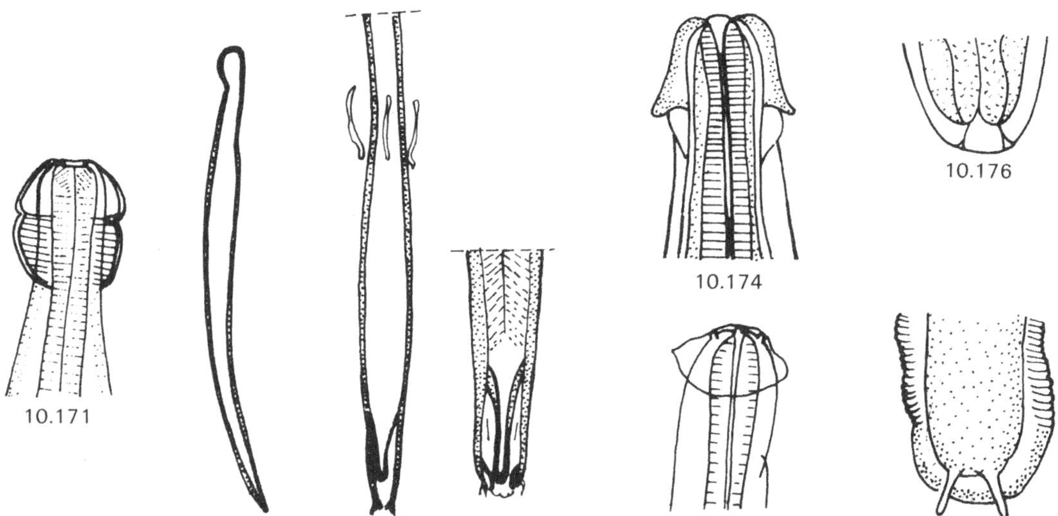

10.171 10.172 10.173 10.174 10.175 10.176 10.177

FIG. 10.171. *Tricholeiperia poeyi* Barus & Valle, 1967, cephalic extremity (after Barus & Valle, 1967).

FIG. 10.172. *Anoplostrongylus* sp., spicule (original).

FIG. 10.173. *Tricholeiperia leiperi* Travassos, 1935, ends of spicules (after Travassos, 1937).

FIG. 10.174. *Torrestrongylus torrei* Vigueras, 1935, cephalic extremity, median view (after Vigueras, 1935).

FIG. 10.175. *Cheiropteronema globocephala* Sandground, 1929, cephalic extremity, lateral view (after Durette-Desset & Tcheprakoff, 1977).

FIG. 10.176. *Cheiropteronema globocephala* Sandground, 1929, female tail, ventral view (after Durette-Desset & Tcheprakoff, 1977).

FIG. 10.177. *Bidigiticauda vivipara* Chitwood, 1938, female tail, ventral view (after Chitwood, 1938).

18– (7) Ray 6 longer than ray 5 (Fig. 10.166).
Vulva close to anus (Fig. 10.165).
Prodelphic.
Parasites of Dasypodidae (also reported in a carnivore).
Trichohelix Ortlepp, 1922

19– (6) Monodelphic or one branch of female reproductive tract atrophied.

20–(21) Didelphic, posterior branch of reproductive tract atrophied (Fig. 10.167).
Parasites of Dasypodidae and Myrmecophagidae.
Moennigia Travassos, 1935
(= *Adolpholutzia* Travassos, 1935; = *Dasypostrongylus* Travassos, 1935; = *Pintonema* Travassos, 1935; = *Pulchrostrongylus* Travassos, 1935)

21–(20) Monodelphic (Fig. 10.168).
Parasites of Myrmecophagidae.
Trifurcata Schulz, 1926

22– (5) Cephalic vesicle absent.
Body markedly slender anteriorly (Fig. 10.169).
Parasites of Myrmecophagidae.
Filicapitis Travassos, 1949

23– (4) Female tail with tubercles (except *Cheiropteronema*).
Rays 2 and 3 generally smaller than, or as long as, rays 5 and 6.
Extremities of rays 5 and 6 widely separated (Fig. 10.170).
Parasites of New World bats.

24–(33) Cephalic extremity lacking cuticular spines.

25–(30) Cephalic extremity with cephalic vesicle in two portions (Fig. 10.171).

26–(29) Posterior portion of cephalic vesicle as long as, or longer than, anterior portion (Fig. 10.171).

27–(28) Spicules not divided at distal end (Fig. 10.172).
Branches of vestibule equal.
Parasites of Mollossidae.
Anoplostrongylus Boulenger, 1926

28–(27) Spicules divided at distal end (Fig. 10.173).
Branches of vestibule unequal.
Parasites of insectivorous Phyllostomidae, Natalidae, Molossidae and Noctilionidae.
Tricholeiperia Travassos, 1935

29–(26) Posterior portion of cephalic vesicle shorter than anterior portion (Fig. 10.174).
Parasites of insectivorous Phyllostomidae.
Torrestrongylus Vigueras, 1935

30–(25) Cephalic extremity with simple cephalic vesicle (Fig. 10.175).
Parasites of nectar feeding Phyllostomidae.

31–(32) Tail of female lacking caudal appendages (Fig. 10.176).
Cheiropteronema Sandground, 1929

32–(31) Tail of female with two caudal appendages (Fig. 10.177).
Bidigiticauda Chitwood, 1938
(= *Didactyluris* Freitas & Dobbin, 1960)

33–(24) Cephalic extremity with cuticular spines (Figs. 10.154, 10.155).

34–(35) Two ventral cephalic spines present (Fig. 10.154).
Parasites of blood feeding Phyllostomidae and Natalidae.
Biacantha Wolfgang, 1954

35–(34) Eight cephalic spines symmetrically disposed around cephalic extremity (Fig. 10.155).

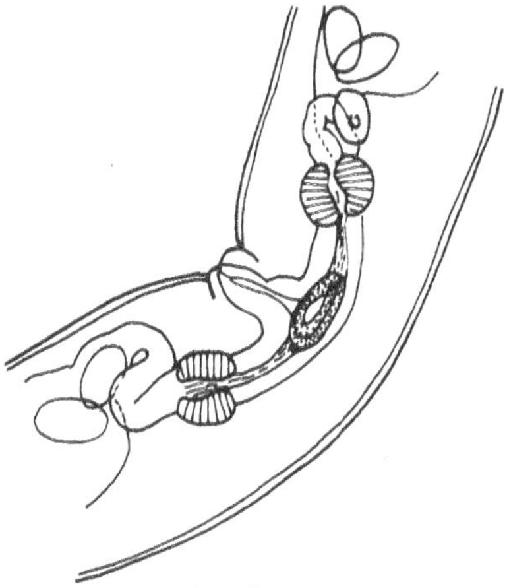

10.178

10.180

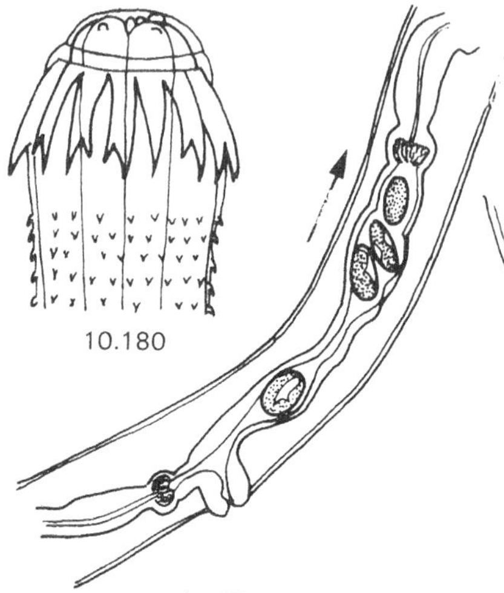

10.179

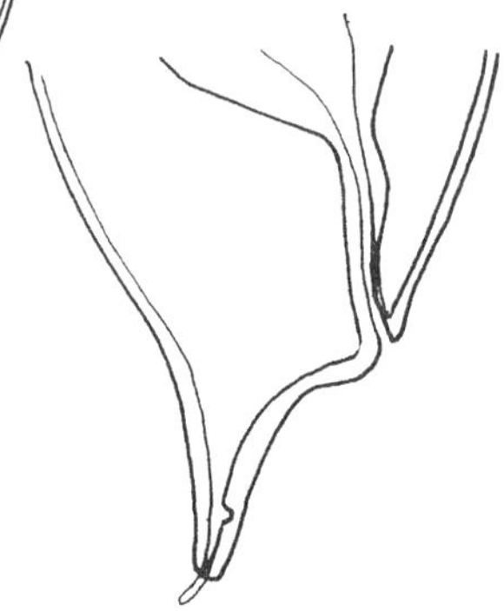

10.181

FIG. 10.178. *Parahistiostrongylus octacanthus* Lent & Freitas, 1940, ovejector, lateral view (after Lent & Freitas, 1940).
FIG. 10.179. *Histiostrongylus coronatus*, Molin, 1861, ovejector, lateral view (after Vigueras, 1941).
FIG. 10.180. *Spinostrongylus spinosus* Boulenger, 1926, cephalic extremity (after Boulenger, 1926).
FIG. 10.181. *Spinostrongylus spinosus* Boulenger, 1926, female tail end, lateral view (after Boulenger, 1926).

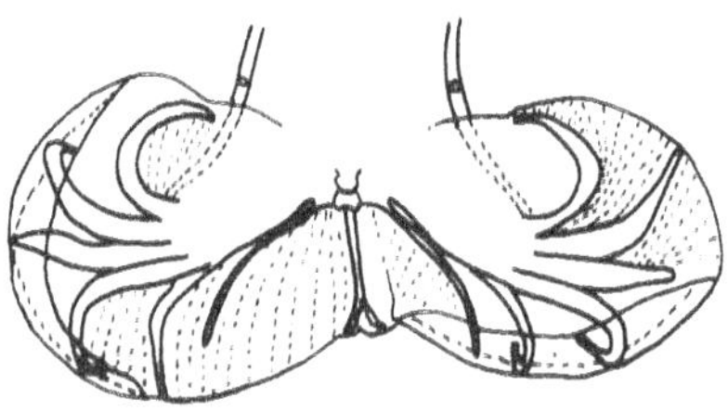

10.182

10.183

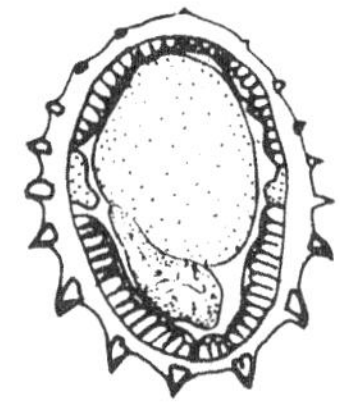

10.184

10.185

FIG. 10.182. *Lamanema chavezi* Becklund, 1963, bursa (after Becklund, 1963).
FIG. 10.183. *Rauschia triangularis* (Boughton, 1932), synlophe (after Durette-Desset, 1979).
FIG. 10.184. *Nematodiroides zembrae* (Bernard, 1965), synlophe (after Durette-Desset, 1979).
FIG. 10.185. *Nematodirus tortuosus* Tucker, 1942, synlophe (after Durette-Desset, 1979).

36–(37) Branches of vestibule equal (Fig. 10.178).
Parasites of insectivorous Phyllostomatidae.
Parahistiostrongylus Vigueras, 1941

37–(36) Anterior branch of vestibule longer than posterior branch (Fig. 10.179).
Parasites of insectivorous and frugivorous Vespertilionidae.
Histiostrongylus Molin, 1861

38– (3) Cephalic extremity with collar of 16 simple or bifurcated hooks.
Cervical region with spines (Fig. 10.180).
Female tail short and sharply pointed, without tubercles (Fig. 10.181).
Parasites of Old World bats.

Spinostrongylus Travassos, 1935

Subfamily **Nematodirinae** Skrjabin & Orloff, 1934

Key to genera

1– (2) Dorsal ray not divided at its base (Fig. 10.182).
Parasites of *Lama*.

Lamanema Becklund, 1963

2– (1) Dorsal ray divided at its base into two projections (Fig. 10.114).

3– (8) Synlophe with ridges not oriented perpendicular to body.
Parasites of lagomorphs.

4– (7) Didelphic.
Coronata radiata present.
Parasites of Leporidae, rarely Ochotonidae.

5– (6) Both branches of reproductive tract functional.
Ventral ridges oriented ventrolaterally (Fig. 10.183).

Rauschia Durette-Desset, 1979

6– (5) Posterior branch of reproductive tract non-functional.
Ventral ridges oriented lateroventrally (Fig. 10.184).

Nematodiroides (Bernard, 1965) Bernard, 1967

7– (4) Monodelphic.
Corona radiata absent.
Parasites of Ochotonidae.

Murielus Dikmans, 1939

8– (3) Synlophe with ridges oriented perpendicular to body (Fig. 10.185).
Parasites of ruminants and rodents.

9–(10) Both branches of reproductive tract functional.

Nematodirus Ransom, 1907

10– (9) Anterior branch of reproductive tract non-functional.

Nematodirella Yorke & Maplestone, 1926

Family ***MACKERRASTRONGYLIDAE*** (Inglis, 1968 subfam.) Durette-Desset & Chabaud, 1981

Key to subfamilies

1– (2) Buccal capsule (Fig. 10.25) or *corona radiata* (Fig. 10.186) present.
Extra-dorsal rays usually absent.
Parasites of monotremes.

Tasmanematinae

2– (1) Buccal capsule and *corona radiata* absent.
Extra-dorsal rays usually present (Fig. 10.187).
Parasites of Dasyuroidea and Perameloidea.

Mackerrastrongylinae

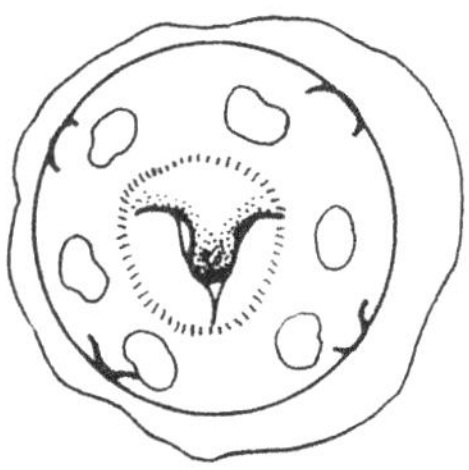

10.186

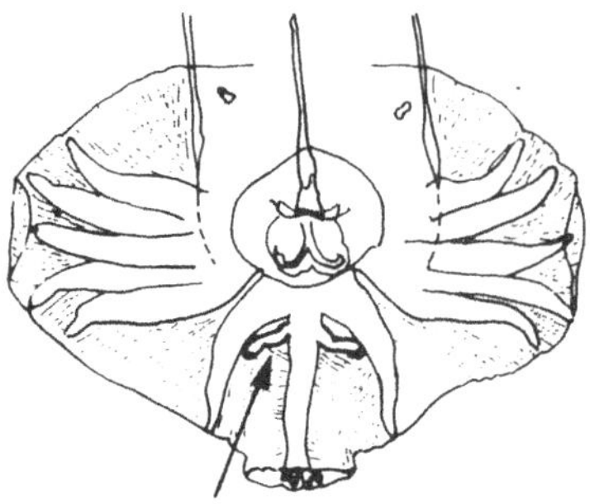

10.187

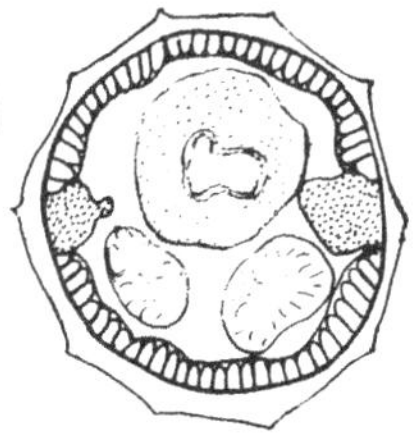

10.188

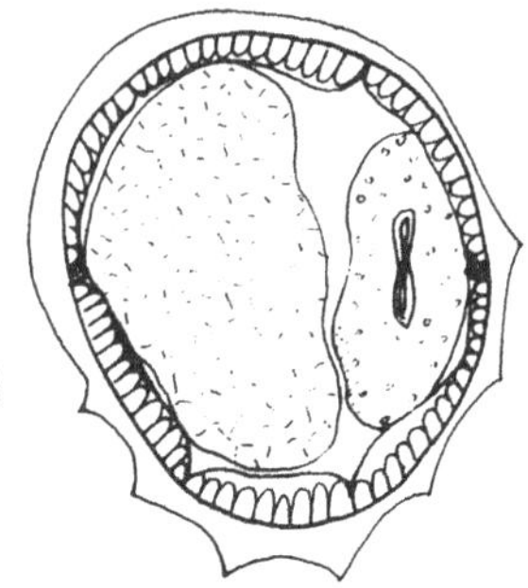

10.189

FIG. 10.186. *Tasmanema mundayi* (Mawson, 1973), cephalic extremity, apical view (after Durette-Desset & Cassone, 1983).

FIG. 10.187. *Mackerrastrongylus isoodon* Durette-Desset & Cassone, ·1981, bursa (after Durette-Desset & Cassone, 1981).

FIG. 10.188. *Mackerrastrongylus isoodon* Durette-Desset & Cassone, 1981, synlophe (after Durette-Desset & Cassone, 1981).

FIG. 10.189. *Sprattellus woolleyae* Durette-Desset & Cassone, 1981, synlophe (after Durette-Desset & Cassone, 1981).

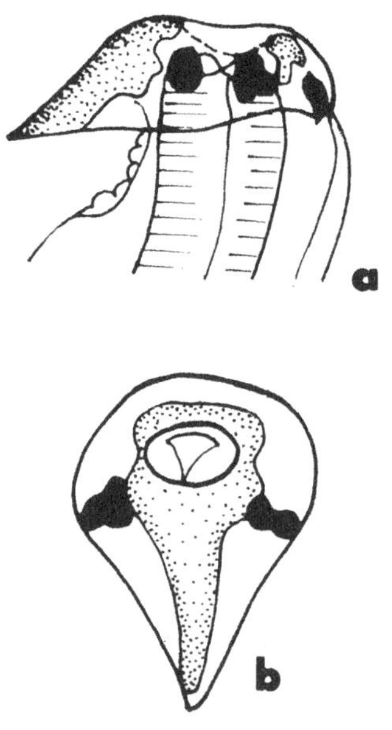

10.190

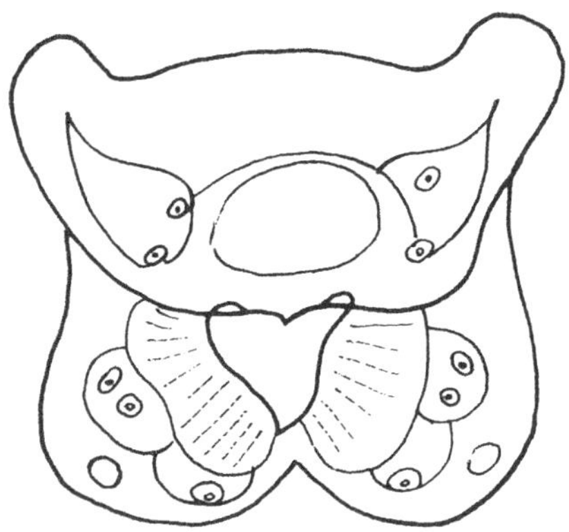

10.191

FIG. 10.190. *Asymmetracantha tasmaniensis* Mawson, 1960, cephalic extremity, (a) lateral view, (b) apical view (after Durette-Desset & Beveridge, 1982).

FIG. 10.191. *Tetrabothriostrongylus mackerrasae* Mawson, 1960, cephalic extremity, apical view (after Durette-Desset, 1979).

Subfamily **Tasmanematinae** Durette-Desset & Chabaud, 1981

Key to genera

1– (4) Cervical groove absent.
Caudal spine present on female tail.
Branches of vestibule equal.
Parasites of monotremes in Australia.

2– (3) *Corona radiata* present (Fig. 10.186).
Ray 4 shorter than rays 3 and 5.
Tasmanema Durette-Desset & Cassone, 1983

3– (2) *Corona radiata* absent (Fig. 10.25).
Ray 4 as long as or longer than rays 3 and 5.
Tachynema Durette-Desset & Cassone, 1983

4– (1) Cervical groove present.
Caudal spine absent on female tail.
Branches of vestibule unequal.
Parasites of monotremes in New Guinea.
Zaglonema Durette-Desset & Beveridge, 1981

Subfamily **Mackerrastrongylinae** Inglis, 1968

Key to genera

1– (4) Cephalic extremity lacking holdfast apparatus.

2– (3) Dorsal oesophageal tooth present.
Cervical groove present.
Synlophe with ridges oriented perpendicular to body and present over entire body surface (Fig. 10.188).
Parasites of Australian Perameloidea.
Mackerrastrongylus Mawson, 1960

3– (2) Dorsal oesophageal tooth absent.
Cervical groove absent.
Synlophe with ridges oriented from right to left; dorsal side lacking ridges (Fig. 10.189).
Parasites of Australian Dasyuroidea.
Sprattellus Durette-Desset & Cassone, 1981

4– (1) Cephalic extremity with holdfast apparatus.

5– (6) Holdfast apparatus in form of large ventral hook (Fig. 10.190).
Parasites of Australian Perameloidea.
Asymmetracantha Mawson, 1960

6– (5) Holdfast apparatus in form of 4 plate-like thickenings (Fig. 10.191).
Parasites of Australian Dasyuroidea.
Tetrabothriostrongylus Mawson, 1960

Family ***NICOLLINIDAE*** (Skrjabin & Schulz, 1937, tribe) Durette-Desset & Chabaud, 1981

Key to genera

1– (4) Dorsal oesophageal tooth present.
Didelphic.
Spicules relatively long and slender (Fig. 10.192).
Parasites of amphibians and monotremes.

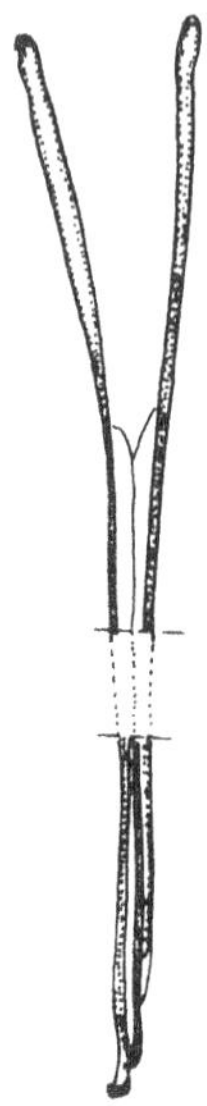

10.192

FIG. 10.192. *Nicollina cameroni* Thomas, 1959, spicules (after Durette-Desset & Cassone, 1983).

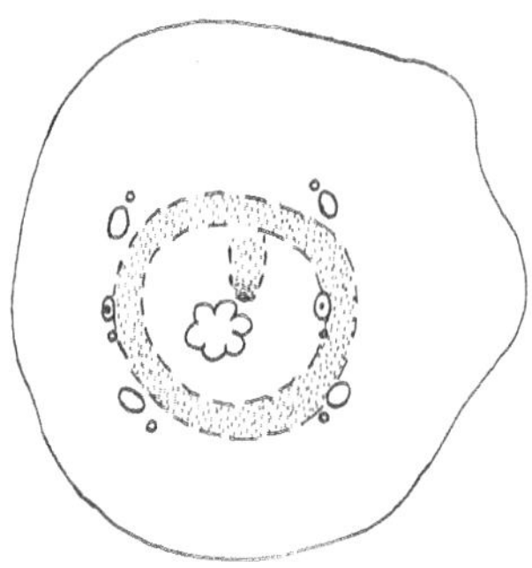

10.193

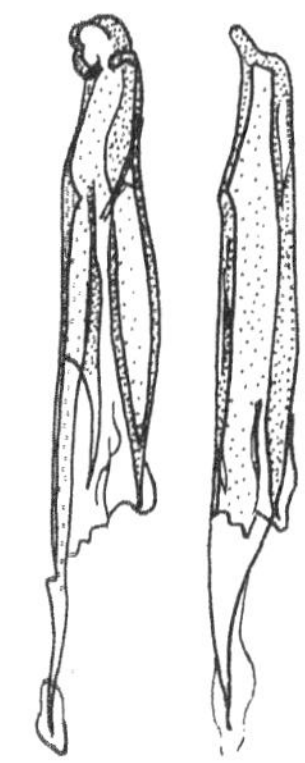

10.194

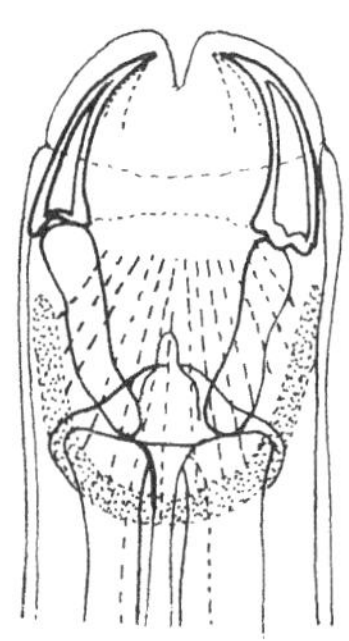

10.195

10.196

FIG. 10.193. *Batrachonema synaptospicula* Yuen, 1965, cephalic extremity, apical view (after Baker, original).
FIG. 10.194. *Copemania obendorfi* Durette-Desset & Beveridge, 1981, spicules (after Durette-Desset & Beveridge, 1981).
FIG. 10.195. *Amphicephaloides thylogale* Beveridge, 1979, cephalic extremity, dorsal view (after Beveridge, 1979).
FIG. 10.196. *Herpetostrongylus pythonis* Baylis, 1931, cephalic extremity, apical view (after Humphery-Smith, 1981).

2– (3) Six lips present (Fig. 10.27).
Gubernaculum present (Fig. 10.33).
Parasites of monotremes.

Nicollina Baylis, 1930
(= *Nicollia* Baylis, 1930)

3– (2) Lips absent (Fig. 10.193).
Gubernaculum absent.
Parasites of Malaysian amphibians.

Batrachonema Yuen, 1965

4– (1) Dorsal oesophageal tooth absent.
Monodelphic.
Spicules relatively short and stout (Fig. 10.194).
Parasites of australasian Dasyuroidea.

Copemania Durette-Desset & Beveridge, 1981

Family *HERPETOSTRONGYLIDAE* (Skrjabin & Schulz, 1937 subfam.) Durette-Desset & Chabaud, 1981

Key to subfamilies

1– (2) Oral opening oval to slit-like, bounded by two lateral jaw-like structures (Figs. 10.29, 10.195).
Parasites of australasian Macropodidae.

Globocephaloidinae

2– (1) Oral opening round or hexagonal.
Six lips present (Fig. 10.196).
Parasites of Malaysian and Indian reptiles and australasian reptiles and marsupials.

Herpetostrongylinae

Subfamily **Globocephaloidinae** Inglis, 1968

Key to genera

1– (2) Buccal capsule not subdivided, approximately as wide as long.
Single dorsal and two small subventral teeth present (Figs. 10.28, 10.29).
Gubernaculum absent.

Globocephaloides Yorke & Maplestone, 1926

2– (1) Buccal capsule elongated, subdivided.
Dorsal tooth present, subventral teeth absent (Fig. 10.195).
Gubernaculum present.

Amphicephaloides Beveridge, 1979

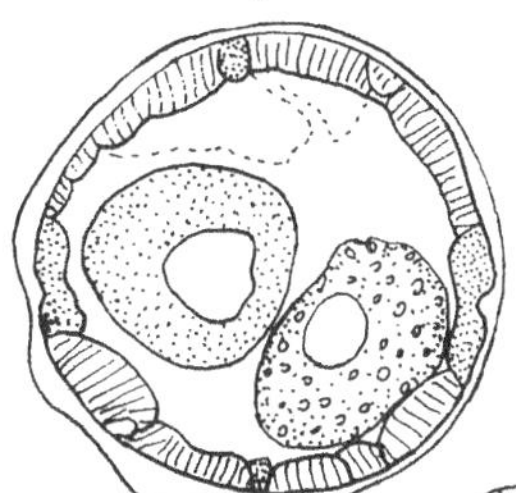

10.197

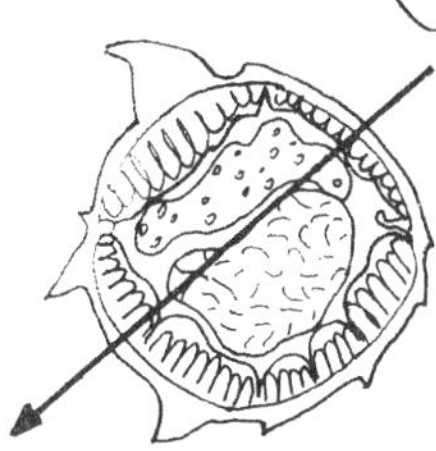

10.198

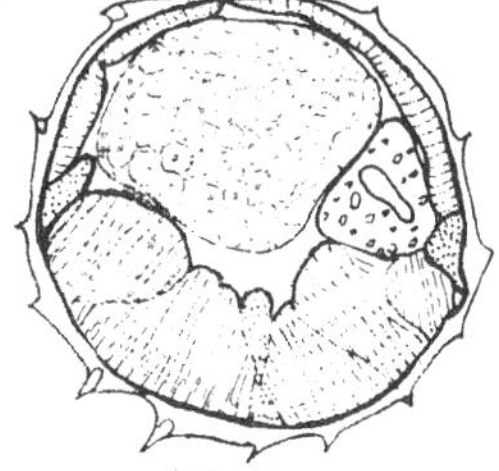

10.199

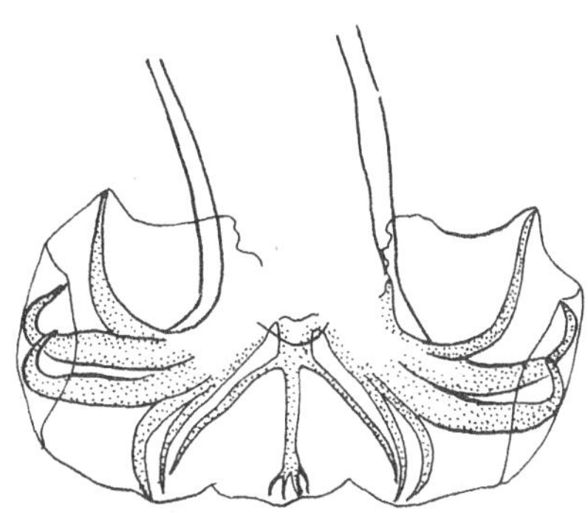

10.200

FIG. 10.197. *Herpetostrongylus pythonis* Baylis, 1931, synlophe (after Humphery-Smith, 1981).

FIG. 10.198. *Vaucherus vaucheri* Durette-Desset, 1980, synlophe (after Durette-Desset, 1980).

FIG. 10.199. *Herpetostrongylus varani* Baylis, 1931, synlophe (after Humphery-Smith, 1981).

FIG. 10.200. *Herpetostrongylus pythonis* Baylis, 1931, bursa (after Baylis, 1931).

Subfamily **Herpetostrongylinae** Skrjabin & Schulz, 1937

Key to genera

1–(16) Two small cuticular dilations present on left side of body (Fig. 10.197) or synlophe with ridges not oriented perpendicular to body (Fig. 10.198).
Parasites of digestive tract.

2– (5) Synlophe formed by two cuticular dilations on left side of body (Fig. 10.197) or by more than ten cuticular ridges oriented from right, ventral-right towards left, dorsal-left (Fig. 10.199), or by less than ten cuticular ridges oriented from dorsal-right towards ventral-left (Fig. 10.198).
Parasites of oriental and australasian reptiles.

3– (4) Synlophe formed by two cuticular dilations on left side of body (Fig. 10.197) or by more than ten cuticular ridges oriented from right, ventral-right towards left, dorsal-left (Fig. 10.199).
Ray 8 not crossing over dorsal side of ray 6 (Fig. 10.200).
Parasites of australasian reptiles.
Herpetostrongylus Baylis, 1931

4– (3) Synlophe formed by fewer than ten cuticular ridges oriented from dorsal-right towards ventral-left (Fig. 10.198).
Ray 8 crossing over dorsal side of ray 6 (Fig. 10.35).
Parasites of Malaysian reptiles.
Vaucherus Durette-Desset, 1980

5– (2) Synlophe formed by fewer than ten cuticular ridges never oriented from dorsal-right to ventral-left (Fig. 10.201).
Parasites of australasian marsupials.

6– (7) Cuticular ridges restricted to ventral side of body (Fig. 10.201).
Parasites of Dasyuridae, Thylacinidae and rarely Muridae.
Woolleya Mawson, 1973

7– (6) Synlophe with ventral and dorsal ridges (Fig. 10.202).

8–(11) Cuticular ridges with oblique axis of orientation (Fig. 10.202).

9–(10) Didelphic.
Female tail with caudal spine.
Spicules divided into two distal branches.
Parasites of Myrmecobiidae and Peramelidae.
Beveridgiella Humphery-Smith, 1981

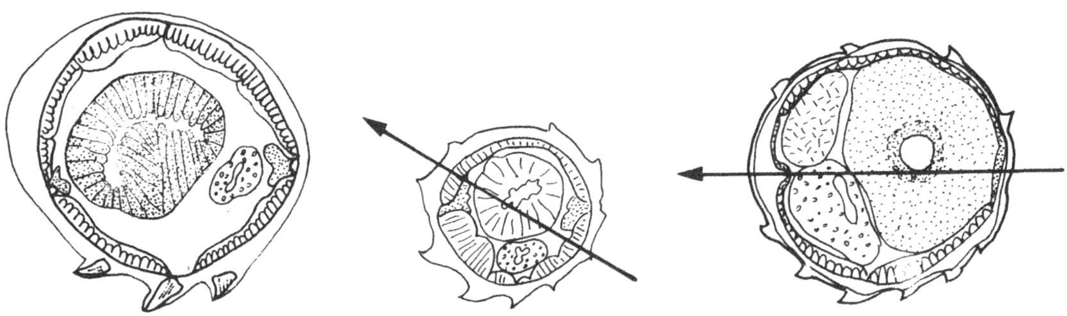

10.201 10.202 10.203

FIG. 10.201. *Woolleya sprenti* Mawson, 1973, synlophe (after Humphery-Smith & Durette-Desset, 1981).
FIG. 10.202. *Beveridgiella pearsoni* Humphery-Smith, 1980, synlophe (after Humphery-Smith, 1980).
FIG. 10.203. *Patricialina hickmani* (Mawson, 1973), synlophe (after Humphery-Smith & Durette-Desset, 1981).

10– (9) Monodelphic.
Female tail lacking caudal spine.
Spicules divided into three distal branches.
Parasites of Dasyuridae.

Dessetostrongylus Humphery-Smith, 1981

11– (8) Cuticular ridges with frontal axis of orientation (Fig. 10.203).

12–(13) Body cuticle on lateral sides not markedly swollen.
Parasites of Dasyuridae.

Patricialina Inglis, 1968

13–(12) Body cuticle on lateral sides markedly swollen (Fig. 10.204).
Parasites of Phalangeroidea.

14–(15) Didelphic.
Genital cone not markedly sclerotized.

Austrostrongylus Chandler, 1924

15–(14) Monodelphic.
Genital cone markedly sclerotized (Fig. 10.205).

Paraustrostrongylus Mawson, 1973

16– (1) Cuticular dilations absent and synlophe with ridges oriented perpendicular to body (Fig. 10.206).
Parasites of nasal cavity of *Antechinus* in Australia.

Nasistrongylus Durette-Desset & Beveridge, 1982.

Family ***VIANNAIIDAE*** (Neveu-Lemaire, 1944, subfam.)
Durette-Desset & Chabaud, 1981

Key to genera

1– (8) Caudal bursa slightly asymmetrical.
Dorsal ray divided into two branches (Figs. 10.38, 10.39).

2– (5) Synlophe formed by three ventral ridges (Fig. 10.40).

3– (4) Caudal bursa heart-shaped.
Dorsal ray usually long (Fig. 10.38).
Parasites of American marsupials.

Viannaia Travassos, 1914
(= *Philostrongylus* Wolfgang, 1951)

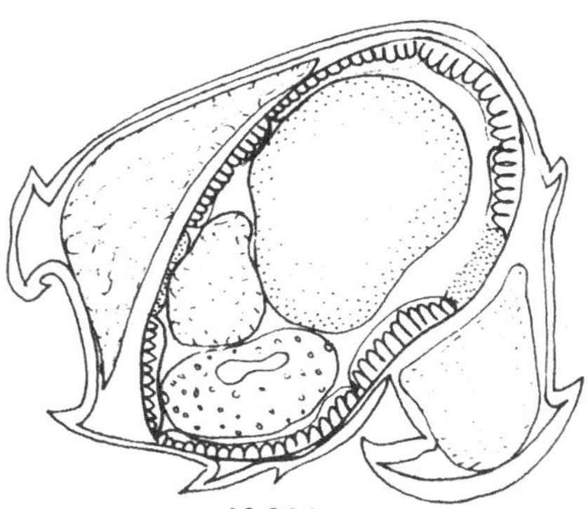

10.204

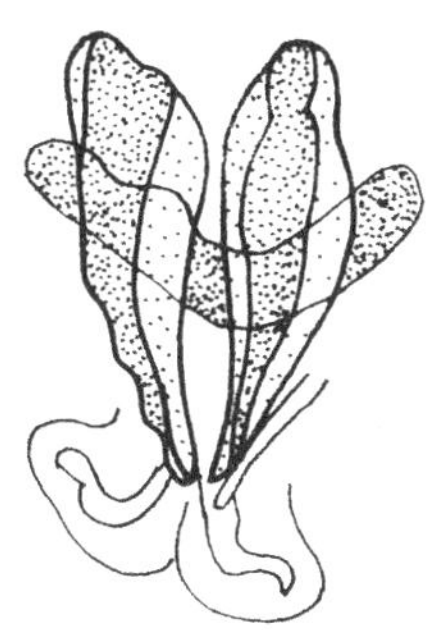

10.205

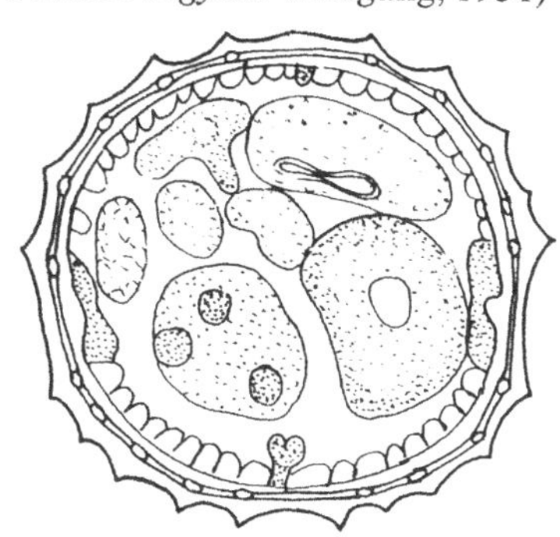

10.206

FIG. 10.204. *Austrostrongylus aggregatus* Johnston & Mawson, 1940, synlophe (after Durette-Desset, 1980).

FIG. 10.205. *Paraustrostrongylus gymnobelideus* Humphery-Smith, 1981, genital cone, ventral view (after Humphery-Smith, 1981).

FIG. 10.206. *Nasistrongylus antechini* (Beveridge & Baker, 1975), synlophe (after Durette-Desset & Beveridge, 1982).

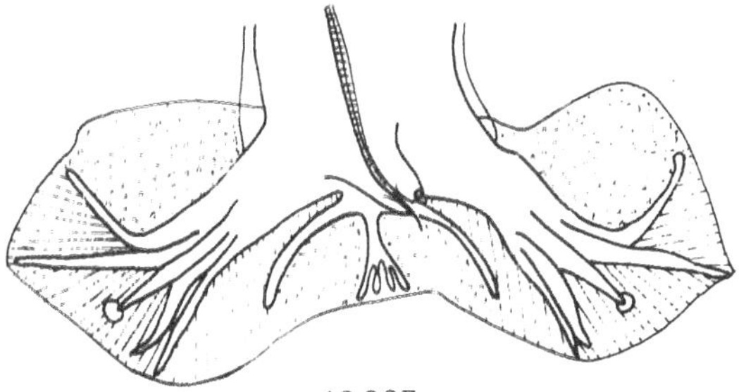

10.207

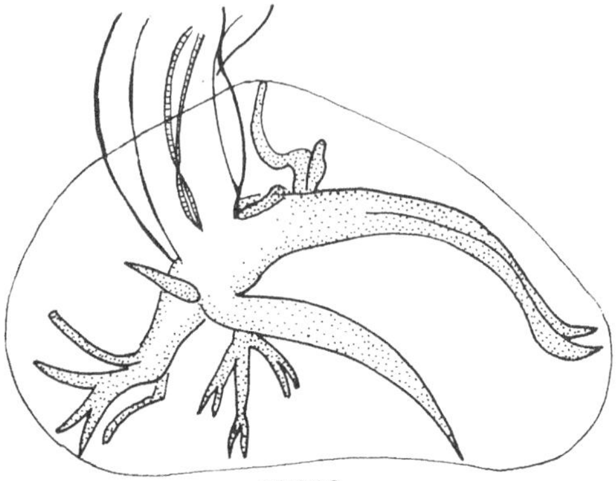

10.208

FIG. 10.207. *Hoineffia cayennensis* Diaw, 1976, bursa (after Diaw, 1976).

FIG. 10.208. *Oswaldonema cruzi* Travassos, 1927, bursa (after Travassos, 1927).

4– (3) Caudal bursa elongated transversely.
Dorsal ray short (Fig. 10.207).
Parasites of American marsupials.

Hoineffia Diaw, 1976

5– (2) Synlophe not as above (Fig. 10.41).

6– (7) Cuticular ridges with frontal axis of orientation (Fig. 10.42).
Didelphic.
Parasites of American marsupials.

Travassostrongylus Orloff, 1933
(= *Camerostrongylus* Wolfgang, 1951)

7– (6) Cuticular ridges with oblique axis of orientation (Fig. 10.41).
Monodelphic.
Parasites of neotropical caviomorphs (Hydrochoeridae, Caviidae, Chinchillidae, Cuniculidae) and primates (Cebidae).

Viannella Travassos, 1918
[= *Avellaria* Freitas & Lent, 1934;
= *Longistriata* (*Brevispiculoides*) Ortlepp, 1939]

8– (1) Caudal bursa markedly asymmetrical with well developed left lobe.
Dorsal ray divided into three branches (Fig. 10.208).
Parasites of *Agouti* in Brazil.

Oswaldonema Travassos, 1927
(= *Heligmoskrjabinia* Freitas & Lent, 1937)

Family ***ORNITHOSTRONGYLIDAE*** (Travassos, 1937, subfam.) Durette-Desset & Chabaud, 1981

Key to subfamilies

1– (2) Buccal capsule and lips present (Figs. 10.36, 10.37).
Parasites of Cricetidae of Chile.

Inglamidinae

2– (1) Buccal capsule and lips absent.
Parasites of birds (cosmopolitan), New World bats, rarely Old World bats, Tupaiidae of Borneo and geomyoid rodents of New World.

Ornithostrongylinae

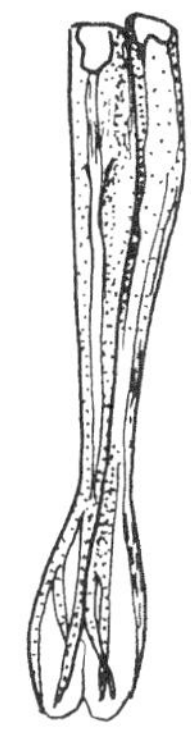

10.209

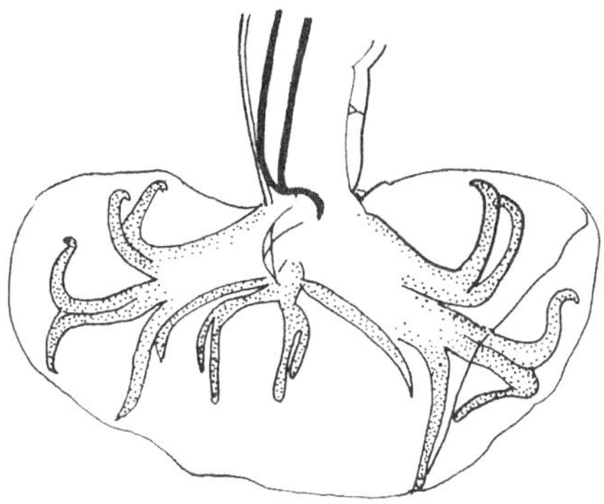

10.210

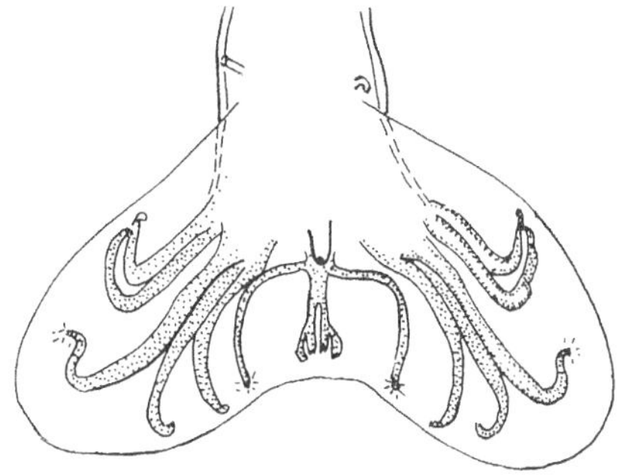

10.211

FIG. 10.209. *Ornithostrongylus crami* Vigueras, 1934, spicules (after Barus, 1968).
FIG. 10.210. *Oswaldostrongylus cruzi* Lent & Freitas, 1934, bursa (after Lent & Freitas, 1934).
FIG. 10.211. *Columbostrongylus shoutedeni* Puylaert, 1968, bursa (after Puylaert, 1968).

Subfamily **Inglamidinae** Durette-Desset, Diaw & Murua, 1976

One genus (Figs. 10.36, 10.37).

Inglamidum Durette-Desset, Diaw & Murua, 1976

Subfamily **Ornithostrongylinae** Travassos, 1914

Key to genera

1–(12) Didelphic.
Cuticular ridges almost equal in size and lacking cuticular supports (Fig. 10.44).
2– (9) Ray 4 longer than ray 5 (Fig. 10.43).
3– (8) Dorsal ray short.
4– (5) Spicules usually short and thick, divided into several distal branches (Fig. 10.209).
Parasites of neotropical birds.

Ornithostrongylus Travassos, 1914
(= *Cephalostrongylus* Irwin-Smith, 1920)

5– (4) Spicules slender, not divided into distal branches (Fig. 10.210).
6– (7) Dorsal ray divided at its base.
Ray 8 short and attached to common base of rays 4 to 6 (Fig. 10.210).
Parasites of neotropical birds.

Oswaldostrongylus Lent & Freitas, 1934

7– (6) Dorsal ray divided in middle.
Ray 8 long and not attached to common base of rays 4 to 6 (Fig. 10.211).
Parasites of ethiopian birds.

Columbostrongylus Puylaert, 1968

8– (3) Dorsal ray long (Fig. 10.212).
Parasites of neotropical birds.

Ornithonema Travassos, 1935

9– (2) Ray 4 shorter than, or only as long as, ray 5 (Figs. 10.213, 10.214).
10–(11) Ray 8 well separated from ray 6 (Fig. 10.213).
Spicules filiform.
Parasites of neotropical birds.

Lutznema Lent & Freitas, 1934

10.212

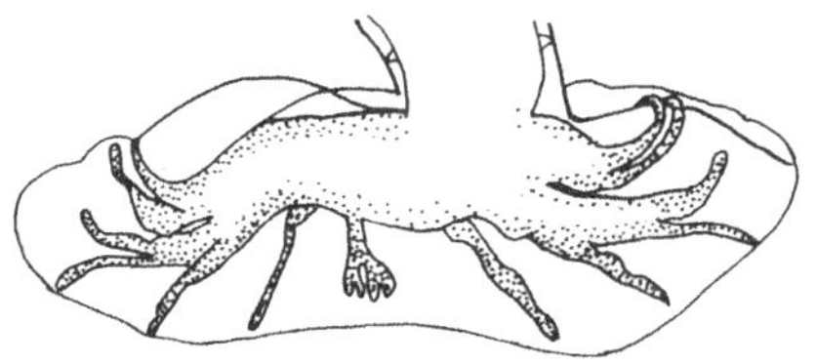

10.213

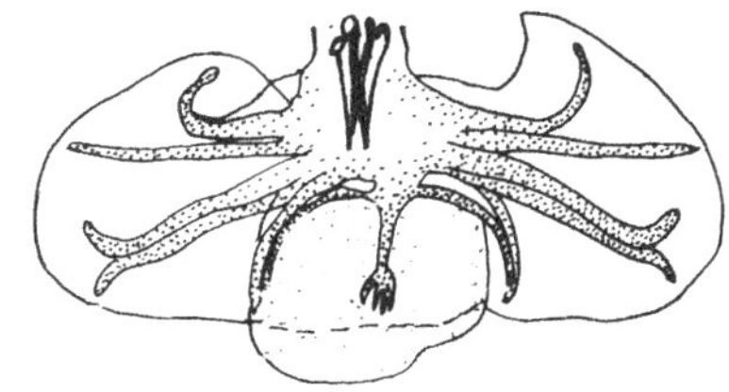

10.214

FIG. 10.212. *Ornithonema mensoris* Travassos, 1935, bursa (after Travassos, 1935).

FIG. 10.213. *Lutznema lutzi* Lent & Freitas, 1934, bursa (after Lent & Freitas, 1934).

FIG. 10.214. *Allintoshius nycticeius* Chitwood, 1937, bursa (after Durette-Desset, 1978).

11–(10) Ray 8 close to ray 6 (Fig. 10.214).
Spicules conical (Fig. 10.214).
Parasites of American bats, rarely Old World bats and *Tupaia* in Borneo.
Allintoshius Chitwood, 1937
(= *Parallintoshius* Araujo, 1940)

12– (1) Monodelphic.
Cuticular ridges markedly unequal in size and with cuticular supports (Fig. 10.215).
Parasites of American Geomyoidea.
Vexillata (Travassos, 1937) Durette-Desset, 1971

Family ***HELIGMOSOMIDAE*** (Travassos, 1914, subfam.) Cram, 1927

Key to genera

1– (4) Each branch of vestibule no longer than infundibulum (Fig. 10.216).
Deirids in form of short blunt projections.
Parasites of Soricoidea.

2– (3) Synlophe formed by three ventral ridges (Fig. 10.45).
Didelphic.
Parasites of ethiopian and oriental Soricoidea.
Suncinema Durette-Desset, 1973

3– (2) Synlophe formed by ventral and dorsal ridges (Fig. 10.217).
Monodelphic.
Parasites of holarctic Soricoidea.
Longistriata Schulz, 1926
(= *Longistrioides* Yeh, 1958; = ? *Neoheligmonoides* Sadovskaja, 1952)

4– (1) Each branch of vestibule longer than infundibulum (Fig. 10.218).
Deirids setiform.
Parasites of lagomorphs and sciuromorph and myomorph rodents.

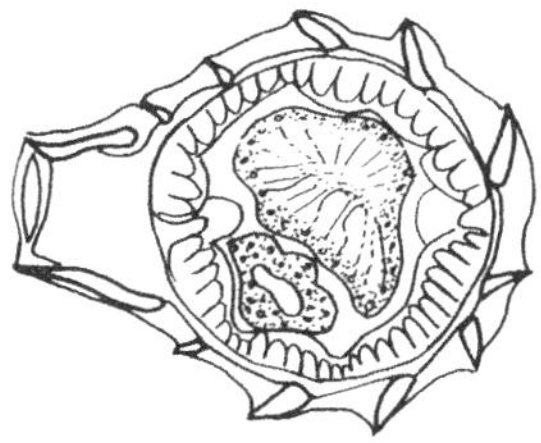

10.215

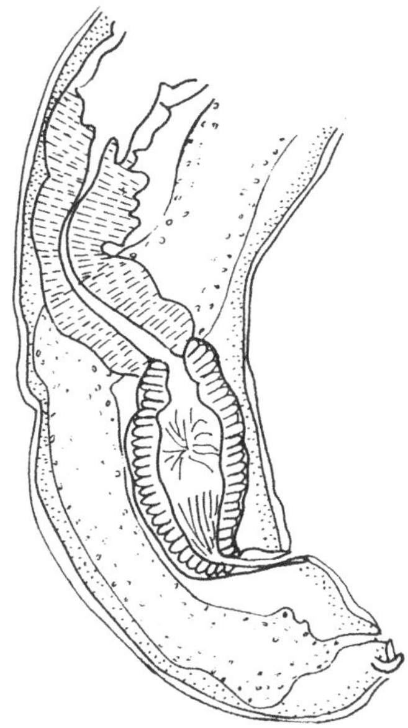

10.216

10.217

FIG. 10.215. *Vexillata legallae* Denké, 1977, synlophe (after Denké, 1977).
FIG. 10.216. *Longistriata ohbayashii* Durette-Desset, 1973, ovejector, lateral view (after Durette-Desset, 1973).
FIG. 10.217. *Longistriata ohbayashii* Durette-Desset, 1973, synlophe (after Durette-Desset, 1973).

5– (8) Didelphic.
6– (7) Spicule shaft divided into two parts in mid-portion (Fig. 10.219).
Parasites of holarctic sciuromorph rodents.

Citellinema Hall, 1916
(= *Warrenius* Hall, 1916)

7– (6) Spicule shaft not divided into two parts in mid-portion.
Parasites of nearctic myomorph rodents.

Citellinoides Dikmans, 1939

8– (5) Monodelphic.
9–(10) Synlophe dorso-ventrally symmetrical (i.e. same number, size and orientation of ridges with respect to frontal axis) (Fig. 10.220).
Parasites of palearctic lagomorphs.

Ohbayashinema Durette-Desset, 1974

10– (9) Synlophe not dorso-ventrally symmetrical.
Parasites of holarctic rodents.
11–(16) Synlophe composed of longitudinal ridges (Fig. 10.221), usually most developed left-ventrally (Fig. 10.46).
12–(15) Left ventral comarête absent.
13–(14) Bursa nearly symmetrical, if asymmetrical rays 8 of same size.
Prebursal rays small (Fig. 10.47).
Parasites of holarctic Arvicolidae and Muridae and palearctic Spalacidae and Cricetidae.

Heligmosomoides Hall, 1916
(= *Nematospira* Walton, 1923; = *Nematospiroides* Baylis, 1926;
= *Paranematospira* Sprehn, 1935; = *Sincosta* Roé, 1929).

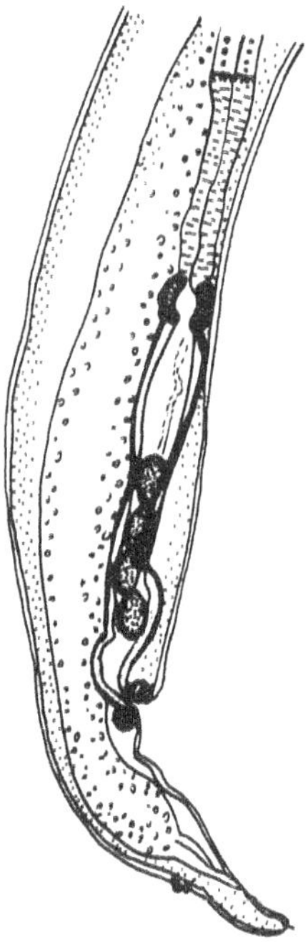

10.218

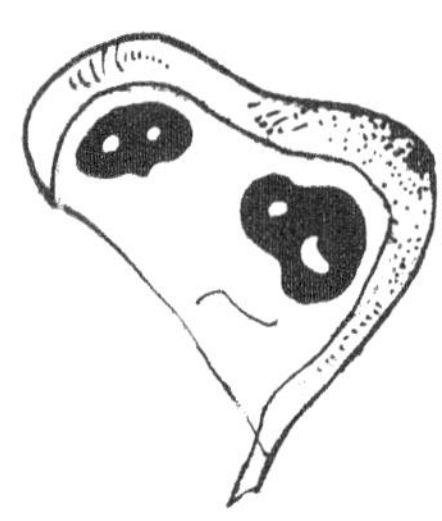

10.219

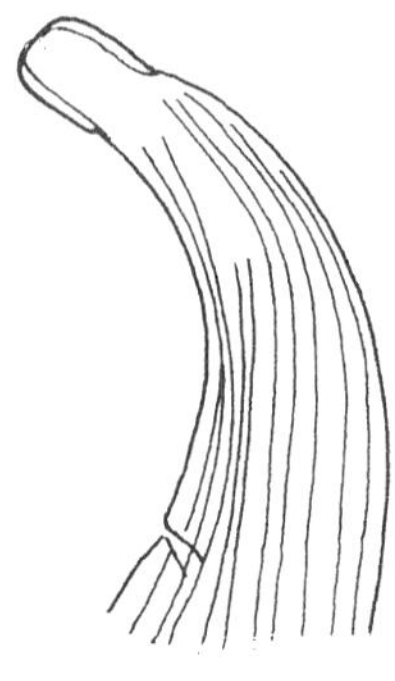

10.221

10.220

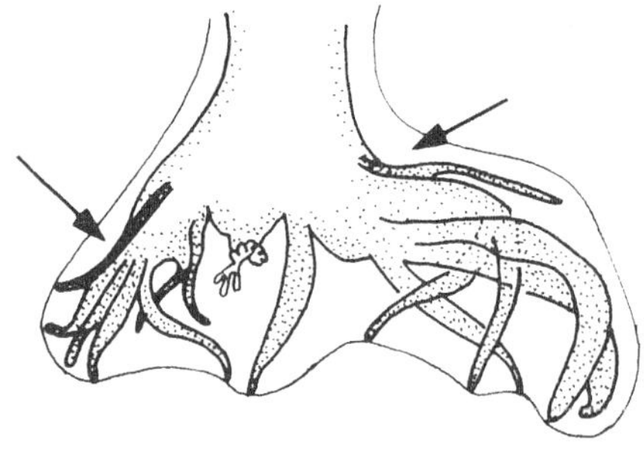

10.222

FIG. 10.218. *Heligmosomoides desportesi* (Chabaud, Rausch & Desset, 1963), ovejector, lateral view (after Chabaud, Rausch & Desset, 1963).

FIG. 10.219. *Citellinema* sp. Durette-Desset, 1969, section through both spicules near mid-portion (after Durette-Desset, 1969).

FIG. 10.220. *Ohbayashinema ochotoni* Durette-Desset, 1974, synlophe (after Durette-Desset, 1974).

FIG. 10.221. *Heligmosomoides*, cuticular ridges in anterior part of body, diagrammatic (after Durette-Desset, 1971).

FIG. 10.222. *Heligmoptera myospalaxi* (Nadtochyi, 1970), bursa (after Nadtochyi, 1976).

14–(13) Bursa markedly asymmetrical; rays 8 of different sizes.
Prebursal ray strongly developed (Fig. 10.222).
Parasites of palearctic Spalacidae.

Heligmoptera Nadtochyi, 1976

15–(12) Left ventral comarête present (Fig. 10.226).
Parasites of palearctic Spalacidae.

Dessetia Genov & Janchev, 1981

16–(11) Synlophe with oblique ridges (Fig. 10.223) or with longitudinal ridges on ventral side and oblique ridges on dorsal side (Fig. 10.224); left-ventral ridges not more developed than dorsal (Fig. 10.225).
Parasites of holarctic Arvicolidae.

Heligmosomum Railliet & Henry, 1909

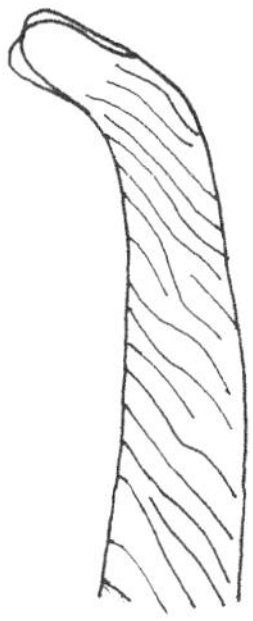

10.223

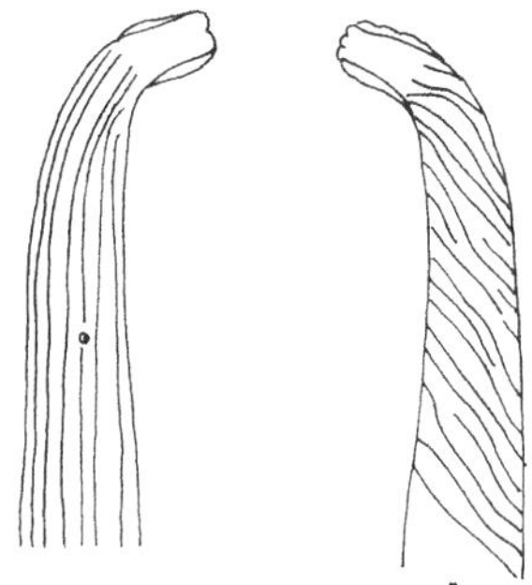

a b

10.224

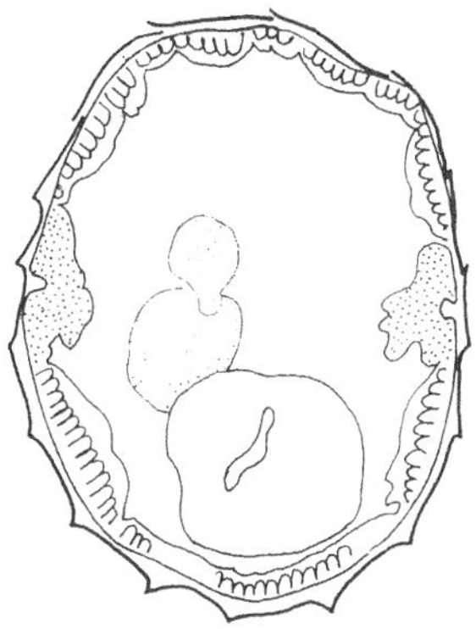

10.225

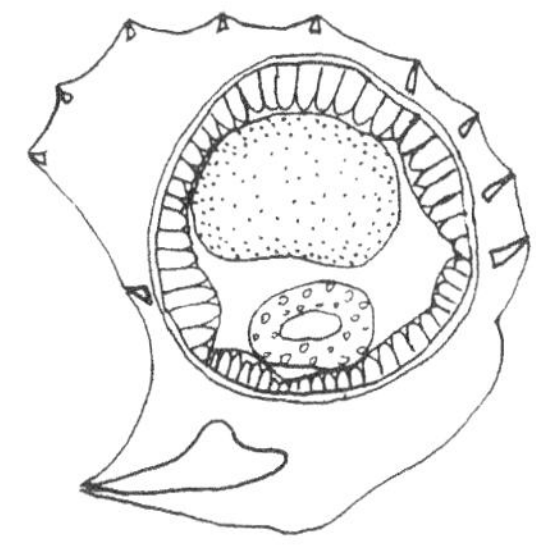

10.226

FIG. 10.223. *Heligmosomum*, cuticular ridges in anterior part of body, diagrammatic (after Durette-Desset, 1971).

FIG. 10.224. *Heligmosomum mixtum* Schulz, 1926, cuticular ridges in anterior part of body, (a) ventral view, (b) dorsal view (after Durette-Desset, 1968).

FIG. 10.225. *Heligmosomum mixtum* Schulz, 1926, synlophe (after Durette-Desset, 1968).

FIG. 10.226. *Dessetia moldovensis* (Andrejko, 1963), synlophe (after Genov & Janchev, 1981).

Family *HELIGMONELLIDAE* (Skrjabin & Schikhobalova, 1952, tribe) Durette-Desset & Chabaud, 1977

Key to subfamilies

1– (2) Axis of orientation of ridges inclined no more than 45° from sagittal axis (Fig. 10.227A).
Carene absent.
Gradient in size of ridges usually from left to right on ventral side and from right to left on dorsal side (Fig. 10.228).
Parasites of Talpidae, lagomorphs and archaic rodents.

Heligmonellinae

2– (1) Axis of orientation of ridges inclined more than 45° from sagittal axis (Fig. 10.227B,C).
Carene present or absent.
Gradient in size of ridges variable.
Parasites of various mammals (except insectivores).

3– (6) Axis of orientation of ridges inclined 67–90° from sagittal axis (Fig. 10.227C).
Gradient in size of ridges lacking or from right to left on dorsal side (Fig. 10.48) or medio-lateral (Fig. 10.229).
Ridges continuous or discontinuous.
Parasites of rodents.

4– (5) Carene present (Fig. 10.230) and dorsal ray deeply divided (Fig. 10.231).
Parasites of caviomorphs, i.e. Chinchilloidea (Dasyproctidae), Octodontoidea (Echimyidae, Capromyidae, Myocastoridae), Erethizontoidea (Erethizontidae) and Sciuridae, rarely Cricetidae and lagomorphs. New World distribution.

Pudicinae

5– (4) Carene absent (Fig. 10.229) or, if present, dorsal ray divided in its posterior half (Fig. 10.232).
Parasites of archaic rodents, Hystricidae, Sciuridae, Gliridae, rarely Tragulidae and Muridae. Old World distribution.

Brevistriatinae

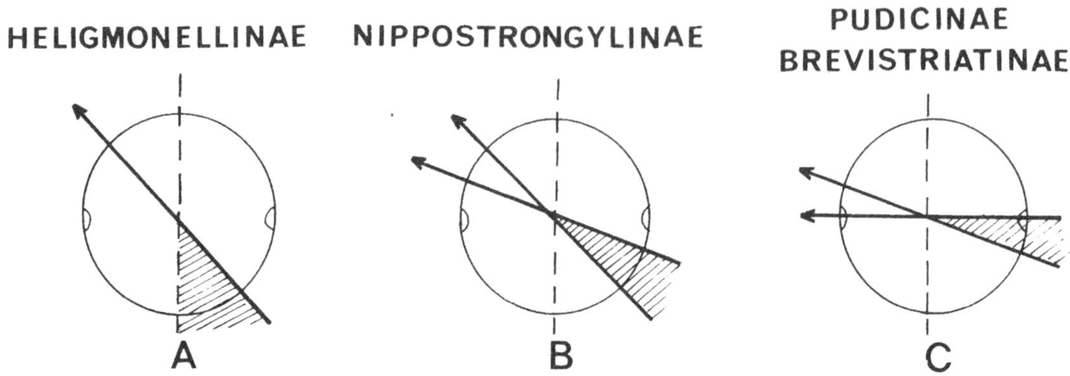

10.227

FIG. 10.227. Schematic diagram showing degree of inclination of axis of orientation with respect to the sagittal axis in Heligmonellidae.
A: up to 45°: Heligmonellinae.
B: 45° to 67°: Nippostrongylinae.
C: 67° to 90°: Pudicinae—Brevistriatinae.

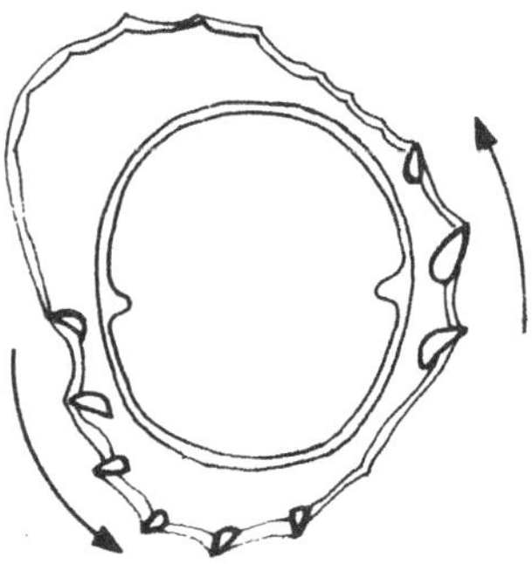

10.228

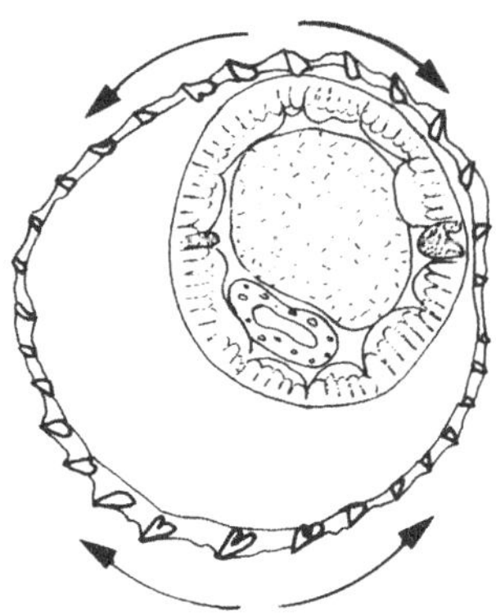

10.229

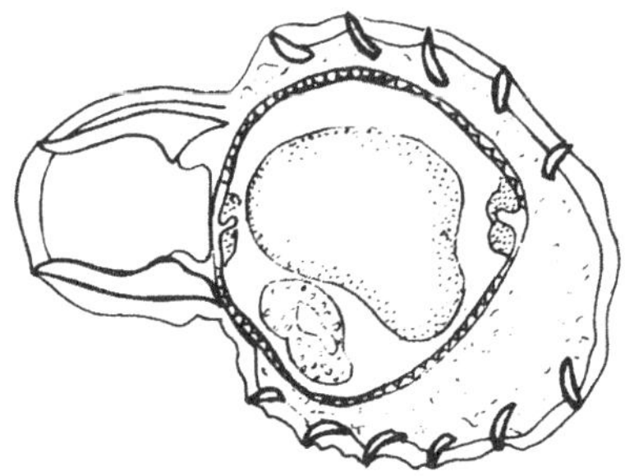

10.230

FIG. 10.228. *Tricholinstowia linstowi* (Travassos, 1918), synlophe (after Durette-Desset, & Vaucher, 1974).
FIG. 10.229. *Kuala chaii* Durette-Desset & Krishnasamy, 1976, synlophe (after Durette-Desset & Krishnasamy, 1976).
FIG. 10.230. *Heligmostrongylus elegans* (Travassos, 1921), synlophe (after Durette-Desset, 1968).

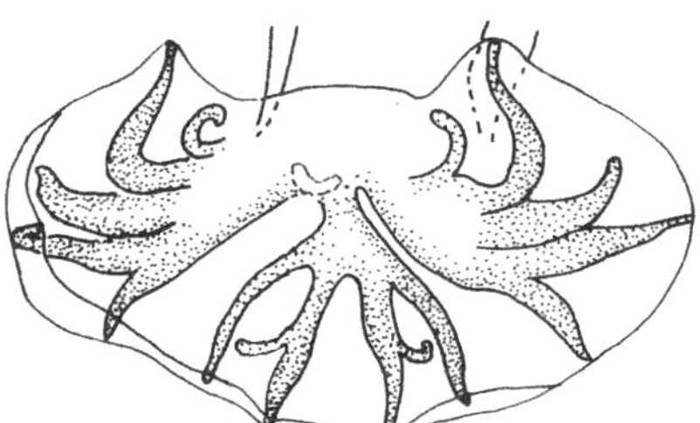

10.231

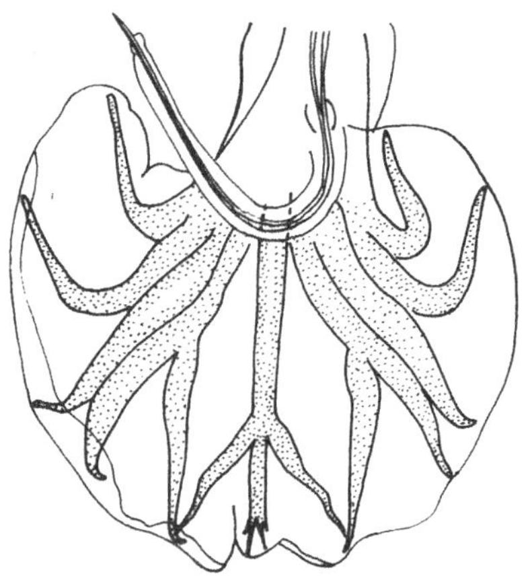

10.232

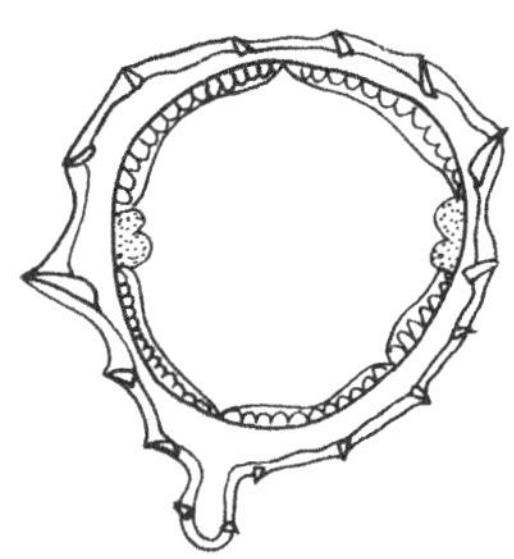

10.233

FIG. 10.231. *Heligmostrongylus elegans* (Travassos, 1921), bursa (after Travassos, 1921).
FIG. 10.232. *Calypsostrongylus titasuthi* Kliks & Durette-Desset, 1976, bursa (after Kliks & Durette-Desset, 1976).
FIG. 10.233. *Carolinensis dikmansi* (Durette-Desset, 1974) n. comb., synlophe (after Durette-Desset, 1974).

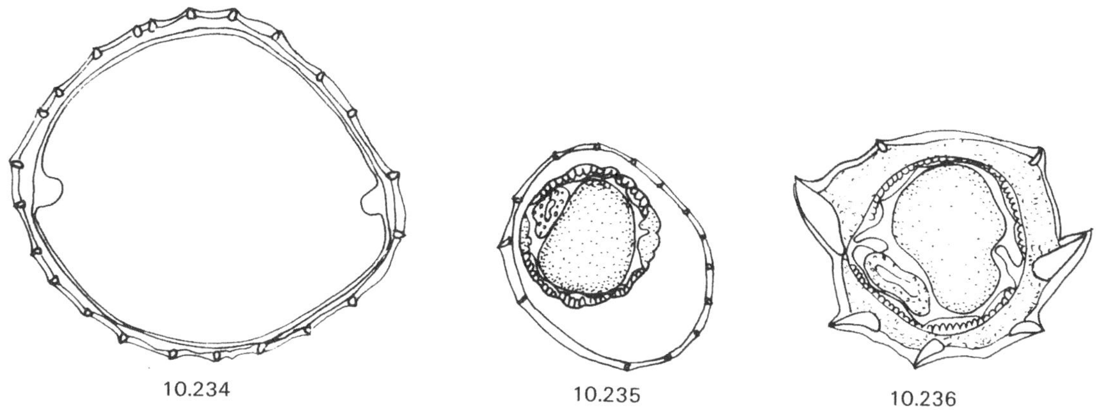

FIG. 10.234. *Mammanidula melomyos* (Mawson, 1961), synlophe (after Durette-Desset, 1969).

FIG. 10.235. *Hypocristata thomasomysi* Durette-Desset, 1970, synlophe (after Durette-Desset, 1970).

FIG. 10.236. *Heligmonella spira* Mönnig, 1927, synlophe (after Durette-Desset, 1969).

6– (3) Axis of orientation of ridges inclined 45–67° from sagittal axis (Fig. 10.227B).
Gradient in size of ridges usually latero median (Fig. 10.233) or ridges very numerous and lacking gradient in size (Fig. 10.234) or ridges barely extending above surface of cuticle (Fig. 10.235).
Ridges continuous.
Parasites of recent rodents, i.e. Gerbillidae, Muridae, Arvicolidae throughout the world and New World Cricetidae, rarely lagomorphs, Dermoptera and other small mammals.
Nippostrongylinae

Subfamily **Heligmonellinae** (Skrjabin & Schikhobalova, 1952, tribe)
Durette-Desset & Chabaud, 1977

Key to genera

1– (2) Dorsal side of body cuticle lacking ridges (Fig. 10.228).
Parasites of Talpoidea.
Tricholinstowia Travassos, 1937
(= *Morganiella* Travassos, 1937; = *Morganostrongylus* Fahmy, 1956)

2– (1) All sides of body cuticle with ridges.

3– (8) Ridges continuous over entire body.

4– (7) Left and right ridges much more developed than adjacent dorsal ridges.
Carene absent (Fig. 10.236).

5– (6) Ray 6 equal to, or shorter than, ray 5.
Bursa lacking separate dorsal lobe (Fig. 10.237).
Parasites of *Thryonomys* (rodent), palearctic and oriental lagomorphs, ethiopian and oriental Hystricidae.
Heligmonella Mönnig, 1927

6– (5) Ray 6 markedly longer than ray 5.
Bursa with separate dorsal lobe (Fig. 10.238).
Parasites of oriental Sciuridae.
Sciuricola n. g.*

---

* See definition in Introduction, p. 54.

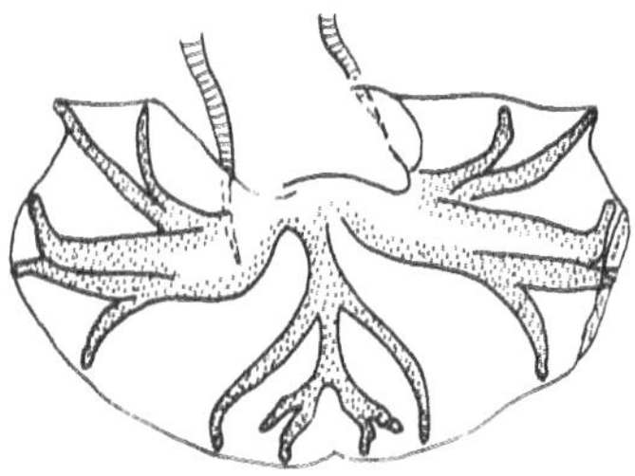

10.237

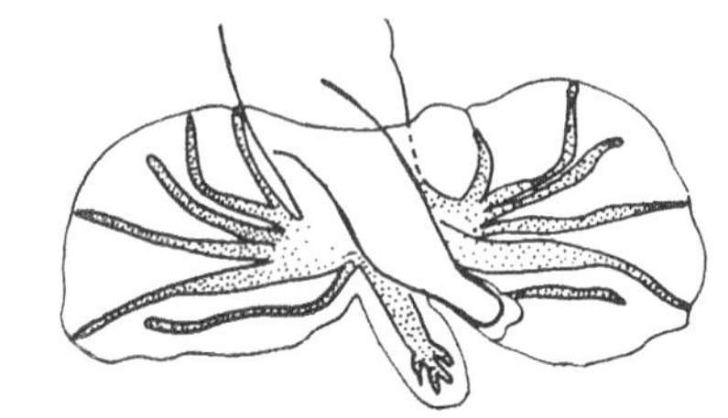

10.238

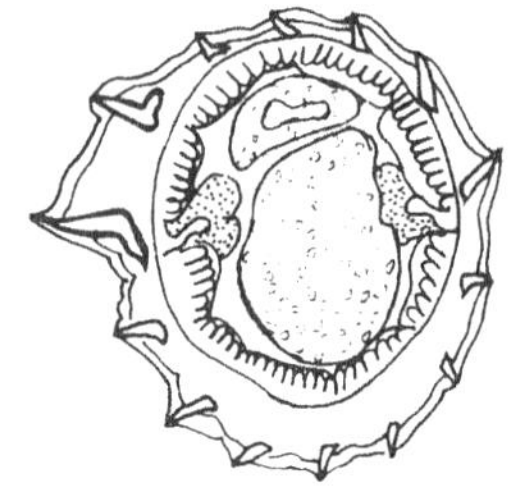

10.239

FIG. 10.237. *Heligmonella spira* Mönnig, 1927, bursa (after Durette-Desset, 1969).

FIG. 10.238. *Sciuricola moreli* (Durette-Desset, 1974), n. comb., bursa (after Durette-Desset, 1974).

FIG. 10.239. *Xericola marocanus* Durette-Desset, 1974, synlophe (after Durette-Desset, 1974).

7– (4) Left and right ridges equal in size to adjacent dorsal ridges.
Small carene present (Fig. 10.239).
Parasites of Moroccan Sciuridae.
Xericola Durette-Desset, 1974

8– (3) At least four ridges discontinuous.

9–(10) Four dorsal ridges discontinuous.
Gradient in size of ridges right to left on dorsal side and left to right on ventral side (Fig. 10.240).
Parasites of neotropical Echimyidae and lagomorphs.
Paraheligmonella Durette-Desset, 1971

10– (9) All ridges discontinuous.
Gradient in size of ridges absent (Fig. 10.241).
Parasites of caviomorphs (Echimyidae).
Trichotravassosia Lent & Freitas, 1938

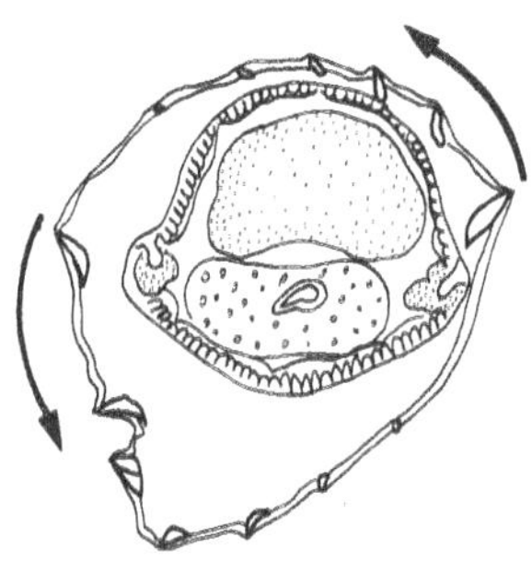

10.240

10.241

10.242

FIG. 10.240. *Paraheligmonella cubaensis* (Vigueras, 1943), synlophe (after Durette-Desset, 1971).

FIG. 10.241. *Trichotravassosia travassosi* Lent & Freitas, 1938, synlophe (after Durette-Desset & Magalhaes Pinto, 1977).

FIG. 10.242. *Sciurodendrium oliverai* Lent & Freitas, 1938, synlophe (after Durette-Desset, 1968).

Subfamily **Pudicinae** (Skrjabin & Schikhobalova, 1952, tribe) Durette-Desset, 1971

Key to genera

1– (8) Carene supported by two left ridges larger than other ridges (Fig. 10.230).

2– (5) Ray 4 shorter than ray 5 (Fig. 10.231).
Comarêtes and ventral spines absent.

3– (4) Synlophe with five dorsal ridges and five or six ventral ridges (Fig. 10.230).
Parasites of caviomorphs (Dasypodidae, Echimyidae, Myocastoridae and Erethizontidae).
Heligmostrongylus Travassos, 1917
(= *Fuellebornema* Travassos & Darriba, 1929; = *Heligmodendrium* Travassos, 1937; = *Squamastrongylus* Travassos, 1937; = *Stunkardionema* Arnold, 1941)

4– (3) Synlophe usually with more than five dorsal ridges and always with more than six ventral ridges (Fig. 10.242).
Parasites of American Sciuridae, rarely Cricetidae.
Sciurodendrium Durette-Desset, 1971

5– (2) Ray 4 as long as or longer than ray 5.
Comarêtes present at least in anterior part of body (Fig. 10.243).
Ventral spines present or absent.

6– (7) Spines on ventral right side of body absent.
Parasites of caviomorphs (Dasyproctidae and Echimyidae).
Pudica Travassos & Darriba, 1929
(= *Evandroia* Travassos, 1937)

7– (6) Two groups of spines present on ventral right side of body (Fig. 10.244).
Parasites of caviomorphs (Echimyidae).
Acanthostrongylus Travassos, 1937

8– (1) Carene supported ventrally by hypertrophied ridge and dorsally by small ridge (Fig. 10.245).
Parasites of caviomorphs (Echimyidae of Cuba).
Pseudoheligmosomum Travassos, 1937

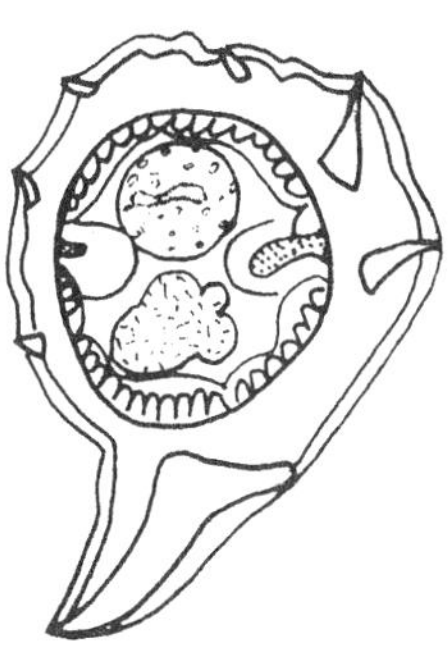

10.243

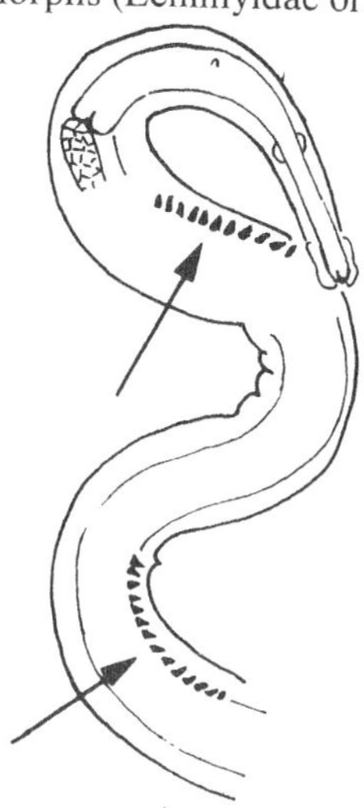

10.244

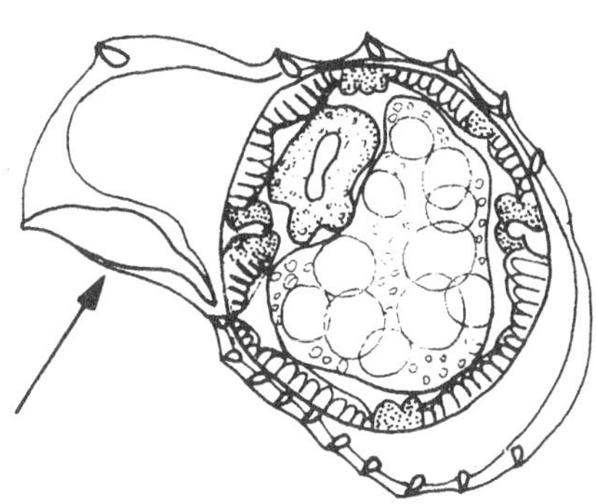

10.245

FIG. 10.243. *Pudica petterae* Durette-Desset, 1970, synlophe (after Durette-Desset, 1970).

FIG. 10.244. *Acanthostrongylus acanthostrongylus* Travassos, 1937, ventral spines (after Durette-Desset & Tcheprakoff, 1983).

FIG. 10.245. *Pseudoheligmosomum howelli* (Vigueras, 1934), synlophe (after Durette-Desset, 1972).

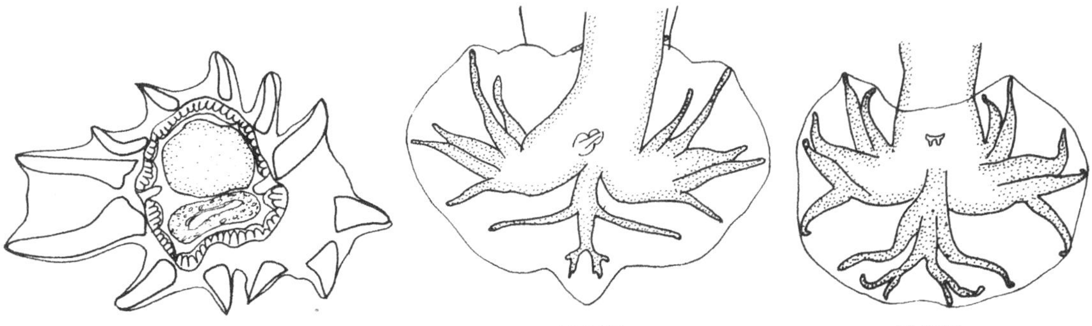

FIG. 10.246. *Srivastavanema yapi* Durette-Desset & Lim-Boo-Liat, 1975, synlophe (after Durette-Desset & Lim-Boo-Liat, 1975).

FIG. 10.247. *Paraheligmonina posterior* Durette-Desset, 1970, bursa (after Durette-Desset, 1970).

FIG. 10.248. *Metheligmonella petri* (Durette-Desset, 1970), bursa (after Durette-Desset, 1970).

Subfamily **Brevistriatinae** Durette-Desset, 1971

Key to genera

1–(10) Carene present.
2– (9) Ridges continuous.
3– (4) Ventral and dorsal ridges large (Fig. 10.246).
Genital cone large.
Parasites of oriental Petauristinae.
Srivastavanema (Singh, 1962) Durette-Desset & Lim Boo Liat, 1975
4– (3) Ventral and dorsal ridges small (Fig. 10.48).
Genital cone small.
5– (8) Dorsal lobe of bursa not hypertrophied.
6– (7) Dorsal ray divided at posterior end (Fig. 10.247).
Parasites of ethiopian Hystricidae and Sciuridae.
Paraheligmonina (Ortlepp, 1939) Durette-Desset, 1971
(= *Heligmobaylisia* Mawson, 1961)
7– (6) Dorsal ray divided in anterior third (Fig. 10.248).
Parasites of ethiopian Muridae.
Metheligmonella Durette-Desset, 1971
8– (5) Dorsal lobe of bursa hypertrophied (Fig. 10.249).
Parasites of oriental Hystricidae.
Cordicauda Durette-Desset, 1971
9– (2) Ridges discontinuous, distributed in alternating lines except those supporting carene (Fig. 10.250).
Parasites of oriental Sciuridae.
Calypsostrongylus Schmidt, Myers & Kuntz, 1967
10– (1) Carene absent.
11–(12) Ridges continuous over entire body surface.
Gradient in size of ridges medio-lateral (Fig. 10.229).
Dorsal ray divided at posterior end.
Parasites of oriental Sciuridae and Tragulidae.
Kuala Durette-Desset & Krishnasamy, 1976

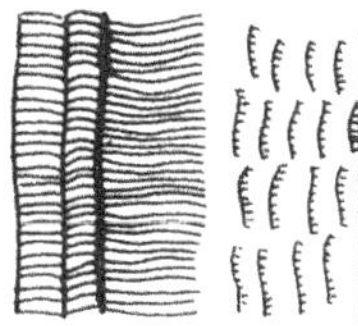

10.250

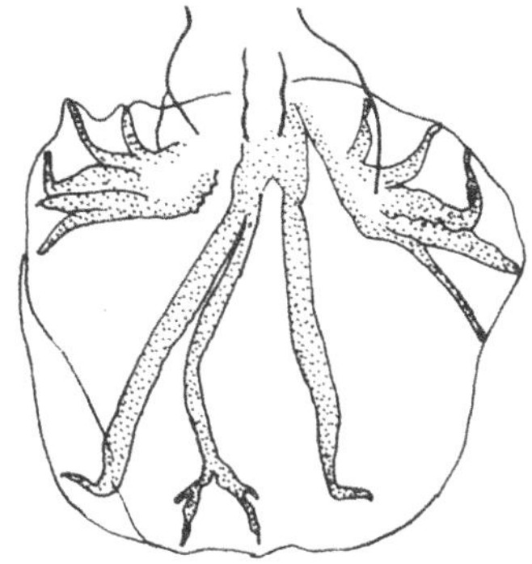

10.249

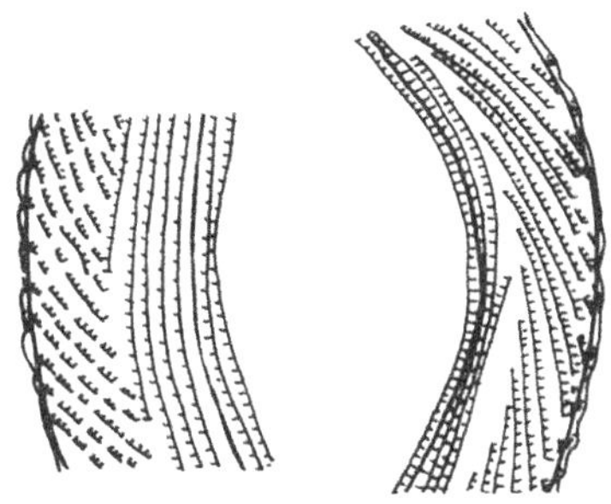

a b

10.251

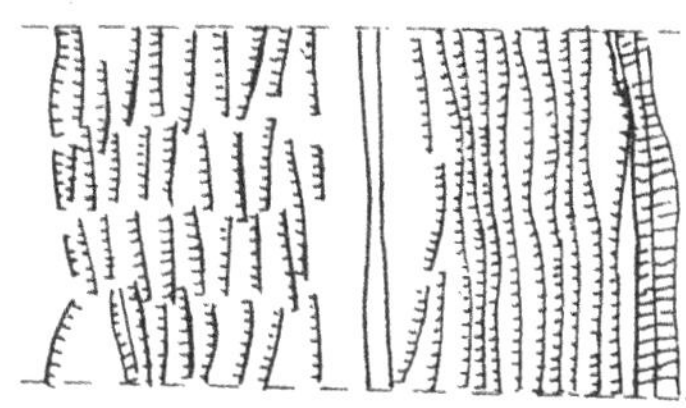

10.252

FIG. 10.249. *Cordicauda levanhoai* (Durette-Desset, 1966), bursa (after Durette-Desset, 1966).

FIG. 10.250. *Calypsostrongylus titasuthi* Kliks & Durette-Desset, 1976, synlophe with ridges disposed in alternating discontinuous lines except carene (after Kliks & Durette-Desset, 1976).

FIG. 10.251. *Quentinstrongylus graphiuri* Durette-Desset, 1969, cuticular ridges along body, (a) right lateral view, (b) left lateral view (after Durette-Desset, 1969).

FIG. 10.252. *Brevistriata skrjabini* (Schulz & Lubimov, 1932), cuticular ridges along body, left lateral view (after Durette-Desset, 1971).

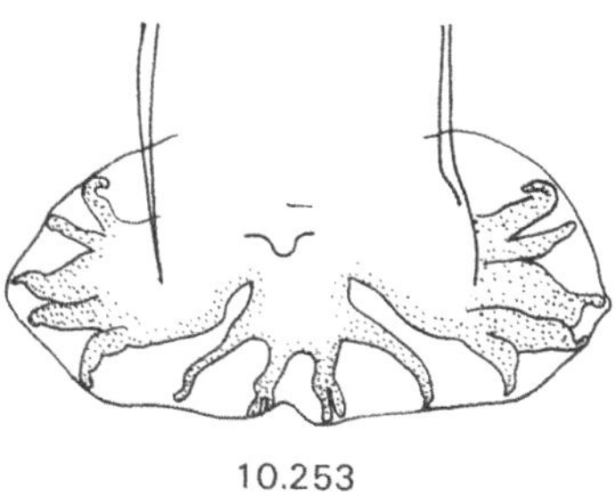

10.253

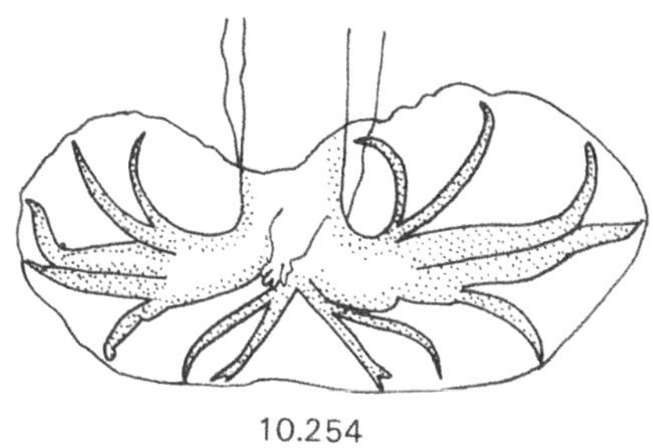

10.254

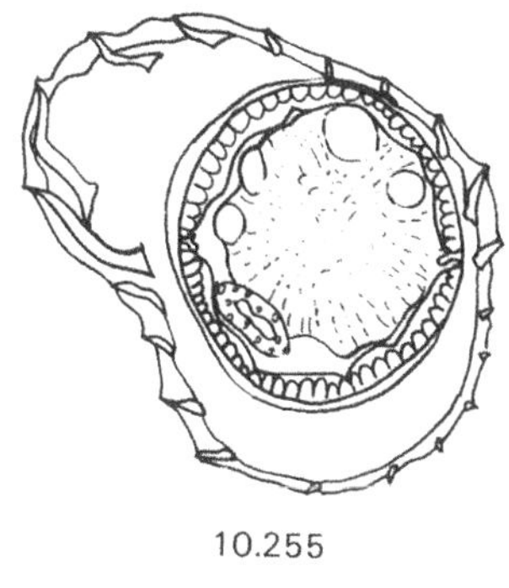

10.255

FIG. 10.253. *Orientostrongylus krishnasamyi* Durette-Desset & Lim-Boo-Liat, 1974, bursa (after Durette-Desset & Lim-Boo-Liat, 1974).

FIG. 10.254. *Heligmonoides josephi* (Wertheim & Durette-Desset, 1976) n. comb., bursa (after Wertheim & Durette-Desset, 1976).

FIG. 10.255. *Heligmonoides josephi* (Wertheim & Durette-Desset, 1976) n. comb., synlophe (after Wertheim & Durette-Desset, 1976).

12–(11) Ridges discontinuous at least on one side of body.
Gradient in size of ridges and dorsal ray not as above.

13–(14) Ridges discontinuous and directed obliquely on dorsal side of body.
Ridges continuous and directed longitudinally on ventral side (Fig. 10.251).
Parasites of ethiopian glirids.
Quentinstrongylus Durette-Desset, 1969

14–(13) Ridges discontinuous and directed longitudinally at least on right side and ventral left quadrant of body (Fig. 10.252).

15–(16) Ridges continuous on dorsal left quadrant of body (Fig. 10.252).
Dorsal ray divided at posterior end.
Parasites of palearctic Sciuridae.
Brevistriata Travassos, 1937

16–(15) Ridges discontinuous on all sides of body.
Dorsal ray deeply divided (Fig. 10.50).
Parasites of oriental Sciuridae, rarely Muridae and Tragulidae.
Fissicauda Durette-Desset & Krishnasamy, 1976

Subfamily **Nippostrongylinae** Durette-Desset, 1971

Key to genera

1–(20) Synlophe well developed with conspicuous, sharply pointed ridges (Fig. 10.49).

2– (7) Carene absent or of type B (supported by two or more pairs of ridges larger than other) (Fig. 10.255).

3– (4) Dorsal ray with thick trunk, extremity in form of inverted "U" (Fig. 10.253).
Bursa symmetrical or subsymmetrical.
Parasites mainly of Muridae, rarely Tupaiidae, Sciuridae and Cricetidae.
Distribution mainly oriental.
Orientostrongylus Durette-Desset, 1970

4– (3) Dorsal ray not as above, extremity in form of inverted "V" (Fig. 10.254).
Bursa usually asymmetrical.

5– (6) Carene absent.
Right lobe of bursa generally larger than left (Fig. 10.256).
Parasites of ethiopian lagomorphs.
Paraheligmonelloides Fukumoto, Kamiya & Suzuki, 1980

6– (5) Carene of type B.
Left lobe of bursa generally larger than right.
Parasites of Old World Muridae (one species in Gerbillidae) (Fig. 10.254).
Heligmonoides Baylis, 1928
(= *Tenorastrongylus* Durette-Desset, 1970)

7– (2) Carene of type A (supported by a hypertrophied left lateral ridge; adjacent ridge on left dorsal side often hypertrophied) (Figs. 10.257, 10.258) or type C (with ridges equal in size to those on ventral left and dorsal right side of body) (Fig. 10.49).

8–(17) Dorsal left ridges smaller than lateral left ridge (Fig. 10.258).
Mainly parasites of Old World rodents.

9–(16) Lateral left ridge not strongly hypertrophied.

10–(13) Bursa symmetrical or sub-symmetrical and ridges of synlophe unequal.

11–(12) Ten to thirteen ridges present: if thirteen, dorsal ray divided above its posterior third (Fig. 10.259).
Parasites of ethiopian Muridae (one species in Gerbillidae).
Neoheligmonella Durette-Desset, 1971

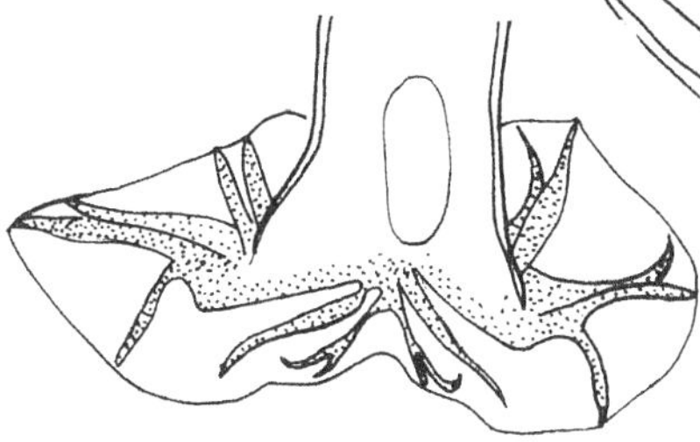

10.256

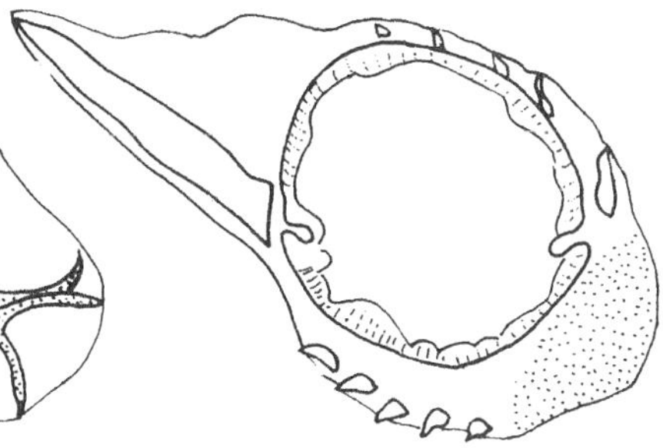

10.257

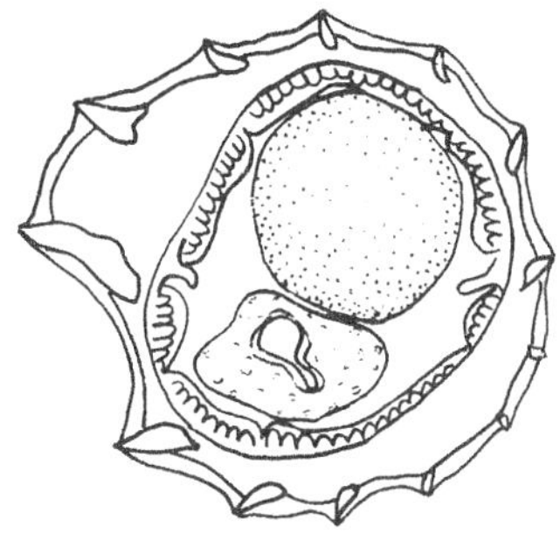

10.258

FIG. 10.256. *Paraheligmonelloides kenyensis* Fukumoto, Kamiya & Suzuki, 1980, bursa (after Fukumoto, Kamiya & Suzuki, 1980).

FIG. 10.257. *Heligmonina dupuisi* (Desset, 1964), synlophe (after Desset, 1964).

FIG. 10.258. *Neoheligmonella bainae* Durette-Desset, 1970, synlophe (after Durette-Desset, 1970).

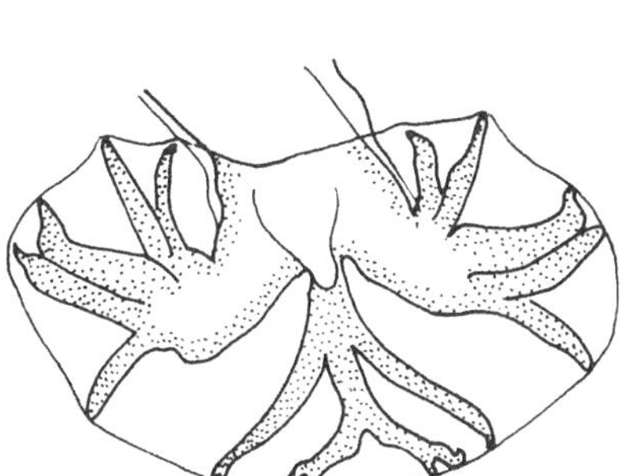

10.259

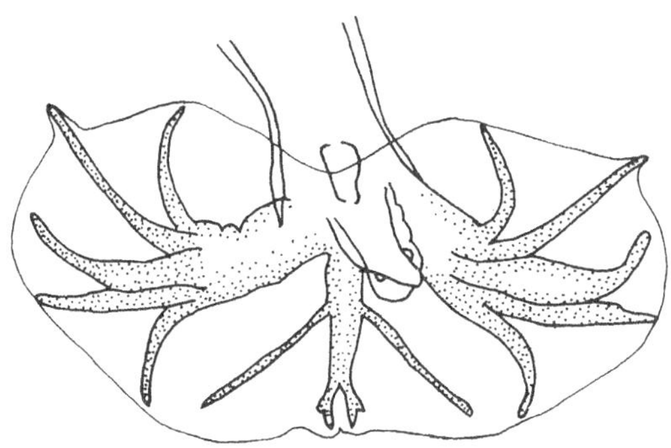

10.260

FIG. 10.259. *Neoheligmonella bainae* Durette-Desset, 1970, bursa (after Durette-Desset, 1970).

FIG. 10.260. *Carolinensis kinsellai* (Durette-Desset, 1969) n. comb., bursa (after Durette-Desset, 1969).

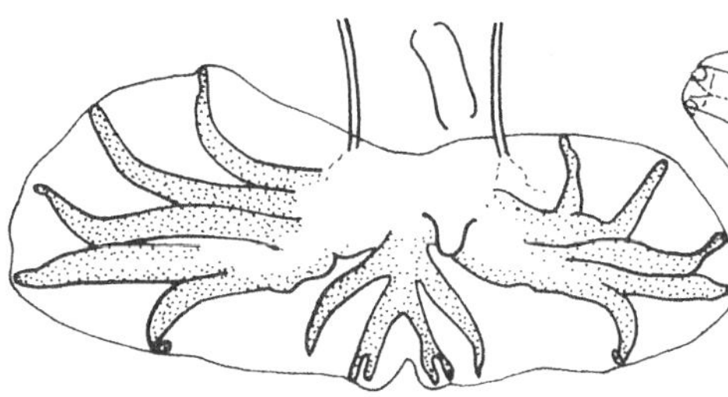

10.261

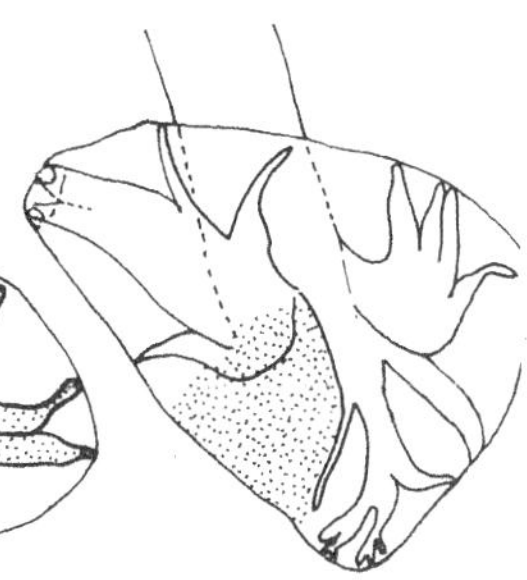

10.262

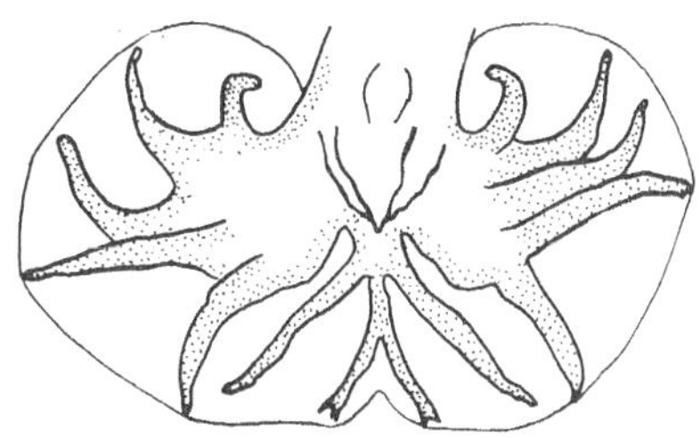

10.263

FIG. 10.261. *Odilia mawsonae* (Durette-Desset, 1969), bursa (after Durette-Desset, 1969).

FIG. 10.262. *Nippostrongylus typicus* (Mawson, 1961), bursa (after Mawson, 1961).

FIG. 10.263. *Hassalstrongylus echalieri* Diaw, 1976, bursa (after Diaw, 1976).

12–(11) Thirteen to sixteen ridges present: if thirteen, dorsal ray divided at its extremity (Fig. 10.260).
Parasites of holarctic Arvicolidae and Gerbillidae.
Carolinensis (Travassos, 1937, subgen.)
(= *Boreostrongylus* Durette-Desset, 1971)

13–(10) Bursa asymmetrical or, if symmetrical, ridges of synlophe of same size.

14–(15) Dorsal ray divided in anterior half (Fig. 10.261).
Rays 6 equal in size.
Parasites of australasian Muridae.
Odilia Durette-Desset, 1973
(= *Austrostrongylus sensu* Durette-Desset, 1971, *nec* Chandler, 1927)

15–(14) Dorsal ray divided in posterior half (Fig. 10.262).
Left ray 6 larger than right ray 6 (Fig. 10.262).
Parasites of oriental and Australian Muridae and oriental Dermoptera.
Cosmopolitan in domiciliated rats.
Nippostrongylus Lane, 1923
(= *Austroheligmonema* Mawson, 1961)

16– (9) Left lateral ridge strongly hypertrophied (Fig. 10.257).
Parasites of ethiopian Muridae especially Guinea-Congo forest zone.
Heligmonina Baylis, 1928
(= *Trichobaylisia* Travassos, 1937; = *Heligmospiroides* Ortlepp, 1939)

17– (8) Dorsal left ridges of same size as left lateral ridge (Fig. 10.49).
Parasites of New World rodents.

18–(19) Synlophe with ridges unequal in size.
Bursa usually symmetrical.
Genital cone not hypertrophied (Fig. 10.263).
Parasites mainly of Cricetidae.
Hassalstrongylus Durette-Desset, 1971

19–(18) Synlophe with ridges almost equal in size (Fig. 10.264).
Bursa usually asymmetrical.
Genital cone hypertrophied (Fig. 10.265).
Parasites mainly of neotropical Cricetidae.
Stilestrongylus Freitas, Lent & Almeida, 1937
(= *Mirandaia* Travassos, 1937)

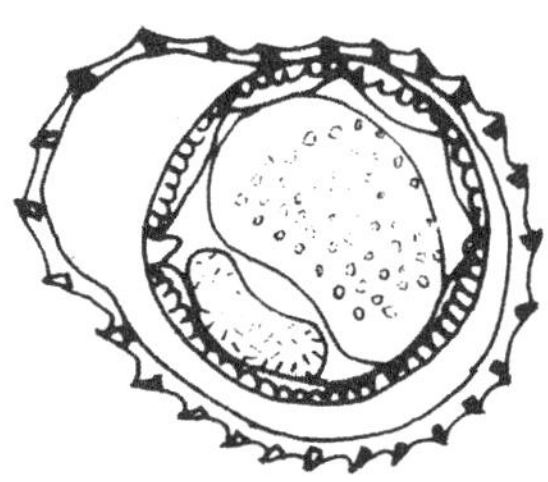

10.264

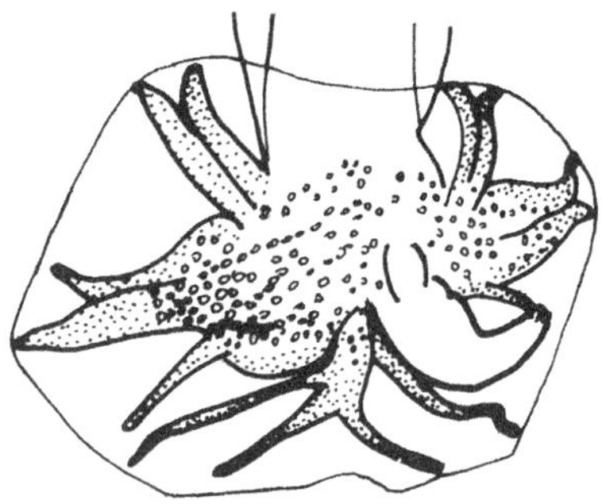

10.265

FIG. 10.264. *Stilestrongylus moreli* Diaw, 1976, synlophe (after Diaw, 1976).
FIG. 10.265. *Stilestrongylus freitasi* Durette-Desset, 1969, bursa (after Durette-Desset, 1969).

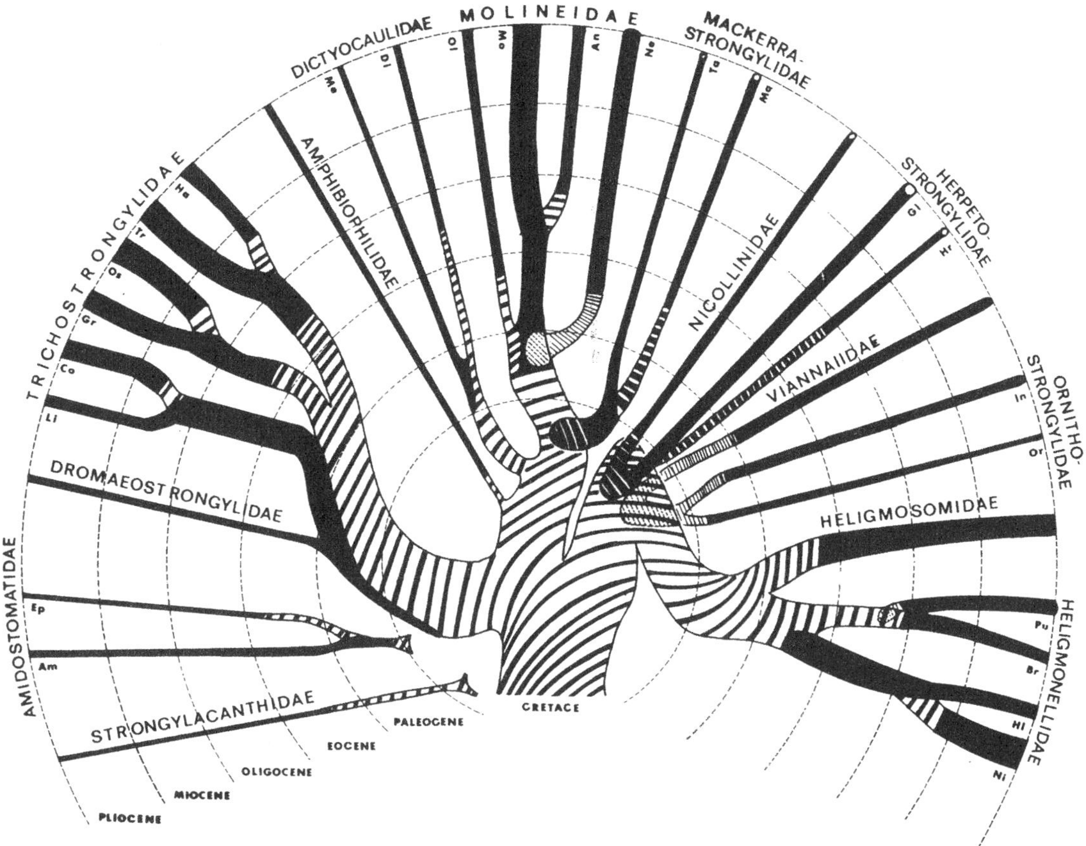

10.266

FIG. 10.266. Hypothetical phylogeny of the Trichostrongyloidea showing proposed period of origin of each of the families and subfamilies.

With the exception of the Strongylacanthidae and the Amidostomatidae, the group is considered monophyletic and to be composed of three principal branches: A, trichostrongylids, B, molineids, C, heligmosomids.

The branch of each family or subfamily is shaded in black starting at the geological period in which the groups originated. Typically Australian groups are depicted as arising from the underside whereas typically neotropical groups are depicted as arising from the upper side of the principal branches. (After Durette-Desset & Chabaud, 1981.)

20– (1) Synlophe reduced, with blunt ridges projecting only slightly above body cuticle.

21–(22) Body relatively thick.
Oesophagus not very long.
Synlophe with many small equal ridges (Fig. 10.234).
Parasites of mammary glands of Old World insectivores and rodents.

Mammanidula Sadovskaja, 1952
(= *Mammolongistriata* Dubinin, 1953; = *Mastonema* Mawson, 1961;
= *Mammaniduloides* Ohbayashi & Fujimaki, 1968)

22–(21) Body relatively slender.
Oesophagus very long.
Synlophe with small number of unequal ridges (Fig. 10.235) or ridges only on left side.
Parasites of neotropical Cricetidae.

Hypocristata Durette-Desset, 1971

REFERENCES

ALI, S. M. & DESHPANDE, G. T. 1969. A new strongylid *Shikhobalovia* gen. nov., from *Hipposideros speoris* from Marathwada, India. *Marathwada Univ. J. Sc.*, **8**, 151–155.

ANDERSON, R. C. 1978. Keys to genera of the Superfamily Metastrongyloidea. *CIH Keys to the Nematode Parasites of Vertebrates* No. 5. Anderson, R. C., Chabaud, A. G. & Willmott, S. (editors), Farnham Royal, Bucks, England: Commonwealth Agricultural Bureaux, 40 pp.

BARUS, V. & VALLE, M. T. 1967. Systematic survey of nematodes parasitizing bats (Chiroptera) in Cuba. *Folia parasit. Praha*, **14**, 121–140.

BAYLIS, H. A. 1928. On a collection of nematodes from Nigerian mammals (chiefly rodents). *Parasitology*, **20**, 280–304.

BRUYN, W. M. DE. 1933. Beitrage zür Kenntnis von *Strongylus circumlitus* Railliet aus den Lungen des Seehundes die neue Gattung *Otostrongylus*. *Zool. Anz.*, **103**, 142–153.

CAMERON, T. W. M. 1935. Studies on the endoparasitic fauna of Trinidad mammals. I. Some parasites of Trinidad deer. *Can. J. Res.*, **13**D, 89–96.

CHABAUD, A. G. 1959. Remarques sur la systématique des nématodes *Trichostrongyloidea*. *Bull. Soc. Zool. France*, **84**, 473–483.

———, PUYLAERT, F., BAIN, O., PETTER, A. J. & DURETTE-DESSET, M. C. 1970. Remarques sur l'homologie entre les papilles cloacales des rhabditides et les côtes dorsales des Strongylida. *C. R. Acad. Sci.* Paris, **271**, 1771–1774.

DIKMANS, G. 1935. New nematodes of the genus *Longistriata* in rodents. *J. Wash. Acad. Sci.*, **25**, 78–81.

DURETTE-DESSET, M. C. 1970. *Tenorastrongylus* n. gen. (nématode—héligmosomatidé) parasite de muridés. *Annls Parasit. hum. comp.*, **45**, 823–828.

———. 1971. Essai de classification des nématodes héligmosomes. Corrélations avec la paléobiogéographie des hôtes. *Mém. Mus. natn. Hist. nat.*, nlle sér., sér. A, Zool., **49**, 126 pp.

———. 1974a. Description de deux nouveaux nématodes héligmosomes chez un sciuridé du Népal. *Bull. Mus. natn. Hist. nat.*, 3e sér., n° 232, Zool. **156**, 819–825.

———. 1974b. Nippostrongylinae (Nematoda: Heligmosomidae) néarctiques. *Annls Parasit. hum. comp.*, **49**, 435–450.

———. 1982a. Sur les divisions génériques des nématodes Ostertagiinae. *Annls Parasit. hum. comp.* **57**, 375–381.

———. 1982b. Sur les divisions génériques des Nématodes Cooperiinae (Trichostrongylidae). *Annls Parasit. hum. comp.* **57**, 383–387.

——— & CHABAUD, A. G. 1977. Essai de classification des nématodes Trichostrongyloidea. *Annls Parasit. hum. comp.*, **52**, 539–558.

——— & ———. 1981a. Nouvel essai de classification des nématodes Trichostrongyloidea. *Annls Parasit. hum. comp.*, **56** (3), 297–312.

——— & ———. 1981b. Sur les Molineinae parasites de mammifères. *Annls Parasit. hum. comp.*, **56**, 489–502.

FREITAS, J. F. TEIXEIRA DE & DOBBIN, J. E. 1960. Nota previa sobre novo nematodeo Strongyloidea parasito de quireptero. *Atas Soc. Biol. Rio de J.*, **4**(4), 56–58.

——— & ———. 1962. Contribuiçao ao conhecimento da fauna helminthologica de quiropteros no estado de Pernambuco, Brasil. *Anais. Fac. Farm., Univ. Recife*, **5**, 53–83.

——— & MENDONCA, J. MACHADO DE, 1960. Nota prévia sobre um novo gênero de nematodeo tricostrongilideo da subfamilia Graphidiinae Travassos, 1937. *Atas Soc. Biol. Rio de J.*, **4**(4), 47–50.

FUKUMOTO, S. I., KAMIYA, M. & SUZUKI, H. 1980. Three trichostrongylid nematodes from the red rock rabbits *Pronolagus* sp. in Kenya. *Jap. J. Vet. Res.*, **28**, 129–136.

GENOV, T. & JANCHEV, J. 1981. *Dessetia* gen. n. (Heligmosomidae) and new data on the morphology and taxonomy of *Dessetia moldovensis* (Andrejko, 1963) comb. n. *C.R. Acad. bulg. Sci.*, **34**, 1157–1160.

GIBBONS, L. M., DURETTE-DESSET, M.-C. & DAYNES, P. 1977. A review of the genus *Impalaia* Mönnig, 1923 (Nematoda: Trichostrongyloidea). *Annls Parasit. hum. comp.*, **52**, 435–446.

GIBBONS, L. M. & KHALIL, L. F. 1982. A key for the identification of genera of the nematode family Trichostrongylidae Leiper, 1912. *J. Helminth.*, **56**, 185–233.

GUTIERREZ, R. O. 1945. *Contribucion al conocimiento de los nematodos parasitos de anfibios argentinos.* Thesis. La Plata No. 8, 37 pp.

INGLIS, W. G. 1968. The geographical and evolutionary relationships of Australian trichostrongyloid parasites and their hosts. *J. Linn. Soc. (Zool.)*, **47**, 327–347.

LOPEZ-NEYRA, C. R. 1946. *Parahistiostrongylus viguerasi* n. sp. Trichostrongylidae nuevo de quiropteros en Espana. *Revta ibér. Parasit.*, **6**, 245–256.

MAWSON, P. M. 1954. *Ichthyostrongylus clelandi* n. gen., n. sp., from an Australian shark. *Trans. R. Soc. S. Aust.*, **77**, 162–163.

———. 1961. Trichostrongyles from rodents in Queensland, with comments on the genus *Longistriata* (Nematoda: Heligmosomatidae). *Aust. J. Zool.*, **9**, 791–826.

———. 1973. Amidostomatinae (Nematoda: Trichostrongyloidea) from Australian marsupials and monotremes. *Trans. R. Soc. S. Aust.*, **97**, 257–279.

NADTOCHYI, E. V. 1970. [Helminth fauna of rodents in the Far East]. *Uchenye Zapiski Dal'nevostochnyi Gosudarstvennyi Universitet (Parasitologicheskie i Zoologischeskie issledovaniya na Dal'nem Vostoke*, **16**, 62–80.

———. 1976. [The erection of a new genus *Heligmoptera* n. g. (Nematoda: Heligmosomatidae) from the Manchurian zokor (*Myospalax psilurus*) from the Primorsk region] [Abstract], 134–135. [*In Third National Conference of Parasitology, Albena, Bulgaria, 12–14 Oct. 1977. Summaries.*]

NIKOLAEVA, V. M. 1969. *Ichthyostrongylus thunni* n. sp. from tunnifish in the Gulf of Mexico. *Mater. nauch. Konf. vses. Obshch. Gel'mint.*, part I, 207–210.

SADOVSKAJA, N. P. 1952, *in* SKRJABIN, K. I., SCHIKHOBALOWA, N. P. & SCHULZ, R. S. 1954. *Osnovi Nematodologii. III. [Trichostrongylids of animals and man], 683 pp. [In Russian].*

SCHULZ, R. S. 1926. Zur Kenntnis der Helminthenfauna der Nagetiere der Union S.S.R. 1. Subordo *Strongylata:* 1. Fam. Trichostrongylidae Leiper, 1912. *Trudy gos. Inst. éksp. Vet.*, **4**, 5–32.

SKRJABIN, K. I., SCHIKHOBALOWA, N. P. & SCHULZ, R. S. 1954. *Osnovi Nematodologii.* III. [Trichostrongylids of animals and man.] 683 pp. [In Russian].

———, ———, & ———. 1954. *Osnovi Nematodologii.* IV. [Dictyocaulidae, Heligmosomatidae and Ollulanidae of Animals.], 323 pp. [In Russian].

SOOD, M. L. & PARSHAD, V. R. 1974. *Gobindonema boodugi* gen. n., sp. n. (Nematoda, Trichostrongylidae) from rodents in Ludhiana (Punjab) India. *Acta Parasit. pol.*, **22**, 93–96.

TANTALEAN, V. & NAUPAY, I. A. 1974. *Parabatrachostrongylus lumbrerasi* n. g., n. sp. (Nematoda, Trichostrongylidae) parasito de anfibio de Arequipa, Peru. *Biota, Peru,* **10**(79), 159–165.

TRAVASSOS, L. 1935. Alguns novos generos e especies de Trichostrongylidae. *Revta méd. cirurg. Bras.*, **43**, 345–361.

———. 1937. Revisão da familia Trichostrongylidae Leiper, 1912. *Monographias do Instituto Oswaldo Cruz,* No. 1, 512 pp.

——— & DARRIBA, A. R. 1929. Notas sobre *Heligmosominae. Sci. med., Rio de J.*, **7**(9), 432–438.

VUYLSTEKE, C. 1956. Note sur quelques nématodes parasites avec description de neuf espèces nouvelles. *Revue Zool. Bot. afr.*, **53**, 441–477.

WILLMOTT, S. 1974. Glossary of terms. *In*: Anderson, R. C., Chabaud, A. G. & Willmott, S. (Editors), *CIH Keys to the Nematode Parasites of Vertebrates.* No 1, pp. 1–5. Farnham Royal, Bucks, England: Commonwealth Agricultural Bureaux.

YEN, W. C. 1973. Helminths of birds and wild animals from Lin-tsan Prefecture, Yunnam Province, China. II. Parasitic nematodes of mammals. *Acta zool. sinica,* **19**, 354–364.

7 STRONGYLIDA

METASTRONGYLOIDEA

by ROY C. ANDERSON*

INTRODUCTION

The metastrongyloids are restricted to certain families of mammals. The group is economically and medically important because many species are significant pathogens of domestic and game animals as well as man. The classification of the lungworms has been subject to two contrasting tendencies even though there is now considerable agreement as to the major groups which should be recognized. Dougherty and others recognized a single family, the Metastrongylidae, which was divided into four subfamilies. On the other hand, Soviet specialists have recognized six families and some fifteen subfamilies. The present classification is an attempt to take the best of both systems, namely, the economy of Dougherty's system and the wealth of sound and detailed morphological information contained in the latest system proposed by Soviet helminthologists (Boev, 1975; Kontrimavichus *et al.*, 1976).

We believe the Metastrongyloidea consists of seven clearly defined families most of which would be regarded as subfamilies in Dougherty's system. The use of several families places the classification of the metastrongyloids more in accord with that of the other bursate nematodes. We do not use, however, the subfamilies proposed for some families which often contain a limited number of genera.

The family Metastrongylidae is biologically unique if confined to *Metastrongylus* of swine. Similarly *Skrjabingylus* of the frontal sinuses of mustelids is unique and requires its own family. The Protostrongylidae contains a homogeneous group of genera found in ruminants and lagomorphs. The Crenosomatidae seems to be a clearly defined but small family found mainly in carnivores and insectivores. The Pseudaliidae is an archaic group restricted generally to the toothed whales. The most controversial proposal in the current work is the separation of the Filaroididae and the Angiostrongylidae. We believe it will be helpful to restrict the former family to abursate forms with the vulva near the tail end, even though genera intermediate between the two families can be shown to exist. Both families require much systematic work and a large number of the smaller species are inadequately described. Undoubtedly many species are still undiscovered. It is hoped that the recognition of the two families will permit the development of the systematics of these groups in a more satisfactory manner.

METASTRONGYLIDAE

Dougherty (1949, 1951) placed *Heterostrongylus* Travassos, 1925 with *Metastrongylus* in the Metastrongylinae, a system followed later by Chabaud (1965). Dougherty was apparently impressed by the lips reported in Travassos's genus. Skrjabin *et al.* (1952) created the subfamily Heterostrongylinae for *Heterostrongylus* and, most recently, Kontrimavichus *et al.* (1976) added *Madafilaroides*, *Madangiostrongylus*, *Marsupostrongylus* and *Plectostrongylus* to this subfamily which, along with the Metastrongylinae, was placed in the Metastrongylidae.

*Department of Zoology, University of Guelph, Canada.

We believe the above systems are in error and that Metastrongylidae should be restricted to *Metastrongylus* of swine, the genus being characterized by a pair of large, lateral, trilobed labia, thick-shelled sculptured eggs, an atypical bursa and earthworm intermediate hosts.

Heterostrongylus heterostrongylus Travassos, 1925 was very inadequately described. We have studied this species from *Didelphis marsupialis* from Cali, Colombia,* as well as the types. There are six prominent lips, a large rounded, unbranched dorsal ray and a fairly complex gubernaculum. We believe that the six lips found in *Heterostrongylus* and other genera (e.g. *Didelphostrongylus*) are homologous to the six flattened structures (perityls) found in many lungworms and that they are morphologically distinct from the large trilobed labia characteristic of *Metastrongylus*. For this reason, Heterostrongylinae is suppressed and *Heterostrongylus* is transferred to the Angiostrongylidae where it is related to *Didelphostrongylus* and other genera with six prominent lips, a well-developed gubernaculum and short spicules. Other genera which were placed in Heterostrongylinae are readily accommodated in Angiostrongylidae.

PROTOSTRONGYLIDAE

This family includes common and important parasites of ruminants and lagomorphs. Its members are transmitted through the agency of terrestrial gastropods. The group is distinguished by a highly developed bursa and complex accessory structures, especially the telamon and gubernaculum (Figs. 5.8, 5.9, 5.15, 5.20, 5.21); spicules are also generally highly developed. We adopt genera and subgenera recognized by Boev (1975) in his excellent monograph. Boev, however, divided the group into six subfamilies mainly on the basis of the structure of the dorsal ray, gubernaculum and telamon. Since the family consists of only 12 genera and three subgenera the number of subfamilies proposed seems excessive. We do not utilize subfamilies but the following key takes into account modifications and synonymies proposed by Boev (1975).

CRENOSOMATIDAE

This family includes a number of species from the bronchi, frontal sinuses and veins of insectivores and carnivores. Gastropods serve as intermediate hosts. The species are characterized by a median vulva and a typical and highly developed bursa. The family includes *Crenosoma*, *Paracrenosoma*, *Troglostrongylus* (= *Bronchostrongylus*) and *Otostrongylus*. *T. delicatus* Travassos, 1946 from *Didelphis* is distinct from *Troglostrongylus* and we propose the following new genus for it:

Prestwoodia n.g.

Generic Diagnosis: Metastrongyloidea, Crenosomatidae. Large worms with highly developed bursa joined anterior to anus and not incised posteriorly. Ventral rays in form of long stalk split at apices. Externolateral ray independent of other lateral rays. Mediolateral and externolateral rays in form of long stalk split at apices. Externodorsal ray well defined, arising independently of other rays. Dorsal ray in form of two mounds, each with small papilla and two long branches. Spicules long and slender. Gubernaculum present. Ovoviviparous. Parasites of nasal sinuses of marsupials (*Didelphis*). Type and only species: *P. delicata* (Travassos, 1946) n. comb. (= *Troglostrongylus delicatus*). The new genus is named in honour of Dr. A. K. Prestwood, University of Georgia.

Unlike Kontrimavichus *et al.* (1976) we do not recognize subfamilies or the validity of *Bronchostrongylus*. These authors also apparently exclude *Otostrongylus*, which in Soviet works is generally placed in the Dictyocaulinae along with *Dictyocaulus* and *Bronchonema*. We follow Chabaud (1965) in associating it with *Crenosoma* and its allies.

*We express our gratitude to Dr. M. D. Little who provided these specimens as well as specimens of *Troglostrongylus delicatus*.

ANGIOSTRONGYLIDAE*

The Angiostrongylinae Boehm & Gebauer, 1934 is elevated to family rank. This family embraces a variety of lungworms with a typical bursa and a posterior vulva. Life-cycles studied involve gastropod intermediate hosts. The group is well represented in marsupials, insectivores, carnivores and rodents. As indicated earlier, the family requires considerable systematic study. It is inherently a difficult group. Many species are extremely small and the bursa is frequently difficult to orientate and study and often subject to distortion upon fixation. Thus, the possibilities for misinterpretation are substantial. Variability in bursa characters has rarely been given sufficient consideration. For these reasons we avoid phylogenetic speculation and, in some instances, recognize subgenera where some authors might prefer genera.

We believe that *Mustelivingylus* is a synonym of *Sobolevingylus* and include in the latter *S. petrovi* and *S. skrjabini* (Romanov & Kontrimavichus, 1962) n. comb. (= *Mustelivingylus skrjabini*). *Sobolevingylus microti* is transferred to *Stefanskostrongylus*, i.e. *S. microti* (Rausch & Rausch, 1969) n. comb. *Sobolevingylus coloradoensis* (Olsen, 1952) Craig & Anderson, 1972 and *S. rodentius* are tıansferred to *Aelurostrongylus* (*Perostrongylus*) resulting in the following: *A.* (*P.*) *coloradoensis* (Olsen, 1952) n. comb. and *A.* (*P.*) *rodentius* (Gubanov & Fedorov, 1965) n. comb. *Glirovingylus* Kontrimavichus, 1967 is placed tentatively as a synonym of *Aelurostrongylus* (*Perostrongylus*). We follow Spratt (1978) in regarding *Plectostrongylus* as a synonym of *Marsupostrongylus*.

We believe that *Perostrongylus* should be recognized, if only as a subgenus of *Aelurostrongylus*, because ovoviviparity and oviparity are easily recognized characters in a group where systematic characters are scarce. *Angiostrongylus* is divided into the subgenera *Angiostrongylus* and *Parastrongylus* as proposed by Dróżdż, 1970; some authors may prefer to regard these subgenera as genera (Chabaud, 1972).

FILAROIDIDAE†

Filaroides and its allies, from primates and carnivores, include delicate worms which are difficult to remove from the tissues of the host and difficult to study morphologically. Life-cycles are extremely varied; in some instances they are direct, in others gastropods or vertebrate intermediate hosts are necessary for transmission. Such variability may indicate the group is far more heterogeneous than generally believed. Numerous species are poorly described and undoubtedly many, especially in carnivores, await discovery. A bursa is lacking in these worms and we believe this justifies assigning them to a distinct family. The more primitive species (e.g. *asper* van Thiel, 1926) which are placed in the genus *Filariopsis* by Webster (1978), align the family to the Angiostrongylidae because in them remnants of lateral and ventral rays are well defined. Following Seneviratna (1959), *Oslerus* is reinstated for species in which the vulva and anus are terminal. This genus is divided into the subgenera *Oslerus* for *osleri* (Cobbold, 1879) and *Anafilaroides* for *rostratus* (Gerichter, 1949) n. comb.

Filaroides is divided into the subgenera *Filaroides* and *Parafilaroides*. The latter in pinnipeds is separated from the former by its reputed lack of caudal papillae and its small spicules. We question the accuracy of the former character as the various species of *Parafilaroides* have been inadequately studied. It may be prudent to emphasize the affinities of this marine group to species in terrestrial carnivores. *Filaroides* includes *martis* (Werner, 1782), *canadensis* Anderson, 1963, *mephitis* Webster, 1967, *hirthi* Georgi & Anderson, 1975, *milksi* Whitlock, 1956, *myonaxi* Sandground, 1937 (= *myonactis*) and *F. orientalis* Kontrimavichus, 1966. The subgenus *Parafilaroides* is reserved for a few poorly described species in pinnipeds, i.e. *F.* (*P.*) *gymnurus* (Railliet,

*We are particularly grateful to Dr. D. Spratt who redescribed species from Australian marsupials and critically examined the following classification for the placement of Australian genera.

†We are grateful for the comments of Mr. W. A. Webster on this group.

1899) n. comb., *F.* (*P.*) *decorus* (Dougherty & Herman, 1947) n. comb., *F.* (*P.*) *nanus* (Dougherty & Herman, 1947) n. comb., *F.* (*P.*) *prolificus* (Dougherty & Herman, 1947) n. comb. The following are regarded as *species inquirendae*: *F. kreisi* Dougherty, 1943, *F. barretoi* (Travassos, 1921) and *F. gordius* (Travassos, 1921).

SKRJABINGYLIDAE

The genus *Skrjabingylus* of the frontal sinuses of mustelids has been a source of controversy for many years. Dougherty (1949) placed *Skrjabingylus* with *Crenosoma*, *Bronchostrongylus*, *Otostrongylus*, *Troglostrongylus* and *Dictyocaulus* in the Skrjabingylinae which he assigned to the Trichostrongyloidea. Skrjabin *et al.* (1952) placed *Skrjabingylus* and *Skrjabinalius* in the Pseudaliidae. Most recently Kontrimavichus *et al.* (1976) have placed the genus in its own family, distinct from the pseudaliids and the crenosomatids. We agree with this decision because of the median vulva and the uniquely modified bursa found in *Skrjabingylus*. We do not believe the genus has close affinities with the pseudaliids or the filaroidids and it certainly belongs in the Metastrongyloidea since its life-cycle is typical of the group (Lankester & Anderson, 1971).

PSEUDALIIDAE

Three subfamilies are recognized in this group which is, except for *Stenuroides* of the mongoose, restricted to the Odontoceti. In this we follow Kontrimavichus *et al.* (1976). However, we follow Arnold & Gaskin (1975) in recognizing *Pharurus* with the synonyms *Otophocaenurus* and *Pseudostenurus*. *Irukanema* seems to be a synonym of *Torynurus* and *Delamurella* a synonym of *Halocercus*. We see little to be gained by recognizing subgenera in *Halocercus* (see Skrjabin, 1942). Life-cycles are unknown.

METASTRONGYLOIDEA

Key to families

1-(2) Cephalic extremity with massive, lateral trilobed lips (Fig. 5.1).
Spicules extremely long and filiform.
Bursa divided into broad lateral lobes by deep terminal incision (Fig. 5.2).
Dorsal ray much reduced; other rays atypical, some absent.
Vulva near anus, prevulvar swelling often prominent and enclosed by provagina (Fig. 5.3).
Eggs with thick sculptured shells containing larvae, passed in faeces of host.
Intermediate hosts earthworms.
Parasites of bronchioles of wild and domesticated swine.
Metastrongylidae

2-(1) Cephalic extremity with six prominent elevations or flat perityls, or smooth.

3-(4) Gubernaculum generally highly developed, in several parts, often paired: telamon often highly developed and forming supporting structure.
Bursa generally highly developed, often bilobed.
Vulva near caudal end; provagina often present.
Oviparous.*

**Oviparous* females deposit unsegmented eggs which subsequently embryonate in the host's tissues. *Ovoviviparous* females deposit first-stage larvae or thin-shelled eggs containing larvae. In both cases, free first-stage larvae are passed in the faeces of the host. *Metastrongylus* is exceptional in the metastrongyloids in that thick-shelled eggs are produced.

First-stage larva often with dorsal spine on tail.
Intermediate hosts terrestrial gastropods.
Parasites of the lungs, musculature or central nervous system of ungulates and lagomorphs.

Protostrongylidae

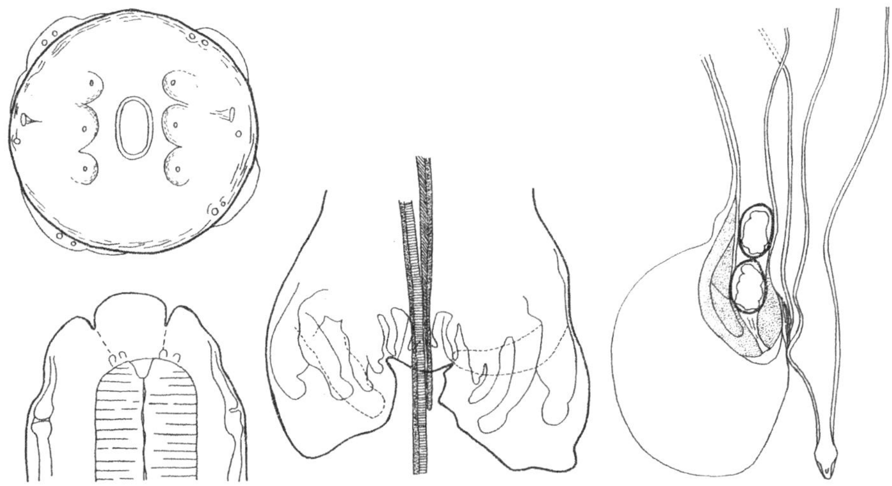

FIG. 5.1. *Metastrongylus madagascariensis* Chabaud & Grétillat, 1956, cephalic extremity female, apical view (after Chabaud & Grétillat, 1956) and *Metastrongylus* sp. from swine, cephalic end female, lateral view. (Original.)
FIG. 5.2. *Metastrongylus madagascariensis* Chabaud & Grétillat, 1956, caudal end male, ventral view. (After Chabaud & Grétillat, 1956.)
FIG. 5.3. *Metastrongylus pudendotectus* Vostokov, 1905, caudal end female, lateral view. (After Dougherty, 1944.)

4-(3) Gubernaculum generally weakly developed or absent; if present unpaired: telamon rarely developed.
5-(8) Bursa typical in form.
6-(7) Vulva in midregion of body.
Amphidelphic, ovejector with prominent sphincter.
Bursa highly developed; rays generally long; dorsal ray large.
Spicules long or short.
Ovoviviparous.*
Intermediate hosts gastropods.
Parasites of the bronchi, sinuses and veins of insectivores and carnivores (including pinnipeds), rarely marsupials.

Crenosomatidae

7-(6) Vulva near anus.
Prodelphic.
Ovejector generally without prominent sphincter.
Bursa highly developed or reduced but typical rays usually always clearly defined.
Oviparous or ovoviviparous.
Parasites of marsupials, insectivores, carnivores and rodents.

Angiostrongylidae

8-(5) Bursa absent or atypical in form.
9-(10) Bursa absent, rays reduced to papillae.
Spicules small, simple in form.
Gubernaculum absent or poorly defined.
Vulva near anus, both sometimes terminal.
Ovejector sometimes muscular with well developed sphincter.
Ovoviviparous.
Intermediate hosts gastropods, fish, or unnecessary.
Parasites of carnivores (including pinnipeds), primates and marsupials (*Antechinus*).
Filaroididae

10-(9) Bursa present but atypical in form.
11-(12) Vulva in midregion of body (Fig. 5.4).
Amphidelphic, ovejector with prominent sphincter.
Bursa modified to form lateral fleshy lobes with six papillae (Fig. 5.5).
Spicules slender.
Gubernaculum greatly reduced.
Ovoviviparous.
Intermediate hosts gastropods.
Parasites of frontal sinuses of carnivores.

Skrjabingylidae

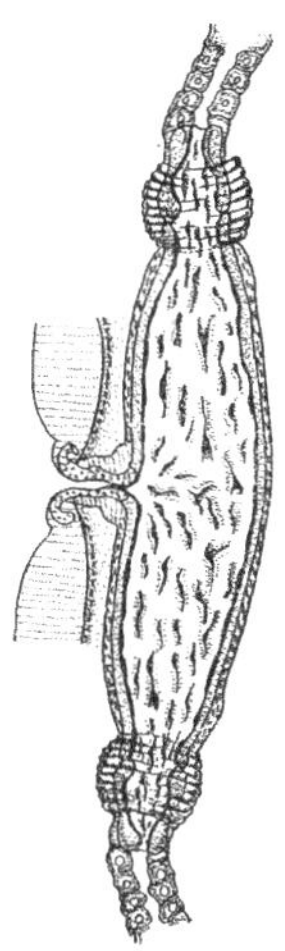
4

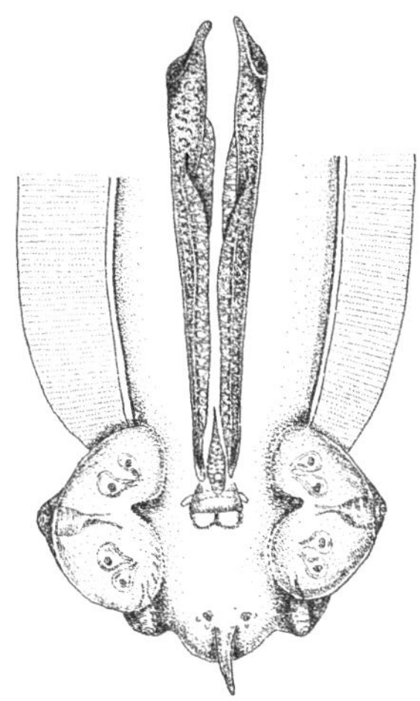
5

FIG. 5.4. *Skrjabingylus nasicola* (Rudolphi, 1842), vulvar region. (After M. Lankester, 1970, University of Guelph, Ph.D. Thesis.)
FIG. 5.5. *Skrjabingylus nasicola* (Rudolphi, 1842), caudal end male, ventral view. (After Lankester, 1970, University of Guelph, Ph.D. Thesis.)

12-(11) Vulva near anus.
Prodelphic.
Ovejector without prominent sphincter.
Bursa reduced, some rays fused to form trunk-like structures, others absent.
Gubernaculum weakly developed, simple.

Ovoviviparous.
Intermediate hosts unknown.
Parasites of lungs, circulatory system and cranial sinuses of toothed cetaceans (one species in the mongoose (*Herpestis*)).

Pseudaliidae

Family *METASTRONGYLIDAE* Leiper, 1908

One genus
Parasites of Suidae

Metastrongylus Molin, 1861
(= *Choerostrongylus* Gedoelst, 1923)

Family *PROTOSTRONGYLIDAE* Leiper, 1926.

Key to genera

1-(2) Bursa reduced in size, ventral and lateral rays short (Fig. 5.6).
Dorsal ray highly developed, trifid.
Telamon consisting of simple basal and transverse plates.
Spicules split 1/3 to 1/2 of length (Fig. 5.7).
Gubernaculum in form of two simple plates.
Provagina weakly developed.
First-stage larva with dorsal spine on tail.
Parasites of air passages of sheep and mouflon (*Ovis*), goats (*Capra*) and chamois (*Rupicapra*).

Muellerius Cameron, 1927

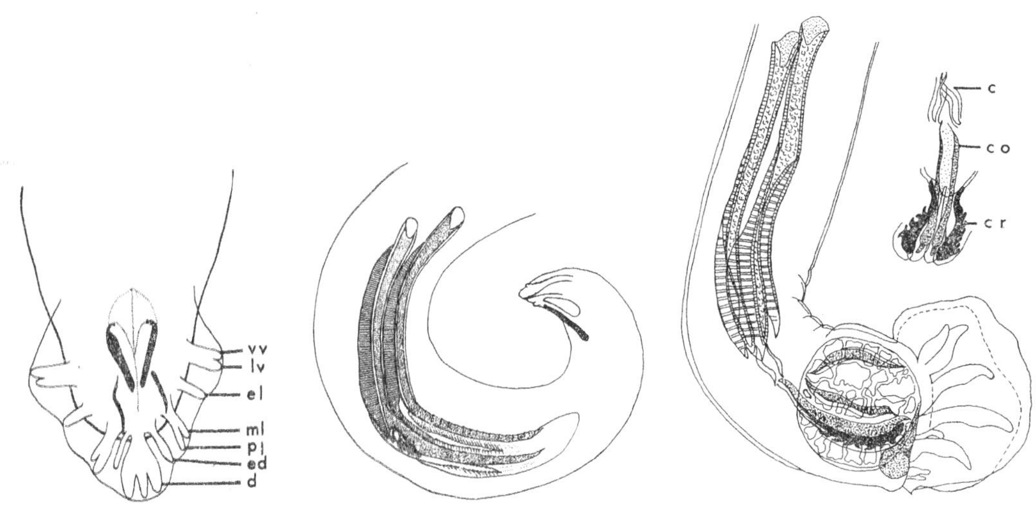

FIG. 5.6. *Muellerius capillaris* (Mueller, 1889), caudal end male, ventral view. (After Gerichter, 1951.) (vv = ventroventral, lv = lateroventral, el = externolateral, ml = mediolateral, pl = posterolateral, ed = externodorsal, d = dorsal.)
FIG. 5.7. *Muellerius capillaris* (Mueller, 1889), caudal end male, lateral view. (After Gerichter, 1951.)
FIG. 5.8. *Pneumocaulus kadenazii* Schulz & Andreeva, 1948, caudal end of male, lateral view, and gubernaculum ventral view. (After Schulz & Andreeva, 1948.) (c = capitulum, co = corpus, cr = crura.)

2*

2-(1) Bursa well developed, ventral and lateral rays normal in size.
3-(4) Pair of large oval gland-like swellings present in body immediately anterior to bursa (Fig. 5.8).
Bursa bilobed, with terminal incision.
Telamon forming complex supporting apparatus.
Dorsal ray short with two bifurcated branches.
Spicules stout.
Gubernaculum present, consisting of capitulum, corpus and crura with serrated edges.
Provagina present.
First-stage larva with dorsal spine on tail.
Parasites of bronchi of musk deer (*Moschus*).

Pneumocaulus Schulz & Andreeva, 1948

4-(3) Oval gland-like swellings not present in body immediately anterior to bursa.
5-(16) Crura if present, stout, much shorter than corpus; capitulum usually absent.
Telamon present or absent.
6-(7) Telamon forming complex supporting structure in male (Fig. 5.9).
Spicules markedly dissimilar in size, thin and long.
Bursa not bilobed.
Dorsal ray spherical, with papillae.
Crura poorly developed.
Provagina small.
Parasites of lungs of goats (*Capra*) and chamois (*Rupicapra*).

Neostrongylus Gebauer, 1932
(= *Neometastrongylus* Kreis, 1944)

7-(6) Complex supporting structure in male absent.
Spicules similar in size.
8-(11) Bursa divided by terminal cleft into two equal lobes.
9-(10) Dorsal ray long with three terminal papillae and one median digitiform papilla (Fig. 5.10).
Ventral rays arising from long common stalk.
Externolateral ray separated from other lateral rays.
Externodorsal ray long and slender.
Spicules short and stout.
Gubernaculum with oval crura, long corpus and delicate capitulum.
Provagina absent.
Parasites of lungs of roe-deer (*Capreolus*), foxes (*Vulpes*) and cats (*Felis*).

Skrjabinocaulus Boev & Sulimov, 1963

10-(9) Dorsal ray not long, usually with several small papillae (Fig. 5.11).
Ventral rays on long common stalk.
Externolateral rays well separated from other lateral rays which arise from common stalk.
Spicules stout, with lancet-shaped lamina.

Crura long, claw-like; capitulum absent; corpus well developed (Fig. 5.12).
Cuticle in vulva region forming sleeve-like structures.
First-stage larva with dorsal spine on tail.
Parasites of lungs of African antelope (*Aepyceros*, *Gazella*, *Damaliscus*, *Alcelaphus* and *Connochaetus*).

Pneumostrongylus Mönnig, 1932

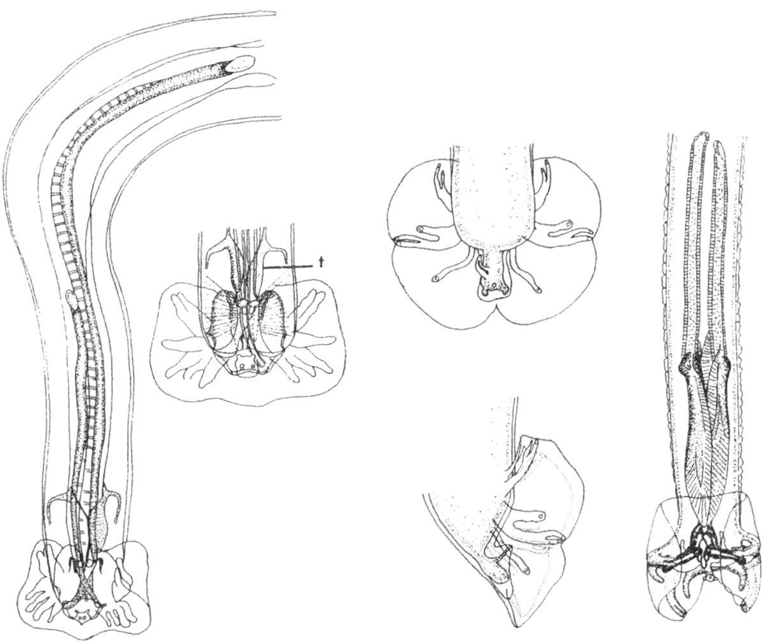

9 10 11

FIG. 5.9. *Neostrongylus linearis* (Marotel, 1913), caudal end male, lateral view. (After Schulz, Kadenazii & Andreeva, 1949 and Boev, 1949.) (t = telamon.)

FIG. 5.10. *Skrjabinocaulus sofievi* Boev & Sulimov, 1963, caudal end male, ventral and lateral views. (After Boev & Sulimov, 1963.)

FIG. 5.11. *Pneumostrongylus calcaratus* Mönnig, 1932, caudal end male, ventral view. (After Boev, 1975.)

11-(8) Bursa not divided by terminal cleft into lobes.

12-(13) Crura absent; corpus boat-shaped; capitulum absent (Fig. 5.13).
Spicules short and stout.
Dorsal ray mound-like, well developed, usually with two or three branches (but variable).
Provagina absent.
First-stage larva with dorsal spine on tail.
Parasites of skeletal musculature and central nervous system of cervids (*Cervus*, *Rangifer* and *Alces*).

Elaphostrongylus Cameron, 1931
(= *Protostrongyloides* Baudet & Verwey, 1951)

13-(12) Crura present.

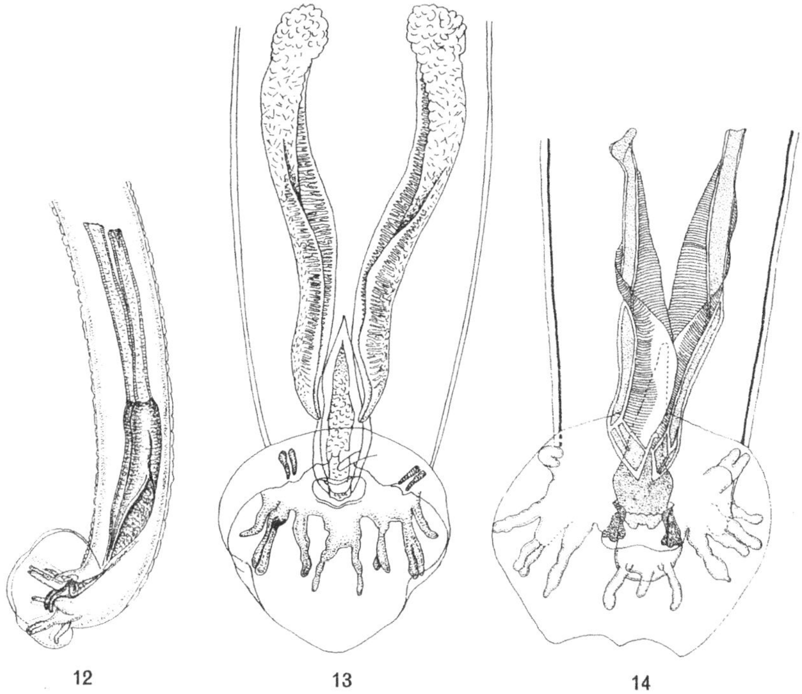

FIG. 5.12. *Pneumostrongylus calcaratus* Mönnig, 1932, caudal end male, lateral view. (After Boev, 1975.)
FIG. 5.13. *Elaphostrongylus cervi* Cameron, 1931, caudal end male, ventral view. (After Lankester & Northcott, 1978.)
FIG. 5.14. *Parelaphostrongylus tenuis* Dougherty, 1945, caudal end male, ventral view. (After Dougherty, 1945.)

14-(15) Dorsal ray well developed, mound-like with 2 to 4 branches and small papillae (Figs. 5.14, 5.15).
Corpus wedge-shaped, tapering anteriorly; crura usually well defined, sometimes fused to corpus; capitulum absent.
Spicules short, stout, lamina well developed.
Provagina absent.
First-stage larva with dorsal spine on tail.
Parasites of skeletal musculature and central nervous system of Cervidae (*Odocoileus*).
Parelaphostrongylus Boev & Schulz, 1950
(= *Odocoileostrongylus* Schulz, 1951;
= *Neurofilaria* Whitlock, 1952)

15-(14) Dorsal ray much reduced.
Crura with serrated edges; corpus long and thin; capitulum present or absent (Fig. 5.16).
Telamon forming supporting structure in form of simple arch.
Spicules moderately long, massive.
Female tail pointed.
Provagina present, moderately developed.
First-stage larva with dorsal spine on tail.

Parasites of lungs of sheep (*Ovis*), goats (*Capra*), deer (*Capreolus*, *Odocoileus*, *Alces*, *Cervus* and *Dama*).

Varestrongylus Bhalerao, 1932
(= *Bicaulus* Schulz & Boev, 1940;
Leptostrongylus Dougherty & Goble, 1946;
Capreocaulus Schulz & Kadenatsii, 1948)

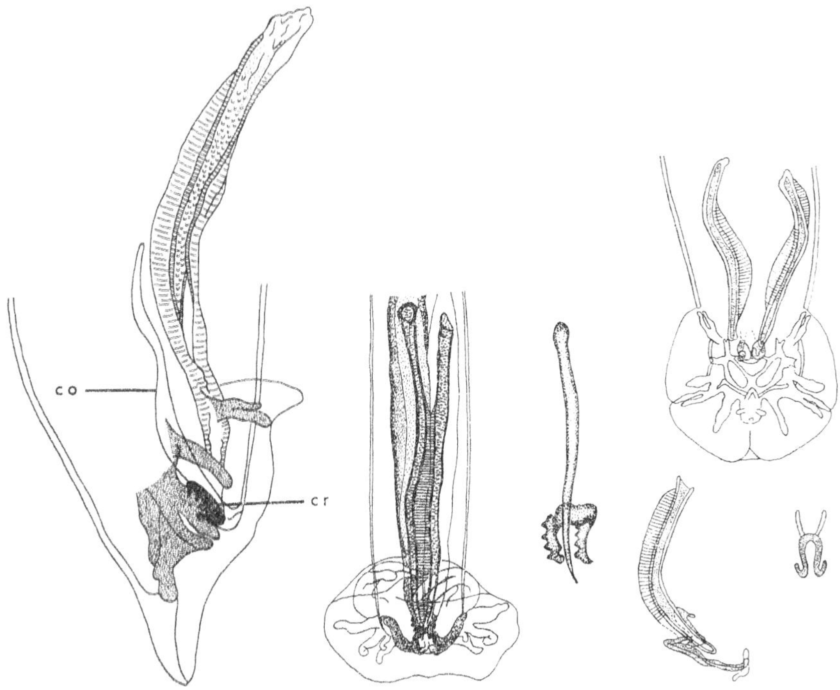

15 16 17

FIG. 5.15. *Parelaphostrongylus tenuis* Dougherty, 1945, caudal end male, lateral view. (After Anderson, 1963.) (co = corpus, cr = crura.)

FIG. 5.16. *Varestrongylus pneumonicus* Bhalerao, 1932, caudal end male and gubernaculum, ventral views. (After Boev, 1957.)

FIG. 5.17. *Orthostrongylus macrotis* (Dikmans, 1931), caudal end male, ventral view; spicules and gubernaculum lateral view; crura ventral view. (After Dikmans, 1931.)

16-(5) Crura thin, as long or almost as long as corpus; capitulum and telamon supporting structure always present.

17-(18) Telamon supporting structure complex (Fig. 5.17).
Capitulum reduced; crura well developed.
Bursa with terminal incision, highly developed.
Ventral rays widely separated from lateral rays.
Externolateral ray independent of other lateral rays.
Dorsal ray spherical with two to three papillae.
Spicules short.
Provagina absent.
Parasites of lungs of deer (*Odocoileus* and *Cervus*) and pronghorn antelope (*Antilocapra*).

Orthostrongylus Dougherty & Goble, 1946

18-(17) Telamon supporting structure in form of simple arches or plates.
Capitulum generally well developed.

19-(20) Dorsal ray long and trifid (Fig. 5.18).
Ventral rays joined to long common stalk.
Externolateral ray independent of other lateral rays which are joined to common stalk.
Externodorsal ray highly developed.
Telamon in form of transverse plate.
Spicules massive, lamina clearly separated from shaft.
Crura with dorsal claw-like extensions; corpus well developed.
Provagina well developed.
First-stage larva with dorsal spine on tail.
Parasites of lungs of sheep (*Ovis*) and goats (*Capra*).
Cystocaulus Schulz, Orlov & Kutass, 1933

20-(19) Dorsal ray mound-like, with papillae.

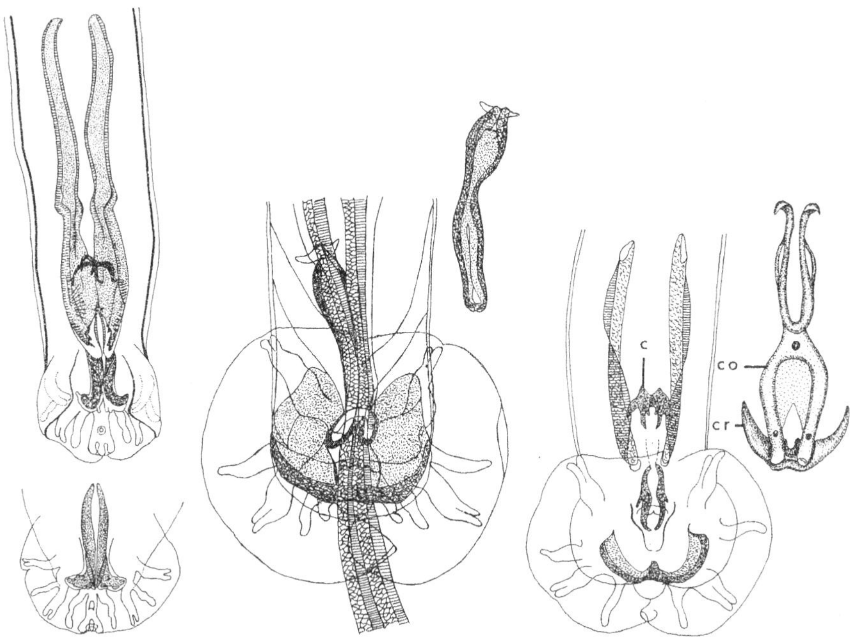

18 19 20

FIG. 5.18. *Cystocaulus ocreatus* (Railliet & Henry, 1907), caudal end male, ventral views. (After Gerichter, 1951.)
FIG. 5.19. *Spiculocaulus leuckarti* Schulz, Orlov & Kutass, 1933, caudal end male and gubernaculum, ventral views. (After Boev, 1952 in Boev, 1975.)
FIG. 5.20. *Protostrongylus* (*Pulmostrongylus*) *pulmonalis* (Frölich, 1802), caudal end male and gubernaculum, ventral views. (After Boev, 1975.) (c = capitulum, co = corpus, cr = crura.)

21-(22) Spicules very long, filiform (Fig. 5.19).
Bursa with terminal incision.
Ventral rays joined to long common stalk.
Externolateral ray independent of other lateral rays which are joined to common stalk.

Telamon forming supporting plates and arches.
Crura well developed; corpus paired, reduced; capitulum usually present and well developed.
Provagina short.
Parasites of lungs of sheep (*Ovis*), goats (*Capra*), roe-deer (*Capreolus*) and chamois (*Rupicapra*).

Spiculocaulus Schulz, Orlov & Kutass, 1933
(= *Spiculostrongylus* Cameron, 1934)

22-(21) Spicules relatively short and stout.
Bursa with or without terminal incision.
Rays well developed, externolateral ray independent of other lateral rays which are joined to common stalk.
Telamon usually forming simple plates and arches.
Gubernaculum with capitulum (usually), corpus and paired crura.
Provagina usually present.
First-stage larva with pointed tail and no dorsal spine.
Parasites of lungs of sheep (*Ovis*), goats (*Capra* and *Oreamnos*), chamois (*Rupicapra*), antelope (*Damaliscus*, *Alcelaphus*, *Connochaetes*, *Alpyceros* and *Gazella*), cervids (*Moschus* and *Odocoileus*), goral (*Naemorhedus*) and lagomorphs (*Lepus*, *Oryctolagus* and *Sylvilagus*).

Protostrongylus Kamensky, 1905
(= *Gelanocaulus* Asadov, 1958;
= *Synthetocaulus* Railliet & Henry, 1907)

23-(24) Edges of crura smooth (Fig. 5.20).

Protostrongylus (*Pulmostrongylus*) Boev, 1975

24-(23) Edges of crura nodulated or toothed (Figs. 5.21, 5.22)

25-(26) Provagina well developed (Fig. 5.23).

Protostrongylus (*Protostrongylus*) Kamensky, 1905

26-(25) Provagina weakly developed or absent.

Protostrongylus (*Davtianostrongylus*) Boev, 1950

Family *CRENOSOMATIDAE* Schulz, 1951

Key to genera

1-(2) Cuticle of body thrown into series of longitudinally striated crenated folds, especially at anterior end (Fig. 5.24).
Spicules moderate in length, delicate and slender, generally with dorsal appendage.
Bursa well developed, each lateral lobe divided into three sub-lobes (Fig. 5.25).
Dorsal ray undivided except for papilla-like structures.
Gubernaculum present.
Vulva often on slight protuberance with pair of cuticular plates.
Ovoviviparous.
Development in gastropods.
Parasites of bronchi of insectivores and carnivores.

Crenosoma Molin, 1861

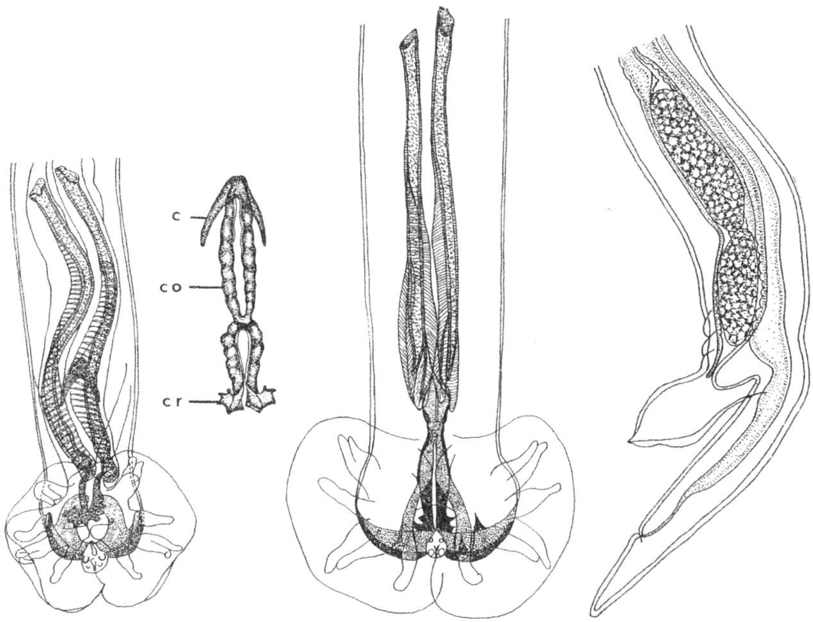

FIG. 5.21. *Protostrongylus* (*Protostrongylus*) *rufescens* (Leuckart, 1865), caudal end male and gubernaculum, ventral views. (After Boev, 1975.)
FIG. 5.22. *Protostrongylus* (*Davtianostrongylus*) *davtiani* (Savina, 1940), caudal end male, ventral view. (After Boev, 1975.)
FIG. 5.23. *Protostrongylus* (*Davtianostrongylus*) *davtiani* (Savina, 1940), caudal end female, lateral view. (After Boev, 1975.)

2-(1) Cuticle of body without longitudinally striated crenated folds.
3-(4) Dorsal ray long, divided into four terminal appendages and bearing externodorsal rays midway along its length (Fig. 5.26).
Lips absent.
Spicules moderately long and slender.
Gubernaculum weakly developed.
Ventroventral and lateroventral rays with long common stalk.
Mediolateral and posterolateral rays with common stalk.
Ovoviviparous.
Parasites of lungs of Soricidae.

Paracrenosoma Yun & Kontrimavichus, 1963

4-(3) Dorsal ray not in form of long stalk divided into four terminal appendages.
Externodorsal rays independent of dorsal ray.
5-(8) Spicules long and filiform (Fig. 5.27).

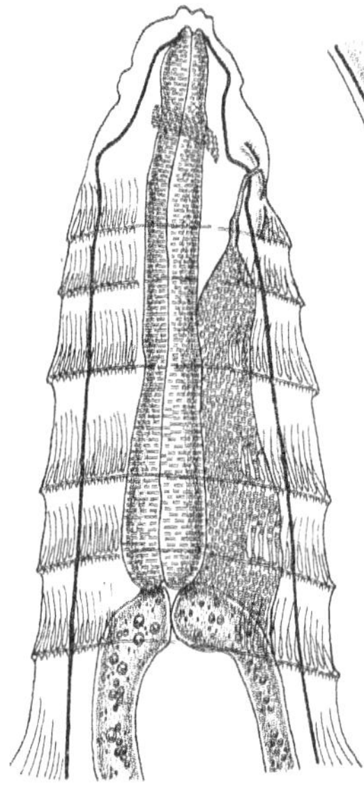

24

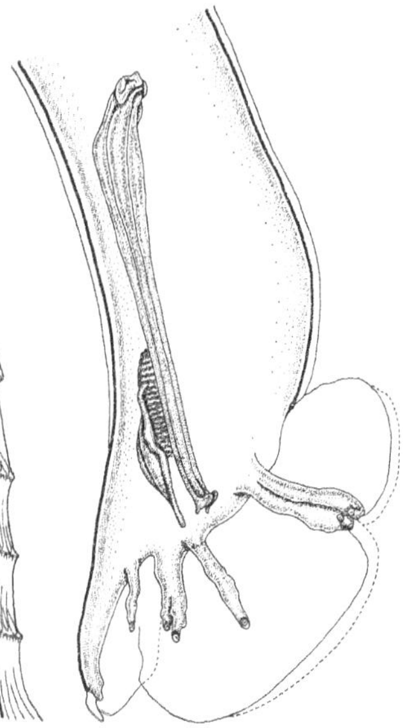

25

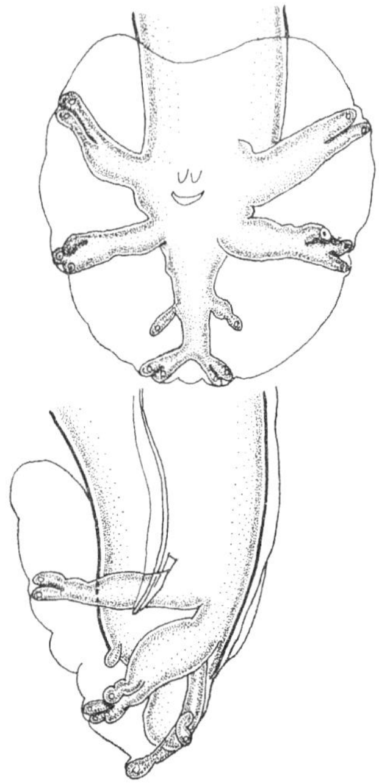

26

FIG. 5.24. *Crenosoma mephitidis* Hobmaier, 1941, anterior end female, lateral view. (After R. Craig, 1972, University of Guelph, M.Sc. Thesis.)

FIG. 5.25. *Crenosoma vulpis* (Dujardin, 1844) caudal end male, lateral view. (After Craig & Anderson, 1972.)

FIG. 5.26. *Paracrenosoma abei* Chabaud, 1973, caudal end male, ventral and lateral views. (After Chabaud, 1973.)

6-(7) Dorsal ray long and broad, terminating in 2 to 4 tiny papillae.
Ventroventral and lateroventral rays with common stalk (Fig. 5.28).
Gubernaculum usually present.
Ovoviviparous.
Development in gastropods.
Parasites of Felidae.

Troglostrongylus Vevers, 1923
(= *Bronchostrongylus* Cameron, 1934;
= *Lynxrufus* Stough, 1953)

7-(6) Dorsal ray in form of two mounds each with a single papilla and two long branches (Fig. 5.29).
Ventral rays on long common stalk.
Externolateral rays well separated from other lateral rays.
Medio- and posterolateral rays on long common stalk.
Externodorsal ray independent of other rays.
Gubernaculum present.
Ovoviviparous.
Parasites of nasal sinuses of *Didelphis*.

Prestwoodia n.g.*

*For generic diagnosis see p. 2.

8-(5) Spicules short and stout with rounded tips and dorsal appendages (Fig. 5.30).
Gubernaculum present, small and oval.
Rays stout, ventroventral and lateroventral rays, and mediolateral and posterolateral rays arising from common stalks.
Parasites of bronchi and veins of pinnipeds (Phocidae).

Otostrongylus de Bruyn, 1933
(= *Kutassicaulus* Skrjabin, 1933)

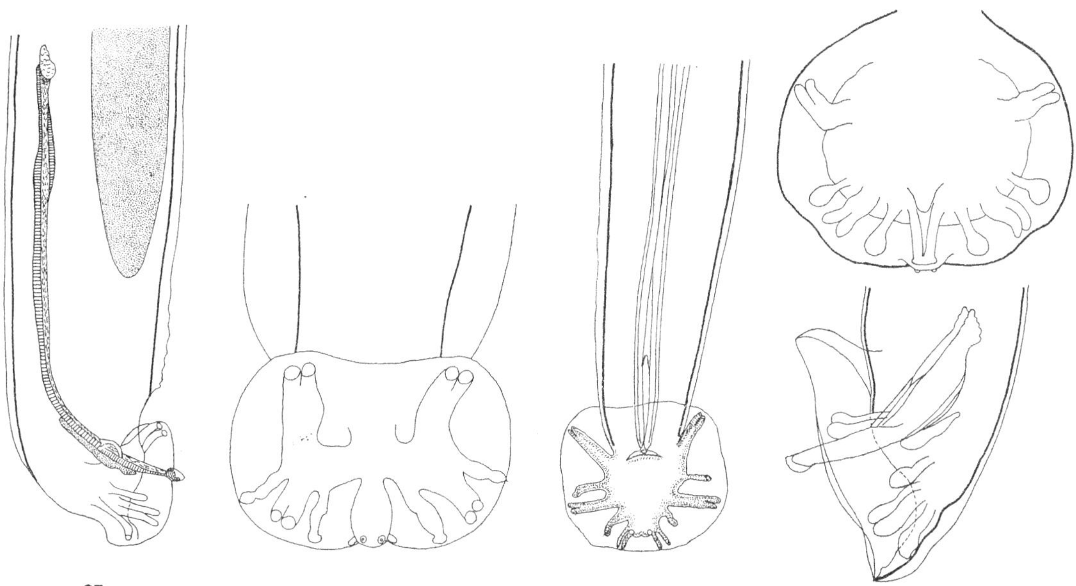

27 28 29 30

FIG. 5.27. *Troglostrongylus brevior* Gerichter, 1949, caudal end male, lateral view. (After Gerichter, 1949.)
FIG. 5.28. *Troglostrongylus brevior* Gerichter, 1949, caudal end male, ventral view. (After Gerichter, 1949.)
FIG. 5.29. *Prestwoodia delicata* (Travassos, 1925) n. comb. from *Didelphis marsupialis*, Colombia, caudal end male, lateral view. (Original.)
FIG. 5.30. *Otostrongylus andreewoi* Skrjabin, 1933, caudal end male, ventral and lateral views. (After Skrjabin, 1933.)

Family *ANGIOSTRONGYLIDAE*

(Boehm & Gebauer, 1934 subfam.) n. fam.*

Key to genera and subgenera

1-(2) Bursa divided into 3 pairs of lobes.
Dorsal ray long and digitiform (Fig. 5.31).
Ventral, lateral and externodorsal rays stout.
Spicules long and slender.
Gubernaculum tiny, triangular.
Tail of female pointed; anus and vulva well separated, not terminal.
Ovoviviparous.
Development in gastropods.
Parasites of Mustelidae (*Martes*).

Trilobostrongylus Anderson, 1963

2-(1) Bursa not divided into 3 pairs of lobes.
Dorsal ray not long and digitiform.

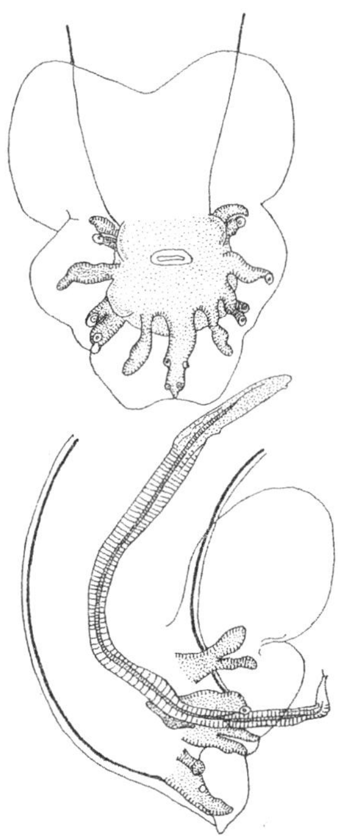

31

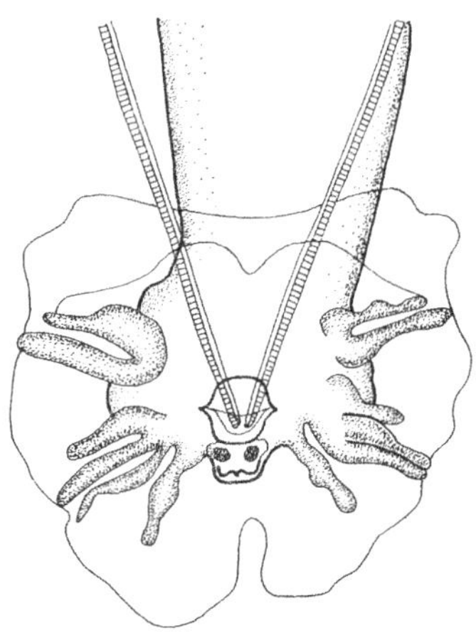

32

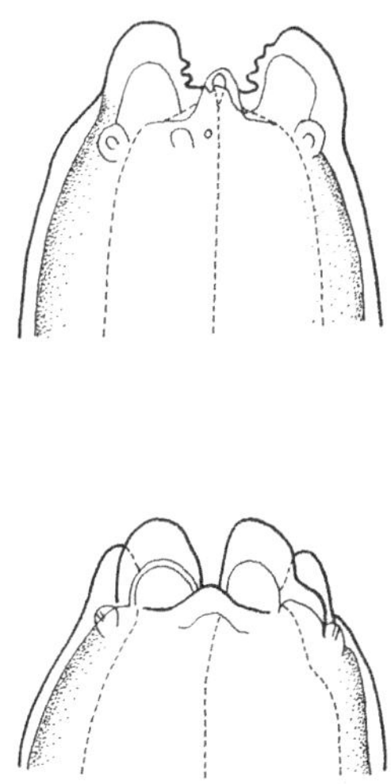

33

FIG. 5.31. *Trilobostrongylus bioccai* Anderson, 1963, caudal end male, ventral and lateral views. (After Anderson, 1963.)

FIG. 5.32. *Madangiostrongylus schulzi* Chabaud & Brygoo, 1960, caudal end male, ventral view. (After Chabaud & Brygoo, 1960.)

FIG. 5.33. *Madangiostrongylus schulzi* Chabaud & Brygoo, 1960, cephalic extremity, lateral and dorsoventral views. (After Chabaud & Brygoo, 1960.)

*See p. 3.

3-(8) Six prominent lips present.
4-(5) Spicules long, slender.
Dorsal ray reduced, mound-like, with papillae (Fig. 5.32).
Bursa with terminal incision.
Ventral and lateral rays and externodorsal rays long and independent.
Gubernaculum long, enveloping spicules.
Lateral lips less prominent than subdorsal and subventral lips (Fig. 5.33).
Vulva and anus not terminal.
Provagina weakly developed.
Ovejector highly muscular.
Parasites of bronchi of insectivores (*Tenrec*).
Madangiostrongylus Chabaud & Brygoo, 1960
5-(4) Spicules short and stout.

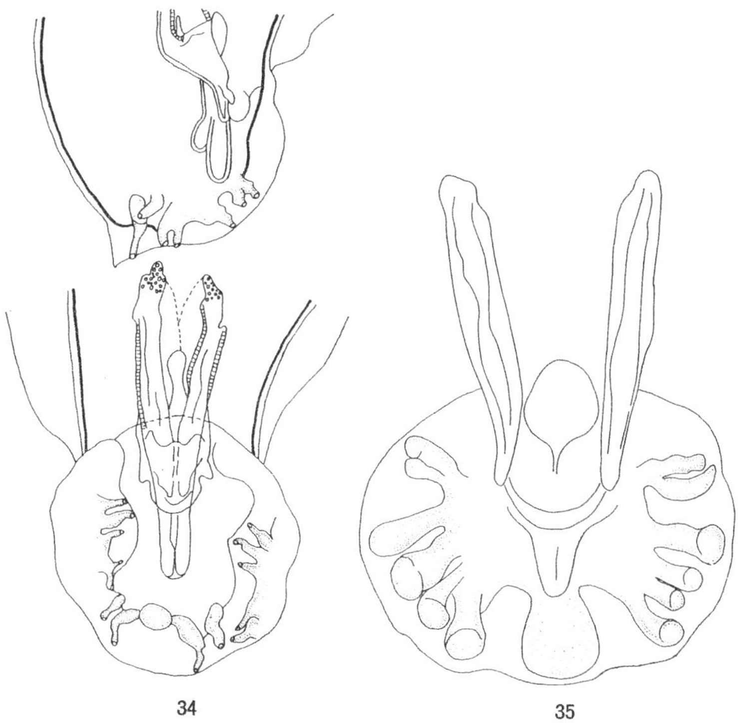

FIG. 5.34. *Didelphostrongylus hayesi* Prestwood, 1976, caudal end male, lateral and ventral views. (Original.)
FIG. 5.35. *Heterostrongylus* sp. from *Didelphis marsupialis*, Colombia, caudal end male, ventral view. (Original.)

6-(7) Dorsal ray mound-like with two long branches (Fig. 5.34).
Gubernaculum complex with terminal tongue-like extension from anus.
Lips set off from body often by narrow constriction.
Teguminal sheath present.

Lateral and ventral rays each joined to common stalk.
Anus and vulva in female not terminal.
Ovejector well developed.
Ovoviviparous.
Development in gastropods.
Parasites of lungs of opossum (*Didelphis*).

Didelphostrongylus Prestwood, 1976

7-(6) Dorsal ray large, rounded, unbranched (Fig. 5.35).
Gubernaculum well developed with terminal tongue-like extension from anus.
Lips set off from body by narrow constriction (Fig. 5.36).
Teguminal sheath present.
Externodorsal ray associated with lateral rays.
Ventral rays joined to short common stalk.
Tail in female short, finger-like.
Ovejector weakly developed.
Ovoviviparous.
Parasites of lungs of opossum (*Didelphis*).

Heterostrongylus Travassos, 1925

8-(3) Six prominent lips not present.
9-(26) Dorsal ray well developed, as large as or larger than most other rays.

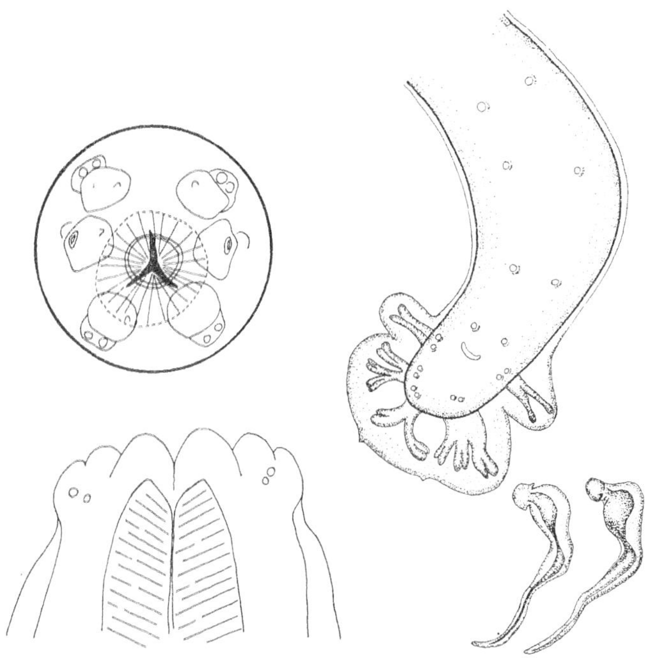

36 37

FIG. 5.36. *Heterostrongylus* sp. from *Didelphis marsupialis*, Colombia, anterior extremity female, en face and lateral views. (Original.)
FIG. 5.37. *Sobolevingylus petrowi* Romanov, 1952, caudal end male, ventral view, and spicules. (After Romanov, 1952.)

4*

10-(11) Lateral rays split at apices (Fig. 5.37).
Dorsal ray divided into 2 main stalks.
Externodorsal rays absent.
Ventral rays separate.
Spicules short, tapering.
Gubernaculum much reduced.
Anus and vulva not terminal.
Parasites of lungs of Mustelidae (*Martes* and *Mustela*).

Sobolevingylus Romanov, 1952
(= *Mustelivingylus* Romanov & Kontrimavichus, 1962)

11-(10) Lateral rays not split at apices.
12-(13) Female tail long and pointed (Fig. 5.38).
Buccal cavity well developed.
Dorsal ray in two branches each split at apices (Fig. 5.39).
Externodorsal rays highly developed.
Ventral rays joined to long common stalk.
Externolateral ray separate from other lateral rays which are on common stalk.
Gubernaculum minute.
Spicules short and stout.
Parasites of bronchi of mongoose (*Herpestes*).

Pulmostrongylus Hsü, 1935
(= *Herpestostrongylus* Khera, 1954)

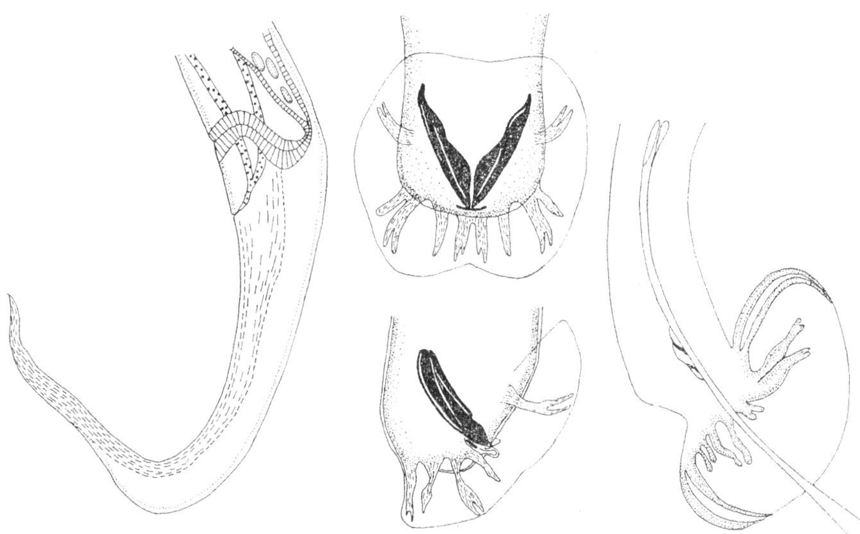

38 39 40

FIG. 5.38. *Pulmostrongylus fengi* Hsü, 1935, caudal end female, lateral view. (After Hsü, 1935.)
FIG. 5.39. *Pulmostrongylus fengi* Hsü, 1935, caudal end male, ventral and lateral views. (After Hsü, 1935.)
FIG. 5.40. *Gurltia paralysans* Wollfhügel, 1933, caudal end male, lateral view. (After Wollfhügel, 1933.)

13-(12) Female tail short and rounded.
Buccal cavity usually weakly developed.
14-(15) Bursa extremely large, with terminal incision (Fig. 5.40).
Spicules long and filiform.
Gubernaculum minute.
Vulva and anus not terminal.
Parasites of veins of leptomeninges of Felidae (*Felis*).
Gurltia Wolffhügel, 1933

15-(14) Bursa not extremely large, without terminal incision.
16-(17) Telamon forming arch-like cuticularized support (Fig. 5.41).
Bursa well developed, asymmetrical.
Dorsal ray with two short branches.
Lateral and ventral rays each with common stalk.
Spicules filiform.
Gubernaculum simple, pointed proximally.
Vulva and anus not terminal, former some distance from latter.
Parasites of pulmonary artery of rodents (*Ondatra*).
Rodentocaulus Schulz, Orlov & Kutass, 1938
17-(16) Arch-like cuticularized support in bursa absent.

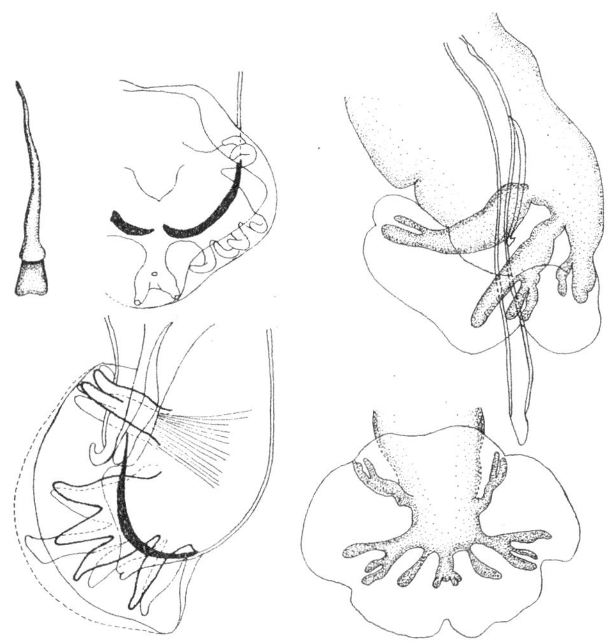

41 42

FIG. 5.41. *Rodentocaulus ondatrae* Schulz, Orlov & Kutass, 1933, caudal end male, ventral and lateral views and gubernaculum. (After Schulz, Orlov & Kutass, 1933.)
FIG. 5.42. *Stefanskostrongylus dubosti* Chabaud, 1972, caudal end male, lateral and ventral views. (After Chabaud, 1972.)

18-(19) Ventral rays and lateral rays each joined to long common stalks (Fig, 5.42).
Dorsal ray usually long with two branches bifurcated at tips.
Spicules and gubernaculum delicate.
Anus and vulva not terminal.
Oviparous.
Parasites of lungs of insectivores and lemurs.
Stefanskostrongylus Dróżdż, 1970

19-(18) Ventral rays and lateral rays not joined to long common stalks.

20-(23) Bursa well developed (Figs. 5.43, 5.44).
Dorsal ray divided into two branches.
Gubernaculum present.

21-(22) Each branch of dorsal ray split at apices (Fig. 5.43).
Gubernaculum well developed.
Teguminal sheath prominent.
Lateral rays arising from common stalk.
Spicules stout, arcuate (Fig. 5.44).
Three subdivided lips present.
Anus and vulva in female not terminal.
Ovoviviparous.
Development in gastropods.
Parasites of lungs of Australian marsupials (*Perameles* and *Isoodon*).
Filostrongylus Mackerras, 1955

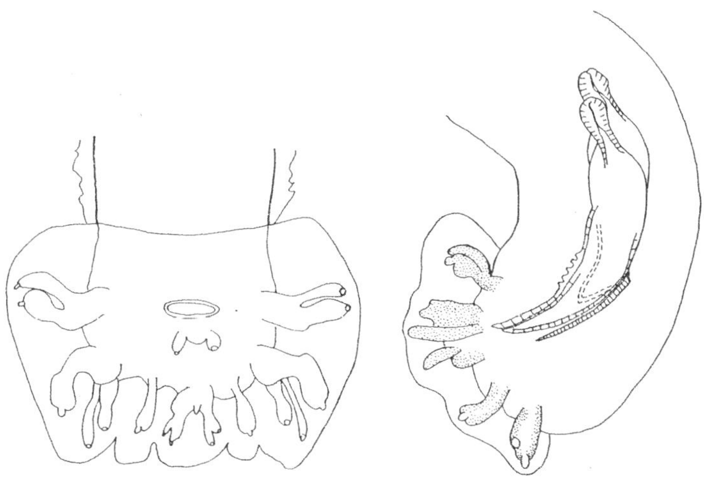

43 44

FIG. 5.43. *Filostrongylus peramelis* Mackerras, 1955, caudal end male, ventral view. (After Spratt, 1978.)
FIG. 5.44. *Filostrongylus peramelis* Mackerras, 1955, caudal end male, lateral view. (After Spratt, 1978.)

22-(21) Each branch of dorsal ray not split (Fig. 5.45).
Gubernaculum weakly developed.
Ventral rays on stout common stalk.
Medio- and posterolateral rays on common stalk.
Spicules delicate, arcuate.
Anus and vulva in female not terminal, tail bluntly pointed.
Oviparous.
Parasites of lungs of rodents.

Gallegostrongylus Mas-Coma, 1978

23-(20) Bursa considerably reduced (Figs. 5.46, 5.47).
Dorsal ray not divided into two branches.
Gubernaculum absent or weakly developed.
Spicules rather stout, often arcuate.
Vulva and anus not terminal.
Parasites of lungs of felids, mustelids and rodents.

Aelurostrongylus Cameron, 1927

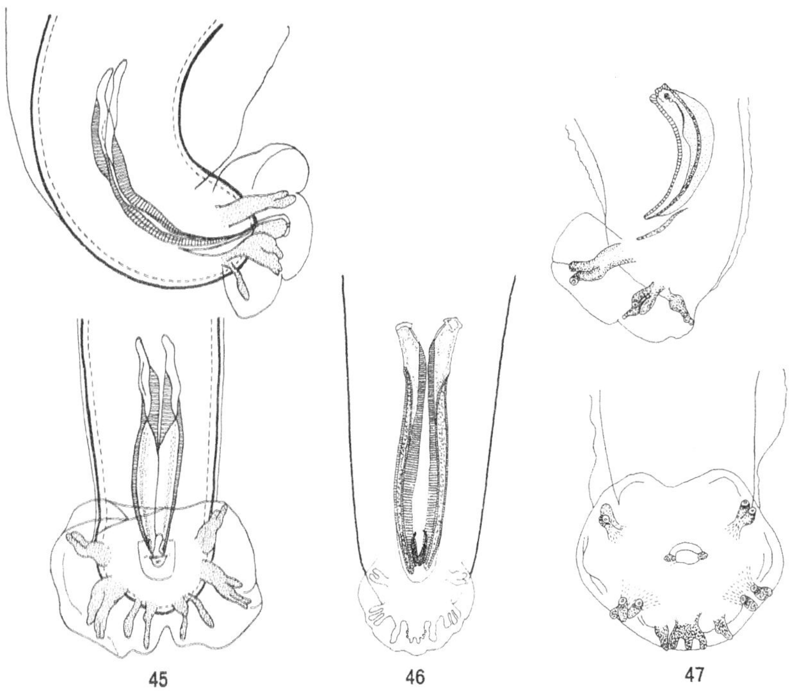

45 46 47

FIG. 5.45. *Gallegostrongylus ibicensis* Mas-Coma, 1977, caudal end male, lateral and ventral views. (After Mas-Coma, 1977.)

FIG. 5.46. *Aelurostrongylus abstrusus* (Railliet, 1898), caudal end male, ventral view. (After Gerichter, 1949.)

FIG. 5.47. *Aelurostrongylus (Perostrongylus) pridhami* Anderson, 1962, caudal end male, lateral and ventral views. (After Anderson, 1962.)

24-(25) Oviparous.

Aelurostrongylus (*Aelurostrongylus*) n. subg.

25-(24) Ovoviviparous.

Aelurostrongylus (*Perostrongylus*) (Schlegel, 1933) n. subg.
(= *Glirovingylus* Kontrimavichus, 1967, genus)

26-(9) Dorsal ray much reduced or absent.

27-(30) Dorsal ray present, often with tiny terminal branches.
Bursa well developed and set off from body.
Terminal bursal incision generally present.
Ventral rays generally on long common stalk.
Externodorsal ray well developed.
Spicules long or short.
Gubernaculum weakly developed or absent.
Vulva and anus not terminal, tail rounded.
Papillae often present near anus.
Parasites of the lungs and blood vessels of insectivores, rodents, felids and canids.

Angiostrongylus Kamensky, 1905

28-(29) Externolateral ray separate from other lateral rays (Fig. 5.48).

Angiostrongylus (*Angiostrongylus*) Dróżdż, 1970
(= *Haemostrongylus* Railliet & Henry, 1907;
= *Cardionema* Yamaguti, 1941; = *Angiocaulus* Schulz, 1951)

29-(28) Externolateral ray joined to common stalk with other lateral rays (Fig. 5.49).

Angiostrongylus (*Parastrongylus*) (Baylis, 1928, genus)
(= *Pulmonema* Chen, 1935; = *Rattostrongylus* Schulz, 1951;
= *Morerastrongylus* Chabaud, 1972)

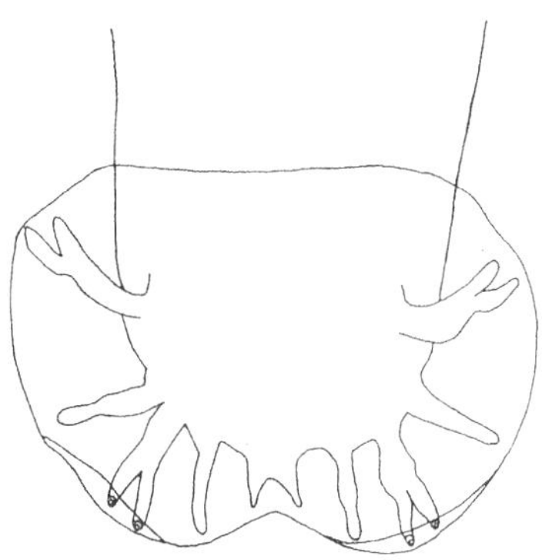

48

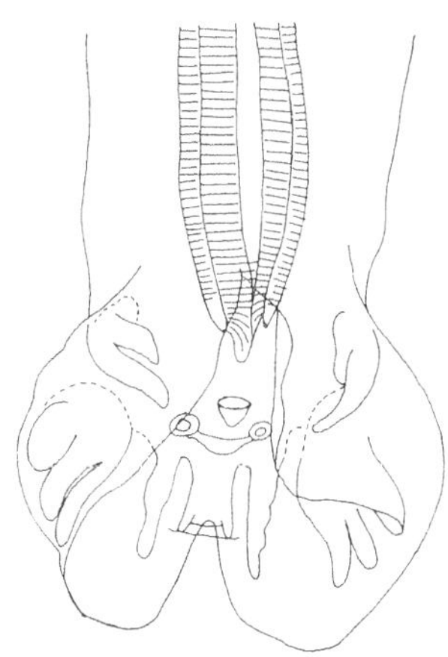

49

FIG. 5.48. *Angiostrongylus* (*Angiostrongylus*) *vasorum* (Baillet, 1866), caudal end male, ventral view. (After Dróżdż, 1970.)

FIG. 5.49 *Angiostrongylus* (*Parastrongylus*) *tateronae* (Baylis, 1928), caudal end male, ventral view. (After Baylis, 1928.)

30-(27) Dorsal ray absent or represented by papillae only.

31-(32) Bursa well developed, clearly set off from body; rays well developed.
Ventral rays rising from common stalk (Fig. 5.50).
Lateral rays rising from common swelling.
Externodorsal ray prominent.
Dorsal ray absent or represented by pair of laterally situated papillae.
Gubernaculum variable, often complex.
Spicules short arcuate (Fig. 5.51).
Anus and vulva not terminal, sometimes close to each other.
Ovoviviparous or oviparous.
Parasites of bronchi of Australian marsupials (*Isoodon*, *Antechinus*, *Perameles*, *Trichosurus*, *Vombatus*, *Schoinobates*, *Thylogale* and *Wallabia*).

Marsupostrongylus Mackerras & Sandars, 1953
(= *Plectostrongylus* Mackerras & Sandars, 1953)

32-(31) Bursa much reduced, not clearly set off from body; rays short.

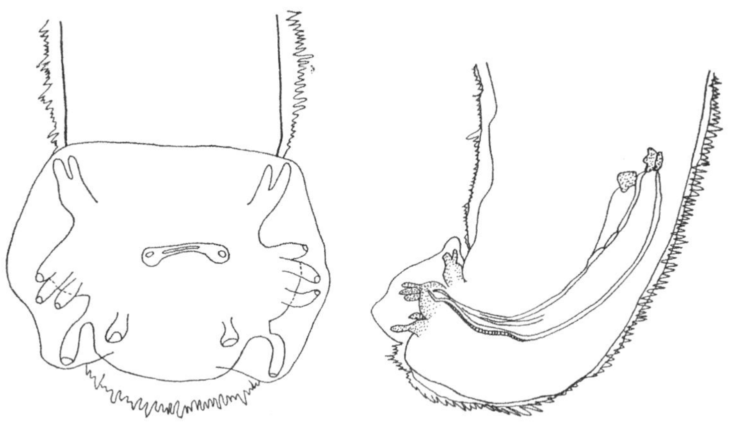

50 51

FIG. 5.50. *Marsupostrongylus bronchialis* Mackerras & Sandars, 1953, caudal end male, ventral view. (After Spratt, 1978.)

FIG. 5.51. *Marsupostrongylus bronchialis* Mackerras & Sandars, 1953, caudal end male, lateral view. (After Spratt, 1978.)

33-(34) Anus and vulva terminal (Fig. 5.52).
Bursa round, ventral in position; rays short and stubby (Fig. 5.53).
Spicules small, thin, subequal.
Gubernaculum minute.
Ovejector muscular, long.
Parasites of bronchi of insectivores (*Tenrec*).

Madafilaroides Chabaud & Brygoo, 1960

34-(33) Anus and vulva in female some distance from caudal extremity (Fig. 5.54).
Bursa reduced to lateral wing-like cuticular expansions (Fig. 5.55).
Spicules arcuate.
Gubernaculum tear-drop shaped.
Teguminal sheath present.
Ovejector with large terminal sphincter.
Ovoviviparous.
Parasites of lungs of Mustelidae (*Mephitis*).

Andersonstrongylus Webster, 1978

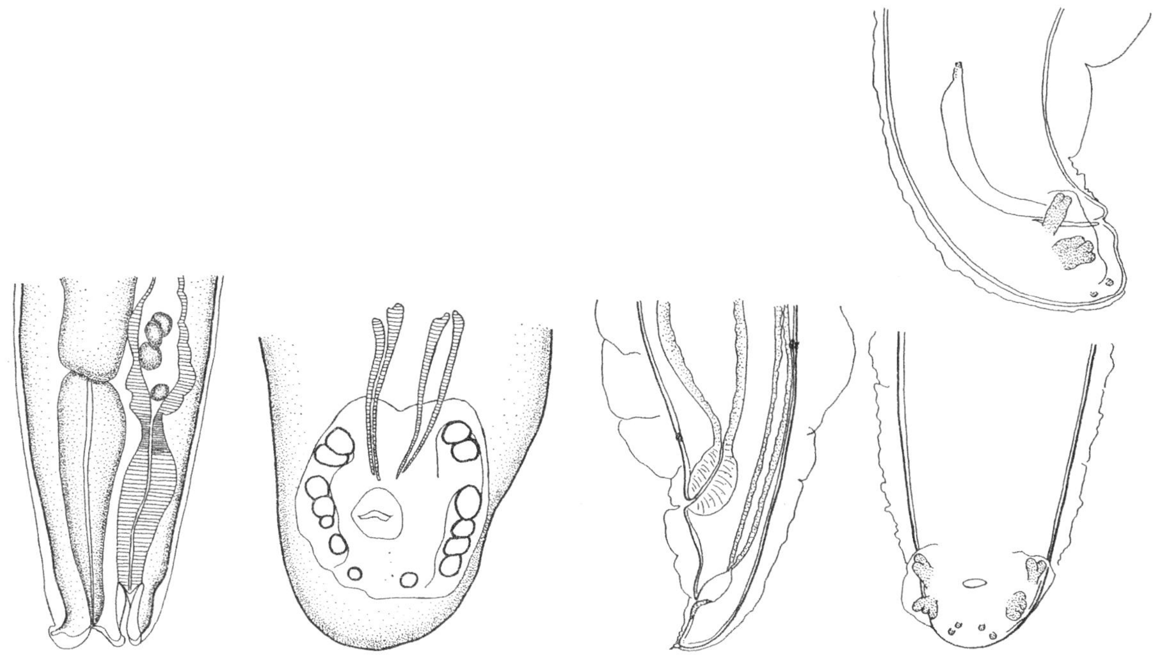

52 53 54 55

FIG. 5.52 *Madafilaroides doughertyi* Chabaud & Brygoo, 1960, caudal end female, lateral view. (After Chabaud & Brygoo, 1960.)

FIG. 5.53. *Madafilaroides doughertyi* Chabaud & Brygoo, 1960, caudal end male, ventral view. (After Chabaud & Brygoo, 1960.)

FIG. 5.54. *Andersonstrongylus captivensis* Webster, 1978, caudal end female, lateral view. (After Webster, 1978.)

FIG. 5.55. *Andersonstrongylus captivensis* Webster, 1978, caudal end male, lateral and ventral views. (After Webster, 1978.)

Family *FILAROIDIDAE* Schulz, 1951

Key to genera and subgenera

1-(2) Vestiges of lateral and ventral rays clearly defined, generally reduced to stubby double or triple appendages or papillae (Fig. 5.56).
Teguminal sheath well developed.
Spicules short, arcuate.
Gubernaculum much reduced or absent.
Anus and vulva not terminal.
Ovejector well developed.
Ovoviviparous.
Parasites of primates.
Filariopsis van Thiel, 1926

2-(1) Vestiges of lateral and ventral rays not clearly defined, papillae, if present, single and sessile.

3-(6) Female tail bluntly rounded with anus and vulva subterminal or terminal in position (Figs. 5.57, 5.58).
Teguminal sheath well developed.
Male tail with several pairs of ventrolateral sessile papillae.
Spicules short and stout, slightly arcuate.
Gubernaculum present.
Parasites of canids and felids.
Oslerus Hall, 1921

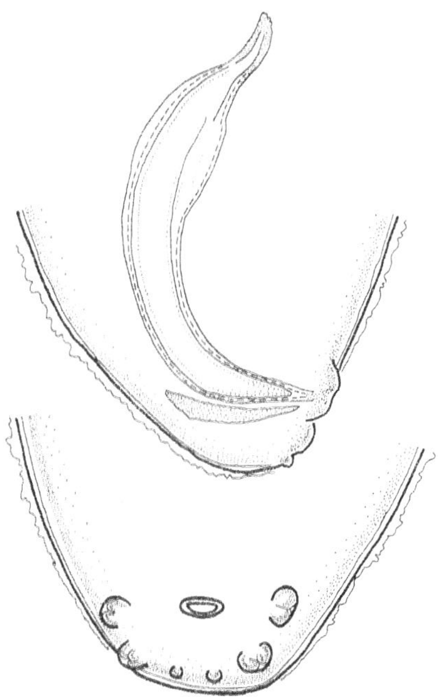

56

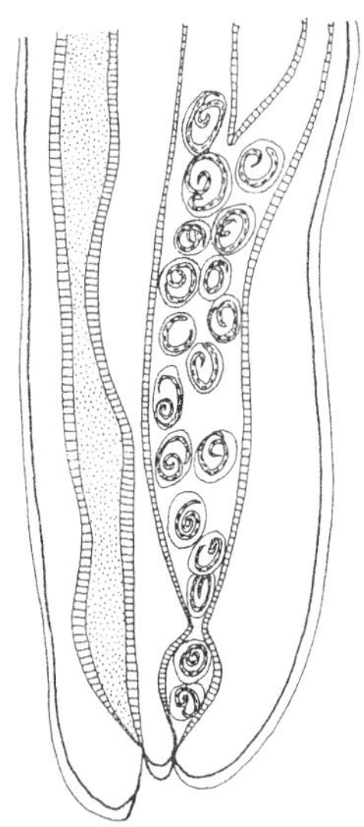

57

FIG. 5.56. *Filariopsis arator* Chandler, 1931, caudal end male, lateral and ventral views. (After Webster, 1978.)
FIG. 5.57. *Oslerus* (*Oslerus*) *osleri* (Cobbold, 1879), caudal end female, lateral view. (After Seneviratna, 1959.)

4-(5) Vaginal sphincter absent or inconspicuous (Fig. 5.57).
Oslerus (*Oslerus*) (Hall, genus) n. subg.

5-(4) Vaginal sphincter highly developed (Fig. 5.58).
Oslerus (*Anafilaroides*) (Gerichter 1948, genus) n. subg.

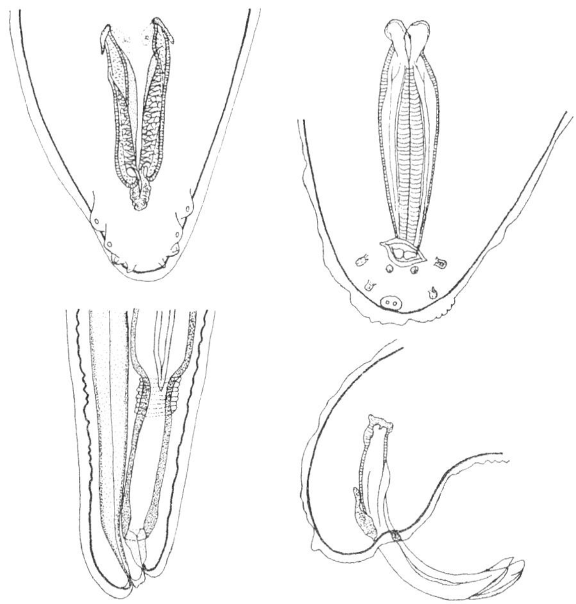

58 59

FIG. 5.58. *Oslerus* (*Anafilaroides*) *rostratus* (Gerichter, 1948), caudal end female, lateral view. (After Gerichter, 1948.)

FIG. 5.59. *Filaroides* (*Filaroides*) *martis* (Werner, 1782), caudal end male, ventral and lateral views. (After Anderson, 1962.)

6-(3) Female tail bluntly rounded with anus and vulva some distance from caudal extremity, never terminal or subterminal.
Caudal papillae in male well developed, absent, or inconspicuous.
Teguminal sheath generally well developed.
Spicules small to medium sized, generally arcuate.
Gubernaculum, if present, weakly developed.

Ovoviviparous.
Intermediate hosts gastropods, fish, or unnecessary.
Parasites of lung parenchyma (often in nodules beside bronchi and bronchioles) of carnivores (including pinnipeds) and, rarely, primates.

Filaroides van Beneden, 1858

7-(8) Caudal papillae in male well developed (Fig. 5.59).
Spicules well developed.
Intermediate hosts gastropods or unnecessary.
Parasites of carnivores, primates and Australian marsupials (*Antechinus*).

Filaroides (*Filaroides*) (van Beneden, genus) n. subg.
(= *Pseudostrongylus* Cameron, 1927)

8-(7) Caudal papillae in male absent or inconspicuous (Fig. 5.60).
Spicules minute.
Gubernaculum weakly developed.
Intermediate hosts coprophagous fish.
Parasites of lung parenchyma of Pinnipedia.

Filaroides (*Parafilaroides*) (Dougherty, 1946, genus) n. subg.

Family *SKRJABINGYLIDAE*

(Skrjabin, 1933 subfam.) Delamure & Kontrimavichus, 1976

One genus.
Parasites of Mustelidae.

Skrjabingylus Petrov, 1927

Family *PSEUDALIIDAE* Railliet & Henry, 1909

Key to subfamilies

1-(4) Lateral rays fused to form pair of prominent, lateral trunk-like structures.
Ventral rays minute and fused.

2-(3) Dorsal ray much reduced or absent (Fig. 5.61); dorsal lobe of bursa absent.
Buccal cavity much reduced.
Perityls absent.
Lateral trunk-like extensions massive, bent ventrally, united by dorsal cuticular flange, extensions at right angles to body, with terminal papillae; stalked papillae near base and median swelling between base and tip.
Ventral rays fused and reduced to small appendages with enlarged tip with two papillae.
Spicules stout with broad blades.
Gubernaculum absent.
Tail of first-stage larva with bend near terminal end.
Parasites of lungs and circulatory system of *Phocoena phocoena*.

Pseudaliinae

3-(2) Dorsal ray well developed; dorsal lobe of bursa usually present.

Stenurinae

4-(1) Lateral rays not fused to form pair of prominent trunk-like structures.
Fused ventral rays more prominent than lateral rays.
Rays often papilliform.

Halocercinae

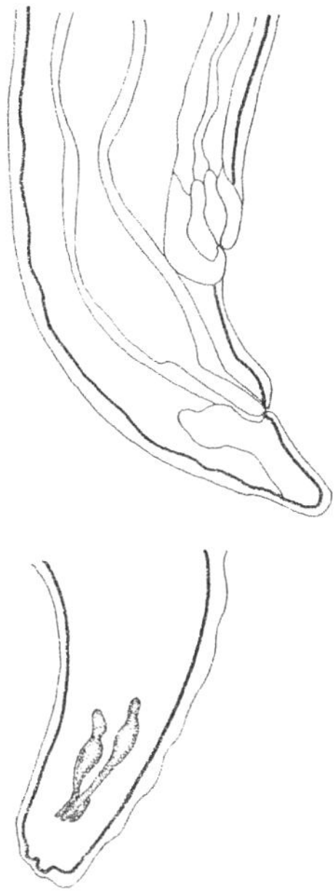

60

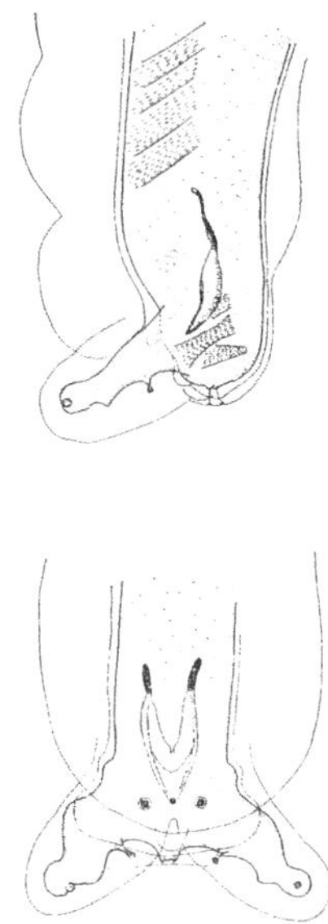

61

FIG. 5.60. *Filaroides* (*Parafilaroides*) *decorus* (Dougherty & Herman, 1947), caudal ends female and male, lateral and ventral views respectively. (After Dougherty & Herman, 1947.)

FIG. 5.61 *Pseudalius inflexus* (Rudolphi, 1808), caudal end male, lateral and ventral views. (After Arnold & Gaskin, 1975.)

Subfamily *Pseudaliinae* Railliet & Henry, 1909

One genus.

Pseudalius Dujardin, 1845

Subfamily **Stenurinae** Skrjabin, 1942

Key to genera

1-(2) Pair of prebursal muscle pads present (Fig. 5.62).
Buccal cavity well developed with distinctly cuticularized walls.
Perityls weakly developed.
Bursa bilobed, dorsal lobe absent.
Dorsal ray reduced but broad with pair of papillae near tip.
Ventral rays fused to form pair of single digitiform appendages.
Lateral trunk-likc cxtensions each with median papillae and three terminal digitiform papillae.
Spicules slender.
Gubernaculum present.
Prebursal cuticular inflation present.
Tail of first-stage larva with terminal spike set off from rest of tail by circular groove.
Parasites of cranial sinuses of Phocoenidae.

Torynurus Baylis & Daubney, 1925
(= *Irukanema* Yamaguti, 1951)

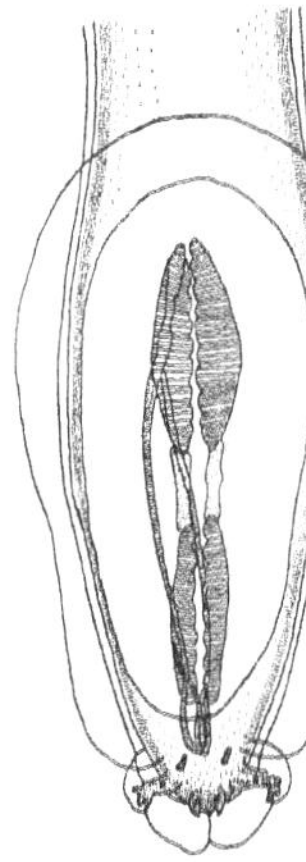

62

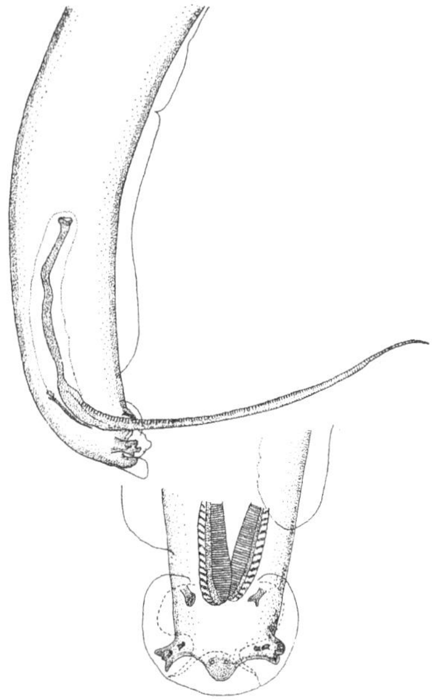

63

FIG. 5.62. *Torynurus convolutus* (Kühn, 1829), caudal end male, ventral view. (After Arnold & Gaskin, 1975.)
FIG. 5.63. *Pharurus alatus* (Leuckart, 1948), caudal end male, lateral and ventral views. (After Arnold & Gaskin, 1975.)

2-(1) Prebursal muscle pads absent.

3-(4) Spicules long and slender (Fig. 5.63).
Buccal cavity well developed with distinctly cuticularized walls.
Perityls weakly developed.
Bursa trilobed, with dorsal and two lateral lobes.
Dorsal ray short and truncate with two ventrolateral papillae at tip.
Ventral rays reduced to pair of small terminally bifid appendages.
Lateral trunk-like structures stout, terminally bifid or trifid, each with pair of digitiform appendages.
Spicules slender, blade with transverse striae.
Gubernaculum much reduced or absent.
Ventrolateral prebursal cuticular inflations present.
Parasites of cranial sinuses of Monodontidae.

Pharurus Leuckart, 1848
(= *Otophocaenurus* Skrjabin, 1942;
= *Pseudostenurus* Yamaguti, 1951)

4-(3) Spicules short and broad, lancet-shaped.

5-(6) Buccal cavity highly developed with thick walls.
Perityls present.
Lateral lobes of bursa united near posterior end of body.
Dorsal ray highly developed with two terminal papillae (Fig. 5.64).
Ventral rays fused to form pair of small appendages with dilated or bifid tips.
Lateral trunk-like structures often trifid terminally, additional digitiform papilla often present near base.
Shaft of spicules rudimentary.
Gubernaculum weakly developed.
Tail of first-stage larva with terminal spike set off from rest of tail by circular groove.
Parasites of lungs and cranial sinuses of Delphinidae, Phocoenidae and Monodontidae.
Stenurus Dujardin, 1845

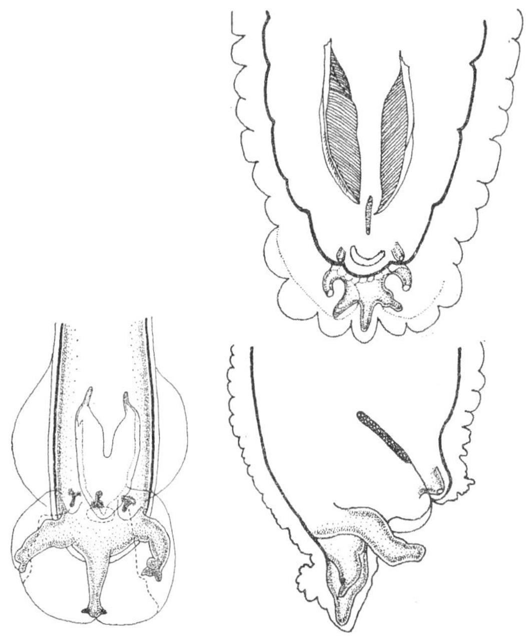

FIG. 5.64. *Stenurus minor* (Kühn, 1829), caudal end male, ventral view. (After Arnold & Gaskin, 1975.)
FIG. 5.65. *Stenuroides herpestis* Gerichter, 1951, caudal end male, ventral and lateral views. (After Gerichter, 1951.)

6-(5) Buccal cavity inconspicuous or absent.
Three shallow lips present.
Oesophagus pear-shaped.
Bursa much reduced, lobes not clearly defined from body cuticle (Fig. 5.65).
Dorsal ray highly developed, trifid.

Ventral rays fused and reduced to pair of tiny digitiform papillae.
Lateral trunk-like structures digitiform and devoid of extensions or papillae.
Gubernaculum much reduced.
First-stage larva with uniformly tapered tail.
Anus and vulva terminal.
Parasites of lungs of *Herpestes ichneumon*.

Stenuroides Gerichter, 1951

Subfamily **Halocercinae** Delamure, 1952

Key to genera

1-(2) Buccal cavity well developed with thick walls.
Spicules long and slender.
Lateral lobes of bursa clearly defined and independent of greatly reduced dorsal lobe.
Dorsal ray broad and undivided.
Ventral rays fused to form pair of large digitiform papillae (Fig, 5.66).
Lateral rays fused to form pair of stubby trifid appendages.
Gubernaculum well developed.
Parasites of *Delphinus delphis*.

Skrjabinalius Delyamure, 1942

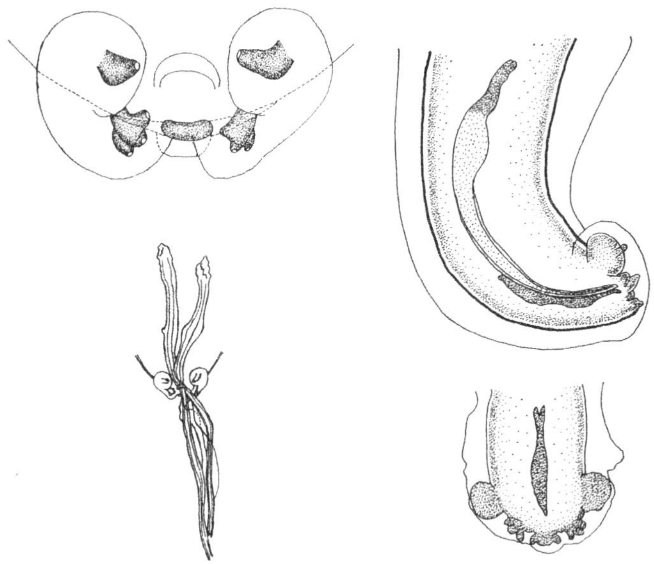

66 67

FIG. 5.66. *Skrjabinalius cryptocephalus* Delyamure, 1942, caudal end male, ventral views. (After Delyamure in Skrjabin, 1932.)

FIG. 5.67. *Halocercus taurica* Delyamure (in Skrjabin, 1942), caudal end male, lateral and ventral views. (After Arnold & Gaskin, 1975.)

2-(1) Buccal cavity weakly developed with thin walls.
Spicules short and arcuate.
Perityls absent.
Bursa generally much reduced with lateral lobes not independent of dorsal lobe (Fig. 5.67).
Dorsal ray weakly developed, sometimes bifid or with pair of papillae.
Ventral rays fused to form pair of large rounded papillae.
Lateral rays fused to form pair of stubby trifid appendages.
Gubernaculum long and slender.
Parasites of lungs of Delphininae.

Halocercus Baylis & Daubney, 1925
(= *Delamurella* Gubanov, 1952)

Acknowledgement

The contribution of Mrs. Uta Strelive in preparing the drawings for these keys and assisting with the manuscript is acknowledged with gratitude.

REFERENCES

ALICATA, J. E. 1968. *Angiostrongylus sandarsae* sp.n. (Nematoda: Metastrongyloidea), a lungworm of rodents in Mozambique, East Africa. *J. Parasit.*, **54,** 896-899.

——— 1968. The gubernaculum of *Angiostrongylus cantonensis* (Chen). *J. Parasit.*, **54,** 1193.

ANDERSON, R. C. 1956. *Elaphostrongylus odocoilei* Hobmaier & Hobmaier, 1934 in the cranial case of *Odocoileus virginianus borealis* Miller. *Can. J. Zool.*, **34,** 167-173.

——— 1962. The systematics and transmission of new and previously described metastrongyles (Nematoda: Metastrongylidae) from *Mustela vison*. *Can. J. Zool.*, **40,** 893-920.

——— 1963. The incidence, development and experimental transmission of *Pneumostrongylus tenuis* Dougherty (Metastrongyloidea: Protostrongylidae) of the white-tailed deer (*Odocoileus virginianus borealis*) in Ontario. *Can. J. Zool.*, **41,** 775-792.

——— 1963. Further studies on the taxonomy of metastrongyles (Nematoda: Metastrongyloidea) of Mustelidae in Ontario. *Can. J. Zool.*, **41,** 801-809.

——— 1968. The pathogenesis and transmission of neurotropic and accidental nematode parasites of the central nervous system of mammals and birds. *Helminth. Abstr.*, **37,** 191-210.

ARNOLD, P. W. & GASKIN, D. E. 1975. Lungworms (Metastrongyloidea: Pseudaliidae) of harbour porpoise *Phocoena phocoena* (L. 1758). *Can. J. Zool.*, **53,** 713-735.

ASH, L. R. 1967. *Angiostrongylus michiganensis* sp.n. (Nematoda: Metastrongyloidea), a lungworm occurring in the shrew, *Sorex cinereus cinereus*, in Michigan. *J. Parasit.*, **53,** 625-629.

BABERO, B. & THOMAS, L. 1960. A record of *Pharurus oserskaiae* (Skrjabin, 1942) in an Alaskan whale. *J. Parasit.*, **46,** 726.

BABOS, S. 1961. Zur Kenntnis der Protostrongylosen der Leporiden, unter besonderer Berücksichtigung der in Ungarn vorkommenden *Protostrongylus*-Arten. *Helminthologia*, **3,** 13-37.

BARUŠ, V. & BLAŽEK, K. 1973. Report on the finding of larval nematodes—*Elaphostrongylus cervi* (Protostrongylidae) in the cranial cavity of a stag. *Folia Parasit.*, **20,** 279-280.

——— & PROKOPIČ, I. 1972. The systematic position and the distribution of nematodes of the genus *Crenosoma* Molin, 1861 parasites in the hedgehog (*Erinaceus europaeus*). *Věst. čsl. Spol. zool.*, **36,** 6-11.

BAYLIS, H. A. 1928. On a collection of nematodes from Nigerian mammals (chiefly rodents). *Parasitology*, **20,** 280-304.

——— & DAUBNEY, R. 1925. A revision of the lungworms of Cetacea. *Parasitology*, **17,** 313-334.

BHAIBULAYA, M. 1968. A new species of *Angiostrongylus* in an Australian rat, *Rattus fuscipes*. *Parasitology*, **58,** 789-799.

——— & CROSS, J. H. 1971. *Angiostrongylus malaysiensis* (Nematoda: Metastrongylidae), a new species of rat-lungworm from Malaysia. *Southeast Asian J. trop. Med. publ. Hlth*, **2,** 527-533.

BIOCCA, E. 1957. *Angiostrongylus chabaudi* n.sp., parassita del cuore e dei vasi polmonari del gatto selvatico (*Felis silvestris*). *Atti Accad. naz. Lincei Rc.*, **22,** 526-532.

BOEV, S. N. 1949. [Characters of the copulatory organs of the lungworm of mountain goats—*Neostrongylus zretkovi* n.sp.] *Dokl. Akad. Nauk SSSR.*, **67,** 759-761. [In Russian.]

——— 1950. [On the systematics of lung nematodes of the genus *Protostrongylus* Kamensky, 1905.] *Trudy gel'mint. Lab.*, **4,** 64-67. [In Russian.]

——— 1957. Lung nematodes of hoofed animals in Kazakhstan. Alma-Ata: Izdatelstvo Akad. Nauk Kazakhskoi SSR. (Translated for the National Science Foundation and the Department of Agriculture, Washington, D.C. by the Israel Program for Scientific Translations, Jerusalem, 1962.)

——— 1958. [On the systematic position of some protostrongylids.] [Collected papers on helminthology presented to Prof. R. S. Shults on his 60th birthday.] Alma-Ata: Kazakhskoe Gosudarstvennoe Izdatelstvo pp. 89-99. [In Russian.]

——— 1960. [The systematic position of *Gelanocaulus boevi* Asadov, 1958.] *Trudy Inst. Zool. Alma-Ata.*, **14,** 43-46. [In Russian.]

——— 1963. [A new nematode from the lungs of *Cervus elaphus*.] *Trudy Inst. Zool., Alma-Ata*, **19,** 89-92. [In Russian.]

——— 1968. [Revision of the subfamily Capreocaulinae.] [Papers on helminthology presented to Academician K. I. Skrjabin on his 90th birthday.] Moscow: Izdat. Akad. Nauk. SSSR, pp. 101-108. [In Russian.]

——— 1968. [Revision of the subfamily Elaphostrongylinae, brain parasites of deer.] *Parazitologiya*, **2**, 312-316. [In Russian.]

——— 1970. Genus *Varestrongylus* Bhalerao, 1932 (Nematoda: Metastrongyloidea) and its composition. In: Singh, K. S. & Tandan, B. K. (Editors), *H. D. Srivastava commemoration volume*. Izatnagar, U.P., Indian Vet. Res. Inst., pp. 263-269.

——— 1971. [An attempt to construct a multivariate key for the Protostrongylidae.] *Trudy Inst. Zool., Alma-Ata*, **31**, 5-13. [In Russian.]

——— 1975. [*Principles of nematology*, edited by K. M. Ryzhikov, Vol. XXV. *Protostrongylidae*.] Moscow, USSR, Izdatel'stvo "Nauka". [In Russian.]

——— & SCHULZ, R. S. 1950. [Reconstruction of the systematics of the nematode family Protostrongylidae Leiper, 1926.] *Dokl. Akad. Nauk. SSSR.*, **70**, 355-358. [In Russian.]

——— & SULIMOV, A. D. 1962. [*Protostrongylus moschi* n.sp. from *Moschus moschiferus*.] *Trudy Inst. Zool., Alma-Ata*, **16**, 42-45. [In Russian.]

——— & SULIMOV, A. D. 1963. [*Skrjabinocaulus sofievi* n.g., n.sp., a new nematode from the lungs of *Capreolus capreolus*.] *Helminthologia*, **4**, 109-114. [In Russian.]

——— & SULIMOV, A. D. 1963. [A new nematode from the lungs of wild carnivores.] [In: *Helminths of man, animals and plants and their control.* Papers on helminthology presented to Academician K. I. Skrjabin on his 85th birthday.] Moscow: Izdatel'stvo Akad. Nauk. SSSR, pp. 174-177. [In Russian.]

BÖHM, L. K. & GEBAUER, O. 1934. Zur System der Familie der Metastrongylidae Leiper, 1908. *Zool. Anz.*, **105**, 287-294.

BRUNETTI, O. D. 1969. Redescription of *Parelaphostrongylus* Boev & Shulz, 1950 in California deer, with studies on its life history and pathology. *Calif. Fish Game*, **55**, 307-316.

CAMERON, T. W. M. 1927. Studies on three new genera and some little known species of the nematode family Protostrongylidae Leiper, 1926. *J. Helminth.*, **5**, 1-24.

CHABAUD, A. G. 1965. Superfamille des Metastrongyloidea. In: Grassé, P. P. (Editor), *Traité de Zoologie*, Masson et Cie, Paris, pp. 921-931.

——— 1972. Description de *Stefanskostrongylus dubosti* n.sp. parasite du potamogale et essai de classification des nématodes Angiostrongylinae. *Annls Parasit. hum. comp.*, **47**, 735-744.

——— 1973. Nouveaux nématodes Metastrongyloidea parasites d'insectivores du Nepal. *Bull. Mus. natn. Hist. nat., Paris, Ser. 3*, No. 136, *Zoologie*, **100**, 751-774.

——— & BAIN, O. 1965. *Aelurostrongylus pottoi* n.sp. métastrongylide parasite de primates. Remarques sur les affinités entre les nématodes de carnivores, d'insectivores et de lemuriens. *Annls Parasit. hum. comp.*, **40**, 569-573.

——— & BRYGOO, E. R. 1960. Deux nouveaux métastrongylides parasites du *Tenrec*. *Mém. Inst. scient. Madagascar*, **14**, 161-172.

———, BRYGOO, E. R. & PETTER, A. J. 1965. Nématodes pulmonaires du limnogale (Insectivores-Tenrecidae). *Annls Parasit. hum. comp.*, **40**, 467-475.

——— & GRÉTILLAT, S. 1956. *Metastrongylus madagascariensis* n.sp. quatrième espèce de strongyle pulmonaire chez le porc domestique. *Annls Parasit. hum. comp.*, **31**, 572-577.

CHANDLER, A. C. 1931. New genera and species of nematode worms. *Proc. U.S. nat. Mus.*, **78** (Article 23), pp. 1-11.

CHEN, H. T. 1935. Un nouveau nématode pulmonaire, *Pulmonema cantonensis* n.g., n.sp. des rats de Canton. *Annls Parasit. hum. comp.*, **13**, 312-317.

COWAN, D. F. 1967. Helminth parasites of the pilot whale *Globicephala melaena* (Traill, 1809). *J. Parasit.*, **53**, 166-167.

CRAIG, R. E. 1972. *Lungworms (Nematoda: Metastrongyloidea) of striped skunk* (Mephitis mephitis). M.Sc. Thesis, University of Guelph, Ontario, Canada.

——— & ANDERSON, R. C. 1972. The genus *Crenosoma* (Nematoda: Metastrongyloidea) in New World mammals. *Can. J. Zool.*, **50**, 1555-1561.

DAILEY, M. D. 1972. Distribution of helminths in the Dall porpoise (*Phocaenides dalli* True). *J. Parasit.*, **57**, 1348.

——— & BROWNELL, R. 1972. The checklist of marine mammal parasites. In: S. H. Ridgway (Editor), *Mammals of the sea*. Springfield, Ill.: C. C. Thomas Publ., pp. 528-589.

DEBRUYN, W. M. 1933. Beiträge zur Kenntnis von *Strongylus circumlitus* Railliet aus den Lungen des Seehundes: die neue Gattung *Otostrongylus*. *Zool. Anz.*, **103**, 142-153.

DELYAMURE, S. L. 1955. Helminthofauna of marine mammals (ecology and phylogeny). [Jerusalem: Israel Program for Scientific Translations (London: H. A. Humphrey Ltd.). Translated from the Russian edition in 1968.]

——— & ALEKSEEV, E. V. 1966. [*Parafilaroides arcticus* n.sp. from *Pusa hispida* in the Chukotsk Sea.] *Problemy Parazit.*, No. 6, pp. 11-15. [In Russian.]

——— & KLEINENBERG, S. E. 1958. [New data on the helminth fauna of the white whale.] *Byull. Mosk. O-Va Ispyt. Prir. Otd. Biol.*, **63**, 25-32. [In Russian.]

——— & SKRYABIN, A. S. 1972. [Diseases, pests and parasites.] *In*: Yablokov, A. V., Bel'kovich, V. M. & Borisov. V. I. [*Whales and Dolphins*]. Moscow: Izdatel'stvo Nauka, pp. 380-409.

DIKMANS, G. 1931. Two new lungworms from North American ruminants and a note on the lungworms of sheep in the United States. *Proc. U.S. natn. Mus.*, **79**, 1-4.

——— 1935. Two new lungworms, *Protostronglyus coburni* n.sp. and *Pneumostrongylus alpenae* from the deer, *Odocoileus virginianus*, in Michigan. *Trans. Am. microsc. Soc.*, **54**, 138-144.

——— 1943. The lungworm, *Protostrongylus rushi* Dikmans, 1937, of the mountain sheep, *Ovis canadensis*. *Proc. helminth. Soc. Wash.*, **10**, 8-9.

——— 1957. A note on the specific identity of *Protostrongylus frosti* Honess, 1942. *Proc. helminth. Soc. Wash.*, **24**, 116-120.

——— & MAPES, C. R. 1950. The lungworm, *Protostrongylus rufescens*, found in domestic sheep, *Ovis aries*, in the United States. *Proc. helminth. Soc. Wash.*, **7**, 16-24.

DINNIK, J. A. & SACHS, R. 1968. A gigantic *Protostrongylus*, *P. africanus* sp.nov., and other lung nematodes of antelopes in the Serengeti, Tanzania. *Parasitology*, **58**, 819-829.

DOUGHERTY, E. C. 1943. The genus *Filaroides* van Beneden, 1858, and its relatives. Preliminary note. *Proc. helminth Soc. Wash.*, **10**, 69-74.

——— 1943. Notes on the lungworms of porpoises and their occurrence on the California coast. *Proc. helminth. Soc. Wash.*, **10**, 74-75.

——— 1944. The lungworms (Nematoda: Pseudaliidae) of the Odontoceti. Part 1. *Parasitology*, **36**, 80-94.

——— 1944. The genus *Metastrongylus* Molin, 1861 (Nematoda: Metastrongylidae). *Proc. helminth. Soc. Wash.*, **11**, 66-73.

——— 1945. A review of the genus *Crenosoma* Molin, 1861 (Nematoda: Trichostrongylidae): its history, taxonomy, adult morphology, and distribution. *Proc. helminth. Soc. Wash.*, **12**, 44-61.

——— 1946. The genus *Aelurostrongylus* Cameron, 1927 (Nematoda: Metastrongylidae), and its relatives; with descriptions of *Parafilaroides* gen.nov., and *Angiostrongylus gubernaculatus* sp.nov. *Proc. helminth. Soc. Wash.*, **13**, 16-26.

——— 1949. The phylogeny of the nematode family Metastrongylidae Leiper, [1909]: a correlation of host and symbiote evolution. *Parasitology*, **39**, 222-234.

——— 1951. A further revision in the classification of the family Metastrongylidae Leiper [1909] (Phylum Nematoda). *Parasitology*, **41**, 91-96.

——— 1951. Evolution of zooparasitic groups in the phylum Nematoda, with special reference to host-distribution. *J. Parasit.*, **37**, 353-378.

——— & GOBLE, F. C. 1946. The genus *Protostrongylus* Kamenskii, 1905 (Nematoda: Metastrongylidae), and its relatives: preliminary note. *J. Parasit.*, **32**, 7-16.

——— & HERMAN, C. M. 1947. New species of the genus *Parafilaroides* Dougherty, 1946 (Nematoda: Metastrongylidae), from sea-lions, with a list of the lungworms of the Pinnipedia. *Proc. helminth. Soc. Wash.*, **14**, 77-87.

DRÓŻDŻ, J. 1970. Révision de la systématique du genre *Angiostrongylus* Kamensky, 1905 (Nematoda: Metastrongyloidea). *Annls Parasit. hum. comp.*, **45**, 597-603.

——— & DOBY, J. M. 1970. *Angiostrongylus dujardini* sp.n. (Nematoda: Metastrongyloidea), parasite de rongeurs (*Apodemus sylvaticus* et *Clethrionomys glareolus*). *Bull. Soc. zool. Fr.*, **95**, 659-668.

FLORES-BARROETA, L. 1955. Helmintos de los perros *Canis familiaris* y gatos *Felis catus* en la Ciudad de Mexico. *An. Esc. nac. Cienc. biol., Méx.*, **7**, 159-202.

GAGARIN, V. G. 1958. [The helminth fauna of wild Carnivora in southern Kirgiz, SSR.] [Collected papers on helminthology presented to Prof. R. S. Shults on his 60th birthday.] Alma-Ata: Kazakhskoe Gosudarstvennoe Izdatel'stvo, pp. 116-121. [In Russian.]

GEBAUER, O. 1933. Beitrag zur Kenntnis von Nematoden aus Affenlungen. *Z. ParasitKde*, **5**, 724-734.

GEORGI, J. R. & ANDERSON, R. C. 1975. *Filaroides hirthi* sp.n. (Nematoda: Metastrongyloidea) from the lung of the dog. *J. Parasit.*, **61**, 337-339.

GERICHTER, C. B. 1949. Studies on the nematodes parasitic in the lungs of Felidae in Palestine. *Parasitology*, **39**, 251-262.

——— 1951. Two new lung nematodes from near-east mammals. *Parasitology*, **41**, 184-188.

——— 1951. Studies on the lung nematodes of sheep and goats in the Levant. *Parasitology*, **41**, 166-183.

GOBLE, F. C. & DOUGHERTY, E. C. 1943. Notes on the lungworms (genus *Protostrongylus*) of varying hares (*Lepus americanus*) in eastern North America. *J. Parasit.*, **29**, 397-404.

GRISI, L. 1971. Ocorrência de *Angiocaulus raillieti* (Travassos, 1927) comb. n. em *Canis familiaris* L. (Nematoda, Protostrongylidae). *Revta bras. Biol.*, **31**, 27-32.

GUBANOV, N. M. & FEDOROV, K. P. 1965. [A new species of parasitic nematode from the lungs of rodents in the Yakutsk region.] [In: *Parasitic worms of domestic and wild animals:* Papers on helminthology presented to Prof. A. A. Sobolev on the 40th anniversary of his scientific and teaching activity.] Vladivostok: Dalnevostochnii Gosudarstvennie Universitet, pp. 78-80. [In Russian.]

HSÜ, H. F. 1935. A study of some Strongyloidea and Spiruroidea from French Indo-China and of *Thelazia chungkingensis* Hsü, 1933 from China. *Z. ParasitenKde*, **7**, 579-600.

KHERA, S. 1954. Nematode parasites of some Indian vertebrates. *Indian J. Helminth.*, **6**, 27-133.

KINSELLA, J. M. 1971. *Angiostrongylus schmidti* sp.n. (Nematoda: Metastrongyloidea) from the rice rat, *Oryzomys palustris*, in Florida, with a key to the species of *Angiostrongylus* Kamensky, 1905. *J. Parasit.*, **57**, 495-497.

KONTRIMAVICHUS, V. L. 1961. [The helminth fauna of *Martes flavigula* and a description of *Skrjabingylus ryjikovi* n.sp. (Pseudaliidae: Nematoda).] *Helminthologia*, **3**, 168-173. [In Russian.]

——— 1967. [Position of the subfamilies Skrjabingylinae Skrjabin, 1933 and Sobolevingylinae Romanov, 1952 in the classification of nematodes.] *Mater. nauch. Konf. vses. Obshch. Gel'mint.*, Year 1966, Pt. 5, pp. 195-202. [In Russian.]

———, DELYAMURE, S. L. & BOEV, S. N. 1976. [*Principles of nematology*, edited by K. M. Ryzhikov. Vol. XXVI. *Metastrongyloidea of domestic and wild animals*.] Moscow, USSR, Izdatel'stvo "Nauka". [In Russian.]

——— & KAZAKOV, B. E. 1966. [Systematic position of nematodes of the genus *Filaroides* Beneden, 1858 from *Mustela sibirica* in the Far East.] *Mater. nauch. Konf. vses. Obshch. Gel'mint.*, Year 1966, Pt. 3, pp. 137-141. [In Russian.]

KREIS, H. A. 1938. Beitrag zur Kenntnis parasitischer Nematoden. VI. Parasitische Nematoden aus dem Zoologischen Garten in Basel. *Zbl. Bakt.*, **141**, 279-304.

LANKESTER, M. W. 1970. *The biology of* Skrjabingylus *spp. (Metastrongyloidea: Pseudaliidae) in mustelids (Mustelidae)*. Ph.D. Thesis, University of Guelph, Ontario, Canada.

——— & ANDERSON, R. C. 1971. The route of migration and pathogenesis of *Skrjabingylus* spp. (Nematoda: Metastrongyloidea) in mustelids. *Can. J. Zool.*, **49**, 1283-1293.

——— & CRICHTON, V. J. 1972. *Skrjabingylus lutrae* n.sp. (Nematoda: Metastrongyloidea) from otter (*Lutra canadensis*). *Can. J. Zool.*, **50**, 337-340.

——— & NORTHCOTT, T. H. 1978. *Elaphostrongylus cervi* Cameron, 1931 (Nematoda: Metastrongyloidea) in caribou (*Rangifer tarandus*) of Newfoundland. *Can. J. Zool.* (in press.)

LEVINE, N. D., IVENS, V., REILLY, J. R. & SIMON, J. 1965. *Filaroides milksi* (Nematoda: Filaroididae) in the lungs of a striped skunk, *Mephitis mephitis*. *J. Parasit.*, **51**, 628-630.

LIU, S. 1965. *Filaroides cebuellae* sp.n. (Nematoda: Metastrongyloidea) from the lung of a pygmy marmoset, *Cebuella pygmaea* (Spix, 1823). *J. Helminth.*, **39**, 225-228.

MACKERRAS, M. J. 1955. A new lungworm from Australian marsupials (Nematoda: Metastrongylidae). *Proc. R. Soc. Qd.*, **66**, 77-81.

——— & SANDARS, D. F. 1953. Two new metastrongyle lungworms from Australian marsupials. *Proc. Roy. Soc. Qd.*, **63**, 71-76.

MAS-COMA, S. 1978. *Gallegostrongylus ibicensis* n.gen., n.sp. (Nematoda: Metastrongylidae) parasite pulmonaire de *Mus musculus* Linnaeus, 1758 (Rodentia: Muridae) à Ibiza (Baléares). *Annls Parasit. hum. comp.*, **52**, 637–642.

——— 1978. Metastrongylidés parasites des Soricidés d'Europe. Description de *Paracrenosoma combesi* n.sp. de *Crocidura russula* Hermann, 1780. *Annls Parasit. hum. comp.*, **52**, 447–456.

MAWSON, P. M. 1953. Parasitic Nematoda collected by the Australian National Antarctic Research Expedition: Heard Island and Macquarie Island, 1948–1951. *Parasitology*, **43**, 291-297.

MÖNNIG, H. O. 1932. New strongylid nematodes of antelopes (Preliminary notes). *Jl S. Afr. vet. med. Ass.*, **3**, 171-175.

MORERA, P. & CESPEDES, R. 1971. *Angiostrongylus costaricensis* n.sp. (Nematoda: Metastrongyloidea), a new lungworm occurring in man in Costa Rica. *Revta Biol. trop.*, **18**, 173-185.

MUMINOV, P. 1964. [*Troglostrongylus badanini* n.sp., a new nematode from the bronchi of *Felis* (*Chaus*) *chaus chaus* Gueldenstadt, 1776.] *Dokl. Akad. Nauk uzbek SSR*, **21**, 55-58. [In Russian.]

NODA, R. 1973. A new species of *Metastrongylus* (Nematoda) from a wild boar with remarks on other species. *Bull. Univ. Osaka Prefect. Ser. B.*, **25**, 21-29.

OGREN, R. E. 1954. Lungworm, *Angiostrongylus blarini* n.sp. from the short-tailed shrew, with observations on the histopathology and life cycle. *J. Parasit.*, **40**, 681-684.

OHBAYASHI, M. & UENO, H. 1974. A new lungworm, *Protostrongylus* (*Davtianostrongylus*) *shiozawai* n.sp. from the Japanese serow, *Capricornis crispus* (Temminck). *Jap. J. vet. Res.*, **22**, 111-115.

OLSEN, O. W. 1952. *Crenosoma coloradoensis* n.sp. (Nematoda: Metastrongylidae) from the lungs of martens, *Martes caurina origenes* (Rhoads). *J. Parasit.*, **38**, 207-209.

ORTLEPP, R. I. 1962. Lungworms from South African antelopes. *Onderstepoort J. vet. Res.*, **29**, 173-181.

PETTER, A. J. 1972. Description d'une nouvelle espèce d'*Aelurostrongylus* parasite de rongeur africain. *Annls Parasit. hum. comp.*, **47**, 131-137.

PRESTWOOD, A. K. 1972. *Parelaphostrongylus andersoni* n.sp. (Metastrongyloidea: Protostrongylidae) from the musculature of white-tailed deer (*Odocoileus virginianus*). *J. Parasit.*, **58**, 897-902.

——— 1976. *Didelphostrongylus hayesi* gen. et sp.n. (Metastrongyloidea: Filaroididae) from the opossum, *Didelphis marsupialis. J. Parasit.*, **62**, 272-275.

PRYADKO, E. I. & BOEV, S. N. 1971. [Systematics, phylogeny and evolution of Elaphostrongylinae—nematodes of deer.] *Izv. Akad. Nauk kazakh. SSR, Ser. Biol.*, No. 5. pp. 41-48. [In Russian.]

RAUSCH, R. L. & RAUSCH, V. R. 1969. Studies on the helminth fauna of Alaska. XLVII. *Sobolevingylus microti* sp. nov. (Nematoda: Pseudaliidae), a lungworm of rodents. *Can. J. Zool.*, **47**, 443-447.

ROMANOV, I. V. 1952. [New species of helminths from *Martes zibellina.*] *Trudy gel'mint. Lab.*, **6**, 323-330. [In Russian.]

——— & KONTRIMAVICHUS, V. L. 1962. [*Mustelivingylus skrjabini* n.g., n.sp. from Mustelidae.] *Trudy gel'mint. Lab.*, **12**, 94-97. [In Russian.]

SANDGROUND, J. A. 1937. Three new parasitic nematodes from the Belgian Congo. *Rev. Zool. Bot. afr.*, **29**, 230-236.

SENEVIRATNA, P. 1959. Studies on the family Filaroididae Schulz, 1951. *J. Helminth.*, **33**, 123-144.

SCHULZ, R. E. 1951. [Phylogenesis of strongylid nematodes and reconstruction of the Metastrongyloidea.] *Dokl. Akad. Nauk SSSR.*, **80**, 293-296. [In Russian.]

——— & ANDREEVA, N. K. 1948. [Morphology and biology of a new nematode from the musk-deer.] *Dokl. Akad. Nauk SSSR.*, **62**, 841-843. [In Russian.]

——— & KADENATSII, A. N. 1948. [Morphology of the copulatory organs in the new nematode genus *Capreocaulus* from roe-deer.] *Dokl. Akad. Nauk SSSR*, **63**, 341-344. [In Russian.]

———, ——— & ANDREEVA, N. K. 1949. [Anatomical structure of the male genitalia in the nematode genus *Neostrongylus* Gebauer, 1932.] *Dokl. Akad. Nauk SSSR*, **67**, 763-765. [In Russian.]

———, ORLOV, I. V. & KUTASS, A. J. 1933. Zur Systematik der Subfamilie Synthetocaulinae Skrj. nebst Beschreibung einiger neuer Gattungen und Arten. Vorläufige Mitteilungen. *Zool. Anz.*, **102**, 303-310.

SKRJABIN, K. I. 1933. Sur la position systématique des nématodes du genre *Oslerus* Hall, 1921. *Bull. Soc. zool. Fr.*, **58**, 87-89.

——— 1940. [On the position of the genus *Filariopsis* van Thiel, 1926 in the system of nematodes.] *Dokl. Akad. Nauk SSSR*, **27**, 1048-1051.

———, SHIKHOBALOVA, N. P., SCHULZ, R. S., POPOVA, T. I., BOEV, S. N. & DELYAMURE, S. L. 1952. Key to parasitic nematodes. Vol. III. Moscow Izdatelstvo Akademii Nauk SSSR. 890 pp. (Translated for the National Science Foundation and the Department of Agriculture, Washington, D.C., by the Israel Program for Scientific Translations, Jerusalem, 1961.)

SMITH, F. W. & THRELFALL, W. 1973. Helminths of some mammals from Newfoundland. *Am. Midl. Nat.*, **90**, 215-218.

SOŁTYS, A. 1954. [Helminthofauna of Soricidae in the Białowieża National Park.] *Acta parasit. pol.*, **1**, 353-402. [In Russian.]

——— 1962. Helminth parasites of Mustelidae of Lublin Palatinate. *Acta parasit. pol.*, **10**, 73-76.

SPRATT, D. 1978. A taxonomic revision of the lungworms (Nematoda: Metastrongyloidea) of Australian marsupials. *Aust. J. Zool.* (in press.)

TARZHIMANOVA, R. A. & CHERTKOVA, A. N. 1969. [*Rattostrongylus petrovi* n.sp., a new nematode from *Dryomys nitedula*.] *Trudy azerb. nauch.-issled. Inst. Medit. Parazit. Trop. Medit. S. M. Kirova*, *7*, 307-310. [In Russian.]

TRAVASSOS, L. 1921. Nematodeos novos. I. *Bras.-méd.*, **35**, 367-368.

——— 1925. Un nouveau type de Metastrongylidae. *C.r. Séanc. Soc. Biol.*, **93**, 1259-1262.

THIEL, P. H. VAN. 1926. On some filariae parasitic in Surinam mammals with the description of *Filariopsis asper* n.g., n.sp. *Parasitology*, **18**, 128-136.

VOGEL, H. 1928. Ueber die Nematodengattung *Oslerus* Hall, 1921 und zwei neue Arten derselben *Oslerus cynopitheci* und *Oslerus felis*. *Zentbl. Bakt. ParasitKde.*, **109**, 430-444.

WEBSTER, W. A. 1964. *Crenosoma canadensis* n.sp. (Nematoda: Metastrongylidae) from the lungs of skunk. *Can. J. Zool.*, **42**, 807-809.

——— 1965. *Skrjabingylus magnus* n.sp. (Nematoda: Trichostrongylidae) from the skunk, *Mephitis mephitis*. *Can. J. Zool.*, **43**, 229-231.

——— 1967. *Filaroides mephitis* n.sp. (Metastrongyloidea: Filaroididae) from the lungs of eastern Canadian skunks. *Can. J. Zool.*, **45**, 145-147.

——— 1978. *Andersonstrongylus captivensis* gen. et sp.n. (Metastrongyloidea: Filaroididae) from the lungs of the striped skunk, *Mephitis mephitis*. *J. Parasit.* (in press.)

——— 1978. The resurrection of *Filariopsis* (Van Thiel, 1926) (Metastrongyloidea: Filaroididae) for filaroidid lungworms from primates. *Can. J. Zool.*, **56**, 369-373.

———, NEUFELD, I. L. & MCNEILL, A. C. 1973. *Halocercus monoceris* n.sp. (Nematoda: Metastrongyloidea) from the narwhal, *Monodon monoceros*. *Proc. helminth. Soc. Wash.*, **40**, 255-258.

WESENBERG-LUND, E. 1947. On three parasitic nematodes from Cetacea. *Vidensk. Meddr dansk natur. Foren.*, **110**, 17-30.

WETZEL, R. 1938. Zur Biologie und systematischen Stellung des Dachslungenwurmes. *Libro Jubilar do Prof. Lauro Travassos*, pp. 531-536.

WHITLOCK, J. H. 1956. A description of a new dog lungworm *Filaroides milksi* n.sp. (Nematoda: Metastrongyloidea). *Wien. tierärztl. Mschr.* **43**, 730-738.

WOLFFHUGEL, K. 1934. Paraplegia cruralis parasiteria felis durch *Gurltia paralysans* nov. gen. nov. sp. Nematoda. *Z. InfektKrankh. parasit. Krankh. Hyg. Haustiere*, **46**, 28-47.

YAMAGUTI, S. 1951. Studies on the helminth fauna of Japan. Part 46. Nematodes of marine mammals. *Arb. med. Fak. Okayama*, **7**, 295-306.

YUN, LYAN & KONTRIMAVICHUS, V. L. 1963. [The systematic position of *Crenosoma skrjabini* Polozhentsev, 1935 from *Sorex* spp.] *Trudy gel'mint. Lab.*, **13**, 52-55. [In Russian.]

YURAKHNO, M. V. & SKRJABIN, A. S. 1971. [*Parafilaroides krascheninikova* n.sp. from the lungs of *Pusa hispida krascheninikova*.] *Vest. Zool.*, **5**, 32-36. [In Russian.]

YUSHKOV, V. F. 1971. [*Angiocaulus ryjikova*) (Nematoda: Strongylata), a parasite of *Clethrionomys rutilus* from the northern Ural mountains.] *Parazitologiya*, **5**, 344-346. [In Russian.]

ZAM, S., CALDWELL, D. & CALDWELL, M. 1971. Some endoparasites from small odontocete cetaceans collected in Florida and Georgia. *Cetology*, **2**, 1-11.

8 OXYURIDA

OXYUROIDEA

by ANNIE J. PETTER* and JEAN-CLAUDE QUENTIN*

Oxyuroidea from vertebrates can be grouped into 3 families, the Pharyngodonidae, Oxyuridae and Heteroxynematidae.

PHARYNGODONIDAE

The Pharyngodonidae, parasites of lower vertebrates, is classically distinguished from the Thelastomatidae, parasites of invertebrates, by the number of cephalic papillae, i.e. four papillae or eight joined in pairs in the Pharyngodonidae, and eight separate papillae in the Thelastomatidae. In fact, however, eight separate papillae can be observed in several species of Pharyngodonidae and some Thelastomatidae have only four papillae. Some genera of Pharyngodonidae, like *Synodontisia* of fish, *Parathelandros* and *Batracholandros* of amphibians, and some species of *Skrjabinodon* and *Thelandros* of carnivorous reptiles have great similarities with some Thelastomatidae, particularly in the male genital structures. The separation of the two families is, therefore, arbitrary and possibly the Pharyngodonidae is polyphyletic and has arisen at various times from the oxyuroids of insects.

The structure of the male caudal extremity of the Pharyngodonidae seems to show best the phylogenetic relationships of the group. The genera parasitic in fish and amphibians are especially close to some Thelastomatidae and they appear to have evolved least since becoming parasites of vertebrates. The genera parasitic in reptiles are usually the furthest removed from the Thelastomatidae and in these genera two lines can be distinguished by the position of the vulva and the disposition of the genital papillae. In the first, which evolved in carnivorous reptiles, the tail remains long and tapering and there is a progressive reduction of the peduncles and the size of papillae, and a progressive disappearance of caudal alae. The second line evolved in herbivorous reptiles (*Uromastix*, herbivorous Iguanidae and Testudinidae). The papillae lose their peduncles but remain large and there is a progressive shortening of the caudal extremity and a progressive disappearance of the caudal alae.

Thelandros, parasitic in both carnivorous and herbivorous reptiles, shows affinities sometimes with the line in herbivores, sometimes with the line in carnivores, as well as with species of *Batracholandros* of amphibians. It would be desirable to place *Thelandros* at the base of the two evolutionary lines but the reduction in thickness of the caudal appendix characteristic of the genus seems to be highly evolved and we believe it may, in fact, be a character showing convergence.

Two main modifications are made in earlier classifications, notably that of Chabaud (1965):

(i) The presence or absence of a gubernaculum has been used to differentiate parasites in herbivores, and parasites in carnivores and the genus *Thelandros*. The gubernaculum in the Pharyngodaeonid is a V-shaped sclerotized piece supporting the posterior anal lip which is in the

* Muséum National d'Histoire Naturelle, Paris.

form of a truncated cone; the V-shaped piece is highly developed in herbivore parasites but we have confirmed its presence in some species of *Thelandros* and it has been noted, more or less well developed, in numerous species from carnivores. This character is, therefore, of no use for the differentiation of thc two main groups.

(ii) The presence or absence of lateral alae has also been used to separate some genera. However, lateral alae may be present or absent in a species, depending on the sample examined (Chabaud & Golvan, 1957), and we will not, therefore use this character. This leads us to consider *Parapharyngodon* a synonym of *Thelandros* (see Baylis, 1936) and *Pseudalaeuris* a synonym of *Alaeuris*.

Neopharyngodon, placed by Chabaud in the Cosmocercidae, is synonymized with *Pharyngodon* since the two "feebly chitinized" spicules and the "pair of small, highly chitinized and closely apposed accessory pieces" reported in this genus are, in our opinion, the sclerotized piece which supports the genital cone in some species of *Pharyngodon*.

Following Petter & Douglas (1975), *Thelastomoides* is considered to be a synonym of *Alaeuris*. *Travassozolaimus* is also placed in the synonymy of *Alaeuris*. We examined specimens of *T. travassosi* and we found it to be identical with *Alaeuris* as far as the number of genital papillae and the structure of the V-shaped piece were concerned. The only differences were in the development of this V-shaped piece, which is highly sclerotized in *Travassozolaimus* so that the branches of the V are prolonged ventrally. *Mammillomacracis* is placed in the synonymy of *Paralaeuris*. Baruš *et al.* (1969) have redescribed the type species *M. cyclurae*, which has all the characters of the latter genus.

Ozolaimus and *Alaeuris* are separated on the basis of apical structure. This character seems the only valid one of the three classically used to distinguish the two genera. We found no differences in the structure of the male genitalia. Inglis *et al.* (1960), in synonymizing *Macracis* and *Ozolaimus*, showed that the oesophageal swelling had no generic value. Apical structures of the *Ozolaimus* type appear to group species with genuine affinities, although these structures cannot usually be used as generic characters in the Pharyngodonidae because of their extreme diversity (Petter, 1966). It is preferable, however, to maintain the two genera since the synonymy would suppress *Alaeuris*, used by all authors, and the generotype for present species of both *Alaeuris* and *Ozolaimus* would be *O. megatyphlon*, an atypical species with a dilated oesophagus. We do not accept the synonymy of *Spauligodon* with the subgenus *Neyrapharyngodon* (see Freitas & Ibañez, 1963) since the type species of the latter has caudal alae supported by the three pairs of papillae, a character which is not found in *Spauligodon*.

Several genera and sub-genera created during the last decade have been included in the key, namely *Ichthyouris*, *Cithariniella*, *Ortleppnema*, *Skrjabinodon*, *Synodontisia*, *Thelandros* (*Archithelandros*) and *Batracholandros*. The definition of the last-named genus is expanded to include species parasitic in North and South American amphibians attributed to the genus *Thelandros*, with the exception of *T. minutus* Read & Amrein, 1952 which is insufficiently described. Baruš (1973) has noted that these species form a natural group.

Following Inglis (1961) and Quentin (1973), *Callistoura* and *Ctenodactylina* of mammals are placed in the Pharyngodonidae.

Propharyngodon Biswas & Chakravarty, 1963 is insufficiently known and is excluded from the key. If one ignores the differences in hosts, there is nothing in the description of the type species of *P. ranae* to differentiate it from *Skrjabinodon*. If further study confirms this, *Skrjabinodon* would be placed in the synonymy of *Propharyngodon*.

Cosmoxynema Travassos, 1948 and *Cosmoxynemoides* Travassos, 1948, the males of which are unknown, are considered of doubtful status and excluded from the key.

Neoprotozoophaga Biswas & Chakravarty, 1963 is considered to belong in the Cosmocercidae. The single spicule described by the authors is apparently the gubernaculum.

OXYURIDAE

The evolution of the Oxyuridae can be interpreted on the basis of genital characters, such as the reduction in number of genital papillae and the differentiation of the cuticularized genital pieces, and on the basis of cephalic characters, such as the reduction of pharyngeal teeth and the development of labial and interlabial structures.

Several changes are made in the systematics proposed by Chabaud (1965).

Callistoura and *Ctenodactylina* are transferred to the Pharyngodonidae (see Inglis, 1961; Quentin, 1973).

Hoplodontophorus was placed in the Oxyuridae by Inglis, 1959.

Paraustroxyuris, Macropoxyuris and *Potoroxyuris* of marsupials are added to the family.

The most important changes concern Oxyuridae of primates and rodents.

Oxyuridae of primates occur in *Enterobius, Trypanoxyuris* and *Lemuricola.* The last-named is sub-divided into 5 sub-genera, namely *Lemuricola, Protenterobius, Ingloxyuris, Madoxyuris* and *Rodentoxyuris. Biguetius* was made synonymous with *Lemuricola* by Chabaud *et al.* (1965) who thought that this genus might become a subgenus of *Lemuricola* when its male became known.

Paraoxyuronema Artigas, 1937 of *Ateles,* a neo-tropical primate, is considered of doubtful validity by Inglis & Cosgrove, 1964.

Of Oxyuridae in rodents, *Wellcomia* Sambon, 1907 is considered provisionally as a *genus inquirendus* as its type species is known only by secondary characters. The other species of *Wellcomia* are distributed among *Evaginuris, Sypharista, Hilgertia* and *Heteromyoxyuris. Carolodelatorella* was placed in the synonymy of *Helminthoxys* by Baruš (1972).

Syphaciuris, parasitic in anomalures, is a synonym of *Acanthoxyurus* which is removed from the Heteroxynematidae and placed in the Oxyuridae (cf. Quentin, 1974).

Among rodent Oxyuridae, there is a close correlation between the evolution of cephalic structures and the differentiation of the male sexual characters. Because of this, *Zenkoxyuris* with its highly differentiated head is placed in the Oxyuridae possessing a gubernaculum in the male, even though the male is unknown.*

Finally *Syphacia* is subdivided into the subgenera *Syphacia, Syphatineria* and *Syphabulea.*

HETEROXYNEMATIDAE

The classification of the Heteroxynematidae appears more natural if one relies on the arrangement of the male genital papillae, the anatomy of the ovejector and the structure of the egg-shell in the female, and on oesophageal, pharyngeal and labial structures rather than the more obvious cuticular cephalic characters and caudal ornamentation. The former characters seem to be inter-related and they allow the establishment of correlations between the distribution of genera and species and the phylogenetic relationships of their hosts.

The most important modifications from the classification of Skrjabin & Schikhobalova (1948) are as follows:

(i) *Acanthoxyurus* is transferred to Oxyuridae on the basis of the male genital structures. It follows that Acanthoxyurinae used by Chabaud (1965) must be replaced by Heteroxynematinae.

(ii) *Eugenuris, Smirnoviella* (=*Smirnovia* Schulz & Andrejeva, 1950 preoccupied) and *Pikaeuris* are placed in the synonymy of *Labiostomum.*

* The only members of Oxyuridae without a gubernaculum which parasitize rodents are *Passalurus* of Leporidae, and *Lemuricola* (*Rodentoxyuris*) *sciuri* (Cameron, 1932) and *Enterobius parallela* (Linstow, 1908) of Sciuridae. The absence of a gubernaculum is associated with very primitive cephalic structures in *Passalurus. L. sciuri* and *E. parallela* are derived from oxyurids normally found in primates.

(iii) The Heteroxynematidae is separated into two distinct groups on the basis of genital and anatomical characters. The first, more primitive, is represented by the Heteroxynematinae. Four genera are restricted to certain bird and rodent families, namely *Syphaciella* from birds of the Pteroclidae, *Eudromoxyura* from birds of the Tinamidae, *Fastigiuris* from Ochotonidae and *Rauschoxyuris* Quentin, 1975 from Aplodontidae. Four genera make up an independent line which evolved in Sciuromorpha, Caviomorpha and Myomorpha, namely, *Heteroxynema*, a primitive genus divided into three subgenera, *Heteroxynema* from neoarctic Sciuridae, *Proxyuronema* from palaearctic Sciuridae and *Cavioxyura* from neo-tropical Caviomorpha. From *Heteroxynema* the other oxyurids of Sciuridae and Muroidea seem to have evolved, i.e. *Dermatopallarya* from palaearctic Sciuridae, *Dentostomella* from neoarctic Sciuridae and palaearctic Gerbillidae and *Aspiculuris* from Muroidea. The second group, the Labiostomatinae, encompasses more specialized forms which evolved in Lagomorpha. This subfamily includes *Dermatoxys* from Leporidae, and *Labiostomum* (with the subgenera *Labiostomum* and *Eugenuris*) and *Cephaluris* from Ochotonidae.

OXYUROIDEA
KEY TO FAMILIES

1-(2) Amphids on peduncle closely adherent to cephalic surface or jutting forward (Figs. 4.19, 4.25, 4.33).
Genital cone often supported by a V-shaped sclerotized structure.
2-(1) Parasites of cold-blooded vertebrates, rarely of archaic mammals.
Pharyngodonidae

2-(1) Amphids non-pedunculate (Figs. 4.35, 4.107).
Genital cone without sclerotized supporting structure (Figs. 4.42, 4.98).
Parasites of mammals and rarely of birds.
3-(4) Tail of male irregular in shape, sometimes bluntly truncate with or without dorsal point, sometimes with large digitiform papillae extending into caudal alae (Figs. 4.48, 4.53).
Oxyuridae

4-(3) Tail of male regular in shape, conical or flattened dorso-ventrally, not bluntly truncate, without long protruding papillae extending into cuticular alae (Fig. 4.81).
Genital papillae mainly concentrated in perianal region (Figs. 4.85, 4.95).
Heteroxynematidae

Family *PHARYNGODONIDAE* Travassos, 1919
KEY TO GENERA

1-(6) Oesophageal bulb without sclerotized apparatus.
2-(3) Oesophagus divided into 2 almost equal parts, each dilated into bulb at its extremity (Fig. 4.1).
Parasites of fish.
Travnema Pereira, 1938

3-(2) Oesophagus not divided into 2 almost equal parts and only dilated into bulb at posterior extremity (Fig. 4.2).
Posterior lip of anus prolonged into a truncated cone-like structure and supported by a sclerotized V-shaped piece (Figs. 4.4, 4.5).
Parasites of mammals.
4-(5) Anal region of male provided with cuticular blades and complex digitiform appendages (Fig. 4.4).
Parasites of lemurs.
Callistoura Chabaud & Petter, 1958

5-(4) Anal region of male without cuticular blades and complex digitiform appendages (Fig. 4.5).
Parasites of rodents (Ctenodactylidae).
Ctenodactylina Bernard, 1969

6-(1) Oesophageal bulb with sclerotized apparatus (Fig. 4.3).
7-(14) Vulva anterior, near excretory pore.
Tail usually long and tapering in both sexes.
8-(13) Genital papillae not clearly separated into anterior group around anus and posterior pair, the latter usually only a short distance from others, papillae of this latter pair well separated from each other and not in shape of rosette.
Protruding genital cone supported by sclerotized structure may be present, but two anterior pairs of papillae not situated on cone (Figs. 4.6, 4.7, 4.8).
Spicule often absent.
Female tail often with spines.
Parasites of carnivorous reptiles.

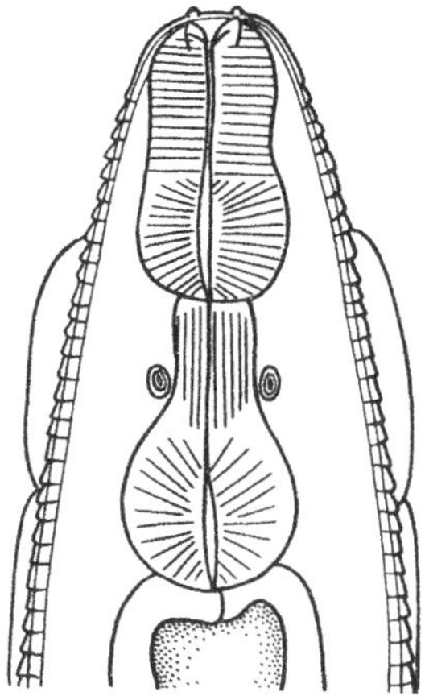

4.1

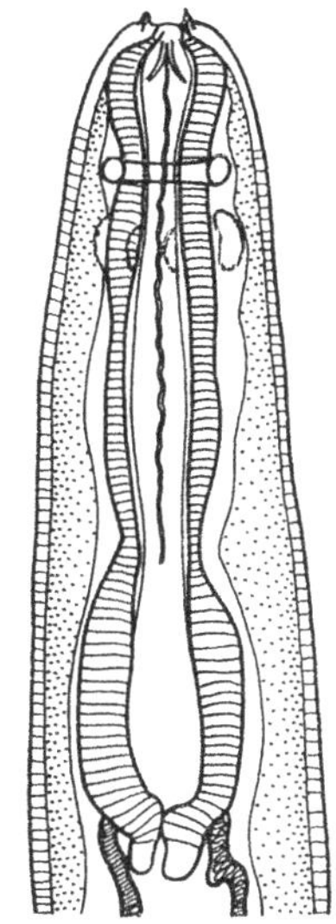

4.2

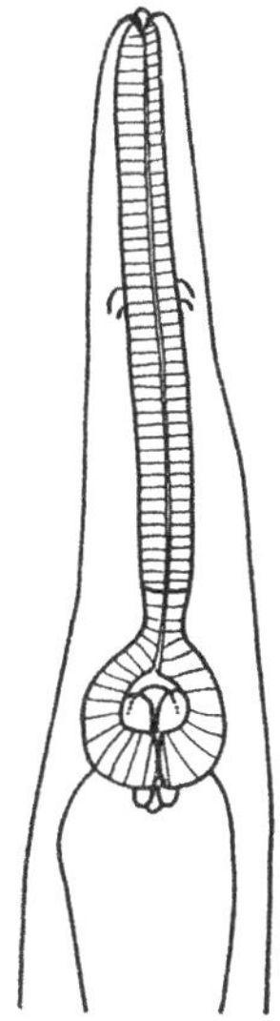

4.3

FIG. 4.1. *Travnema travnema*, oesophageal region. (After Pereira, 1938.)
FIG. 4.2. *Ctenodactylina tunetae*, oesophageal region. (After Quentin, 1973.)
FIG. 4.3. *Batracholandros spectatus*, oesophageal region. (After Freitas & Ibáñez, 1962.)

9-(12) Caudal alae present.
Genital papillae voluminous, the two posterior pairs pedunculate.
10-(11) Caudal alae supported by the 3 pairs of genital papillae (Fig. 4.6).
Pharyngodon Diesing, 1861
(=*Neopharyngodon* Chakravarty & Bhaduri, 1948)
11-(10) Caudal alae not supported by last pair of genital papillae (Fig. 4.7).
Spauligodon Skrjabin, Schikhobalova & Lagodovskaja, 1960
12-(9) Caudal alae absent.
Genital papillae sessile and often reduced (Fig. 4.8).
Skrjabinodon Inglis, 1968

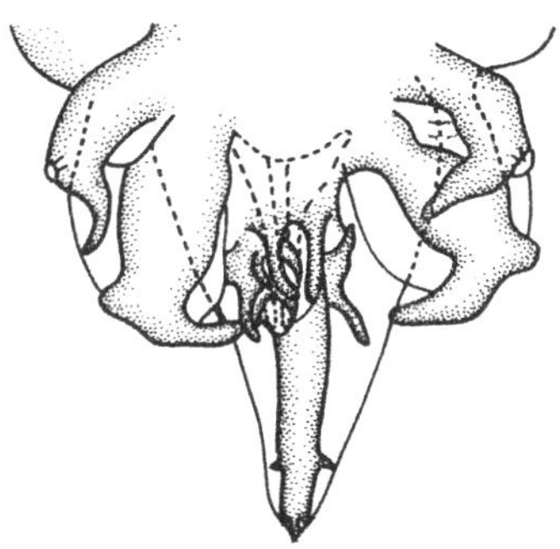

4.4

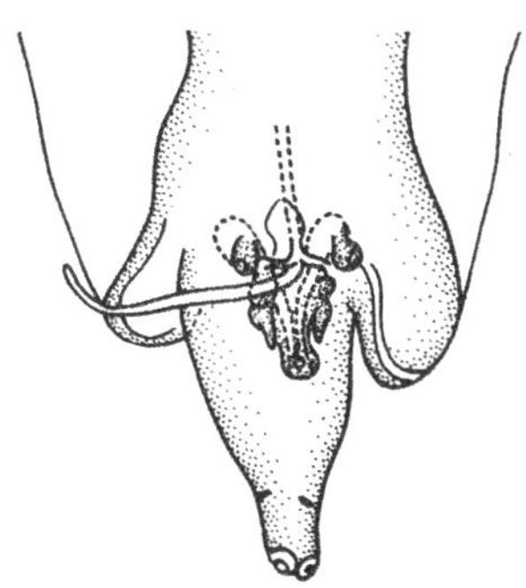

4.5

FIG. 4.4. *Callistoura brygooi*, male, ventral view of posterior extremity. (After Chabaud & Petter, 1958.)
FIG. 4.5. *Ctenodactylina tunetae*, male, ventral view of posterior extremity. (After Quentin, 1973.)

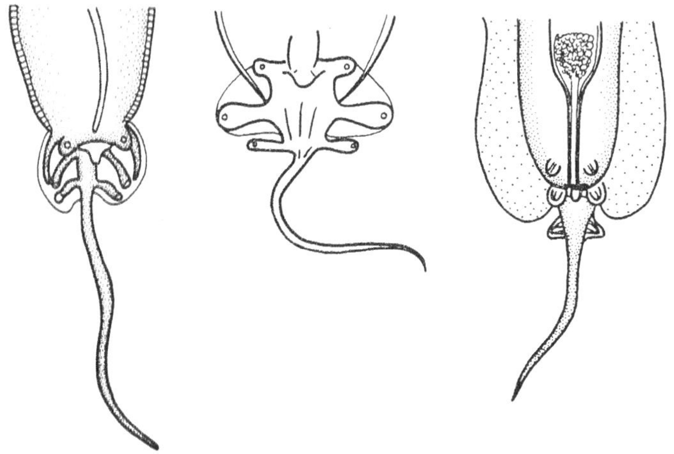

4.6 4.7 4.8

Fig. 4.6. *Pharyngodon spinicauda*, male, ventral view of posterior extremity. (After Seurat, 1914.)
Fig. 4.7. *Spauligodon tarentolae*, male, ventral view of posterior extremity. (After Spaul, 1926.)
Fig. 4.8. *Skrjabinodon mabuyae*, male, ventral view of posterior extremity. (After Sandground, 1936.)

13-(8) Caudal alae absent.
Genital papillae clearly separated into anterior group on protruding genital cone and one posterior pair; each papilla of latter in shape of rosette and pair often sharing common base. (Fig. 4.9).
Parasites of Australian amphibians.
Parathelandros Baylis, 1930

14-(7) Vulva in middle region of body (excluding tail) or in posterior region (except *Thelandros maculatus* Caballero, 1968).

15-(16) Genital papillae grouped around anus and caudal appendage without papillae.
Pair of strongly cuticularized plate-like structures present, lateral and posterior to anus.
Caudal alae present (Fig. 4.10).
Parasites of fish.
Ichthyouris Inglis, 1962

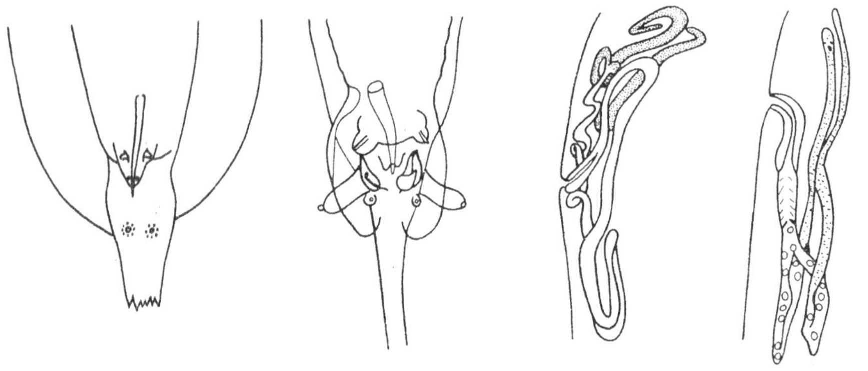

4.9 4.10 4.11 4.12

Fig. 4.9. *Parathelandros mastigurus*, male, ventral view of cloacal region. (After Inglis, 1968.)
Fig. 4.10. *Ichthyouris ro*, male, ventral view of cloacal region. (After Inglis, 1962.)
Fig. 4.11. *Veversia tuberculata*, female genital apparatus. (After Thapar, 1925.)
Fig. 4.12. *Thelandros bulbosus*, female genital apparatus. (After Chabaud & Golvan, 1957.)

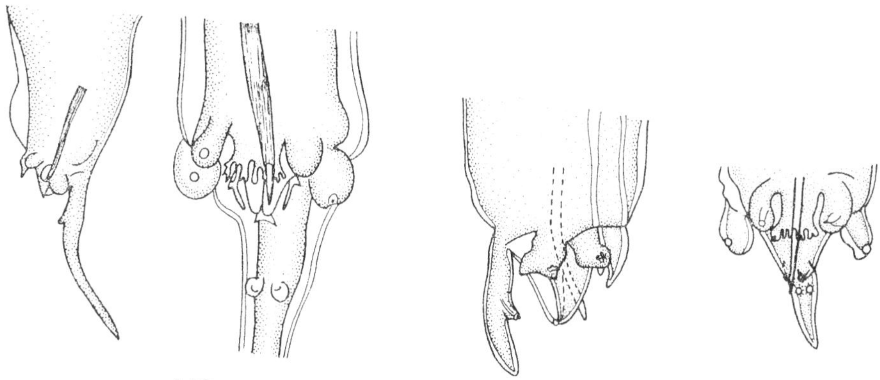

4.13 4.14

FIG. 4.13. *Thelandros* (*Archithelandros*) *pyxis*, male, lateral and ventral views of cloacal region. (After Petter, 1966.)

FIG. 4.14. *Thelandros* (*Thelandros*) *alatus*, male, lateral and ventral views of caudal extremity. (After Lucker, 1952.)

16-(15) Genital papillae consisting of anterior group around anus and posterior pair on caudal appendage.
Anal region without a pair of strongly cuticularized plate-like structures.
Caudal alae present or absent.

17-(22) Male tail reduced to a slim appendage, usually inserted dorsally and directed obliquely to longitudinal axis of body (Figs 4.13, 4.14).
V-shaped sclerotized piece supporting posterior anal lip often present.

18-(19) Monodelphic (Fig. 4.11).
Caudal alae absent.
Parasites of carnivorous reptiles.

Veversia Thapar, 1925

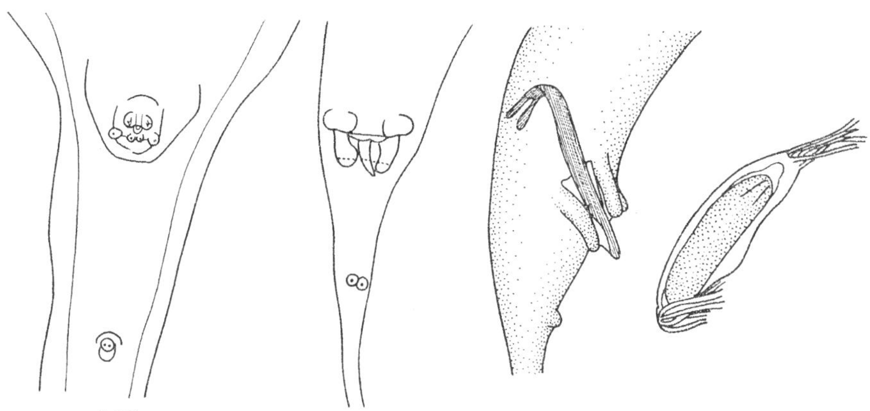

4.15 4.16 4.17 4.18

FIG. 4.15. *Synodontisia thelastomoides*, male, ventral view of cloacal region. (After Petter, Vassiliades & Troncy, 1972.)

FIG. 4.16. *Cithariniella khalili*, male, ventral view of cloacal region. (After Petter, Vassiliades & Troncy, 1972.)

FIG. 4.17. *Laurotravassoxyuris travassosi*, male, lateral view of cloacal region. (After Vigueras, 1938.)

FIG. 4.18. *Cithariniella khalili*, egg. (After Petter, Vassiliades & Troncy, 1972.)

19-(18) Didelphic (Fig. 4.12).
Anterior anal lip often covered by well developed, fringed membrane (Figs. 4.13, 4.14).
Parasites of carnivorous and herbivorous reptiles.
Thelandros Wedl, 1862
(= *Parapharyngodon* Chatterji, 1933 = *Pseudothelandros* Read, Amrein & Walton, 1952)

20-(21) Caudal alae present (Fig. 4.13).
Thelandros (*Archithelandros*) Petter, 1966

21-(20) Caudal alae absent (Fig 4.14).
Thelandros (*Thelandros*)

22-(17) Male tail not reduced to a slim appendage, at least up to level of posterior pair of genital papillae, and directed parallel to longitudinal axis of body.

23-(30) V-shaped sclerotized structure supporting posterior lip of anus absent or poorly developed.
Genital papillae of posterior pair fused or closely joined.
Caudal alae absent.
Tail long and tapering in both sexes (Figs. 4.15, 4.16, 4.17, 4.21, 4.22).
Parasites of fish and amphibians.

24-(25) Three pairs of small round genital papillae present on summit of protruding genital cone (Fig. 4.15).
Parasites of fish.
Synodontisia Petter, Vassiliades & Troncy, 1972

25-(24) Genital papillae not on summit of protruding genital cone.

26-(29) Two pairs of large subventral papillae around anus (Fig. 4.16).
Spicule with dorsally bent capitulum (Fig. 4.17).
Eggs with filaments (Fig. 4.18).
Parasites of fish.

27-(28) Cephalic extremity with four large sub-median and two narrow lateral flat lips.
Buccal cavity with three sclerotized, striated pieces (Fig. 4.19).
Citharinella Khalil, 1964

28-(27) Cephalic extremity with six rectangular lamellae raised above apical surface and originating from depth of buccal cavity (Fig. 4.20).
Laurotravassoxyuris Vigueras, 1938

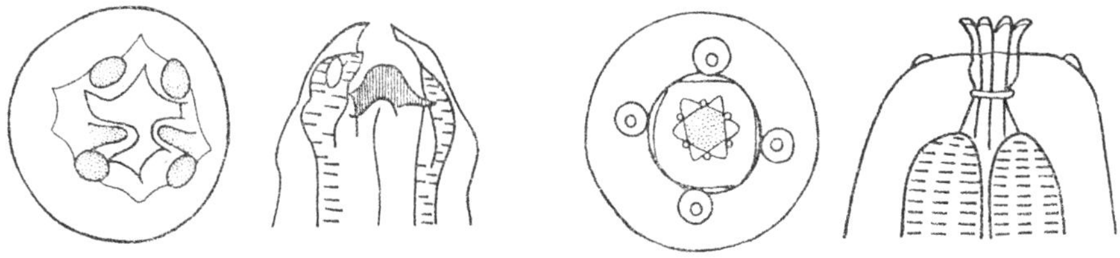

4.19 4.20

FIG. 4.19. *Citharinella citharini*, apical and median views of anterior extremity. (After Petter, Vassiliades & Troncy, 1972.)

FIG. 4.20. *Laurotravassoxyuris travassosi*, apical and median views of anterior extremity. (After Vigueras, 1938.)

29-(26) Two to three pairs of small papillae around anus, one of which is situated laterally.
Posterior anal lip in form of conical papilla-like appendix.
Spicule simple (Figs. 4.21, 4.22).
Eggs without filaments.
Parasites of amphibians.

Batracholandros Freitas & Ibañez, 1965

30-(23) Posterior lip of anus prolonged into a truncated cone supported by tip of a V-shaped sclerotized piece.
Parasites of herbivorous reptiles, rarely of carnivorous reptiles.

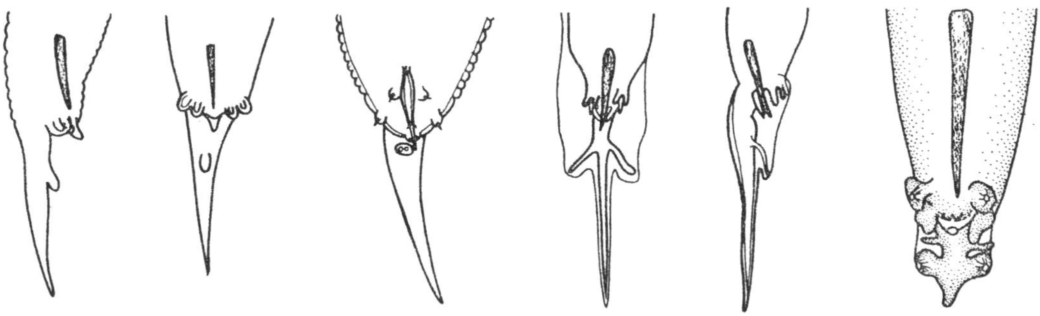

4.21 4.22 4.23 4.24

FIG. 4.21. *Batracholandros spectatus*, male, lateral and median views of caudal extremity. (After Freitas & Ibáñez, 1965.)

FIG. 4.22. *Batracholandros salamandrae*, male, ventral view of posterior extremity. (After Schad, 1960.)

FIG. 4.23. *Paralaeuris dorochila*, male, lateral and ventral views of posterior extremity. (After Walton, 1942.)

FIG. 4.24. *Ortleppnema possompesi*, male, ventral view of posterior extremity. (After Petter, 1966.)

31-(42) Oesophageal isthmus very short in relation to corpus.
32-(39) Caudal alae present.
33-(36) Caudal alae supported by posterior pair of genital papillae.
34-(35) Posterior pair of genital papillae long and slender (Fig. 4.23).
Parasites of herbivorous Iguanidae.

Paralaeuris Cuckler, 1938
(= *Mammillomacracis* Dosse, 1939)

35-(34) Posterior pair of genital papillae large and short (Fig. 4.24).
Parasites of Testudinidae in Madagascar.

Ortleppnema Petter, 1966

36-(33) Caudal alae not supported by posterior pair of genital papillae.
37-(38) Oral opening with raised lateral, cuticular elevations penetrated by amphidial ducts (Fig. 4.25).
Oesophageal corpus sometimes with more or less pronounced anterior swelling (Fig. 4.27).
Male tail, short and rounded.
Posterior pair of genital papillae subterminal (Fig. 4.26).
Parasites of herbivorous Iguanidae.

Ozolaimus Dujardin, 1845
(= *Macracis* Gedoelst, 1916)

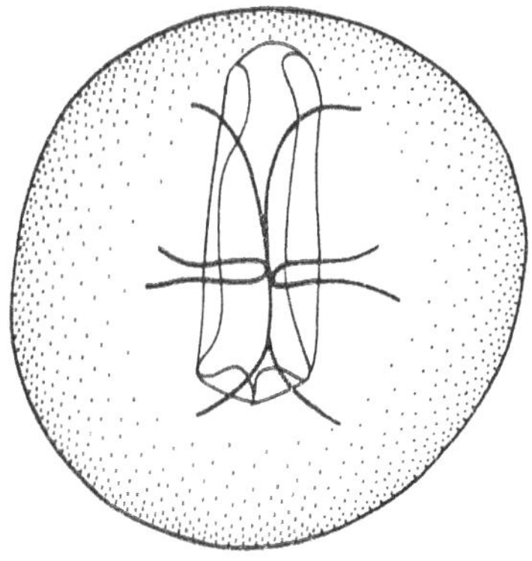
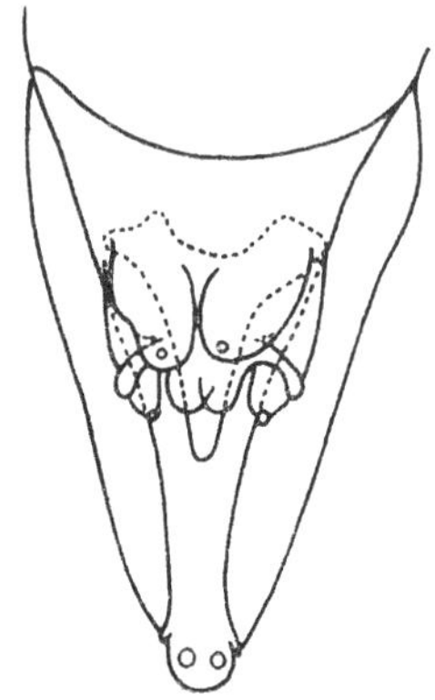

4.25 4.26

FIG. 4.25. *Ozolaimus megatyphlon*, apical view. (Original drawing.)

FIG. 4.26. *Ozolaimus megatyphlon*, male, ventral view of posterior extremity. (After Inglis, Diaz-Ungría & Coles, 1960.)

38-(37) Oral opening without raised lateral cuticular elevations penetrated by amphidial ducts.
Oesophageal corpus cylindrical.
Male tail sometimes tesminated by long tapering point or short and rounded with subterminal papillae (Fig. 4.28).
Parasites of herbivorous Testudinidae and Iguanidae, rarely carnivorous reptiles.
Alaeuris Thapar, 1925
(= *Pseudalaeuris* Walton,1942 = *Travassozolaimus* Vigueras, 1938 = *Thelastomoides* Walton, 1927)

39-(32) Caudal alae absent.
Genital papillae of posterior pair very large, clearly separated from one another and situated laterally (Figs. 4.30, 4.31).

40-(41) Male tail ending in point (Fig. 4.30).
Parasites of Testudinidae.
Mehdiella Seurat, 1918

41-(40) Male tail truncated at extremity (Fig. 4.31).
Parasites of Testudinidae and *Uromastix*.
Tachygonetria Wedl, 1862

42-(31) Oesophageal isthmus greatly elongated, dividing oesophagus into two almost equal parts (Fig. 4.32).
Amphids in form of small rigid rods raised above the apical surface (Fig. 4.33).
Parasites of Testudinidae.
Thaparia Ortlepp, 1933

Family *OXYURIDAE* Cobbold, 1864

Key to genera and subgenera

1-(26) Four or five pairs of genital papillae present, clearly separated from each other, of which two pairs at least are pedunculate (Fig. 4.34) (except subgenus *Ingloxyuris*).
Gubernaculum absent.
Parasites of primates and Australian marsupials (secondarily ungulates and Sciuridae).

4.27

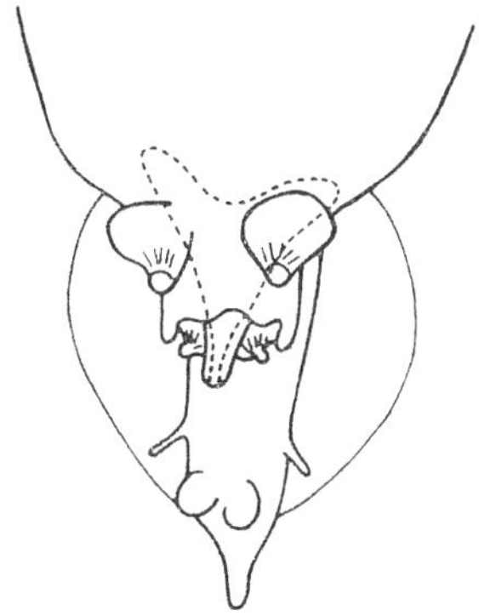

4.28

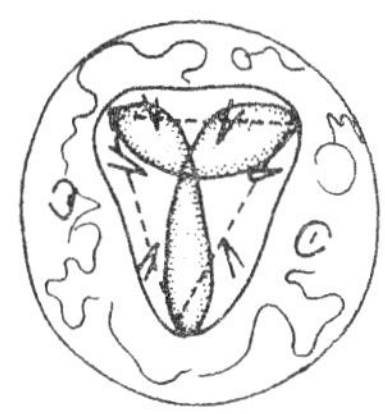

4.29

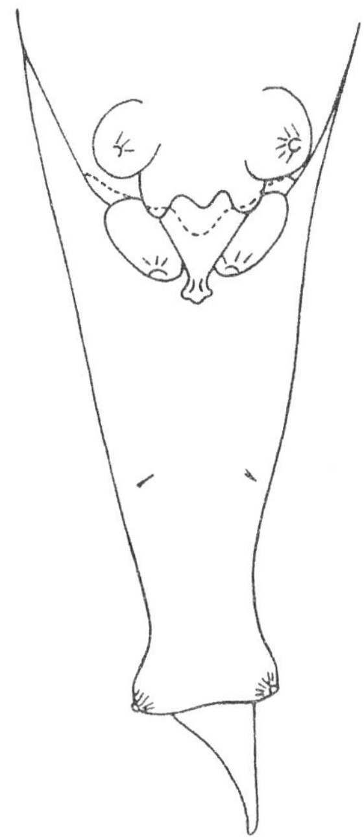

4.30

FIG. 4.27. *Ozolaimus cirratus*, oesophagus. (After Lent & Freitas, 1948.)
FIG. 4.28. *Alaeuris numidica*, male, ventral view of posterior extremity. (Original drawing.)
FIG. 4.29. *Alaeuris numidica*, apical view. (After Petter, 1966.)
FIG. 4.30. *Mehdiella microstoma*, male, ventral view of posterior extremity. (Original drawing.)

2-(25) Buccal cavity simple without complicated armature with spines (Fig. 4.35) but sometimes with interradial blades.
Cephalic papillae, small in size, never differentiated.
Parasites of primates and Australian marsupials (secondarily Sciuridae).

3-(18) Buccal cavity without interradial blades between three pharyngeal teeth (Fig. 4.35).
Eggs not operculated.
Parasites of primates (secondarily Sciuridae).

4-(15) Three simple lips or six labial lobes equal in size and simple cephalic vesicle present (Fig. 4.35).
Including Old World species and one nearctic species.

5-(14) Sclerotized elements reinforcing one to two pairs of post-anal papillae present (Fig. 4.36).
External circle with four or eight papillae; internal circle of papillae sometimes present.
Male with or without cuticular ornamentation along midventral line and with or without caudal point.
Parasites of lemurs and holarctic Sciuridae.

Lemuricola Chabaud & Petter, 1959

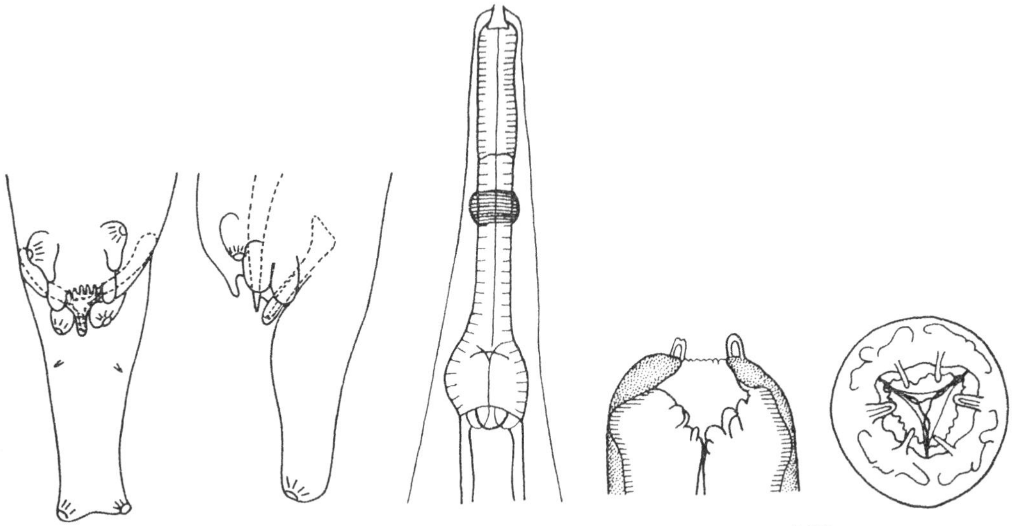

4.31 4.32 4.33

FIG. 4.31. *Tachygonetria longicollis*, male, lateral and ventral views of posterior extremity. (Original drawing.)

FIG. 4.32. *Thaparia thapari*, oesophageal region. (Original drawing.)

FIG. 4.33. *Thaparia domerguei*, median and apical views of anterior extremity. (After Petter, 1966.)

6-(11) Ventral cuticular fringe present on male (Fig. 4.37).

7-(10) With 3 lips (Fig. 4.35).
Oesophageal isthmus short.

8-(9) Post-anal plate strongly sclerotized.
Male with long caudal point (Figs. 4.36, 4.38).
Parasites of lemurs.

Lemuricola (*Lemuricola*)

9-(8) Post-anal plate only slightly sclerotized.
Male without caudal point (Fig. 4.34).
Parasites of lemurs.

Lemuricola (*Madoxyuris*) Chabaud, Brygoo & Petter, 1965

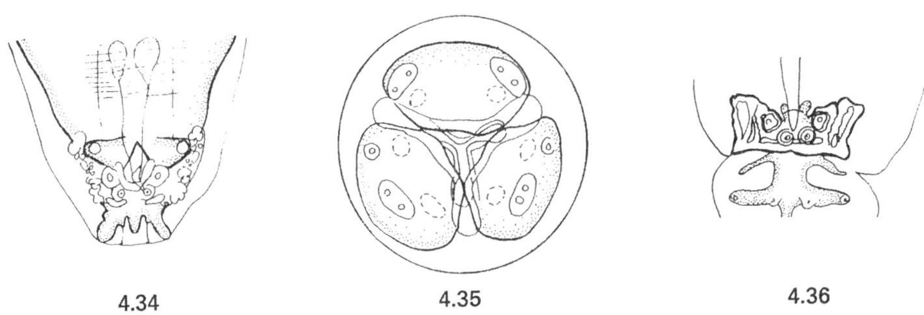

4.34 4.35 4.36

FIG. 4.34. *Lemuricola* (*Madoxyuris*), cloacal papillae. (After Petter *et al.*, 1972.)

FIG. 4.35. *Lemuricola* (*Rodentoxyuris*), apical view. (After Quentin & Tenora, 1974.)

FIG. 4.36. *Lemuricola* (*Lemuricola*), chitinoid elements. (After Chabaud & Petter, 1959.)

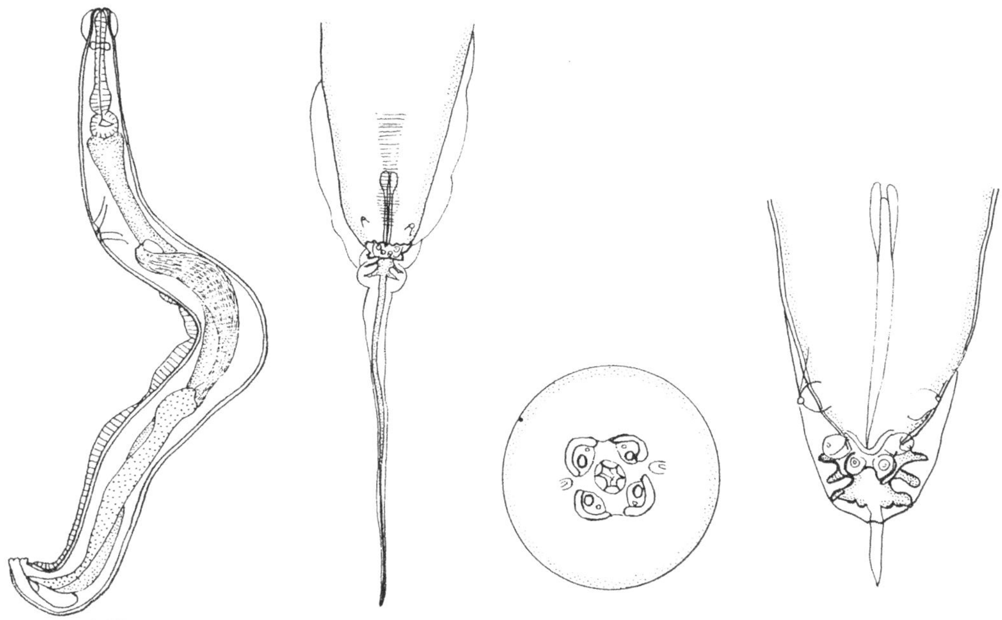

FIG. 4.37. *Lemuricola*, male, cuticular fringe. (After Chabaud, Brygoo & Petter, 1965.)
FIG. 4.38. *Lemuricola* (*Lemuricola*), male, tail tip. (After Chabaud & Petter, 1959.)
FIG. 4.39. *Lemuricola* (*Ingloxyuris*), apical view. (After Chabaud, Petter & Golvan, 1961.)
FIG. 4.40. *Lemuricola* (*Protenterobius*), caudal bursa. (After Inglis & Dunn, 1963.)

10-(7) Six labial lobes present (Fig. 4.39).
Oesophageal isthmus long.
Parasites of lemurs.

Lemuricola (*Ingloxyuris*) Chabaud, Petter, & Golvan, 1961

11-(6) Ventral cuticular fringe on male absent.

12-(13) Oesophageal isthmus long.
Highly sclerotized post-anal plate present (Fig. 4.40).
Parasites of lorises.

Lemuricola (*Protenterobius*) Inglis, 1961

13-(12) Oesophageal isthmus short.
Post-anal sclerotized plate reduced to cuticular ring around first pair of post-anal papillae (Fig. 4.41).
Parasites of Sciuridae.

Lemuricola (*Rodentoxyuris*) Quentin, 1974

14-(5) Sclerotized elements reinforcing 2 post-anal pairs of papillae absent (Fig. 4.42).
Four cephalic papillae present.
Male without midventral cuticular ornamentation or caudal point.
Parasites of catarrhine primates and Ethiopian Sciuridae.

Enterobius Leach, 1853

15-(4) Lips usually complex; when simple (e.g. *T. sceleratus*) cephalic vesicle deeply striated or divided into lobes (Fig. 4.43).
Neo-tropical in distribution.
Parasites of platyrrhine primates.

Trypanoxyuris Vevers, 1923.

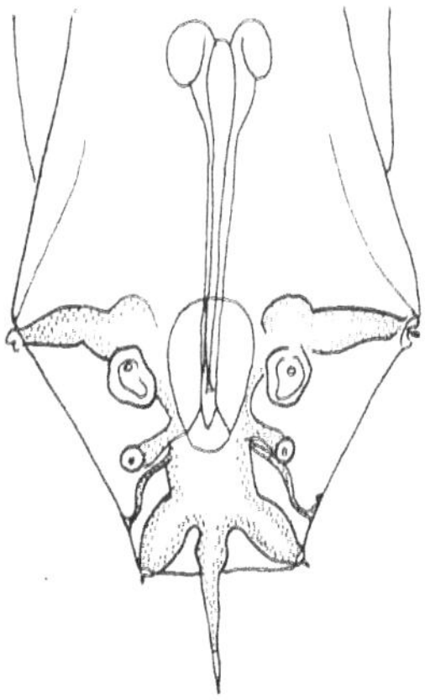

4.41

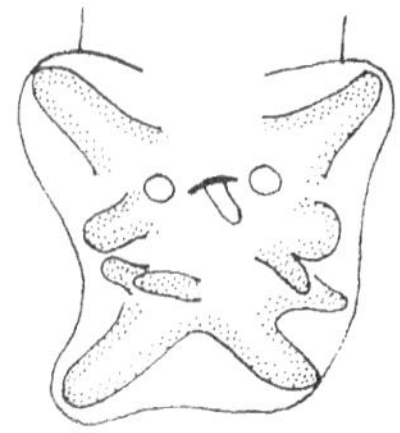

4.42

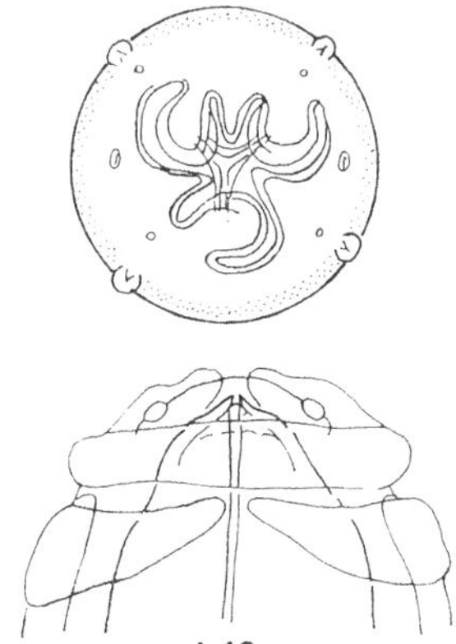

4.43

FIG. 4.41. *Lemuricola* (*Rodentoxyuris*), caudal bursa. (After Quentin & Tenora, 1974.)
FIG. 4.42. *Enterobius*, caudal bursa. (After Sandosham, 1950.)
FIG. 4.43. *Trypanoxyuris*, apical view and cephalic vesicle. (After Inglis & Diaz-Ungría, 1959 and Inglis, 1961.)

16-(17) Oesophagus with long isthmus.
Parasites of Hapalidae.
Trypanoxyuris (*Hapaloxyuris*) Inglis & Cosgrove, 1964

17-(16) Oesophagus with short isthmus.
Parasites of Cebidae.
Trypanoxyuris (*Trypanoxyuris*)

18-(3) Buccal cavity with interradial blades between 3 pharyngeal teeth (Fig. 4.44).
Eggs operculate at one extremity.
Parasites of Australian marsupials.

19-(20) External surface of wall of buccal cavity with annular, sclerotized reinforcement (Fig. 4.45).
Austroxyuris Johnson & Mawson, 1938

20-(19) External surface of wall of buccal cavity without annular sclerotized reinforcement (Fig. 4.46).

21-(24) Pharyngeal teeth not protruding from oral opening (Fig. 4.46).

22-(23) Teeth without tubercles (Fig. 4.47).
Pre-anal cuticular differentiation absent (Fig. 4.48).
Paraustroxyuris Mawson, 1964

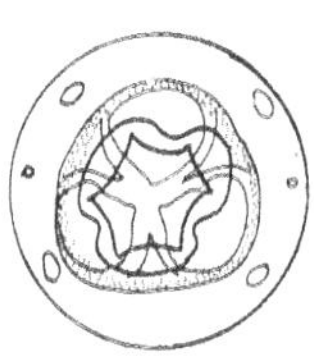

4.44

4.45

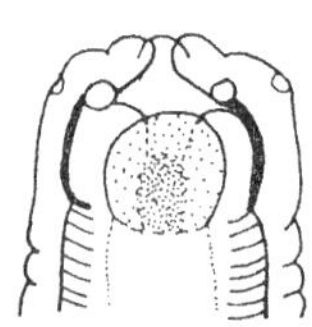

4.46

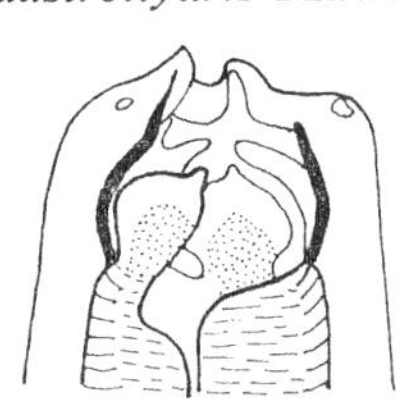

4.47

FIG. 4.44. *Paraustroxyuris*, apical view. (After Mawson, 1964.)
FIG. 4.45. *Austroxyuris*, buccal capsule. (After Mawson, 1964.)
FIG. 4.46. *Paraustroxyuris*, head, dorsal view. (After Mawson, 1964.)
FIG. 4.47. *Paraustroxyuris*, head, longitudinal section. (After Mawson, 1964.)

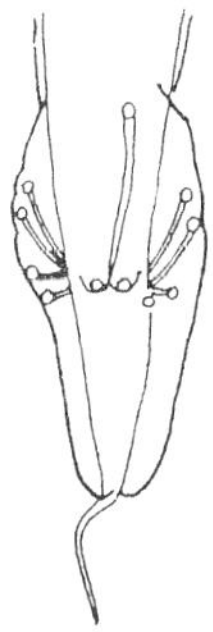
4.48

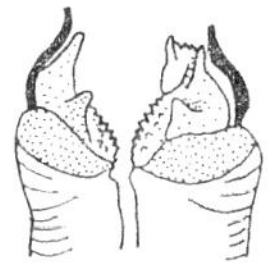
4.49

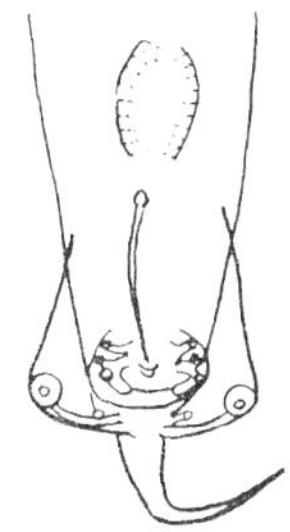
4.50

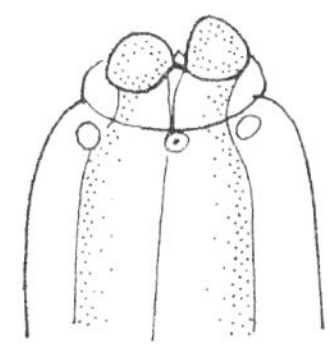
4.51

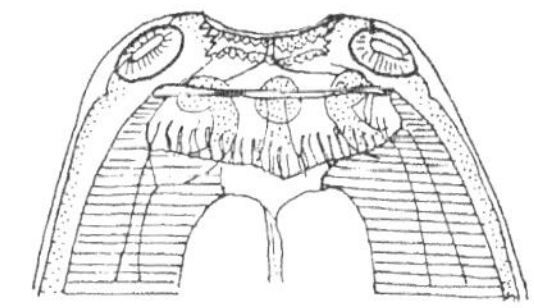
4.52

FIG. 4.48. *Paraustroxyuris*, caudal bursa. (After Mawson, 1964.)
FIG. 4.49. *Macropoxyuris*, buccal teeth. (After Mawson, 1964.)
FIG. 4.50. *Macropoxyuris*, caudal bursa. (After Mawson, 1964.)
FIG. 4.51. *Potoroxyuris*, head, lateral view. (After Mawson, 1964.)
FIG. 4.52. *Oxyuris*, head, ventral view. (After Skrjabin & Shikhobalova, 1949.)

23-(22) Teeth with numerous tubercles (Fig. 4.49).
Pre-anal cuticular differentiation present (Fig. 4.50).
Macropoxyuris Mawson, 1964

24-(21) Pharyngeal teeth protruding from oral opening (Fig. 4.51).
Potoroxyuris Mawson, 1964

25-(2) Buccal cavity with complex armature with spines (Fig. 4.52).
Cephalic papillae large, simple or differentiated into cuticular network.
Parasites of ungulates.
Oxyuris Rudolphi, 1803

26-(1) Four or less pairs of well-separated genital papillae present, of which only one pair pedunculate (Fig. 4.53).
Gubernaculum present or absent.
Parasites of rodents, Dermoptera, ruminants and hyracoids.

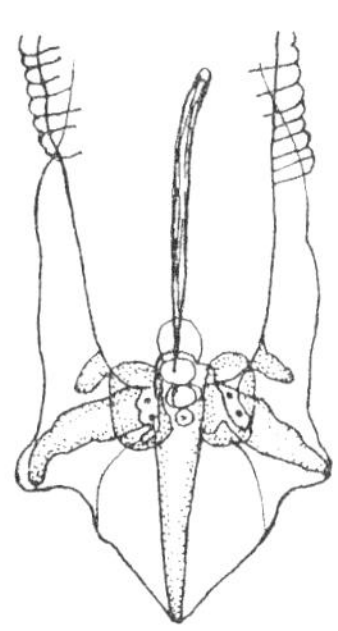
4.53

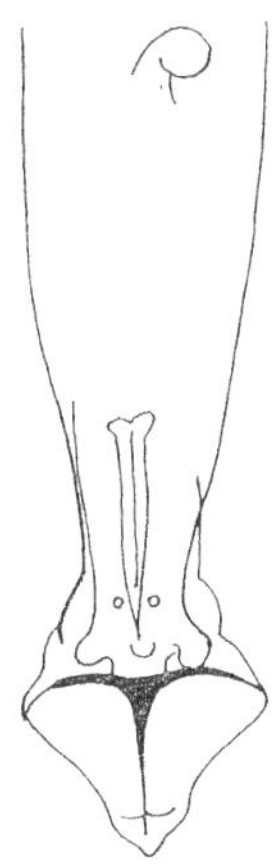
4.54

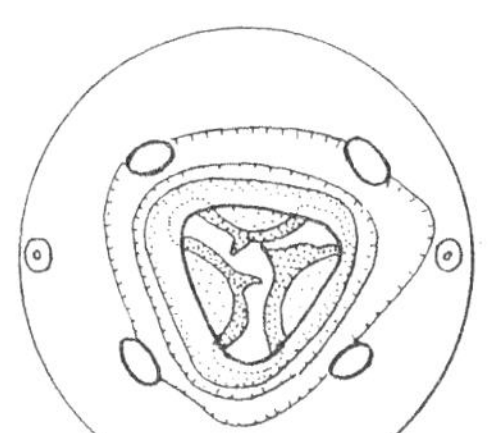
4.55

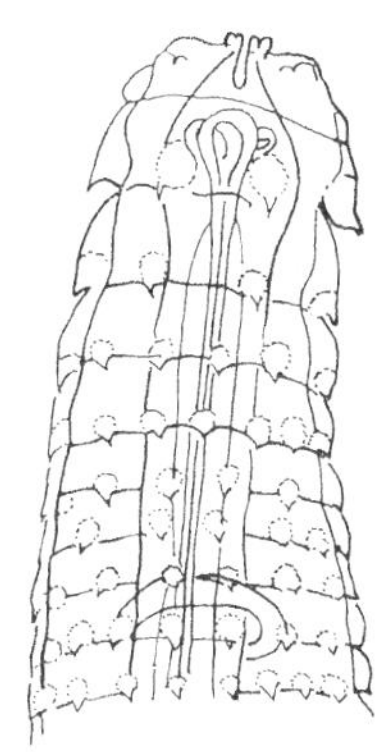
4.56

FIG. 4.53. *Citellina*, caudal bursa. (After Merkuscheva, 1955.)
FIG. 4.54. *Auchenacantha*, caudal bursa. (After Baylis, 1929.)
FIG. 4.55. *Passalurus*, apical view. (After Mawson, 1964.)
FIG. 4.56. *Auchenacantha*, cuticle. (After Baylis, 1929.)

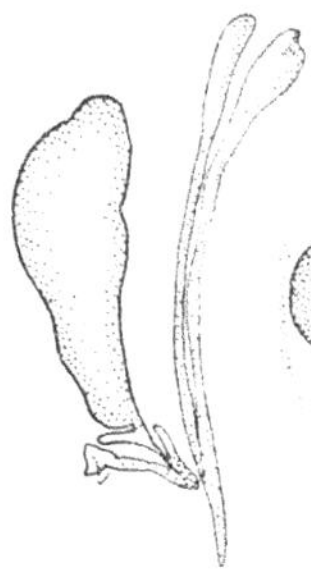

4.57

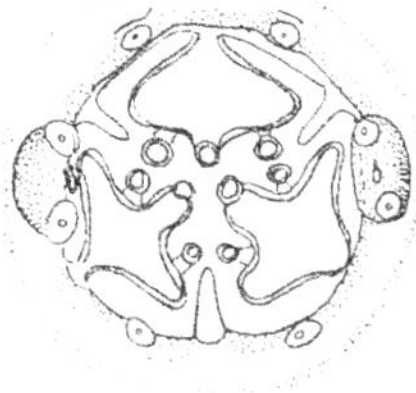

4.58

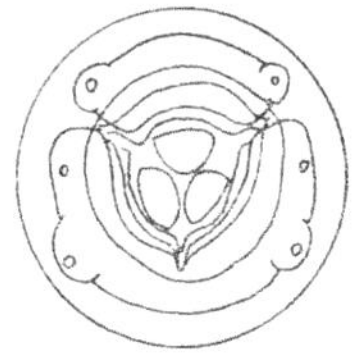

4.59

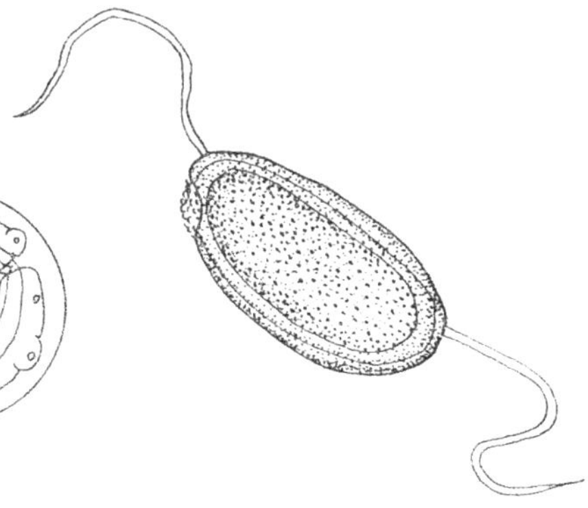

4.60

FIG. 4.57. *Sypharista*, spicule and gubernaculum. (After Quentin, 1971b.)
FIG. 4.58. *Zenkoxyuris*, pharyngeal teeth. (After Quentin, 1974.)
FIG. 4.59. *Citellina*, apical view. (After Inglis & Ogden, 1965.)
FIG. 4.60. *Citellina*, egg. (After Prendel, 1928.)

27-(30) Gubernaculum absent (Fig. 4.54).
28-(29) Cuticle unarmed.
Caudal alae absent.
Lips absent (Fig. 4.55).
Male pre-anal papillae sessile; double vestigial post-anal papilla present.
Caudal appendix highly developed.
Female tail with moniliform constrictions in old individuals.
Parasites of Leporidae.
Passalurus Dujardin, 1845

29-(28) Cuticle with many spines (Fig. 4.56).
Caudal alae supported by rib-like papillae and caudal appendage (Fig. 4.54).
Male with three lips or six labial lobes.
Female with six labial lobes.
Male with a pre-anal cuticular appendix.
Tail of female smooth.
Parasites of Dermoptera.
Auchenacantha Baylis, 1929

30-(27) Gubernaculum present (Fig. 4.57).
31-(32) Pharyngeal teeth complex, each with three cylindrical diverticula (Fig. 4.58).
Parasites of anomaluridean rodents.
Zenkoxyuris Quentin, 1975

32-(31) Pharyngeal teeth simple without cylindrical diverticula (Fig. 4.59).
33-(36) Caudal bursa terminating in large alae with rib-like papillae reminiscent of bursa of strongyles (Fig. 4.53).
34-(35) Eggs with two polar filaments (Fig. 4.60).
Cuticulae expansions anterior and posterior to vulva present.
Parasites of Sciuridae.
Citellina Prendel, 1928

35-(34) Eggs without polar filaments (Fig. 4.62).
Vulva without cuticular expansions.
Parasites of ruminants.
Skrjabinema Werestchajin, 1926

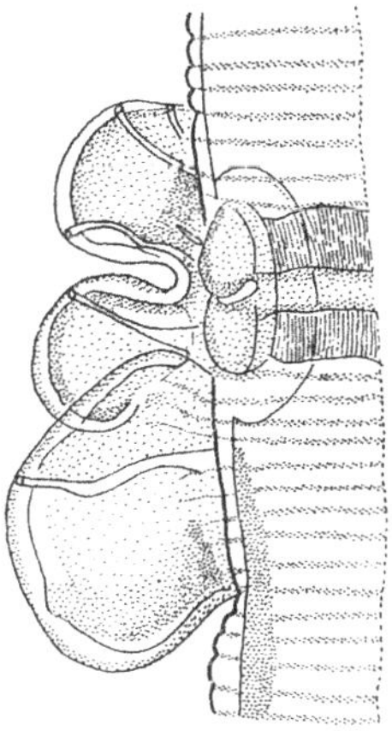

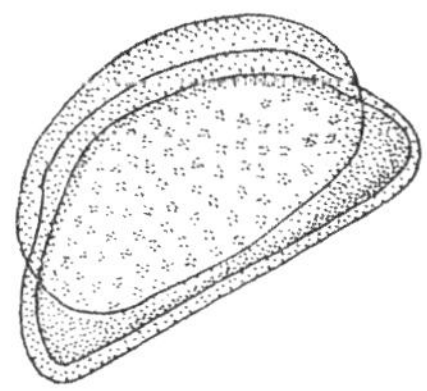

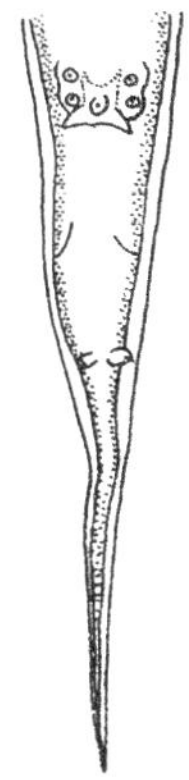

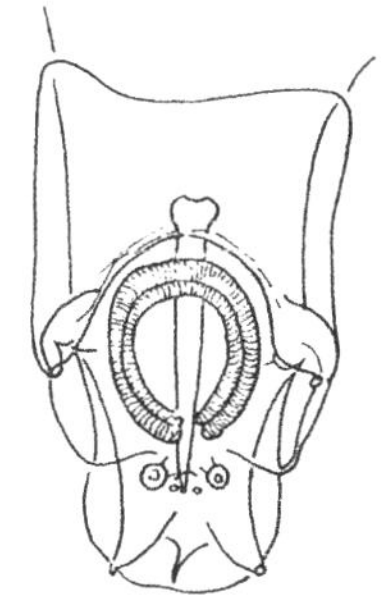

4.61 4.62 4.63 4.64

FIG. 4.61. *Citellina*, vulva. (After Kornev, 1951.)
FIG. 4.62. *Skrjabinema*, egg. (After Skrjabin & Mizkevitsch, 1930.)
FIG. 4.63. *Sypharista*, caudal bursa. (After Quentin, 1971b.)
FIG. 4.64. *Hoplodontophorus*, caudal bursa. (After Inglis, 1959.)

36-(33) Caudal bursa without lateral alae reminiscent of strongyles (Fig. 4.63).
37-(38) Horse-shoe shaped pre-anal sucker present (Fig. 4.64).
Oesophagus short.
Parasites of hyracoids.

Hoplodontophorus Turner, 1921

38-(37) Pre-anal sucker absent (Fig. 4.63).
Oesophagus long.
Parasites of rodents.
39-(50) Oral opening circular without expansions of buccal capsule and not surrounded by three lips or six labial lobes (Fig. 4.65).
40-(41) Accessory hook on gubernaculum absent (Fig. 4.66).
Vagina more or less extroverted, with muscular base.
Parasites of African Bathyergidae and Ctenodactylidae.

Hilgertia Quentin, 1973

41-(40) Accessory hook on gubernaculum present (Fig. 4.57).
Vagina rarely extroverted, without muscular base.
42-(49) Cephalic extremity of female without cuticular shield (Fig. 4.65).
Cephalic vesicle often present.
Eggs operculate.
43-(44) Sexual dimorphism present in cephalic structures of male and female (Fig. 4.67).
Cephalic papillae arranged in square, never laterally close to amphids.
Cuticular swellings present or absent on ventral surface of male; if present, one to three in number.
Parasites of Asian Petauristinae.

Sypharista Quentin, 1971

44-(43) Cephalic structures identical in both sexes.
Cephalic papillae not arranged in square, more or less laterally close to amphids (Fig. 4.65).

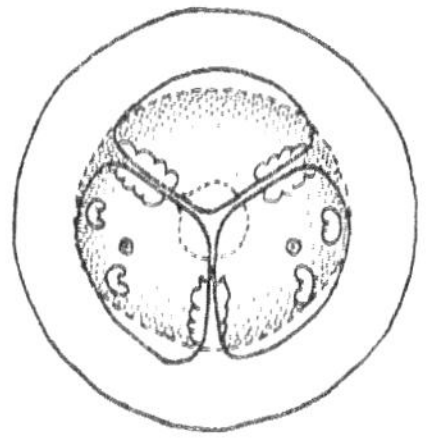
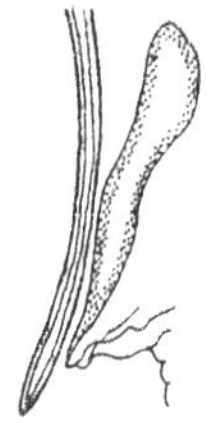
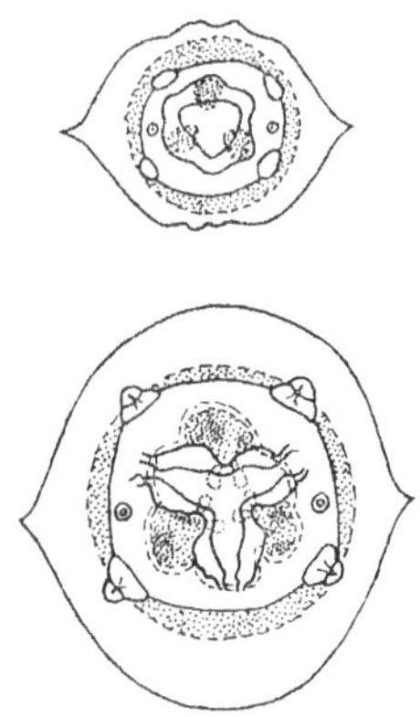
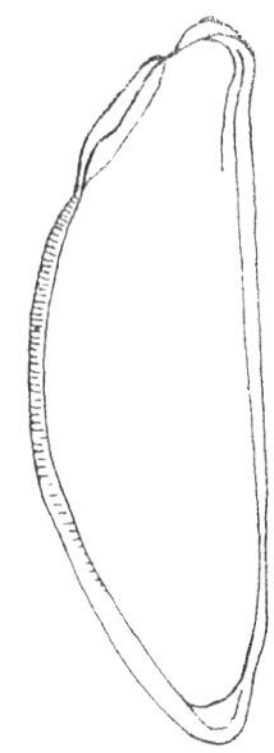

4.65 4.66 4.67 4.68

FIG. 4.65. *Syphacia*, apical view. (After Quentin, 1971a.)
FIG. 4.66. *Hilgertia*, male, spicule and gubernaculum. (After Quentin, 1973.)
FIG. 67. *Sypharista*, heads of male and female. (After Quentin, 1971b.)
FIG. 4.68. *Syphacia*, egg. (After Quentin, 1971a.)

Two to four ventral cuticular swellings present in the male.
Parasites of Sciuridae and Muroidea.
Syphacia Seurat, 1916

45-(48) Eggs with small operculum at one end (Fig. 4.68).
Vagina never protuberant in gravid females.
Two to three ventral cuticular swellings present in male.

46-(47) Male with two ventral cuticular swellings (Fig. 4.69).
Cephalic papillae widely spaced.
Parasites of Sciuridae.
Syphacia (*Syphatineria*) Chabaud & Biocca, 1955

47-(46) Male with 3 ventral cuticular swellings (Fig. 4.70).
Cephalic papillae near amphids.
Parasites of Sciuridae, Cricetidae, Gerbillidae, Muridae and Microtidae.
Syphacia (*Syphacia*)

48-(45) Eggs with large operculum occupying greater part of convex side of shell (Fig. 4.72).
Vagina usually protuberant (Fig. 4.77) in gravid females.
Three to four ventral cuticular swellings present in male. (Figs. 4.70, 4.71).
Parasites of Petauristinae, less commonly of Sciuridae and Muridae.
Syphacia (*Syphabulea*) Gubanov, 1964

49-(42) Female cephalic extremity asymmetrical, with cuticular shield reinforced dorsally by hook-like or lobe-like thickenings (Fig. 4.73).
Cephalic vesicule absent.
Eggs non-operculate.
Parasites of African Anomalurinae, Petromyscidae and Petromyidae.
Acanthoxyurus Sandground, 1928

50-(39) Oral opening surrounded by six expansions of buccal capsule or by 3 pseudolabia extending along interlabial axes (Fig. 4.74).

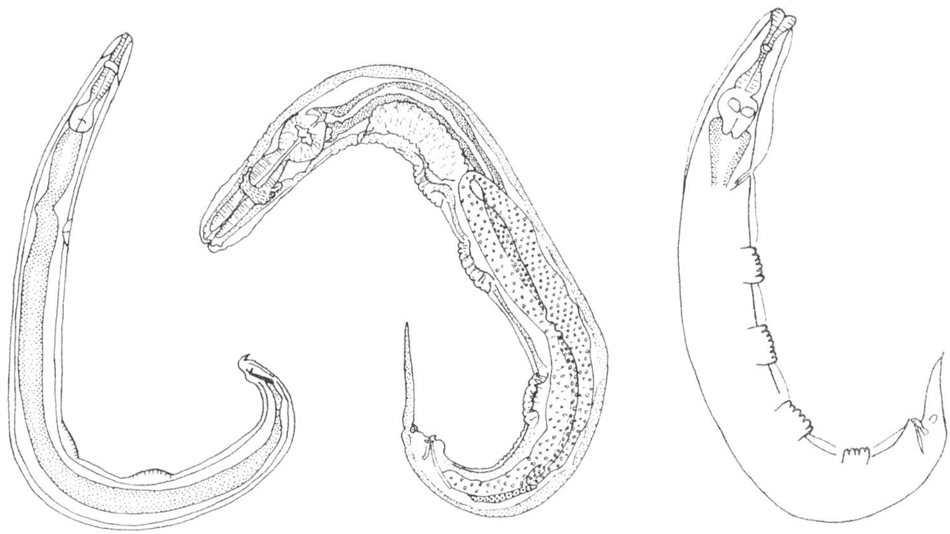

4.69 4.70 4.71

FIG. 4.69. *Syphatineria*, male, lateral view. (After Tiner, 1948.)
FIG. 4.70. *Syphacia*, male, lateral view. (After Quentin, 1969.)
FIG. 4.71. *Syphabulea*, male, lateral view. (After Gubanov, 1964.)

51-(58) Cephalic extremity without cuticular shields.
52-(53) Protruding labial papillae present.
Preanal ornamentation in male absent (Fig. 4.75).
Parasites of neoarctic Heteromyidae.

Heteromyoxyuris Quentin, 1973

53-(52) Labial papillae absent (Fig. 4.74).
Preanal ornamentation usually present (Fig. 4.76).
54-(55) Oral opening with six expansions. Pedunculated cephalic papillae present (Fig. 4.75 bis).
Parasites of Caviidae.

Protozoophaga Travassos, 1923

55-(54) Oral opening with three pseudolabia. Cephalic papillae not pedunculated.
56-(57) Buccal cavity thick, with sclerotized expansions (Fig. 4.74).
Stout pharyngeal teeth present.
Pseudolabia reduced.
Preanal cuticular swelling present or absent, if present one.
Cervical alae absent.
Gubernaculum without accessory hook.
Vagina more or less protuberant (Fig. 4.77).
Spiral caudal cuticular ornamentation of female present or absent.
Parasites of Hystricidae, Erethizontidae and Dinomyidae.

Evaginuris Skrjabin & Schikhobalova, 1951

57-(56) Buccal cavity reduced.
Pharyngeal teeth reduced.
Pseudolabia well developed (Fig. 4.78).

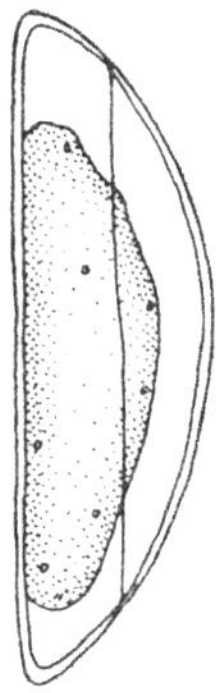

4.72

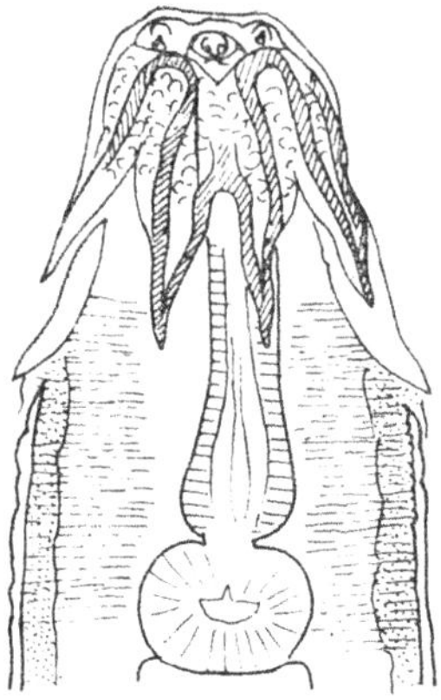

4.73

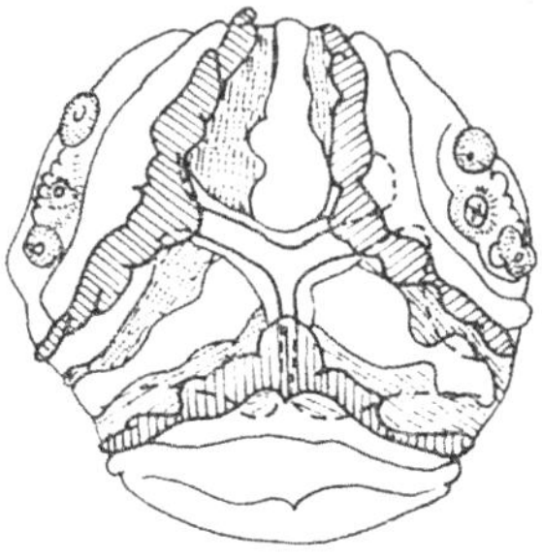

4.74

FIG. 4.72. *Syphacia* (*Syphabulea*), egg. (After Schmidt & Kuntz, 1968.)
FIG. 4.73. *Acanthoxyurus*, anterior extremity. (After Quentin, 1974.)
FIG. 4.74. *Evaginuris*, apical view. (After Quentin, 1973.)

Two preanal cuticular swellings present.
Cervical alae well developed.
Gubernaculum with accessory hook (Fig. 4.79).
Vagina not protuberant.
Tail smooth.
Parasites of neotropical Caviomorpha.

Helminthoxys Freitas, Lent & Almeida, 1937

58-(51) Cephalic extremity with dorsal cuticular shield (Fig. 4.80) ending in 2 lobes.
Eggs not operculate.

Octodonthoxys Quentin, Courtin & Fontecilla, 1975

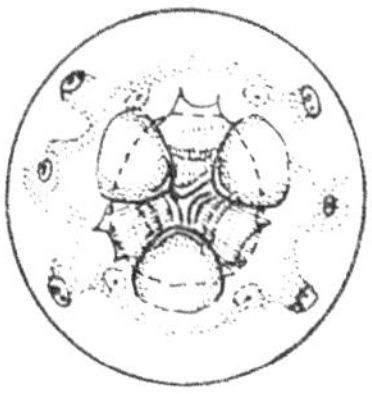

4.75

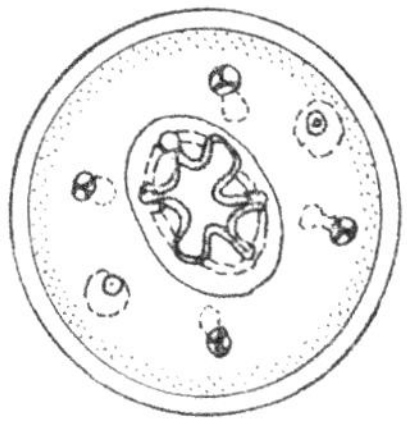

4.75 bis

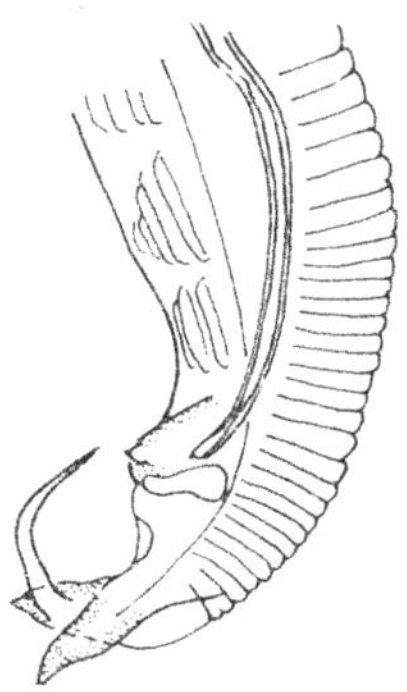

4.76

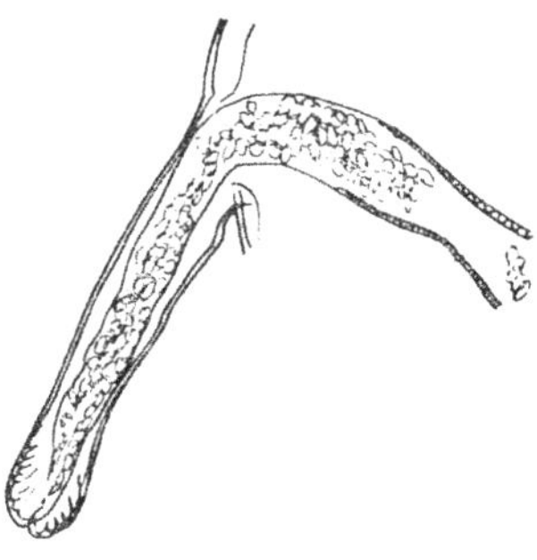

4.77

FIG. 4.75. *Heteromyoxyuris*, apical view. (After Quentin, 1973.)
FIG. 4.75 bis. *Protozoophaga*, male, apical view. (After Diaz-Ungría & Quentin, 1976.)
FIG. 4.76. *Evaginuris*, caudal bursa. (After Quentin, 1973.)
FIG. 4.77. *Evaginuris*, vagina. (After Quentin, 1973.)

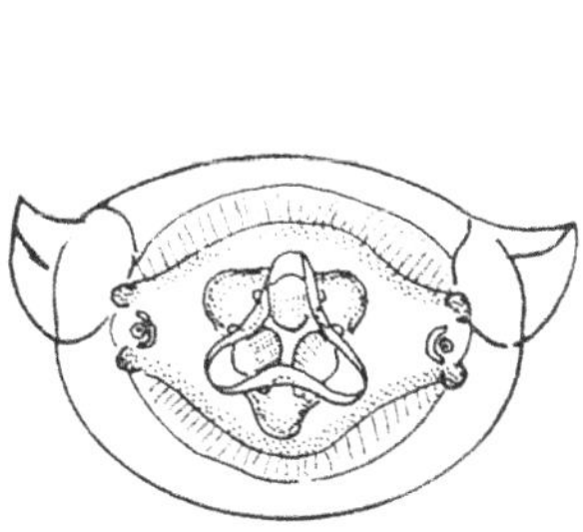

4.78

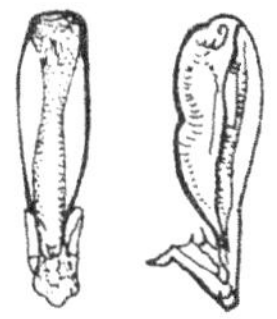

4.79

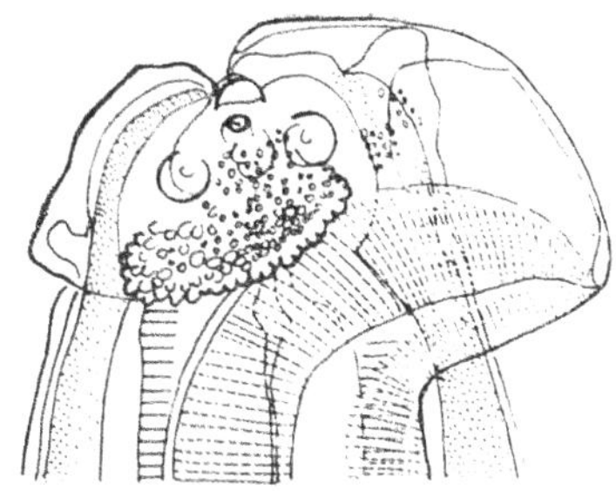

4.80

FIG. 4.78. *Helminthoxys*, apical view. (After Quentin, 1973.)
FIG. 4.79. *Helminthoxys*, gubernaculum. (After Quentin, 1973.)
FIG. 4.80. *Octodonthoxys*, head. (After Quentin, Courtin Lyon & Fontecilla Gallardo, 1975.)

Family *HETEROXYNEMATIDAE* (Skrjabin & Shikhobalova, 1948)

Key to subfamilies

1-(2) Last pair of male postanal papillae isolated near extremity of caudal bursa (Fig. 4.81) (except in *Fastigiurus prudhoei*).
Glandular part of ovejector never in form of enlarged ring at distal end of muscular *vagina vera* (Fig. 4.82).
Eggs without operculum (Fig. 4.83).
Pharyngeal and buccal structures simple (Fig. 4.84).
Parasites of birds and rodents (one genus in Lagomorpha).

Heteroxynematinae

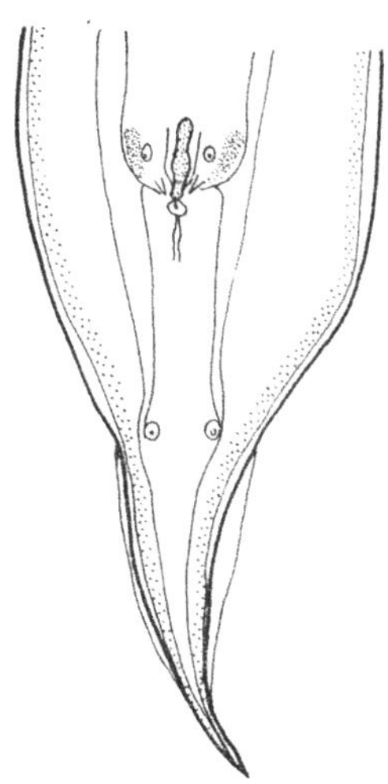

4.81

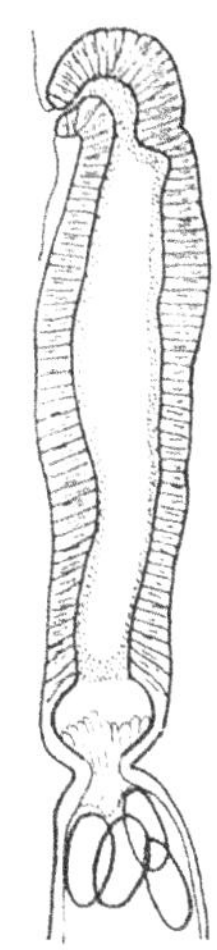

4.82

FIG. 4.81. *Syphaciella*, male caudal bursa, ventral view. (After Vassiliades, 1970.)
FIG. 4.82. *Syphaciella*, ovejector. (After Vassiliades, 1970.)

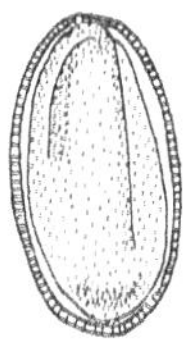

4.83

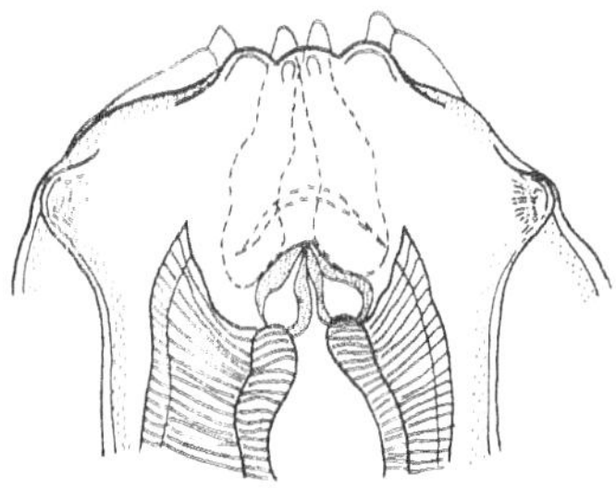

4.84

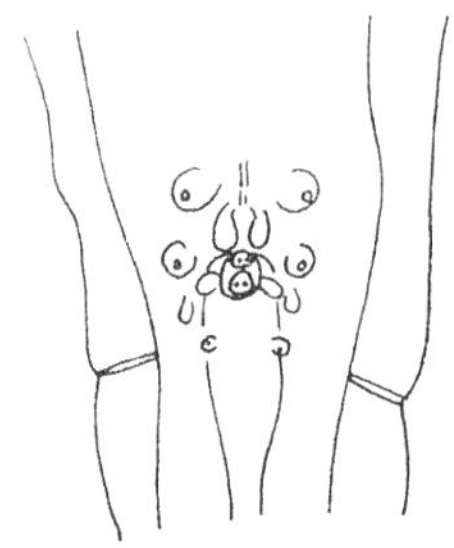

4.85

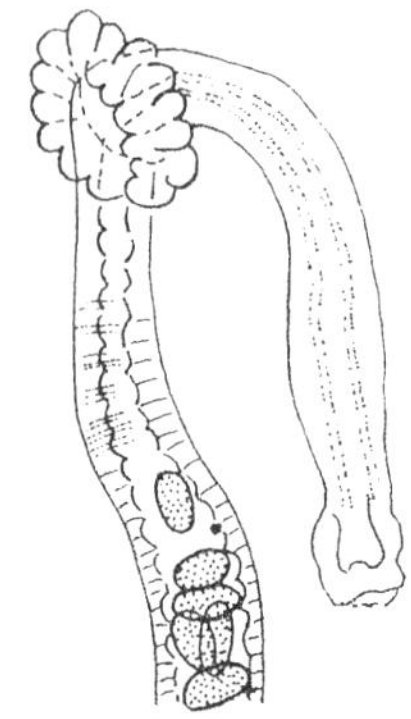

4.86

FIG. 4.83. *Aspiculuris*, egg. (After Quentin, 1966.)
FIG. 4.84. *Syphaciella*, pharyngeal structures. (After Vassiliades, 1970.)
FIG. 4.85. *Labiostomum*, cloacal papillae. (After Inglis, 1959.)
FIG. 4.86. *Dermatoxys*, glandular ring of ovejector. (After Quentin, 1975.)

2-(1) Anal papillae grouped or fused together around anus (Fig. 4.85).
Ovejector with glandular ring around distal end of muscular *vagina vera* (Fig. 4.86).
Eggs with thick double shells and thick operculum (Fig. 4.87).
Complex pharyngeal and buccal structures with lamellar extensions present (Fig. 4.88).
Parasites of Lagomorpha.
Labiostomatinae Akhtar, 1956 sub fam.

Subfamily **Heteroxynematinae** (Skrjabin & Schikhobalova, 1948 fam.)

Key to genera and subgenera

1-(2) Gubernaculum present (Fig. 4.89).
Spicule double.
Parasites of birds of the Pteroclidae.
Syphaciella Mönnig, 1924

2-(1) Gubernaculum absent.
Spicule single or absent (Fig. 4.90).

3-(4) Intestinal caecum present behind oesophagus (Fig. 4.91).
Parasites of birds of the Tinamidae.
Eudromoxyura Anderson & Prestwood, 1972

4-(3) Intestinal caecum absent (Fig. 4.92).
Parasites of rodents or Lagomorpha.

5-(16) Eggs with single thin shell (Fig. 4.83).
Parasites of sciuromorph, caviomorph and myomorph rodents.

6-(13) Pharyngeal cavity highly developed with massive teeth (Fig. 4.93).
Pre-anal cuticular ornamentation in shape of crest or comb often present (Fig. 4.94).
Parasites of Sciuromorpha and Caviomorpha.

7-(12) Cuticular swellings on ventral surface of males absent.
Parasites of holarctic Sciuromorpha and neotropical Caviomorpha.
Heteroxynema Hall, 1916

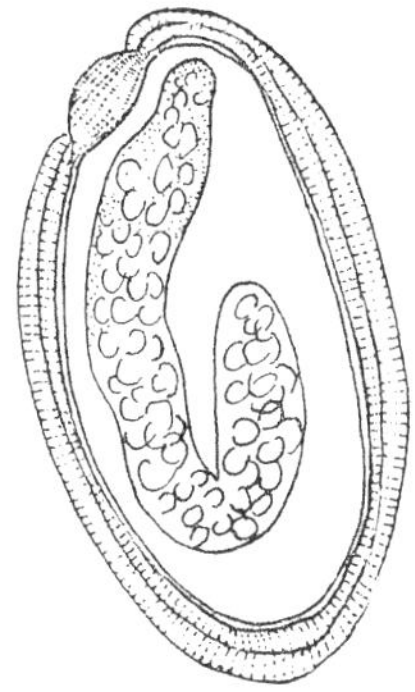

4.87

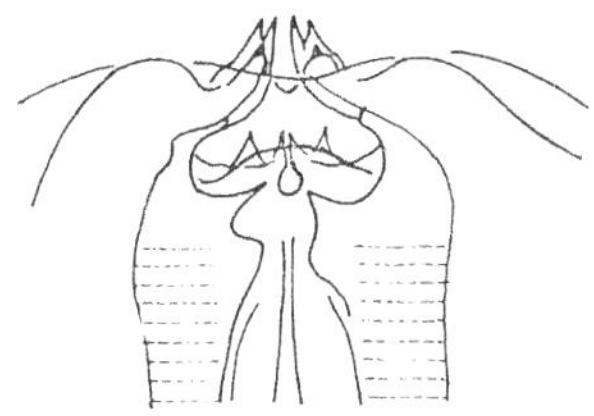

4.88

FIG. 4.87. *Labiostomum*, egg. (After Quentin, 1975.)
FIG. 4.88. *Labiostomum*, pharyngeal structures. (After Quentin, 1975.)

8-(11) Single pair of papillae present near end of caudal bursa of male (Fig. 4.95).
Sexual dimorphism of labial structures absent.
Parasites of holarctic Sciuridae.

9-(10) *Vagina vera* elongated, U-shaped, with thick muscular wall near vulva (Fig. 4.96).
Oesophagus elongated with spherical bulb with strongly developed valves (Fig. 4.92).
Parasites of palaearctic Sciuridae.

Heteroxynema (*Proxyuronema*) Quentin, 1975

10-(9) *Vagina vera* short, not U-shaped (Fig. 4.97) with uniformly muscular wall.
Oeosophagus with non-spherical bulb and reduced valves.
Parasites of nearctic Sciuridae.

Heteroxynema (*Heteroxynema*)

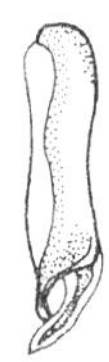

4.89

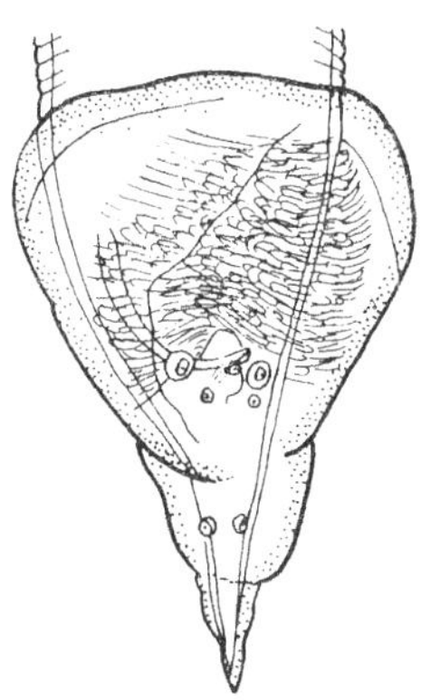

4.90

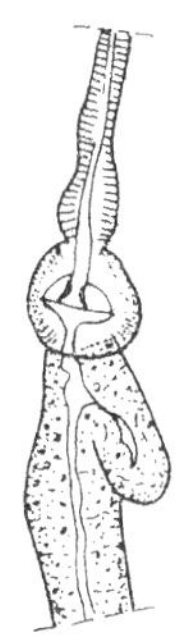

4.91

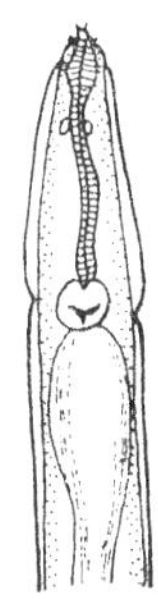

4.92

FIG. 4.89. *Syphaciella*, gubernaculum. (After Vassiliades, 1970.)
FIG. 4.90. *Dentostomella*, male, caudal bursa. (After Schulz & Krepkogorskaja, 1932.)
FIG. 4.91. *Eudromoxyura*, intestinal caecum. (After Anderson & Prestwood, 1972.)
FIG. 4.92. *Heteroxynema* (*Proxyuronema*), oesophagus. (After Biocca & Chabaud, 1955.)

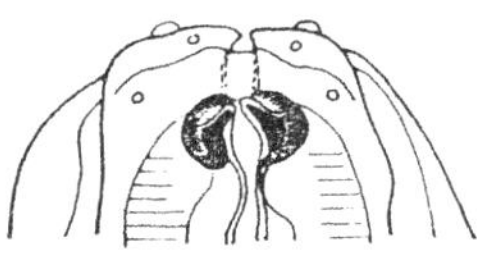

4.93

4.94

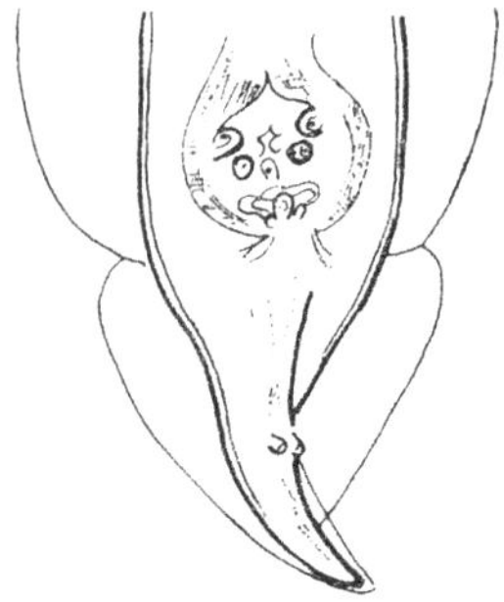

4.95

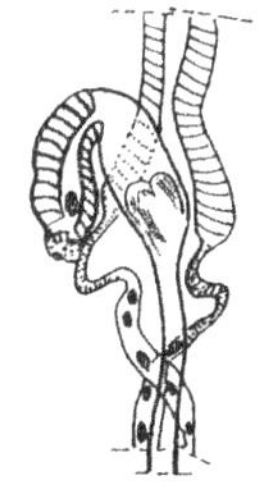

4.96

FIG. 4.93. *Heteroxynema* (*Cavioxyura*), teeth. (After Freitas & Almeida, 1936.)

FIG. 4.94. *Proxyuronema*, comb-like precloacal cuticular ornamentation. (After Seurat, 1915.)

FIG. 4.95. *Heteroxynema* (*Proxyuronema*), caudal bursa. (After Biocca & Chabaud, 1955.)

FIG. 4.96. *Heteroxynema* (*Proxyuronema*), ovejector. (After Biocca & Chabaud, 1955.)

11-(8) Two pairs of isolated papillae present well behind anus, anterior pair sometimes fused (Fig. 4.98).
Sexual dimorphism of labial structures present or absent.
Glandular part of ovejector elongated.
Parasites of neotropical Caviomorpha.
Heteroxynema (*Cavioxyura*) Quentin, 1975

12-(7) Two cuticular swellings present on ventral side of male (Fig. 4.99).
Parasites of palaearctic Sciuridae.
Dermatopallarya Skrjabin, 1924

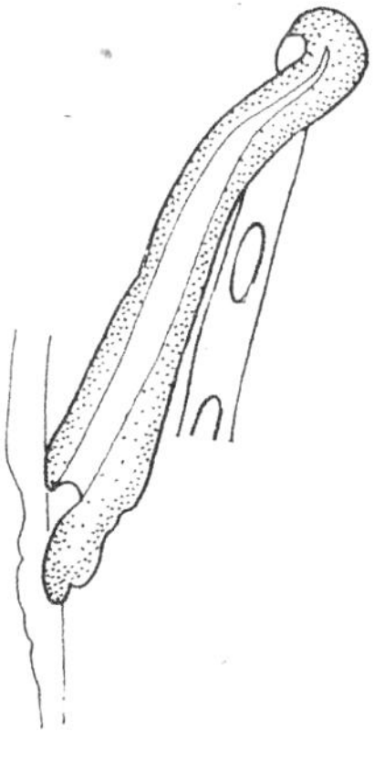

4.97

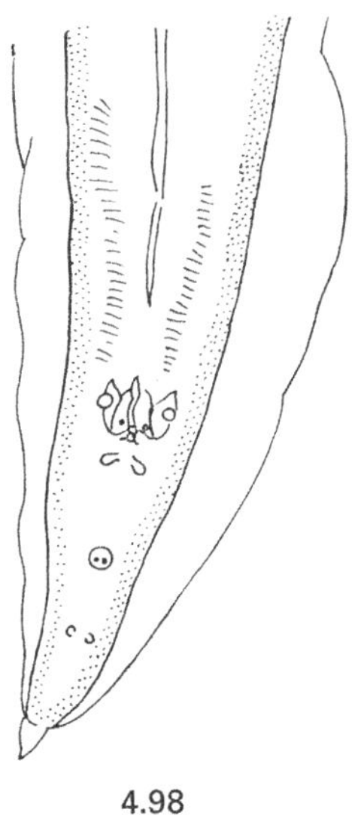

4.98

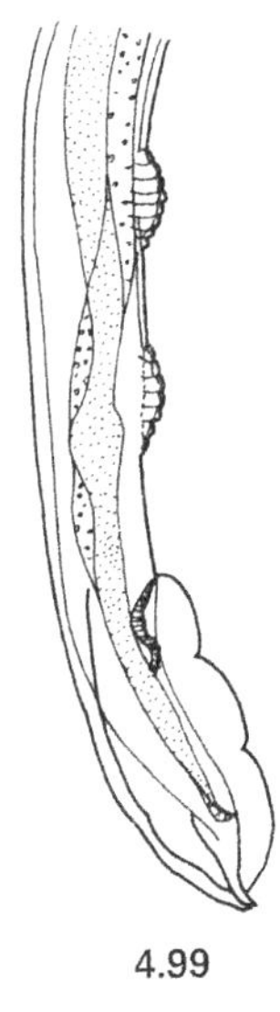

4.99

FIG. 4.97. *Heteroxynema* (*Heteroxynema*), ovejector. (After Hall, 1916.)

FIG. 4.98. *Heteroxynema* (*Cavioxyura*), caudal bursa. (After Quentin, 1975.)

FIG. 4.99. *Dermatopallarya*, cuticular mamelons. (After Skrjabin, 1924.)

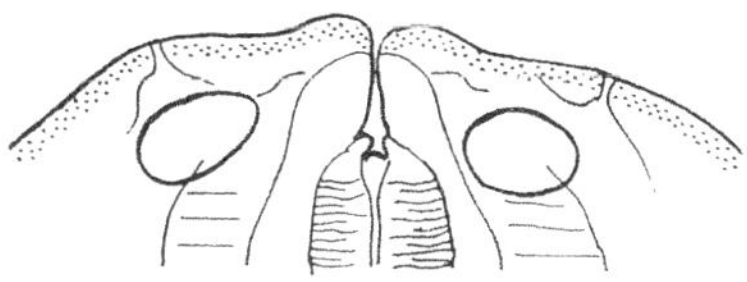

4.100

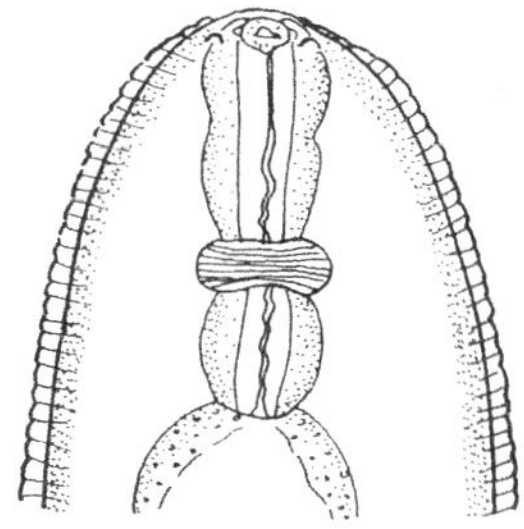

4.101

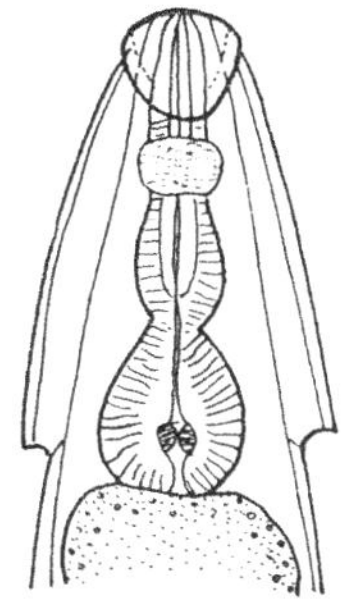

4.102

FIG. 4.100. *Aspiculuris*, ventral view. (After Quentin, 1975.)

FIG. 4.101. *Dentostomella*, anterior extremity. (After Schultz & Krepkogorskaja, 1932.)

FIG. 4.102. *Aspiculuris*, anterior extremity. (After Schulz, 1924.)

13-(6) Pharyngeal cavity reduced or absent.
Buccal teeth reduced, sometimes with tubercles (Fig. 4.100).
Preanal cuticular ornamentation absent.

14-(15) Cuticular striations marked.
Cervical alae absent.
Cephalic vesicle absent (Fig. 4.101).
Parasites of nearctic Sciuridae and Gerbillidae.

Dentostomella Schulz & Krepkogorskaja, 1932

15-(14) Cuticular striations poorly marked.
Cervical alae well developed, extending into cephalic vesicle (Fig. 4.102).
Parasites of Muroidea.

Aspiculuris Schulz, 1927

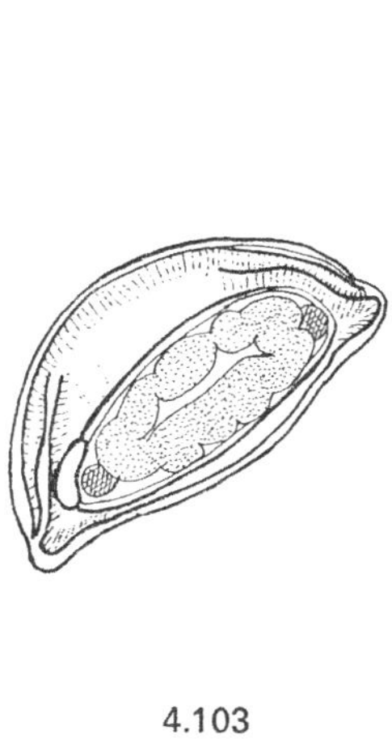

4.103

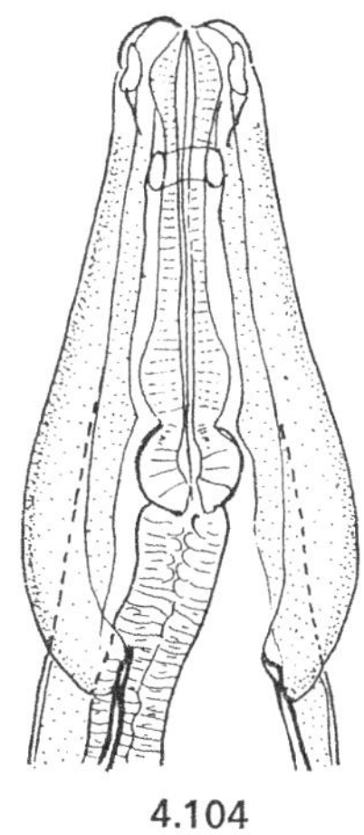

4.104

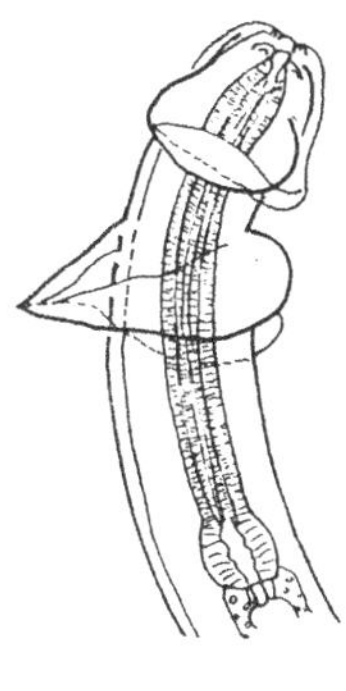

4.105

FIG. 4.103. *Rauschoxyuris*, egg. (After Quentin, 1975.)

FIG. 4.104. *Rauschoxyuris*, anterior extremity. (After Quentin, 1975.)

FIG. 4.105. *Fastigiuris*, anterior extremity. (After Babaev, 1966.)

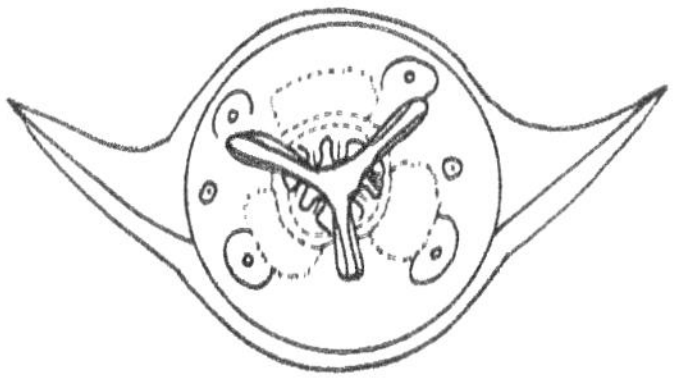

4.106

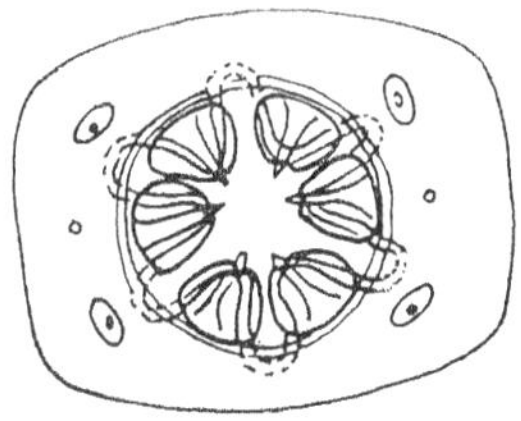

4.107

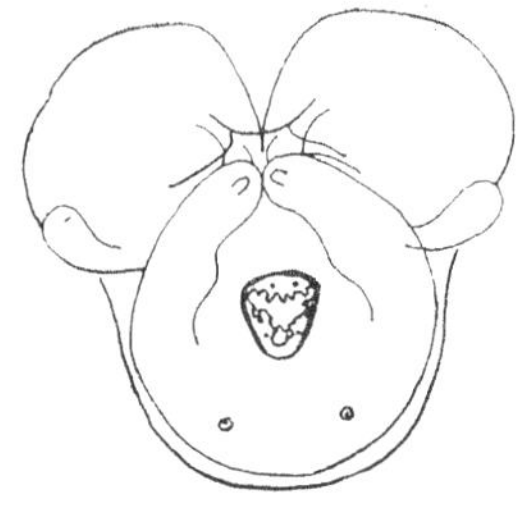

4.108

FIG. 4.106. *Dermatoxys*, apical view. (After Vicente, 1969.)
FIG. 4.107. *Labiostomum*, apical view. (After Inglis, 1959.)
FIG. 4.108. *Cephaluris*, apical view. (After Akhtar, 1947.)

16-(5) Eggs with thick or double shells (Fig. 4.103).
Parasites of aplodontid rodents and Lagomorpha.

17-(18) Oesophagus short with well-defined spherical bulb.
Cervical alae long and thin (Fig. 4.104).
Parasites of aplodontid rodents.
Rauschoxyuris Quentin, 1975

18-(17) Oesophagus elongated with reduced bulb or no bulb.
Cervical alae short, thick, markedly bent on ventral region (Fig. 4.105).
Parasites of Ochotonidae.
Fastigiuris Babaev, 1966

Subfamily **Labiostomatinae** Akhtar, 1956

Key to genera and subgenera

1-(6) Head without dorsal cuticular shield.
Preanal sucker absent.

2-(3) Oral opening surrounded by three lips in male and female (Fig. 4.106).
Three preanal crests present.
Parasites of Leporidae.
Dermatoxys Schneider, 1866

3-(2) Oral opening surrounded by three lips or six labial lobes in male and female (Fig. 4.107).
One preanal crest present.
Parasites of Ochotonidae.
Labiostomum Akhtar, 1941

4-(5) Cephalic vesicle present.
Labiostomum (*Labiostomum*)

5-(4) Cephalic vesicle absent.
Labiostomum (*Eugenuris*) (Schulz, 1948)

6-(1) Cephalic extremity with dorsal cuticular shield terminated by two lobes (Fig. 4.108).
Preanal sucker present.
Parasites of Ochotonidae.
Cephaluris Akhtar, 1947

REFERENCES*

ANDERSON, R. C. & PRESTWOOD, A. K. 1972. *Eudromoxyura elonbyrdi* n.g., n.sp. (Nematoda: Oxyuroidea) from crested tinamou (*Eudromia elegans*). *Can. J. Zool.*, **50** (3): 297-300. [En, fr].

BABAEV, YU. 1966. *Fastigiuris devexus* gen. et sp. nov. (Nematoda: Heteroxyncmatidae), parasite du rongeur *Ochotona rufescens* Gray. *Izv. Akad. Nauk turkmen. SSR, Ser biol.*, No. 1: pp. 94-96. [Ru].

BARUŠ, V. 1972. Remarks on the Cuban species of the genus *Helminthoxys* (Nematoda: Syphaciidae). *Folia Parasit.*, **19** (2): 105-111. [En, ru].

BARUŠ, V. 1973. Some remarks on the neotropical species of the genera *Parapharyngodon* and *Batracholandros* (Oxyuridae). *Folia Parasit.*, **20** (2): 131-139. [En, ru].

BARUŠ, V., COY OTERO, A. & GARRIDO, O. H. 1969. Helmintofauna de *Cyclura macleayi* Gray (Sauria, Iguanidae) en Cuba. *Torreia*, n.s. No. 8, pp. 2-20. [Es, en].

BAYLIS, H. A. 1936. The fauna of British India, including Ceylon and Burma. Nematoda. Vol. I. (Ascaroidea and Strongyloidea). *London: Taylor & Francis*, xxxvi+408 pp.

BERNARD, J. 1969. Quelques nématodes parasites nouveaux ou non encore signalés en Tunisie. *Archs Inst. Pasteur Tunis*, **46** (3): 397-411.

BISWAS, P. K. & CHAKRAVARTY, G. K. 1963.The systematic studies of the zoo-parasitic oxyuroid nematodes. *Z. ParasitKde*, **23** (5): 411-428.

CABALLERO R., G. 1968. Contribution à la connaissance des nématodes de sauriens malgaches. *Annls Parasit. hum. comp.*, **43** (2): 149-200; (3): 353-379. [Fr, en].

CHABAUD, A. G. 1965. Superfamille des Oxyuroidea. *In:* Grassé, P. P., Traité de zoologie. Tome IV, Fasc. III. *Paris: Masson et Cie*, pp. 956-988.

CHABAUD, A. G., BRYGOO, E. R. & PETTER, A. J. 1965. Les nématodes parasites de lémuriens malgaches. VI. Description de six espèces nouvelles et conclusions générales. *Annls Parasit. hum. comp.*, **40** (2): 181-214.

CHABAUD, A. G. & GOLVAN, Y. J. 1957. Miscellanea helminthologica maroccana XXIV. Nématodes parasites de lézards de la forêt de Nefifik. *Archs Inst. Pasteur Maroc*, **5** (7): 447-469.

CHABAUD, A. G. & PETTER, A. J. 1958. Les nématodes parasites de lémuriens malgaches. *Mem Inst. scient. Madagascar*, Sér, A. **12**: 139-158.

CHABAUD, A. G. & PETTER, A. J. 1959. Les nématodes parasites de lémuriens malgaches. II. Un nouvel oxyure: *Lemuricola contagiosus. Mem. Inst. scient. Madagascar*, Sér. A, **13**: 127-132.

CHABAUD, A. G., PETTER, A. J. & GOLVAN, Y. J. 1961. Les nématodes parasites de lémuriens malgaches. III. Collection récoltée par M. et Mme Francis Petter. *Annls Parasit hum. comp.*, **36** (12): 113-126.

DIAZ-UNGRÍA, C. & QUENTIN, J. C. 1976. Morphologie et position systématique de l'oxyure *Protozoophaga obesa* (Diesing, 1851) Travassos, 1923. *Excerta Parasitológica en Memoria del Doctor Eduardo Caballero y Caballero*. [In press.]

FREITAS, J. F. TEIXEIRA DE & IBÁÑEZ H., N. 1963. Fauna helmintológica del Perú. Notas sobre Pharyngodonidae Travassos, 1920 y descripción de *Pharyngodon micrurus* sp. n. (Nematoda, Oxyuroidea) *Revta Univ. Lima*, No. 23/24: pp. 3-10.

FREITAS, J. F. TEIXEIRA DE & IBÁÑEZ H., N. 1965. Fauna helmintológica do Peru: alguns nematodeos parasitos de *Bufo spinulosus limensis* (Werner). *Papéis Dep. Zool. S Paulo*, **17** (18): 229-240.

INGLIS, W. G. 1959a. Some oxyurid parasites (Nematoda) from *Ochotona rufescens vizier* (Mammalia: Lagomorpha) in Iran. *Bull. Soc. zool. France*, **84** (2/3): 178-187.

INGLIS, W. G. 1959b. The systematic position of the nematode genus *Hoplodontophorus. Revue Zool. Bot. afr.*, **59** (3/4): 316-325.

INGLIS, W. G. 1961. The oxyurid parasites (Nematoda) of primates. *Proc. zool. Soc. Lond.*, **136** (1): 103-122.

INGLIS, W. G. 1962. *Ichthyouris ro* gen. et sp. nov. (Nematoda): an oxyurid from a freshwater fish. *J. Helminth.*, **36** (1/2): 45-50.

INGLIS, W. G. 1968. Nematodes parasitic in Western Australian frogs. *Bull. Br. Mus. nat. Hist. (Zool.)*, **16** (4): 163-183.

INGLIS, W. G. & COSGROVE, G. E. 1965. The pin-worm parasites (Nematoda: Oxyuridae) of the Hapalidae (Mammalia: primates). *Parasitology*, **55** (4): 731-737.

INGLIS, W. G. & DIAZ-UNGRÍA, C. 1959. Nematodes de Venezuela, III. Nematodes parasitos vertebrados venezolanos, I. Una revisión del género *Trypanoxyuris* (Ascaridina: Oxyuridae). *Mems Soc. Cienc. nat. "La Salle"*, **19** (54): 176-212. [Es, en].

INGLIS, W. G., DIAZ-UNGRÍA, C. & COLES, J. W. 1960. Nematodes de Venezuela, IV. Nematodes parasitos de vertebrados venezolanos, II. El género *Ozolaimus* (Oxyuridae: Pharyngodoninae). *Acta biol. venez.*, **3** (1): 1-24. [Es, en].

INGLIS, W. G. & DUNN, F. L. 1963. The occurrence of *Lemuricola* (Nematoda: Oxyurinae) in Malaya: with the description of a new species. *Z. ParasitKde*, **23** (4): 354-359.

INGLIS, W. G. & DUNN, F. L. 1964. Some oxyurids (Nematoda) from neotropical primates. *Z. ParasitKde*, **24** (1): 83-87.

INGLIS, W. G. & OGDEN, C. G. 1965. Observations on the nematode genus *Citellina:* with the description of a new species, *Citellina himalensis*. *J. Helminth.*, **39** (1): 11-18.

JOHNSON, S. 1967. A new nematode of the genus *Syphacia* (Oxyuroidea) from the squirrel, *Funambulus pennanti*, from Rajasthan, India. *Proc. zool. Soc. Calcutta*, **20** (1): 83-85.

KHALIL, L. F. 1964. *Cithariniella citharini* gen. et sp. nov. (Nematoda): an oxyurid from a freshwater fish, *Citharinus citharus* in the Sudan. *J. Helminth.*, **38** (1/2): 41-46.

MAWSON, P. M. 1964. Some Nematoda (Strongylina and Oxyurina) from kangaroos (*Macropus* spp.) from eastern Australia. *Parasitology*, **54** (2): 237-262.

PETTER, A. J. 1966. Equilibre des espèces dans les populations de nématodes parasites du côlon des tortues terrestres. *Mém. Mus. natn. Hist. nat., Paris*, Sér. A. Zool., **39** (1): 1-252.

PETTER, A. J., CHABAUD, A. G., DELAVENAY, R. & BRYGOO, E. R. 1972. Une nouvelle espece de nématode du genre *Lemuricola*, parasite de *Daubentonia madagascariensis* Gmelin, et considérations sur le genre *Lemuricola*. *Annls Parasit. hum. comp.*, **47** (3): 391-398. [Fr, en].

PETTER, A. J. & DOUGLASS, J. 1976. Etude des populations d'oxyures du côlon des *Gopherus* (Testudinidae). *Bull. Mus. natn. Hist. nat., Paris* [In preparation].

PETTER, A. J., VASSILIADES, G. & TRONCY, P. M. 1972. Trois espèces d'oxyures parasites de poissons en Afrique. *Annls Parasit. hum. comp.*, **47** (4): 569-579. [Fr, cn].

QUENTIN, J. C. 1966. Oxyures de Muridae africains. *Annls Parasit. hum. comp.*, **41** (5): 443-452.

QUENTIN, J. C. 1971a. Morphologie comparée des structures céphaliques et génitales des oxyures du genre *Syphacia*. *Annls Parasit. hum. comp.*, **46** (1): 15-60. [Fr, en].

QUENTIN, J. C. 1971b. Description d'un nouvel Oxyurinae: *Sypharista kamegaii* n.gen., n.sp., parasite d'un écureuil volant du Japon. *Bull. Mus. natn. Hist. Nat., Paris*, 2ème sér., Year 1970, **42** (5): 989-995. [Fr, en].

QUENTIN, J. C. 1973. Les Oxyurinae de rongeurs. *Bull. Mus. natn. Hist. nat., Paris*, 3e ser., No. 167, Zool. 112: pp. 1045-1096. [Fr, en].

QUENTIN, J. C. 1974, Publ. 1975. Sur les oxyures d'anomalures. *Bull. Mus. natn. Hist. nat., Paris*, 3e sér., No. 256, Zool. 178, pp. 1507-1523. [Fr, en].

QUENTIN, J. C. 1975. Essai de classification des oxyures Heteroxynematinae. *Mém. Mus. natn. Hist. nat. Paris, Sér. A. Zool.*, **94:** 51-96.

QUENTIN, J. C., COURTIN LYON, S. & FONTECILLA GALLARDO, J. 1975. *Octodonthoxys gigantea* n.gen., n.sp., nuevo nematodo Oxyurinae parasito de un roedor caviomorpho de Chile. *Boln chil. Parasit.*, **30** (1/2): 21-25.

QUENTIN, J. C. & KRISHNASAMY, 1975. Spéciations des oxyures parasites de rongeurs Petauristinae en Malaisie. *Mem. Mus. natn. Hist. nat. Paris, Sér. A. Zool.*, **94**: 1-50.

QUENTIN, J. C. & TENORA, F. 1974. Publ. 1975. Morphologie et position systématique de *Lemuricola* (*Rodentoxyuris*) *sciuri* (Cameron, 1932) nov.comb., n.subgen., et *Syphacia* (*Syphatineria*) *funambuli* Johnson, 1967, oxyures (Nematoda) parasites de rongeurs sciurides. *Bull. Mus. natn. Hist. nat., Paris*, 3e sér., No. 256, Zool. 178, pp. 1525-1535. [Fr, en].

SCHAD, G. A. 1960. The genus *Thelandros* (Nematoda: Oxyuroidea) in North American salamanders, including a description of *Thelandros salamandrae* n.sp. *Can. J. Zool.*, **38** (1): 115-120.

SCHMIDT, G. D. & KUNTZ, R. 1968. Nematode parasites of Oceanica. IV. Oxyurids of mammals of Palawan, P.I., with descriptions of four new species of *Syphacia*. *Parasitology*, **58** (4): 845-854.

SIMON VICENTE, F. 1969. On *Dermatoxys hispaniensis* n.sp. (Nematoda: Oxyuroidea) from *Oryctolagus cuniculus* and *Lepus timidus* of Spain. *J. Helminth.*, **43** (3/4): 417-426.

SKRJABIN, K. I., SHIKHOBALOVA, N. P. & LAGODOVSKAYA, E. A. 1960. Oxyurata of animals and man. Part 1. Oxyuroidea. Principles of Nematology, Vol. VIII, edited by Academician K.I.Skrjabin. *Moscow: Izdatel'stvo Akademii Nauk SSSR*, 557 pp. Translated from the Russian by Israel Program for Scientific Translations, Jerusalem, 1974.

VASSILIADES, G. 1970. Nématodes parasites d'oiseaux malgaches. *Annls Parasit. hum. comp.*, **45** (1): 47-88. [Fr, en].

YAMAGUTI, S. 1961. Systema helminthum. Volume III. The nematodes of vertebrates. *New York: Interscience Publishers, Inc,*. Part II, pp. 681-1261.

*For most papers published before 1958 please refer to bibliography in YAMAGUTI (1961) and CHABAUD (*in* GRASSÉ, 1965).

9 ASCARIDIDA

COSMOCERCOIDEA, SEURATOIDEA AND HETERAKOIDEA

by ALAIN G. CHABAUD*

COSMOCERCOIDEA

The over-all classification and the division of the superfamily into the Cosmocercidae, Kathlaniidae and Atractidae follows Chabaud (1965), but many modifications and additions are included in the latest system.

Cosmocercidae

Cosmocercinae

The classification adopted is similar to that of Travassos (1931). Changes which have been proposed and new genera which have been added since the work of Travassos are not generally followed because they seem to be based on artificial characters. For example, the presence or apparent absence of a gubernaculum is often difficult to determine and is a character without generic value and one which is liable to lead to many errors.

In many species, spicules are more or less completely atrophied and some authors have mistaken the gubernaculum for a spicule. For this reason *Paracosmocerca* Kung & Wu, 1945 is considered a synonym of *Cosmocerca.*

Neoprotozoophaga Biswas & Chakravarty, 1963 is a member of the Cosmocercidae and not an oxyuroid (see Petter & Quentin, 1976). The description of the type species is not sufficient to determine the exact synonymy but it may be a species of *Maxvachonia.*

In 1961 Le Van Hoa redescribed, from African material, *Oxyuris preputialis* Skrjabin, 1916, the type species of *Neoraillietnema* Ballesteros Márquez, 1945. This species has a large gubernaculum and two spicules which are almost atrophied and falls into the genus *Aplectana.* Semenov's (1929) redescription, based on material from the U.S.S.R., is clearly of another species and we believe that *Neoraillietnema* is a synonym of *Aplectana.*

Aplectana mexicana (Caballero, 1933) lacks special characters and the genus *Neoxysomatoides* Yamaguti, 1961 proposed for it is unjustified.

Pseudaplectana Yamaguti, 1961 was proposed for *Oxyuris* (s.l.) *papillocauda* Walton, 1940. Walton correctly left his species as *incertae sedis* since the female was unknown. It is not known if this species is a peculiar member of the Cosmocercidae, a *Meteterakis* with a reduced preanal sucker, or a member of some other group. The genus *Pseudaplectana* is not, therefore, considered in the following keys.

The original description of *Aplectana pharyngeodentata* Belle, 1957 does not mention a ventral sucker although the structure is clearly represented in a drawing. This species is a member

* Muséum National d'Histoire Naturelle, Paris, France.

of the Strongylurinae and *Bellaplectana*, proposed for it by Skrjabin *et al.* in 1961, falls into the synonymy of *Moaciria* (see Inglis, 1967).

Following Petter & Quentin (1976) we think that the "pair of small, highly chitinized and closely apposed accessory pieces" which characterizes *Neopharyngodon* Chakravarty & Bhaduri (1948) are really supports for the genital cone and that this genus is a synonym of *Pharyngodon*.

In *Aplectana* the gubernaculum is often much reduced and *Neyraplectana* Ballesteros Márquez, 1955, which was based on the supposed absence of a gubernaculum, is regarded as a synonym of *Aplectana*.

Alaplectana Azim, 1931 and *Latibuccana* Patwardhan, 1935 seem to belong to the Subuluroidea and not to the Cosmocercoidea.

Luzonema Oliveira Rodriguez *et al.*, 1973, known from a single male specimen, cannot be classified. It may be a seuratoid or an atractid but it is not, as suggested by the authors, a member of the Cosmocercinae since it has two small lips.

Neoxysomatium Ballesteros Márquez, 1945 was distinguished from *Oxysomatium* by the presence of a gubernaculum. Hartwich (1975) studied the type species and showed that this character was false. However, he retained *Neoxysomatium* and stressed the absence (in *Oxysomatium*) or presence (in *Neoxysomatium*) of somatic papillae and lateral alae. We follow Hartwich.

Cosmocercella was redefined by Baker & Adamson in 1977. The type species lacks the vesiculated "bursa", often mentioned as characteristic of the genus, and *Cosmocercella*, which resembles *Raillietnema*, is characterized by its vesiculated rosette caudal papillae.

Walton (1940, 1941) studied the cosmocercid *Aplectana hamatospicula* in detail and verified that Chitwood & Chitwood (1937) were, according to the nomenclature of the time, correct in placing the cosmocercids among the Ascaridoidea and not the Oxyuroidea. It is not known if this species is amphidelphic or prodelphic and it is difficult to determine whether it belongs to *Aplectana* or *Oxysomatium* but the species has, as Walton showed, all the characteristics of a typical member of the Cosmocercinae. *Aplecturis*, which was proposed by Skrjabin & Schikhobalova in 1951 for this species and placed near *Labiduris* seems, therefore, to have been based on an error in interpretation.

Maxvachoniinae

The male of *Maxvachonia* Chabaud & Brygoo, 1960 is a typical cosmocercid but the female has very abberant characters. The chief interest of the genus lies in the fact that it links the cosmocercids with certain seuratoids (Skrjabinalaziinae). Mawson (1972) discovered numerous species in Australia and showed that *Austracerca* Inglis, 1968 is a synonym of *Maxvachonia*.

Typhlonema Kreis, 1958 cannot be classified because the male is unknown. The egg suggests that the genus may be close to *Maxvachonia*. Kreis did not see the anus; if it should prove to be markedly anterior in position one might suggest that the two genera are related.

Kathlaniidae

It is difficult to delimit the Kathlaniidae because the family is intermediate between the Cosmocercidae and certain Seuratoidea (Quimperiidae, Schneidernematidae) in which there is an evolutionary trend towards the Heterakoidea. It is a family in transition, with inevitably arbitrary limits, but nevertheless part of a clearly defined evolutionary line. It seems undesirable, therefore, to create for this group two superfamilies and numerous families and subfamilies, as was done by Skrjabin *et al.* (1964).

To define the limits of Kathlaniidae, we include in it those Cosmocercoidea in which the oesophageal isthmus is expanded into a bulb and in which the ventral musculature of the male tail is specialized to form oblique bundles which lead ultimately to a more or less completely formed preanal sucker. Variations in oesophageal and in male caudal structures nevertheless cause difficulties in classifying the Kathlaniidae. For example, although *Falcaustra* usually has a typical oesophagus with a spherical isthmus, the latter may become cylindrical in certain species when these have been fixed in an extended position.

In *Megalobatrachonema* the isthmus is cylindrical and the bulb has lost its valves and become glandular: in *Chabaudgolvania* the posterior extremity of the oesophagus is not differentiated into an isthmus and a bulb. Apart from the evolutionary changes in these characters, other morphological characters are quite homogeneous and it is important to keep these various genera together in a single taxonomic group. We retain, therefore, only three subfamilies, i.e. Kathlaniinae, Oxyascaridinae and Cruziinae.

Inglis (1957) reviewed and redefined *Kathlania* and *Tonaudia.* The name *Falcaustra* Lane, 1915 is preferred over *Spironoura* Leidy, 1856 as the type species of the latter is poorly defined (see Freitas & Lent, 1941). *Dibulbiger* Caballero, 1935 and *Zanclophorus* Baylis & Daubney, 1922 are not retained as the limits between them and *Falcaustra* seem imprecise (see Chabaud & Golvan, 1957). Inglis (1959) showed that *Nematoxynema* Skrjabin & Schikhobalova, 1951 is a synonym of *Falcaustra,* as is also *Velariocephalus* Singh, 1958 which had been related to *Nematoxynema* (see Chabaud, 1965).

Chabaudgolvania Freitas, 1958 and *Megalobatrachonema* Yamaguti, 1941 differ only by the more or less complete atrophy of the oesophageal bulb and Hartwich (1960) synonymizes the former with the latter.

Amblyonema, from an Australian ceratodontid, was redescribed by Chabaud (1959) and a closely related genus, *Chabaudinema,* was discovered by Díaz-Ungría (1968) in a characinid in Venezuela.

Pteroxyascaris, which is very similar to *Oxyascaris,* was described by Freitas (1958). *Paraoxyascaris* Oliveira Rodrigues & Rodrigues, 1971 is distinguished from *Oxyascaris* only by the presence of a gubernaculum. We believe the former to be a synonym of the latter.

Subulascaris Freitas & Dobbin, 1957 is a member of the Seuratoidea very close to *Gendria* and will be treated under that superfamily.

Freitasoxyascaris Gomes & Motta, 1967 seems to be an *Aplectana.*

Neocruzia Yamaguti, 1961 proposed for *Cruzia morleyi,* is based on the very brief description given by Pearse (1936). Caballero Deloya (1974) recently redescribed this species in detail and none of the morphological details seems to necessitate a special genus for this species.

Pseudocruzia Wolfgang, 1953 and *Schizobucca* Schuurmans Stekhoven, 1950 were created because of their peculiar cephalic structures.

Atractidae

The interpretation and the limits of the family are those proposed earlier by Chabaud & Petter (1960) but a number of new genera have since been erected.

Fitzsimmonsnema Petter, 1966 was proposed for *Probstmayria reptiliae* Fitzsimmons, 1958 in which the male copulatory structures resemble those of *Atractis* and the didelphic uteri constitute a line between *Raillietnema* of the Cosmocercidae and the monodelphic Atractidae.

Maracaya Díaz-Ungría, 1963, which is close to *Ibrahimia,* is characterized by the unique cuticular structure of its pharynx.

Schrankiana Strand, 1942 has priority over *Schranknema* Travassos, 1949. The latter was intended to replace *Schrankia* Travassos, 1925 which was preoccupied at the time it was erected. Freitas (1959) has described *Schrankianella* for a species with a highly modified oesophagus.

In *Labeonema* Puylaert, 1970, which is close to *Schrankiana,* the posterior genital branch is replaced by a larval reservoir and the ovary and oviduct have disappeared. This genus connects didelphic and monodelphic Atractidae.

Cyrtosomum has been reviewed in detail by Bowie & Franz (1974) and Coy Otero & Barus (1975).

Paratractis Sarmiento, 1959 is similar to *Atractis* but has a highly characteristic cuticular ornamentation.

In *Proatractis* G. Caballero R., 1971, which is also close to *Atractis*, the cephalic end has a complex ornamentation.

Neoatractis Gupta & Aggerwal, 1974 is too poorly defined to be retained.

Pseudatractis Yamaguti, 1961 was proposed for *Monhysterides testudinicola* Baylis, 1933 on the grounds that it has a small gubernaculum. This genus is not worth retaining, particularly as this species is very similar to the type species of *Monhysterides*.

Acanthostephanocephalus Kreis, 1960 was described before the author was aware of *Grassenema* Petter, 1959 and must fall into the synonymy of the latter.

Orientatractis Petter, 1966 resembles *Grassenema* in that the four submedian lips are supported by small cuticularized pieces and, as in *Crossocephalus*, the lips appear capable of being more or less completely everted.

Nouvelnema Petter, 1959 has been redescribed by Puylaert (personal communication). The material is in good condition and Puylaert has been able to see that the mouth is elongated in the lateral and not the median axis and that the genus is amphidelphic with the two ovaries being evident in the young females. The male, if it exists, is unknown.

A monograph on the various species of Cosmocercoidea with a bibliography is found in volumes 10, 15 and 19 of Osnovi Nematodologii: Skrjabin, Schikhobalova and Lagodovskaya (1961, 1964) and Skrjabin, Sobolev and Ivaschkin (1967).

COSMOCERCOIDEA

Key to families

1-(4) Oviparous, or if viviparous, larvae laid in the first stage.
Didelphic.
2-(3) Oesophageal isthmus elongated, not spherical.
Male without a preanal sucker.
Cosmocercidae

3-(2) Oesophageal isthmus not elongated, generally spherical (Fig. 6.23).
Male generally with one or several preanal suckers (Figs. 6.25, 6.35).
Kathlaniidae

4-(1) Viviparous with larvae laid in an advanced stage of development and capable of endogenous development.
Generally monodelphic.
Atractidae

Family *COSMOCERCIDAE*

(Railliet, 1916, subfam.) Travassos, 1925

Key to subfamilies

1-(4) Genital structure not highly modified from that of a typical nematode.
2-(3) Vulva generally behind oesophagus.
Female tail not highly modified.
Eggs without filaments.
Parasites mainly of amphibians, rarely of reptiles.
Cosmocercinae

3-(2) Vulva in region of oesophagus.
Female tail very long (1/4 length of body) (Fig. 6.14).
Eggs with filaments (Fig. 6.13).
Parasites mainly of reptiles, rarely of amphibians.
Maxvachoniinae

4-(1) One uterus transformed into a sac containing males (Fig. 6.15).
Parasites of tadpoles.
Gyrinicolinae

Subfamily **Cosmocercinae** Railliet, 1916

Key to genera

1-(6) Ventral cuticle of male ornamented with plectanes (cuticular supports around caudal papillae) (Fig. 6.1) or rosette papillae (papillae surrounded by punctations) (Fig. 6.2).
2-(5) Caudal rosette papillae present, but not on plectanes (Fig. 6.2).

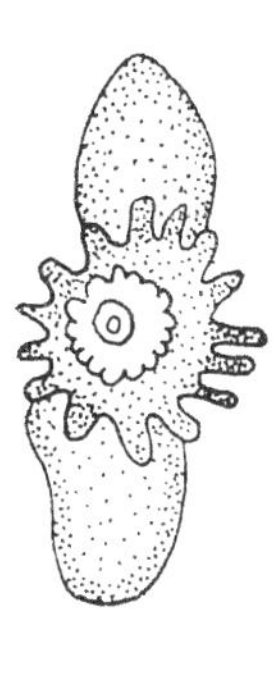

6.1

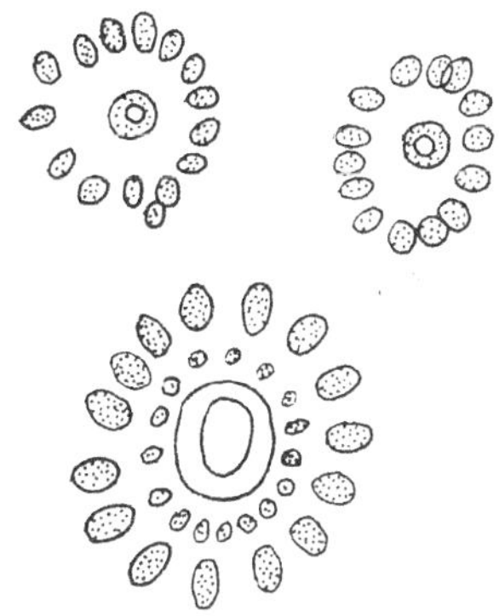

6.2

FIG. 6.1. plectanes.
FIG. 6.2. rosette papillae.

3-(4) Caudal rosette papillae not raised above surface of cuticle (Fig. 6.3).
Eggs numerous, of normal size (less than 100 μm) (Figs. 6.4, 6.5).
Cosmocercoides Wilkie, 1930
(= *Trionchonema* Kreis, 1932)

4-(3) Caudal rosette papillae raised on surface of clear vesicle (Fig. 6.7).
Eggs few in number and large (more than 100 μm) (Fig. 6.6).
Cosmocercella Steiner, 1924

5-(2) Certain caudal papillae on plectanes (Figs. 6.1, 6.8).
Cosmocerca Diesing, 1861
(= *Paracosmocerca* Kung & Wu, 1945)

6-(1) Tail of male without rosette papillae or plectanes.

7-(12) Amphidelphic (Figs. 6.4, 6.6).
Parasites of amphibians.

8-(9) Uterus with few unusually large eggs (more than 100 μm).
Ovaries short (Fig. 6.6).
Raillietnema Travassos, 1927

9-(8) Uterus with numerous eggs of normal size (less than 100 μm).
Ovaries long (Fig. 6.4).

10-(11) Cuticle with eight longitudinal lines of somatic papillae (Fig. 6.9).
Lateral alae present.
Neoxysomatium Ballesteros-Márquez, 1945
sensu Hartwich, 1975

11-(10) Cuticle without somatic papillae.
Lateral alae absent.
Oxysomatium Railliet & Henry, 1916

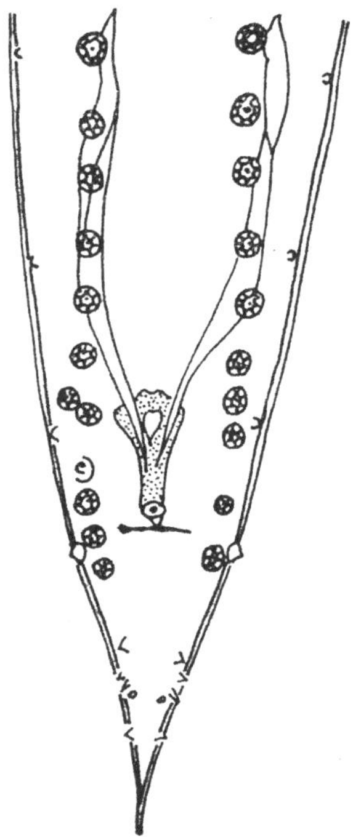

6.3

6.4

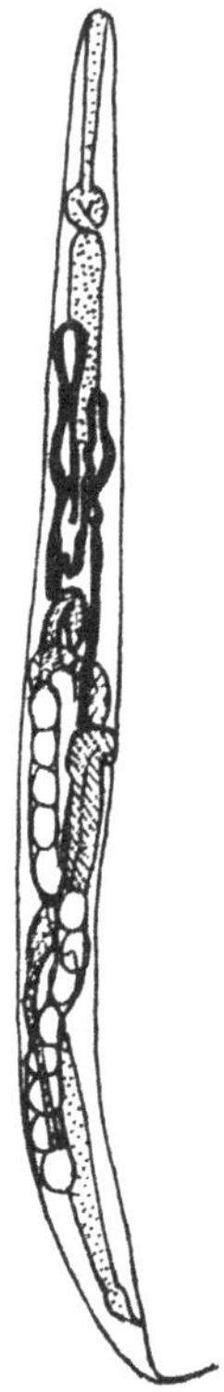

6.5

FIG. 6.3. *Cosmocercoides*, posterior extremity male, ventral view. (After Anderson, 1960.)
FIG. 6.4. *Neoxysomatium*, female, amphidelphic, eggs small and numerous. (After Travassos, 1931.)
FIG. 6.5. *Cosmocerca*, female, prodelphic, eggs small and numerous. (After Chabaud & Campana, 1955.)

12-(7) Prodelphic (Fig. 6.5).
Parasites of amphibians and reptiles.

Aplectana Railliet & Henry, 1916 (Fig. 6.10).
(= *Neoraillietnema* Ballesteros-Márquez, 1945;
= *Neyraplectana* Ballesteros-Márquez, 1945;
= *Neoxysomatoides* Yamaguti, 1961)

Subfamily **Maxvachoniinae** Chabaud & Brygoo, 1960

One genus (Figs. 6.11-6.14).

Maxvachonia Chabaud & Brygoo, 1960
(= *Austracerca* Inglis, 1968)

Subfamily **Gyrinicolinae** (Yamaguti, 1938 fam.)

One genus (Fig. 6.15).

Gyrinicola Yamaguti, 1938

2*

6.6

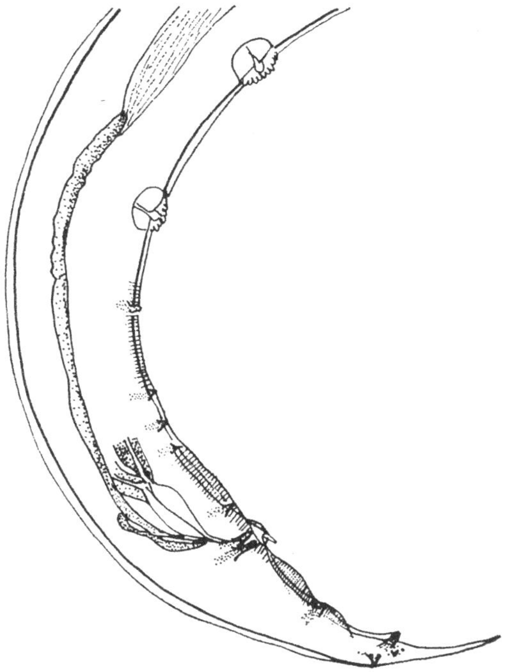

6.7

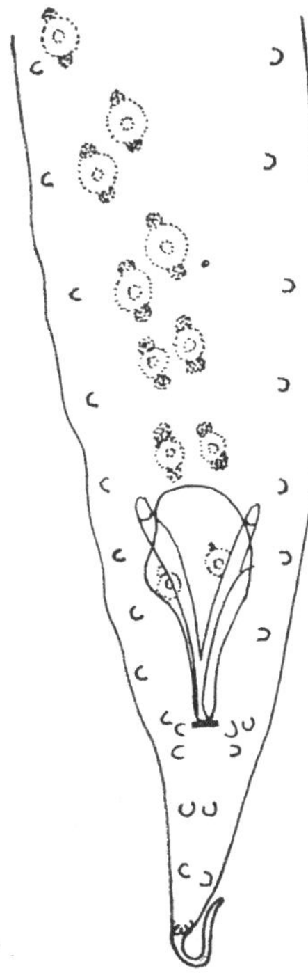

6.8

FIG. 6.6. *Raillietnema*, female, amphidelphic, eggs large and few in number. (After Chabaud & Brygoo, 1962.)
FIG. 6.7. *Cosmocercella*, caudal end male, lateral view. (After Baker & Adamson, 1978.)
FIG. 6.8. *Cosmocerca*, caudal end male, ventral view. (After Lent & Freitas, 1948.)

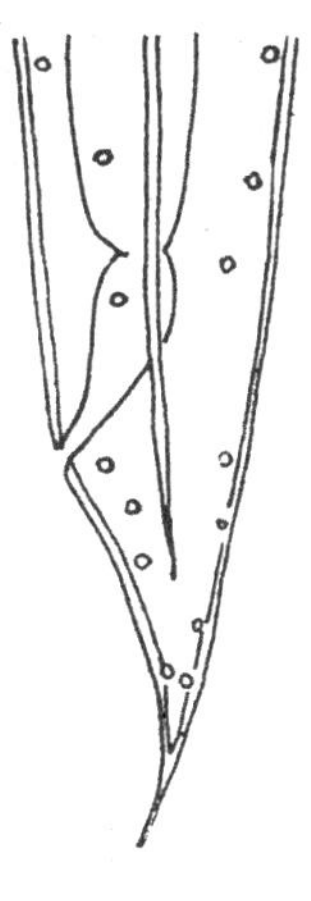

6.9

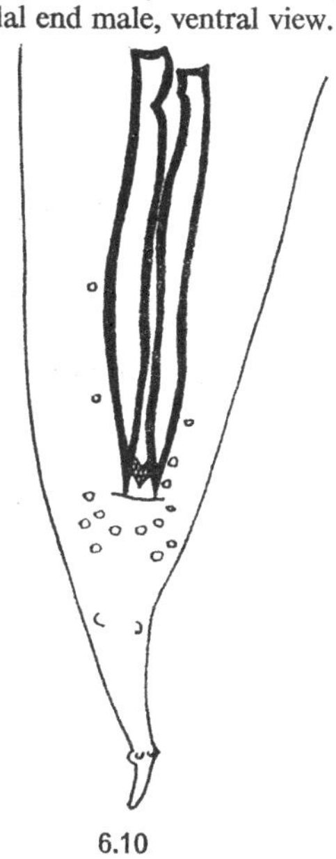

6.10

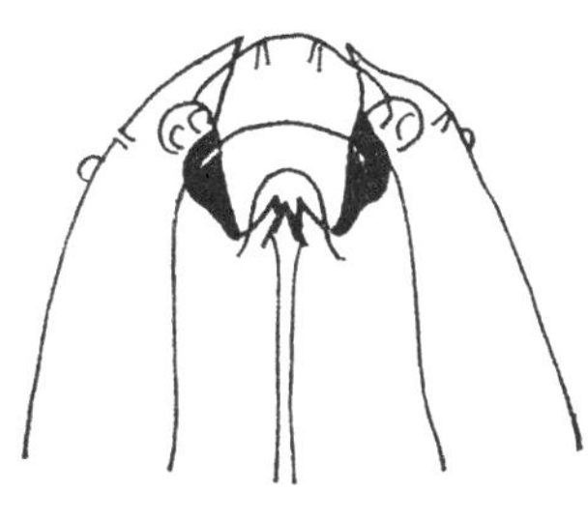

6.11

FIG. 6.9. *Neoxysomatium*, caudal end female, lateral view showing somatic papillae. (After Hartwich, 1975.)
FIG. 6.10. *Aplectana*, posterior extremity male, ventral view. (After Travassos, 1931.)
FIG. 6.11. *Maxvachonia*, cephalic extremity, dorsal view showing detail of buccal cavity. (After Inglis, 1968.)

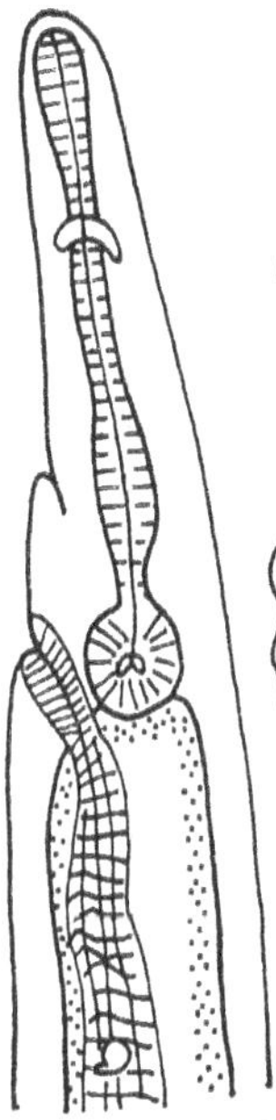

6.12

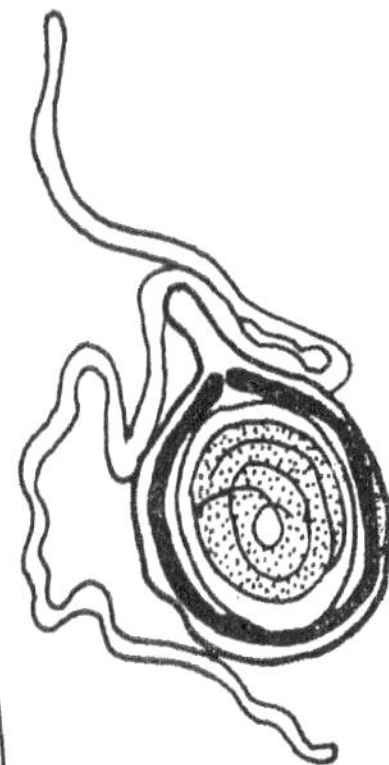

6.13

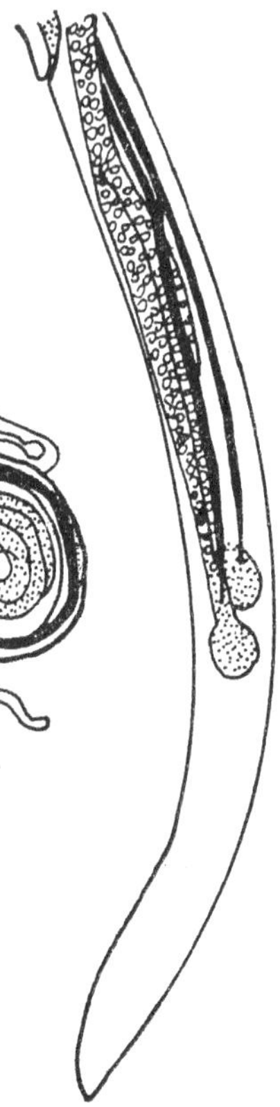

6.14

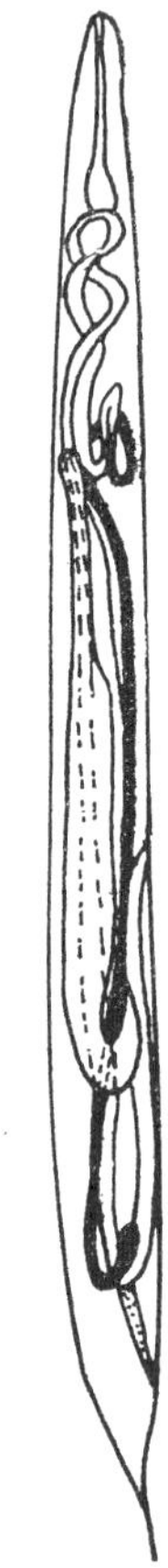

6.15

FIG. 6.12. *Maxvachonia*, anterior extremity female, lateral view. (After Chabaud & Brygoo, 1960.)
FIG. 6.13. *Maxvachonia*, thick-shelled egg. (After Chabaud & Brygoo, 1960.)
FIG. 6.14. *Maxvachonia*, caudal end female, lateral view. (After Chabaud & Brygoo, 1960.)
FIG. 6.15. *Gyrinicola*, female, lateral view, showing genitalia diagrammatically. (After Yamaguti, 1938.)

Family *KATHLANIIDAE*
(Lane, 1914 subfam.) Travassos, 1918

Key to subfamilies

1-(4) Pharyngeal part of oesophagus without longitudinal rows of denticles.
Intestinal diverticulum absent.

2-(3) Oesophageal bulb generally muscular and with valves (Fig. 6.23).
Preanal ventral sucker present (Figs. 6.18, 6.21, 6.35), sometimes multiple (Fig. 6.25).
Kathlaniinae

3-(2) Oesophageal bulb transformed into glandular ventricle (Figs. 6.36, 6.38).
Preanal sucker absent.

Oxyascaridinae

4-(1) Pharyngeal part of oesophagus with three longitudinal rows of denticles (Fig. 6.41).
Intestinal diverticulum present (Fig. 6.42).

Cruziinae

Subfamily **Kathlaniinae** Lane, 1914

Key to genera

1-(14) Dorsal lip little different from subventral lips.
2-(5) Cephalic extremity complex with three main lips separated from each other by subsidiary lobes (Figs. 6.16, 6.19).
3-(4) Buccal cavity triangular in cross section (Fig. 6.16).
Oesophastome with unequal teeth (Fig. 6.17).

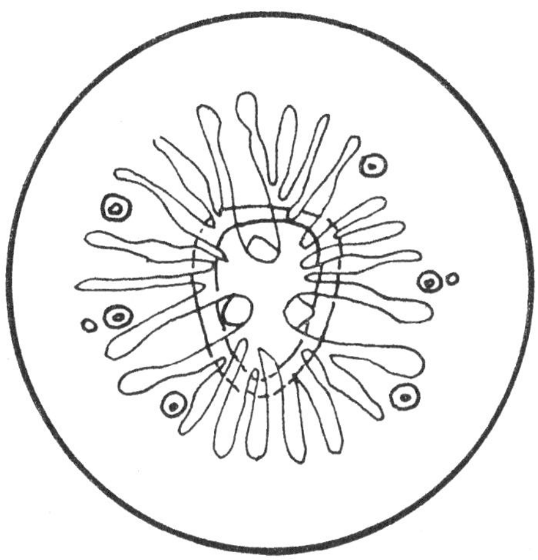

6.16

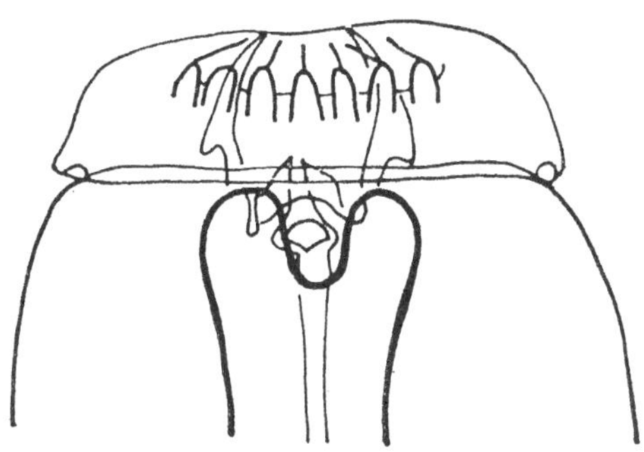

6.17

FIG. 6.16. *Kathlania*, cephalic extremity, apical view. (After Inglis, 1957.)
FIG. 6.17. *Kathlania*, cephalic extremity, ventral view. (After Inglis, 1957.)

Spicules and gubernaculum complex (Fig. 6.18).
Parasites of marine turtles.

Kathlania Lane, 1914
(= *Pseudoheterakis* Travassos, 1917)

4-(3) Buccal cavity hexagonal in cross section (Fig. 6.19).
Oesophastome with teeth of equal size (Fig. 6.20).
Spicules and gubernaculum simple in form (Fig. 6.21).
Parasites of marine turtles and selachians.

Tonaudia Travassos, 1918

5-(2) Cephalic extremity simple with three or six lips (Figs. 6.24, 6.26).
6-(11) Pharyngeal part of oesophagus simple or armed with three small teeth (Fig. 6.26).

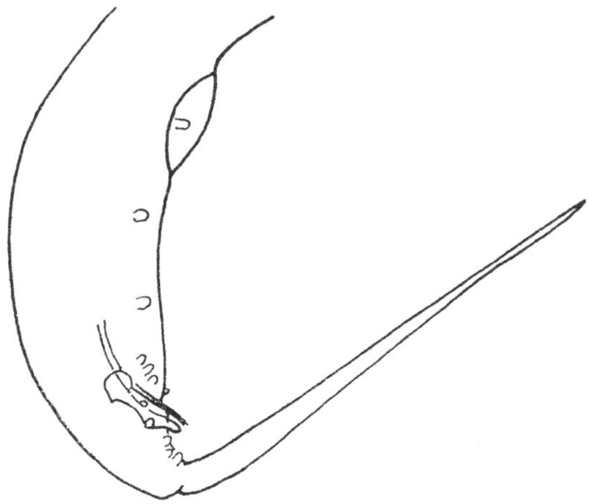

6.18

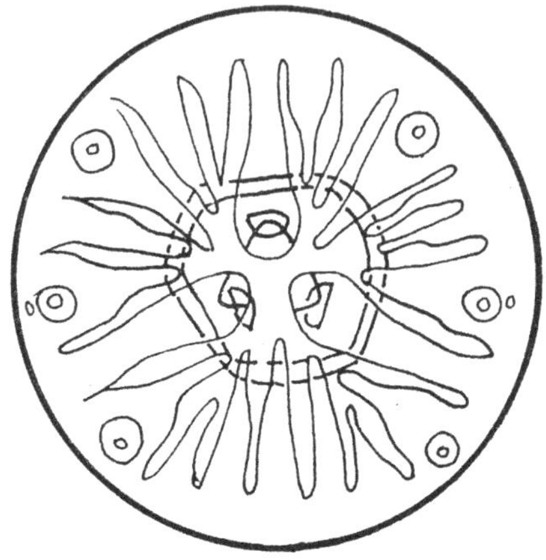

6.19

FIG. 6.18. *Kathlania*, caudal extremity male, lateral view. (After Inglis, 1957.)
FIG. 6.19. *Tonaudia*, cephalic extremity, apical view. (After Inglis, 1957.)

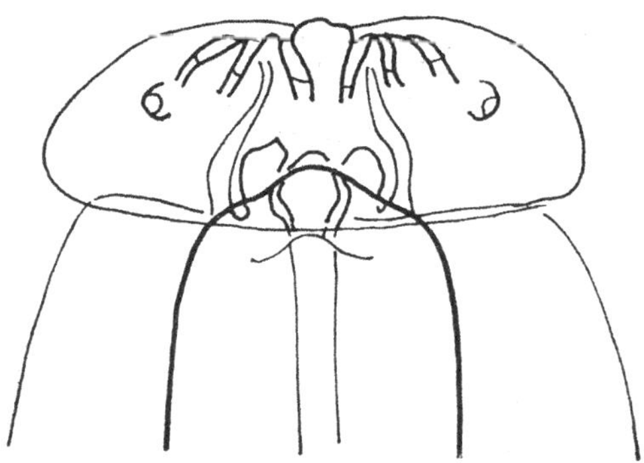

6.20

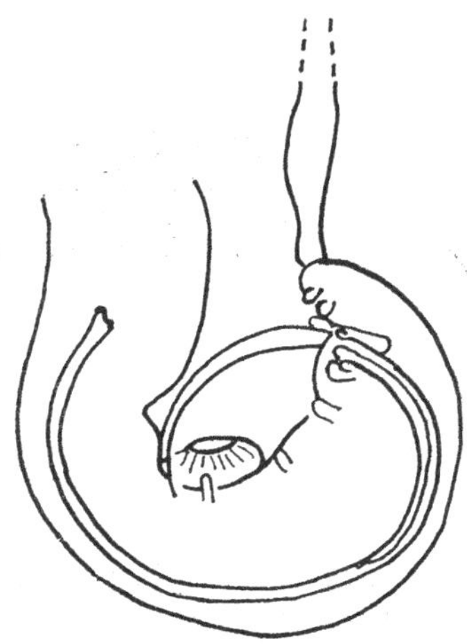

6.21

FIG. 6.20. *Tonaudia*, cephalic extremity, ventral view. (After Inglis, 1957.)
FIG. 6.21. *Tonaudia*, caudal extremity male, lateral view. (After Inglis, 1957.)

7-(8) Anterior extremity of oesophagus simple, not differentiated into a pharyngeal portion (Fig. 6.22).
Parasites of neotropical fish.

Spectatus Travassos, 1923

8-(7) Anterior extremity of oesophagus differentiated into pharyngeal portion (Figs. 6.23, 6.27).

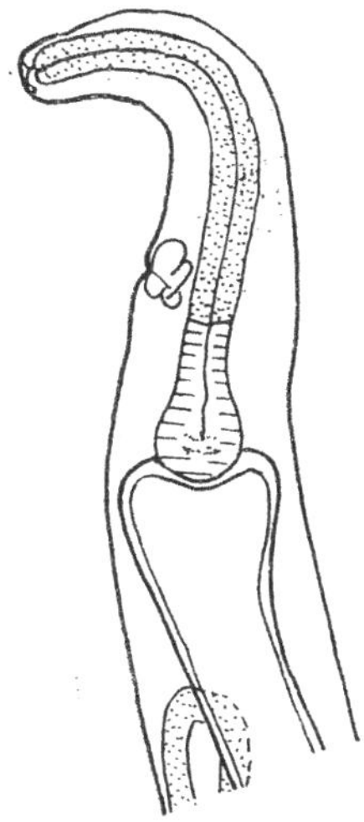

6.22

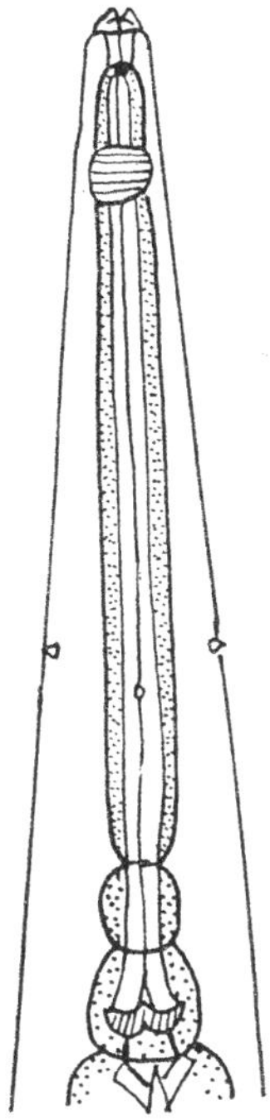

6.23

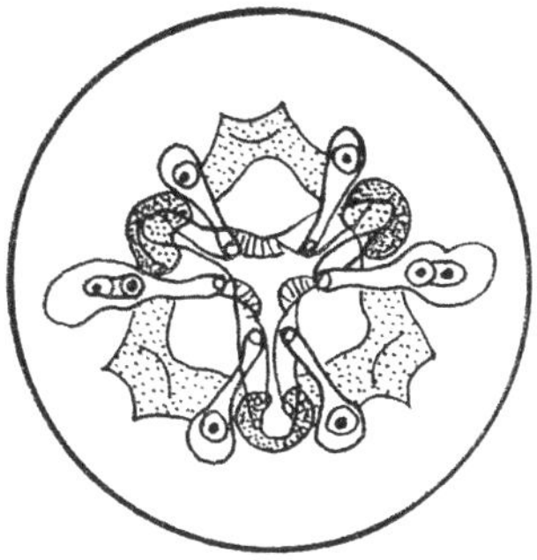

6.24

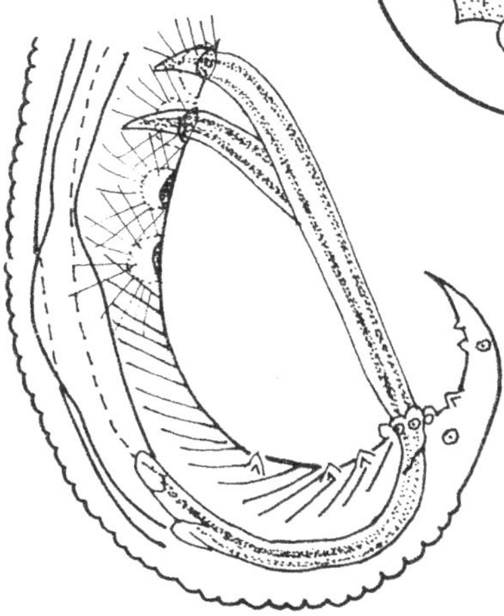

6.25

FIG. 6.22. *Spectatus*, anterior extremity. (After Travassos, Artigas & Pereira, 1928.)
FIG. 6.23. *Falcaustra*, anterior extremity. (After Massino, 1924.)
FIG. 6.24. *Falcaustra*, cephalic extremity, apical view. (After Walton, 1929.)
FIG. 6.25. *Falcaustra*, caudal end male, lateral view. (After Massino, 1924.)

9-(10) Three or six well developed lips present (Fig. 6.24).
Oesophageal isthmus generally spherical.
Oesophageal bulb well differentiated, with well developed valves (Fig. 6.23).
Parasites of fish, amphibians and reptiles.

Falcaustra Lane, 1915 (Fig. 6.25)
(= *Spironoura* Leidy, 1856; = *Florencioia* Travassos, 1920;
= *Dibulbiger* Caballero 1935; = *Zanclophorus* Baylis & Daubney, 1922;
= *Nematoxynema* Skrjabin & Schikhobalova, 1951;
= *Velariocephalus* Singh, 1958)

10-(9) Lips poorly developed (Fig. 6.26).
Oesophageal isthmus and bulb atrophied or much reduced (Fig. 6.27).
Parasites of amphibians.

Megalobatrachonema Yamaguti, 1941
(= *Chabaudgolvania* Freitas, 1958)

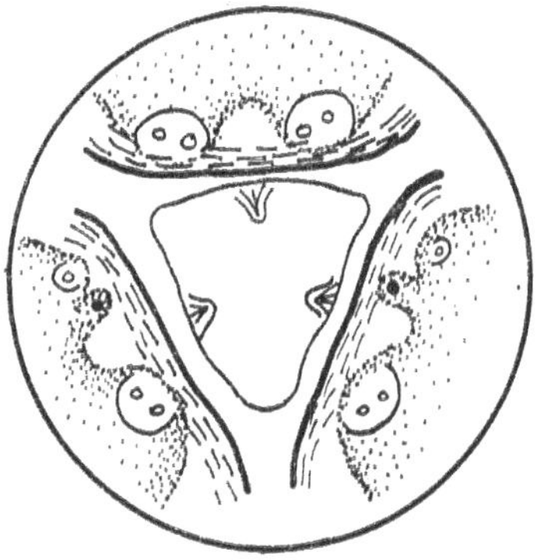

6.26

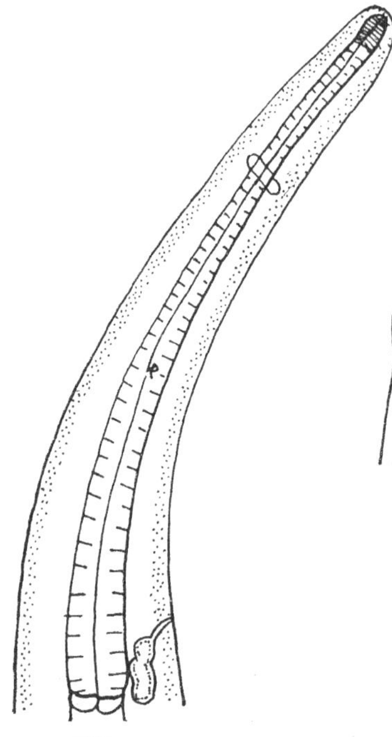

6.27

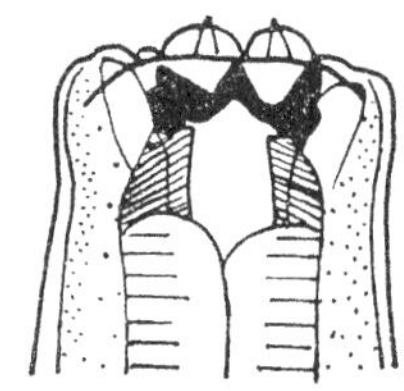

6.28

FIG. 6.26. *Megalobatrachonema*, cephalic extremity, apical view. (After Hartwich, 1960.)
FIG. 6.27. *Megalobatrachonema*, anterior extremity. (After Chabaud & Golvan, 1957.)
FIG. 6.28. *Chabaudinema*, cephalic extremity, lateral view. (After Díaz-Ungría, 1968.)

11-(6) Pharyngeal part of oesophagus armed with complex cuticularized formations (Figs. 6.28, 6.30, 6.31).

12-(13) Pharyngeal part of oesophagus with anterior cuticularized ring (Fig. 6.28).
Parasites of neotropical fish.
Chabaudinema Díaz-Ungría, 1968

13-(12) Pharyngeal part of oesophagus with three large anteriorly directed tricuspid teeth (Figs. 6.30, 6.31).
Parasites of Australian ceratodiform fish.
Amblyonema Linstow, 1898 (Figs. 6.29-6.32).

14-(1) Dorsal lip reduced to cuticularized trident (Fig. 6.33).
Ventrolateral lips with numerous internal lamellae (Fig. 6.34).
Parasites of herbivorous reptiles.
Cissophyllus Railliet & Henry, 1922 (Fig. 6.35)

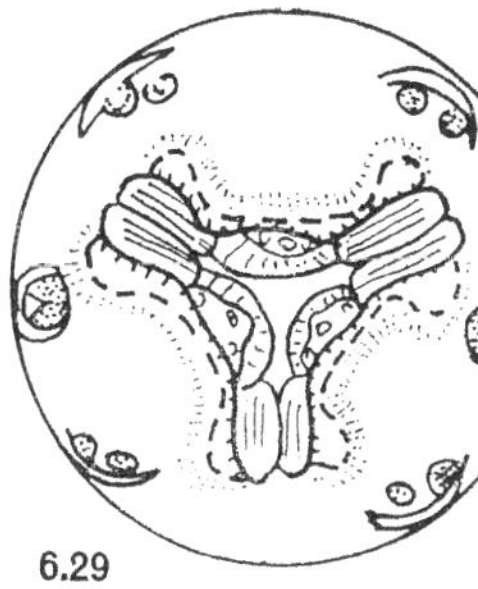

6.29

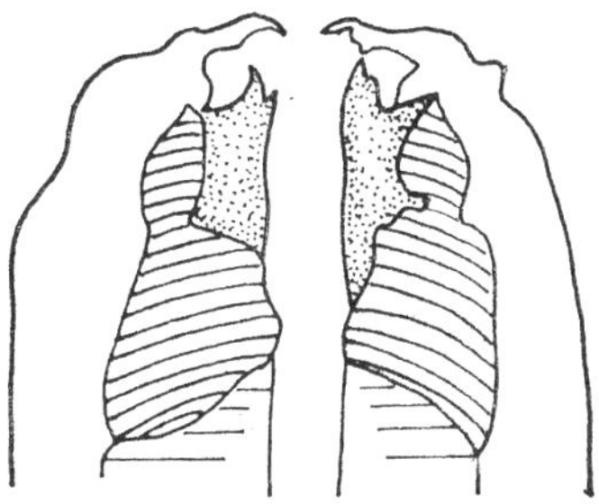

6.30

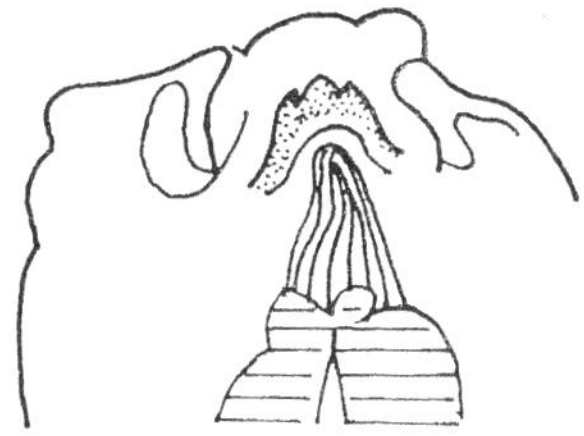

6.31

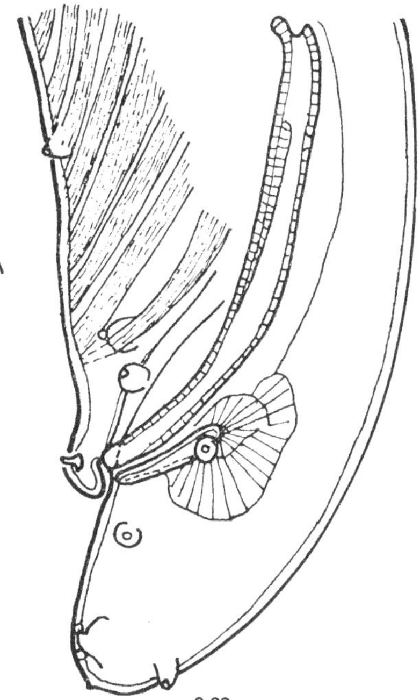

6.32

FIG. 6.29. *Amblyonema*, cephalic extremity, apical view. (After Chabaud, 1959.)
FIG. 6.30. *Amblyonema*, cephalic extremity, optical section. (After Chabaud, 1959.)
FIG. 6.31. *Amblyonema*, pharyngeal tooth. (After Chabaud, 1959.)
FIG. 6.32. *Amblyonema*, caudal end male, lateral view. (After Chabaud, 1959.)

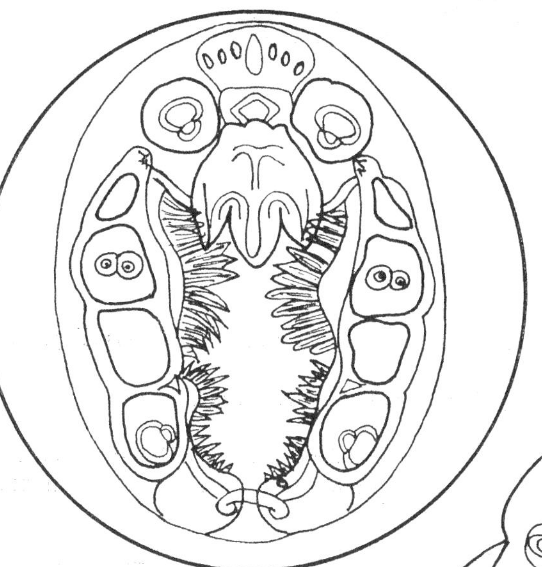

6.33

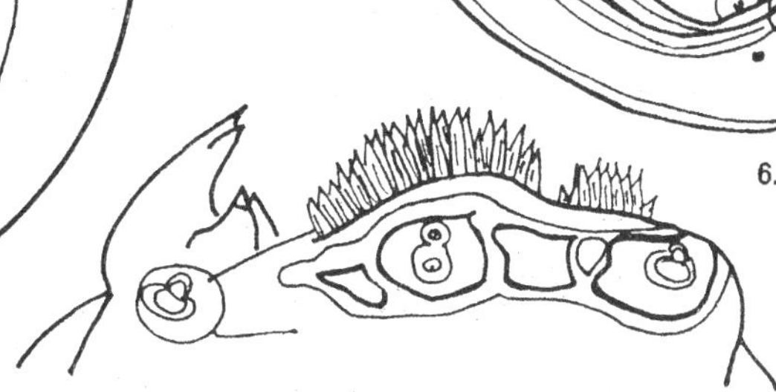

6.34

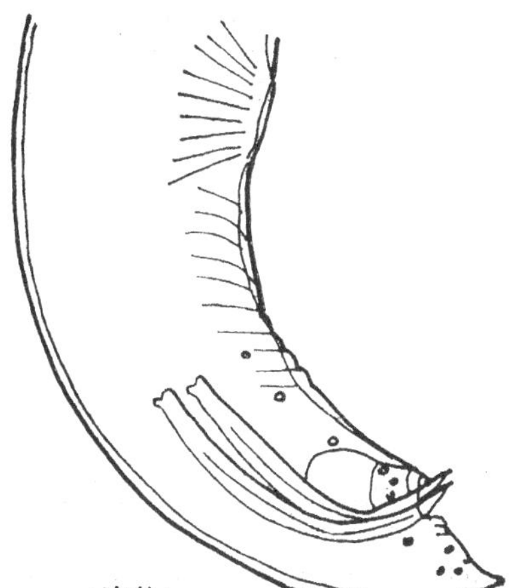

6.35

FIG. 6.33. *Cissophyllus*, cephalic extremity, apical view. (After Railliet & Henry, 1912.)
FIG. 6.34. *Cissophyllus*, lateral lip, lateral view. (After Railliet & Henry, 1912.)
FIG. 6.35. *Cissophyllus*, caudal end male, lateral view. (After Railliet & Henry, 1912.)

Subfamily **Oxyascaridinae** Freitas, 1958

Key to genera

1-(2) Lateral alae absent (Fig. 6.36).
Caudal papillae not numerous (Fig. 6.37).
Parasites of neotropical amphibians and reptiles.

Oxyascaris Travassos, 1920
(= *Paraoxyascaris* Oliveira Rodrigues & Sodre Rodrigues, 1971)

2-(1) Lateral alae present (Fig. 6.38).
Caudal papillae numerous (Fig. 6.39).
Parasites of neotropical amphibians.

Pteroxyascaris Freitas, 1958

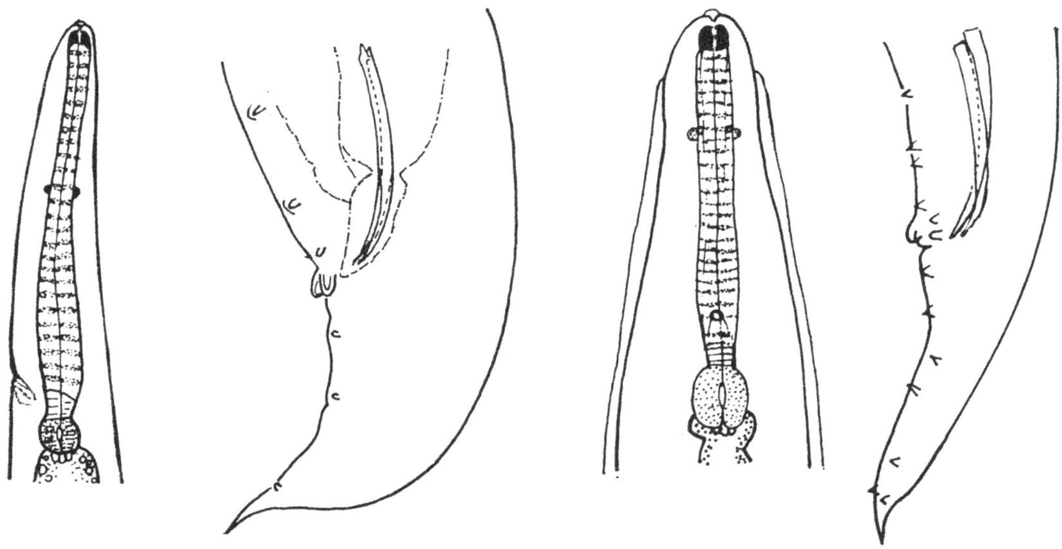

6.36 6.37 6.38 6.39

FIG. 6.36. *Oxyascaris*, anterior extremity. (After Freitas, 1958.)
FIG. 6.37. *Oxyascaris*, caudal end male, lateral view. (After Freitas, 1958.)
FIG. 6.38. *Pteroxyascaris*, anterior extremity. (After Freitas, 1958.)
FIG. 6.39. *Pteroxyascaris*, caudal end male, lateral view. (After Freitas, 1958.)

Subfamily **Cruziinae** (Travassos, 1917 fam.) Ortlepp, 1924

Key to genera

1-(4) Lips not covered by a widely inflated cuticle.
Cephalic extremity not separated from body by constriction (Figs. 6.41, 6.42).
Parasites of American vertebrates.

3*

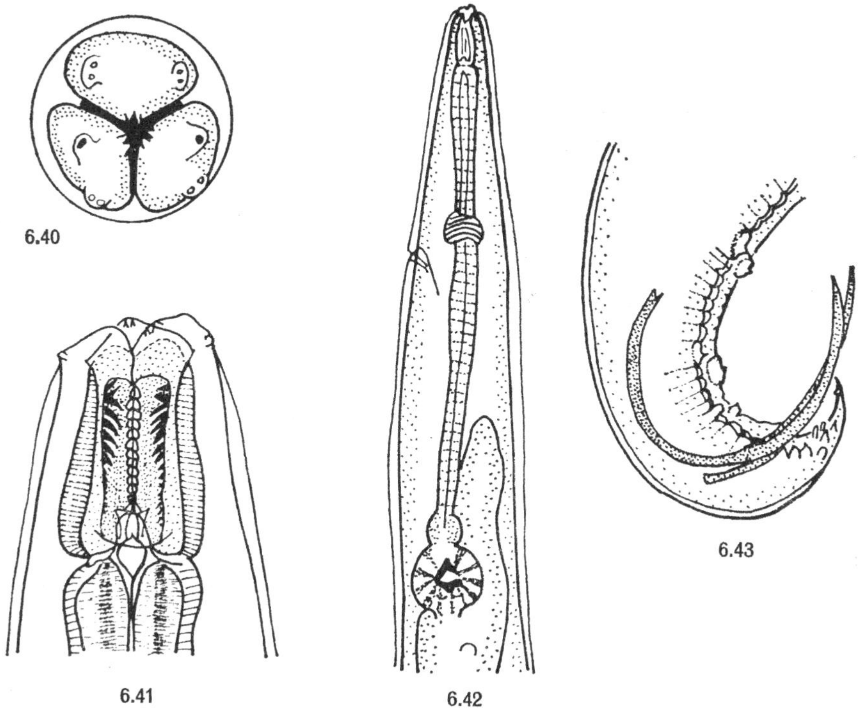

FIG. 6.40. *Cruzia*, cephalic extremity, apical view. (After Harwood, 1932.)
FIG. 6.41. *Cruzia*, cephalic extremity, lateral view. (After Ruiz, 1947.)
FIG. 6.42. *Cruzia*, anterior extremity. (After Wolfgang, 1951.)
FIG. 6.43. *Cruzia*, caudal end male, lateral view. (After Wolfgang, 1951.)

2-(3) Three triangular lips jointed at apex (Fig. 6.40).
Denticles of pharyngeal part of the oesophagus well defined (Fig. 6.41).
Parasites of American amphibians, reptiles, marsupials and *Xenarthra*.
Cruzia Travassos, 1917 (Figs. 6.42, 6.43)
(= *Neocruzia* Yamaguti, 1961)

3-(2) Lips not jointed at apex (Fig. 6.44).
Denticles of pharyngeal part of oesophagus inconspicuous (Fig. 6.45).
Parasites of American lizards.
Schizobucca Schuurmans-Stekhoven, 1950

4-(1) Lips covered with widely inflated cuticle (Figs. 6.46, 6.47).
Cephalic extremity separated from body by constriction (Fig. 6.47).
Parasites of Asian Suidae.
Pseudocruzia Wolfgang, 1953

6.44

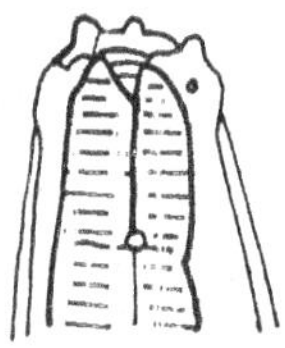

6.45

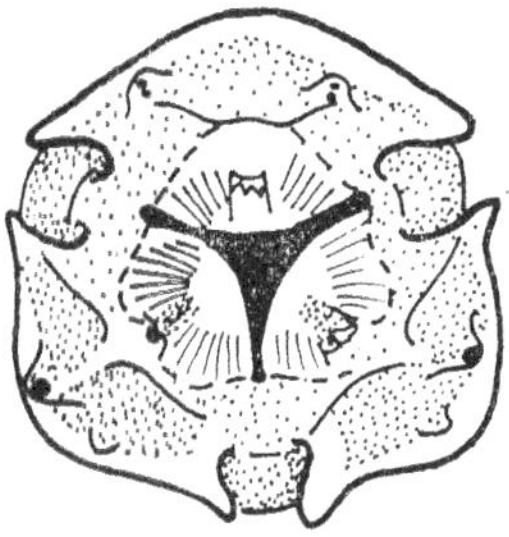

6.46

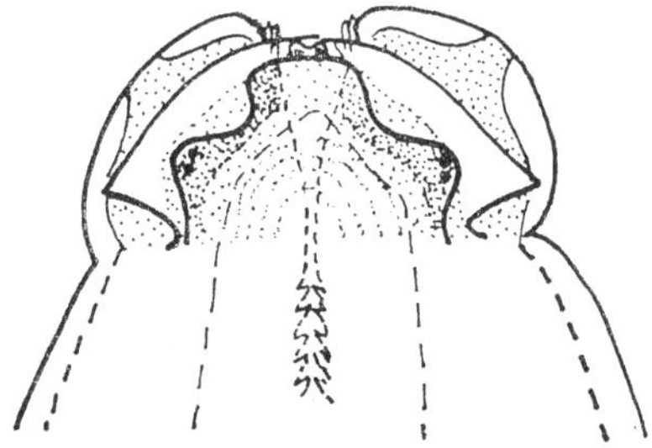

6.47

FIG. 6.44. *Schizobucca*, cephalic extremity, apical view. (After Schuurmans Stekhoven, 1950.)
FIG. 6.45. *Schizobucca*, cephalic extremity, lateral view. (After Schuurmans Stekhoven, 1950.)
FIG. 6.46. *Pseudocruzia*, cephalic extremity, apical view. (After Wolfgang, 1953.)
FIG. 6.47. *Pseudocruzia*, cephalic extremity, dorsal view. (After Wolfgang, 1953.)

Family **Atractidae**

(Railliet, 1917 subfam.) Travassos, 1919

Key to genera

1-(8) Didelphic.
2-(7) Cephalic extremity with three or six lips.
Pharyngeal part of oesophagus muscular and typical in form.
3-(6) Both genital branches functional in female.
4-(5) Tail of male rather short (Fig. 6.48).
Parasites of reptiles.

Fitzsimmonsnema Petter, 1966

5-(4) Tail of male terminated by long filament (Fig. 6.49).
Parasites of Equidae and primates.

Probstmayria Ransom, 1917

6-(3) Females with posterior genital branch transformed into reservoir for larvae and without ovary and oviduct (Fig. 6.50).
Parasites of fish.

Labeonema Puylaert, 1970

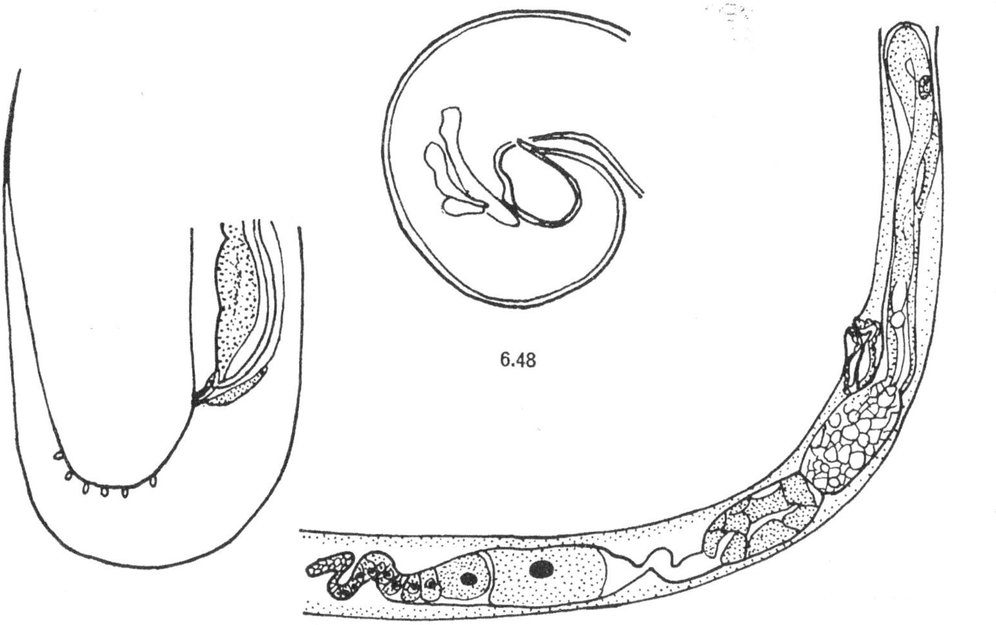
6.48

6.49

6.50

FIG. 6.48. *Fitzsimmonsnema*, caudal end male, lateral view. (After Fitzsimmons, 1958.)
FIG. 6.49. *Probstmayria*, caudal end male, lateral view. (After Ransom, 1907.)
FIG. 6.50. *Labeonema*, female genital system. (After Puylaert, 1970.)

7-(2) Oral opening elongated in lateral axis (Fig. 6.56).
Pharyngeal part of oesophagus forming small buccal cavity (Fig. 6.57).
Parasites of Hyracoidea.

Nouvelnema Petter, 1959

8-(1) Monodelphic.

9-(28) Buccal and pharyngeal structures not elaborate, without specialized cuticularized formations.

10-(15) Oesophagus of cosmocercid form, with pharynx, corpus, isthmus and valved bulb (Fig. 6.51).

11-(14) Spicules short, not much longer than gubernaculum (Fig. 6.52).
Isthmus well marked (Fig. 6.51).
Parasites of amphibians.

12-(13) Oesophageal corpus muscular in appearance and typical in form.
Pharyngeal part of oesophagus typical (Fig. 6.51).

Schrankiana Strand, 1942
(= *Schrankia* Travassos, 1925; = *Schranknema* Travassos, 1949)

13-(12) Oesophageal corpus divided into short muscular procorpus and very long, glandular metacorpus.
Pharyngeal part of oesophagus long (Fig. 6.53).

Schrankianella Freitas, 1959

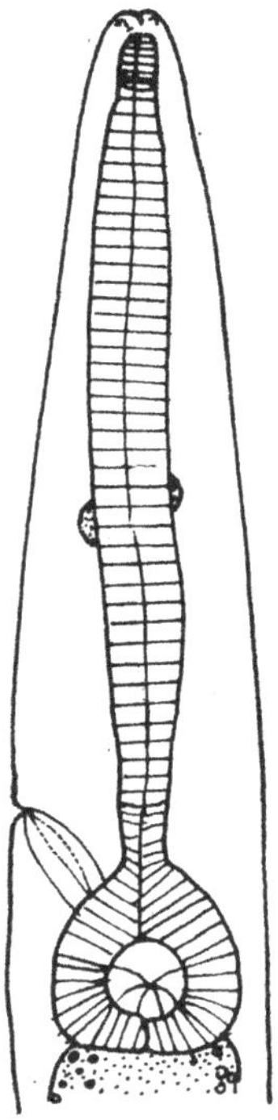

6.51

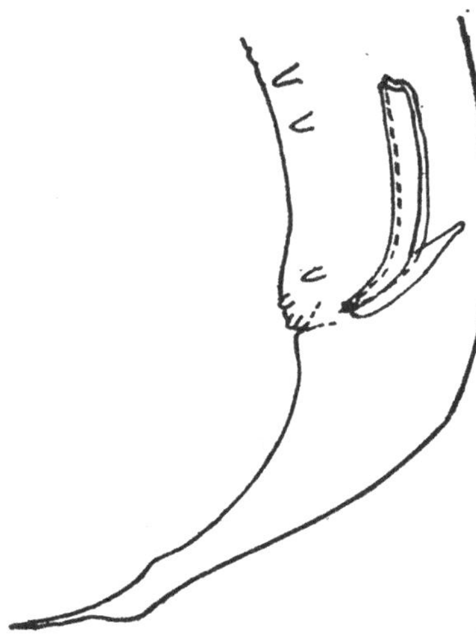

6.52

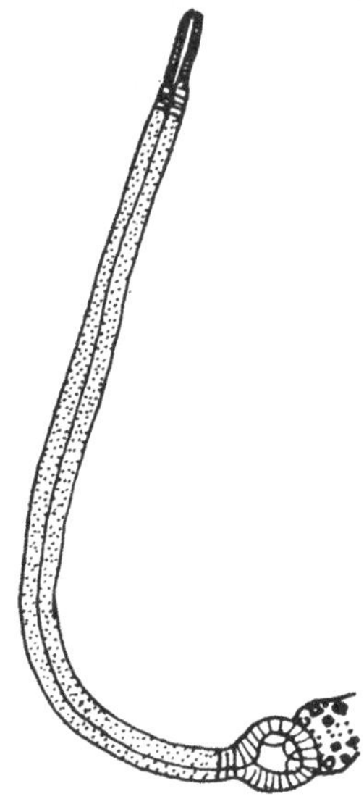

6.35

FIG. 6.51. *Schrankiana*, oesophagus, lateral view. (After Freitas, 1959.)
FIG. 6.52. *Schrankiana*, caudal end male, lateral view. (After Freitas, 1959.)
FIG. 6.53. *Schrankianella*, outline of oesophagus. (After Freitas, 1959.)

14-(11) Spicules much longer than gubernaculum (Fig. 6.55).
Isthmus not prominent.
Parasites of reptiles.

Ibrahimia Khalil, 1932

15-(10) Oesophagus not cosmocercid in form (Figs. 6.58, 6.63, 6.66, 6.85).

16-(21) Oesophageal corpus dilated with lumen armed with cuticularized rods.
Bulb with valves (Fig. 6.58).

17-(18) Vulva opening into rectum forming cloaca (Fig. 6.59).
Parasites of fish.

Rondonia Travassos, 1919

18-(17) Vulva separate from rectum and anus.
Parasites of reptiles and amphibians.

19-(20) Cuticle without spines.

Atractis Dujardin, 1945 (Fig. 6.60)

20-(19) Cuticle covered with numerous scale-like cuticular projections alternating in diagonal rows (Figs. 6.61, 6.62).

Paratractis Sarmiento, 1959

21-(16) Oesophageal corpus not dilated, without cuticularized rods.
Bulb with or without valves (Figs. 6.63, 6.66, 6.67).

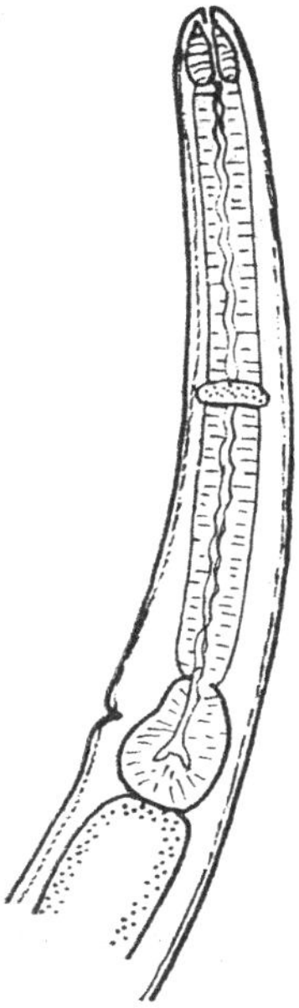

6.54

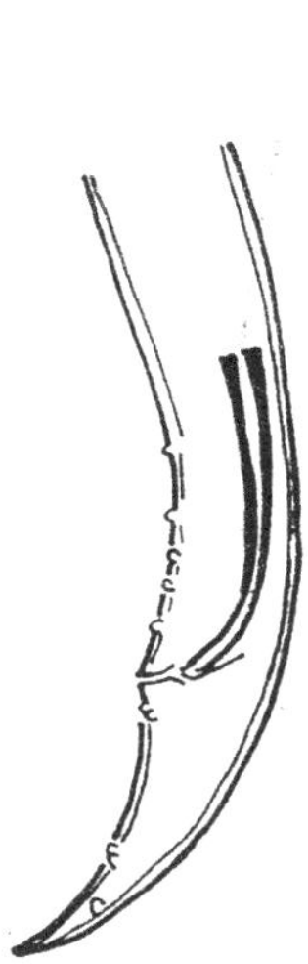

6.55

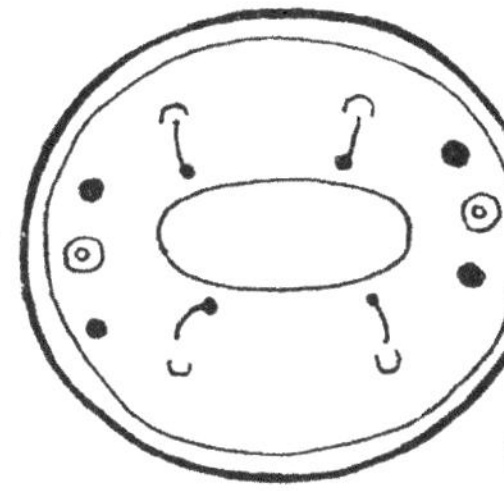

6.56

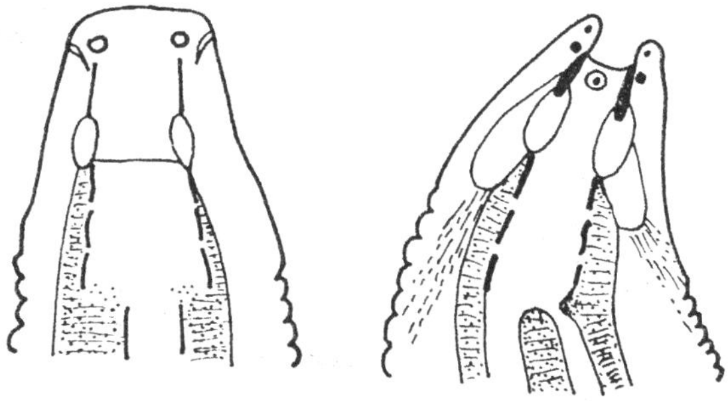

6.57

FIG. 6.54. *Ibrahimia*, oesophagus, lateral view. (After Khalil, 1932.)
FIG. 6.55. *Ibrahimia*, caudal end male, lateral view. (After Khalil, 1932.)
FIG. 6.56. *Nouvelnema*, cephalic extremity, apical view. (After Puylaert, unpublished data.)
FIG. 6.57. *Nouvelnema*, cephalic extremity, median and lateral view. (After Puylaert, unpublished data.)

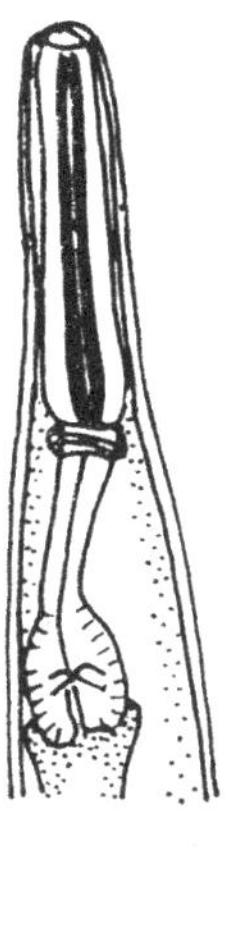

6.58

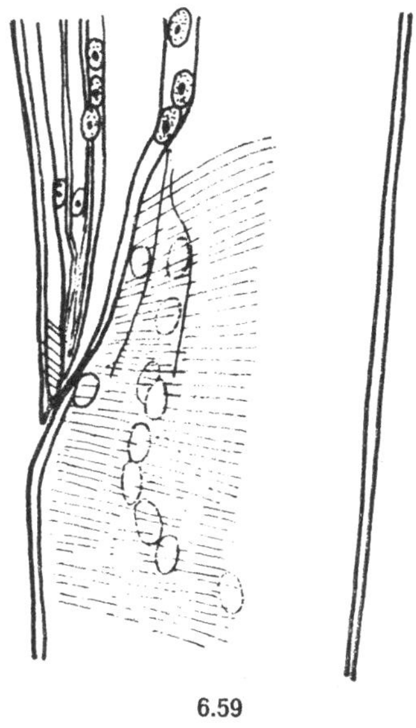

6.59

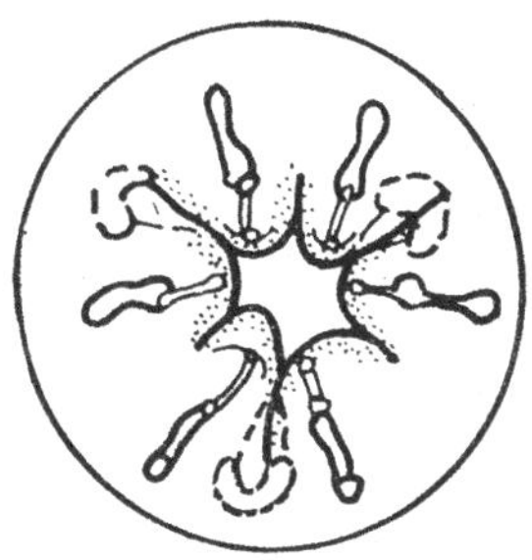

6.60

FIG. 6.58. *Atractis*, anterior extremity, lateral view. (After Petter, 1966.)
FIG. 6.59. *Rondonia*, vulva and anus. (After Baylis, 1936.)
FIG. 6.60. *Atractis*, cephalic extremity, apical view. (After Petter, 1966.)

6.61

6.62

FIG. 6.61. *Paratractis*, anterior extremity, ventral view. (After Sarmiento, 1959.)
FIG. 6.62. *Paratractis*, caudal end male, lateral view. (After Sarmiento, 1959.)

22-(27) Cephalic extremity without lateral projections.

23-(26) Anterior part of oesophagus long, forming more than 1/4 of whole (Fig. 6.63).

24-(25) Tail rather short (Fig. 6.64).
Parasites of American reptiles.

Cyrtosomum Gedoelst, 1919

25-(24) Tail long and attenuated in both sexes (Fig. 6.65).
Parasites of fish and reptiles.

Monhysterides Baylis & Daubney, 1922
(= *Pseudatractis* Yamaguti, 1961)

26-(23) Anterior part of oesophagus short, forming about 1/4 of whole (Fig. 6.66).
Parasites of proboscidians.

Leiperenia Khalil, 1922

27-(22) Anterior extremity with two large lateral pointed elevations (Fig. 6.67).
Parasites of hippopotami.

Cobboldina Leiper, 1911

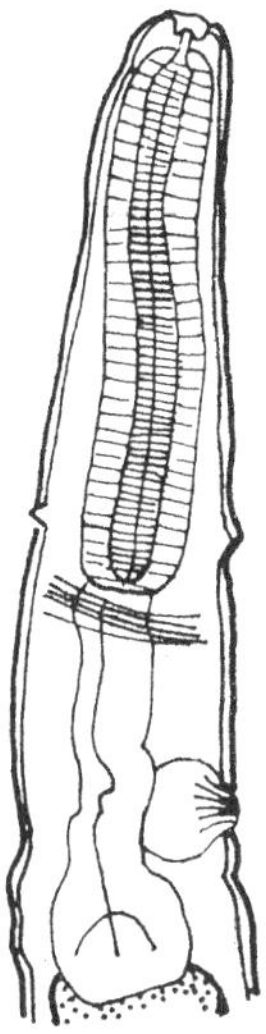

6.63

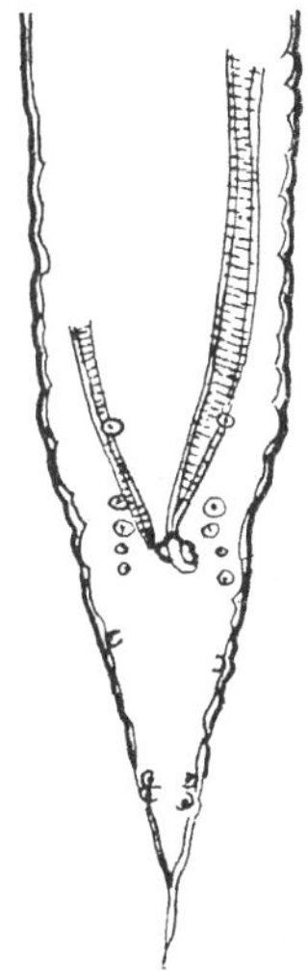

6.64

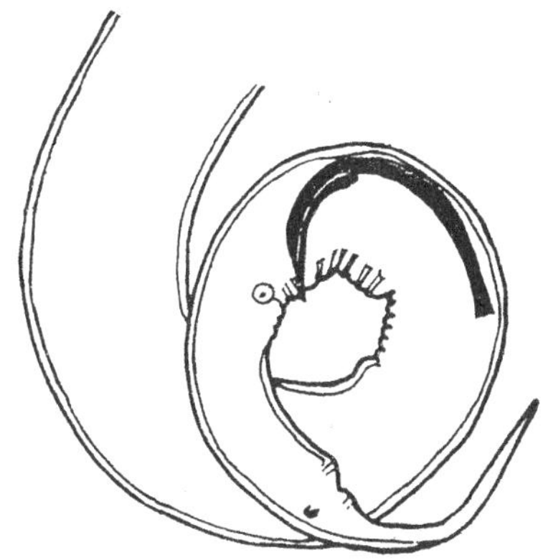

6.65

FIG. 6.63. *Cyrtosomum*, anterior extremity, lateral view. (After Bowie & Franz, 1976.)
FIG. 6.64. *Cyrtosomum*, caudal end male, ventral view. (After Bowie & Franz, 1976.)
FIG. 6.65. *Monhysterides*, caudal end male, lateral view. (After Baylis & Daubney, 1922.)

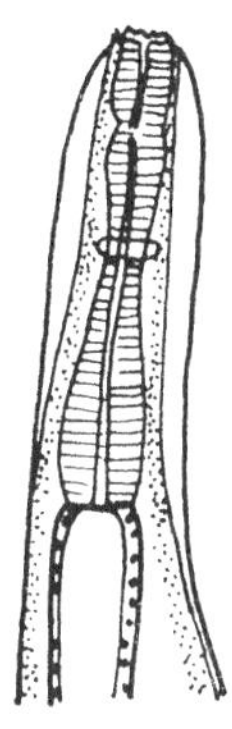

6.66

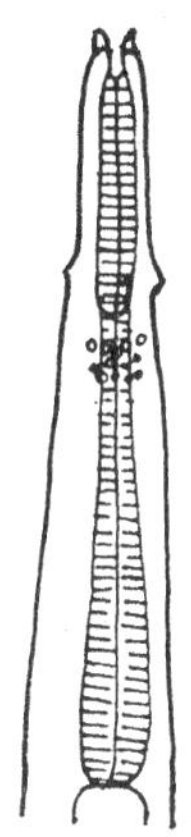

6.67

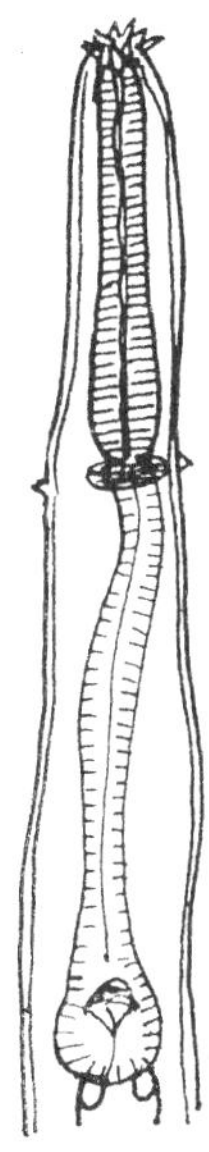

6.69

6.68

6.71

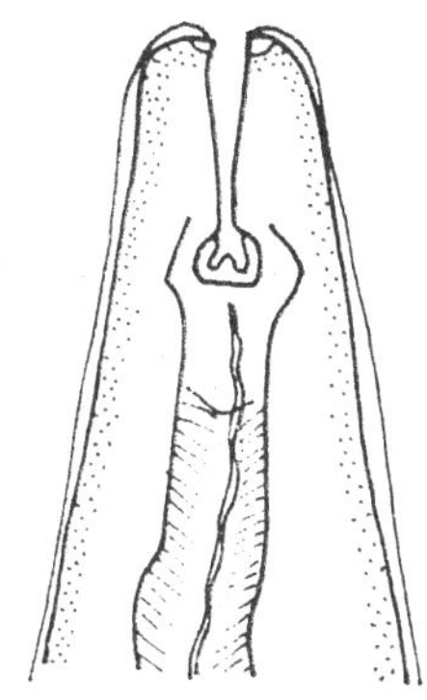

6.70

FIG. 6.66. *Leiperenia*, anterior extremity. (After Khalil, 1922.)
FIG. 6.67. *Cobboldina*, anterior extremity, ventral view. (After Yorke & Maplestone, 1926.)
FIG. 6.68. *Proatractis*, cephalic extremity, lateral view. (After G. Caballero R., 1971.)
FIG. 6.69. *Proatractis*, anterior extremity, dorsal view. (After G. Caballero R., 1971.)
FIG. 6.70. *Maracaya*, optical view of body at level of base of lips. (After Díaz-Ungría, 1964.)
FIG. 6.71. *Maracaya*, cephalic extremity, ventral optical view. (After Díaz-Ungría, 1964.)

6.72

6.73

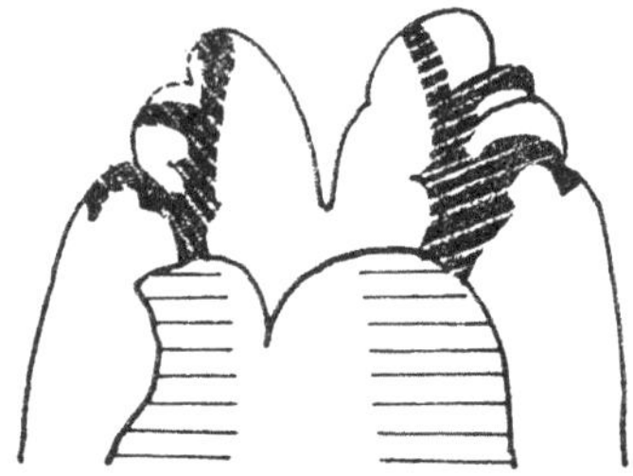

6.74

FIG. 6.72. *Orientatractis*, cephalic extremity, apical view. (After Petter, 1966.)
FIG. 6.73. *Orientatractis*, cephalic extremity apical and optical view. (After Petter, 1966.)
FIG. 6.74. *Orientatractis*, head, lateral optical view. (After Petter, 1966.)

28-(9) Buccal and pharyngeal structures with specialized cuticularized formations (Figs. 6.68, 6.70, 6.72, 6.75, 6.78, 6.82).

29-(38) Oral opening apical in position.

30-(31) Lips with small cuticular tongue-like expansions surrounding oral opening (Figs. 6.68, 6.69).
Parasites of tortoises.
Proatractis G. Caballero R., 1971

31-(30) Lips without small tongue-like cuticular expansions surrounding oral opening.

32-(37) Pharynx lacking powerful eversible jaws.

33-(34) Three large lips present.
Cuticle between lips thickened to form H-shaped supports (one ventral, two dorsolateral) (Figs. 6.70, 6.71).
Parasites of saurians.
Maracaya Díaz-Ungría, 1964

34-(33) Lips reduced.
Cuticularized supports of buccal cavity on four or eight axes (Figs. 6.73, 6.75).

35-(36) Four cuticularized supports present in buccal cavity, on submedian axes (Figs. 6.72-6.74).
Pharynx without laminated structure.
Parasites of tortoises.
Orientatractis Petter, 1966

36-(35) Twelve cuticularized supports present in buccal cavity, eight on submedian axes and four on sublateral axes (Figs. 6.75-6.77).
Dorsal wall of pharynx with laminated structure (Figs. 6.76, 6.77).
Parasites of Hyracoidea.
Grassenema Petter, 1959
(= *Acanthostephanocephalus* Kreis, 1960)

37-(32) Pharynx eversible and armed with powerful jaws in form of pectinate blades (Figs. 6.78-6.81).
Parasites of equids and rhinoceros.
Crossocephalus Railliet, 1909

6.75

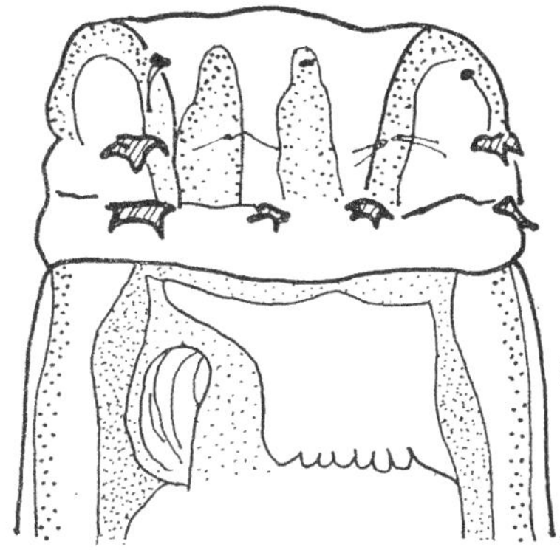

6.76

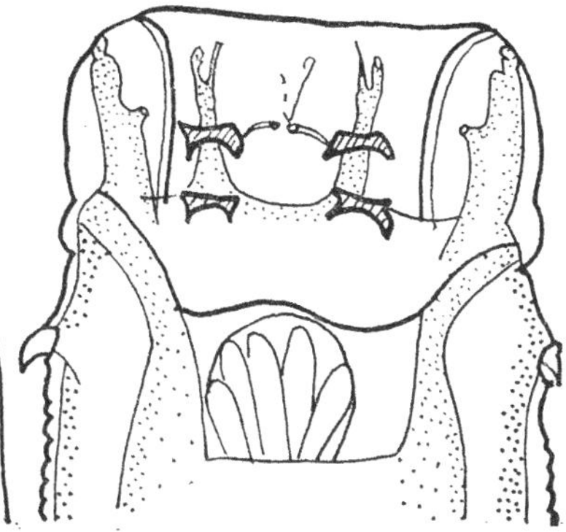

6.77

FIG. 6.75. *Grassenema*, cephalic extremity, apical view. (After Petter, 1959.)
FIG. 6.76. *Grassenema*, cephalic extremity, lateral view. (After Petter, 1959.)
FIG. 6.77. *Grassenema*, cephalic extremity, median view. (After Petter, 1959.)

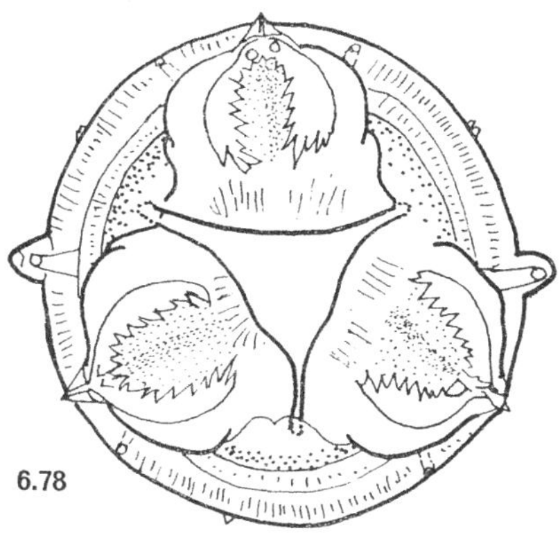

6.78

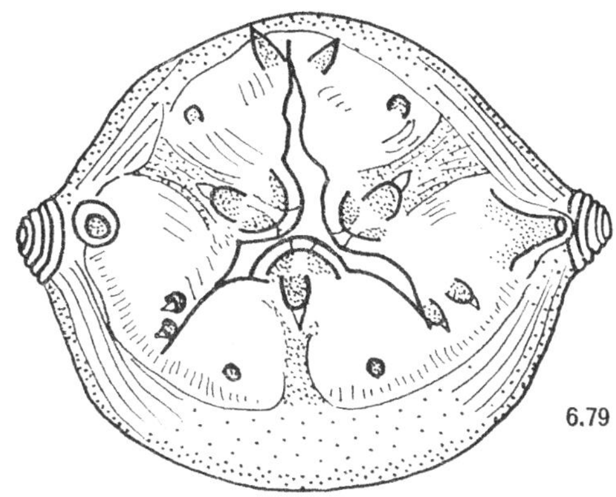

6.79

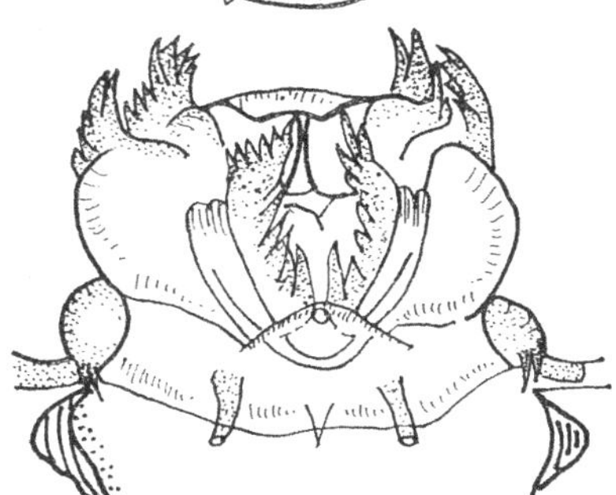

6.80

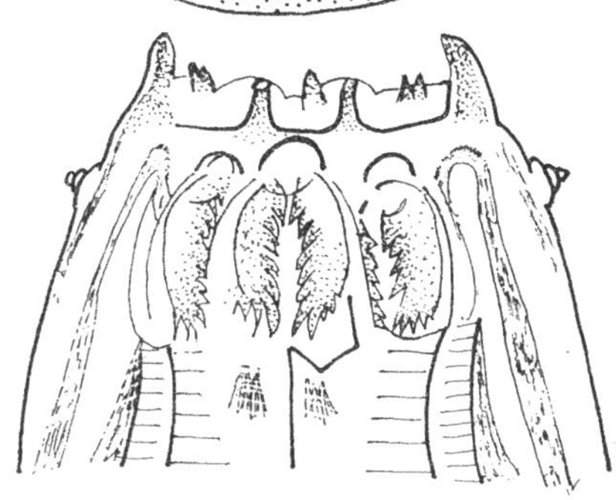

6.81

FIG. 6.78. *Crossocephalus*, cephalic extremity, apical view, evaginated head. (After Le Van Hoa, 1962.)
FIG. 6.79. *Crossocephalus*, cephalic extremity, apical view, invaginated head. (After Le Van Hoa, 1962.)
FIG. 6.80. *Crossocephalus*, cephalic extremity, dorsal view, evaginated head. (After Le Van Hoa, 1962.)
FIG. 6.81. *Crossocephalus*, cephalic extremity, dorsal view, invaginated head. (After Le Van Hoa, 1962.)

38-(29) Oral opening displaced ventrally, between two hypertrophied lobes of subventral lips (Figs. 6.82, 6.83, 6.85).
Lining of buccal cavity in form of pectinate lobes projecting through ventral part of oral opening (Figs. 6.83, 6.84).
Parasites of tortoises.

Labiduris Schneider, 1866

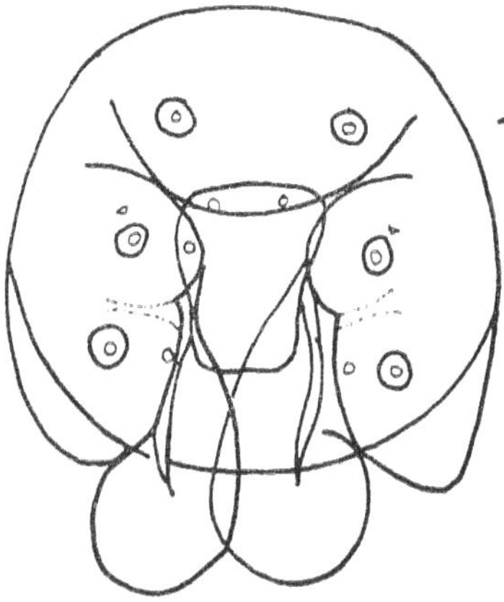

6.82

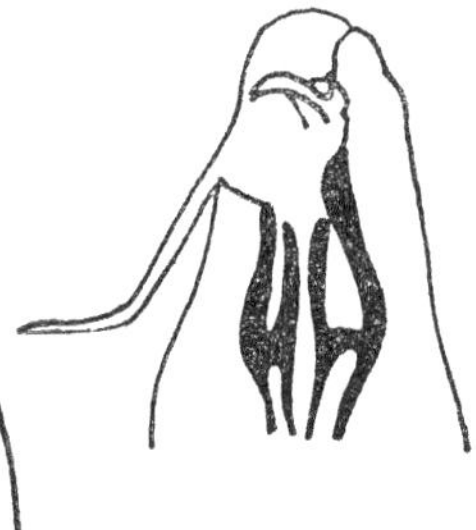

6.83

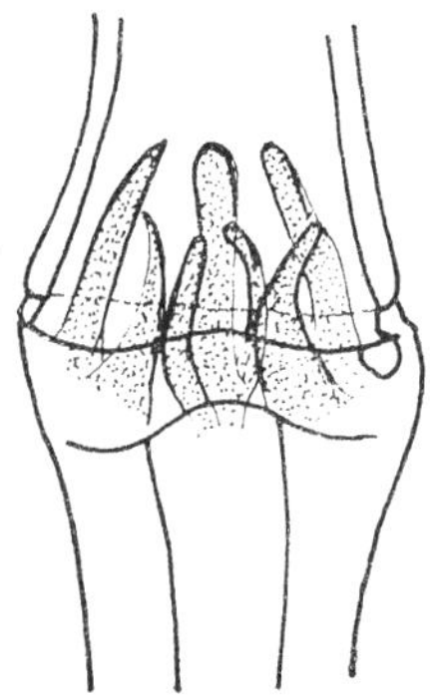

6.84

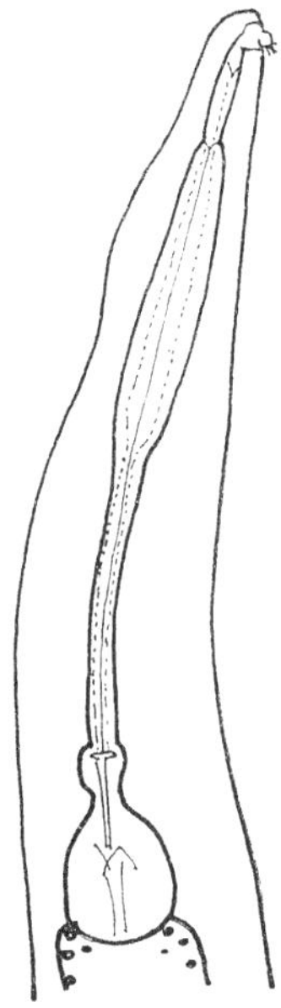

6.85

FIG. 6.82. *Labiduris*, cephalic extremity, apical view. (After Inglis & Díaz-Ungría, 1963.)
FIG. 6.83. *Labiduris*, cephalic extremity, lateral view, optical section. (After Petter, 1966.)
FIG. 6.84. *Labiduris*, pharyngeal teeth. (After Inglis & Díaz-Ungría, 1963.)
FIG. 6.85. *Labiduris*, anterior extremity, lateral view. (After Inglis & Díaz-Ungría, 1963.)

REFERENCES

BAKER, M. R. & ADAMSON, M. 1977. The genus *Cosmocercella* Steiner, 1924 (Nematoda: Cosmocercoidea). *Can. J. Zool.*, **55**, 1644-1649.

BALLESTEROS MÁRQUEZ, A. 1945. Revisión de la familia Cosmocercidae Travassos 1925. *Revta ibér. Parasit.*, Tomo Extraordinario, pp. 150-180.

BAYLIS, H. A. 1933. On a collection of nematodes from Malayan reptiles. *Ann. Mag. nat. Hist.*, Sér. **10**, **11**, 615-633.

BELLE, E. A. 1957. Helminth parasites of reptiles, birds and mammals in Egypt. IV. Four new species of oxyurid parasites from reptiles. *Can. J. Zool.*, **35**, 163-169.

BISWAS, P. K. & CHAKRAVARTY, G. K. 1963. The systematic studies of the zoo-parasitic oxyuroid nematodes. *Z. ParasitKde*, **23**, 411-428.

BOWIE, A. & FRANZ, R. 1974. *Cyrtosomum mega* sp.n. (Nematoda: Oxyuroidea) from the lizard, *Cyclura carinata*. *J. Parasit.*, **60**, 628-631.

CABALLERO y C., E. 1933. Nemátodos parásitos de los batracios de México. II. *Oxysomatium mexicanum* nov.espec. *An. Inst. Biol. Univ. Méx.*, **4**, 187-190.

CABALLERO DELOYA, J. 1974. Estudio helmintológico de los animales silvestres de la Estación de Biología Tropical 'Los Tuxtlas', Veracruz. Nematoda I. Algunos nematodos parasitos de *Bufo horribilis* Wiegmann, 1833. *An. Inst. Biol. Univ. Mex.*, **45**, Ser. Zool. (1), 45-50.

CABALLERO R., G. 1971. Contribución al conocimiento de los nemátodos que parasitan a los reptiles de México. I. Descripción de *Proatractis parvicapiticoronata* n.g., n.sp., parásito de Testudines. *Revta Biol. trop.*, **18**, 149-154.

CHABAUD, A. G. 1959. Redescription d'*Amblyonema terdentatum* Linstow, 1898, nématode parasite du dipneuste australien *Neoceratodus*. *Bull. Soc. zool. Fr.*, **84**, 188-194.

——— 1965. Superfamille des Cosmocercoidea. In: Grassé, P. P. *Traite de Zoologie, Némathelminthes*, Tome IV, fasc. 3. Paris: Masson et Cie, pp. 937-949.

——— & BRYGOO, E. R. 1960. Nématodes parasites de caméléons malgaches. *Mém. Inst. scient. Madagascar*, Sér. A, **14**, 125-159.

——— & GOLVAN, Y. J. 1957. *Megalobatrachonema campanae* n.sp. (Nematoda Kathlaniidae) parasite de tritons de la région parisienne. *Annls Parasit. hum. comp.*, **32**, 243-263.

——— & PETTER, A. J. 1960. Sur les nématodes atractides. *Libro Homenaje al Dr. Eduardo Caballero y Caballero, Jubileo 1930-1960*, pp. 465-470.

CHAKRAVARTY, G. K. & BHADURI, N. V. 1948. An oxyurid nematode *Neopharyngodon gekko*, gen. et sp. nov. from the Indian lizard *Gekko gecko* (Linn.). *Proc. zool. Soc. Bengal*, **1**, 103-107.

COY OTERO, A. & BARUŠ, V. 1973. Notes on nematodes of the genus *Cyrtosomum* (Atractidae) parasitic in Cuban lizards (Sauria). *Folia Parasit. (Praha)*, **20**, 297-305.

DÍAZ-UNGRÍA, C. 1963. Nématodes parasites, nouveaux ou intéressants, du Vénézuéla. *Annls Parasit. hum. comp.*, **38**, 893-914.

——— 1968. Helmintos de peces de Venezuela, con descripción de un género y tres especies nuevas. *Boln Soc. venez. Cienc. nat.*, **27**, 537-549.

FITZSIMMONS, W. M. 1958. On *Probstmayria reptiliae* n.sp., from *Homopus femoralis* and some notes on the genus *Probstmayria*. *J. Helminth.*, **32**, 211-218.

GOMES, D. C. & MOTTA, C. de SILVA. 1967. Sôbre um nôvo gênero e uma nova espécie de Oxyascarididae (Travassos, 1920) Freitas, 1958 (Nematoda, Subuluroidea). *Mems Inst. Oswaldo Cruz*, **65**, 29-31.

GUPTA, N. K. & AGGERWAL, S. C. 1974. On the nematode parasites from some fresh-water chelonians in India. *Third Internat. Cong. Parasit., Munich, Aug. 25th-31st, 1973. Proc.* Vol. 3. Vienna, Austria, FACTA Publ., pp. 1656-1657.

HARTWICH, G. 1960. Über *Megalobatrachonema terdentatum* (Linstow, 1890) nov. comb. und die Stellung von *Megalobatrachonema* Yamaguti, 1941 im System der Ascaridina (Nematoda). *Z. ParasitKde*, **19**, 606-616.

——— 1975. Schlauchwürmer, Nemathelminthes Rund- oder Fadenwürmer, Nematoda Parasitische Rundwürmer von Wirbeltreren. I. Rhabditida und Ascaridida. *Die Tierwelt Deutschlands.* 62 Teil. Jena: Gustav Fischer Verlag, 256 pp.

INGLIS, W. G. 1957. A revision of the nematode genera *Kathlania* and *Tonaudia*. *Ann. Mag. nat. Hist., Ser.* 12, **10**, 785-800.

——— 1959. The systematic position of *Nematoxys piscicola* Linstow, 1907 (Nematoda). *Z. ParasitKde*, **19**, 100.

——— 1968. Nematodes parasitic in Western Australian frogs. *Bull. Br. Mus. nat. Hist. Zool.*, **16**, 161-183.

——— & DÍAZ-UNGRÍA, C. 1963. Sobre el género *Labiduris* (Ascaridata, Kathlaniidae) con una discusión sobre el desarrollo de la cabeza. *Boln Soc. venez. Cienc. nat.*, **25**, 126-154.

KREIS, H. A. 1938. Beiträge zur Kenntnis parasitischer Nematoden. VIII. Neue parasitische Nematoden aus dem Naturhistorischen Museum Basel. *Zentbl. Bakt. Parasit.*, **142**, 329-352.

KUNG, C. C. & WU, H. W. 1945. Parasitic nematodes of amphibians from Pehpei, Szechwan, China. *Sinensia*, **16**, 73-83.

LE VAN HOA. 1961. Nématodes parasites de mammifères, reptiles et amphibiens du Congo. *Parc National de l'Upemba, Mission G. F. de Witte*, fasc. 65, pp. 3-58.

MAWSON, P. M. 1972. The nematode genus *Maxvachonia* (Oxyurata: Cosmocercidae) in Australian reptiles and frogs. *Trans. R. Soc. S. Aust.*, **96**, 101-108.

OLIVEIRA RODRIGUES, H. de & RODRIGUES, S. S. 1971. Sôbre um nôvo gênero e nova espécie de subfamília Oxyascaridinae Freitas, 1958 (Nematoda: Subuluroidea). *Atas Soc. Biol. R. de J.*, **15**, 15-17.

———, VARELA, M. C., RODRIGUES, S. S. & CRISTÓFARO, R. 1973. Alguns nematódeos de peixes do Oceano Atlântico. Costa Continental Portuguêsa e Costa do Norte da Africa. *Mems Inst. Oswaldo Cruz*, **71**, 247-259.

PEARSE, A. J. 1959. VII. Parasites from Yucatan. *Publs Carnegie Instn*, No. 457, 45-59.

PETTER, A. J. 1959. Deux nouveaux genres de nématodes atractides parasites du daman des rochers *Procavia ruficeps* (Ehrenberg). *Bull. Soc. zool. Fr.*, **84**, 195-204.

——— 1966. Equilibre des espèces dans les populations de nematodes parasites du colon des tortues terrestres. *Mem. Mus. natn Hist. nat., Paris, Sér. A. Zool.*, **39**, 252 pp.

——— & QUENTIN, J. C. 1976. Keys to genera of the Oxyuroidea. *CIH Keys to the Nematode Parasites of Vertebrates*, No. 4. Farnham Royal, Bucks, UK: Commonwealth Agricultural Bureaux, 30 pp.

PUYLAERT, F. 1970. Description de *Labeonema intermedia* gen.n., sp.n., parasite d'un *Labeo* (Pisces-Cyprinidae) (Atractidae-Nematoda-Vermes). *Revue Zool. Bot. afr.*, **82**, 370-376.

SARMIENTO, L. 1959. Description of *Paratractis hystrix* (Diesing, 1851) gen. nov. (Nematoda: Atractidae) from *Podocnemis dumeriliana*. *J. Parasit.*, **45**, 65-68.

SCHUURMANS-STEKHOVEN, J. H., Jr. 1950. Nemátodos parasitarios del Chaco Paraguayo y de Argentina del Museo de Estocolmo. *Acta zool. lilloana*, **9**, 325-345.

SEMENOV, V. D. 1929. Eine neue Nematodenart—*Rhabdias microoris* n.sp.—aus den Lungen von Amphibien. *Zentbl. Bakt. ParasitKde*, I. Orig., **114**, 533-539.

SINGH, S. N. 1958. On an interesting new nematode *Velariocephalus trilokiae* gen. et sp. nov. from an Indian frog and new subfamily Velariocephalinae (Cosmocercidae). *J. Helminth.*, **32**, 259-266.

SKRJABIN, K. I., SCHIKHOBALOVA, N. P. & LAGODOVSKAYA, E. A. 1961. *Oxyurata of animals and man*. Osnovi Nematodologii. Vol. 10. Moscow: Izdatel'stvo Akademii Nauk SSSR, 499 pp. (In Russian).

———, ———, ——— 1964. *Oxyurata of animals and man*. Osnovi Nematodologii. Vol. 13. Moscow: Izdatel'stvo Akademii Nauk SSSR, 468 pp. (In Russian).

———, SOBOLEV, A. A. & IVASCHKIN, V. M. 1967. *Spirurata*. Osnovi Nematodologii. Vol. 19. Moscow: Izdatel'stvo Nauka, 239 pp. (In Russian).

TEIXEIRA de FREITAS, J. F. 1958. Estudos sôbre 'Oxyascarididae' (Travassos, 1920) (Nematoda, Subuluroidea). *Mems Inst. Oswaldo Cruz*, **56**, 489-559.

——— 1959. Estudos sôbre Schrankianidae fam.nov. (Nematoda, Subuluroidea). *Archos Mus. nac. Rio de J.*, **49**, 9-68.

——— & DOBBIN, J. E., Jr. 1957. Novo nematódeo parasito de '*Rana palmipes*' Spix: '*Subulascaris falcaustriformis*' g.n., sp.n. (Nematoda, Ascaridiformes). *Revta bras. Biol.*, **17**, 245-248.

——— & LENT, H. 1941. Contribuição ao conhecimento da sub-familia Kathlaniinae (Nematoda: Subuluroidea). *Archos Zool. Est. S Paulo*, **3**, 13-42.

TRAVASSOS, L. 1931. Pesquizas helminthologicas realizadas em Hamburgo. IX. Ensaio monographico de familia Cosmocercidae Trav., 1925 (Nematoda). *Mems Inst. Oswaldo Cruz*, **25**, 237-298.

WALTON, A. C. 1940. Some nematodes from Tennessee amphibia. *J. Tenn. Acad. Sci.*, **15**, 402-405.

WOLFGANG, R. W. 1953. *Pseudocruzia* (Oxyuroidea: Kathlaniidae) a new genus of nematode from domestic swine in India. *Can. J. Zool.*, **31**, 16-19.

YAMAGUTI, S. 1961. *Systema Helminthum*. Vol. III. *The Nematodes of Vertebrates*. New York: Interscience Publishers, Pt. II, pp. 681-1261.

SEURATOIDEA

The superfamily Seuratoidea was created in order to group the numerous archaic genera which originated from the Cosmocercoidea and to connect the latter superfamily with the advanced Ascaridida or the Spirurida (see Chabaud *et al.*, 1960). Le Van Hoa & Pham-Ngoc-Khue (1971) found some genera which closely link some Seuratoidea to the Cucullanidae and it is now necessary to follow Inglis (1967) and include the Cucullanidae in the Seuratoidea.*

Skrjabin & Ivashkin (1968) believed that the Seuratoidea was unnecessary and they distributed the various genera among their four suborders the "Cucullanata, Ascaridata, Spirurata and Oxyurata", assuming that the Cucullanata is the link between the Ascaridata and the Spirurata. The two conceptions of the group are, in fact, fundamentally similar: the various main branches which ramify from a common origin in the Cosmocercoidea are individualized systematically at somewhat different levels in the two systems, a low level in the Soviet system, a little higher in our system (Fig. 1.68).

We prefer to retain the Seuratoidea for a number of reasons. The Cucullanidae has few genera but many species and is characterized, as shown by Inglis, by an unique evolution—the formation of a buccal cavity from an extension of the anterior end of the oesophagus—the *oesophastome* (Inglis, 1967). It is, therefore, not possible theoretically to regard the cucullanids as a connecting link between the Ascaridata and the Spirurata. In fact, the Seuratoidea, when taken as a whole, clearly indicate the archaic generic ancestors of the Heterakoidea, Subuluroidea, Cucullanidae, Thelazioidea, Rictularioidea and Camallanoidea, and how they relate to each other and derive from the Cosmocercoidea.

From a practical point of view, suppression of the Seuratoidea would not result in fewer taxa since a suborder would be necessary for the Cucullanidae. Finally, because of their smaller lips and relatively simple oesophagus, seuratoids are easy to separate from the other nematode groups. Since the Seuratoidea contains all the more primitive and least characteristic genera, the definition of the Heterakoidea, Subuluroidea, Cucullanidae, Thelazioidea, Rictularioidea and Camallanoidea is much simplified.

The Seuratoidea was originally made up of one family divided into seven subfamilies. Inglis (1969) improved the systematics by recognizing four families, namely, (i) Seuratidae with Seuratinae and Skrjabinelaziinae, (ii) Schneidernematidae with Schneidernematinae and Echinonematinae, (iii) Quimperiidae with Quimperiinae and Omeiinae, and (iv) Cucullanidae. Le Van Hoa & Pham-Ngoc-Khue (1967) discovered the subfamily Campanarougetiinae† which established clearly the affinities between the Quimperiidae and Cucullanidae. These affinities were further substantiated by the discovery of the Oceanicucullanidae (= Chitwoodiidae) and the Chitwoodchabaudiidae Puylaert, 1970. Finally, in 1968, Mawson discovered the subfamily Inglisonematinae and, in 1971, Quentin showed that the Echinonematinae should be placed in the Seuratidae and not in the Schneidernematidae.

The classification established herein is in accord with Inglis (1967) and includes the following additions and modifications: (a) the Echinonematinae is placed in the Seuratidae; (b) the Inglisonematinae is added to the Schneidernematidae; (c) the Gendriinae is placed in the Quimperiidae (see Boyce, 1961); (d) the Campanarougetiinae is added to the Cucullanidae (we believe Oceanicucullanidae can be included in Campanarougetiinae); (e) the Chitwoodchabaudiidae is added to the superfamily. Since 1961, the biology of a number of genera of the Seuratoidea has been investigated: *Seuratum* (see Quentin, 1970), *Cucullanus*

*Since the oldest suprageneric taxon is the Cucullanidae Cobbold, 1864, the superfamily should be called the Cucullanoidea. However, since the Soviet authors elevated the only group of the Cucullanidae to superfamily and even to suborder level, it seems preferable, as shown by Le Van Hoa *et al.*, to preserve the taxon Seuratoidea in order to avoid confusion.

†The correct spelling of this subfamily, based on *Campanarougetia* Le Van Hoa & Pham-Ngoc-Khue, 1967, is Campanarougetiinae not Campanarougetinae as given in the original paper.

(see Le Van Hoa & Pham-Ngoc-Khue, 1967a; Gibson, 1972); *Truttaedacnitis* (see Khromova, 1975; Pybus *et al.*, 1978); *Paraquimperia* (see Moravec, 1974); *Quimperia* (see Bain & Philippon, 1969; Vassiliades, 1972). Numerous new genera have been described and lesser known forms redescribed.

The principal innovations in the classification of the superfamily since Grassé's "Traité de Zoologie" are outlined below.

SEURATIDAE

Skrjabinelaziinae

Chabaud (1973) showed that *Salobrella* Freitas, 1949 is a synonym of *Skrjabinelazia* Sypliaxov, 1930; *Rabbium* Chitwood, 1960 can be placed close to the latter. The importance of the subfamily is that it links the Cosmocercoidea (see *Maxvachonia*) and the Thelazioidea, (see Chabaud *et al.*, 1965).

Echinonematinae

The type and only species of *Echinonema* was redescribed by Inglis (1967). After studying the larvae of a species of *Seuratum*, Quentin (1971) was able to show the affinities of this genus to *Echinonema* and he placed the Echinonematinae in the Seuratidae.

Seuratinae

The differentiation of *Seuratinema* Johnston & Mawson, 1941 from *Skrjabinura* Gnedina, 1933 by Inglis in 1967 seems to be based on an error in the description of the type species of *Skrjabinura* and we prefer to accept the synonymy of the two genera proposed by Mawson (1960).

The characters that distinguish *Limonnema* Guerrero, 1971 from *Skrjabinura* do not seem important enough to warrant the creation of a special genus and the former is placed in the synonymy of the latter.

Rictularina Johnston & Mawson, 1941a and *Denticulospirura* Johnston & Mawson, 1941b are known only from immature females and they cannot be included in the keys.

SCHNEIDERNEMATIDAE

Schneidernematinae

The genus *Ascaroterakis* Vicente, 1965 is said by some authors to relate the Schneidernematinae to the Heterakoidea. This is not recommended herein because *Hatterianema*, an extremely archaic genus of Heterakoidea, has three tiny labia and an oesophagus with a valved bulb. This suggests that basically the heterakoids represent a line distinct from that of the Schneidernematinae.

Inglisonematinae

This subfamily, proposed by Mawson (1968), is increased by the addition of *Madelinema* Schmidt & Kuntz, 1971.

QUIMPERIIDAE

Arthur & Margolis (1974) have shown recently that *Haplonema*, *Cottocomephoronema* and *Ichthyobronema* are synonyms and we propose to regard *Paragendria*, *Metaquimperia*, and *Neometaquimperia* as synonymous. Kloss (1966) has completed our knowledge of *Paraseuratum*.

Moravec (1966) has clarified the relationships between *Haplonema* and *Paraquimperia*. *Chabaudus* Inglis & Ogden, 1965 and *Ezonema* Boyce, 1971 have been added to the subfamily.

The genus *Auchmeronema* Puylaert, 1970 was placed in the Subulascaridae but a recent revision by Puylaert (personal communication) leads to the conclusion that *Auchmeronema* is a synonym of *Chabaudus*; the type species has, in addition to a dorsal tooth, two much reduced ventrolateral oesophageal teeth.

The genus *Chabaudus* has a hexagonal oral opening with three bi-lobed lips whereas in *Gendria* the oral opening is rounded and without lips. Some species currently classified with *Gendria* belong in *Chabaudus*. In addition, *Subulascaris* Freitas & Dobbin, 1957, which is closely related to *Chabaudus*, is a seuratoid and not an oxyascaridid.

CHITWOODCHABAUDIIDAE

Puylaert placed this new family in the Cosmocercoidea mainly because of its affinities with the Kathlaniidae. In view of the dentate oesophagostome and, especially, the muscular and glandular oesophagus, we prefer to place it in the Seuratoidea. This also helps avoid complications in the preparation of dichotomous keys. This alteration is of little theoretical importance since the genus *Chitwoodchabaudia* is very synthetic.* This genus is of considerable interest as it represents the relic of a group which evolved towards the Camallanina (see Puylaert, 1970).

CUCULLANIDAE

Campanarougetiinae

The studies of Le Van Hoa & Pham-Ngoc-Khue (1967, 1971), Schmidt & Kuntz (1969) and Le Van Hoa (1974) have resulted in the formation of this important subfamily which links the Quimperiidae and the Cucullanidae. *Paracucullanellus* Agrawal, 1965 may also belong to this group but the description of the cephalic structures does not correspond with the figures given and we prefer to regard this form as a *species inquirenda*.

Cucullaninae

A number of detailed studies based on Campana-Rouget (1957) have brought some order into this difficult group (Rasheed, 1968; Berland, 1970; Maggenti, 1971). Petter (1975) correlated the evolution of the principal morphological characters with the phylogenetic positions of the hosts and her taxonomic conclusions have been adopted herein.

Neocucullanus is retained but with only its type species and not in the sense of Campana-Rouget who placed in it all species with multiple pre-anal papillae; such over-development is found in various groups of Cucullaninae and it lacks phylogenetic significance.

Cucullanus interrogativus Travassos *et al.*, 1928 is characterized by unusual cephalic structures and has been difficult to classify. The problem was resolved with the recent discovery of the cephalic structures of the 4th larval stage by Berland in 1970 and Gibson in 1972. *C. interrogativus* is, as reported by the Brazilian authors, a juvenile form of the genus *Cucullanus*.

Pseudocucullanus Kalyankar, 1971 which is insufficiently described, is not considered.

Amphisakis Chakravarty & Majumdar, 1960 is not a member of the Cucullanidae. It may be a larval ascaridoid.

Ivaschkhin & Khromova (1976) have published a monograph on the "Cucullanata".

*A non-specialized genus with a mixture of characters some of which can be observed in various more highly evolved groups.

SEURATOIDEA

Key to families

1-(6) Buccal cavity absent or, if present, derived from cheilostome; cuticle constituting walls of cheilostome having same structure and staining reactions as external body cuticle (Fig. 6.120).
2-(3) Eggs embryonated, viviparity sometimes present.
Rows of cuticular spines sometimes present (Figs. 6.90, 6.94).
Tail short or very short (Fig. 6.92).
Preanal sucker absent.
Gubernaculum large.
Spicules simple, barely longer than gubernaculum (Fig. 6.92) (except in *Echinonema*).
Parasites of various vertebrates, excluding fish.
Seuratidae

3-(2) Eggs not embryonated.
Rows of cuticular spines absent.
Preanal sucker present or absent.
Gubernaculum present or absent.
Spicules longer than gubernaculum.
Parasites of fish, amphibians, birds or mammals.
4-(5) Circular preanal sucker with clearly defined cuticular rim (Fig. 6.100).
Parasites of mammals and birds.
Schneidernematidae
5-(4) Preanal sucker, when present, elongate, without cuticular rim (Fig. 6.108).
Parasites of fresh-water fish or Amphibia.
Quimperiidae
6-(1) Buccal cavity formed from modifications of anterior end of oesophagus (oesophastome); walls of oesophastome surrounded by oesophageal tissue (Figs. 6.124, 6.129-6.131).
Parasites of fish, rarely chelonians.
7-(8) Oesophagus divided into anterior muscular part and posterior glandular part (Fig. 6.132).
Parasites of amphibians.
Chitwoodchabaudiidae

8-(7) Oesophagus not divided.
Parasites of fish, rarely chelonians.
Cucullanidae

Family *SEURATIDAE* (Hall, 1916) Railliet, 1906

Key to subfamilies

1-(2) Vulva opening near middle of oesophagus (Figs. 6.87, 6.89).
Parasites of reptiles.
Skrjabinelaziinae

2-(1) Vulva opening near middle of body.
Parasites of birds and mammals.
3-(4) Anterior end of body swollen into cephalic bulb bearing large spines (Fig. 6.90).
Spicules elongated (Fig. 6.91).
Parasites of Australian marsupials.
Echinonematinae

5*

4-(3) Anterior end of body not swollen into cephalic bulb bearing large spines.
Spicules short, smaller than gubernaculum (Figs. 6.92, 6.93).
Parasites of birds, Chiroptera and rodents.

Seuratinae

Subfamily **Skrjabinelaziinae**

Chabaud, Campana-Rouget & Brygoo, 1959

(= Salobrellidae Freitas, 1940)

Key to genera

1-(2) Oral opening triangular with delicate leaf crown (Fig. 6.86).
Oesophagus cylindrical (Fig. 6.87).

Skrjabinelazia Sypliaxov, 1930
(= *Salobrella* Freitas, 1940)

2-(1) Oral opening elongated dorsoventrally (Fig. 6.88).
Oesophagus with short muscular anterior part (Fig. 6.89).

Rabbium Chitwood, 1960

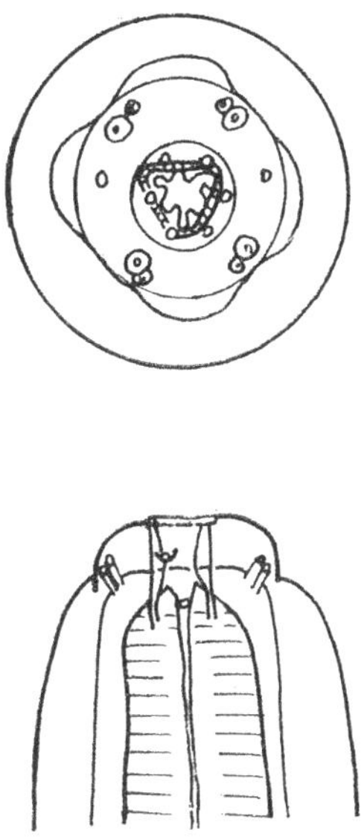

6.86

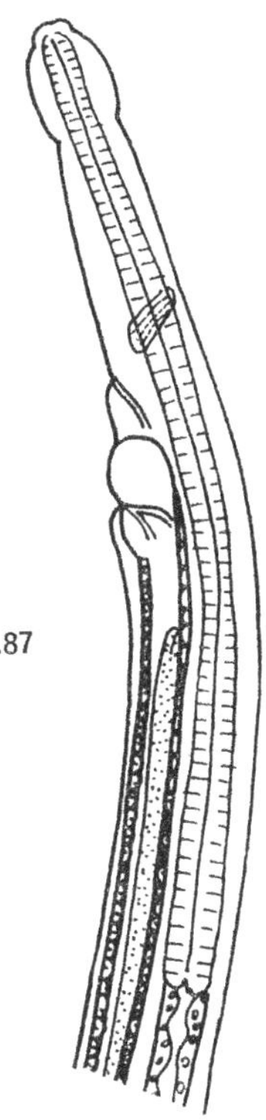

6.87

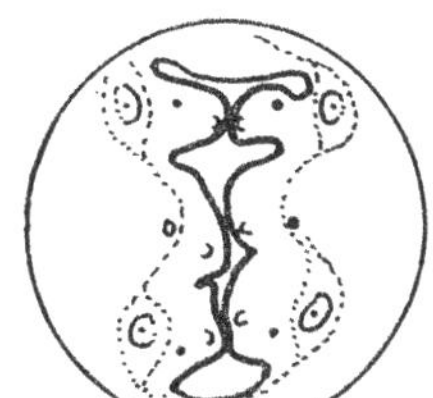

6.88

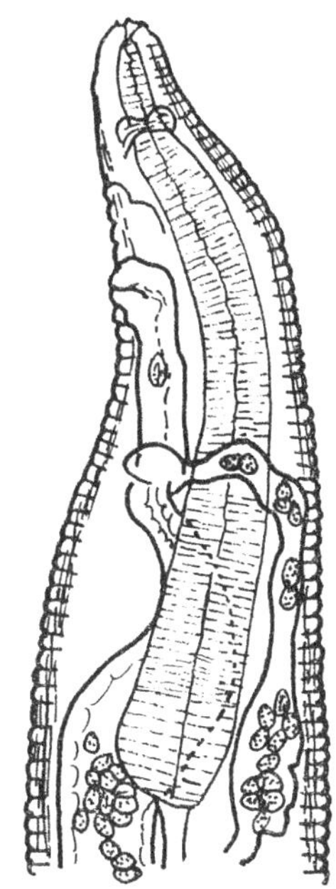

6.89

FIG. 6.86. *Skrjabinelazia*, cephalic extremity, apical and lateral views. (After Chabaud, 1973.)
FIG. 6.87. *Skrjabinelazia*, anterior extremity female, lateral view. (After Chabaud, 1973.)
FIG. 6.88. *Rabbium*, cephalic extremity, apical view. (After Chitwood, 1960.)
FIG. 6.89. *Rabbium*, anterior extremity female, lateral view. (After Chitwood, 1960.)

Subfamily **Echinonematinae** Inglis, 1967

One genus (Figs. 6.90, 6.91).

Echinonema Linstow, 1898

Subfamily **Seuratinae** Hall, 1916

Key to genera

1-(2) Male tail very short, with rounded tip (Fig. 6.92).
Cuticle without rows of spines or special appendages.
Didelphic.
Parasites of birds.

Skrjabinura Gnedina, 1933
(= *Seuratinema* Johnston & Mawson, 1941;
= *Limonnema* Guerrero, 1971)

2-(1) Tail of male slightly longer than wide at level of anus (Fig. 6.93).
Cuticle with rows of spines or special appendages (Figs. 6.94, 6.95).
Monodelphic or didelphic.
Parasites of mammals.

3-(4) Didelphic.
Cuticle with longitudinal rows of spines (Fig. 6.94).
Parasites of Chiroptera and rodents.

Seuratum Hall, 1916

6.90

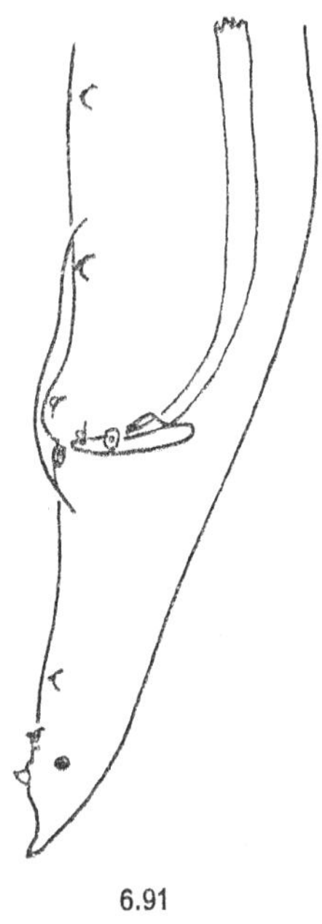

6.91

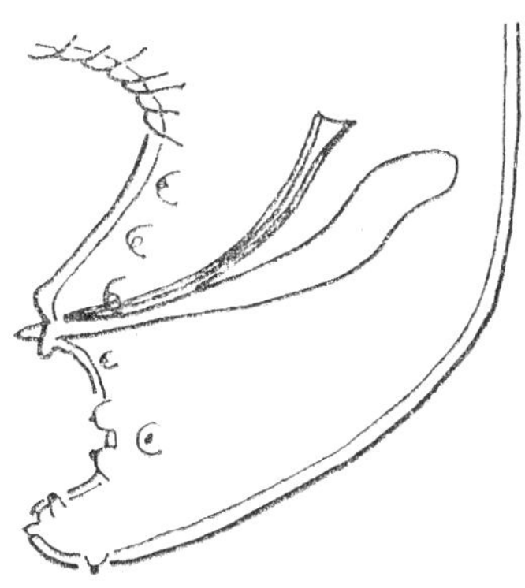

6.92

Fig. 6.90. *Echinonema*, anterior extremity. (After Inglis, 1967.)
Fig. 6.91. *Echinonema*, caudal extremity male, lateral view. (After Inglis, 1967.)
Fig. 6.92. *Skrjabinura*, caudal extremity male, lateral view. (After Inglis, 1967.)

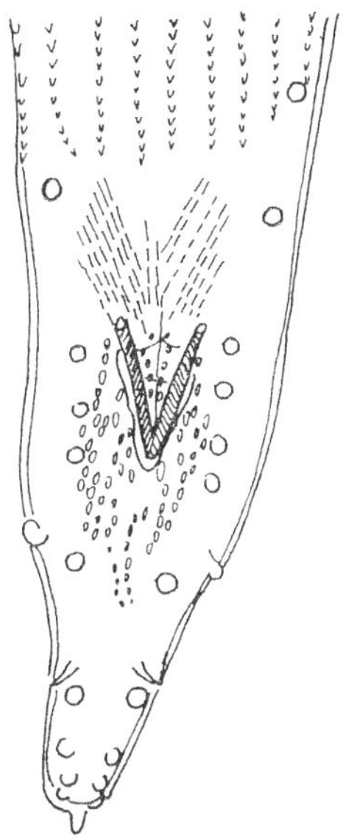

6.93

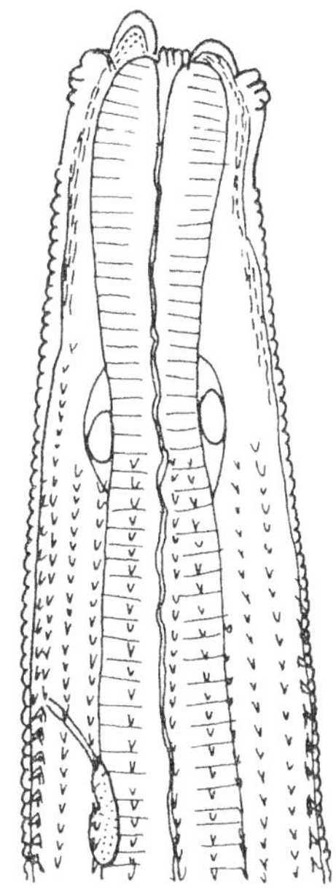

6.94

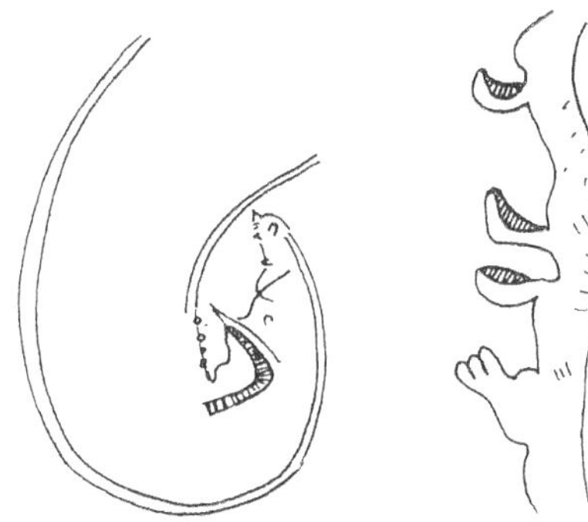

6.95

FIG. 6.93. *Seuratum*, caudal extremity male, ventral view. (After Biocca & Chabaud, 1951.)
FIG. 6.94. *Seuratum*, anterior extremity, lateral view. (After Biocca & Chabaud, 1951.)
FIG. 6.95. *Monovaria*, caudal extremity male, lateral view and cup-shaped appendages. (After Khera, 1953.)

4-(3) Monodelphic.
Cuticle without rows of longitudinal spines but with four pairs of special appendages anterior to preanal male papillae (Fig. 6.95).
Parasites of Chiroptera.

Monovaria Khera, 1953

Family *SCHNEIDERNEMATIDAE* Freitas, 1956

Key to subfamilies

1-(2) Oesophagus swollen anteriorly (Fig. 6.96).
Gubernaculum present.
Spicules either equal or unequal.
Parasites of birds.

Inglisonematinae

2-(1) Oesophagus swollen posteriorly (Fig. 6.99).
Gubernaculum absent.
Spicules equal.
Parasites of mammals.

Schneidernematinae

Subfamily **Inglisonematinae** Mawson, 1968

Key to genera

1-(2) Spicules unequal.
Caudal alae asymmetrical (Fig. 6.97).
Inglisonema Mawson, 1968

2-(1) Spicules equal.
Caudal alae symmetrical (Fig. 6.98).
Madelinema Schmidt & Kuntz, 1971

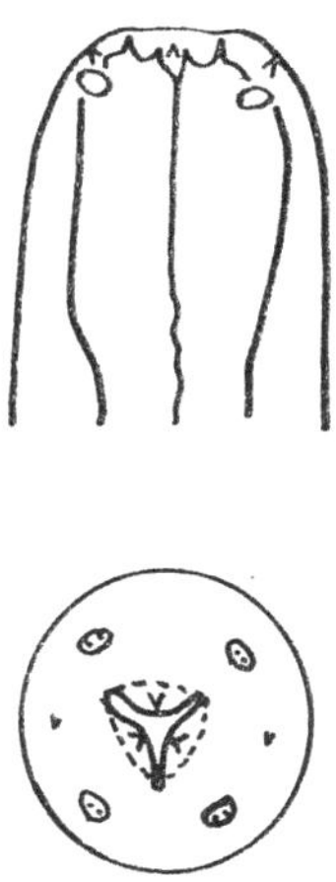

6.96

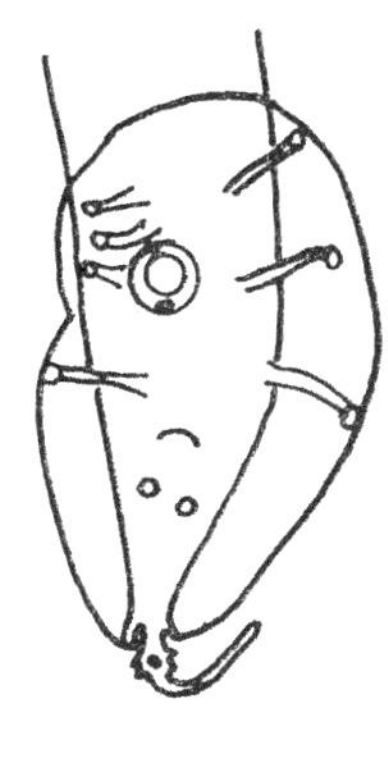

6.97

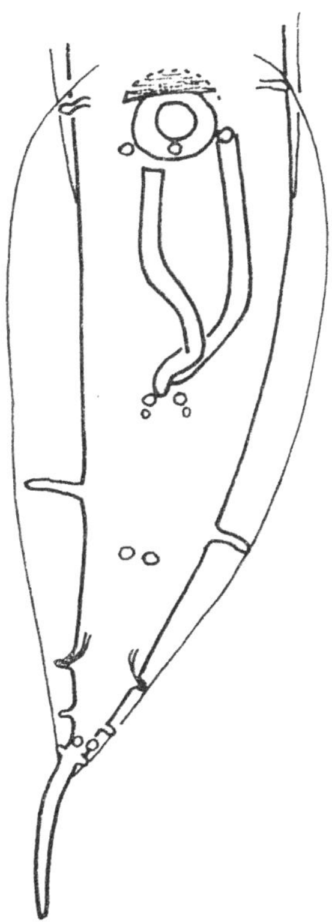

6.98

FIG. 6.96. *Inglisonema*, cephalic extremity, ventral and apical views. (After Mawson, 1968.)
FIG. 6.97. *Inglisonema*, caudal extremity male, ventral view. (After Mawson, 1968.)
FIG. 6.98. *Madelinema*, caudal extremity male, ventral view. (After Schmidt & Kuntz, 1971.)

Subfamily **Schneidernematinae** Freitas, 1956

Key to genera

1-(2) Caudal alae well developed (Fig. 6.100).
Parasites of *Xenarthra*.
Ascaroterakis Vicente, 1965

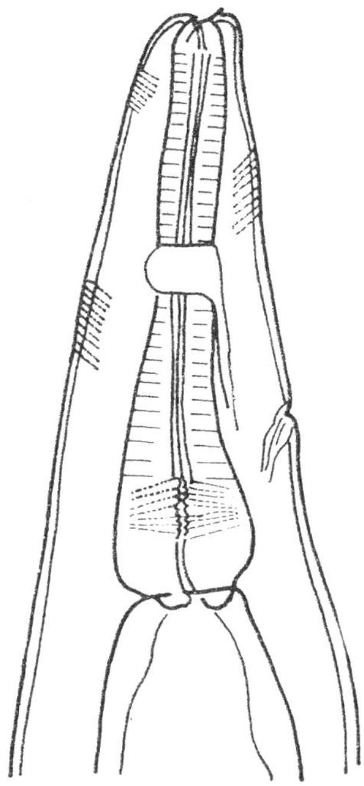

6.99

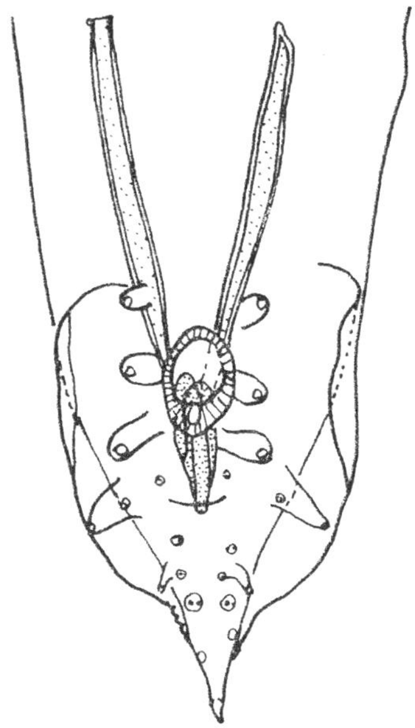

6.100

FIG. 6.99. *Morgascaridia*, anterior extremity, lateral view. (After Inglis, 1958.)
FIG. 6.100. *Ascaroterakis*, caudal end male, ventral view. (After Vicente, 1965.)

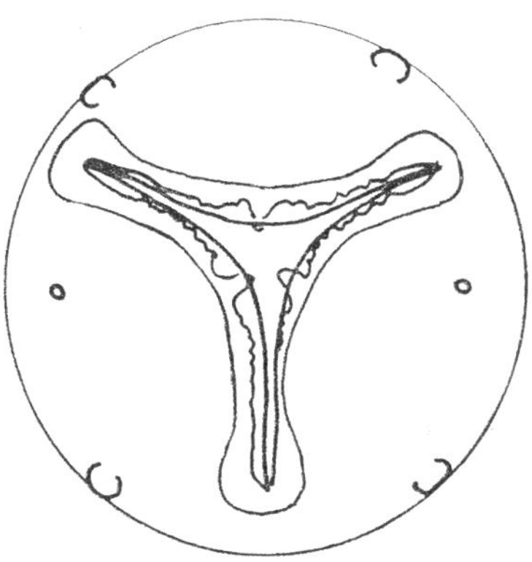

6.101

6.103

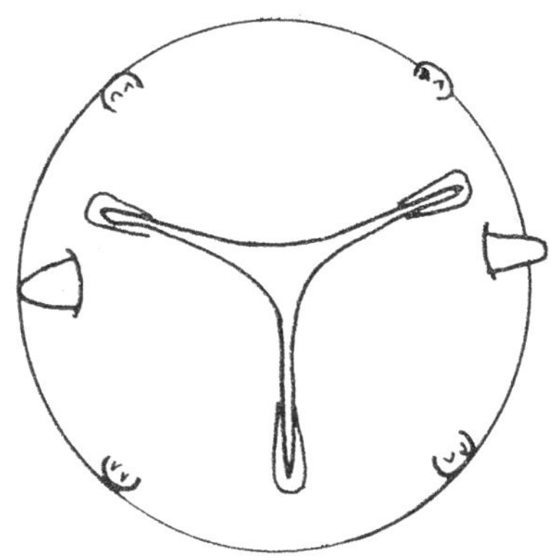

6.102

FIG. 6.101. *Schneidernema*, cephalic extremity, apical view. (After Inglis & Chabaud, 1960.)
FIG. 6.102. *Morgascaridia*, cephalic extremity, apical view. (After Inglis & Chabaud, 1960.)
FIG. 6.103. *Paraseuratum*, cephalic extremity female. (After Kloss, 1966.)

2-(1) Caudal alae absent.
3-(4) Denticles present on inner surface of lips.
Amphids flat (Fig. 6.101).
Parasites of *Xenarthra*.

Schneidernema Travassos, 1927

4-(3) Denticles not present on inner surface of labia.
Amphids protruding anteriorly (Fig. 6.102).
Parasites of African Sciuridae.

Morgascaridia Inglis, 1958

Family *QUIMPERIIDAE* (Gendre, 1928) Baylis, 1930

Key to subfamilies

1-(2) Intestinal caecum absent.
Buccal capsule small or absent.
Gubernaculum small or absent.
Parasites of fish, rarely amphibians.

Quimperiinae

2-(1) Intestinal caecum very large (Fig. 6.121).
Buccal capsule well developed (Fig. 6.120).
Gubernaculum large.
Parasites of batrachians.

Omeiinae

Subfamily **Quimperiinae** Gendre, 1928

Key to genera

1-(6) Preanal sucker absent or weakly developed.
2-(3) Anterior part of body without lateral alae.
Deirids barely visible.
Oesophagus with six teeth protruding into buccal cavity (Fig. 6.103).

Paraseuratum Johnston & Mawson, 1940

3-(2) Anterior part of body with large lateral alae (Fig. 6.104).
Deirids large.
4-(5) Oblique muscle bands in preanal region of male absent (Fig. 6.105).
Lips absent.
Oesophagus without pharynx and without teeth protruding into buccal cavity (Fig. 6.104).
Parasites of fresh-water fish.

Haplonema Ward & Magath, 1917
(= *Cottocomephoronema* Layman, 1933;
= *Ichthyobronema* Gnedina & Savina, 1930)

5-(4) Distinct oblique muscle bands in preanal region of male present.
Oral opening surrounded by three small lips.
Oesophagus with distinct pharyngeal portion and teeth protruding into buccal cavity (Fig. 6.107).
Parasites of eels.

Paraquimperia Baylis, 1934

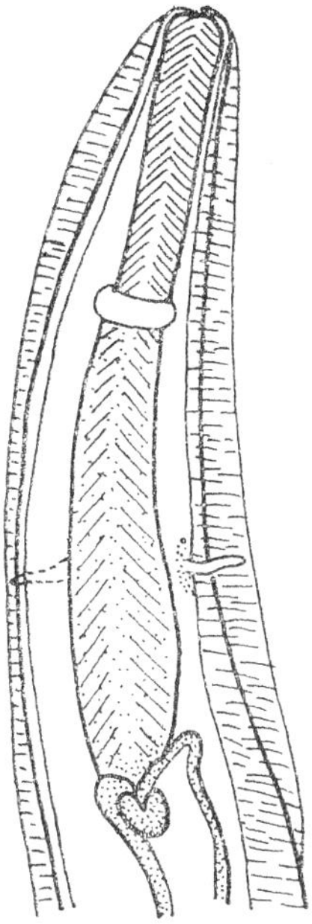

6.104

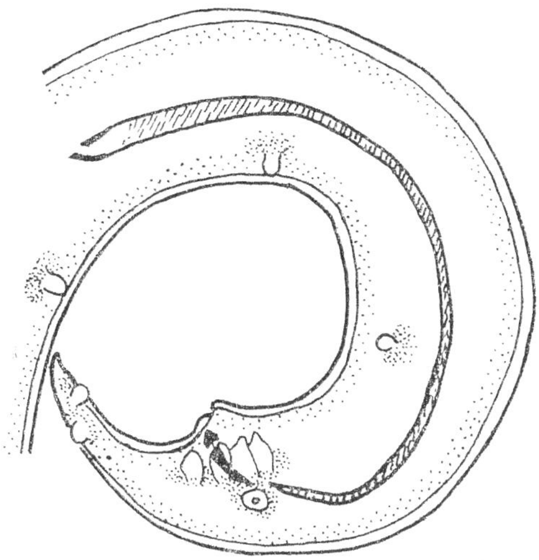

6.105

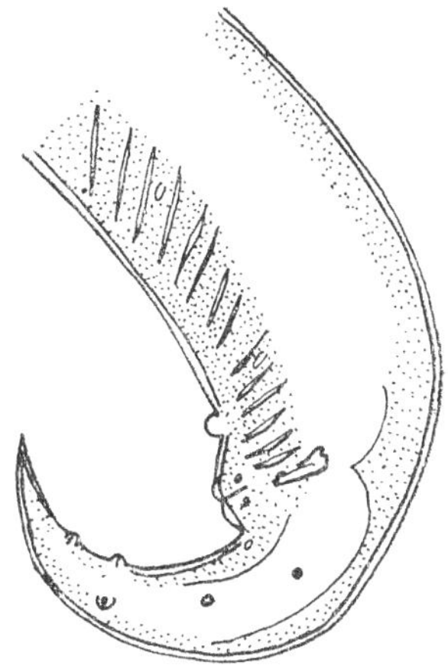

6.106

FIG. 6.104. *Haplonema*, anterior extremity, ventral view. (After Arthur & Margolis, 1975.)
FIG. 6.105. *Haplonema*, caudal extremity male, lateral view. (After Arthur & Margolis, 1975.)
FIG. 6.106. *Paraquimperia*, caudal extremity male, lateral view. (After Moravec, 1966.)

6-(1) Preanal sucker present (Fig. 6.108).
7-(14) Cephalic vesicle absent.
Cervical alae well developed (Fig. 6.109).
8-(9) Anterior third of oesophagus narrow, posterior two thirds much thickened (Fig. 6.110).
Three small lips present.
Buccal capsule absent.
Parasites of African fish.

Quimperia Gendre, 1928

9-(8) Oesophagus uniformly cylindrical or gradually swelling posteriorly.
10-(11) Oesophagus gradually swelling posteriorly (Fig. 6.111).
Buccal cavity present but weakly developed.
Parasites of siluriform fish in India.

Paragendria Baylis, 1939
(= *Metaquimperia* Karve, 1941;
= *Neometaquimperia* Agrawal, 1965)

11-(10) Oesophagus cylindrical (Fig. 6.112).
Buccal cavity absent.
12-(13) Deirids very large and prominent forming two post-oesophageal horns.
Oesophagus without pharynx (Fig. 6.112).
Parasites of fish in Japan.

Ezonema Boyce, 1971

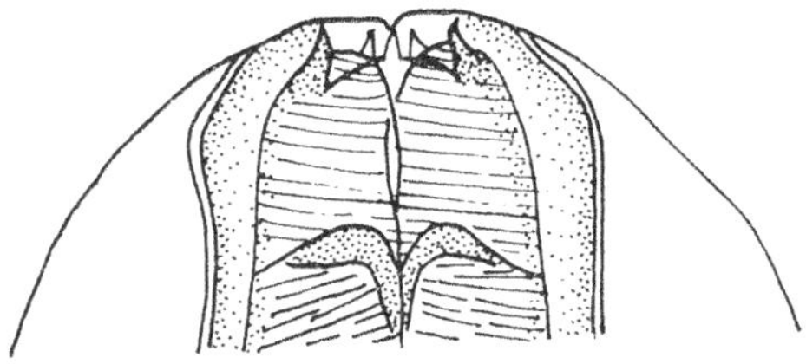

6.107

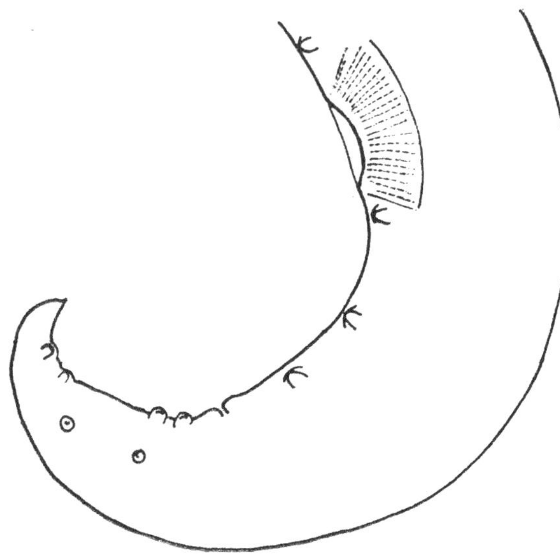

6.108

FIG. 6.107. *Paraquimperia*, cephalic extremity female. (After Moravec, 1966.)
FIG. 6.108. *Quimperia*, caudal end male, lateral view. (After Gendre, 1926.)

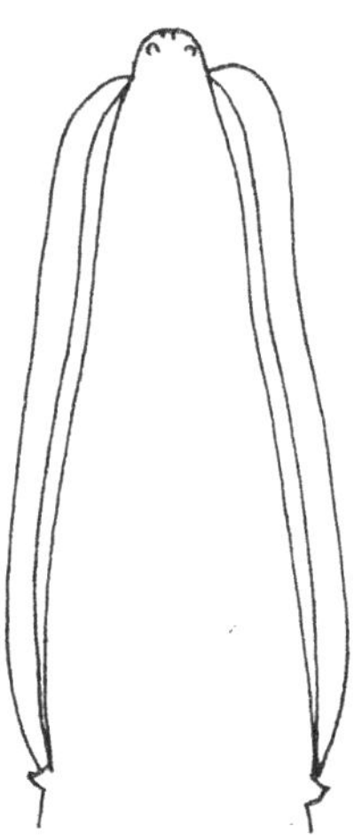

6.109

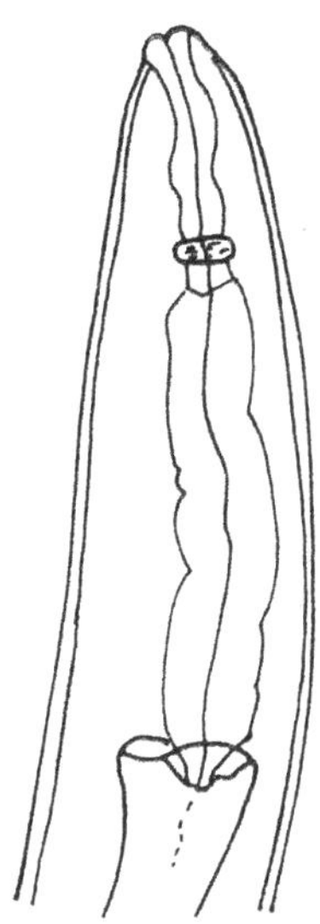

6.110

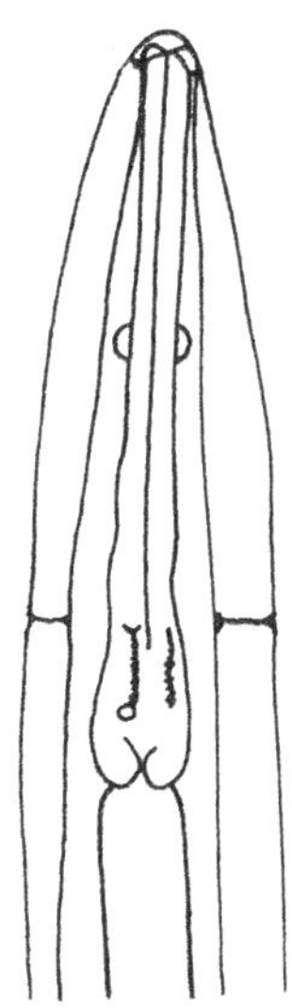

6.111

FIG. 6.109. *Quimperia*, anterior extremity, dorsal view. (After Gendre, 1926.)
FIG. 6.110. *Quimperia*, oesophagus. (After Gendre, 1926.)
FIG. 6.111. *Paragendria*, anterior extremity, median view. (After Stewart, 1914.)

13-(12) Deirids not prominent.
Oesophagus with anterior extremity forming pharynx (Fig. 6.113).
Parasites of fish in China.

Pingus Hsü, 1933

14-(7) Cephalic vesicle present (Fig. 6.114).
Cervical alae usually absent.

15-(20) Caudal papillae numerous.
Cuticle without longitudinal striations.
Parasites of fish and amphibians of the Old World.

16-(19) Buccal cavity with three teeth, the two ventrolaterals sometimes reduced.

17-(18) Oral opening triangular with three bilobed lips giving hexagonal appearance in apical view (Fig. 6.115).

Chabaudus Inglis & Ogden, 1965
(= *Auchmeronema* Puylaert, 1970)

18-(17) Oral opening, rounded, without lips (Fig. 6.116).

Gendria Baylis, 1930

19-(16) Buccal cavity with six teeth (Figs. 6.117, 6.118).
Oesophagus swollen posteriorly (Fig. 6.119).

Buckleynema Ali & Singh, 1954

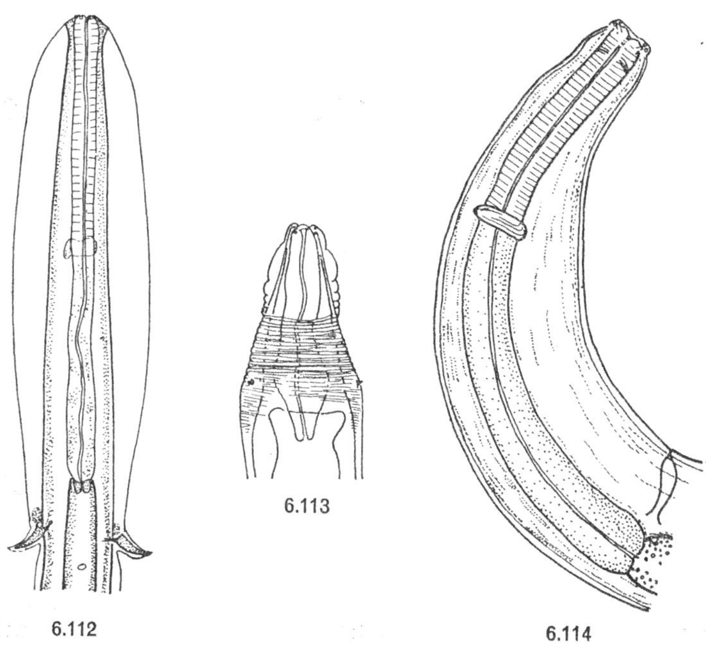

FIG. 6.112. *Ezonema*, anterior extremity, ventral view. (After Boyce, 1971.)
FIG. 6.113. *Pingus*, anterior extremity, lateral view. (After Hsü, 1933.)
FIG. 6.114. *Chabaudus*, anterior extremity, median view. (After Inglis & Ogden, 1965.)

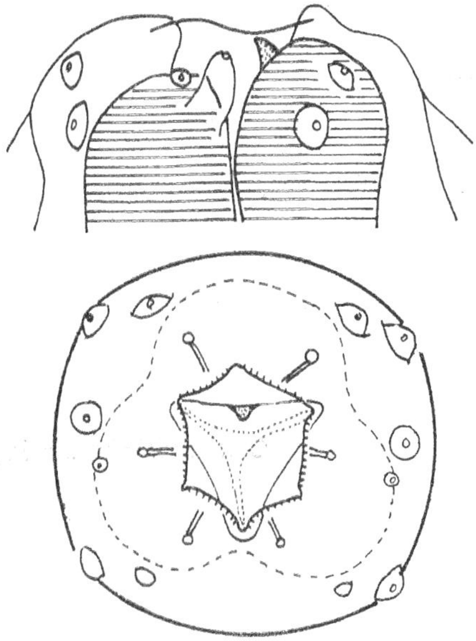

6.115

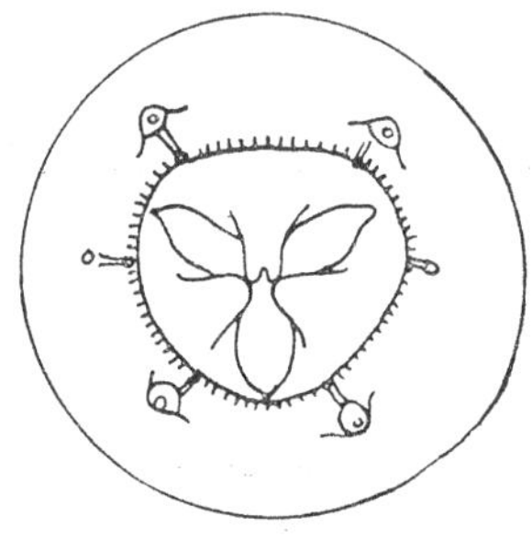

6.116

FIG. 6.115. *Chabaudus*, cephalic extremity, lateral view (with dorsal surface to right) and apical view. (After Inglis & Ogden, 1965.)

FIG. 6.116. *Gendria*, cephalic extremity, apical view. (After Inglis, 1967.)

6.117

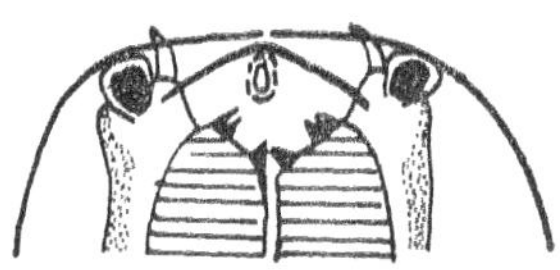

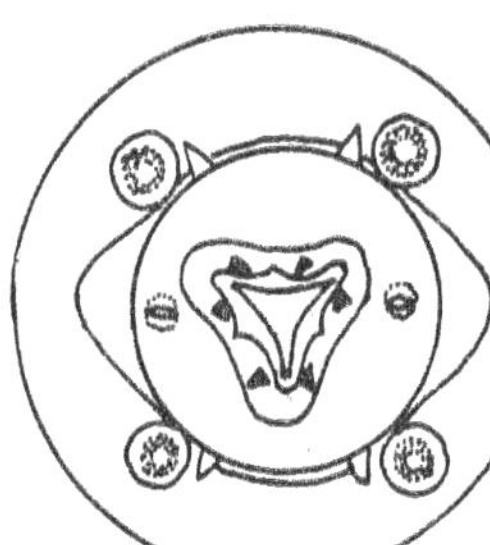

6.118

6.120

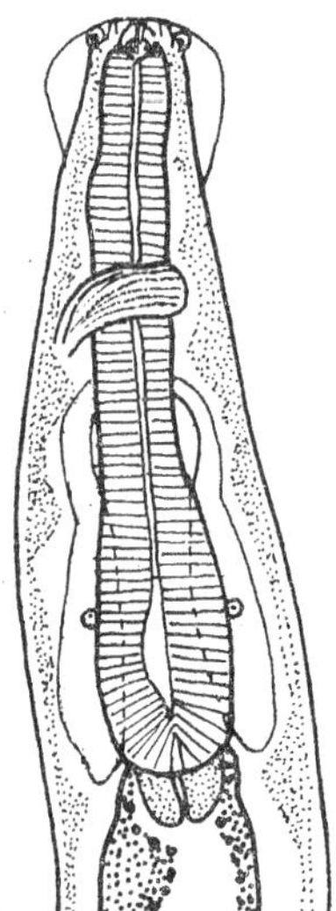

6.119

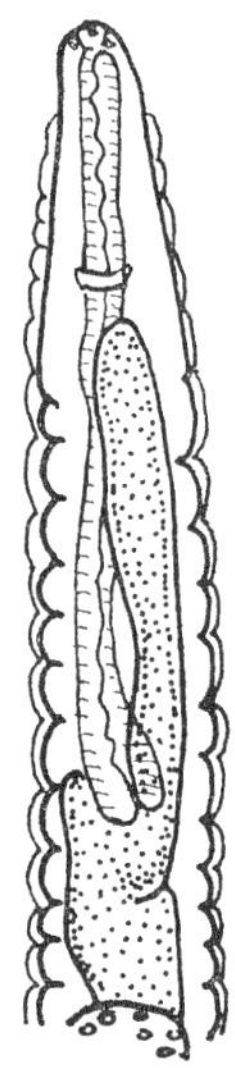

6.121

FIG. 6.117. *Buckleynema*, cephalic extremity, lateral view. (After Ali & Singh, 1954.)

FIG. 6.118. *Buckleynema*, cephalic extremity, apical view. (After Ali & Singh, 1954.)

FIG. 6.119. *Buckleynema*, anterior extremity. (After Ali & Singh, 1954.)

FIG. 6.120. *Omeia*, cephalic extremity, dorsal view. (After Chabaud & Brygoo, 1957.)

FIG. 6.121. *Omeia*, anterior extremity, dorsal view. (After Chabaud & Brygoo, 1957.)

20-(15) Caudal papillae not numerous.
Cuticle with longitudinal striations.
Parasites of amphibians of the New World.

Subulascaris Freitas & Dobbin, 1957

Subfamily **Omeiinae** Sobolev, 1949

One genus (Figs. 6.120, 6.121).

Omeia Hsü, 1933
(= *Harentinema* Chabaud & Brygoo, 1957)

Family *CHITWOODCHABAUDIIDAE* Puylaert, 1970

One genus (Figs. 6.122–6.125).

Chitwoodchabaudia Puylaert, 1970

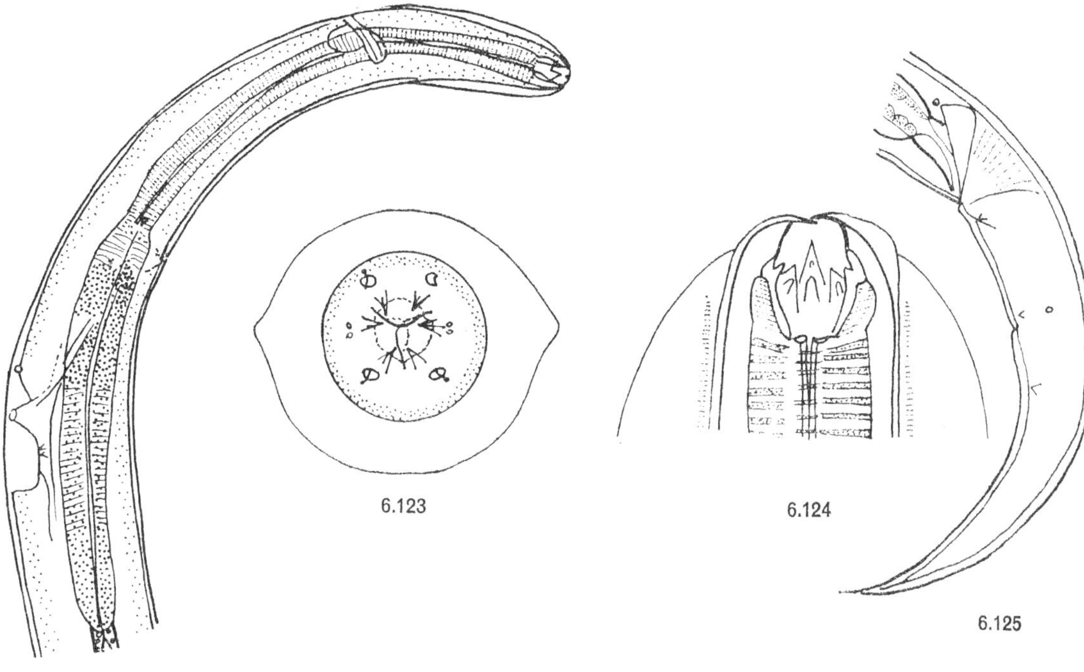

6.122 6.123 6.124 6.125

FIG. 6.122. *Chitwoodchabaudia*, anterior extremity, lateral view. (After Puylaert, 1970.)
FIG. 6.123. *Chitwoodchabaudia*, cephalic extremity. (After Puylaert, 1970.)
FIG. 6.124. *Chitwoodchabaudia*, cephalic extremity, ventral view. (After Puylaert, 1970.)
FIG. 6.125. *Chitwoodchabaudia*, caudal extremity male, lateral view. (After Puylaert, 1970.)

Family *CUCULLANIDAE* Cobbold, 1864

Key to subfamilies

1-(2) Oral opening triangular (Figs. 6.126, 6.128).
Oesophastome triangular.
Lumen of oesophagus without pronounced cuticularized armature.
Parasites of freshwater fish.

Campanarougetiinae

2-(1) Oral opening elongated dorsoventrally (Fig. 6.132).
Oesophastome with more or less pronounced bilateral symmetry.
Lumen of oesophagus reinforced with cuticularized rods.
Parasites of fish and chelonians.

Cucullaninae

Subfamily **Campanarougetiinae** Le Van Hoa & Pham-Ngoc-Khue, 1971

(= Chitwoodiidae Le Van Hoa & Pham-Ngoc-Khue, 1971;
= Oceanicucullanidae Le Van Hoa, 1973)

Key to genera

1-(2) Mouth with three reduced lips (Fig. 6.126).
Oesophastome with three small teeth.

Oceanicucullanus Schmidt & Kuntz, 1969 (Fig. 6.127)
(= *Chitwoodia* Le Van Hoa & Pham-Ngoc-Khue, 1971)

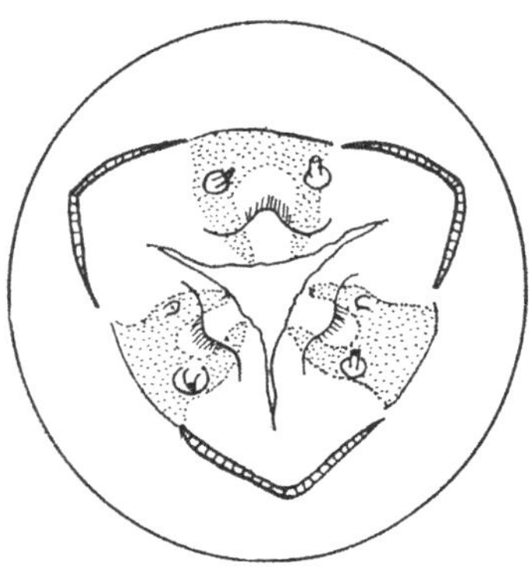

6.126

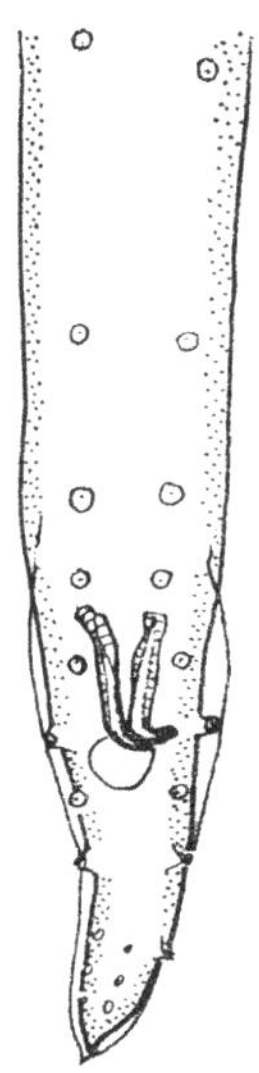

6.127

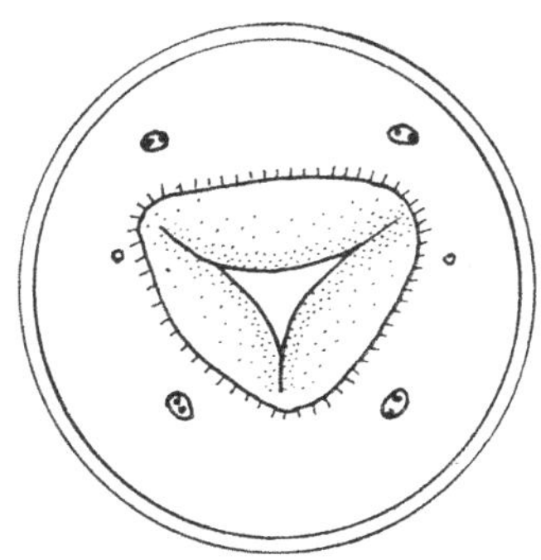

6.128

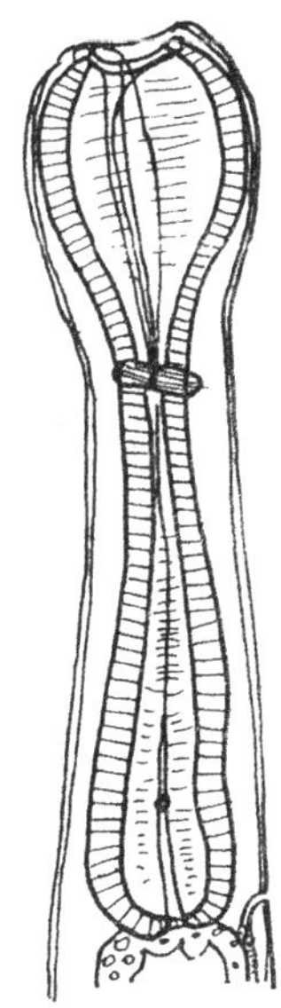

6.129

FIG. 6.126. *Oceanicucullanus*, cephalic extremity, apical view. (After Le Van Hoa & Pham-Ngoc-Khue, 1971.)
FIG. 6.127. *Oceanicucullanus*, caudal extremity male, ventral view. (After Le Van Hoa & Pham-Ngoc-Khue, 1971.)
FIG. 6.128. *Campanarougetia*, cephalic extremity, apical view. (After Le Van Hoa & Pham-Ngoc-Khue, 1967.)
FIG. 6.129. *Campanarougetia*, anterior extremity, lateral view. (After Le Van Hoa & Pham-Ngoc-Khue, 1967.)

2-(1) Mouth without lips (Fig. 6.128).
Oesophastome without teeth (Fig. 6.129).
Campanarougetia Le Van Hoa & Pham-Ngoc-Khue, 1967

Subfamily **Cucullaninae** (Cobbold, 1864)

Key to genera and subgenera

1-(2) Mouth markedly inclined dorsally (Fig. 6.130).
Cuticular lining of oesophastome consisting of complex set of thickened cuticularized pieces separated by sutures.
Cuticularized frame around oral opening absent.
Parasites of cyclostomes, selachians, acipenseriforms and salmoniforms.
Truttaedacnitis Petter, 1974
(= *Dacnitis* Tornquist, 1931 *nec* Dujardin, 1845;
= *Bulbodacnitis* Maggenti, 1971 *nec* Lane, 1916)

2-(1) Oral opening perpendicular to body axis or slightly inclined dorsally (Fig. 6.131).
Cuticularized plates of oesophastome few in number and separated by simple Y-shaped suture (Fig. 6.131).
Oral opening supported by peribuccal cuticularized frame to which are joined some cuticularized pieces inserted into adjacent oesophageal muscles (Fig. 6.132).

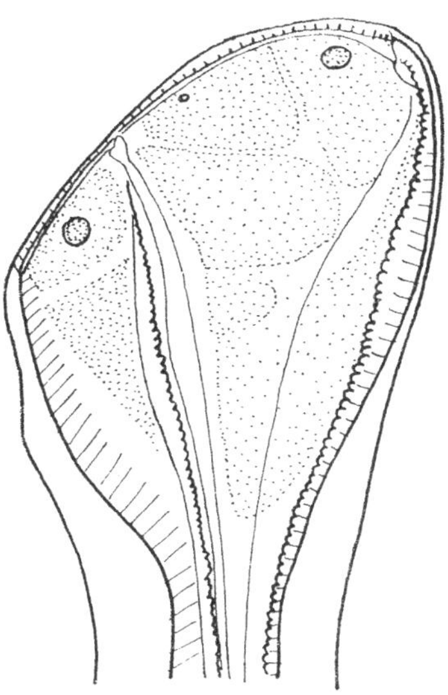

6.130

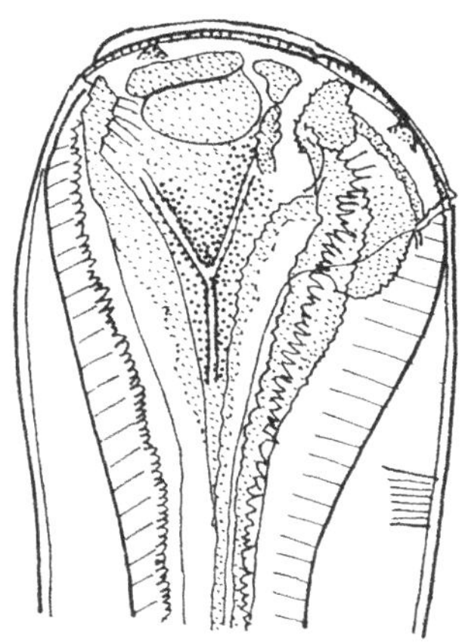

6.131

FIG. 6.130. *Truttaedacnitis*, cephalic extremity, lateral view. (After Berland, 1970.)
FIG. 6.131. *Cucullanus*, cephalic extremity, lateral view. (After Berland, 1970.)

3-(6) Intestinal caecum absent.
Usually medium or large worms with thin cuticle.
Parasites of teleosts (except salmoniforms) and sometimes chelonians.

4-(5) Male tail usually conical and without small terminal caudal alae.
Preanal papillae usually three in number.
Oesophastome well developed.
Cucullanus Mueller, 1777
(= *Indocucullanus* Ali, 1956)

5-(4) Male tail short and rounded with small caudal alae (Fig. 6.133), reminiscent of *Strongyluris*.
Numerous preanal papillae present (seven pairs).
Oesophastome reduced.
Neocucullanus Travassos, Artigas & Pereira, 1928

6-(3) Intestinal caecum present (Fig. 6.134).
Usually small worms with thick cuticle.
Parasites mainly of acanthopterygians (especially perciforms), less commonly of other fish and chelonians.
Dichelyne Jägerskiöld, 1902

7-(10) Eleven pairs of caudal papillae present of which three pairs are preanal.

8-(9) Preanal sucker present.
Dichelyne (*Cucullanellus*) (Tornquist, 1931 gen.)

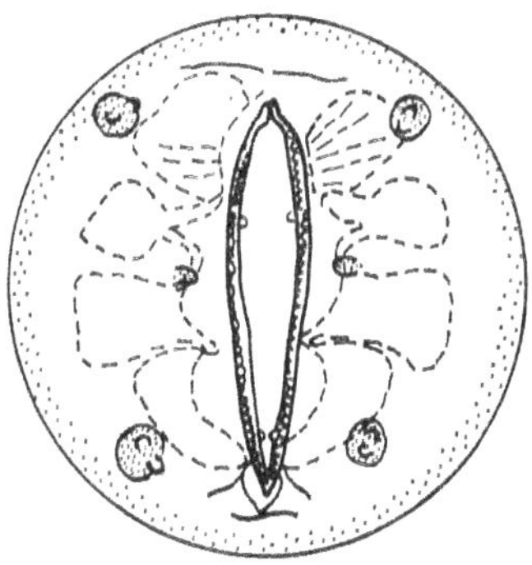

6.132

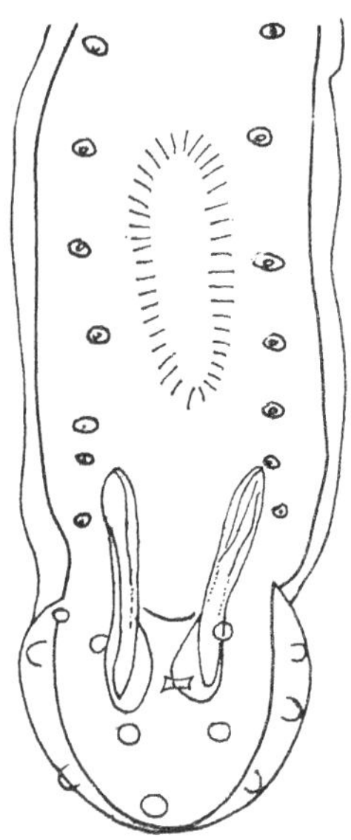

6.133

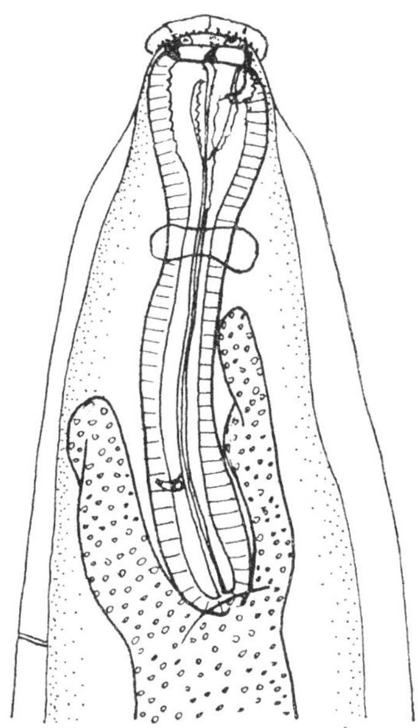

6.134

FIG. 6.132. *Cucullanus*, cephalic extremity, apical view. (After Petter, 1975.)
FIG. 6.133. *Neocucullanus*, caudal end male, ventral view. (After Travassos, Artigas & Pereira, 1928.)
FIG. 6.134. *Dichelyne*, anterior extremity, lateral view. (After Petter, 1975.)

9-(8) Preanal sucker absent.

Dichelyne (*Dichelyne*) Jägerskiöld, 1902

10-(7) More than 11 pairs of caudal papillae present of which more than three pairs are preanal.

Dichelyne (*Neocucullanellus*) (Yamaguti, 1941 gen.)

REFERENCES

AGRAWAL, V. 1965. Two new spirurid nematode parasites from freshwater fishes of India. *Proc. helminth. Soc. Wash.*, **32**, 246-249.

ARTHUR, J. R. & MARGOLIS, L. 1975. Revision of the genus *Haplonema* Ward and Magath, 1917 (Nematoda: Seuratoidea). *Can. J. Zool.*, **53**, 736-747.

BAIN, O. & PHILIPPON, B. 1969. Recherches sur des larves de nématodes Ascaridida trouvées chez *Simulium damnosum*. *Annls Parasit. hum. comp.*, **44**, 147-156.

BERLAND, B. 1970. On the morphology of the head in four species of the Cucullanidae (Nematoda). *Sarsia*, **43**, 15-63.

BOYCE, N. P. 1971. *Ezonema bicornis* gen. et sp.n. (Nematoda: Seuratidae) from freshwater fishes of Hokkaido, Japan. *J. Parasit.*, **57**, 1175-1179.

CAMPANA-ROUGET, Y. 1957. Parasites de poissons de mer ouest-africains récoltés par J. Cadenat. Nématodes (4ème note). Sur quelques espèces de Cucullanidae. Révision de la sous-famille. *Bull. Inst. fr. Afr. noire*, ser. A, **19**, 417-465.

CHABAUD, A. G. 1965. In: Grassé, P. P. *Traité de Zoologie. Nemathelminthes*, Tome IV, fasc. 3. Paris: Masson et Cie, pp. 732-1497.

——— 1973. *Skrjabinelazia* Sypliaxov, 1930 (= *Salobrella* Freitas, 1940), genre de transition entre cosmocercidés et spirurides. *Annls Parasit. hum. comp.*, **48**, 329-334.

———, CAMPANA-ROUGET, Y. & BRYGOO, E. R. 1960. Les nématodes Seuratoidea. *Annls Parasit. hum. comp.*, **35**, 316-346.

———, CABALLERO, R. G. & BRYGOO, E. R. 1965. Affinités entre les genres *Skrjabinelazia* (Ascaridida, Seuratoidea) et *Maxvachonia* (Ascaridida, Cosmocercoidea). *Bull. Mus. natn. Hist. nat.*, 2e sér., **36**, 844-848.

CHAKRAVARTY, G. K. & MAJUMDAR, G. 1960. On two new parasites of the nematode families Ascaridae and Cucullanidae. *Proc. zool. Soc. Calcutta*, **12**, 115-120.

CHITWOOD, M. B. 1960. A new spiruroid nematode, *Rabbium caballeroi*, from the stomach of *Leiocephalus carinatus*, from the Bahama Islands. *Libro Homenaje al Dr. Eduardo Caballero y Caballero, Jubileo 1930-1960*, pp. 471-473.

GIBSON, D. I. 1972. Contributions to the life-histories and development of *Cucullanus minutus* Rudolphi, 1819 and *C. heterochrous* Rudolphi, 1802 (Nematoda: Ascaridida). *Bull. Br. Mus. nat. Hist., Zool.*, **22**, 153-170.

GUERRERO, R. 1971. Helmintos de la hacienda "El Limon", D. F. Venezuela—Nematodes de Vertebrados. I. *Mems Soc. Cienc. nat. La Salle*, **31**, 175-230.

INGLIS, W. G. 1967. The relationship of the nematode superfamily Seuratoidea. *J. Helminth.*, **41**, 115-136.

——— & OGDEN, C. 1965. *Chabaudus chabaudi* gen. et sp. nov. from a freshwater fish in Sierra Leone (Nematoda: Seuratoidea). *Revue Zool. Bot. afr.*, **71**, 171-176.

IVASCHKIN, V. M. & KHROMOVA, L. A. 1976. *Principles of Nematology*, edited by K. I. Skrjabin. Vol. XXVII. *Cucullanata and Gnathostomata of animals and man, and the diseases caused by them.* Moscow: Izdatel'stvo "Nauka". 436 pp. (In Russian).

KALYANKAR, S. D. 1971. On a remarkable new nematode *Pseudocucullanus tachysuri* gen. et sp. nov. (Cucullanidae: Cucullaninae) from marine fish of India and *Neocucullanus yamaguti* n.sp. *Riv. Parassit.*, **32**, 37-42.

KHERA, S. 1953. *Monovaria rhinolophensia* n.g., n.sp. (subfamily Seuratinae Hall, 1916: family Cucullanidae Cobbold, 1964: Nematoda) from the bat, *Rhinolophus affinis*. *Indian J. Helminth.*, **5**, 109-114.

KHROMOVA, L. A. 1975. Development of *Dacnitis sphaerocephalus caspicus* (Nematoda: Cucullanidae). *Zool. Zh.*, **54**, 449-452. (In Russian).

KLOSS, G. R. 1966. Helmintos parasitos de espécies simpátricas de *Astyanax* (Pisces, Characidae). *Papeis Dep. Zool. S Paulo*, **18**, 189-219.

LE VAN HOA. 1974. Mise en synonymie du genre *Chitwoodia* Le Van Hoa *et al.*, 1971 avec *Oceanicucullanus* Schmidt *et al.*, 1969. *Annls Parasit. hum. comp.*, **48**, 699-702.

——— & PHAM-NGOC-KHUE. 1967a. Morphologie et cycle évolutif de *Cucullanus chabaudi* n.sp., parasite de poissons *Pangasius pangasius* H. B. (*P. buchanani*) du Sud-Viet-Nam. Note Préliminaire. *Bull. Soc. Path. exot.*, **60**, 315-318.

———, ——— 1967b. Relation entre les Seuratoidea et les Cucullanidae par l'intermédiaire d'un nouveau nématode *Campanarougetia campanarougetae* n.g., n.sp., parasite des anguilles du Sud-Viet-Nam. *Bull. Soc. Path. exot.*, **60**, 393-398.

———, ——— 1971. Relation taxonomique entre *Gendria* Baylis, 1930 et *Cucullanus* Mueller, 1777 (Remarques sur la superfamille Seuratoidea Chabaud, Campana-Rouget et Brygoo, 1959). *Annls Parasit. hum. comp.*, **46**, 595-604.

MAGGENTI, A. R. 1971. A review of the family Cucullanidae Cobbold, 1864 and the genus *Bulbodacnitis* Lane, 1916 with a description of *Bulbodacnitis ampullastoma* sp.n. (Nematoda: Cucullanidae) from *Salmo gairdnerii* Richardson. *Proc. helminth. Soc. Wash.*, **38**, 80-85.

MAWSON, P. M. 1960. *Seuratinema* Johnston et Mawson 1941, synonyme de *Skrjabinema* Gnedina 1933. *Annls Parasit. hum. comp.*, **35**, 430-431.

——— 1968. *Inglisonema typos* gen. nov. sp. nov. (Nematoda: Seuratoidea: Inglisonematinae, subfam. nov.) from an Australian bird. *Parasitology*, **58**, 71-73.

MORAVEC, F. 1967. On the systematics of the genus *Paraquimperia* Baylis, 1934 (Nematoda: Quimperiidae). *Helminthologia*, Year 1966, **7**, 353-364.

——— 1974. Some remarks on the development of *Paraquimperia tenerrima* Linstow, 1879 (Nematoda: Quimperiidae). *Scripta Fac. Sci. Nat. Ujep. Brunensis, Biol.*, **4**, 135-142.

PETTER, A. J. 1974. Deux nouvelles espèces de Cucullanidae parasites de poissons en Guyane. *Bull. Mus. natn. Hist. nat., Paris*, Sér. 3, No. 255, Zool. **177**, 1459-1467.

PUYLAERT, F. A. 1970. Description d'*Auchmeronema thysi* gen.sp.n. parasite d'*Auchenoglanis punctatus* Blgr. (Pisces) et d'*Auchmeronema williamsi* sp.n. parasite de *Petropedetes natator* Blgr. (Amphibia). (Subulascarididae-Nematoda-Vermes). *Revue Zool. Bot. afr.*, **81**, 82-94.

——— 1970. Description de *Chitwoodchabaudia skryabini* g.n., sp.n. (Chitwoodchabaudiidae fam.nov.), parasite de *Xenopus laevis victorianus* Ahl. (Cosmocercoidea-Nematoda-Vermes). *Revue Zool. Bot. afr.*, **81**, 369-382.

——— 1977. *Personal communication.*

PYBUS, M. J., ANDERSON, R. C. & UHAZY, L. S. 1978. Redescription of *Truttaedacnitis stelmioides* (Vessichelli, 1910) (Nematoda: Cucullanidae) from brook lamprey (*Lampetra lamottenii*) (Lesueur, 1827). *Proc. helminth. Soc. Wash.* (In press).

———, UHAZY, L. S. & ANDERSON, R. C. 1978. Life cycle of *Truttaedacnitis stelmioides* (Vessichelli, 1910) (Nematoda: Cucullanidae) in American brook lamprey (*Lampetra lamottenii*) (Lesueur, 1827). *Can. J. Zool.*, **56**, 1420-1429.

QUENTIN, J. C. 1970. Sur le cycle évolutif de *Seuratum cadarachense* Desportes, 1947 et ses affinités avec ceux des nématodes subulures (Ascaridida) et rictulaires (Spirurida). *Annls Parasit. hum. comp.*, **45**, 605-628.

——— 1971. Sur les modalités d'évolution chez quelques lignées d'helminthes de rongeurs Muroidea. Thèse Doctorat d'Etat, Fac. Sci. Orsay. *Cah. O.R.S.T.O.M., sér. Ent. méd. Parasitol.*, **9**, 103-176.

RASHEED, S. 1968. The nematodes of the genus *Cucullanus* Mueller, 1777 from the marine fish of the Karachi coast. *Anais. Esc. nac. Cienc. biol., Mex.*, **15**, 23-59.

SCHMIDT, G. D. & KUNTZ, R. E. 1969. Nematode parasites of Oceanica. V. Four new species from fishes of Palawan, P.I., with a proposal for *Oceanicucullanus* gen. nov. *Parasitology*, **59**, 389-396.

——— ——— 1971. Nematode parasites of Oceanica. XI. *Madelinema angelae* gen. et sp.n., and *Inglisonema mawsonae* sp.n. (Heterakidae: Inglisonematidae) from birds. *J. Parasit.*, **57**, 479-484.

SKRJABIN, K. I. & IVASCHKIN, V. M. 1968. [On the elimination of the superfamily Seuratoidea Chabaud, Campana-Rouget et Brygoo, 1959, from the nematode classification.] *Izv. Akad. Nauk. SSSR, Ser. Biol.*, **6**, 789-792. (In Russian).

TEIXEIRA de FREITAS, J. F. & DOBBIN, J. E., Jr. 1957. Novo nematódeo parasito de *Rana palmipes* Spix: *Subulascaris falcaustriformis* g.n., sp.n. *Revta bras. Biol.*, **17**, 245-248.

TRAVASSOS, L., ARTIGAS, P. & PEREIRA, C. 1928. Fauna helminthologica dos peixes de ague doce do **Brasil**. *Archos Inst. biol., S Paulo*, **1,** 5-68.

VASSILIADES, G. 1972. Nématodes parasites des poissons de la région de Sangalkam (Sénégal). *Bull. I.F.A.N., Sér. A*, **34,** 529-533.

VICENTE, J. J. 1965. Sôbre um nôvo gênero de Scheidernematidae Freitas, 1956 (Nematoda: Ascaridoidea). *Mems Inst. Oswaldo Cruz*, **63,** 121-125.

HETERAKOIDEA

This superfamily was reviewed and revised comprehensively by Inglis (1967) after a restudy of the type specimens of many genera (Inglis, 1957, 1958). The conclusions of this author are adopted and two new genera discovered since his study are added, namely *Hatterianema* Chabaud & Dollfus, 1966, from the relict reptile *Rhynchocephalus*, which corresponds almost perfectly to the primitive species hypothesized by Inglis, and *Pseudostrongyluris* Guerrero, 1971 which resembles *Cruzia* in that the pharynx is armed with teeth.

Inglis (1967) placed the following into synonymy: *Preterakis* Freitas, 1956; *Heterakoides* Freitas, 1956; *Pareterakis* Freitas, 1956; *Pseudaspidoderoides* Freitas, 1956; *Pseudaspidoderina* Freitas, 1956; *Raillietakis* Freitas, 1956; *Sexansodera* Skrjabin & Schikhobalova, 1947; *Bellaplectana* Skrjabin, Schikhobalova & Lagodovskaja, 1961; *Ganguleterakis* Lane, 1914; *Cheloniheterakis* Yamaguti, 1961; *Spinaspidodera* Skrjabin & Schikhobalova, 1947.

We propose to place into synonymy a number of more recently proposed genera. *Parapseudaspidodera* Johnson, 1967 and *Pseudaspidoderella* Ali & Deshmukh, 1969 are considered as synonyms of *Pseudaspidodera*. As in the acuarids, the unnecessary proliferation of genera based on the complexity of the cordons leads inevitably to an unuseable classification and Johnson himself (1974) proposed reducing genera to the level of subgenera. Moreover, the labial teeth by which Ali & Deshmukh differentiated their genus seem to correspond to structures Inglis (1957) described in *Pseudaspidodera*.

Inglisakis Freitas, Vicente & Santos, 1969 is separated from *Heterakis* in the same way as *Preterakis* Freitas, 1956 is separated from *Africana*. Thus, in both cases, the caudal alae seem to join in front of the preanal sucker. Inglis has shown, however, that in *Preterakis* this character is actually a fixation artefact and that this genus is a synonym of *Africana*. We believe that the same is true in *Inglisakis* and that this genus is a synonym of *Heterakis*.

Proencaia Gomes & Pereira, 1970 is very similar to *Lauroia*. The absence of a cuticularized ring around the preanal sucker does not seem to us to have much generic value since *Lauroia* is itself characterized by a more or less reduced sucker. *Paraheterakis* Nama & Jain, 1974 is a *Subulura* that the authors mistook for a heterakid. *Cotylascaris* Sprent, 1971 was created by error and placed into the synonymy of *Ascaridia* the following year (Sprent, 1972). *Cometeterakis* Cruz & Ching, 1975 is treated herein as a synonym of *Meteterakis* as it is distinguished only by the marked difference in the length of its spicules.

A monograph on the species and a corresponding bibliography have been published by Mozgovoi (1953) and Skrjabin *et al.* (1961).

HETERAKOIDEA

Key to families

1-(4) Oesophagus with tri-valved posterior bulb (Fig. 6.154).

2-(3) Lips rounded (Figs. 6.139, 6.145, 6.149, 6.150, 6.153, 6.156, 6.159), not connected by lateral lobes (Figs. 6.151, 6.157).
Cordons when present, poorly developed and not remaining the same width throughout their length (Fig. 6.156).
Oesophagus relatively short and stout (Fig. 6.154).
Cosmopolitan.

Heterakidae

3-(2) Lips square (Figs. 6.160, 6.164, 6.165, 6.167), connected by lateral lobes (Fig. 6.161).
Cuticle at anterior end of body thickened to form cephalic cap.

Oesophagus long and narrow with small posterior oesophageal bulb (Fig. 6.162).
Cordons, when present, of the same width throughout their lengths (Figs. 6.160, 6.164).
South American.

Aspidoderidae

4-(1) Oesophagus club-shaped, without posterior bulb (Fig. 6.170).
Lips large and stout.
Distinct anterior cuticular flange absent (Fig. 6.169).
Cosmopolitan.

Ascaridiidae

Family *HETERAKIDAE* Railliet & Henry, 1912

Key to subfamilies

1-(2) Lips off-set from body (Figs. 6.139, 6.145).
Caudal alae, when present, either not supported by papillae (Fig. 6.140) or supported by many short, stout papillae (Figs. 6.142, 6.144).
Parasites of amphibians and reptiles.

Spinicaudinae

2-(1) Lips not off-set from body.
Caudal alae present, supported by three or four stout papillae (Fig. 6.148) or by long, thin papillae (Fig. 6.155).

3-(4) Head with three rounded lips, without interlabia (Figs. 6.149, 6.150).
Caudal alae supported by three or four pairs of large fleshy papillae (Fig. 6.148).
Numerous small sessile papillae present on tail (Figs. 6.147, 6.148).
Parasites of amphibians, reptiles and mammals in the Orient.

Meteterakinae

4-(3) Head with interlabia or their homologues (Fig. 6.156).
Caudal alae broad, supported by long, narrow papillae (Figs. 6.155, 6.158).
Few sessile papillae present on tail.
Parasites of birds and, rarely, mammals.

Heterakinae

Subfamily **Spinicaudinae** Travassos, 1920

Key to genera

1-(6) Male tail pointed and sucker directed ventrally (Fig. 6.135).
Caudal alae absent or not supported by papillae (Fig. 6.136).

2-(5) Male tail without caudal alae (Figs. 6.136, 6.138).

3-(4) Cuticular flange projecting beyond main mass of lips absent (Fig. 6.137).
Many (24 to 28) small sessile papillae present on tail (Fig. 6.138).
Parasites of rhynchocephalid reptiles (New Zealand).

Hatterianema Chabaud & Dollfus, 1966

4-(3) Anterior cuticular flange projecting slightly beyond the main mass of the lips (Fig. 6.139).
Few (less than 20) sessile papillae present on tail (Figs. 6.135, 6.136).
Parasites of tropical and subtropical reptiles.

Spinicauda Travassos, 1920

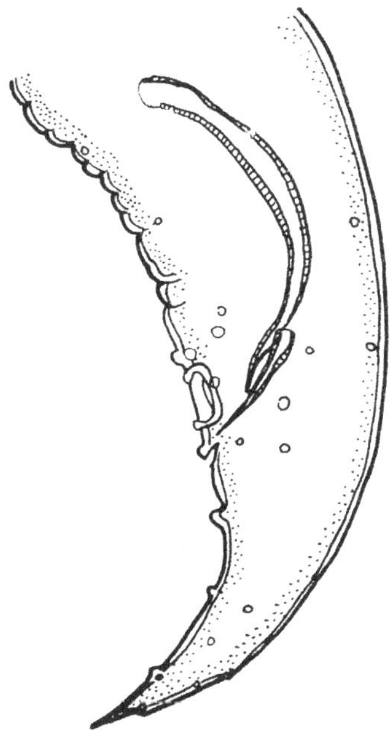

6.135

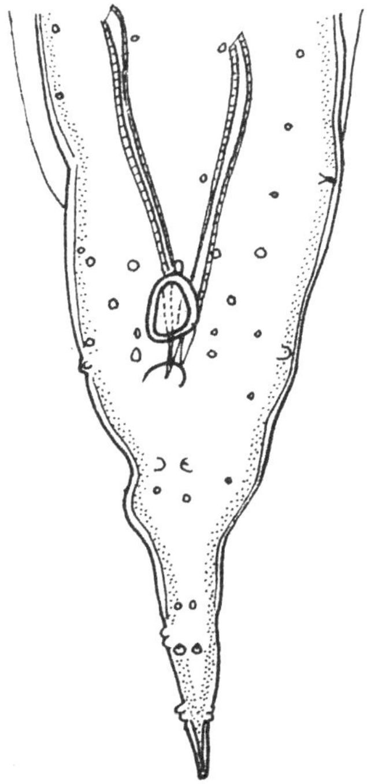

6.136

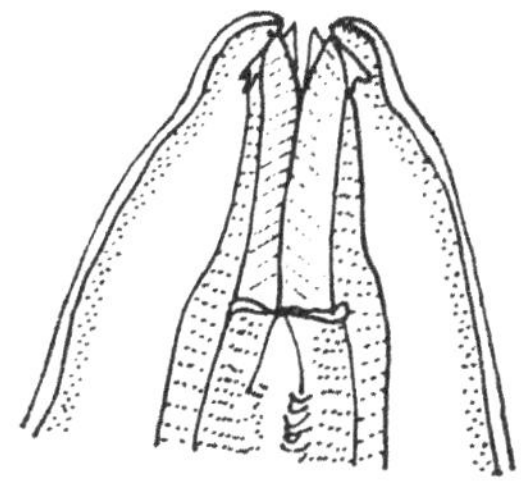

6.137

FIG. 6.135. *Spinicauda*, male tail, lateral view. (After Chabaud & Brygoo, 1960.)
FIG. 6.136. *Spinicauda*, male tail, ventral view. (After Chabaud & Brygoo, 1960.)
FIG. 6.137. *Hatterianema*, anterior extremity, dorsal view. (After Chabaud & Dollfus, 1966.)

5-(2) Male tail with caudal alae not supported by papillae (Fig. 6.140).
Parasites of African and Indian amphibians and reptiles.

Africana Travassos, 1920
(= *Preterakis* Freitas, 1956)

6-(1) Mail tail obliquely truncate, sucker directed posteriorly.
Caudal alae supported by stout papillae (Figs. 6.141-6.144).

7-(10) Rows of cuticularized teeth not present in pharyngeal canal.

8-(9) Male tail with long terminal spike (Figs. 6.141, 6.142).
Anterior flange not major part of lip.
Parasites of reptiles.

Moaciria Freitas, 1956
(= *Bellaplectana* Skrjabin, Schikhobalova & Lagodovskaja, 1961)

9-(8) Male tail without long terminal spike (Figs. 6.143, 6.144).
Anterior cuticular flange major part of lip (Fig. 6.145).
Parasites of reptiles, rarely amphibians.

Strongyluris Mueller, 1894

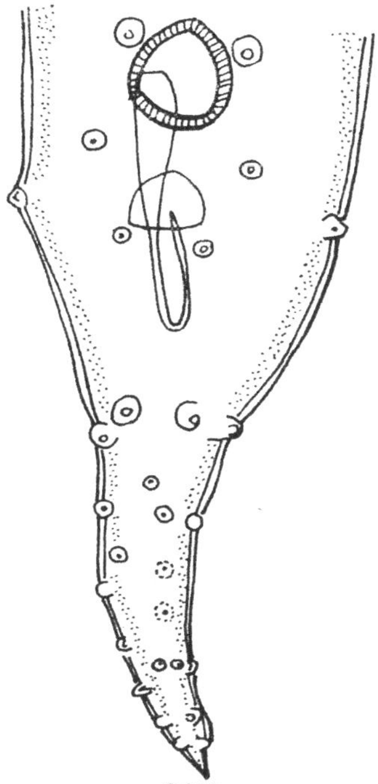

6.138

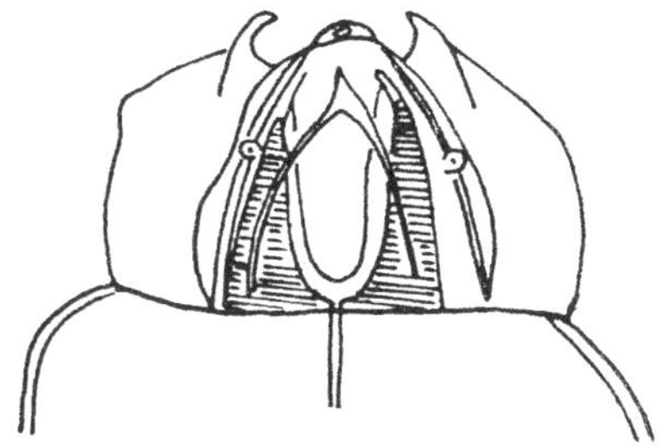

6.139

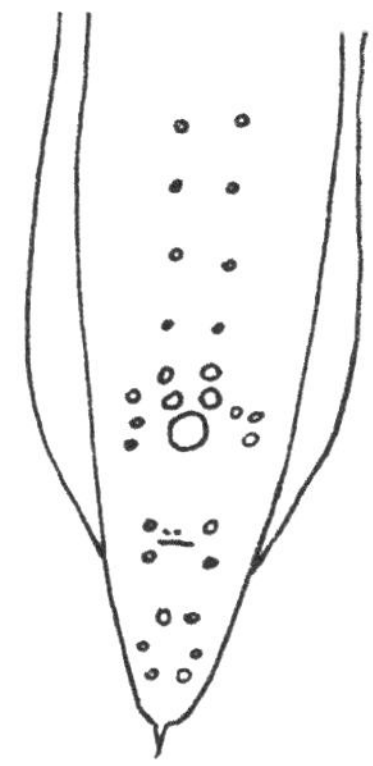

6.140

FIG. 6.138. *Hatterianema*, male tail, ventral view. (After Chabaud & Dollfus, 1966.)
FIG. 6.139. *Spinicauda*, anterior extremity, dorsal view. (After Inglis, 1957.)
FIG. 6.140. *Africana*, male tail, ventral view. (After Gendre, 1925.)

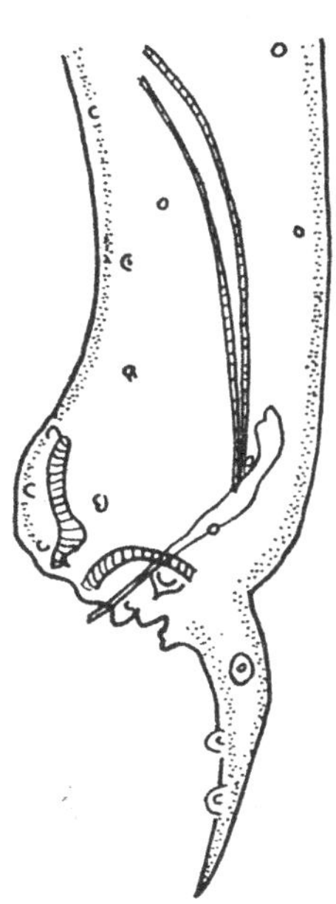

6.141

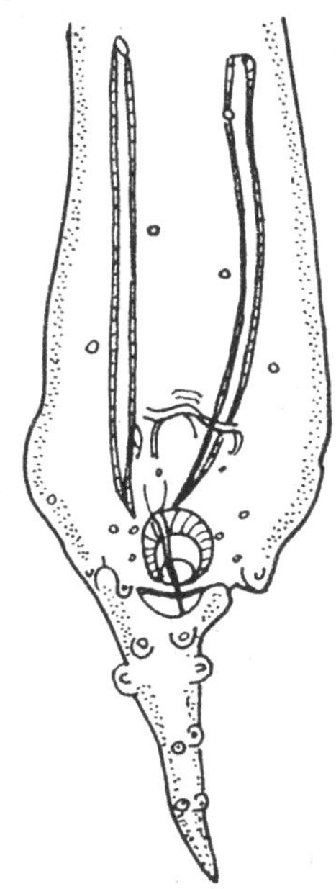

6.142

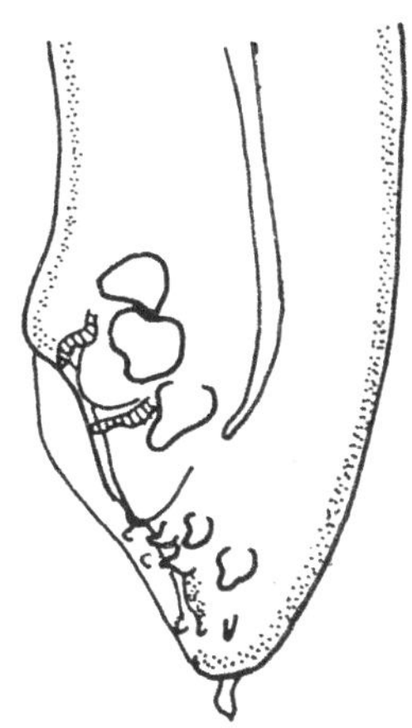

6.143

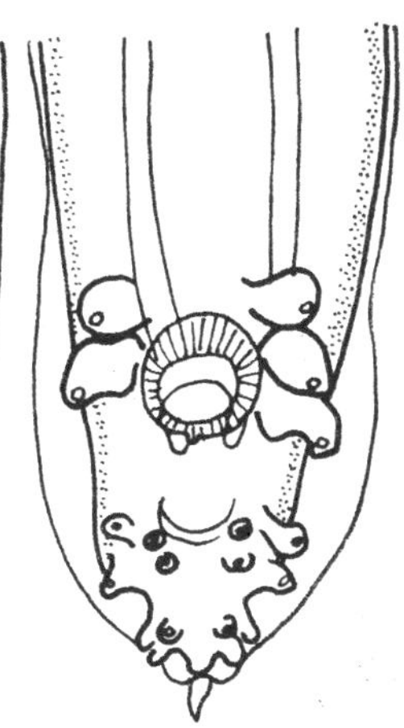

6.144

FIG. 6.141. *Moaciria*, male tail, lateral view. (After Chabaud & Brygoo, 1960.)
FIG. 6.142. *Moaciria*, male tail, ventral view. (After Chabaud & Brygoo, 1960.)
FIG. 6.143. *Strongyluris*, male tail, lateral view. (After Chabaud & Brygoo, 1960.)
FIG. 6.144. *Strongyluris*, male tail, ventral view. (After Chabaud & Brygoo, 1960.)

10-(7) Three rows of cuticularized teeth in pharyngeal canal (Fig. 6.146).
Parasites of South American reptiles.

Pseudostrongyluris Guerrero, 1971

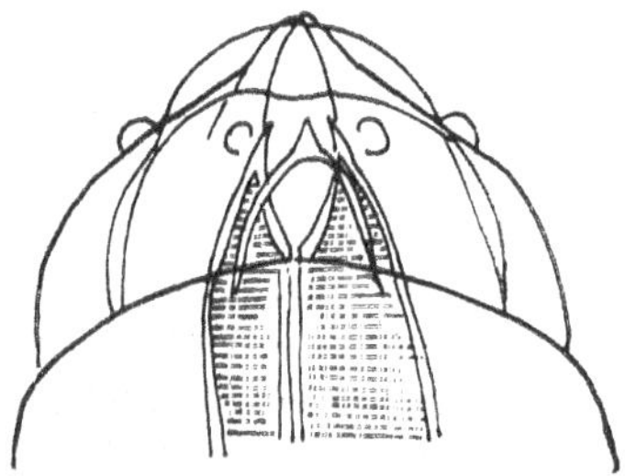

6.145

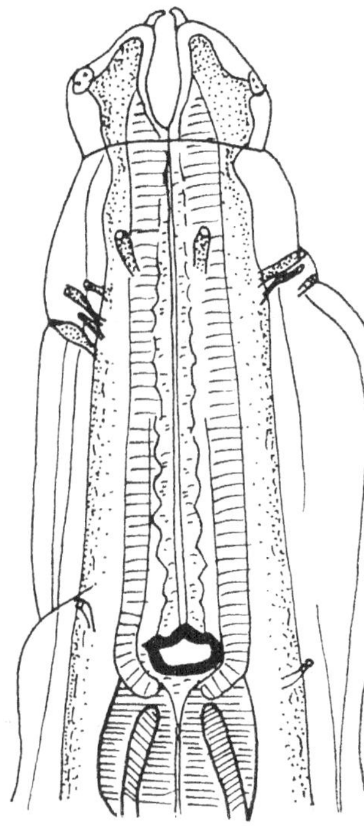

6.146

FIG. 6.145. *Strongyluris*, anterior extremity, dorsal view. (After Inglis, 1957.)
FIG. 6.146. *Pseudostrongyluris*, anterior extremity, ventral view. (After Guerrero, 1971.)

Subfamily **Meteterakinae** Inglis, 1967

Key to genera

1-(2) Head without cordons (Fig. 6.149).
Spicules not complex (Fig. 6.147).
Parasites of oriental amphibians and reptiles.

Meteterakis Karve, 1930
(= *Heterakoides* Freitas, 1956;
= *Pareterakis* Freitas, 1956;
= *Cometeterakis* Cruz & Ching, 1975)

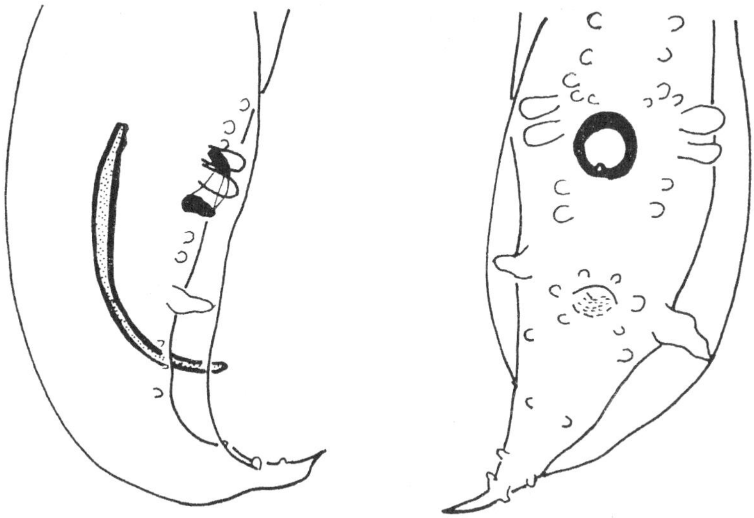

6.147 6.148

FIG. 6.147. *Meteterakis*, male tail, lateral view. (After Inglis, 1958.)
FIG. 6.148. *Meteterakis*, male tail, ventral view. (After Inglis, 1958.)

2-(1) Three straight simple cordons present, one arising from each interlabial space (Figs. 6.150, 6.151).
Spicules extremely complex (Fig. 6.152).
Parasites of *Hystrix* in India.

Gireterakis Lane, 1917

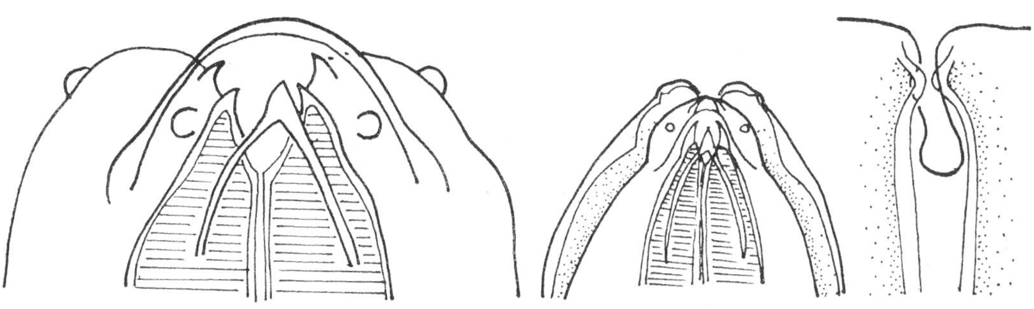

6.149 6.150 6.151

FIG. 6.149. *Meteterakis*, anterior extremity, dorsal view. (After Inglis, 1957.)
FIG. 6.150. *Gireterakis*, anterior extremity, dorsal view. (After Inglis, 1957.)
FIG. 6.151. *Gireterakis*, interlabia, showing formation of cordons. (After Inglis, 1957.)

Subfamily **Heterakinae** Railliet & Henry, 1912

Key to genera

1-(2) Head without cordons or labial grooves (Fig. 6.153).
Parasites mainly of ground-feeding birds (mainly Galliformes), one species in mammals.
Heterakis Dujardin, 1945 (Figs. 6.154, 6.155)
(= *Raillietakis* Freitas, 1956;
= *Ganguleterakis* Lane, 1914;
= *Inglisakis* Freitas, Vicente & Santos, 1969)

2-(1) Head with cordons or labial grooves (Figs. 6.156, 6.157, 6.159).

6.153

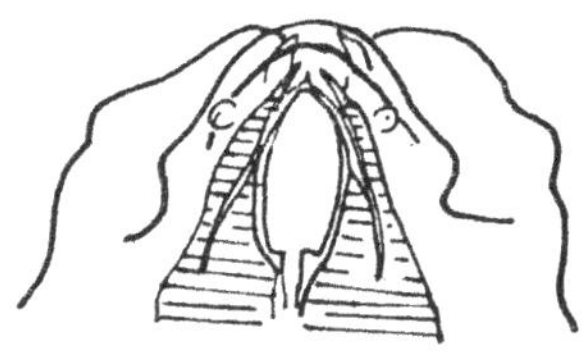

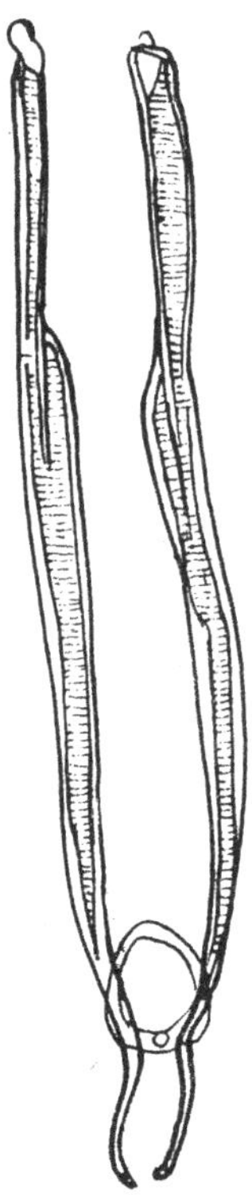

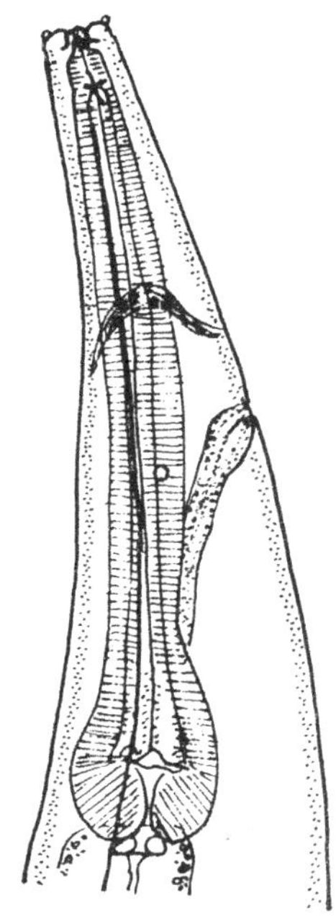

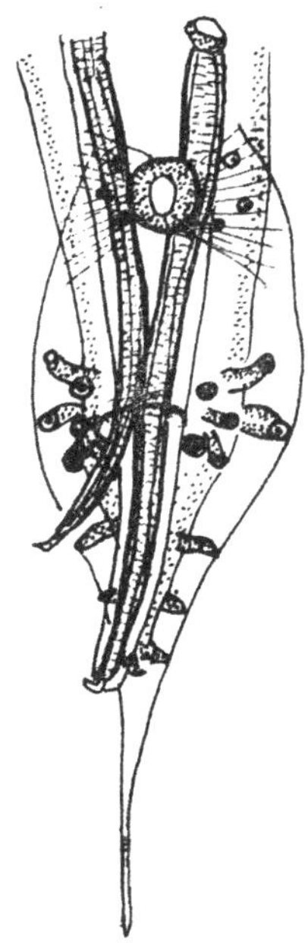

6.152 6.154 6.155

FIG. 6.152. *Gireterakis*, spicules, ventral view. (After Lane, 1917.)
FIG. 6.153. *Heterakis*, anterior extremity, dorsal view. (After Inglis, 1957.)
FIG. 6.154. *Heterakis*, anterior extremity, lateral view. (After Skrjabin & Schikhobalova, 1949.)
FIG. 6.155. *Heterakis*, male tail, ventral view. (After Skrjabin & Schikhobalova, 1949.)

3-(4) Spicules dissimilar, left with broad alae, right slim and needle-like (Fig. 6.158).
Parasites of phasianid birds in India and associated regions.
Pseudaspidodera Baylis & Daubney, 1922 (Figs. 6.156, 6.157)
(= *Pseudaspidoderoides* Freitas, 1956;
= *Pseudaspidoderina* Freitas, 1956;
= *Spinaspidodera* Skrjabin & Schikhobalova, 1947;
= *Parapseudaspidodera* Johnson, 1967;
= *Pseudaspidoderella* Ali & Deshmukh, 1969)

4-(3) Spicules similar, never alate.
Parasites of birds (mainly tinamous) in South America.
Odonterakis Skrjabin & Schikhobalova, 1947 (Fig. 6.159)

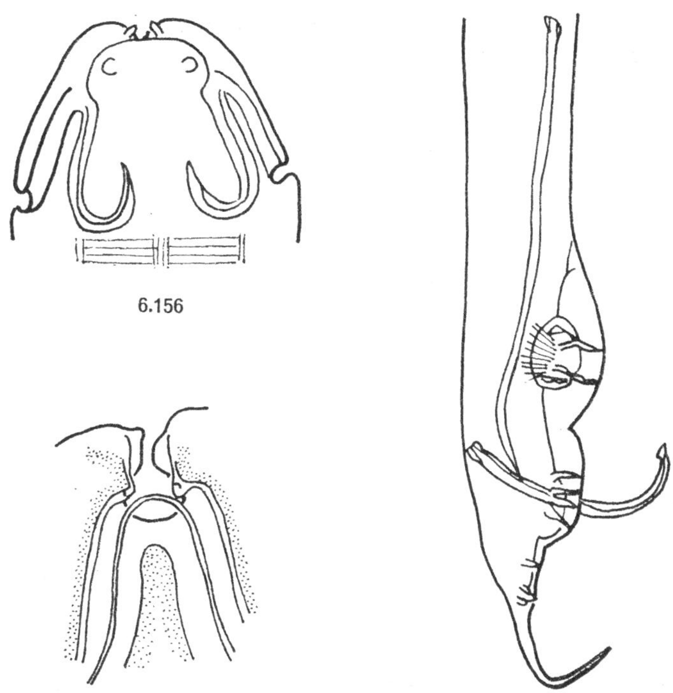

FIG. 6.156. *Pseudaspidodera*, anterior extremity, dorsal view of outer surface. (After Inglis, 1957.)
FIG. 6.157. *Pseudaspidodera*, interlabia showing formation of cordons. (After Inglis, 1957.)
FIG. 6.158. *Pseudaspidodera*, male tail, lateral view. (After Baylis & Daubney, 1922.)

Family *ASPIDODERIDAE* Skrjabin & Schikhobalova, 1947

Key to subfamilies

1-(2) Cephalic cordons present (Figs. 6.160, 6.164).
Lateral lobes of lips complex (Fig. 6.161).
Aspidoderinae

2-(1) Cephalic cordons lacking (Figs. 6.165, 6.167).
Lateral lobes of lips simple.
Lauroiinae

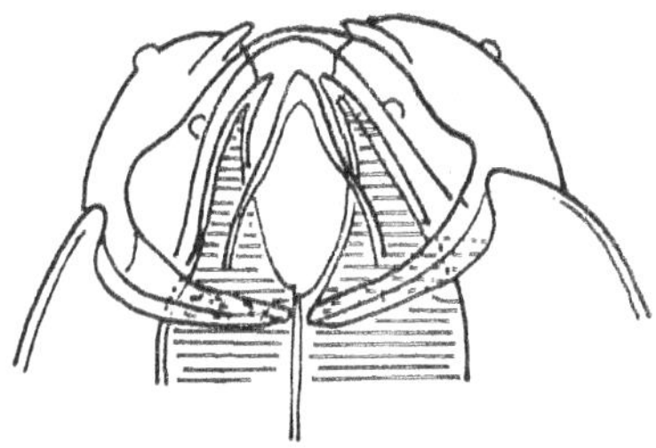

6.159

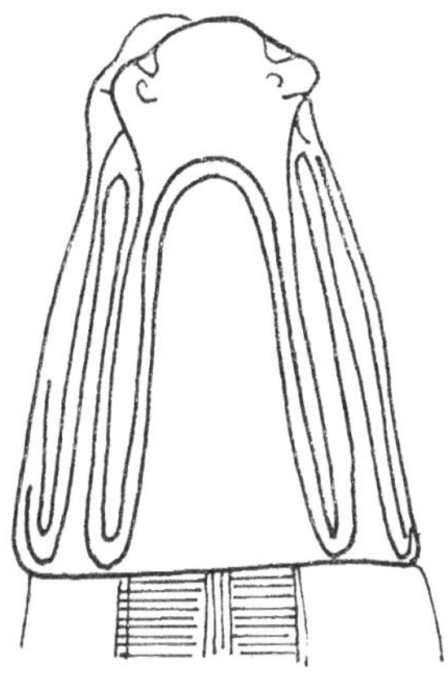

6.160

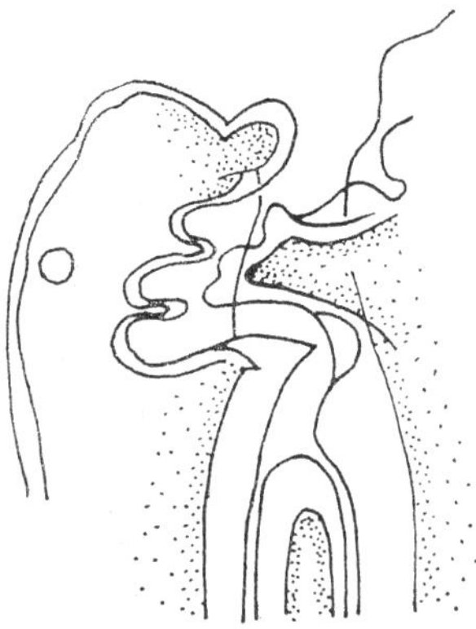

6.161

FIG. 6.159. *Odonterakis*, anterior extremity, dorsal view. (After Inglis, 1957.)
FIG. 6.160. *Aspidodera*, anterior extremity, dorsal view. (After Inglis, 1957.)
FIG. 6.161. *Aspidodera*, interlabia showing formation of cordons. (After Inglis, 1957.)

Subfamily **Aspidoderinae** Skrjabin & Schikhobalova, 1947

Key to genera

1-(2) Cephalic cordons recurrent and anastomosing (Fig. 6.160).
Parasites of marsupials, *Xenarthra* and rodents in South America and related areas.
Aspidodera Railliet & Henry, 1912 (Figs. 6.161-6.163)
(= *Sexansodera* Skrjabin & Schikhobalova, 1947)

2-(1) Cephalic cordons not anastomosing (Fig. 6.164).
Parasites of *Xenarthra.*
Ansiruptodera Skrjabin & Schikhobalova, 1947

Subfamily **Lauroiinae** Skrjabin & Schikhobalova, 1951

Key to genera

1-(2) Cephalic cap modified into three "plates" slightly undercut posteriorly (Fig. 6.165).
Posterior end of male simple, with relatively few papillae and reduced pre-anal sucker (Fig. 6.166).
Parasites of *Xenarthra* and rodents in South America.
Lauroia Proenca, 1938
(= *Proencaia* Gomes & Pereira, 1970)

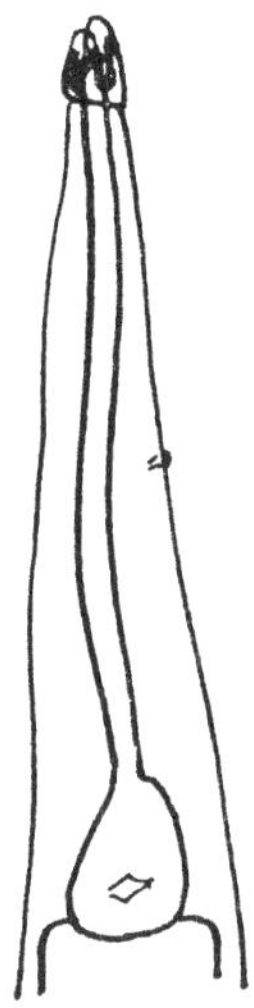

6.162

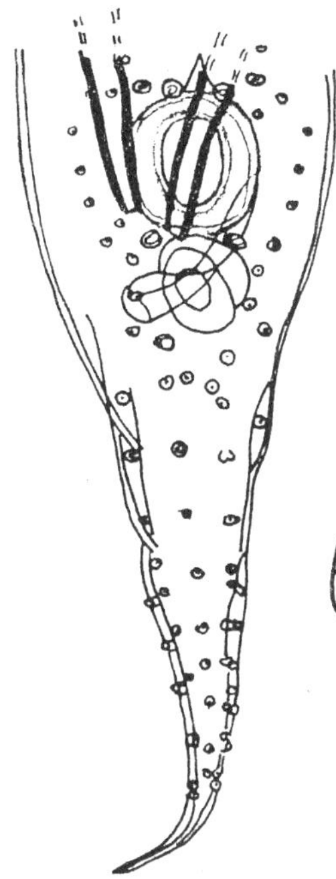

6.163

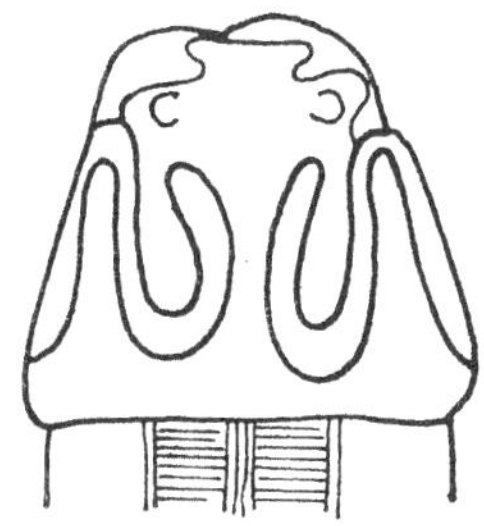

6.164

FIG. 6.162. *Aspidodera*, anterior extremity, lateral view. (After Freitas & Lent, 1937.)
FIG. 6.163. *Aspidodera*, male tail, ventral view. (After Travassos, 1913.)
FIG. 6.164. *Ansiruptodera*, anterior extremity, dorsal view. (After Inglis, 1957.)

2-(1) Cephalic cap not modified into "plates" (Fig. 6.167).
Posterior end of male relatively complex with relatively large number of caudal papillae and well developed pre-anal sucker (Fig. 6.168).
Parasites of South American rodents.

Paraspidodera Travassos, 1914

Family *ASCARIDIIDAE* Travassos, 1919

One genus (Figs. 6.169-6.171).
Parasites of birds and, rarely, mammals.

Ascaridia Dujardin, 1845
(= *Cotylascaris* Sprent, 1971)

6.165

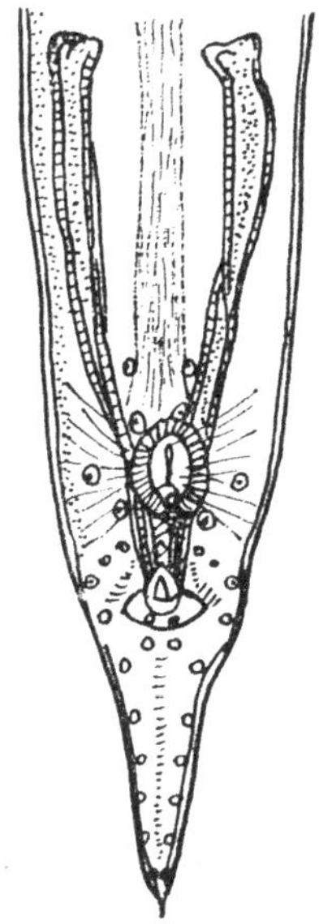

6.168

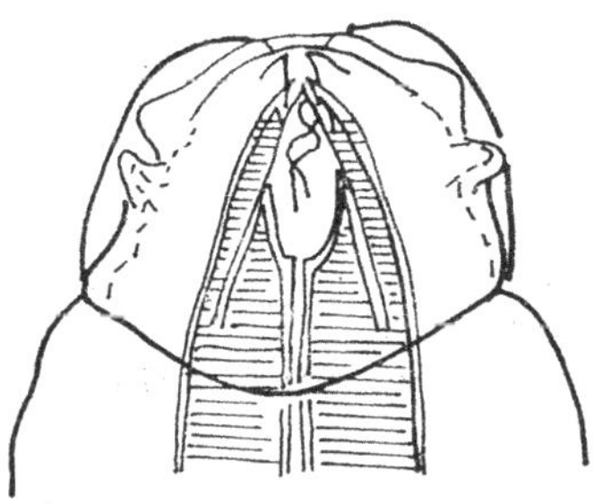

6.167

6.166

FIG. 6.165. *Lauroia*, anterior extremity, dorsal view. (After Proenca, 1938.)
FIG. 6.166. *Lauroia*, male tail, ventral view. (After Proenca, 1938.)
FIG. 6.167. *Paraspidodera*, anterior extremity, dorsal view. (After Inglis, 1957.)
FIG. 6.168. *Paraspidodera*, male tail, ventral view. (After Travassos, 1922.)

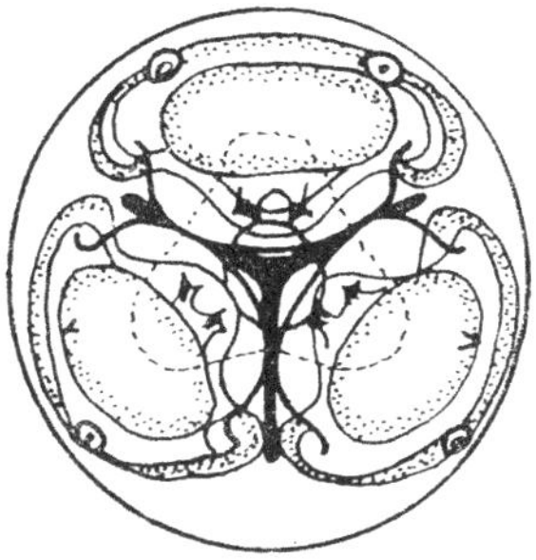

6.169

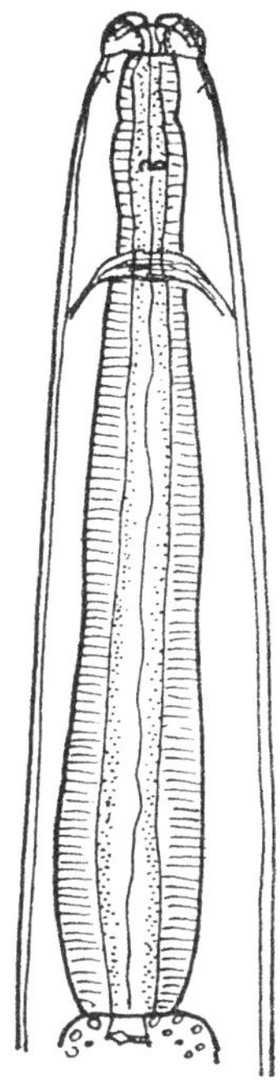

6.170

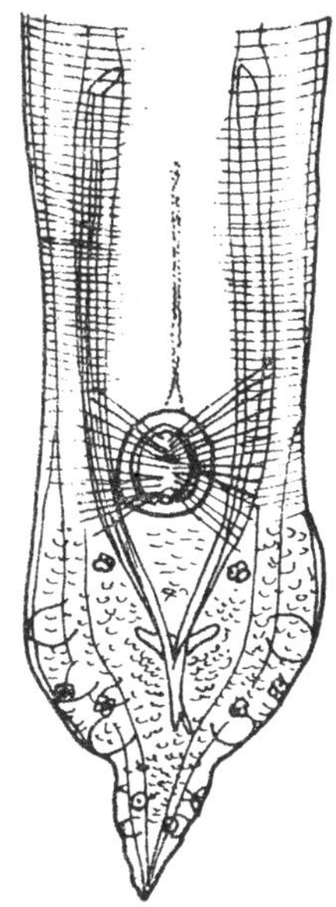

6.171

FIG. 6.169. *Ascaridia*, anterior extremity, apical view. (After Skrjabin & Schulz, 1937.)
FIG. 6.170. *Ascaridia*, anterior extremity, ventral view. (After Skrjabin & Schulz, 1937.)
FIG. 6.171. *Ascaridia*, male tail, ventral view. (After Skrjabin & Schulz, 1937.)

REFERENCES

ALI, S. M. & DESHMUKH, P. G. 1969. *Pseudaspidoderella indica* n.gen. n.sp. from grey partridge *Francolinus pondicerianus*. *Riv. Parassit.*, **30,** 45-48.

CHABAUD, A. G. & DOLLFUS, R.Ph. 1966. *Hatterianema hollandei* n.g., n.sp., nématode hétérakide parasite de rhynchocéphale. *Bull Mus. natn. Hist. nat.*, 2e sér., **37,** 1041-1055.

CRUSZ, H. & CHING, C. C. 1975. Parasites of the relict fauna of Ceylon. VI. More new helminths from amphibians and reptiles, with a new host-record and redescription of *Acanthocephalus serendibensis* Crusz and Mills, 1970. *Annls Parasit. hum. comp.*, **50,** 531-558.

GOMES, D. C. & PEREIRA, R. C. da S. 1970. Sôbre um nôvo genero da sub-familia Lauroiinae Skrjabin & Schikhobalova, 1951 (Nematoda, Subuluroidea). *Atas Soc. Biol. Rio de J.*, **12** (Suppl.), 35-37.

GUERRERO, R. 1971. Helmintos de la hacienda "El Limon", D. F. Venezuela—Nematodos de Vertebrados. I. *Mems Soc. Cienc. nat. La Salle*, **31,** 175-230.

INGLIS, W. G. 1957. The comparative anatomy and systematic significance of the head in the nematode family Heterakidae. *Proc. zool. Soc. Lond.*, **128,** 133-143.

——— 1958. A revision of the nematode genus *Meteterakis* Karve, 1930. *Parasitology*, **48,** 9-31.

——— 1967. The evolution, host relationships and classification of the nematode superfamily Heterakoidea. *Bull. Br. Mus. nat. Hist.*, **15,** 3-28.

JOHNSON, S. 1967. A new heterakid genus from the painted spurfowl in India (Nematoda, Spinaspidoderinae). *Revta bras. Biol.*, **27,** 197-200.

——— 1974. Reconstruction of the nematode superfamily Spinaspidoderinae Freitas, 1956 (Heterakidae). *Indian J. Helminth.*, **24,** 56-62.

MOZGOVOI, A. A. 1953. *Principles of Nematodology*, edited by K. I. Skrjabin. Vol. II. *Ascaridata of animals and man and the diseases caused by them.* Part I. Moscow: Izdatel'stvo Akademii Nauk SSSR, 351 pp. (In Russian).

NAMA, H. S. & JAIN, R. C. 1974. A new oxyurid nematode *Paraheterakis gubernaculi* n.gen., n.sp. from squirrel *Funambulus pennanti*. *Indian J. Helminth.*, **24**, 84-88.

SKRJABIN, K. I., SHIKHOBALOVA, N. P. & LAGODOVSKAJA, E. A. 1961. *Principles of Nematodology*, edited by K. I. Skrjabin. Vol. X. *Oxyurata of animals and man.* Part 2. Moscow: Izdatel'stvo Akademii Nauk SSSR, 499 pp. (In Russian).

SPRENT, J. F. A. 1971. A new genus and species of ascaridoid nematode from the marsupial wolf (*Thylacinus cynocephalus*). *Parasitology*, **63**, 37-43.

——— 1972. *Cotylascaris thylacini:* a synonym of *Ascaridia columbae*. *Parasitology*, **64**, 331-332.

TEIXEIRA de FREITAS, J. F. 1956. Notas sôbre "Heterakidae" Railliet et Henry, 1914 (Nematoda, Subuluroidea). *Revta bras. Biol.*, **16**, 461-482.

———, VICENTE, J. J. & SANTOS, E. dos. 1969. Sôbre um novo genero da familia "Heterakidae" Railliet & Henry, 1914 (Nematoda, Subuluroidea). *Atas Soc. Biol. Rio de J.*, **12**, 285-286.

10 ASCARIDIDA

ASCARIDOIDEA

by GERHARD HARTWICH*

INTRODUCTION

The following key to the genera and higher systematic categories of the superfamily Ascaridoidea follows the system proposed by Chabaud (1965). In contrast to the classifications of Mozgovoi (1951, 1953), Mozgovoi & Shakhmatova (1971) and Yamaguti (1961), this system reflects better the phylogenetic relationships of the ascaridoids in the light of the latest knowledge, mainly provided by the studies of Hartwich (1954, 1957) and Osche (1958). The genera described since Chabaud's (1965) treatise are included in the key. At the same time certain changes are proposed in the taxonomy of the group as outlined below.

Chabaud's (1965) tribe Raphidascaridinea for fish parasites of the Anisakinae is regarded herein as the subfamily Raphidascaridinae *sensu* Hartwich (1954).

Lappetascaridinae Rasheed, 1965 (for *Lappetascaris* Rasheed, 1965) and Aliascaridinae Kalyankar, 1971 (for *Aliascaris* Kalyankar, 1971) are regarded as tribes of the Raphidascaridinae.

Paranisakis Baylis, 1923 and *Paranisakiopsis* Yamaguti, 1941 are placed in the new tribe Paranisakinea of the Raphidascaridinae.

The Anisakinae (*sensu stricto*) is divided into the tribes Anisakinea Chabaud, 1965 and Contracaecinea Mozgovoi & Shakhmatova, 1971. Anisakinea (*sensu stricto*) includes *Anisakis* Dujardin, 1845, *Phocanema* Myers, 1959, *Terranova* Leiper & Atkinson, 1914 and *Sulcascaris* Hartwich, 1957. The tribe Contracaecinea, to which Mozgovoi & Shakhmatova assigned *Contracaecum* Railliet & Henry, 1912, *Heterotyphlum* Spaul, 1929 and *Phocascaris* Höst, 1932, is extended herein to include *Galeiceps* Railliet, 1916 and *Duplicaecum* Majumdar & Chakravarty, 1963. *Heterotyphlum* is, however, transferred to the tribe Raphidascaridinea of the Raphidascaridinae.

Following Campana-Rouget (1960), *Dujardinascaris* Baylis, 1927 is removed from the Toxocarinae and placed with *Hartwichia* Chabaud & Bain, 1966 in the tribe Dujardinascaridinea of the Multicaecinae.

Following Chabaud & Bain (1966), the genus *Typhlophorus* von Linstow, 1906 is transferred from the Heterocheilidae to the Ascarididae where it is placed in the tribe Multicaecinea of the Multicaecinae.

Chabaud's (1965) division of the Toxocarinae into the tribes Toxocarinea and Porrocaecinea is not followed since it is not clear to which group *Paradujardinia* Travassos, 1933 belongs.

The following genera are excluded from the key since they are incompletely described: *Heligmus* Dujardin, 1845 (from a pleuronectiform fish); *Metascaris* Schuurmans-Stekhoven, 1950 (from the body-cavity of a bird); *Paraheterotyphlum* Johnston & Mawson, 1948 (from a hydrophiine snake); *Pseudoterranova* Mozgovoi, 1951 (from a whale).

* Zoologisches Museum der Humboldt-Universität zu Berlin.

KEY TO THE FAMILIES

1-(2) Behind semicircular lips is constriction occupied by collar consisting of double row of fimbriae; these pass as a single row around the base of each lip on its inner surface (Fig. 2.1).
Parasites of hyracoids.
Each lip provided near outer edges of its inner aspect with 7 or 8 groups of combs of pointed teeth. Oesophagus long, without ventriculus, but enlarged posteriorly, its anterior portion containing chitinoid apparatus. One or two (Fig. 2.7) intestinal caeca present. Spicules equal, alate. Gubernaculum present.

Crossophoridae

2-(1) Basal region of lips without collar of fimbriae, postlabial region naked or provided with three tongue-like prolongations of lips united by folded membranes (Fig. 2.2), with cordon-like cuticular thickenings (Fig. 2.4), or with thickenings consisting of transversely striated ribs (Fig. 2.3).

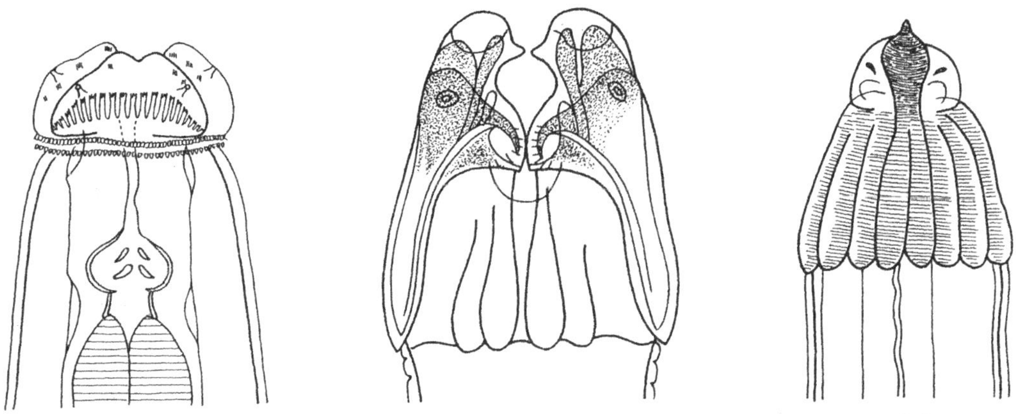

2.1 2.2 2.3

FIG. 2.1. *Crossophorus collaris*, dorsal view of anterior extremity (after Yorke & Maplestone, 1926).
FIG. 2.2. *Heterocheilus tunicatus*, ventral view of anterior extremity (after Drasche, 1883).
FIG. 2.3. *Typhlophorus lamellaris*, dorsal view of anterior extremity (after Linstow, 1906).

3-(4) Each lip prolonged posteriorly, forming tongue-like process; processes united by membranes thrown into four longitudinal folds (Fig. 2.2).
Oesophagus cylindrical, slightly enlarged posteriorly, ventriculus apparently absent. Long intestinal caecum present. Spicules equal, alate. Gubernaculum present. Vulva near anus.
Parasites of sirenians.

Heterocheilidae

4-(3) Postlabial region without cuticular peculiarities, or provided with cordon-like or rib-like structures.

5-(6) Oesophagus with oblong to cylindrical posterior ventriculus with one dorsal and one ventral longitudinal suture-like depression (Fig. 2.13), or with small globular ventriculus giving off posteriorly directed appendix with such depressions (Figs. 2.14, 2.15) or rarely two appendices (Fig. 2.31). Excretory system asymmetrical, confined to left lateral chord. Excretory pore situated between base of subventral lips or near nerve ring. Intestinal caecum present or absent.
Parasites of mammals, birds, reptiles and fishes.

Anisakidae

6-(5) Oesophagus without ventriculus or with globular to ellipsoidal ventriculus without longitudinal sutures (Figs. 2.9 to 2.11); from ventriculus two anteriorly and three posteriorly directed appendices may arise (Fig. 2.12). Excretory system symmetrical, the right lateral canal of which may be somewhat reduced.
Excretory pore situated near nerve ring. Intestinal caecum present or absent.

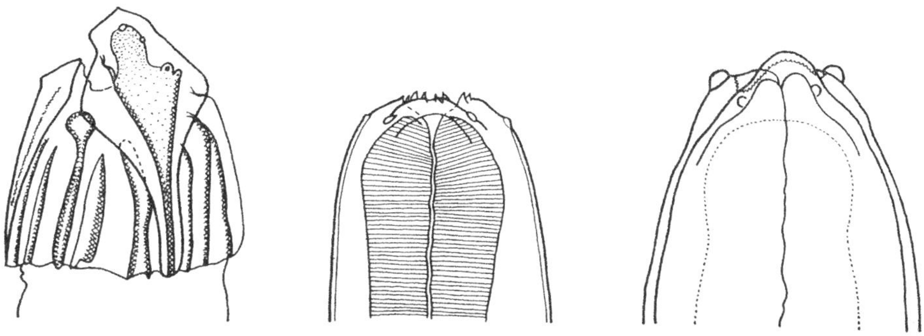

2.4 2.5 2.6

FIG. 2.4. *Hartwichia rousseloti*, lateral view of anterior extremity (after Chabaud & Bain, 1966).
FIG. 2.5. *Acanthocheilus rotundatus*, dorsal view of anterior extremity (after Baylis, 1929).
FIG. 2.6. *Pseudanisakis truncata*, dorsal view of anterior extremity (after Mozgovoi, 1953).

7-(8) Lips small, semicircular, not offset from anterior end; each lip provided on inner surface with two pairs of small pointed teeth (Fig. 2.5) or continuous dentigerous ridge (Fig. 2.6) or two separate transverse dentigerous ridges.
Interlabia absent. Oesophagus with ventriculus. Intestinal caecum absent.
Parasites of elasmobranch fishes.

Acanthocheilidae

8-(7) Lips well-defined, distinctly offset from anterior end; each lip with single, separate dentigerous ridge (Fig. 2.40) on anterior margin of inner surface, or dentigerous ridges absent.
Interlabia present or absent. Oesophagus with or without ventriculus. Intestinal caecum present or absent.
Parasites of mammals, birds, reptiles or amphibians, rarely of fishes.

Ascarididae

KEYS TO SUBFAMILIES, TRIBES AND GENERA

Family *CROSSOPHORIDAE*
(Baylis, 1920, subfam.) Hartwich, 1957

Key to genera

1-(2) Two intestinal caeca present (Fig. 2.7). Tail of female conical, without lateral appendages at tip.
Parasites of hyracoids.

Crossophorus Hemprich & Ehrenberg, 1828

2-(1) One intestinal caecum present. Tail of female with 2 terminal lateral cuticular appendages (Fig. 2.17).
Parasites of hyracoids.

Dartevellenia Ezzat, 1954

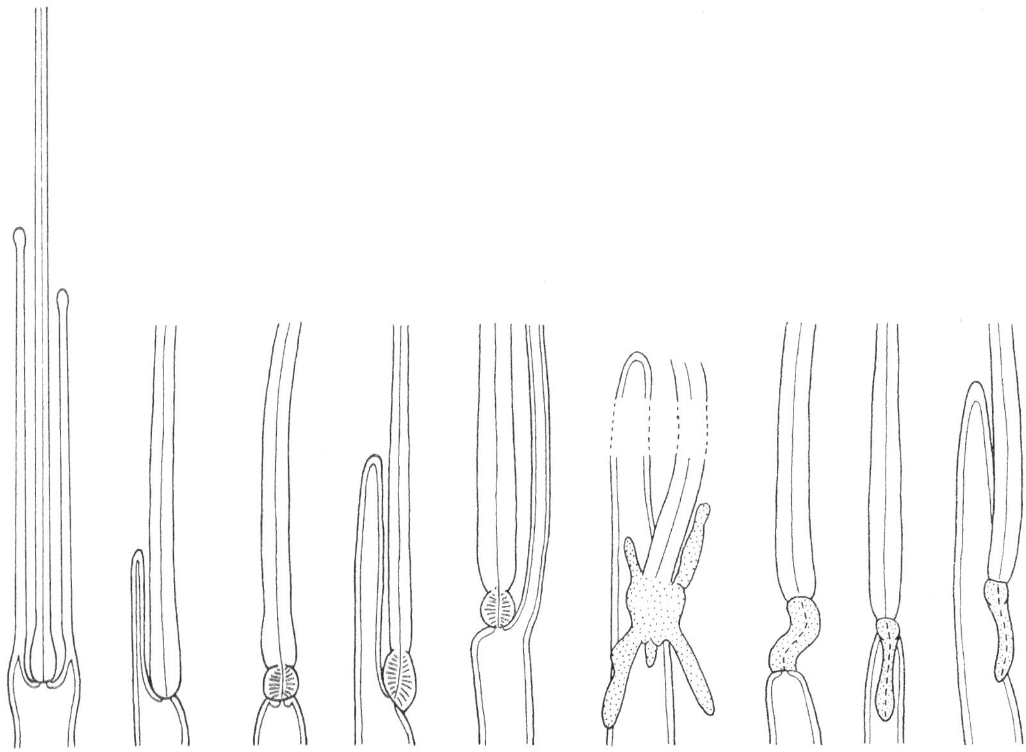

FIG. 2.7 to 2.15. **Oesophagi**; 2.7—*Crossophorus*; 2.8—*Angusticaecum*; 2.9—*Toxocara*; 2.10—*Porrocaecum*; 2.11—*Paradujardinia*; 2.12—*Multicaecum*; 2.13—*Anisakis*; 2.14—*Raphidascaris*; 2.15—*Contracaecum*.

Family *HETEROCHEILIDAE*

(Railliet & Henry, 1912, subfam.) Railliet & Henry, 1915

One genus.
Parasites of sirenians.

Heterocheilus Diesing, 1839
(=*Lobocephalus* Diesing, 1838, nom. nud.; =*Plicatolabia* Mozgovoi, 1951)

Family *ANISAKIDAE*

(Railliet & Henry, 1912, subfam.) Skrjabin & Karokhin, 1945
(=Heterocheilidae Railliet & Henry, 1915, in part; =Stomachidae (Johnston & Mawson, 1945, subfam.) Hartwich, 1957)

Key to Subfamilies

1-(2) Cuticular rings with posterior spines.
Lips flattened and expanded outwards (Fig. 2.18). Interlabia absent. Oesophagus stout, claviform, with small ventriculus giving off a long, single or double appendix (Fig. 2.19). Intestinal caecum present. Spicules subequal. Gubernaculum absent.
Parasites of teleost fishes and, rarely, reptiles.

Goeziinae

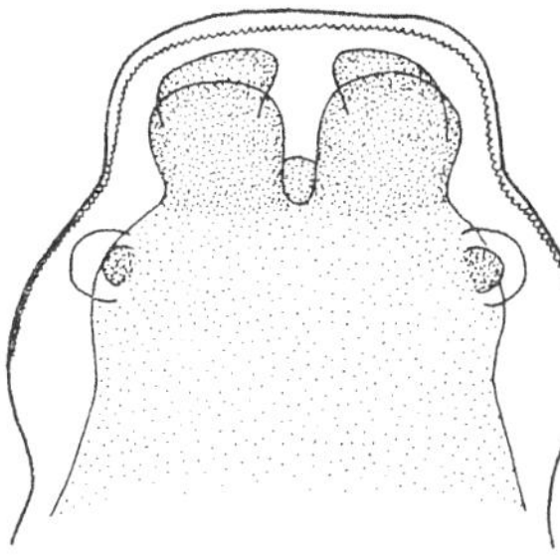
2.16

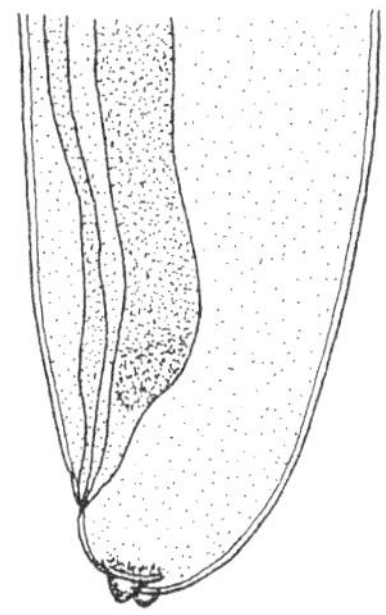
2.17

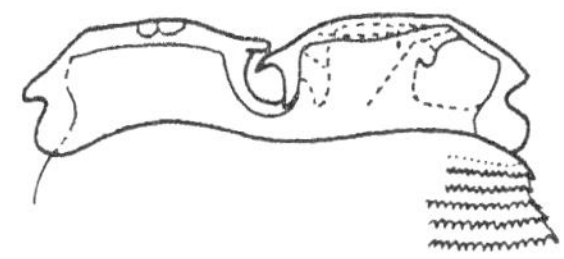
2.18

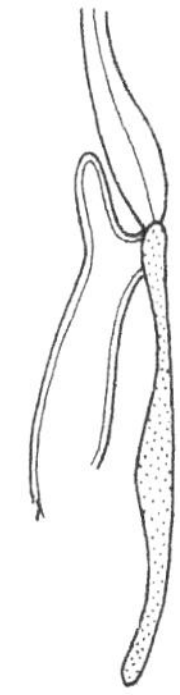
2.19

FIG. 2.16. *Toxocara vitulorum*, dorsal lip (after Baylis, 1936).
FIG. 2.17. *Dartevellenia collaris*, lateral view of posterior extremity of female (after Ezzat, 1954).
FIG. 2.18. *Goezia ascaroides*, anterior extremity (after Goeze, 1782).
FIG. 2.19. *Goezia*, oesophagus.

2-(1) Cuticular rings without spines.

3-(4) Excretory system ribbon-like, extending from left lateral chord into ventral part of body-cavity.
Excretory pore usually between subventral lips or at base of ventral interlabium.
Parasites of mammals, birds, reptiles or elasmobranch fishes.

Anisakinae

4-(3) Excretory system not ribbon-like.
Excretory pore usually near nerve ring.
Parasites of fishes, occasionally of birds.

Raphidascaridinae

Subfamily **Goeziinae** Travassos, 1919

One genus
Parasites of teleost fishes and, perhaps accidentally, of reptiles.

Goezia Zeder, 1800*
(=*Cochlus* Zeder, 1803; =*Cucullanus* Goeze, 1782, *nec* Müller, 1777; =*Lecanocephalus* Diesing, 1839; =*Prionoderma* Rudolphi, 1810, *nec* Cuvier, 1817)

Subfamily **Anisakinae** Railliet & Henry, 1912
(=Capsulariinae Johnston & Mawson, 1943;
=Filocapsulariinae Yamaguti, 1961, in part;
=Stomachinae Johnston & Mawson, 1945)

Key to tribes and genera

1-(8) Oesophagus with oblong to cylindrical ventriculus (Fig. 2.13). Appendix absent.
Intestinal caecum present or absent. Each lip provided with dentigerous ridge.
Parasites of marine mammals, marine turtles, crocodiles or elasmobranch fishes.

ANISAKINEA Chabaud, 1965

* This genus was divided by Mozgovoi (1951) into the following subgenera:
Goezia (*Goezia*) Zeder, 1800—Oesophagus with one appendix;
Goezia (*Pseudogoezia*) Mozgovoi, 1951—Oesophagus with two appendices.

2*

2-(3) Intestinal caecum absent.
Anterior region of lips forming medial, slightly bilobed process (Fig. 2.20). Interlabia absent. Spicules equal or unequal.
Parasites of marine mammals, except sirenians.

Anisakis Dujardin, 1845

(=*Anisakis* (*Skrjabinisakis*) Mozgovoi, 1951; =*Capsularia* Zeder, 1800, *nec* Modeer, 1793; =*Conocephalus* Diesing, 1861, *nec* Thunberg, 1812; =*Filocapsularia* Deslongchamps, 1824, *nom. oblit.*; =*Peritrachelius* Diesing, 1861; =*Stomachus* Goeze in Zeder, 1800, *nom. nud.*)

3-(2) Intestinal caecum present.

4-(7) Interlabia absent. Excretory pore between subventral lips.
Parasites of marine and semi-aquatic mammals, crocodiles or elasmobranch fishes.

5-(6) Anterior region of lips marked by median indentation giving it appearance of two antero-laterally directed, rounded lobes (Fig. 2.21).
Spicules subequal.
Parasites of marine mammals, except sirenians, and occasionally of semiaquatic carnivores.

Phocanema Myers, 1959

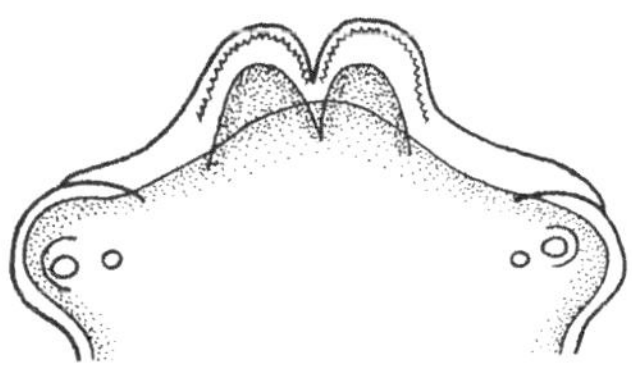

2.20

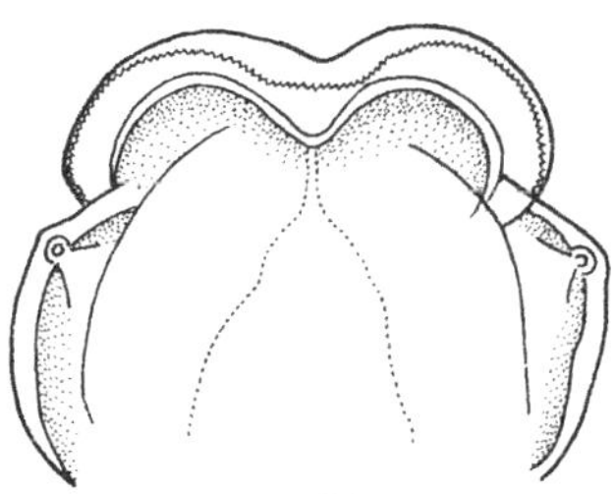

2.21

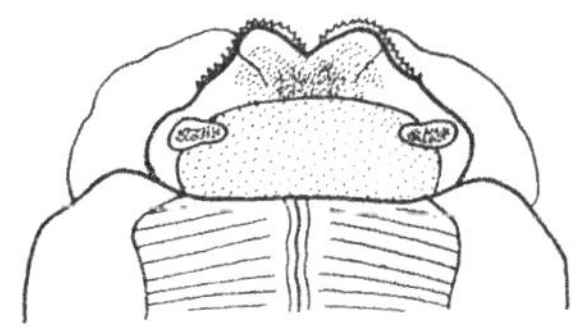

2.22

FIG. 2.20. *Anisakis catodontis*, dorsal lip (after Baylis, 1929).
FIG. 2.21. *Phocanema decipiens*, dorsal lip (after Mozgovoi, 1953).
FIG. 2.22. *Terranova scoliodontis*, anterior extremity (after Baylis, 1931).

6-(5) Anterior region of lips forming medial, nearly triangular process divided into two indistinct lobes (Fig. 2.22).
Spicules equal or unequal.
Parasites of elasmobranch fishes and, rarely, crocodiles.

Terranova Leiper & Atkinson, 1914

(=*Metangusticaecum* Mozgovoi, 1951; ?=*Paracanthocheilus* Kreis, 1952; =*Terranova* (*Sauronema*) Mozgovoi, 1951)

7-(4) Interlabia present. Excretory pore near nerve ring.
Lips rounded quadrangular, with blunt triangular process in, or somewhat in front of, middle of lateral margins (Fig. 2.23).
Spicules equal.
Parasites of turtles.

Sulcascaris Hartwich, 1957

8-(1) Oesophagus with reduced globular ventriculus giving off more or less elongated appendix (Fig. 2.15), rarely two appendices (Fig. 2.31).
Intestinal caecum present. Lips with or without dentigerous ridge.
Parasites of marine, semi-aquatic or terrestrial mammals, or fish-eating birds.

CONTRACAECINEA Mozgovoi & Shakhmatova, 1971

9-(10) Ventriculus with two appendices, one longer and straight, and the other shorter and coiled (Fig. 2.31).
Lips separated from body by groove. Dentigerous ridges absent. Prominent interlabia present. Spicules equal.
Parasites of ciconiiform birds.
Duplicaecum Majumdar & Chakravarty, 1963

10-(9) Ventriculus with one straight appendix (Fig. 2.15).

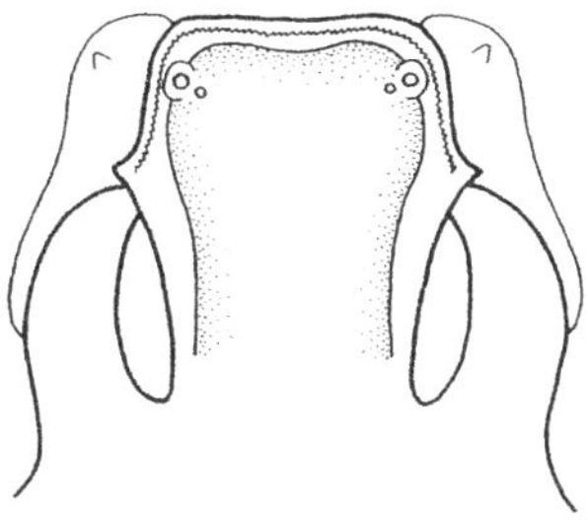

2.23

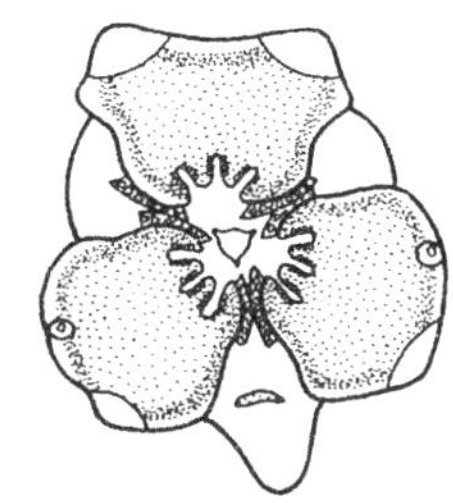

2.24

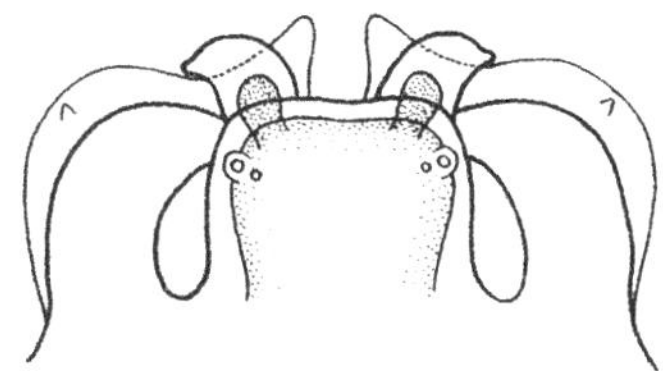

2.25

FIG. 2.23. *Sulcascaris sulcata*, dorsal view of anterior extremity (after Hartwich, 1954).
FIG. 2.24. *Phocascaris phocae*, apical view of anterior extremity (after Delyamure, in Mozgovoi, 1953).
FIG. 2.25. *Contracaecum rudolphii*, dorsal view of anterior extremity (after Hartwich, 1954).

11-(12) Postlabial region surrounded by collar-like fold of cuticle; between fold and bases of lips is area covered with small spines (Fig. 2.26). Inner surface of each lip with pair of large conical teeth. Interlabia absent. Spicules equal.
Parasites of semiaquatic carnivorous mammals, occasionally insectivores and monkeys.
Galeiceps Railliet, 1916
(=*Cloeoascaris* Baylis, 1923)

12-(11) Postlabial region without cuticular collar; bases of lips not spined. Lips without tooth-like structures.

13-(14) Interlabia present.
Lips usually rounded hexagonal, sometimes semi-oval; each lip more or less indented medially and provided with pair of laterally directed pointed processes on its anterior angles (Fig. 2.25). Dentigerous ridges absent. Spicules unequal or subequal.
Parasites of fish-eating birds and marine mammals, except sirenians.
Contracaecum Railliet & Henry, 1912
(=*Amphicaecum* Walton, 1927; =*Cerascaris* Cobb, 1929;
=*Contracaecum* (*Ornitocaecum*) Mozgovoi, 1951;
=*Contracaecum* (*Synthetonema*) Kreis, 1952;
=*Kathleena* Leiper & Atkinson, 1914)

14-(13) Interlabia absent.
Lips nearly quadrangular; anterior margin of each lip with three incisions (Fig. 2.24). Dentigerous ridges present. Spicules subequal.
Phocascaris Höst, 1932

Subfamily **Raphidascaridinae** Hartwich, 1954

Key to tribes and genera

1-(2) Lips surrounded posteriorly by ring with small anteriorly directed curves between lips; behind ring two thick conical processes attached to cuticle of posterior border of ventro-lateral lips (Fig. 2.28).
Lips nearly rectangular, each showing anteriorly two small semilunar plates (Fig. 2.27). Small bulbous ventriculus with very long appendix present. Intestinal caecum present. Excretory pore situated immediately behind nerve ring. Spicules equal.
Parasites of marine teleost fishes.

LAPPETASCARIDINEA (Rasheed, 1965, subfam.)

One genus.

Lappetascaris Rasheed, 1965

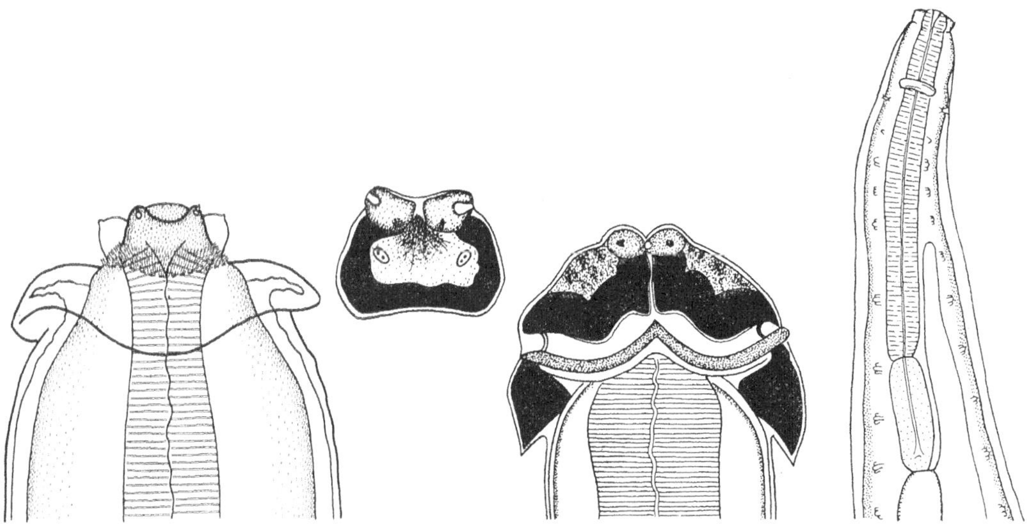

2.26 2.27 2.28 2.29

FIG. 2.26. *Galeiceps spinicollis*, dorsal view of anterior extremity (after Baylis, 1923).
FIG. 2.27. *Lappetascaris lutjani*, dorsal lip (after Rasheed, 1965).
FIG. 2.28. *Lappetascaris lutjani*, ventral view of anterior extremity (after Rasheed, 1965).
FIG. 2.29. *Aliascaris indica*, anterior extremity (after Kalyankar, 1971).

2-(1) Postlabial ring and processes behind ventro-lateral lips absent.

3-(4) Cuticle provided with two lateral longitudinal rows of trilobed papilla-like structures (Fig. 2.29).
Lips distinct. Spheroidal ventriculus and intestinal caecum present. Excretory pore situated near nerve ring. Spicules equal.
Parasites of elasmobranch fishes.

ALIASCARIDINEA (Kalyankar, 1971, subfam.)

One genus.

Aliascaris Kalyankar, 1971

4-(3) Cuticle without longitudinal rows of papilla-like structures.

5-(12) Ventricular appendix present.

RAPHIDASCARIDINEA Chabaud, 1965

6-(9) Intestinal caecum absent.
7-(8) Interlabia absent.
Lips usually not longer than wide (Fig. 2.30).
Parasites of teleost fishes.

Raphidascaris Railliet & Henry, 1915
(=*Hysterothylacium* Ward & Magath, 1916;
=*Ichthyascaris* Wu, 1949; ?=*Neogoezia* Kreis, 1937)

8-(7) Small interlabia present.
Lips markedly longer than wide (Fig. 2.32).
Parasites of teleost fishes.

Raphidascaroides Yamaguti, 1941
(=*Ryjikovascaris* Mozgovoi, 1951)

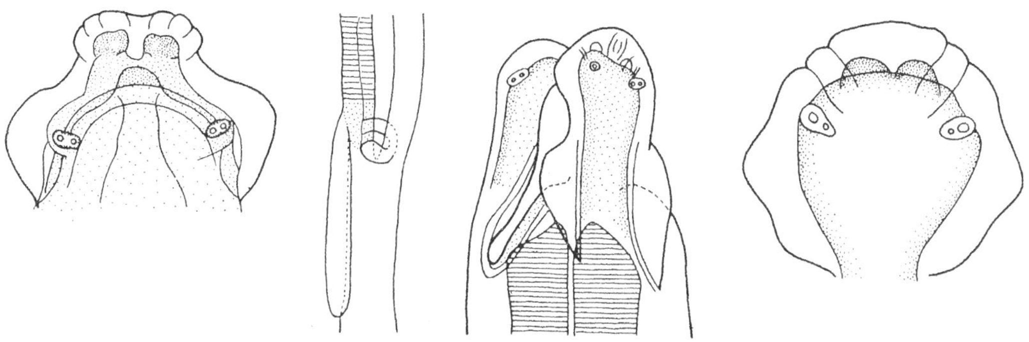

2.30 2.31 2.32 2.33

FIG. 2.30. *Raphidascaris acus*, dorsal lip (after Mozgovoi, 1953).
FIG. 2.31. *Duplicaecum ibisi*, ventricular region of oesophagus (after Majamdar & Chakravarty, 1963).
FIG. 2.32. *Raphidascaroides nipponensis*, anterior extremity (after Yamaguti, 1941).
FIG. 2.33. *Thynnascaris gadi*, dorsal lip (after Mozgovoi, 1953).

9-(6) Intestinal caecum present.
10-(11) Interlabia present. Lips large, with anterior region forming distally broadened process internally supported by pair of longitudinal thickenings (Fig. 2.33).
Parasites of teleost fishes.

Thynnascaris Dollfus, 1933
(=*Contracaecum* (*Thynnascaris*) Dollfus, 1935;
=*Contracaecum* (*Erschowicaecum*) Mozgovoi, 1951;
=*Contracaecum* (*Simplexonema*) Kreis, 1952 *nom. nud.*;
=*Iheringascaris* Pereira, 1935)

11-(10) Interlabia absent. Lips narrow, somewhat asymmetrical, without distinct anterior process and without longitudinal thickenings (Fig. 2.35).
Parasites of teleost fishes, some species found as accidental parasites of fish-eating birds.

Heterotyphlum Spaul, 1927

12-(5) Ventricular appendix absent.

PARANISAKINEA nov. trib.

13-(14) Interlabia prominent, more than half as long as lips (Fig. 2.34). Excretory pore immediately behind base of ventral interlabium.
Parasites of marine teleost fishes.

Paranisakiopsis Yamaguti, 1941

14-(13) Interlabia smaller, not exceeding half the length of lips (Fig. 2.36). Excretory pore near nerve ring.
Parasites of marine fishes.

Paranisakis Baylis, 1923
(=*Ichthyanisakis* Gendre, 1928;
=*Ortoanisakis* (Mozgovoi, 1951, subgen.) Yamaguti, 1961)

Family *ACANTHOCHEILIDAE* Wülker, 1929

Key to genera

1-(2) Each lip provided on its inner surface with two pairs of small, pointed, anteriorly projecting teeth (Fig. 2.5). Dentigerous ridges absent.
Ventriculus nearly spherical. Spicules short, non-alate.
Parasites of elasmobranch fishes, mainly sharks.

Acanthocheilus Molin, 1858

2-(1) Lips without paired projecting teeth but with continuous dentigerous ridge (Fig. 2.6) on inner surface of their margins, or two separate transverse ridges.
Ventriculus oval to ellipsoidal. Spicules long, alate.

3-(4) Lips with two separate transverse dentigerous ridges.
Parasites of elasmobranch fishes, mainly sharks.

Metanisakis Mozgovoi, 1951
(=*Pseudanisakis* Yamaguti, 1941, *nec* Layman & Borovkova, 1926)

4-(3) Lips with one continuous dentigerous ridge.
Parasites of elasmobranch fishes, mainly rays.

Pseudanisakis (Layman & Borovkova, 1926, subgen.) Mozgovoi, 1951
(=*Anacanthocheilus* Wülker, 1930; =*Eustoma* van Beneden, 1870, *nec* Piette, 1855)

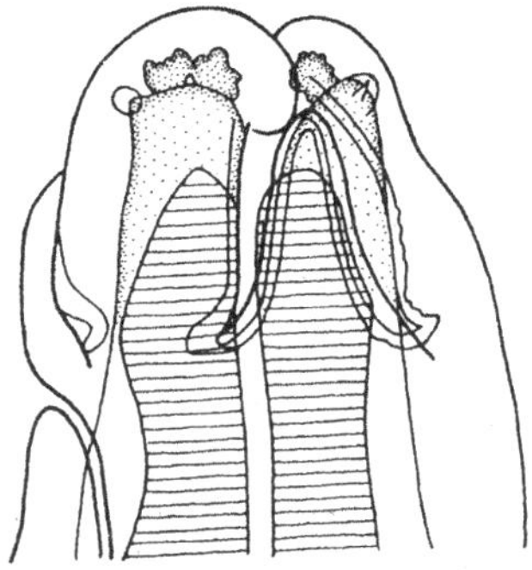

2.34

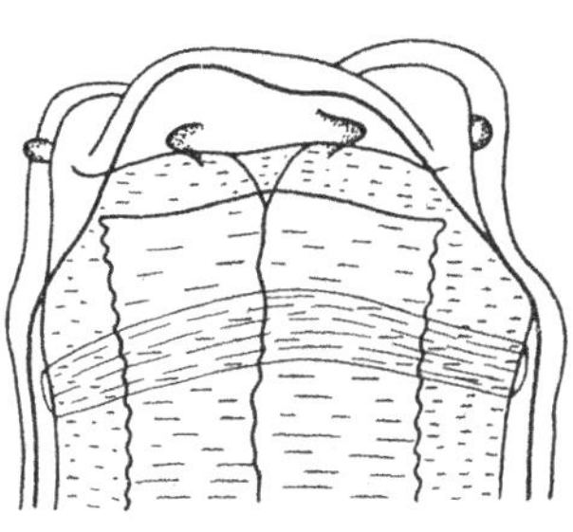

2.35

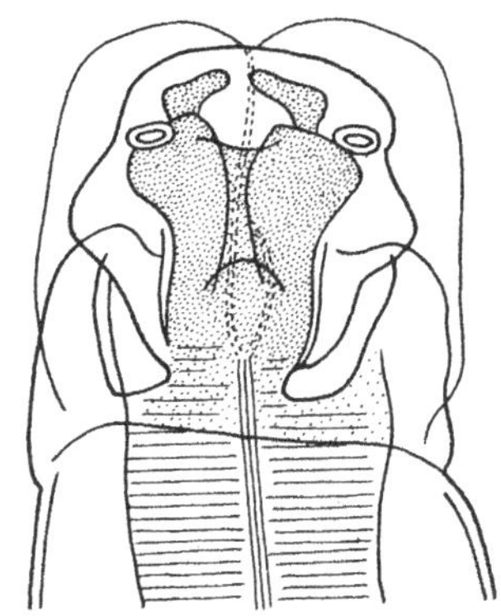

2.36

Fig. 2.34. *Paranisakiopsis coelorhynchi*, subventral view of anterior extremity (after Yamaguti, 1941).
Fig. 2.35. *Heterotyphlum himantolophi*, anterior extremity (after Spaul, in Mozgovoi, 1953).
Fig. 2.36. *Paranisakis squatinae*, dorsal view of anterior extremity (after Baylis, 1923).

Family *ASCARIDIDAE* Baird, 1853

Key to subfamilies

1-(4) Oesophagus with globular to ellipsoidal posterior ventriculus (Figs. 2.9 to 2.11), from which may arise two anteriorly and three posteriorly directed appendices (Fig. 2.12).
Gubernaculum present or absent.

2-(3) Parasites of terrestrial or marine mammals, or of birds.
Ventriculus without appendices. Gubernaculum absent.

Toxocarinae

3-(2) Parasites of crocodilians and, rarely, of fishes.
Ventriculus without appendices or with two anterior and three posterior appendices (Fig. 2.12). Gubernaculum present (not described in *Typhlophorus*).

Multicaecinae

4-(1) Oesophagus simple.
Gubernaculum absent.

5-(6) Parasites of terrestrial mammals.
Lips usually rounded hexagonal, with anterior region distinctly offset from posterior by lateral indentations (Figs. 2.33 to 2.40), or anterior margin of each lip deeply indented in antero-dorsal view (Fig. 2.45). Intestinal caecum absent.

Ascaridinae

6-(5) Parasites of reptiles or amphibians.
Lips nearly quadrangular, not clearly divided into anterior and posterior regions, and without deeply indented anterior margin (Fig. 2.38). Intestinal caecum present or absent.

Angusticaecinae

Subfamily **Toxocarinae**
(Hartwich, 1954, fam.) Osche, 1958

Key to genera

1-(2) Intestinal caecum absent. Interlabia absent.
Parasites of terrestrial mammals.

Toxocara Stiles, 1905
(=*Belascaris* Leiper, 1907; =*Neoascaris* Travassos, 1927)

2-(1) Intestinal caecum present. Interlabia present.

3-(4) Anterior margin of each lip provided on inner surface with dentigerous ridge (Fig. 2.37).
Oesophagus with voluminous globoid to oblong ventriculus. Intestinal caecum as broad or broader than oesophagus (Fig. 2.10).
Parasites of birds.

Porrocaecum Railliet & Henry, 1912
(=*Belanisakis* Maplestone, 1932; =*Porrocaecum* (*Laymanicaecum*) Mozgovoi, 1951, in part)

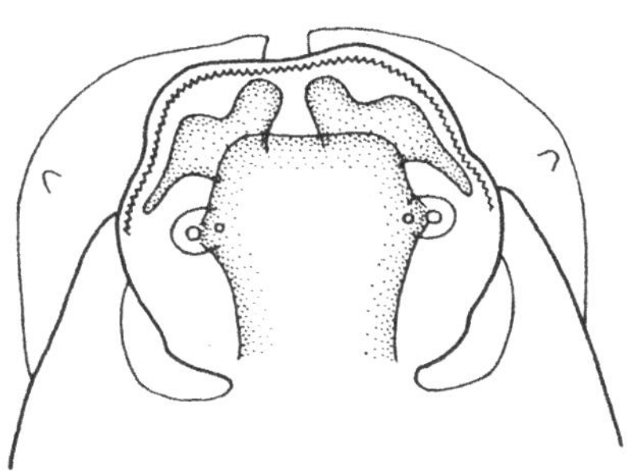

2.37

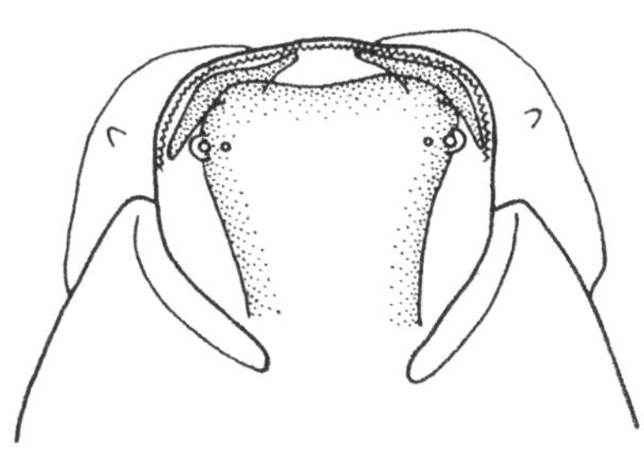

2.38

FIG. 2.37. *Porrocaecum ensicaudatum*, dorsal view of anterior extremity (after Hartwich, 1954).
FIG. 2.38. *Ophidascaris filaria*, dorsal view of anterior extremity (after Hartwich, 1954).

4-(3) Lips without dentigerous ridge.
Oesophagus with small globoid ventriculus. Intestinal caecum usually markedly thinner than oesophagus (Fig. 2.11).
Parasites of sirenians.

Paradujardinia Travassos, 1933
(=*Dujardinia* Gedoelst, 1916, in part, *nec* Quatrefages, 1844)

Subfamily **Multicaecinae**
(Hartwich, 1954, fam.) Campana-Rouget, 1960
Key to tribes and genera

1-(4) Oesophagus with small posterior ventriculus giving off two anteriorly and three posteriorly directed appendices of different sizes (Fig. 2.12).

MULTICAECINEA Campana-Rouget, 1960

2-(3) Cuticular thickening consisting of about sixteen transversely striated ribs behind lips (Fig. 2.3). Interlabia absent.
Parasites of crocodilians.

Typhlophorus von Linstow, 1906

3-(2) No cuticular thickening behind lips. Small interlabia present.
Parasites of crocodilians.

Multicaecum Baylis, 1923
(=*Multicaecum* (*Brevimulticaecum*) Mozgovoi, 1951;
=*Polycaecum* Maplestone, 1930)

4-(1) Ventriculus without appendices.

DUJARDINASCARIDINEA Campana-Rouget, 1960

5-(6) Each lip elongated by two cordons originating on internal surface of each lip and converging and fusing posteriorly; between main cordon-like prolongations of coinciding lips cuticula covered by five additional longitudinal cordons (Fig. 2.4). Interlabia absent.
Parasites of crocodilians.

Hartwichia Chabaud & Bain, 1966

6-(5) Lips not elongated by cordons; additional cordons absent. Interlabia present.
Parasites of crocodilians.

Dujardinascaris Baylis, 1927
(=*Dujardinia* Gedoelst, 1916, in part, *nec* Quatrefages, 1844;
?=*Trispiculascaris* Skrjabin, 1916, *gen. dub.*)

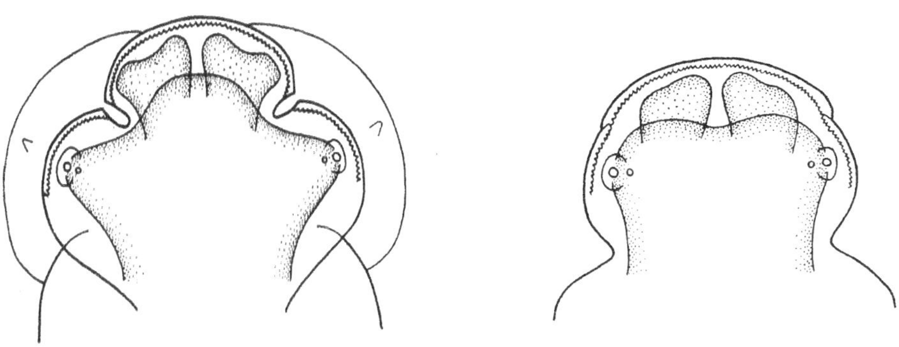

2.39 2.40

FIG. 2.39. *Parascaris equorum*, dorsal view of anterior extremity (after Hartwich, 1954).
FIG. 2.40. *Ascaris lumbricoides*, dorsal lip (after Hartwich, 1954).

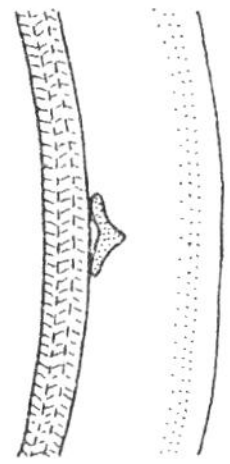

2.41

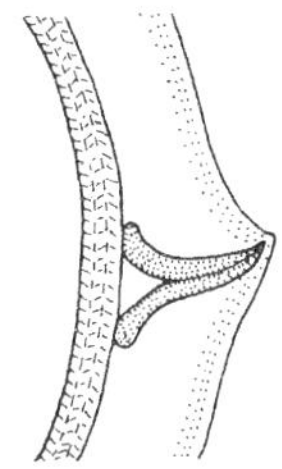

2.42

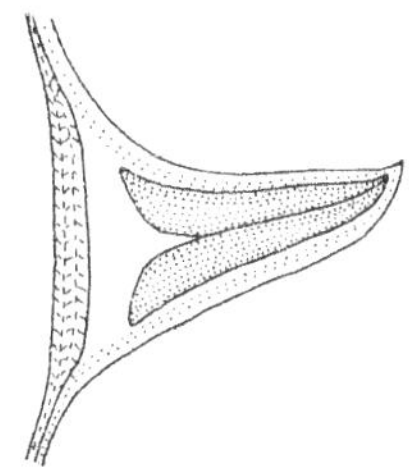

2.43

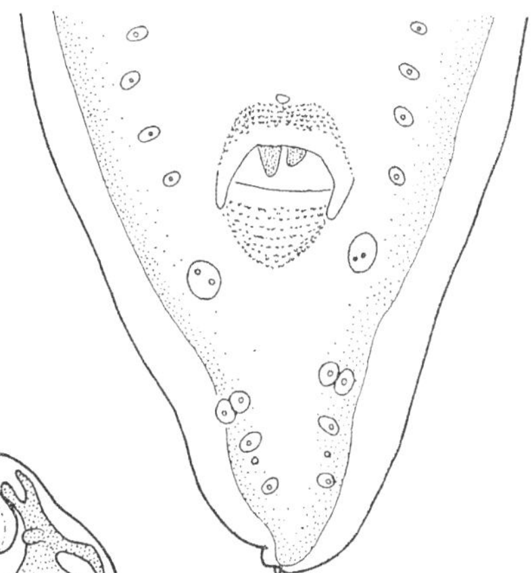

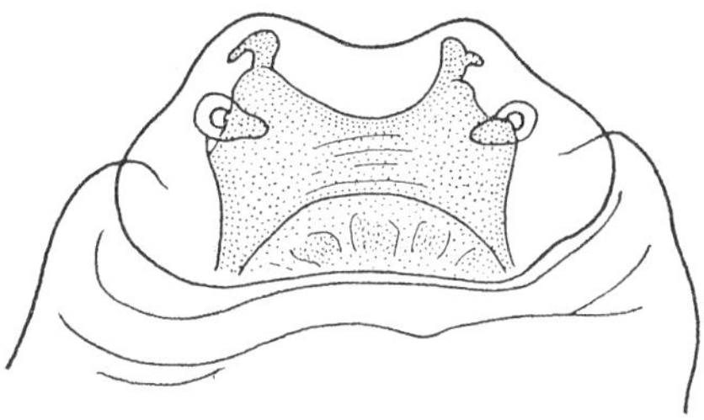

2.44

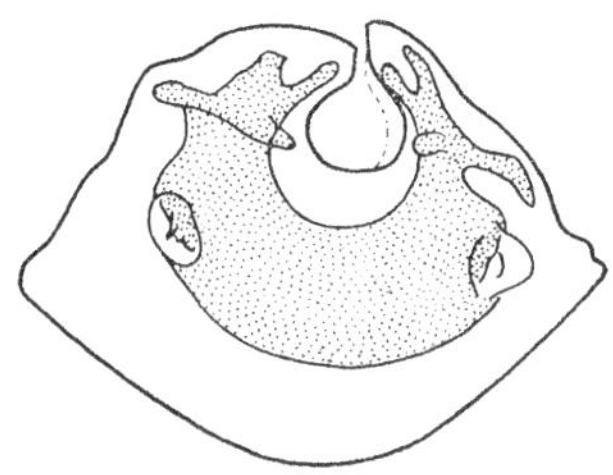

2.45

2.46

FIGS. 2.41 to 2.43. **Cross sections through cuticle showing cuticular bars** (after Hartwich, 1962). 2.41—*Ascaris lumbricoides*; 2.42—*Baylisascaris procyonis*; 2.43—*Toxascaris leonina.*

FIG. 2.44. *Lagochilascaris minor*, dorsal view of dorsal lip (after Sprent, 1971).

FIG. 2.45. *Lagochilascaris minor*, antero-dorsal view of dorsal lip (after Sprent, 1971).

FIG. 2.46. *Baylisascaris procyonis*, ventral view of posterior extremity of male (after Hartwich, 1962).

Subfamily **Ascaridinae** (Baird, 1853)

Key to genera

1-(4) Cervical alae and lateral flanges absent. Longitudinal cuticular bars bracket-like in cross-section, and not reaching surface of cuticle (Fig. 2.41).

2-(3) Lips subdivided into anterior and posterior regions by internal transverse groove forming deep incision on each lateral margin (Fig. 2.39). Small interlabia present.
Parasites of equines.

Parascaris Yorke & Maplestone, 1926

3-(2) Anterior and posterior region of lips separated only by slight lateral indentations (Fig. 2.40). Interlabia absent.
Parasites of primates, artiodactyles and rodents. *Ascaris* L., 1758
(=*Fusaria* Zeder, 1800; =*Lombricoides* Mérat, 1821;
=*Stomachida* Pereboom, 1780)

4-(1) Cervical alae salient or sometimes vestigial, or narrow lateral flanges present. Cuticular bars in region of alae usually funnel-shaped in cross-section, reaching to surface of cuticle (Figs. 2.42, 2.43).

5-(6) Anterior margin of lips deeply indented in antero-dorsal view (Fig. 2.45). Lips separated from body by postlabial groove; behind groove is a cuticular collar from which project distinct interlabia (Fig. 2.44). Narrow lateral alae present, extending length of body, Cervical alae absent.
Parasites of man (in abscesses), felines, and didelphoids.
Lagochilascaris Leiper, 1909*

6-(5) Anterior margin of lips not deeply indented. Continuous postlabial groove, cuticular collar and interlabia absent. Lateral flanges absent. Cervical alae present, but sometimes vestigial.

7-(8) Perianal region of male with rough areas, usually both in front and behind anus (Fig. 2.46). Shell of egg finely pitted.
Parasites of carnivores and rodents.
Baylisascaris Sprent, 1968

8-(7) Perianal region of male without roughened patches. Egg with smooth shell.
Parasites of carnivores.
Toxascaris Leiper, 1907

Subfamily **Angusticaecinae** Skrjabin & Karokhin, 1945

Key to tribes and genera

1-(4) Intestinal caecum present (Fig. 2.8), sometimes paired. Female with two uterine branches.
ANGUSTICAECINEA Chabaud, 1965

2-(3) Interlabia absent.
Parasites of tortoises.
Angusticaecum Baylis, 1920

3-(2) Interlabia present.
Parasites of reptiles and amphibians, except for the type species *Amplicaecum colurum* from a bird†.
Amplicaecum Baylis, 1920 (?=*Orneoascaris* Skrjabin, 1916 *gen. dub.*)

4-(1) Intestinal caecum usually absent, female with four or six uterine branches.
OPHIDASCARIDINEA (Hartwich, 1954; subfam.) Chabaud, 1965
(=Hexametrinae Mozgovoi & Shakhmatova, 1971;
=Polydelphinae Mozgovoi & Shakhmatova, 1971)

5-(6) Interlabia present. Genital tubes in both sexes confined to posterior region of body. Female with two uterine branches.
Parasites of snakes and lizards, occasionally of amphibians.
Ophidascaris Baylis, 1921

6-(5) Interlabia absent. Genital tubes extending through both anterior and posterior regions of body. Female with four or six uterine branches.

7-(8) Female with four uterine branches.
Parasites of snakes and lizards.
Polydelphis Dujardin, 1845

8-(7) Female with six uterine branches.
Parasites of snakes and lizards. *Hexametra* Travassos, 1919

* According to Sprent (1971) this genus belongs in Angusticaecinae because its species bear in some adult and larval characters a close resemblance to reptilian ascaridoids, especially *Ophidascaris*.

† According to Baylis (1947, *J. Linn. Soc., Zool.*, **41**, p. 397) this species, described from an African eagle, was possibly a pseudoparasite, the real host perhaps being a reptile or amphibian.

REFERENCES*

BERLAND, B. 1963. *Phocascaris cystophorae* sp. nov. (Nematoda) from the hooded seal, with an emendation of the genus. *Årbok. Univ. Bergen*, No. 17, 21 pp.

CAMPANA-ROUGET, Y. 1960. Sur la position systématique du genre *Dujardinascaris* Baylis 1947 (Nematoda, Ascaridoidea). *Bull. Soc. zool. Fr.* **85:** 383-388.

CHABAUD, A. G. 1965. Systématique des Nématodes. Sous-classe des Secernentea. Ordre des Ascaridida. Superfamilles des Ascaridoidea, Heterakoidea et Subuluroidea. *In:* Grassé, P. P., Traité de Zoologie, Tome IV, Fasc. III: Némathelminthes (nématodes, gordiacés), rotifères, gastrotriches, kinorhynches. pp. 988-1025. Paris: Masson et Cie.

——— & Bain, O. 1966. Description de *Hartwichia rousseloti* n. gen., n. sp., ascaride parasite de crocodile et remarques sur la famille des Heterocheilidae Railliet et Henry 1912. *Bull. Mus. natn. Hist. nat., Paris.*, Year 1965, **37** (5) : 848-853.

———, ——— & Tcheprakoff, R. 1967. Redescription de *Galeiceps cucullus* (Linstow 1899) et remarques sur l'osmo-régulation des nématodes Anisakides. *Annls Parasit. hum. comp.*, **423:** 321-326.

FREITAS, J. F. TEIXEIRA DE. 1968. Revisão do género *Ophidascaris* Baylis, 1921 (Nematoda, Ascaridoidea). *Mems Inst. Oswaldo Cruz*, **66** (1): 1-83 + 23 pp. plates.

HARTWICH, G. 1954. Die Vorderdarmstrukturen, das Exkretionssystem sowie der Kopfbau der Ascariden und ihre taxonomische Bedeutung. *Wiss. Z. Martin-Luther-Univ. Halle-Wittenb.*, **3** (5): 1171-1211.

——— 1957. Zur Systematik der Nematoden-Superfamilie Ascaridoidea. *Zool. Jb. (Syst.)* **85:** 211-252.

——— 1962. Über den Waschbärenspulwurm *Ascaris procyonis* Stefanski et Żarnowski 1951, und seine Stellung im System der Ascaroidea (Nematoda). *Cslká. Parasit.* **9:** 239-256.

KALYANKAR, S. D. 1971. On a remarkable new nematode, *Aliascaris indica* gen. et sp. nov. (Ascarididea: Heterocheilidae) from a marine fish, *Trygon sephen. Riv. Parassit.* **32** (2): 85-88.

LE-VAN-HOA. 1960. Synonymie des genres *Amplicaecum* Baylis 1920 et *Orneoascaris* Skrjabin 1916. *Annls Parasit. hum. comp.*, **35** (5/6): 760-761.

——— 1962. Nématodes parasites de mammifères, reptiles et amphibiens du Congo. Phasmidiens. *Explor. Parc natn. Upemba Miss. G. F. de Witte*, Fasc. 65: pp. 3-58

MAJUMDAR, G. A. 1964. A revision of the Indian anisakids (Nematoda: Heterocheilidae), with the description of a new species of *Amplicaecum* Baylis, 1920. *Proc. zool. Soc., Bengal*, **17** (1): 111-123

——— & CHAKRAVARTY, G. K. 1963. New nematodes from birds. Part II. *Z. ParasitKde*, **23** (5): 405-410.

MOZGOVOI, A. A. 1951. Ascaridata. *In:* Skryabin, K. I., Shikhobalova, N. P. & Mozgovoi, A. A., [Descriptive catalogue of parasitic nematodes], Vol. II, pp. 405-566. [In Russian]. Moscow: Izdat. Akad. Nauk SSSR.

——— 1953. [Ascaridata of animals and man, and diseases caused by them]. *In:* Skryabin, K. I. [Editor] [Principles of nematodology], Vol. II, Parts 1 and 2. [In Russian]. Moscow: Izdat. Akad. Nauk SSSR.

——— & Shakhmatova, V. I. 1971. [Revision of nematodes of the suborder Ascaridata, Skryabin, 1915]. *Trudy gel'mint. Lab.* **22:** 129-145 [In Russian].

MYERS, B. J. 1959. *Phocanema*, a new genus for the anisakid nematode of seals. *Can. J. Zool.*, **37** (4): 459-465.

OSCHE, G. 1958. Beiträge zur Morphologie, Ökologie und Phylogenie der Ascaridoidea (Nematoda). Parallelen in der Evolution von Parasit und Wirt. *Z. ParasitKde*, **18** (6): 479-572.

RASHEED, S. 1965a. Some parasitic nematodes from the Cameroons (W. Africa). *J. Helminth.*, **39** (1): 67-100.

——— 1965b. On a remarkable new nematode, *Lappetascaris lutjani* gen. et sp. nov. (Anisakidae: Ascaridoidea) from marine fishes of Karachi and an account of *Thynnascaris inquies* (Linton 1901) n. comb. and *Goezia intermedia* n. sp. *J. Helminth.*, **39** (4): 313-342.

SPRENT, J. F. A. 1968. Notes on *Ascaris* and *Toxascaris*, with a definition of *Baylisascaris* gen. nov. *Parasitology*, **58** (1): 185-198.

——— 1971. Speciation and development in the genus *Lagochilascaris*. *Parasitology*, **62** (1): 71-112.

——— & Mines, J. J. 1960. A new species of *Amplicaecum* (Nematoda) from the carpet snake (*Morelia argus variegatus*): with a re-definition and a key for the genus. *Parasitology*, **50** (1/2): 183-198.

WARREN, G. 1970. Studies on the morphology and taxonomy of the genera *Toxocara* Stiles, 1905 and *Neoascaris* Travassos, 1927. *Zool. Anz.* **185** (5/6): 393-442.

WILLIAMS, H. H. & RICHARDS, D. H. H. 1968. Observations on *Pseudanisakis rotundata* (Rudolphi, 1819) Mozgovoi, 1950, a common but little known nematode parasite of *Raia radiata* Donovan in the northern North Sea. *J. Helminth.*, **42** (1/2): 199-220.

YAMAGUTI, S. 1961. Systema Helminthum. Vol. III. The nematodes of vertebrates. Interscience Publishers Inc., New York.

* Important recent publications on taxonomy and nomenclature of the genera and higher taxa of Ascaridoidea not cited in Yamaguti (1961) and Chabaud (1965).

11 ASCARIDIDA

SUBULUROIDEA

by ALAIN G. CHABAUD*

SUBULUROIDEA

The systematics of the Subuluroidea have been revised by Inglis (1958, 1960) after a detailed study of buccal structures. He showed that each of the three oesophageal lobes extends anteriorly and forms a "pharyngeal portion" behind the true buccal cavity. Inglis distinguished the following in the pharyngeal portion of the buccal cavity:

(a) a *peripheral lobe*, ". . . a continuation forwards of the outer wall of the oesophagus and the pharynx which determines the outer limits of the portions. It generally forms a cup internally within which are two lobes which show a great range of variation in structure and degree of cuticularization . . ."; (b) a *radial lobe*, ". . . a continuation forwards of the corresponding sector of the oesophagus which narrows after the separation of the pharyngeal portion to form this wedge-shaped lobe which is, in most cases, attached to the inner side of the peripheral lobe by its broad base with its apex towards the centre-line of the body"; (c) a *chordal lobe* arising ". . . from the right-hand side of the main mass of the pharyngeal portions and is in most cases produced anti-clockwise to over-lie the next portion" (Fig. 6.172).

Quentin (1969, 1971) has stressed the ontogeny of the cephalic extremity and slightly modified and completed Inglis' classification; these modifications are followed herein.

Cyclobulura Quentin, 1977 which has recently been described, is added to the group.

Alaplectana Azim, 1931 was assigned to the Cosmocercidae. Although Azim did not report a preanal sucker we believe the species studied is really a member of the genus *Allodapa* which is found in the Phasianidae. Similarly, *Latibuccana* Patwarhan, 1935 and *Paraheterakis* Nama & Jain, 1974 seem to be synonyms of *Subulura.*

Kaszabospirura Mészáros, 1975, known only from two female specimens, is not a spirurid as suggested by Mészáros. It belongs in the synonymy of *Murisubulura.*

A monograph of the species, published in 1964 by Skrjabin *et al.* was brought up to date by the same authors in 1967.

SUBULUROIDEA

Key to families

1-(2) Oral opening surrounded by 12 leaf-like elements (Fig. 6.173).
Caudal alae broad and composed of a large anterior part and smaller posterior part (Fig. 6.174).

Maupasinidae

2-(1) Oral opening without leaf-like elements.
Caudal alae weakly developed or absent, and not divided into two parts.

Subuluridae

* Muséum National d'Histoire Naturelle, Paris, France.

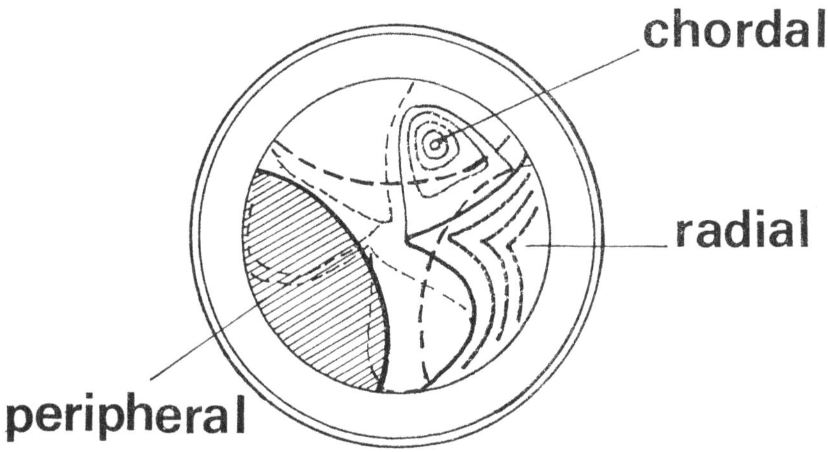

6.172

FIG. 6.172. Arrangement of the lobes of the pharyngeal portions as viewed in apical view (diagrammatic, only the portion on the right is fully drawn). Note how the inner lobes fit into the "cup" formed by the peripheral lobes, which are here drawn overlying the former. (After Inglis, 1958.)

Family *MAUPASINIDAE*
(Lopez-Neyra, 1945 subfam.) Inglis, 1959
(= Dubioxyuridae Ortlepp, 1937)

One genus (Figs. 6.173, 6.174).
Parasites of African Macroscelididae.

Maupasina Seurat, 1913
(= *Maupasiella* Seurat, 1913;
= *Dubioxyuris* Ortlepp, 1937)

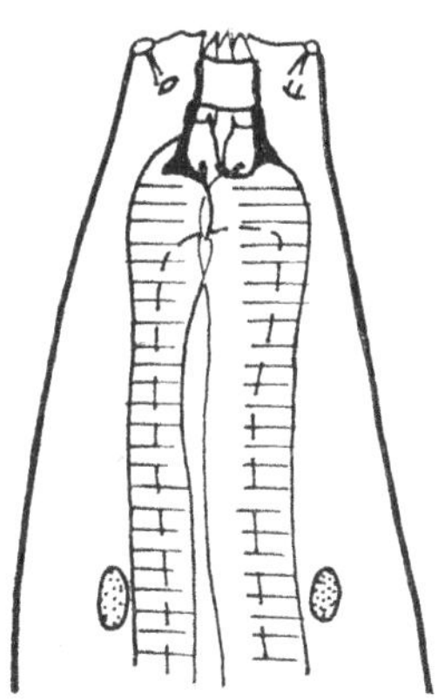

6.173

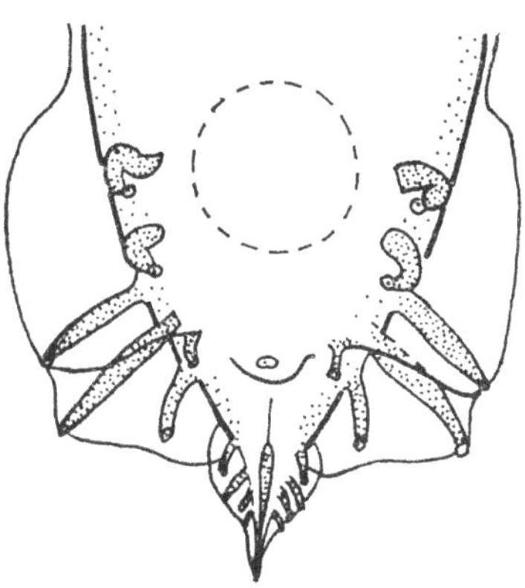

6.174

FIG. 6.173. *Maupasina*, anterior extremity. (After Ortlepp, 1937.)
FIG. 6.174. *Maupasina*, caudal extremity male, ventral view. (After Seurat, 1917.)

Family *SUBULURIDAE* (Travassos, 1914) Yorke & Maplestone, 1926

Key to subfamilies

1-(2) Pharyngeal portions of buccal cavity not lobed or twisted (Figs. 6.175, 6.179) but with elaborate arrangements of ridges (Fig. 6.176).
Parasites of birds and, more rarely, American marsupials.
Allodapinae

2-(1) Pharyngeal portions of buccal cavity more or less twisted (Figs. 6.180, 6.181, 6.184, 6.189, 6.191).
Peripheral, radial and chordal lobes partially or completely distinct.

3-(4) Pharyngeal portions basically cuticular.
Peripheral, radial and chordal lobes independent and elongated vertically (Figs. 6.180, 6.181, 6.184).
Parasites of Australian marsupials, *Xenarthra*, Tupaiidae and Asian Lemuroidea.
Labiobulurinae

4-(3) Pharyngeal portions basically muscular, not elongated vertically (Figs. 6.185, 6.189, 6.191).

5-(6) Peripheral and radial lobes of pharyngeal portions markedly muscular and indistinguishable.
Chordal lobes small, not twisted over neighbouring portions (Fig. 6.185).
Preanal sucker with entire ring of cuticle from which radiate cuticular lamellae (Fig. 6.186).
Parasites of Macroscelididae.
Parasubuluridae

6-(5) Peripheral and radial lobes of pharyngeal portions distinguishable.
Chordal lobes twisted (Figs. 6.188-6.191).
Preanal sucker not elaborated (Fig. 6.187).

7-(8) Peripheral and radial lobes incompletely separated.
Chordal lobes hemispherical, simple, twisted to form helix (Figs. 6.188, 6.189).
Oral opening triangular with three lip lobes off-set from body by grooves (Fig. 6.189).
Parasites of Australian birds.
Leipoanematinae

8-(7) Peripheral and radial lobes completely separated.
Chordal lobes helix in form, covering adjacent portions (Figs. 6.190, 6.191).
Anterior extremity not separated from body by constriction.
Parasites of birds and mammals (primates, carnivores and rodents).
Subulurinae

Subfamily **Allodapinae** Inglis, 1958

Key to genera and subgenera

1-(4) Grooves between lip lobes absent (Fig. 6.175).
Allodapa Diesing, 1861

2-(3) Oral opening simple (Fig. 6.177).
Parasites of various birds and American marsupials.
Allodapa (*Allodapa*) Diesing, 1861
(= ?*Alaplectana* Azim, 1931)

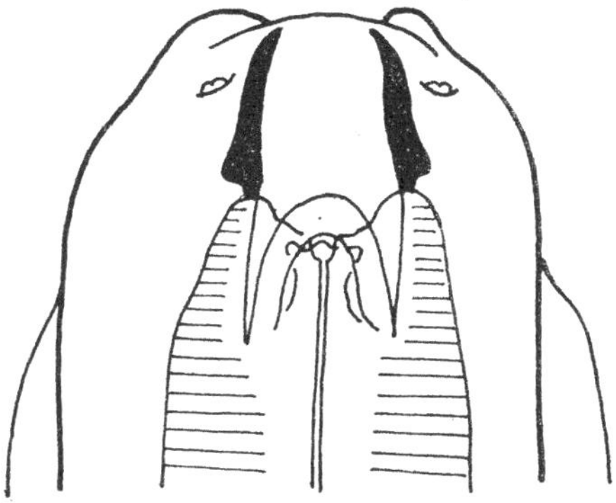

6.175

6.176

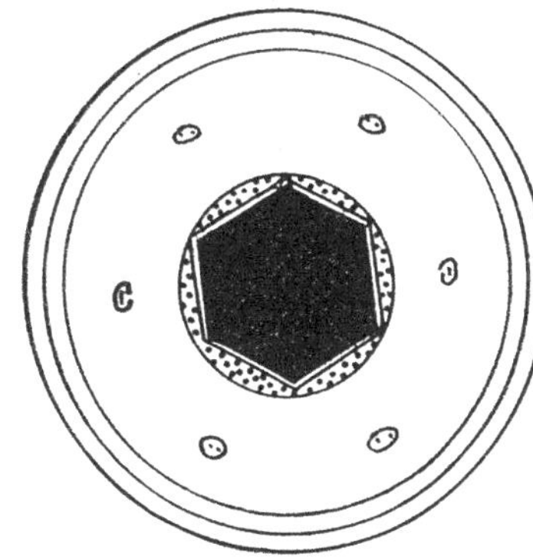

6.177

FIG. 6.175. *Allodapa*, anterior extremity. (After Inglis, 1958.)
FIG. 6.176. *Allodapa*, detail of pharyngeal portions, dorsal view. (After Inglis, 1958.)
FIG. 6.177. *Allodapa (Allodapa)*, apical view. (After Inglis, 1958.)

3-(2) Oral opening surrounded by six labial lobes (Fig. 6.178).
Parasites of galliform, ralliform and passeriform birds.

Allodapa (Labiodapa) Quentin, 1969

4-(1) Grooves between lip lobes present, partly enclosed by lateral expansion of cuticle from lip lobes (Fig. 6.179).
Parasites of Phasianidae.

Aulonocephalus Chandler, 1935

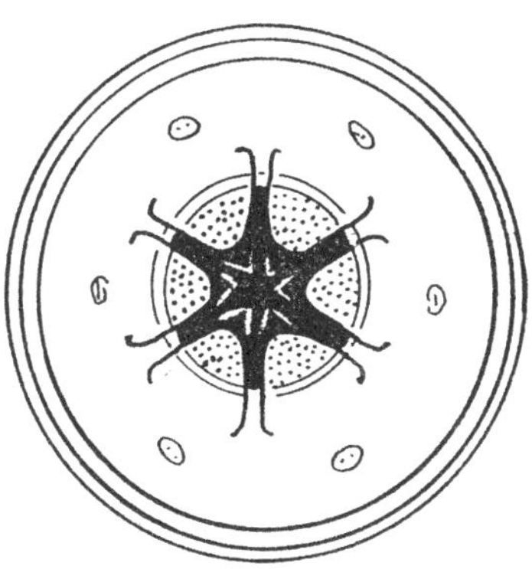

6.178

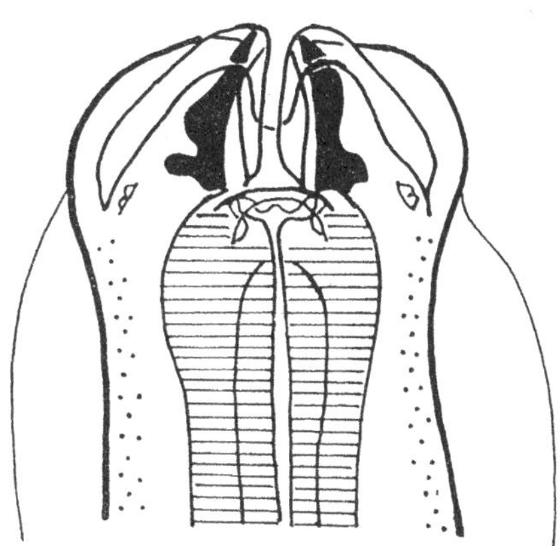

6.179

FIG. 6.178. *Allodapa (Labiodapa)*, apical view. (After Inglis, 1958.)
FIG. 6.179. *Aulonocephalus*, anterior extremity. (After Inglis, 1958.)

Subfamily **Labiobulurinae** Quentin, 1969

Key to genera and subgenera

1-(2) Buccal cavity small (Fig. 6.180), triangular in transverse section.
Oral opening simple without labial lobes.
Parasites of *Xenarthra.*
Cyclobulura Quentin, 1977

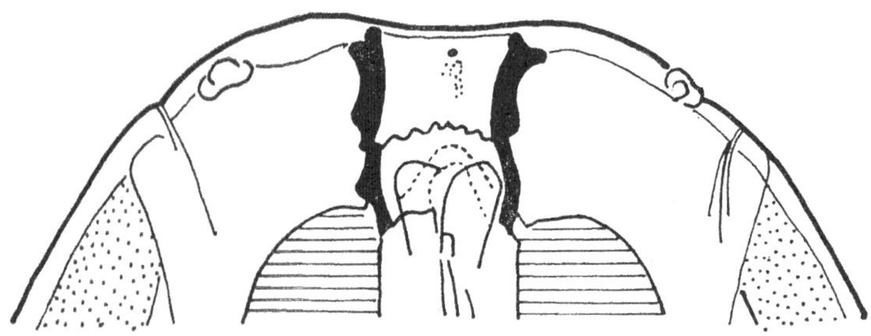

6.180

FIG. 6.180. *Cyclobulura,* anterior extremity. (After Quentin, 1977.)

2-(1) Buccal cavity large (Fig. 6.181), hexagonal or circular in transverse section.
Oral opening bordered by six labial lobes (Figs. 6.182, 6.183).

3-(6) Lobes of pharyngeal portions stout and elaborated (Fig. 6.184).
Parasites of Australian marsupials.
Labiobulura Skrjabin & Schikhobalova, 1948

4-(5) Oral opening surrounded by six simple labial lobes (Fig. 6.182).
Labiobulura (*Archeobulura*) Quentin, 1969

5-(4) Oral opening with six labial lobes separated by interlabia (Fig. 6.183).
Labiobulura (*Labiobulura*) Skrjabin & Schikhobalova, 1948

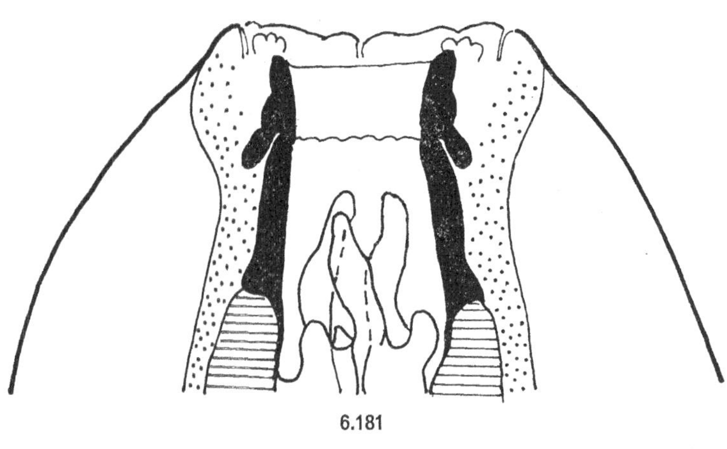

6.181

FIG. 6.181. *Tarsubulura,* anterior extremity. (After Quentin *et al.*, 1977.)

6-(3) Lobes of pharyngeal portions narrow and extending anteriorly into buccal cavity (Fig. 6.181).
Parasites of tarsiers in Asia.

Tarsubulura Inglis, 1958

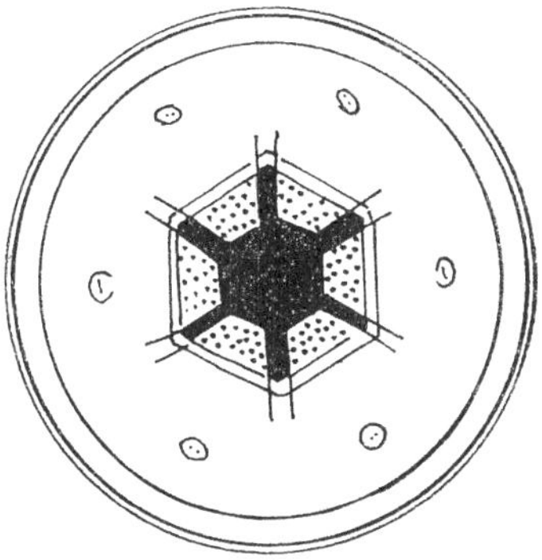

6.182

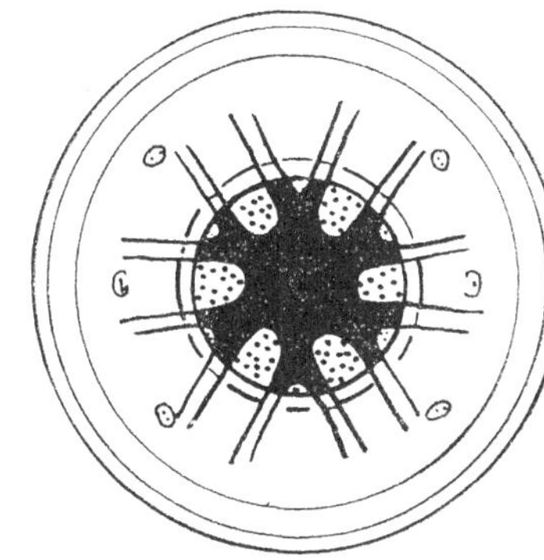

6.183

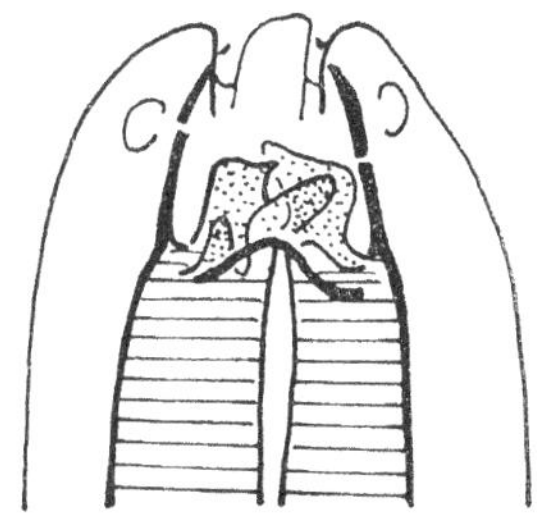

6.184

FIG. 6.182. *Tarsubulura*, apical view. (After Inglis, 1958.)
FIG. 6.183. *Labiobulura* (*Labiobulura*), apical view. (After Inglis, 1958.)
FIG. 6.184. *Labiobulura* (*Labiobulura*), anterior extremity. (After Inglis, 1958.)

Subfamily **Parasubulurinae** (Berghe & Vuylsteke, 1938, fam.)

One genus (Figs. 6.185, 6.186).
Parasites of African Macroscelididae.

Parasubulura Berghe & Vuylsteke, 1938
(= *Baylisnumidica* Lopez-Neyra, 1945;
= *Heterobulura* Skrjabin & Schikhobalova, 1948)

Subfamily **Leipoanematinae** Chabaud, 1957

One genus (Figs. 6.188, 6.189).
Parasites of Australian galliforms.

Leipoanema Johnston & Mawson, 1942

Subfamily **Subulurinae** Travassos, 1914

Key to genera and subgenera

1-(8) Buccal cavity circular or hexagonal in transverse section (Fig. 6.177).
2-(7) Spicules fully cuticularized.
Preanal sucker without cuticular elaboration (Fig. 6.187).
Cervical alae generally present.

Subulura Molin, 1860

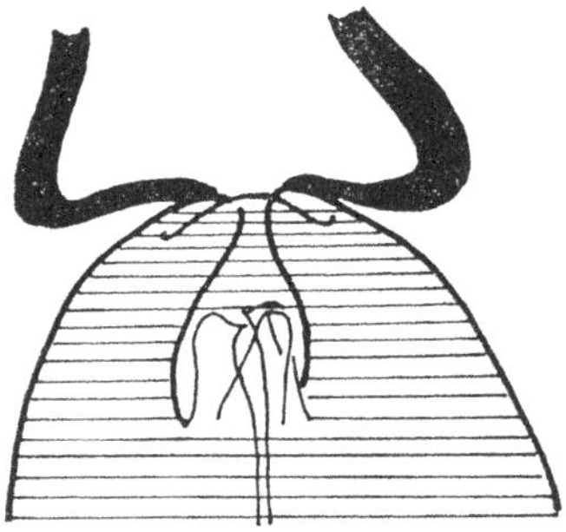

6.185

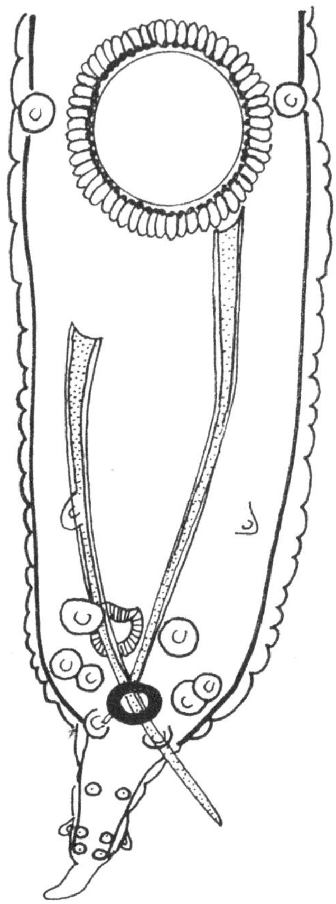

6.186

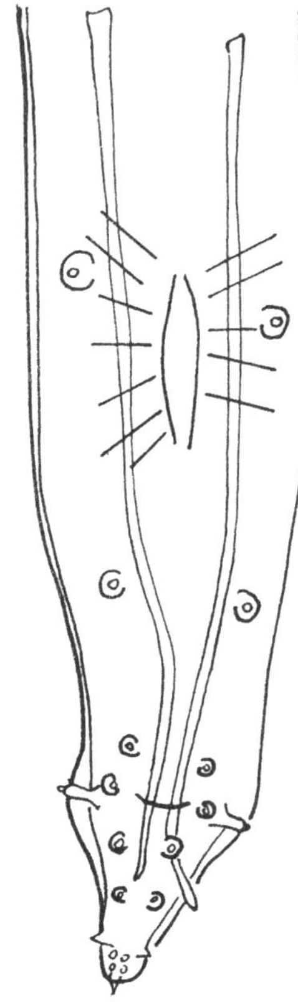

6.187

FIG. 6.185. *Parasubulura*, pharyngeal portions, dorsal view. (After Inglis, 1958.)
FIG. 6.186. *Parasubulura*, caudal extremity male, ventral view. (After Berghe & Vuylsteke, 1938.)
FIG. 6.187. *Subulura*, caudal extremity male, ventral view. (After Inglis, 1960.)

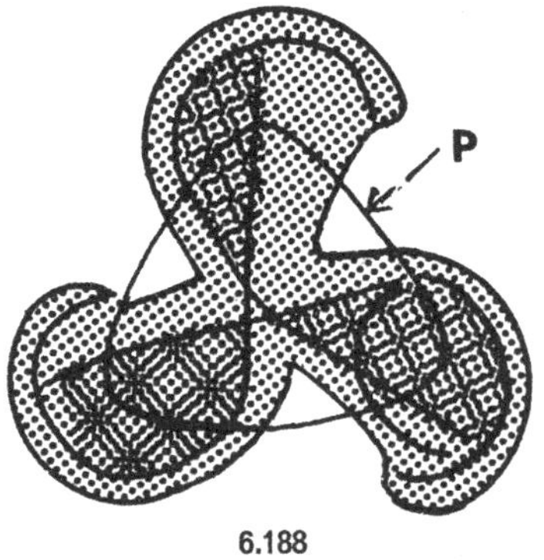

6.188

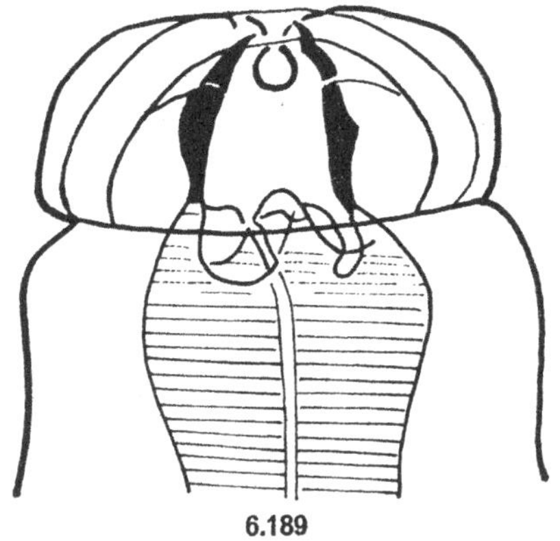

6.189

FIG. 6.188. *Leipoanema*, pharyngeal portions *en face*, showing anterior limits of peripheral lobes. (After Inglis, 1958.)
FIG. 6.189. *Leipoanema*, anterior extremity, dorsal view. (After Inglis, 1958.)

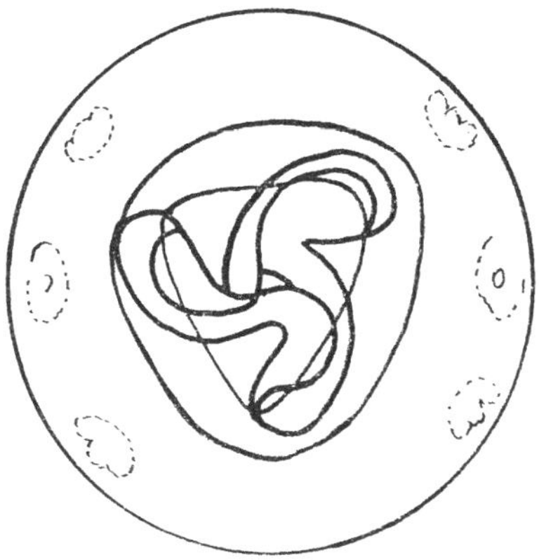

6.190

6.191

FIG. 6.190. *Subulura*, pharyngeal portions, apical view. (After Quentin, 1969.)
FIG. 6.191. *Subulura*, anterior extremity, dorsal view. (After Quentin, 1969.)

3-(4) Oral opening simple, triangular or circular (Fig. 6.190).
Parasites of birds (Cucullidae and Phasianidae), Lemuroidea in Africa and Madagascar, and rodents (Sciuridae and Gerbillidae).

Subulura (*Subulura*) Molin, 1860
(= *Latibuccana* Patwardhan, 1935;
Paraheterakis Nama & Jain, 1974)

4-(3) Oral opening surrounded by six labial lobes (Figs. 6.194, 6.195).
Parasites of rodents (Gerbillidae and Muridae).

5-(6) Labial lobes simple (Fig. 6.194).
Parasites of old-world rodents.

Subulura (*Murisubulura*) Quentin, 1969
(= *Kaszabospirura* Mészáros, 1975)

6-(5) Labial lobes separated by interlabia (Fig. 6.195).
Parasites of new-world rodents.

Subulura (*Tepuinema*) (Díaz-Ungría, 1964, genus)

7-(2) Left spicule weakly cuticularized.
Preanal sucker elongated, without definite rim but surrounded by area with cuticular bosses arranged in regular rows (Fig. 6.193).
Cervical alae generally absent.

Oxynema Linstow, 1899

8-(1) Buccal cavity triangular in transverse section (Fig. 6.192).

Primasubulura Inglis, 1958

9-(10) Oral opening triangular and simple (Fig. 6.196).
Parasites of platyrhinine monkeys.

Primasubulura (*Platysubulura*) Quentin, 1969

10-(9) Oral opening triangular and bordered by lips or labial lobes (Fig. 6.192).
Parasites of catarhinine monkeys.

Primasubulura (*Primasubulura*) Inglis, 1958

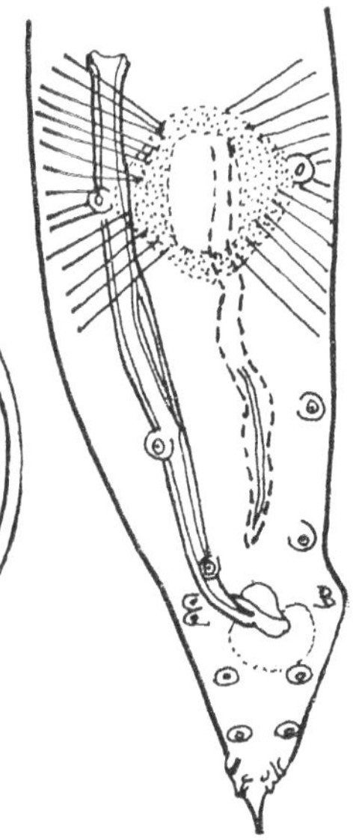

6.192 6.193 6.194

Fig. 6.192. *Primasubulura* (*Primasubulura*), apical view. (After Inglis, 1958.)
Fig. 6.193. *Oxynema*, caudal extremity male, ventral view. (After Baylis, 1930.)
Fig. 6.194. *Subulura* (*Murisubulura*), apical view. (After Quentin, 1965.)

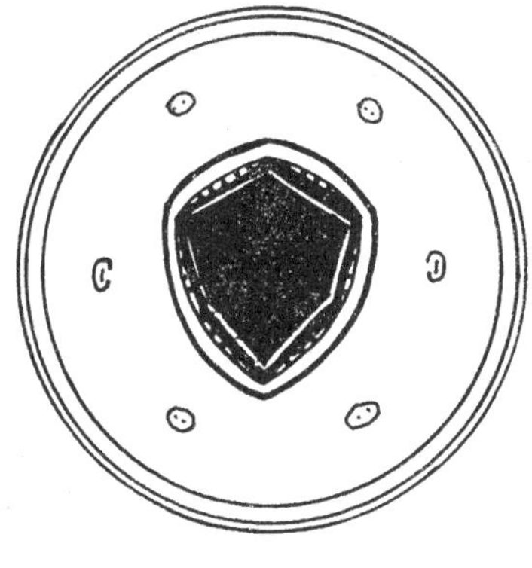

6.195 6.196

Fig 6.195. *Subulura* (*Tepuinema*), outline of oral opening, apical view. (After Quentin, 1971.)
Fig. 6.196. *Primasubulura* (*Platysubulura*), apical view. (After Inglis, 1958.)

REFERENCES*

INGLIS, W. G. 1958. The comparative anatomy of the subulurid head (Nematoda): with a consideration of its systematic importance. *Proc. zool. Soc. Lond.*, **130,** 577-604.

——— 1960. Further observations on the comparative anatomy of the head in the nematode family Subuluridae: with the description of a new species. *Proc. zool. Soc. Lond.*, **135,** 125-136.

MÉSZÁROS, F. 1974. Two spirurids (Nematoda) from *Allactaga sibirica* (Mammalia) in Mongolia. *Acta zool. Hung.*, **21,** 97-100.

QUENTIN, J. C. 1969. Cycle biologique de *Subulura williaminglisi* Quentin, 1965. Ontogénèse des structures céphaliques. Valeur phylogénétique de ce caractère dans la classification des Nématodes Subuluridae. *Annls Parasit. hum. comp.*, **44,** 451-484.

——— 1971. Sur les modalités d'évolution chez quelques lignées d'helminthes de rongeurs Muroidea. *Cah. O.R.S.T.O.M., sér. Ent. méd. Parasitol.*, **9,** 103-176.

——— 1977. *Cyclobulura lainsoni* n.gen. n.sp., Nématode subulure parasite d'un Xénarthre d'Amérique du Sud. *Bull. Mus. natn. Hist. nat.*, 3ème sér., No. 469, Zool., **326,** 771-776.

———, KRISHNASAMY, M. & TCHEPRAKOFF, R. 1977. Cycle biologique de *Tarsubulura perarmata* (Ratzel, 1868). *Annls Parasit. hum. comp.*, **52,** 159-170.

SKRJABIN, K. I., SHIKHOBALOVA, N. P. & LAGODOVSKAYA, E. A. 1964. *Oxyurata of animals and man.* Part 3. *Principles of Nematology*, Vol. XIII, edited by Akad. K. I. Skrjabin, Moscow: Izdatel'stvo Akademii Nauk SSSR, 468 pp. (In Russian).

———, ———, ——— 1967. *Oxyurata of animals and man.* Part 5. *Principles of Nematology*, Vol. XVIII, edited by Akad. K. I. Skrjabin, Moscow: Izdatel'stvo Akademii Nauk SSSR, 243 pp. (In Russian).

*For most papers published before 1969 refer to bibliographies in Inglis (1958), Quentin (1969) and Skrjabin *et al.* (1964 and 1967).

12 SPIRURIDA

CAMALLANOIDEA, DRACUNCULOIDEA, GNATHOSTOMATOIDEA, PHYSALOPTEROIDEA, RICTULAROIDEA AND THELAZIOIDEA

by ALAIN G. CHABAUD*

Nematodes of the order Spirurida possess an anterior extremity which is bilaterally symmetrical and lack lateral, external labial papillae. The caudal papillae are always ventral or ventrolateral in position. The oesophagus is divided, although the division may be indistinct, into an anterior muscular part and a posterior glandular portion, the anterior section being shorter than the posterior section. There is no pre-anal sucker. These nematodes are found in the anterior part of the gut (the oesophagus, stomach or, rarely, the duodenum) or in tissues and tissue spaces. The larval stages, preinfective for the final host, develop entirely within an intermediate host.

There are two sub-orders, the Camallanina and the Spirurina, divided into two and ten superfamilies respectively. These are the Camallanoidea and Dracunculoidea in the Camallanina and the Gnathostomatoidea, Physalopteroidea, Rictularioidea, Thelazioidea, Spiruroidea, Habronematoidea, Acuarioidea, Filarioidea, Aproctoidea and Diplotriaenoidea in the Spirurina. The first six superfamilies are covered here. No. 3, Part 2 will continue keys to the Spiruroidea, Habronematoidea and Acuarioidea and No. 3, Part 3 to the Filarioidea, Aproctoidea and Diplotriaenoidea.

KEY TO GENERA OF THE CAMALLANOIDEA

It now seems clear that the family Cucullanidae lacks fundamental affinities with the Camallanoidea.

Yeh (1960) discovered a character which makes it possible to recognize the species parasitic in reptiles and he created for them the genus *Serpinema*. The genus *Zeylanema* Yeh, 1960 is not accepted here as there are forms which are intermediate between those species which have denticulate longitudinal bands in the cephalic valves and those in which the bands are smooth. The ensemble of species based on this character does not seem to be homogeneous.

Baudin Laurencin (1971) showed that *Trichocephalus gibbosus* Rudolphi, 1819 is a camallanoid in which the female is deformed by parasitism. The position of *Oncophora* Diesing, to which *T. gibbosus* has been assigned, is now clear and *Piscillania* Yeh, 1960 falls as a synonym of this genus.

The genus *Malayocamallanus* has been described by Jothy & Fernando (1971).

The phylogeny of the camallanids is readily determined because, during development, larvae of the highly evolved genera take on cephalic structures found in more primitive genera (Campana-Rouget, 1961) and it is possible, therefore, to base the systematics mainly on the structure of the buccal capsule. It is for this reason the classifications of Ali (1957 and 1960) and Chakravarty, Majumdar & Sain (1960), which stress spicule morphology and the sclerotization of the pieces to which the buccal musculature is attached, are not retained.

* Muséum National d'Histoire Naturelle, Paris, France.

CAMALLANOIDEA

One family

Family *CAMALLANIDAE*

Key to genera

1-(6) Buccal capsule rounded, continuous, not separated into two valves (Fig. 3.1).
2-(3) Buccal capsule with smooth internal walls (Fig. 3.2).

Procamallanus Baylis, 1923
(=*Neocamallanus sensu* Chakravarty *et al.*, 1962, not Ali, 1957;
=*Indocamallanus* Chakravarty *et al.*, 1963)

3-(2) Buccal capsule with walls reinforced by sclerotized bands.
4-(5) Sclerotized bands of the buccal capsule longitudinal.
Bands rather irregular, sometimes armed at the base with denticles.

Malayocamallanus Jothy & Fernando, 1971

5-(4) Sclerotized bands of buccal capsule in form of spiral (Fig. 3.3).

Spirocamallanus Olsen, 1952

6-(1) Buccal capsule divided into two lateral valves (Fig. 3.4).
7-(8) Buccal capsule divided into two levels, a large cavity being present behind the valves (Fig. 3.5).

Paracamallanus Yorke & Maplestone, 1928
(=*Neocamallanus* Ali, 1957)

8-(7) Buccal cavity reduced or absent behind valves.
9-(10) Buccal valves with a thick lateral, longitudinal external band (Fig. 3.6).
Dorsal and ventral tridents each replaced by a sclerotized rod.
Parasites of snakes.

Camallanides Baylis & Daubney, 1922

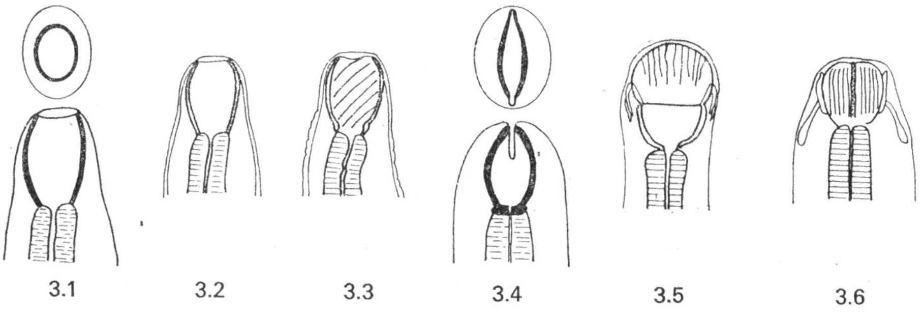

FIG. 3.1. *Procamallanus*, buccal capsule undivided, apical view, ventral view. (After Yeh, 1960.)
Fig. 3.2. *Procamallanus*, buccal capsule with smooth walls. (After Yeh, 1960.)
FIG. 3.3. *Spirocamallanus*, buccal capsule with chitinoid bands arranged in a spiral. (After Yeh, 1960.)
FIG. 3.4. *Camallanus*, buccal capsule divided into two valves, apical view, ventral view. (After Yeh, 1960.)
FIG. 3. 5. *Paracamallanus*. (After Yeh, 1960.)
FIG. 3.6. *Camallanides*. (After Yeh, 1960.)

10-(9) Buccal valves without a thick lateral longitudinal band.

11-(12) Longitudinal bands sustaining the buccal valves transformed posteriorly into rows of spines or a few combs (Fig. 3.7).
Posterior buccal cavity present. Mature female with posterior extremity enlarged as in that of *Trichocephalus*.
Parasites of marine fish.

Oncophora Diesing, 1851
(=*Piscillania* Yeh, 1960)

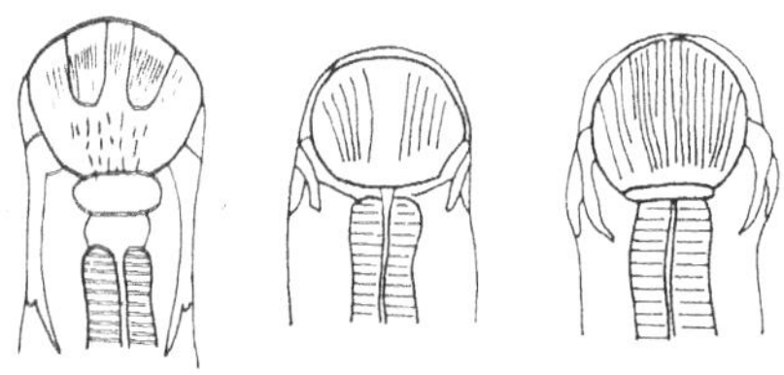

3.7 3.8 3.9

FIG. 3.7. *Oncophora*. (After Yeh, 1960.)
FIG. 3.8. *Serpinema*. (After Yeh, 1960.)
FIG. 3.9. *Camallanus*. (After Yeh, 1960.)

12-(11) Longitudinal bands sustaining the buccal valves continuous.

13-(14) Longitudinal bands sustaining the buccal valves separated into a ventral and a dorsal group (Fig. 3.8).
Parasites of reptiles.

Serpinema Yeh, 1960

14-(13) Longitudinal bands sustaining the buccal valves not separated into ventral and dorsal groups (Fig. 3.9).
Parasites of fishes and amphibians.

Camallanus Railliet & Henry, 1915
(=*Zeylanema* Yeh, 1960)

REFERENCES

ALI, S. M. 1957. Studies on the nematode parasites of fishes and birds found in Hyderabad State. *Indian J. Helminth.*, **8** (1), 1-83.

ALI, S. M. 1960. On two new species of *Procamallanus* Baylis, 1923 from India, with a key to the species. *J. Helminth.*, **34** (1/2), 129-138.

BAUDIN LAURENCIN, F. 1970. *Oncophora melanocephala* (Rud.1819) n.comb., Nématode Camallanidae, parasite du thon albacore *Neothunnus albacares*. *Bull. Mus. natn. Hist. nat.*, **42**, 984-988.

CAMPANA-ROUGET, Y. 1961. Remarques sur le cycle évolutif de *Camallanus lacustris* (Zoega, 1776) et la phylogénie des Camallanidae. *Annls Parasit. hum. comp.*, **36** (3), 425-434.

CHAKRAVARTY, G. K., MAJUMDAR, G. & SAIN, S. K. 1961. On a camallanid nematode *Neocamallanus heteropneusti* n.gen. & sp. with emendation of the family. *Zool. Anz.*, **166** (5/6), 221-224.

JOTHY, A. A. & FERNANDO, C. H. 1971. A new camallanid nematode, *Malayocamallanus intermedius* gen. et sp.nov., from a Malayan freshwater fish, *Fluta alba* (Zuiew) with a key to the genera of the subfamily Procamallaninae. *Helminthologia*, Year 1970, **11** (1/4), 87-91.

YEH, L. S. 1960. On a reconstruction of the genus *Camallanus* Railliet and Henry, 1915. *J. Helminth.*, **34** (1/2), 117-124.

KEY TO GENERA OF THE DRACUNCULOIDEA

Forms are gradually being discovered which relate the dracunculoids, markedly deformed by their adaptation to life within tissues, to species with relatively little specialization and close to the camallanoids. The works of Shigin & Shigina (1958), Molnar (1966) and Moravec (1968) have established the Skrjabillanidae and that of Petter (1975) the Guyanemidae so that the family Anguillicolidae is not as isolated as previously. The recognition of the Philoneminae as a subfamily by Skrjabin *et al.* (1971) neatly allies the Philometridae to the preceding families.

The data provided by Rasheed (1963) on the Philometridae are adopted here since the males of many genera are unknown and it seems necessary, at least provisionally, to use characters provided by the cuticular ornamentation to classify this difficult group.

Mudry & Dailey (1969) have definitely established the affinities of the curious genus *Phlyctainophora* with the Philometridae. On the other hand, *Tetanonema* Steiner, 1937 has not been rediscovered. It is either some adenophorean or dracunculoid and is treated here as *incertae sedis*.

Our classification of the dracunculids falls between the highly restricted approach of Skrjabin *et al.* (1971) and the broad one of Yamaguti (1961). The same species, namely *D. globocephalus*, is considered as a synonym of all the *Dracunculus* of reptiles by the Soviet authors but is used by Yamaguti as the type species of a new genus and of a new subfamily. We do not accept Yamaguti's new taxa but we also do not believe that the *Dracunculus* of reptiles throughout the whole world comprises only a single species (see for example Vaucher & Bain, 1973).

The genus *Pesteria* Tadros, 1966 is a synonym of *Dracunculus* and the type species almost certainly corresponds with badly preserved specimens of *D. oesophageus* (Polonio, 1859).

KEY TO FAMILIES

1-(2) Buccal capsule present. Oesophagus short. Vulva functional. Spicules absent or poorly developed.
Parasites of fishes.

Anguillicolidae

2-(1) Buccal capsule absent or reduced to a peribuccal ring.

3-(4) Vulva anterior, well developed in mature female. Monodelphic. Caudal alae of male large (Fig. 3.10).
Peribuccal ring absent. Anus functional. Oesophagus divided, with long glandular part.
Parasites of fishes.

Guyanemidae

4-(3) Vulva equatorial or posterior, more or less completely atrophied in the mature female. Two ovaries present (except *Phlyctainophora*). Caudal alae in male narrow or absent.

5-(6) Glandular part of oesophagus short (Fig. 3.11) or absent. Tail of male rounded and very short (Fig. 3.12) (except *Philonema*, Fig. 3.13). Peribuccal ring absent.
Parasites of fishes.

Philometridae

6-(5) Glandular part of oesophagus long. Tail of male sharply pointed, of average length.

7-(8) Peribuccal ring absent. Sexual dimorphism moderate.
Anus functional.
Parasites of archaic reptiles (crocodilians and chelonians).

Micropleuridae

8-(7) Peribuccal ring present (Fig. 3.14) (except *Avioserpens*). Sexual dimorphism marked.
Anus atrophied in adult.
Parasites of reptiles, birds, and mammals.

Dracunculidae

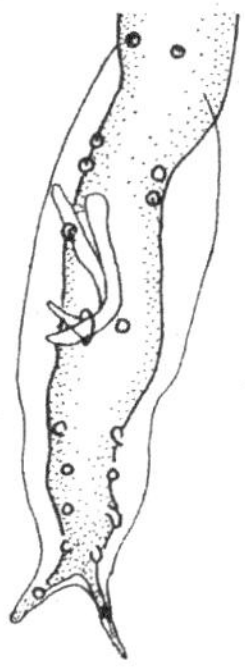

3.10

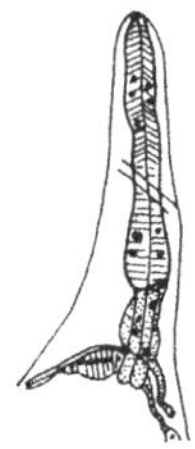

3.11

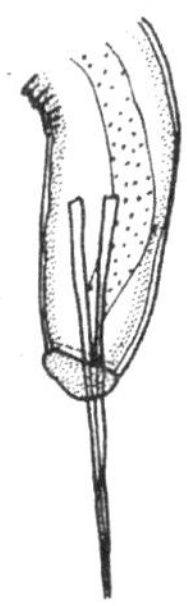

3.12

3.13

3.14

FIG. 3.10. *Guyanema*, caudal alae. (After Petter, 1973.)
FIG. 3.11. *Ichthyofilaria*, oesophagus. (After Yamaguti, 1935.)
FIG. 3.12. *Philometra*, male tail, ventral view. (After Skrjabin *et al.*, 1971.)
FIG. 3.13. *Philonema*, male tail, ventral view. (After Platzer & Adams, 1967.)
FIG. 3.14. *Dracunculus*, head, apical view. (After Chabaud, 1960.)

KEY TO SUBFAMILIES AND GENERA

Family *ANGUILLICOLIDAE* Yamaguti, 1935

Key to subfamilies

1-(2) Oesophagus short and divided (Fig. 3.15). Tail of male without caudal alae.
Spicules and gubernaculum absent. Anus atrophied. Vulva posterior.
Parasites of swim bladder of eels.

Anguillicolinae

2-(1) Oesophagus short, not divided. Tail of male with large caudal alae in the form of a bursa (Fig. 3.16).
Spicules minute. Anus functional. Vulva anterior.
Parasite of the peritoneal cavity of palearctic freshwater fishes.

Skrjabillaninae

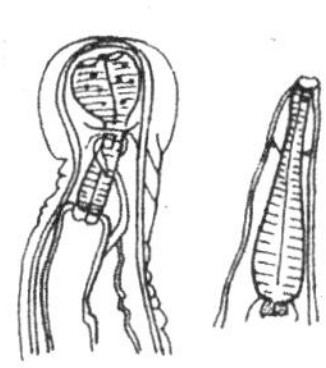

3.15

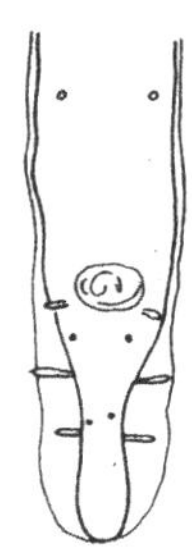

3.16

FIG. 3.15. *Anguillicola*, oesophagi, female on left, male on right. (After Wu, 1956.)
FIG. 3.16. *Skrjabillanus*, male tail, ventral view. (After Shigin & Shigina, 1958.)

Subfamily **Anguillicolinae** (Yamaguti, 1935, fam.)

One genus.
Parasites of the swim bladder of eels.

Anguillicola Yamaguti, 1935

Subfamily **Skrjabillaninae** (Shigin & Shigina, 1958 fam.) Chabaud, 1965

Key to genera

1-(2) Cephalic end broad, with four dome-shaped papillae protruding anteriorly. Buccal cavity reduced to a thick-walled cuticularized ring (Fig. 3.17).
Molnaria Moravec, 1968

2-(1) Cephalic extremity rounded, dome shaped, without protruberances. Oral opening and buccal cavity well developed (Fig. 3.18).
Skrjabillanus Shigin & Shigina, 1958
(=*Agrachanus* Tichomirova *in* Skrjabin *et al.*, 1971)

Family *GUYANEMIDAE* Petter, 1975

One genus
Parasites of body-cavity of neotropical freshwater fishes.

Guyanema Petter, 1975

Family *PHILOMETRIDAE*
Baylis & Daubney, 1926

Key to subfamilies

1-(4) Body elongated, more or less cylindrical. With two ovaries. Cuticle smooth or ornamented with bosses or small excrescences.

2-(3) Female tail attenuated. Male tail spirally coiled and attenuated. Anus some distance from caudal extremity (Fig. 3.13).
Philoneminae

3-(2) Female tail blunt, rounded, or slightly tapering. Male tail blunt and rounded with or without lobes. Cloaca terminal (Fig. 3.12).
Philometrinae

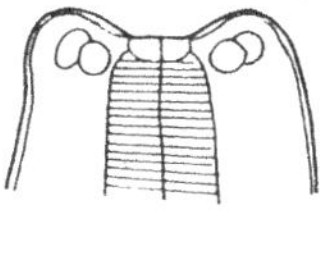

3.17

3.18

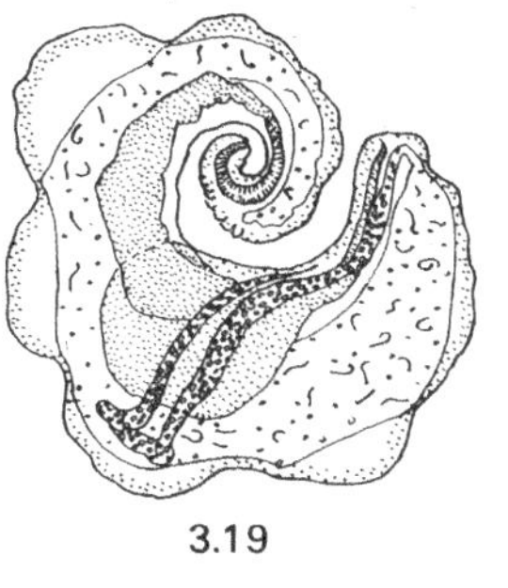

3.19

FIG. 3.17. *Molnaria*, head. (After Moravec, 1968.)
FIG. 3.18. *Skrjabillanus*, head. (After Moravec, 1968.)
FIG. 3.19. *Phylctainophora*, body of female. (After Mudry & Dailey, 1969.)

4-(1) Body of female bent dorsally. One ovary present (Fig. 3.19).
Ventral part of cuticle swollen by large vesicular structures.

Phlyctainophorinae

Subfamily **Philoneminae**
(Skrjabin, Sobolev & Khromova, 1971)

One genus
Parasites of fishes, especially salmonids.

Philonema Kuitunen-Ekbaum, 1933
(=? *Coregonema* Bauer, 1946)

Subfamily **Philometrinae**
(Baylis & Daubney, 1926 fam.)

Key to genera

1-(12) Cuticle smooth.
2-(3) Muscular part of oesophagus elongate with two swellings separated by nerve ring, with free posterior glandular appendix at junction of two portions (Fig. 3.11).
Ichthyofilaria Yamaguti, 1935
3-(2) Oesophagus unlike the above.
4-(7) Four cephalic papillae, large and lobe-like (Fig. 3.20).
5-(6) Oesophagus swollen or widened near mouth.
Female tail blunt or rounded.

Thwaitia Rasheed, 1963

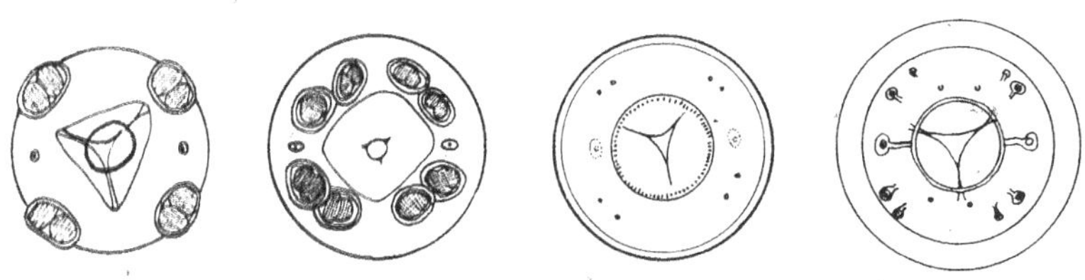

3.20 3.21 3.22 3.23

FIG. 3.20. *Thwaitia*. (After Rasheed, 1963.)
FIG. 3.21. *Philometra* (*Ranjhinema*). (After Rasheed, 1963.)
FIG. 3.22. *Philometra* (*Alinema*). (After Rasheed, 1963.)
FIG. 3.23. *Philometra* (*Philometra*.) (After Rasheed, 1963.)

6-(5) Oesophagus cylindrical, not swollen, near oral opening.
Female tail conical.

Rumai Travassos, 1960

7-(4) More than four cephalic papillae present, may be obscure, small or large.
Philometra Costa, 1845
(=*Sanguinofilaria* Yamaguti, 1935; =? *Clavinema* Yamaguti, 1935)

8-(9) Cephalic papillae of external circle, large, lobe-like and in double pairs (Fig. 3.21).
Philometra (*Ranjhinema*) Rasheed, 1963

9-(8) Cephalic papillae of external circle smaller, often fused but not in double pairs.

2*

10-(11) Rim of oral opening studded by small, cuticularized, bead-like structures. Cephalic papillae minute and indistinct (Fig. 3.22).

Philometra (Alinema) Rasheed, 1963

11-(10) Rim of oral opening simple and unornamented. Cephalic papillae of both circles distinct in apical view (Fig. 3.23).

Philometra (Philometra) Costa, 1845

12-(1) Cuticle ornamented.

13-(14) Cuticle with regular rows of spines (Fig. 3.24).
Body of female coiled.

Spirophilometra Parukhin, 1971

14-(13) Cuticle without regular rows of spines.

15-(16) Cuticle with cone-shaped excrescences (Fig. 3.25).
Oesophagus cylindrical, not swollen near mouth.
Female tail pointed.

Nilonema Khalil, 1960

16-(15) Cuticle without cone-shaped excrescences.
Oesophagus bulbous or somewhat swollen near oral opening.
Female tail rounded or blunt.

17-(18) Cuticle with rod-like structures arranged in patterns (Fig. 3.26).
Cuticularized oesophageal teeth present (Fig. 3.27).
Cephalic papillae large, flat, lobe-like (Fig. 3.27).

Buckleyella Rasheed, 1963

3.24

3.25

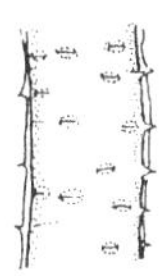
3.26

3.27

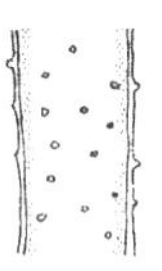
3.28

FIG. 3.24. *Spirophilometra*, body spines. (After Parukhin, 1971.)
FIG. 3.25. *Nilonema*, cuticle. (After Rasheed, 1963.)
FIG. 3.26. *Buckleyella*, cuticle. (After Rasheed, 1963.)
FIG. 3.27. *Buckleyella*, head. (After Rasheed, 1963.)
FIG. 3.28. *Philometroides*, cuticle. (After Rasheed, 1963.)

18-(17) Cuticle with small, irregular bosses (Fig. 3.28).
Oesophageal teeth present or absent.
Cephalic papillae small.

Philometroides Yamaguti, 1935
(=*Pseudophilometroides* Parukhin, 1966)

Subfamily **Phlyctainophorinae** (Roman, 1965, fam.)

One genus

Parasites of subcutaneous tissues of selachians.

Phlyctainophora Steiner, 1921

Family *MICROPLEURIDAE* (Baylis & Daubney, 1926, subfam.) Travassos, 1960

One genus
Parasites of body-cavity of crocodilians and chelonians.

Micropleura Linstow, 1906

Family *DRACUNCULIDAE* (Stiles, 1907 subfam.) Leiper, 1912

Key to genera

1-(2) Peribuccal ring not clearly defined.
Parasites of birds.

Avioserpens Wehr & Chitwood, 1934
(=*Oshimaia* Sugimoto, 1934; =*Petroviprocta* Schachtachtinskaja, 1951)

2-(1) Peribuccal ring thick and clearly defined (Fig. 3.14).
Parasites of reptiles and mammals.

Dracunculus (Reichard, 1759)
(=*Ophiodracunculus* Yamaguti, 1961; =*Chelonidracunculus* Yamaguti, 1961; =*Pesteria* Tadros, 1966)

REFERENCES

KHALIL, L. F. 1960. On a new nematode, *Nilonema gymnarchi* gen. et sp.nov. (Dracunculidae), from a freshwater fish in the Sudan. *J. Helminth.*, **34** (1/2), 55-58.

MOLNÁR, K. 1966. On some little-known and new species of the genera *Philometra* and *Skrjabillanus* from fishes in Hungary. *Acta Vet. hung.*, **16** (2), 143-157.

MORAVEC, F. 1968. A new nematode genus, *Molnaria* gen.n. (Nematoda : Skrjabillanidae). *Folia Parasit.*, **15** (4), 322.

MUDRY, D. R. & DAILEY, M. D. 1969. *Phlyctainophora squali* sp.nov. (Nematoda, Philometridae) from the spiny dogfish, *Squalis acanthias*. *Proc. Helminth. Soc. Wash.*, **36** (2), 280-284.

PARUKHIN, A. M. 1966. [*Pseudophilometroides atropi* n.g., n.sp., a new nematode of the family Dracunculidae Leiper, 1912.] *Zool. Zh.*, **45** (5), 766-767. [In Russian: English summary p. 767.]

PARUKHIN, A. M. 1971. *Spirophilometra eichleri* n.g. n.sp. ein neuer Fisch-Nematode aus dem Indik. *Angew. Parasit.*, **12** (4), 220-224.

PETTER, A. J. 1975. Deux nouvelles espèces de Camallanina parasites de *Hoplerythrinus unitaeniatus* (Characidae, Cypriniformes) en Guyane. *Bull. Mus. natn. Hist. nat.*, 3ème sér., No. 228, 1974, Zool. 156.

PLATZER, E. G. & ADAMS, R. A. 1967. The life-history of a dracunculoid, *Philonema oncorhynchi*, in *Oncorhynchus nerka*. *Canad. J. Zool.*, **45** (1), 31-43.

RASHEED, S. 1963. A revision of the genus *Philometra* Costa, 1845. *J. Helminth.*, **37** (1/2), 89-130.

SHIGIN, A. A. & SHIGINA, N. G. 1958. [A new parasite of tench, *Skrjabillanus tincae* n.g., n.sp. (Nematoda : Camallanata).] [Papers on helminthology presented to Acad. K. I. Skrjabin on his 80th birthday.] Moscow: Izdat. Akad. Nauk SSSR, pp. 395-399. [In Russian.]

SKRJABIN, K. I., SOBOLEV, A. A. & KHROMOVA, A. 1971. Osnovi nematodologii. (Camallanat'). Vol. **22**. Moscow: Izdat. "Nauka", 388 pp. [In Russian.]

TADROS, G. 1966. On *Pesteria inglisi* gen.nov., sp.nov. (Filariidae) from the English grass snake (*Tropidonotus natrix*). *J. Vet. Sci. U.A.R.*, **3** (1), 55-64.

TRAVASSOS, L. 1960. Sôbre nematódeos cavitários de peixes do Rio Amazonas. *Atas Soc. Biol. Rio de Janeiro*, **4** (2), 15-20.

VAUCHER, C. & BAIN, O. 1973. Développement larvaire de *Dracunculus doi* (Nematoda), parasite d'un serpent malgache et description de la femelle. *Annls Parasit. hum. comp.*, **48** (1), 91-104.

YAMAGUTI, S. 1961. Systema Helminthum. Vol. III. The Nematodes of Vertebrates. New York: Interscience Publishers Inc., Pt. I, 679 pp., Pt. II, pp. 681-1261.

KEY TO GENERA OF THE GNATHOSTOMATOIDEA

The arrangement of three subfamilies proposed by Chitwood and confirmed by Ivashkin (1960) is followed. The genus *Ancyracanthus*, previously known only from older descriptions, has been recently redescribed by Gomes & Kohn (1970).

The genus *Piayussunema* Kohn, Gomes & Motta (1968) is not retained because ornaments found on the pseudolabia of certain Spirurida are extremely varied and it seems advisable to limit, as far as possible, the number of genera created on the basis of these structures.

Family *GNATHOSTOMATIDAE* Railliet, 1895

Key to subfamilies

1-(4) Cephalic bulb absent (Figs. 3.29; 3.30).
2-(3) Pseudolabia without posterior appendages (Fig. 3.29). **Spiroxyinae**
3-(2) Pseudolabia with long ramified appendages (Fig. 3.30). **Ancyracanthinae**
4-(1) Cephalic bulb present (Figs. 3.31-3.33). **Gnathostomatinae**

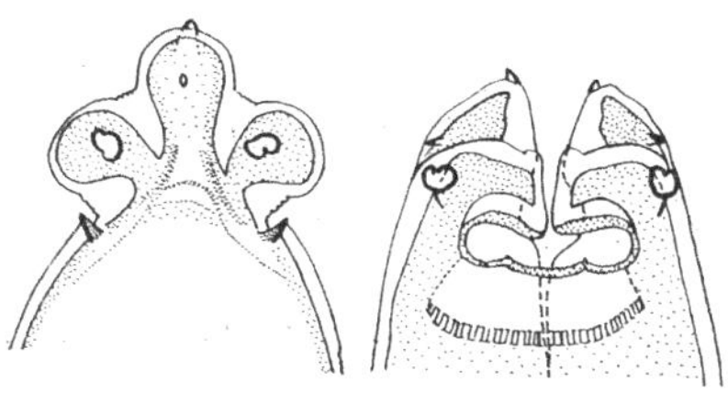
3.29

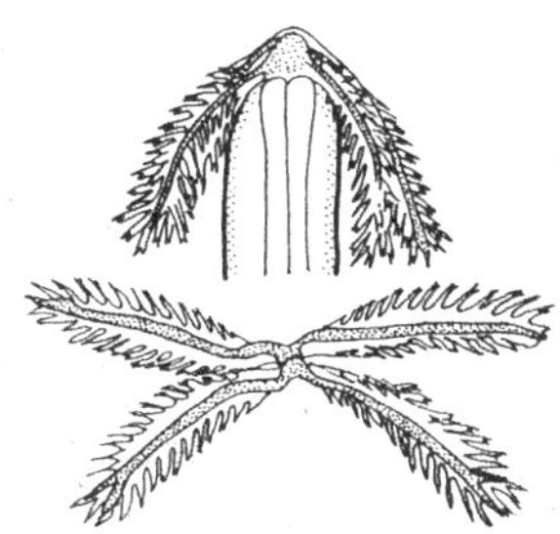
3.30

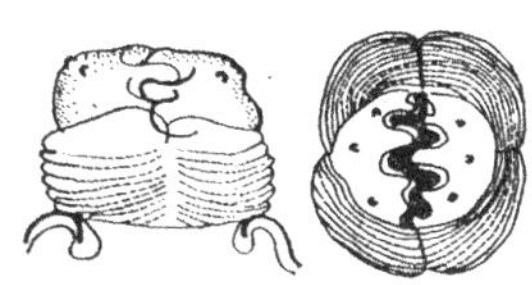
3.31

FIG. 3.29. *Spiroxys*, head, lateral view on left, median view on right. (After Freitas & Dobbin, 1971.)
FIG. 3.30. *Ancyracanthus*, head, lateral view, cephalic appendages, apical view. (After Gomes & Kohn, 1970.)
FIG. 3.31. *Tanqua*, head, median view, apical view. (After Baylis & Lane, 1920 and Mönnig, 1924.)

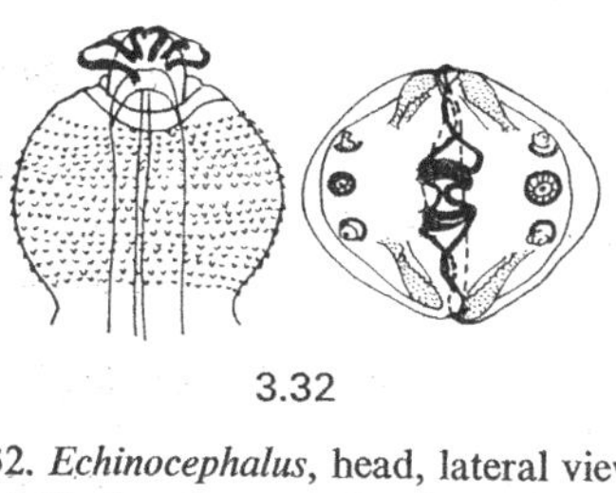
3.32

3.33

FIG. 3.32. *Echinocephalus*, head, lateral view, apical view. (After Troncy, 1969.)
FIG. 3.33. *Gnathostoma*, head, median view. (After Baylis & Lane, 1920.)

Subfamily **Spiroxyinae** Baylis & Lane, 1920

One genus
Parasites of snakes, aquatic tortoises and amphibians.

Spiroxys Schneider, 1866

Subfamily **Ancyracanthinae** Yorke & Maplestone, 1926

One genus
Parasites of fishes and neotropical chelonians.

Ancyracanthus Diesing, 1858
(=*Piayussunema* Kohn, Gomes & Motta, 1968)

Subfamily **Gnathostomatinae** (Railliet, 1895, fam.) Baylis & Lane, 1920

Key to genera

1-(2) Cephalic bulb divided into 2 or 4 lobes and furnished with transverse cuticular ridges with sharp, posteriorly projecting edges but without hooks (Fig. 3.31).
Parasites of aquatic reptiles.

Tanqua Blanchard, 1904

2-(1) Cephalic bulb undivided externally and armed with transverse rows of recurved hooks (Fig. 3.32).

3-(4) Body unarmed.
Parasites of fishes.

Echinocephalus Molin, 1858

4-(3) Body partially or wholly armed with posteriorly directed spines (Fig. 3.33).
Parasites of mammals.

Gnathostoma Owen, 1836

REFERENCES

FREITAS, J. F. TEIXEIRA DE & DOBBIN, JR., J. E. 1971. Contribuicão ao conhecimento da fauna helmintológica de Quelônios no estado de Pernambuco, Brasil. *Mems. Inst. Oswaldo Cruz*, **69** (1), 33-39.

GOMES, D. C. & KOHN, A. 1970. Sôbre a subfamília Ancyracanthinae Yorke & Maplestone, 1926 (Nematoda, Spiruroidea). *Atas Soc. Biol. Rio de Janeiro*, **13** (3/4), 83-88.

IVASHKIN, V. M. 1960. [The position of Spiroxyinae (Baylis & Lane, 1920) and Ancyracanthinae (Yorke & Maplestone, 1926) in the Spirurata.] *Trudy Gel'mint. Lab.*, **10**, 92-93. [In Russian.]

KOHN, A., GOMES, D. C. & SILVA MOTTA, C. DA. 1968. Nota prévia sôbre um nôvo gênero de Ancyracanthinae Yorke & Maplestone, 1926 (Nematoda). *Atas Soc. Biol. Rio de Janeiro*, **12** (1), 27-28.

TRONCY, P. M. 1969. Description de deux nouvelles espèces de nématodes parasites de poissons. *Bull. Mus. natn. Hist. nat.*, 2 ème sér., **41** (2), 598-605.

KEY TO GENERA OF THE PHYSALOPTEROIDEA

The view of the systematics of this superfamily does not differ substantially from author to author (cf. Skrjabin & Sobolev, 1964; Chabaud, 1965). We include here only one family divided into three subfamilies, the Thubunaeinae, Proleptinae and Physalopterinae.

The Thubunaeinae cannot be characterized by the asymmetry of the oral opening since the type species of the type genus, *Thubunaea*, is perfectly symmetrical. We have had to rely, therefore, on characters provided by the cephalic collarette and the caudal bursa.

The Proleptinae, separated from the Thubunaeinae, corresponds exactly with the group which is parasitic in fishes.

In the Physalopterinae we retain genera which seem to correspond to the diversification of a homogeneous line, either in a group of hosts, e.g. *Pseudophysaloptera* in Soricoidea, or in a specific geographical region, e.g. *Turgida* in South America. We have not stressed, on the other hand, divisions on the basis of uterine branches (e.g. *Tridelphynema*, *Tetradelphynema*, *Didelphyoptera*, *Polydelphyoptera* etc.) since multiplication of the uterine branches occurs in species in various hosts and from various regions and can be considered as a convergence.

The cephalic dentition, which corresponds to a great extent with distribution in one or another groups of hosts, is regarded as a valuable character; the nomenclature employed to describe cephalic teeth is indicated in Fig. 3.34.

Pseudabbreviata (Lichtenfels & Quigley, 1968) can possibly, as suggested by the authors, be regarded as an intermediate form between Thubunaeinae and Physalopterinae.

Pseudorictularia Dollfus & Desportes, 1944, *Bancroftinema* Johnston & Mawson, 1941 and *Dogielina* Sobolev, 1949 seem to belong to the Physalopteroidea but, as yet, are not well enough known to be classified.

The problem posed by *Bulbocephalus* and *Cestocephalus* which were discovered by Rasheed (1966) has been resolved by Le Van Hoa *et al.* (1972) who found, under lobes evaginated from the oesophagus, reduced pseudolabia recognizable as those of a physalopteroid. *Bulbocephalus* is, therefore, classified near *Heliconema* and *Cestocephalus* is regarded as a synonym of *Bulbocephalus*.

Family *PHYSALOPTERIDAE* (Railliet, 1893 subfam.) Leiper, 1908

Key to subfamilies

1-(2) Cephalic collarette absent. Oesophageal lobes which can be everted from cephalic extremity absent. Caudal papillae often numerous and not very different from cuticular ornamentation surrounding them (Fig. 3.35).
Parasites of reptiles and rarely amphibians.

Thubunaeinae

2-(1) Cephalic collarette (Fig. 3.36) or oesophageal lobes (Fig. 3.37) which can be everted from cephalic extremity present. Caudal papillae very distinct from cuticular ornamentation (Fig. 3.38).

3-(4) Caudal alae of male fading into lateral borders of body. *Area rugosa* limited to the medioventral zone anterior to anus (Fig. 3.39).
Parasites of fishes.

Proleptinae

4-(3) Caudal alae of male united on ventral surface of body (Fig. 3.38). Caudal bursa ornamented (Fig. 3.38).
Parasites of various vertebrates other than fishes.

Physalopterinae

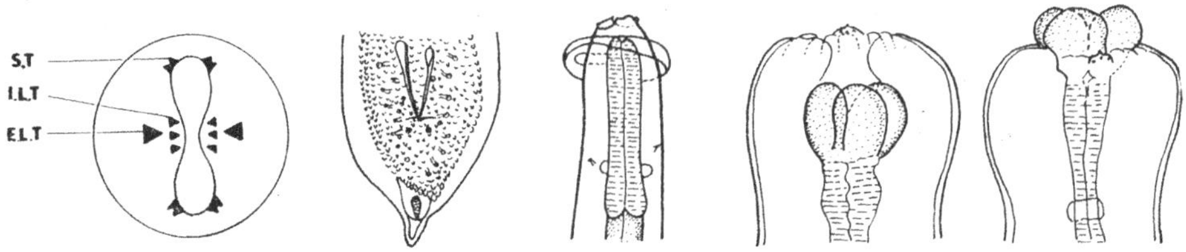

3.34 3.35 3.36 3.37

FIG. 3.34. Nomenclature of the cephalic teeth of physalopterans. E.L.T.: externo lateral tooth; I.L.T.: interno lateral teeth; S.T.: double pairs submedian teeth.

FIG. 3.35. *Thubunaea*, caudal bursa, ventral view. (After Seurat, 1914.)

FIG. 3.36. *Proleptus*, cephalic collarette. (After Lent & Freitas, 1948.)

FIG. 3.37. *Bulbocephalus*, head, lateral view, oesophageal lobes invaginated on left, evaginated on right. (After Le Van Hoa *et al.*, 1972.)

Subfamily **Thubunaeinae** (Sobolev, 1949, tribe)

Key to genera

1-(2) Oral opening symmetrical (Fig. 3.40).
Thubunaea Seurat, 1914

2-(1) Oral opening asymmetrical, teeth on one pseudolabium more pronounced or more numerous than those on the other (Fig. 3.41).
Physalopteroides Wu & Liu, 1940

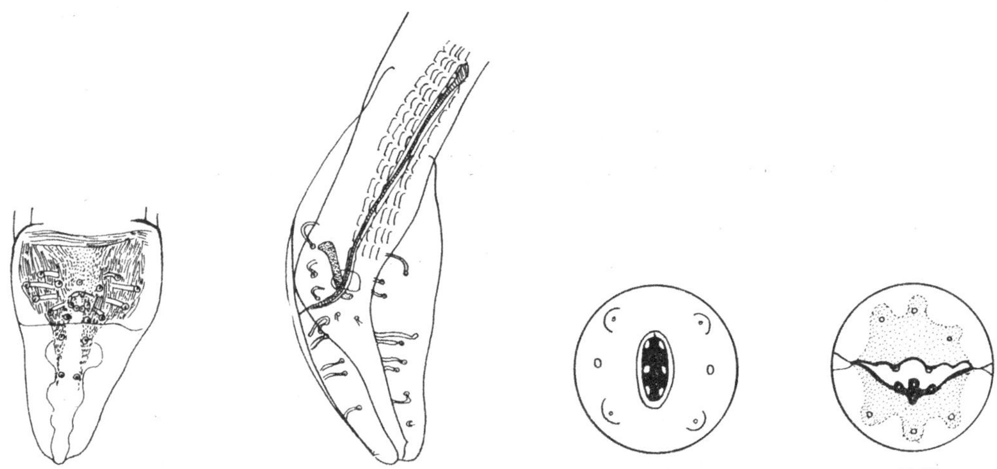

3.38 3.39 3.40 3.41

FIG. 3.38. *Physaloptera*, caudal bursa, ventral view. (After Quentin, 1968.)

FIG. 3.39. *Proleptus*, caudal bursa, ventral view. (After Seurat, 1919.)

FIG. 3.40. *Thubunaea*, head, apical view. (After Chabaud & Golvan, 1957.)

FIG. 3.41. *Physalopteroides*, head, apical view. (After Sandground, 1933.)

Subfamily **Proleptinae** (Schulz, 1927)

Key to genera

1-(2) Vulva near anus.
Spicules unequal. Each pseudolabium with a single interno-lateral tooth (Fig. 3.43). Parasites of selachians, rarely of teleosts, chelonians and even crustaceans.

Proleptus Dujardin, 1845

2-(1) Vulva far from anus.

3-(6) Oesophageal lobes not protrusible.

4-(5) Spicules equal.
Vulva slightly postequatorial. Internal border of each pseudolabium provided with a series of denticles in addition to the internolateral tooth (Fig. 3.42).
Parasites of selachians.

Paraleptus H.W. Wu, 1927

5-(4) Spicules markedly unequal.
Vulva slightly pre-equatorial in position. Internal border of each pseudolabium carrying only an internolateral tooth and, sometimes, a simple tooth at each dorsoventral extremity (Fig. 3.43).
Parasites of anguilliform fishes.

Heliconema Travassos, 1919
(=*Ortleppina* Schulz, 1927)

6-(3) Protrusible oesophageal lobes present (Fig. 3.37).
Pseudolabia reduced. Collarette absent. Vulva in posterior half of body.
Parasites of marine fishes.

Bulbocephalus Rasheed, 1966
(=*Cestocephalus* Rasheed, 1966)

Subfamily **Physalopterinae** Railliet, 1893

Key to genera

1-(2) Buccal cavity well developed (Fig. 3.44). Amphids situated lateral to pseudolabia (Fig. 3.45).
Parasites of reptiles.

Pseudabbreviata Lichtenfels & Quigley, 1968

3.42

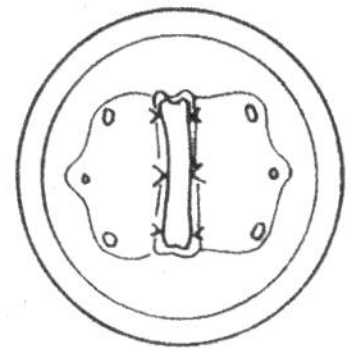

3.43

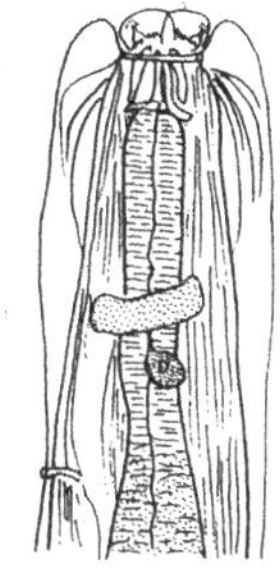

3.44

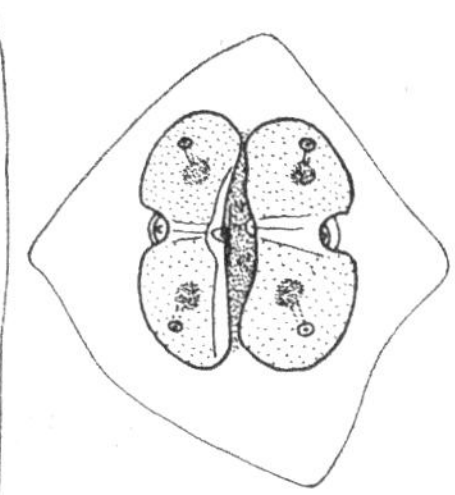

3.45

FIG. 3.42. *Paraleptus*, pseudo-labium, lateral view. (After Wu, 1927.)
FIG. 3.43. *Heliconema*, head, apical view. (After Campana-Rouget, 1956.)
FIG. 3.44. *Pseudabbreviata*, anterior extremity, lateral view. (After Lichtenfels & Quigley, 1968.)
FIG. 3.45. *Pseudabbreviata*, head, apical view. (After Lichtenfels & Quigley, 1968.)

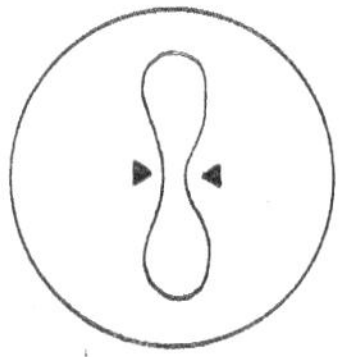

3.46

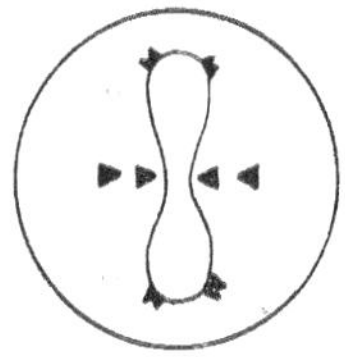

3.47

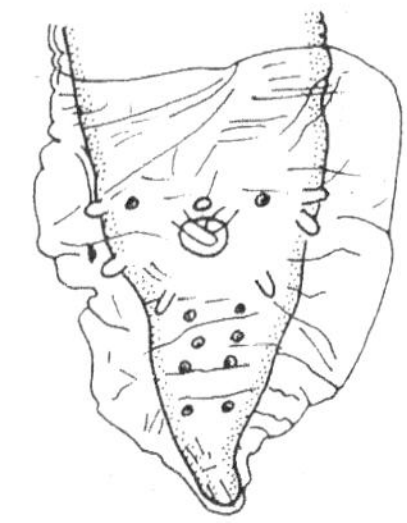

3.48

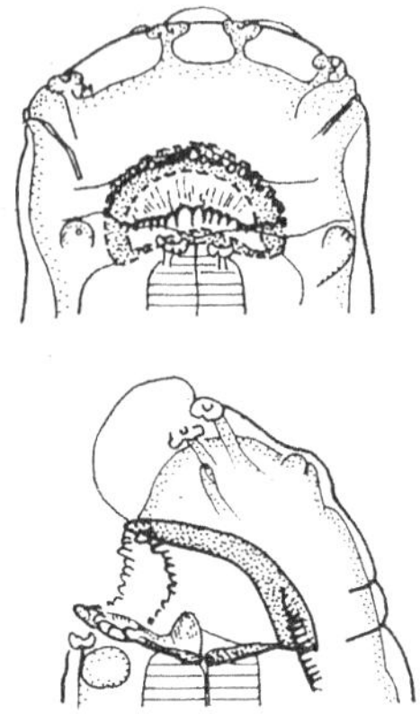

3.49

FIG. 3.46. *Skrjabinoptera*, diagrammatic apical view.
FIG. 3.47. *Abbreviata*, diagrammatic apical view.
FIG. 3.48. *Pseudophysalopteroides*, caudal bursa, ventral view. (After Morosov, 1960.)
FIG. 3.49. *Rictularia*, dorsal view, lateral view. (After Quentin, 1969.)

2-(1) Buccal cavity atrophied. Amphids on pseudolabia.
3-(4) Externolateral tooth absent from each pseudolabium (Fig. 3.46).
Each pseudolabium with a single internolateral tooth (Fig. 3.46).
Parasites of reptiles.
Skrjabinoptera Schulz, 1927

4-(3) Externolateral tooth present on each pseudolabium (Fig. 3.47).
5-(6) A single internolateral tooth and two double pairs of submedian teeth present on each pseudolabium (Fig. 3.47).
Parasites of reptiles (especially saurians), more rarely amphibians, rodents, and primates.
Abbreviata Travassos, 1920

6-(5) Three internolateral teeth present on each pseudolabium (Fig. 3.34).
7-(12) Double pair of submedian teeth absent from each pseudolabium.
8-(9) Caudal papillae scarce, small, sessile or slightly pedunculate (Fig. 3.48).
Spicules rudimentary or absent.
Parasites of insectivores (Soricoidea).
Pseudophysaloptera Baylis, 1934

9-(8) Caudal papillae of usual number and well developed (Fig. 3.38).
Spicules well developed.
10-(11) Two to four uteri present.
Parasites essentially of birds of prey and Carnivora and, incidentally, of snakes and various mammals which ingest flesh or insects.
Physaloptera Rudolphi, 1819

11-(10) More than 4 uteri present.
Cuticle usually inflated
Parasites of neotropical mammals.
Turgida Schulz, 1927

12-(7) Double pair of submedian teeth present on each pseudolabium (Fig. 3.34).
Parasites of palearctic rodents.
Pentadentoptera Schachnasarova, 1949
(=*Physalopteriata* Sobolev, 1949)

REFERENCES

CHABAUD, A. G. 1965. In: Grassé, P. P. Traité de Zoologie. Anatomie, systématique, biologie. Tome IV, fasc. II. Némathelminthes (nématodes). Fasc. III. Némathelminthes (nématodes, gordiacés), rotifères, gastrotriches, kinorhynques. Paris: Masson et Cie, Fasc. II, 731 pp.; fasc. III, pp. 732-1497.

LE VAN HOA, PHAM-NGOC-KHUE & NGUYEN-THI-LIEN, 1972. Etude de deux nouvelles espèces de nématodes du genre *Bulbocephalus* Rasheed, 1966, parasites des poissons de mer du sud Viet Nam. *Bull. Soc. Path. exot.*, **65** (2), 313-322.

LICHTENFELS, J. R. & QUIGLEY, D. 1968. *Pseudabbreviata nudamphida* gen. et sp.n. (Nematoda : Physalopteridae) from an African "lizard". *J. Parasit.*, **54** (6), 1092-1094.

QUENTIN, J. C. 1969. *Physaloptera longispicula* nouvelle espèce de spiruride parasite de *Cercomys cunicularius* Cuvier. *Bull. Mus. natn. Hist. nat.*, Year 1968, 2 ème sér., **40** (5), 1043-1046.

RASHEED, S. 1966. On some interesting nematode parasites of fish from Pakistan. *Parasitology*, **56** (1), 151-160.

SKRJABIN, K. I. & SOBOLEV, A. A. 1964. Osnovi nematodologii. Vol. XII. [Spirurata of animals and man and the diseases caused by them. Part II. Physalopteroidea.] Moscow: Izdat. Akad. Nauk SSSR, 334 pp. [In Russian.]

KEY TO GENERA OF THE RICTULARIOIDEA

The rictularioids, generally classified within the Spiruroidea or Thelazioidea are here given their own superfamily because of Quentin's (1969) conclusions that this group appears to have originated close to the Seuratoidea and is an evolutionary line independent of other spirurids. Quentin's conclusions with respect to genera and subgenera are accepted in their entirety.

Pseudorictularia Dollfus & Desportes, 1945, which is insufficiently known, and *Rictularioides* Hall, 1916 which may be a juvenile *Pterygodermatites*, are excluded from the key.

Diserratosomus Mirza, 1933 corresponds possibly to the subgenus *Mesopectines* Quentin, 1969 but since this is uncertain we retain the taxon *Mesopectines*.

There is one family, the Rictulariidae.

Family *RICTULARIIDAE* (Hall, 1915 subfam.) Railliet, 1916

Key to genera and subgenera

1-(2) Oral opening dorsal in position and transverse. A single oesophageal tooth present (Fig. 3.49).
Caudal papillae pairs 1-4-8* directed laterally (Fig. 3.50). Number of prevulvar spines, 34 pairs or less.
Parasites essentially of rodents (Sciuridae, Gliridae, Muridae, Microtidae) and bats in the holarctic area.
Rictularia Froelich, 1802

2-(1) Oral opening apical in position or displaced somewhat dorsally, never totally dorsal and transverse. 3 oesophageal teeth present (Fig. 3.51).
Number of prevulvar spines, 29-58 pairs.
Pterygodermatites Wedl, 1861

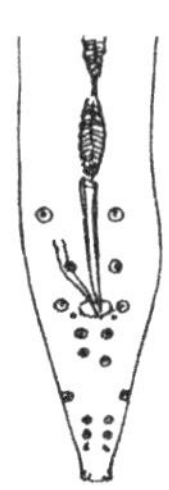
3.50

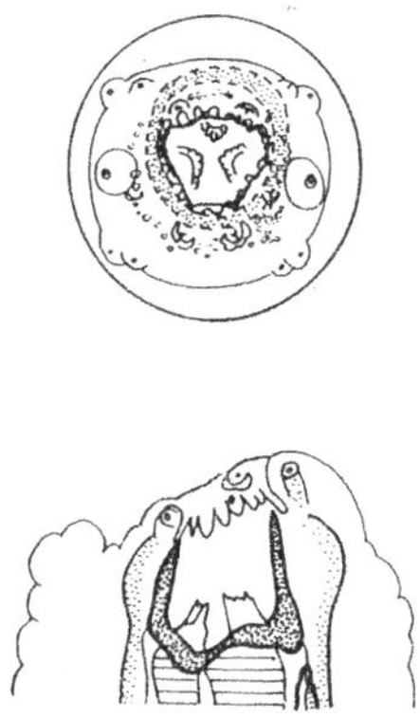
3.51

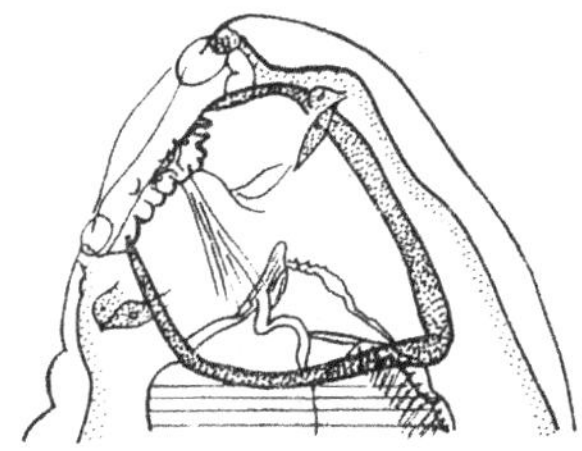
3.52

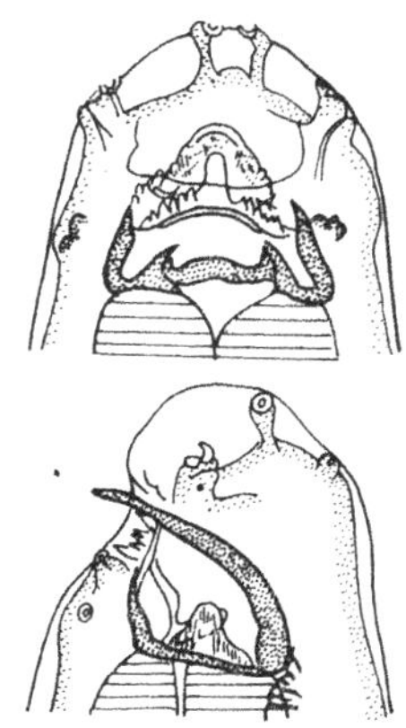
3.53

FIG. 3.50. *Pterygodermatites* (*Paucipectines*), male tail. (After Quentin, 1969.)
FIG. 3.51. *Pterygodermatites* (*Paucipectines*), head, apical view, lateral view. (After Quentin, 1969.)
FIG. 3.52. *Pterygodermatites* (*Neopaucipectines*), head, lateral view. (After Quentin, 1969.)
FIG. 3.53. *Pterygodermatites* (*Pterygodermatites*), head, dorsal view, lateral view. (After Quentin, 1969.)

* For the nomenclature of caudal papillae see Chabaud & Petter, 1961.

3-(4) Oral opening apical in position (Fig. 3.51).
Caudal papillae pairs 1-4-8 directed laterally (Fig. 3.50). With 29-39 pairs of prevulvar spines.
Parasites essentially of holarctic and neotropical Cricetidae, Microtidae and Sciuridae.
Pterygodermatites (*Paucipectines*) Quentin, 1969

4-(3) Oral opening always more or less displaced dorsally (Figs. 3.52-3.55; 3.57).

5-(6) Oral opening slightly inclined dorsally (Fig. 3.52). 34-38 pairs of prevulvar spines present.
Caudal papillae pairs 1-4-8 laterally directed (Fig. 3.50).
Parasites of European bats, Ethiopian rodents, and Malagasy lemurs.
Pterygodermatites (*Neopaucipectines*) Quentin, 1969

6-(5) Oral opening clearly dorsal in position (Figs. 3.53-3.55; 3.57). Pairs of prevulvar spines generally more numerous.

7-(10) Oral opening surrounded by a crown of denticles (Figs. 3.53; 3.54).

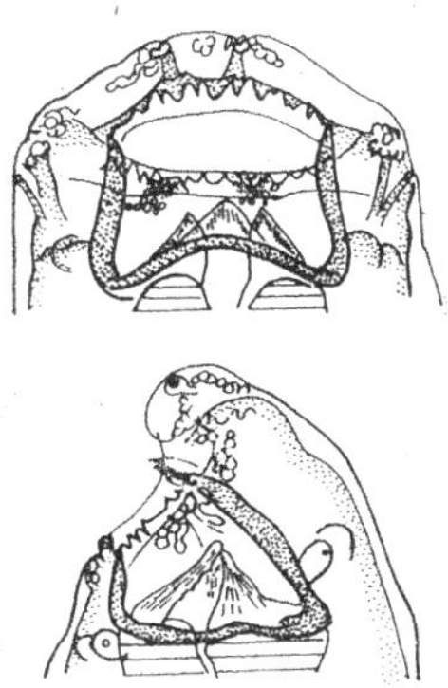

3.54

3.55

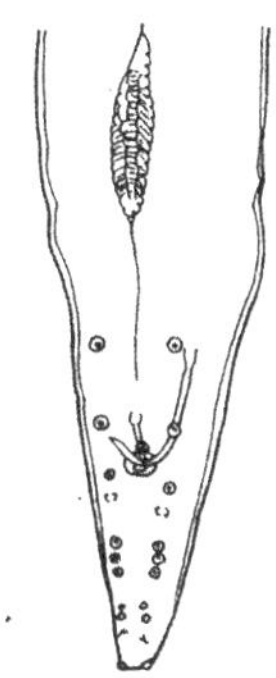

3.56

FIG. 3.54. *Pterygodermatites* (*Mesopectines*) *taterilli*, head, dorsal view, lateral view. (After Quentin, 1969.)
FIG. 3.55. *Pterygodermatites* (*Mesopectines*) *leiperi*, head, dorsal view. (After Quentin, 1969.)
FIG. 3.56. *Pterygodermatites* (*Mesopectines*), male tail, ventral view. (After Quentin, 1969.)

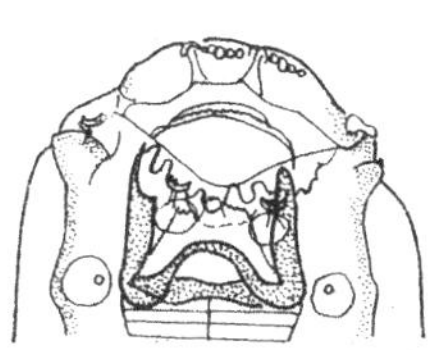

3.57

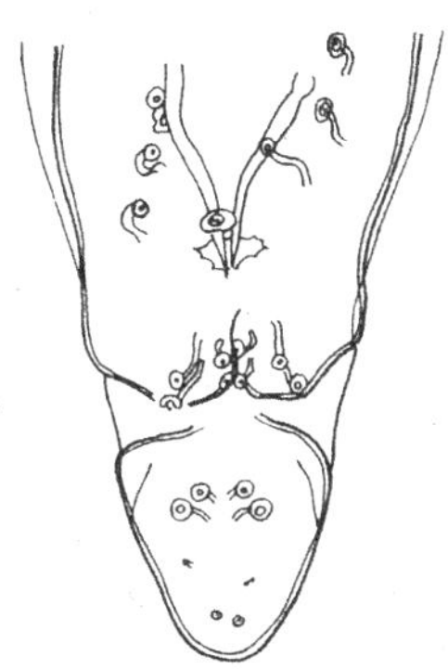

3.58

FIG. 3.57. *Pterygodermatites* (*Multipectines*), head, dorsal view. (After Quentin, 1969.)
FIG. 3.58. *Pterygodermatites* (*Multipectines*), male tail, ventral view. (After Quentin, 1969.)

8-(9) Peribuccal denticles of irregular sizes (Fig. 3.53).
40-46 pairs of prevulvar spines present. (Disposition of caudal papillae unknown.)
Parasites mainly of bats and insectivores, in the Mediterranean area and North Africa.
Pterygodermatites (*Pterygodermatites*) Wedl, 1861

9-(8) Peribuccal denticles in the form of a regular crown (Fig. 3.54), sometimes replaced on the ventral side by one or two semi-lunar apophyses (Fig. 3.55).
37-51 pairs of prevulvar spines present. Caudal papillae in two subventral rows (Fig. 3.56).
Parasites of rodents (Gerbillidae, Muridae), carnivores (Viverridae) and of primates in Africa and Asia.
Pterygodermatites (*Mesopectines*) Quentin, 1969

10-(7) Peribuccal denticles replaced on the ventral border by a sclerotized apophysis (Fig. 3.57).
47-58 pairs of prevulvar combs. Caudal papillae pedunculate and grouped around anus (Fig. 3.58).
Parasites of Mustelidae, Felidae, and Canidae; world-wide in distribution.
Pterygodermatites (*Multipectines*) Quentin, 1969

REFERENCES

CHABAUD, A. G. & PETTER, A. J. 1961. Remarques sur l'évolution des papilles cloacales chez les nématodes phasmidiens parasites de vertebrés. *Parassitologia*, **3** (1/2), 51-70.

QUENTIN, J. C. 1969. Essai de classification des nématodes rictulaires. *Mém. Mus. natn. Hist. nat.*, sér. A, **54** (2), 55-115.

KEY TO GENERA OF THE THELAZIOIDEA

The division of the superfamily into three families corresponds to that adopted by Chabaud (1959, 1965), each of the taxa being raised one level.

In contrast to Soviet authors, we remove from the Thelazioidea the following: *Crassicauda* which seems to belong to the Tetrameridae; *Rictularia* which is regarded as a distinct line originating directly from the Seuratoidea; Gongylonematinae which appears to be allied to the Spiruroidea, and Desmidocercidae which is, for convenience, regarded as a member of the Filarioidea and its allies.

The Thelazioidea includes forms which have a round or hexagonal oral opening which is not compressed laterally. In contrast, the Spiruroidea have either a complex oral opening with invaginated teeth or an oral opening dorsoventrally elongated by the development of lateral neoformations known as pseudolabia. The latter, with some few exceptions, attain their highest development only in the Habronematoidea.

The classification conforms, therefore, to the general ideas of Chitwood & Wehr (1934), ideas verified by all the more recent work on the ontogeny of the cephalic structures of the Spirurida.

THELAZIIDAE

Oxyspirurinae

Oxyspirura has been divided into 3 subgenera for a long time but, more recently, Barus (1963) split it into 5 subgenera. Rodriguez & Freitas (1964) have shown that the type species of *Oxyspirura* has a divided buccal capsule and that *Yorkeispirura*, which was separated from the former genus on the basis of this character, falls into its synonymy. This explains why Yeh (1957), having seen the type species, considered that the division of species into groups with or without a divided buccal cavity was not sound.

The character provided by the structure of the buccal capsule, however, has been almost universally adopted, and we think it useful to use it here. *Yorkeispirura* cannot be used and we prefer to follow the classification of Barus (1963) and propose, for the species that was called *Yorkeispirura*, a new subgenus *Barusispirura* n.subg. with *Oxyspirura* (*Barusispirura*) *rodriguesi* Barus, 1968 as type.

Because of the general homogeneity of *Oxyspirura*, *Molinospirura* Rodriguez, 1964 is also regarded as a subgenus.

Hamulofilaria Chandler, 1924, previously known only from one male, has been classified as a filarioid of dubious status. Soota & Chaturvedi (1971) found both sexes and redescribed the species. But there seems to be nothing which separates it from *Oxyspirura*: the buccal capsule is not clearly divided, the spicules are dissimilar and unequal in length and lateral alae are absent. The species agree with the subgenus *Skrjabinispirura* which must, however, fall into the synonymy of *Hamulofilaria* which now constitutes a subgenus of *Oxyspirura*.

Thelaziinae

Anderson & Diaz-Ungría (1959) rejected *Thelaziella* Travassos as it was then defined (relative lengths of the spicules), but Skrjabin *et al.* (1967), relying on other characters (length of the female tail, egg shells, presence or absence of a gubernaculum), have been able to separate rather well the parasites of birds (*Thelaziella*) from those of mammals (*Thelazia*). *Thelaziella* is retained, therefore, but as a subgenus because the morphological differences between *Thelazia* and *Thelaziella* are slight.

Annulospira Jairajpuri *et al.* (1969), is placed as a synonym of *Ceratospira* Schneider, 1866.

Thylaconema Chandler (1929), is valid but the head structure is not well known, especially in *en face* view. Perhaps it is related to *Ceratospira*.

RHABDOCHONIDAE

Following Chitwood & Wehr (1934) and Campana-Rouget (1955), all Cystidicolinae (Spiruroidea) parasitic in fishes are now excluded from the Rhabdochonidae (Thelazioidea). Contrary to the definition of Moravec (1972), the Rhabdochonidae do not have pseudolabia, even tiny ones. The small lateral reliefs which may be present on some rare species do not impinge on the oral opening which remains round or hexagonal and is never compressed laterally. In the Cystidicolinae, on the other hand, even in species in which the pseudolabia are markedly reduced, lips always impinge on the oral opening which is elongated dorsoventrally (cf. Rasheed, 1965a).

The Rhabdochonidae have 8 cephalic papillae; the Cystidicolinae almost always have only 4. The caudal papillae of Rhabdochonidae are numerous often with pairs 1-4-8 laterally positioned as in the Seuratoidea and unlike the positioning found in the Spiruroidea (Chabaud & Petter, 1961). In contrast, caudal papillae in the Cystidicolinae are like those found in the Spiruroidea.

Since the Cystidicolinae are from every point of view more evolved, it was feasible, until recently, to consider the hypothesis of a type of rhabdochonid progressively modifying in response to the adoption of life within the tissues which is often characteristic of species of Cystidicolinae. However, this theory is no longer tenable because numerous genera of true rhabdochonids which are adapted to living in tissues are now known and these do not resemble the Cystidicolinae. For the Rhabdochonidae the evolution of the cephalic extremity comes about by simple atrophy of the buccal capsule although the pharynx is retained and even increased in length.

We believe, therefore, that the Rhabdochonidae and the Cystidicolinae belong to two very different evolutionary lines and we do not agree with Skrjabin *et al.* (1961) and Yamaguti (1961) who place them together in a single group.

The genus *Rhabdochona* has occasioned some controversy. Certain authors accord great importance to the presence or absence of filaments on the eggs and create for this character (which is common in many parasites of aquatic animals) a subfamily (Filochinae Yamaguti, 1961) or a family (Rhabdochonoididae Janizewska, 1955 publ. 1956). Other authors totally reject such characters. In our view, such a character has no phylogenetic value but is useful to separate the numerous species in the genus and in these keys we follow Moravec (1972), who used polar filaments to separate subgenera.

The *Rhabdochona* line, which remains very stable while it lives in the intestine of freshwater fish, is capable of adapting and diversifying in marine fish, either in the tissues of fish or finally in unexpected hosts ("hôtes de capture"*) such as reptiles and South American primates. Commencing with *Johnstonmawsonia* Campana-Rouget, 1955 (of which Moravec & Puylaert (1970) have shown *Prosungulonema* Roitmann, 1963 to be a synonym) there are a series of recently discovered genera which have evolved by simplifying the cephalic structures, elongating the body (sometimes in particular the pharynx), displacing the vulva, and even sometimes losing one uterus. These are *Vasorhabdochona* Martin & Zam, 1967 in the blood vessels of fish, *Heptochona* Rasheed, 1965b in the body cavity of fish, *Hepatinema* Rasheed, 1964b in the liver of fish, *Freitasia* Barus & Coy Otero, 1968 in the intestine of lizards in Cuba, and finally *Trichospirura* Smith & Chitwood, 1967 in the pancreatic vessels of South American primates. We do not hesitate to place *Trichospirura* in the Rhabdochonidae as the morphological modifications it shows are very similar to those of the genera cited above.

Pontochona Mamaev, 1968 is a synonym of *Heptochona*, the author being apparently unaware at the time of Rasheed's work.

* hôte de capture—a host, distinct from the particular group in which the nematodes have evolved and diversified, to which the nematodes suddenly adapt. This adaptation is almost always accompanied by pronounced specialization and often by unusual localization within the host.

PNEUMOSPIRURIDAE

The classification is based on the studies of Dougherty (1951, 1952), except that the group is considered to belong to the Spirurida and not the Strongylida.

KEY TO FAMILIES

1-(2) Buccal capsule and pharynx* small or absent (Figs. 3.72-3.74).
Six lips sometimes well developed, sometimes atrophied. Cuticle with sheath. Gubernaculum often double.
Parasites of the lungs of mammals.

Pneumospiruridae

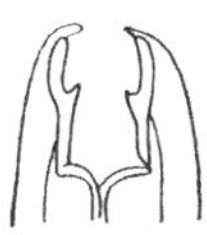

3.59

3.60

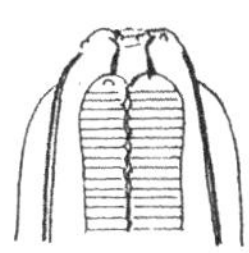

3.61

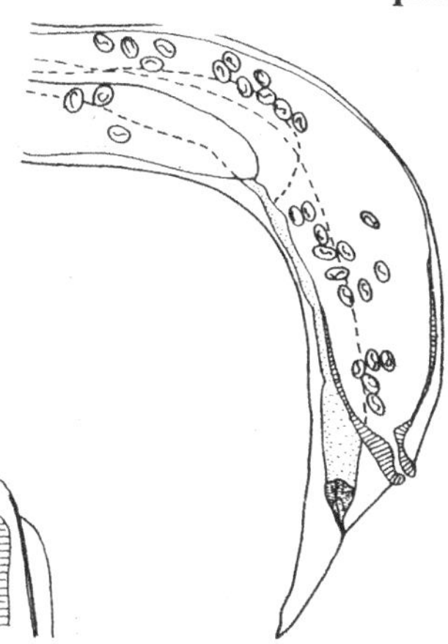

3.62

FIG. 3.59. *Oxyspirura* (*Oxyspirura*), buccal capsule. (After Chitwood & Wehr, 1934.)
FIG. 3.60. *Thelazia* (*Thelaziella*), head. (After Anderson & Diaz Ungría, 1959.)
FIG. 3.61.*Oxyspirura* (*Baruspirura*), head. (After Barus, 1968.)
FIG. 3.62. *Oxyspirura* (*Molinospirura*), ovejector. (After Rodrigues, 1963.)

2-(1) Buccal capsule or pharynx well developed (Fig. 3.59-3.71).
Not parasites of lungs of mammals.

3-(4) Pharynx not markedly elongated (Figs. 3.59-3.63).
Parasites generally of the eyes of birds and mammals.

Thelaziidae

4-(3) Pharynx cylindrical, elongated (Figs. 3.64-3.71).
Parasites of the intestine of fish or of various organs of fish and sometimes other vertebrates.

Rhabdochonidae

Family *THELAZIIDAE* Skrjabin, 1915

Key to subfamilies

1-(2) Inner surface of buccal capsule often armed with teeth (Fig. 3.59). Oesophagus divided.
Vulva near anus. Posterior end of body pointed in both sexes.

Oxyspirurinae

* Here used for the posterior part of the buccal cavity.

2-(1) Inner surface of buccal capsule without teeth (Fig. 3.60). Oesophagus not divided. Vulva in anterior half of body (except *Hempelia*). Posterior end of body often rounded in both sexes.

Thelaziinae

Subfamily **Oxyspirurinae** (Skrjabin, 1916 fam.) Yamaguti, 1961

One genus

Oxyspirura Drasche in Stossich, 1897

Key to subgenera

1-(8) Buccal capsule not clearly divided into anterior and posterior chambers (Figs. 3.61; 3.63).
2-(5) Spicules subequal, little different in morphology.
3-(4) Ovejector dilated into a large chamber for eggs (Fig. 3.62). Oral opening surrounded by sclerotized ring (Fig. 3.63).

Oxyspirura (*Molinospirura*) (Rodrigues, 1964 gen.)

4-(3) Ovejector not dilated.
Oral opening not surrounded by sclerotized ring.

Oxyspirura (*Cramispirura*) Skrjabin, 1931

5-(2) Spicules morphologically dissimilar and unequal in length.
6-(7) Lateral alae present. Gubernaculum absent.

Oxyspirura (*Barusispirura*) n. subg.
[Type: *O.*(*B.*) *rodriguesi* Barus, 1968]

7-(6) Lateral alae absent. Gubernaculum present or absent.

Oxyspirura (*Hamulofilaria*) Chandler, 1924 gen.
(=*Oxyspirura* (*Skrjabinispirura*) Barus, 1963)

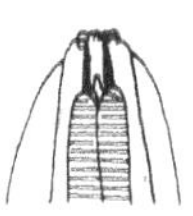

3.63

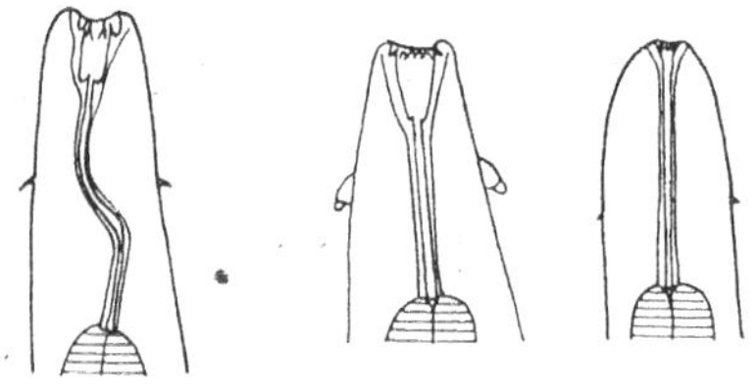

3.64

FIG. 3.63. *Oxyspirura* (*Molinospirura*), head. (After Rodrigues, 1963.)
FIG. 3.64. *Rhabdochona* spp., buccal capsules. (After Moravec, 1972.)

8-(1) Buccal capsule clearly divided into anterior and posterior chambers (Fig. 3.59).
9-(10) Lateral alae present. Gubernaculum present.

Oxyspirura (*Caballeroispirura*) Barus, 1963

10-(9) Lateral alae absent. Gubernaculum present or absent.

Oxyspirura (*Oxyspirura*) Drasche in Stossich, 1897
(=*Oxyspirura* (*Yorkeispirura*) Skrjabin, 1931)

Subfamily **Thelaziinae** (Skrjabin 1915, fam.) Baylis & Daubney, 1926

Key to genera and subgenera

1-(2) Vulva in posterior half of body.

Hempelia Vaz, 1937

2-(1) Vulva in anterior half of body.

3-(4) Male caudal alae well developed. Spicules markedly unequal.

Ceratospira Schneider, 1866
(=*Annulospira* Jairajpuri & Siddiqi, 1969)

4-(3) Male without caudal alae.

Thelazia Bosc, 1819
(=*Annulofilaria* G.N. Hsü, 1957)

5-(6) Tail of female relatively long. Gubernaculum generally present. Eggs with rather thick shells.
Parasites of birds.

Thelazia (*Thelaziella*) Travassos, 1918

6-(5) Tail of female short (less than 100 μ). Gubernaculum absent. Eggs with thin shells.
Parasites of mammals.

Thelazia (*Thelazia*) Bosc, 1819

Family *RHABDOCHONIDAE* (Travassos, Artigas & Pereira, 1928 subfam.) Skrjabin, 1946

Key to genera and subgenera

1-(12) Pharynx dilated anteriorly to form a well defined buccal capsule (Figs. 3.64; 3.68).

2-(7) Buccal capsule armed with longitudinal sclerotized rods projecting anteriorly as teeth (Fig. 3.64).
Preanal papillae numerous.
Parasites of intestine of freshwater fishes.

Rhabdochona Railliet, 1916
(=*Ichthyospirura* Skrjabin, 1917; =*Rhabdochonoides* Janizewska, 1955; =*Pseudorhabdochona* Liu & Wu, 1941; =? *Ichthyobronema* Gnedina & Ssavina, 1930)

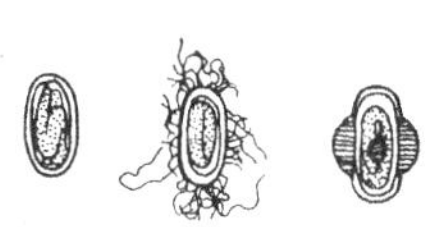

3.65 3.66 3.67

FIG. 3.65. *Rhabdochona* (*Rhabdochona*), egg. (After Moravec, 1972.)
FIG. 3.66. *Rhabdochona* (*Filochona*), egg. (After Moravec, 1972.)
FIG. 3.67. *Rhabdochona* (*Globochona*), egg. (After Moravec, 1972.)

3-(4) Surface of mature eggs smooth or covered with thin ill-defined gelatinous covering (Fig. 3.65).

Rhabdochona (*Rhabdochona*) Railliet, 1916

4-(5) Surface of eggs with filaments or floats.

5-(6) Eggs with filaments (Fig. 3.66).

Rhabdochona (*Filochona*) Saidov, 1953

6-(5) Eggs with special hemispherical floats (Fig. 3.67).
Rhabdochona (*Globochona*) Moravec, 1972

7-(2) Buccal capsule without sclerotized rods. Oral opening unarmed (Fig. 3.68). Preanal papillae not numerous.
Rarely parasites of the intestine of freshwater fishes.

8-(11) Didelphic.

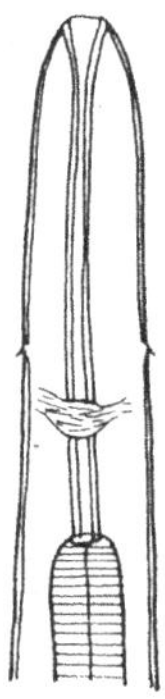

3.68

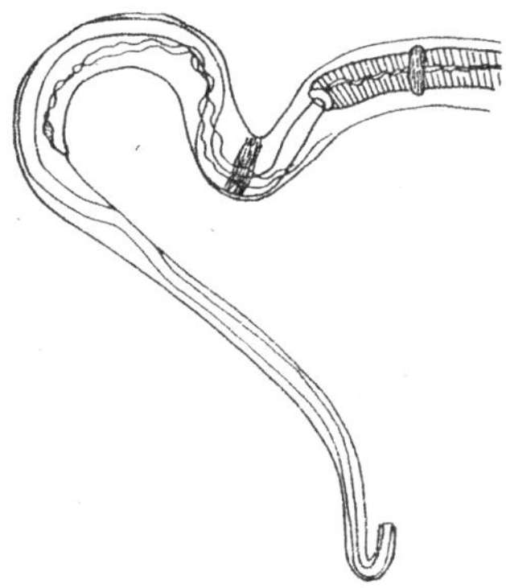

3.69

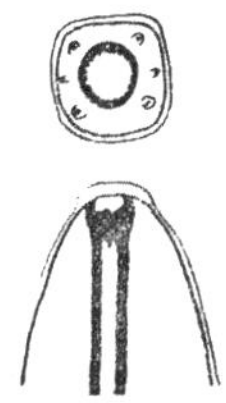

3.70

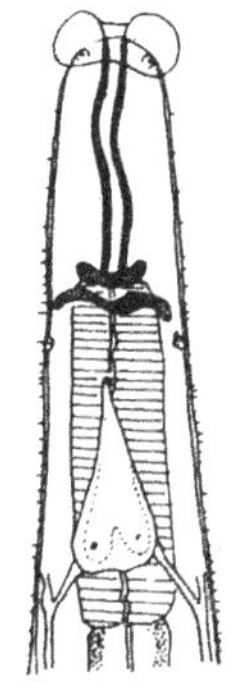

3.71

FIG. 3.68. *Johnstonmawsonia*, anterior extremity. (After Moravec & Puylaert, 1970.)
FIG. 3.69. *Freitasia*, anterior extremity. (After Moravec & Puylaert, 1970.)
FIG. 3.70. *Heptochona*, pharynx and head, apical view. (After Rasheed, 1965.)
FIG. 3.71. *Hepatinema*, anterior extremity. (After Rasheed, 1964.)

9-(10) Vulva near middle of body.
Pharynx typical of Rhabdochonidae (Fig. 3.68).
Parasites of intestine, bile-duct or liver of marine and freshwater fishes.
Johnstonmawsonia Campana-Rouget, 1955
(=*Prosungulonema* Roytman, 1963)

10-(9) Vulva near anus.
Pharynx markedly elongated, much longer than muscular oesophagus (Fig. 3.69).
Parasites of the intestine of reptiles in Cuba.
Freitasia Barus & Coy Otero, 1968

11-(8) Monodelphic.
Vulva slightly postoesophageal in position.
Parasites of blood vessels of marine fishes.
Vasorhabdochona Martin & Zam, 1967

12-(1) Pharynx not dilated or only slightly dilated anteriorly.
Buccal capsule absent or vestigial (Figs. 3.70; 3.71).

13-(14) Sclerotized rods and teeth present in oral opening (Fig. 3.70). Vulva in posterior half of body.
Parasites of the gut or body cavity of marine fishes.
Heptochona Rasheed, 1965
(=*Pontochona* Mamaev, 1968)

14-(13) Sclerotized rods or teeth absent (around oral opening). (Fig. 3.71).
Vulva near anus

15-(16) Monodelphic.
Pharynx of usual length for rhabdochonids.
Parasites of liver of marine fishes.

Hepatinema Rasheed, 1964

16-(15) Didelphic.
Pharynx greatly elongated. Cuticle unornamented. Prodelphic.
Parasites of pancreatic canal of American primates.

Trichospirura Smith & Chitwood, 1967

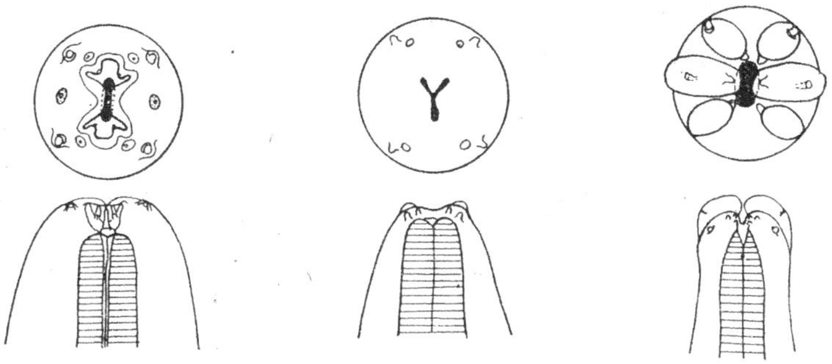

3.72 3.73 3.74

FIG. 3.72. *Pneumospirura*, head and buccal capsule. (After Gerichter, 1948.)
FIG. 3.73. *Metathelazia*, head, apical view, median view. (After Gerichter, 1948.)
FIG. 3.74. *Vogeloides*, head, apical view, median view. (After Gerichter, 1948.)

Family *PNEUMOSPIRURIDAE* Wu & Hu, 1938

Key to genera

1-(2) Buccal capsule present (Fig. 3.72).
Lips reduced. With 6 denticles surrounding oral opening.
Parasites of carnivores.

Pneumospirura Wu & Hu, 1938

2-(1) Buccal capsule not well developed.

3-(4) Lips tiny (Fig. 3.73).
Parasites of carnivores and insectivores.

Metathelazia Skinker, 1931
(=*Oslerus* Hall, 1971 (in part); =*Osleroides* Orlov, Davtian & Lubimov in Skrjabin, 1933; =? *Papillosleru*s Khera, 1944)

4-(3) Six well developed salient lips present (Fig. 3.74).
Parasites of carnivores and primates.

Vogeloides (Orlov, Davtian & Lubimov in Skrjabin, 1933
(=*Osleroides* Orlov, Davtian & Lubimov, 1933

REFERENCES

ANDERSON, R. C. & DÍAZ-UNGRÍA, C. 1959. Revisión preliminar de las especies de *Thelazia* Bosc (Spiruroidea : Thelaziidae), parásitas de aves. *Mems. Soc. Cienc. nat. "La Salle"*, **19** (52), 37-75.

BARUS, V. 1963. Ein Beitrag zur Systematik des Genus *Oxyspirura* Drasche in Stossich, 1897 (Nematoda, Thelaziidae). *Z. Parasitkde*, **22** (6), 545-559.

BARUS, V. 1968. Parasitic nematodes of birds of the family Icteridae (Passeriformes) in Cuba. *Folia Parasit.*, **15** (2), 131-146.

BARUS, V. & COY OTERO, A. 1968. *Freitasia teixeirai* gen.n. et sp.n. and other nematodes parasitizing *Anolis equestris* (Squamata : Iguanidae). *Folia Parasit.*, **15** (1), 41-54.

CAMPANA-ROUGET, Y. 1955. Sur deux nouveaux genres de spirurides parasites de poissons; discussion systématique des genres voisins. *Annls Parasit. hum. comp.*, **30** (4), 346-362.

CHABAUD, A. G. 1959. Phénomène d'évolution régressive des structures céphaliques et classification des nématodes Spiruroidea. *Parassitologia*, **1** (1), 11-20.

CHABAUD, A. G. in GRASSÉ, P. P. 1965. Traité de Zoologie, **4** (fasc. 2 et 3): 1-1497.

CHABAUD, A. G. & PETTER, A. J. 1961. Remarques sur l'évolution des papilles cloacales chez les nématodes phasmidiens parasites de vertébrés. *Parassitologia*, **3** (1/2), 51-70.

CHANDLER, A. C. 1924. New filariae from Indian birds. *Parasitology*, **16**, 398-404.

CHITWOOD, B. G. & WEHR, E. E. 1934. The value of cephalic structures as characters in nematode classification, with special reference to the superfamily Spiruroidea. *Z.ParasitKde*, **7** (3), 273-335.

DOUGHERTY, E. C. 1951. A further revision in the classification of the family Metastrongylidae Leiper [1909] (Phylum Nematoda). *Parasitology*, **41** (1/2), 91-96.

DOUGHERTY, E. C. 1952. A note on the genus *Metathelazia* Skinker, 1931 (Nematoda : Metastrongylidae). *Proc. helminth. Soc. Wash.*, **19** (1), 55-63.

JAIRAJPURI, D. S. & SIDDIQI, A. H. 1969. *Annulospira oculata* gen. et sp.nov. (Nematoda : Thelaziidae) from India. *J. Helminth.*, **43** (3/4), 341-342.

JANIZEWSKA, J. 1955 publ. 1956. *Rhabdochonoides barbi* g.n., sp.n., subfamily Rhabdochonoidinae subfam.n. (fam. Rhabdochonidae Skrjabin), an intestinal parasite in cyprinid fish. *Acta Parasit. Pol.*, 3 (6/12), 233-244.

MAMAEV, YU. L. 1968. [*Pontochona* n.g. (Rhabdochonidae, Spirurata), a new nematode genus from fish in the South China Sea.] In: Skrjabin, K. I. & Mamaev, Yu. L. (Editors), [Helminths of animals of the Pacific Ocean.] Moscow: Izdat. "Nauka", pp. 30-35. [In Russian.]

MARTIN, W. E. & ZAM, S. G. 1967. *Vasorhabdochona cablei*, gen. et sp.n. (Nematoda) from blood vessels of the marine fish, *Gillichthys mirabilis* Cooper. *J. Parasit.*, **53** (2), 389-391.

MORAVEC, F. 1972. General characterization of the nematode genus *Rhabdochona* with a revision of the South American species. *Věst. čsl. Spol. zool.*, **36** (1), 29-46.

MORAVEC, F. & PUYLAERT, F. A. 1970. On *Johnstonmawsonia africana* sp.n. (Nematoda : Rhabdochonidae) from the freshwater fish *Haplochromis schwetzi* of Angola. *Rev. Zool. Bot. afr.*, **82** (3/4), 306-314.

OLIVEIRA RODRIGUES, H. DE. 1964. Considerações sôbre a subfamilia Oxyspirurinae Yamaguti, 1961 (Nematoda, Spiruroidea). *Atas Soc. Biol. Rio de Janeiro*, **8** (5), 43-44.

OLIVEIRA RODRIGUES, H. DE & FREITAS, J. F. TEIXEIRA DE. 1964. Notula helminthologica: sôbre o material original de *Spiroptera cephaloptera* Molin, 1860, atualmente existente uo Museum de Viena (Nematoda). *Atas Soc. Biol. Rio de Janeiro*, **8**, 33-35.

RASHEED, S. 1964. On *Hepatinema karachiensis* gen. et sp.nov. (Thelaziidae Railliet, 1916) from a marine fish of Pakistan, with a note on the genus *Filochona* (Saidov, 1954) Yamaguti, 1961. *J. Helminth.*, **38** (1/2), 63-76.

RASHEED, S. 1965a. Observations on the spiruroid nematodes of fish with a revision of the genus *Metabronema* Yorke & Maplestone, 1926. *Z. zool. Syst. Evolutionsforsch.*, **3** (3/4), 359-387.

RASHEED, S. 1965b. A preliminary review of the genus *Rhabdochona* Railliet, 1916 with a description of a new and related genus. *Acta Parasit. Pol.*, **13** (39/46), 407-424.

ROYTMAN, V. A. 1963. [Nematodes from fishes of the water basin of the River Zeya.] *Trudy Gel'mint. Lab.*, **13**, 253-300. [In Russian.]

SKRJABIN, K. I., SOBOLEV, A. A. & IVASHKIN, V. M. 1967. Osnovi nematodologii. Vol. XVI. [Spirurata of animals and man and the diseases caused by them. Part 4. Thelazioidea.] Moscow: Izdat. "Nauka", 624 pp. [In Russian.]

SMITH, W. N. & CHITWOOD, M. B. 1967. *Trichospirura leptostoma* gen. et sp.n. (Nematoda : Thelazioidea) from the pancreatic ducts of the white-eared marmoset *Callithrix jacchus*. *J. Parasit.*, **53** (6), 1270-1272.

SOOTA, T. D. & CHATURVEDI, Y. 1971. On some filariid nematodes and a description of a new family Hamulofilariidae. *Zool Anz.*, **186** (5/6), 359-367.

YAMAGUTI, S. 1961. Systema Helminthum. Vol. III. The nematodes of vertebrates. New York: Interscience Publishers Inc., Pt. I, 679 pp., Part II, pp. 681-1261.

YEH, L. A. 1957. A collection of helminths from the great bustard, *Otis tarda* from Spain with a description of a new species of *Oxyspirura* (Nematoda). *Proc. zool. Soc. Lond.*, **128** (2), 279-286.

13 SPIRURIDA

SPIRUROIDEA, HABRONEMATOIDEA AND ACUARIOIDEA

by ALAIN G. CHABAUD*

KEY TO GENERA OF THE SPIRUROIDEA

We (Chabaud, 1959 a & b) were obliged to modify the classification of Chitwood & Wehr (1934) after showing that the terrestrial Spirurinae belong to the same group as, and have perhaps given rise to, the Ascaropsinae and the Spirocercinae. It was, therefore, impossible to leave the Spirurinae with the habronemes (which are highly evolved), and the Ascaropsinae and the Spirocercinae with the thelaziids (which are little evolved). The name of the family is derived from the genus *Spirura* and it has been necessary to introduce some important taxonomic changes. In the present work the family Thelaziidae includes only those genera which constitute the Thelaziinae of Chitwood & Wehr. The family Spiruridae corresponds to the four other sub-families of the Thelaziidae in Chitwood & Wehr, together with the terrestrial Spirurinae. The family Hedruridae (now designated as the Habronematoidea) corresponds to the Spiruridae of Chitwood & Wehr, with the exception of the terrestrial Spirurinae which are excluded.

Quentin (1970, 1971), stressing larval cephalic structures and ontogeny, demonstrated the homogeneity of the Spiruridae as defined above (excluding the Rictulariinae) and worked out more precisely the relationships between the various genera. He created 2 new subfamilies, the Mastophorinae and the Hartertiinae.

In the present study the Spiruridae is raised to a superfamily and, following the work of Quentin, four families are recognized, namely, Gongylonematidae, Spiruridae, Spirocercidae and Hartertiidae. The Spirocercidae is divided into the Spirocercinae, Ascaropsinae and Mastophorinae.

Skrjabin *et al.* (1967) adopted our point of view to a great extent and classified the Ascaropsinae and Spirocercinae in the Spiruridae which are separated from the Habronematidae. The difference in the two systems concerns only the place of the aquatic spirurids and the gongylonemes.

Genera recently described which have been incorporated in the key are *Tejeraia* Díaz-Ungría, 1963, *Spiralatus* Chabaud, Brygoo & Durette, 1963, *Texicospirura* M. B. Chitwood & Cordero del Campillo, 1966 and *Alainchabaudia* Mawson, 1968.

The classification of the Spirurida is based mainly on cephalic structures and a certain number of genera not yet studied in apical view cannot be classified; two of these are *Mazzia* Khalil & Vogelsang, 1932 (which is probably a valid genus of Spirocercinae) and *Paraleiuris* Vaz & Pereira, 1929 (probably a member of the Ascaropsinae), both from Xenarthra.

Chlamydoprocta Chandler, 1954 may be a synonym of *Vigisospirura*. *Neospirocerca* Johnson, 1968 is unclassifiable but it may be a juvenile female of *Protospirura*. (*Thaprospirura* Sood & Parshad, 1974 is a synonym of *Protospirura*. We had the opportunity to see one cotype).

* Muséum National d'Histoire Naturelle, Paris, France.

Taphozoia Ali & Lovekar, 1966 is regarded as a synonym of *Spirura* because the 6 labial formations of the latter are weakly developed in some species but highly developed in others (see Quentin, 1973 for *Spirura guianensis*); the type species of *Taphozoia* is only a transitional form.

Skrjabinitectus Majumdar & Banerjee, 1966 is too poorly described to be assigned to any family.

SPIRUROIDEA

Key to families

1-(2) Body covered with large verruciform thickenings, especially anteriorly (Fig. 3.75).
Salient lips or pseudolabia absent.
Oral opening octagonal with medial lobes corresponding to small dorsal and ventral elevations (Fig. 3.76).
Parasites of the mucosa of the anterior part of the gut of birds and mammals.
Gongylonematidae

2-(1) Body without verruciform thickenings.

3-(6) Pseudolabia rarely well developed but, if so, each consists of 3 distinct hypertrophied lobes (Figs. 3.77; 3.78).
Parasites of mammals, rarely birds.

4-(5) Pharynx laterally flattened.
Oral opening dorsoventrally elongated, generally constricted by 2 lateral elevations and 4 small submedian formations (Fig. 3.79).
Buccal cavity without teeth.
Spiruridae

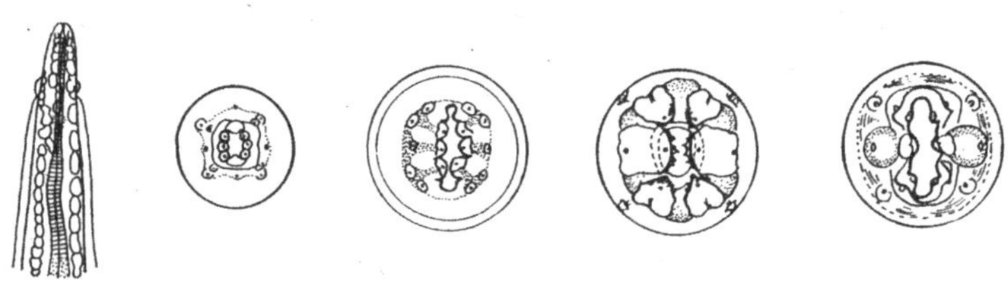

3.75 3.76 3.77 3.78 3.79

FIG. 3.75. *Gongylonema*, anterior extremity. (After Baylis, 1929).
FIG. 3.76. *Gongylonema*, head, apical view. (After Quentin, 1969).
FIG. 3.77. *Physocephalus*, head, apical view. (After Quentin, 1965).
FIG. 3.78. *Mastophorus*, head, apical view. (After Quentin, 1971).
FIG. 3.79. *Spirura*, head, apical view. (After Chabaud, 1954).

5-(4) Pharynx not compressed laterally, sometimes with rugosities (Figs. 3.80; 3.81) or annular thickenings (Figs. 3.82; 3.83).
Oral opening generally hexagonal.
Buccal cavity generally with teeth (Fig. 3.84).
Spirocercidae

6-(3) Pseudolabia large, each forming a mass slightly subdivided into lobes (Fig. 3.85).
Parasites of birds.
Hartertiidae

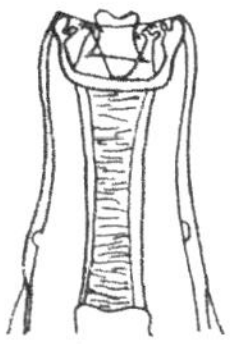

3.80

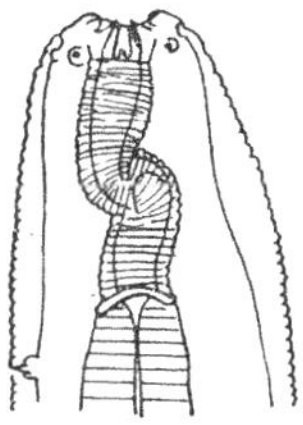

3.81

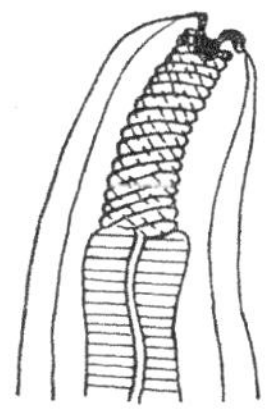

3.82

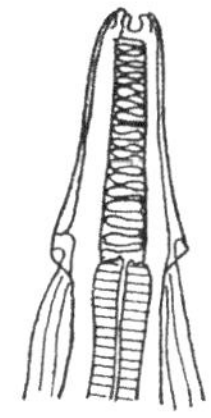

3.83

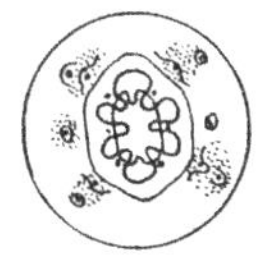

3.84

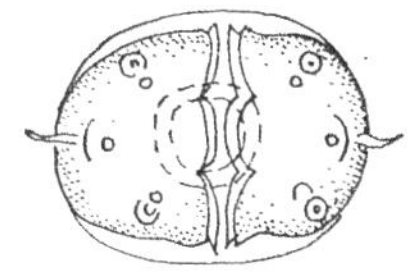

3.85

FIG. 3.80. *Leiuris*, anterior extremity. (After Vaz & Pereira, 1929).
FIG. 3.81. *Streptopharagus*, anterior extremity. (After Baylis, 1923).
FIG. 3.82. *Ascarops*, anterior extremity. (After Foster, 1912).
FIG. 3.83. *Physocephalus*, anterior extremity. (After Quentin, 1965).
FIG. 3.84. *Streptopharagus*, head, apical view. (After Chabaud, 1954).
FIG. 3.85. *Hartertia*, head, apical view. (After Chabaud, 1958).

Family *GONGYLONEMATIDAE* (Hall, 1916, subfam.) Sobolev, 1949

One genus

Gongylonema Molin, 1857

Key to subgenera

1-(2) Verruciform cuticular thickenings generally numerous (Fig. 3.75).
Spicules markedly different in length.
Gubernaculum present.
Parasites of mammals and birds.

Gongylonema (*Gongylonema*) (Molin, 1857)

2-(1) Verruciform cuticular thickenings not numerous (Fig. 3.86).
Spicules unequal but length not markedly different.
Gubernaculum absent.
Parasites of American marsupials.

Gongylonema (*Gongylonemoides*) (Lent & Freitas, 1937, gen.)

Family *SPIRURIDAE* Oerley, 1885

Key to genera

1-(4) Pseudolabia poorly developed, not markedly protruding above oral opening.
2-(3) Oral opening an elongated oval without lateral and submedian bulges (Fig. 3.87) but with fine denticulated border (Fig. 3.87).
Parasites of reptiles.

Paraspirura Sandground, 1936

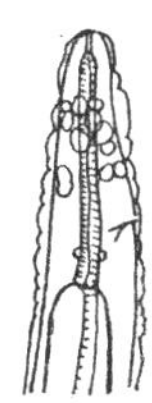

3.86

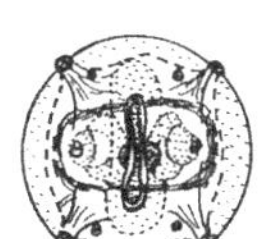

3.87

FIG. 3.86. *Gongylonema* (*Gongylonemoides*), anterior extremity. (After Freitas & Lent, 1937).
FIG. 3.87. *Paraspirura*, head, apical view. (After Sandground, 1936).

3-(2) Oral opening constricted at level of lateral and submedian bulges, without fine denticulate border (Fig. 3.79).
Oesophageal region of body with one or two ventral bosses (Fig. 3.88).
Spirura Blanchard, 1849
(= *Travassospirura* Mönnig, 1958; = *Spiruracerca* Erickson, 1938; = *Taphozoia* Ali & Lovekar, 1967)

4-(1) Pseudolabia highly developed, raised above oral opening, each ornamented with small teeth (Fig. 3.89).
Protospirura Seurat, 1914
(= *Thaprospirura* Sood & Parshad, 1974)

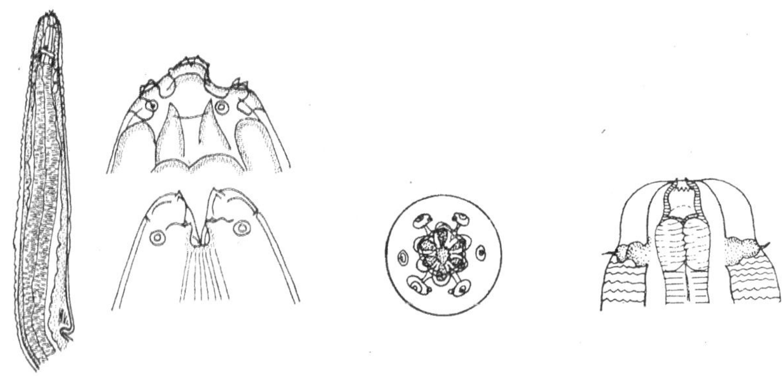

3.88 3.89 3.90 3.91

FIG. 3.88. *Spirura*, anterior extremity. (After Quentin, 1973).
FIG. 3.89. *Protospirura*, head, lateral view, median view. (After Schulz, 1927).
FIG. 3.90. *Cylicospirura*, head, apical view. (After Sandground, 1933).
FIG. 3.91. *Spiralatus*, head, median view. (After Chabaud *et al.*, 1963).

Family *SPIROCERCIDAE* (Chitwood & Wehr, 1932, subfam.)

Key to subfamilies

1-(4) Lips poorly developed, not prominently raised above oral opening.
2-(3) Pharynx without rugous or annular thickenings.
Spirocercinae
3-(2) Pharynx with rugous or annular thickenings (Figs. 3.80-3.83).
Ascaropsinae
4-(1) Lips highly developed, forming 6 denticulate labial lobes, separate over oral opening (Fig. 3.78).
Mastophorinae

Subfamily **Spirocercinae** Chitwood & Wehr, 1932

Key to genera and subgenera

1-(10) Buccal cavity without median lobes.
2-(9) Buccal cavity armed with 6 teeth (Fig. 3.90).
3-(4) Deirids very large and markedly anterior in position.
Lateral alae hypertrophied (Fig. 3.91).
Parasites of birds.
Spiralatus Chabaud, Brygoo & Durette, 1963

4-(3) Deirids small.
Lateral alae reduced or absent.
Parasites of mammals.

Cylicospirura Vevers, 1922
(= *Petrowospirura* Matschulsky, 1952)

5-(8) Vulva in anterior region of body.
6-(7) With 4 pairs of preanal papillae.
Caudal end of male without spines.
Parasites of carnivores.

Cylicospirura (*Cylicospirura*) Vevers, 1922

7-(6) With 7 pairs of preanal papillae.
Caudal end of male with one pair of spines.
Parasites of insectivores.

Cylicospirura (*Gastronodus*) (S. N. Singh, 1934 gen.)

8-(5) Vulva near anus.
Parasites of rodents.

Cylicospirura (*Skrjabinocercina*) (Matschulsky, 1952, gen.)

9-(2) Buccal cavity without teeth (Fig. 3.92).
Parasites of carnivores.

Spirocerca Railliet & Henry, 1917

3.92 3.93 3.94 3.95

FIG. 3.92. *Spirocerca*, head, apical view, lateral view. (After Baylis, 1923).
FIG. 3.93. *Didelphonema*, head, apical view. (After Hill, 1939).
FIG. 3.94. *Cyathospirura*, head, apical view. (After Chabaud, 1959).
FIG. 3.95. *Vigisospirura*, head, apical view. (After Chabaud, 1958).

10-(1) Buccal cavity with one ventral and one dorsal lobe (Fig. 3.93-3.95).
11-(12) Lateral cephalic reliefs well defined (Fig. 3.93).
Caudal end of female with spines.
Parasites of American marsupials.

Didelphonema Wolfgang, 1953

12-(11) Lateral cephalic reliefs not well defined.
Cephalic extremity rounded.
Parasites of carnivores.
13-(14) Buccal cavity with 8 teeth (Fig. 3.94).

Cyathospirura Baylis, 1934

14-(13) Buccal cavity without teeth (Fig. 3.95).

Vigisospirura Petrow & Potekhina, 1953

Subfamily **Ascaropsinae**
Alicata & McIntosh, 1933

Key to genera

1-(12) Lips flat, poorly developed.
Buccal cavity with teeth (Fig. 3.84) or median pharyngeal blades (Fig. 3.96).

2-(7) Buccal cavity with 4 or 6 teeth or groups of teeth, pharyngeal blades absent.

3-(6) Buccal cavity with 6 teeth or 6 groups of teeth (Fig. 3.84).

4-(5) Pharynx straight, wall with spiral thickenings (Fig. 3.82).
Ascarops van Beneden, 1873
(= *Arduenna* Railliet & Henry, 1911)

5-(4) Pharynx with a loop, wall with rugous markings (Fig. 3.81).
Streptopharagus Blanc, 1912

6-(3) Buccal cavity with 4 teeth, 2 median and 2 lateral (Fig. 3.97).
Posterior extremity of body of mature female inflated in form of sphere.
Wall of pharynx with annular thickenings (Fig. 3.83).
Simondsia Cobbold, 1864

7-(2) Buccal cavity with 2 pharyngeal blades, one ventral, the other dorsal (Fig. 3.96).
Parasites of the New World.

8-(9) Buccal border with numerous teeth (Fig. 3.98).
Parasites of Bradypodoidea.
Leiuris Leuckart, 1850
(= *Spirocercella* Thwaite, 1928)

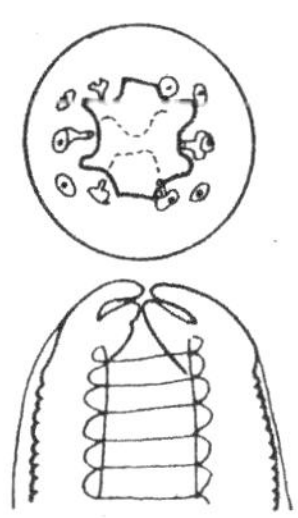

3.96

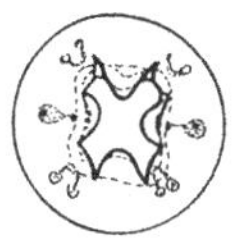

3.97

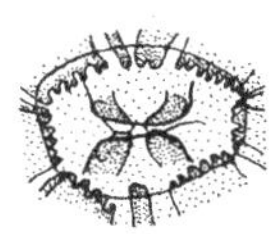

3.98

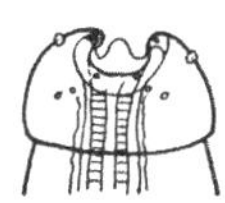

3.99

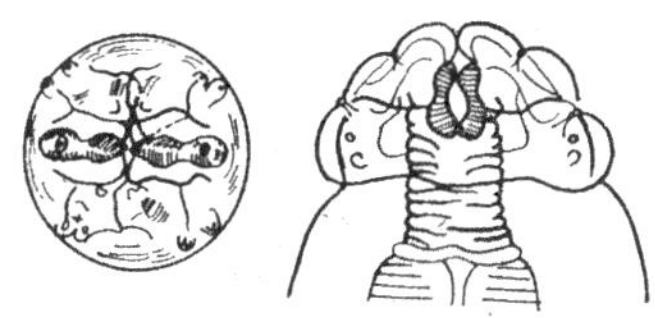

3.100

FIG. 3.96. *Tejeraia*, head, apical view, lateral view. (After Diaz-Ungria, 1963).
FIG. 3.97. *Simondsia*, head, apical view. (After Chitwood & Wehr, 1934).
FIG. 3.98. *Leiuris*, mouth, apical view. (After Vaz & Pereira, 1929).
FIG. 3.99. *Texicospirura*, head, median view. (After Chitwood & Cordero del Campillo, 1966).
FIG. 3.100. *Pygarginema*, head, apical view, median view. (After Chabaud & Rousselot, 1956).

9-(8) Buccal border without teeth.

10-(11) Pharynx with annular thickenings. Cephalic extremity without cephalic vesicle (Fig. 3.96).
Parasites of caviomorphs.
Tejeraia Díaz-Ungría, 1963

11-(10) Pharynx with rugous markings.
Cephalic extremity with cephalic vesicle (Fig. 3.99).
Parasites of Suidae.
Texicospirura M. B. Chitwood & Cordero del Campillo, 1966

12-(1) Lips of great size, markedly elevated above anterior extremity (Figs. 3.77; 3.100).
Buccal cavity unarmed.

13-(14) With 6 independent lips, 2 lateral and 4 submedian.
Pharynx short (Fig. 3.100).
Parasites of Bovidae.

Pygarginema Kadenatsy, 1948

14-(13) With 6 lips joined to form 2 lateral masses (Fig. 3.77).
Pharynx elongated.

Physocephalus Diesing, 1861
(= *Pereiraia* Cuocolo, 1943; = *Physocephaloides* Maplestone, 1932)

Subfamily **Mastophorinae** Quentin, 1970

One genus (Fig. 3.78).
Parasites of Muridae, Microtidae and numerous accidental hosts.

Mastophorus Diesing, 1853

Family *HARTERTIIDAE*
(Quentin, 1970, subfam.)

1-(2) Mouth with 6 teeth.
Buccal capsule with 4 submedian teeth.
Parasites of Australian birds.

Alainchabaudia Mawson, 1968

2-(1) Mouth and buccal capsule without teeth.
Parasites of birds.

Hartertia Seurat, 1914

REFERENCES

ALI, S. M. & LOVEKAR, C. D. 1967. On a new spirurid *Taphozoia aurangabadensis* n.g., n.sp. from a micro-bat *Taphozous kacchensis*. *Indian J. Helminth.*, Year 1966, **18**, Suppl. pp. 68-73.

CHABAUD, A. G. 1959a. Phénomène d'évolution regressive des structures céphaliques et classification des nématodes Spiruroidea. *Parassitologia*, **1** (1), 11-20.

——— 1959b. Sur la systématique des nématodes proches de *Spirocerca lupi* (Rud., 1809). *Parassitologia*, **1** (2), 129-135.

———, BRYGOO, E. R. & DURETTE, M. C. 1963. Spirurides, parasites d'oiseaux malgaches. (Deuxième note). *Annls Parasit. hum. comp.*, **38** (1), 93-108.

CHITWOOD, B. G. & WEHR, E. E. 1934. The value of cephalic structures as characters in nematode classification, with special reference to the superfamily, Spiruroidea. *Z. ParasitKde*, **7** (3), 273-335.

CHITWOOD, M. B. & CORDERO DEL CAMPILLO, M. 1966. *Texicospirura turki* gen. et sp. n. (Nematoda: Spiruroidea) from the stomach of the peccary in the United States and a key to the genera of Ascaropsinae. *J. Parasitol.*, **52**, (2), 307-310.

DÍAZ-UNGRÍA, C. 1963. Nématodes parasites, nouveaux ou intéressants, du Vénézuéla. *Annls Parasit. hum. comp.*, **38**, (6), 893-913.

JOHNSON, S. 1968. A new spirurid genus from the soft-furred field rat, *Rattus meltada* (Nematoda, Spirurinae). *Revta bras. Biol.*, **28** (2), 111-114.

MAJUMDAR, G. & BANERJEE, S. K. 1966. Records on *Skrjabinitectus herpestesi* gen. nov., sp. nov. (Nematoda: Rhabdochonidae) from a carnivorous mammal, *Herpestes mungo* in India. *Věst. čsl. Spol. Zool.*, **30** (4), 319-322.

MAWSON, P. M. 1968. A new genus, *Alainchabaudia* (Nematoda: Spirurinae) from coraciform birds in Australia. *Parasitology*, **58**, 741-744.

QUENTIN, J. C. 1970. Morphogénèse larvaire du spiruride *Mastophorus muris* (Gmelin, 1907). *Annls Parasit. hum. comp.*, **45**, 839-855.

——— 1971. Sur les modalités d'évolution chez quelques lignées d'helminthes de rongeurs Muroidea. *Cah. ORSTOM, Sér. Ent. méd. Parasit.*, **9**, 103-176.

——— 1973. Présence de *Spirura guianensis* (Ortlepp, 1924) chez des marsupiaux néotropicaux. Cycle évolutif. *Annls Parasit. hum. comp.*, **48**, 117-133.

SOOD, M. L. & PARSHAD, V. R. 1974. A new spirurid nematode parasite, *Thaprospirura meltadi* gen.n., sp.n. from rodents in Ludhiana, India. *Acta parasit. pol.*, **22**, 133-138.

KEY TO GENERA OF THE HABRONEMATOIDEA

The superfamily as treated here corresponds almost to the family Hedruridae in Grassé's Traité de Zoologie. The only difference is that we have now included the Tetrameridae in the superfamily.

The name Hedruridae had been chosen to conform to the International Code of Zoological Nomenclature as Hedruridae Railliet, 1916 was the oldest suprageneric taxon. Technically, the superfamily should now be designated Tetrameroidea since Tetrameridae was proposed in 1915. However, both these groups concern genera which are poorly representative of the habronemes in general. In addition, Habronematinae Chitwood & Wehr, 1932 is more recent than Histiocephalinae Gendre, 1922 but, if the latter were given priority, it would cause confusion and entail superficial but annoying changes in the names of taxa used by other authors. It is proposed, therefore, not to follow the International Code of Zoological Nomenclature in this special case and to use the names Habronematoidea and Habronematidae.

The superfamily is divided into 4 families, namely Hedruridae, Habronematidae, Tetrameridae and Cystidicolidae.

HEDRURIDAE

Petter (1971), on the basis of the ontogeny of cephalic structures, concluded that *Hedruris* does not have close affinities with the Habronematidae and should be placed in a distinct family.

HABRONEMATIDAE

Habronematinae

The taxonomy of the Habronematinae has not been greatly modified for some years. The subgenus *Cyrnea* (*Metacyrnea*) was proposed by Chabaud (1960). For convenience the subgenus is now raised to the level of a genus just as *Procyrnea.* Inglis (1965) has shown that the conclusions of S. M. Ali (1961) cannot be sustained and *Aviabronema* Ali falls into the synonymy of *Procyrnea.*

The following genera are regarded as synonyms of *Gendrespira: Chenospirura* Kou, 1958 *nec* Hsü, 1957; *Cheniellospirura* Kou, 1962; and *Sinicaspirura* Skrjabin *et al.*, 1963. *Chenospirura* Hsü, 1957 *nec* Kou, 1958 is considered a synonym of *Cyrnea,* and *Ruschiella* Freitas, 1967 as a synonym of *Procyrnea.*

Draschearia Skrjabin *et al.*, 1965, proposed for the *species inquirenda Spiroptera bullosa* Molin, 1860 which has not been described since Drasche, 1884, cannot be retained. It is probably a synonym of *Procyrnea.*

Histiocephalinae

Since the appearance of Grassé's Traité de Zoologie, in which only the genera *Hadjelia* and *Histiocephalus* were removed from the Acuaridae (Schistorophinae) and reclassified with the habronemes, a whole series of genera has been restudied and the distinction between the two groups clarified (Chabaud, 1960; Inglis, 1965; Bain & Chabaud, 1965). The subfamily Histiocephalinae is now enlarged by the addition of *Viguiera, Torquatoides* (see Chabaud, 1960; Chabaud *et al.*, 1963) and *Stellocaronema* (see Quentin & Wertheim, 1974). On the other hand, *Schistogendra* has been reassigned to the Schistorophinae (Bain & Chabaud, 1965).

Serticeps has been placed in the synonymy of *Viguiera* (see Chabaud, 1960) and the taxon *Torquatella* has been replaced by *Torquatoides* (Williams) and not by *Cathematella* Yamaguti (see Inglis, 1965).

Another new synonymy proposed here is that of *Sobolevicephalus* Parukhin, 1964 with *Hadjelia.* Cephalic "blades" which ornament the pseudolabia are very diverse in size and they can be simple or increasingly indented. As in the Schistorophinae, generic or subgeneric divisions based on such characters can be multiplied without end and this seems undesirable. Also, in our opinion, *Spirocaudata* Sharma, 1957 is a synonym of *Viguiera.*

The genus *Leprirophus* Skrjabin *et al.*, 1965 was created for *Spiroptera bidens* Rudolphi, 1819, a parasite of the gizzard of *Merops apiaster*. Schneider (1866) gave an illustration of the posterior extremity of a male tail with 16 pairs of papillae, which corresponds well with the acuarid type of arrangement. Lepri (1898) assigned to *S. bidens*, specimens from birds of prey, described the anterior end in a fanciful fashion and illustrated the caudal end of the male in a way which clearly suggests the spirurid type of arrangement. In our view, *S. bidens* of Rudolphi is the classical Histiocephalinae of *Merops*, namely, *Torquatoides balanocephala* (Gendre, 1922) and Lepri's specimens represent a different species. *Leprirophus* is placed, therefore, in the synonymy of *Torquatoides*. The genus *Leprirophus* is useless since it is based on inadequate descriptions of worms of different species.

Parabronematinae

This subfamily is separated from the Histiocephalinae because it seems to represent a parallel evolutionary line. *Okapinema* Ivaschkin, 1960 is accepted. Díaz-Ungría (1963), after a study of cephalic structures, confirmed the synonymy of *Squamanema* with *Parabronema*.

TETRAMERIDAE

In 1835 Wiegmann emended the name *Tropisurus* to *Tropidurus*. This accords with Article 33 (a) (i) of the International Code of Zoological Nomenclature (1964 edition) but the latter name was preoccupied (by *Tropidurus* Neuwied, 1824, reptile) and was replaced by the next available taxon, which was *Tetrameres* Creplin, 1846, by Travassos in 1914. It is unfortunate, therefore, that Yamaguti (1961) should have felt it necessary to readopt *Tropisurus*, particularly as *Tetrameres* has been accepted by everyone for many years.

The species of the genera *Tetrameres* and *Microtetrameres* have highly varied cephalic structures, the female being often more specialized than the male. A revision of the genera and their subdivision into subgenera based on cephalic characters would undoubtedly be valuable, but the present division into subgenera (*Gynaecophila* Gubanov, 1950, *Petrowimeres* Chertkova, 1953 and *Gubernacules* Rasheed, 1960) which is based on the presence or absence of somatic spines and a gubernaculum can lead to errors and has little phylogenetic significance. We prefer, therefore, not to retain them.

The genus *Microhadjelia* Jogis, 1965, recently restudied by Quentin & Wertheim (1974) seems to reveal the line between Tetramerinae deformed by parasitism and the spirurid origin of the subfamily. The genus confirms well the common origin and the close parallel evolution between the habronemes and the tetramerids.

CYSTIDICOLIDAE

As in the habronemes of mammals and birds, the evolution of cephalic structures in this group seems to have involved the atrophy and progressive invagination of the lateral pseudolabia; this suggestion is quite compatible with the basic information provided by Dollfus & Campana (1957). The classification adopted here commences with species with large pseudolabia and 8 cephalic papillae (*Cristitectus*) and passes towards species with only 4 papillae and with pseudolabia greatly reduced or almost absent (*Spinitectoides*).

In the family, as in many other spirurids, there is a superficial analogy between primitive Thelazioidea and those which are in fact highly evolved forms (Chabaud, 1959). The cephalic structures which are the basic characters for classification are tiny and can be studied only by experienced workers. There is, therefore, much confusion in the group but recent work by Campana-Rouget (1955), Dollfus & Campana-Rouget (1957), Trofimenko (1962), Rasheed (1965), Moravec (1967), Margolis (1968), and Ko & Anderson (1969) have established many synonymies and permit a more precise and better balanced view of the family.

A number of genera have been discovered recently, namely, *Cristitectus* Petter, 1970, *Salvelinema* Trofimenko, 1962, *Ctenascarophis* Mamaev, 1968 and *Spinitectoides* Petter, 1969. Rasheed (1965) and Moravec (1967) have synonymized *Sterliadochona* Skrjabin, 1946 with *Cystidicoloides*. Characters used by Maggenti & Paxman (1971) to re-establish two genera have no generic value for nematode parasites of vertebrates and it seems important not to reintroduce the confusion which was created.

It was suggested earlier (Chabaud, 1965) that *Pseudoproleptus* Khera, 1954 is a cystidicoline and not a physalopterid and Le Van Hoa *et al.* (1969) have confirmed this.

Notopteroides Chakravarty *et al.*, 1962 had been placed in the synonymy of *Pseudoproleptus* by Johnson & Khera (1966) and *Collarinema* Sey, 1970 is probably also a synonym of this genus.

Margolis & Kabata (1967) have studied carefully the buccal structures of *Salvelinema* Trofimenko, 1962 and shown that *Pseudometabronema* Bogdanova, 1963 is a synonym of this genus.

Having studied *Cyclozone acipenserina* in apical view, we do not share the view of E. C. Skrjabina (1969) who placed this genus in the Physalopteroidea. The truncated apical tooth of the pseudolabia seems characteristic of the Cystidicolidae, not of the physalopterids, and the genus is placed near *Ascarophis*.

Neospinitectus Kalyankar, 1971 which has a peculiar cervical ornamentation, is too poorly known to be classified at present.

KEY TO FAMILIES

1-(2) Median lips (ventral and dorsal) present, extending over lateral pseudolabia.
Cephalic papillae on lateral pseudolabia (Fig. 3.101).
Oesophagus simple.
Tail of female modified as attachment organ (Fig. 3.102).
Parasites of amphibians and reptiles.
Hedruridae

2-(1) Median lips (ventral and dorsal) not extending over lateral pseudolabia.
Cephalic papillae never on lateral pseudolabia.
Oesophagus divided.
Tail of female not in form of attachment organ.

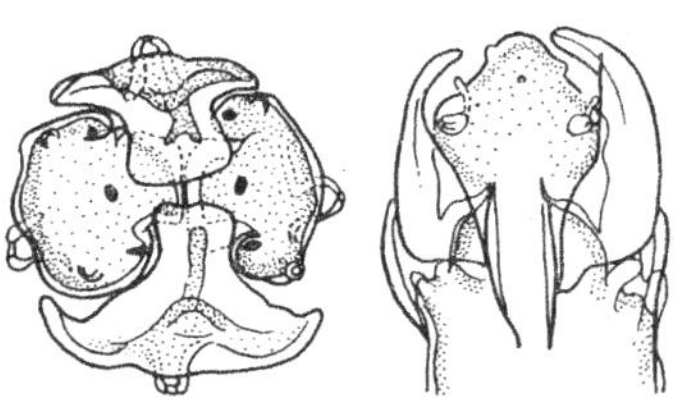

3.101

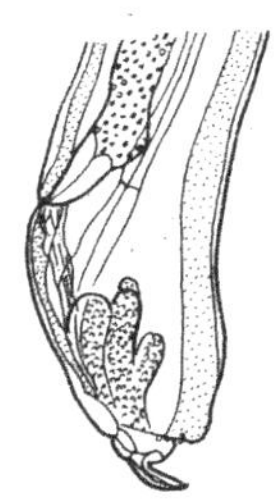

3.102

FIG. 3.101. *Hedruris*, head, apical view, lateral view. (After Petter, 1971).
FIG. 3.102. *Hedruris*, female tail, lateral view. (After Petter, 1971).

3-(4) Body of female spirally coiled or globular in form, or else with the posterior extremity more slender than the remainder of the body, with the distal end inflated. (Figs. 3.103; 3.104).
Sexual dimorphism often marked.
Tissue parasites of marine mammals or forming tumours in the gut wall of birds (except *Microhadjelia*).

Tetrameridae

4-(3) Body of female not spirally coiled, not globular in form, and caudal extremity not inflated.
Sexual dimorphism not marked.

5-(6) Pseudolabia very large, generally covering median lips (Fig. 3.105).
Eight papillae on median lips.
Parasites of birds and mammals.

Habronematidae

6-(5) Pseudolabia small, often reduced to small appendix (Fig. 3.106).
Papillae usually reduced to four at base of pseudolabia.
Parasites of fishes.

Cystidicolidae

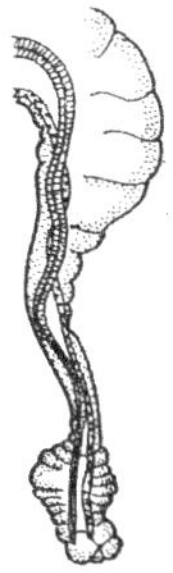
3.103

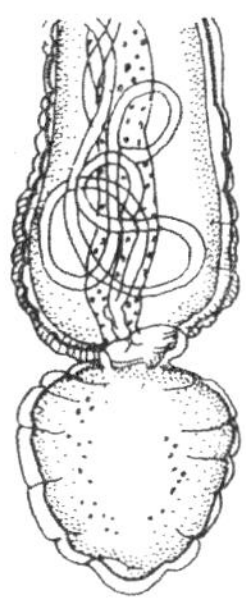
3.104

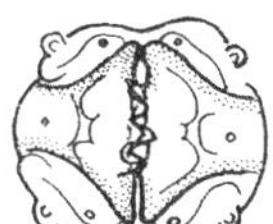
3.105

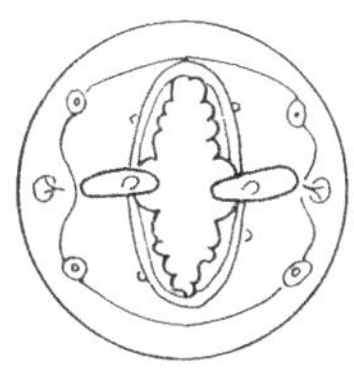
3.106

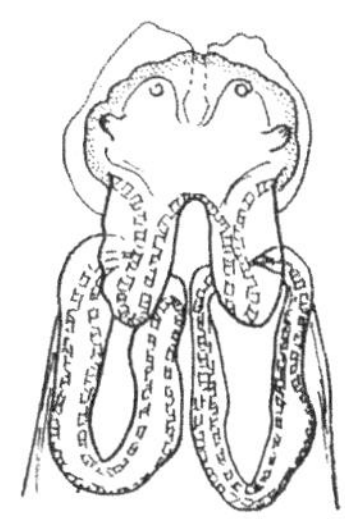
3.107

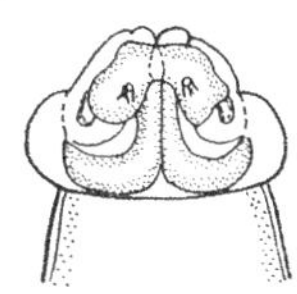
3.108

FIG. 3.103. *Geopetitia*, female tail, lateral view. (After Chabaud, 1951).
FIG. 3.104. *Crassicauda*, female tail. (After Skrjabin & Andreeva, 1934).
FIG. 3.105. *Procyrnea*, head, apical view. (After Chabaud, 1958).
FIG. 3.106. *Cystidicola*, head, apical view. (After Petter, 1969).
FIG. 3.107. *Parabronema*, head, ventral view. (After Chabaud & Mouchet, 1956).
FIG. 3.108. *Torquatoides*, head, ventral view. (After Chabaud, Brygoo & Durette, 1963).

Family *HEDRURIDAE* (Railliet, 1916, subfam.)

One genus (Fig. 3.101).
Parasites of alimentary tract of reptiles and amphibians.

Hedruris Nitzsch, 1812

Family *HABRONEMATIDAE* (Chitwood & Wehr, 1932) Ivaschkin, 1961

Key to subfamilies

1-(2) Posterior border of lips and pseudolabia without ornamentation.

Habronematinae

2-(1) Posterior border of pseudolabia and sometimes of lips with leaf-like structures or shields which extend posteriorly and form attachment organ (Figs. 3.108; 3.109).

3-(4) Attachment organ consisting of cordons arising from convex cuticular shields (Fig. 3.107).
Parasites of mammals.

Parabronematinae

4-(3) Attachment organ consisting of leaf-like structures, simple or denticulate, sometimes also with shields and various ornamentations (Figs. 3.108; 3.109).
Parasites of birds.

Histiocephalinae

Subfamily **Habronematinae** Chitwood & Wehr, 1932

Key to genera

1-(4) Pseudolabia with concave anterior border (Fig. 3.110). Salient lobe behind pseudolabia absent.

2-(3) Anterior border of pseudolabia with well developed teeth (Fig. 3.111).
Spicules equal, narrow and extremely long.
Parasites of birds.

Odontospirura Wehr, 1933
(= *Vaznema* Freitas & Lent, 1947)

3-(2) Anterior border of pseudolabia without 6 well developed teeth (Fig. 3.110).
Spicules unequal.
Parasites of Equidae.

Draschia Chitwood & Wehr, 1934

4-(1) Pseudolabia with convex anterior border (Fig. 3.112), or with projecting lobe behind pseudolabia (Fig. 3.113).

3.109

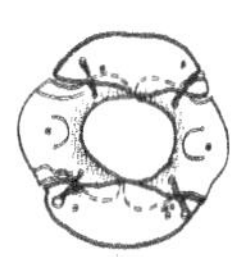
3.110

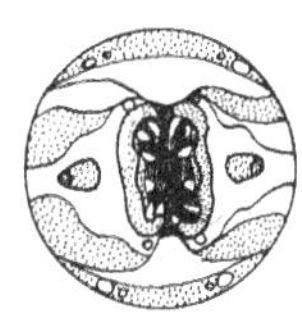
3.111

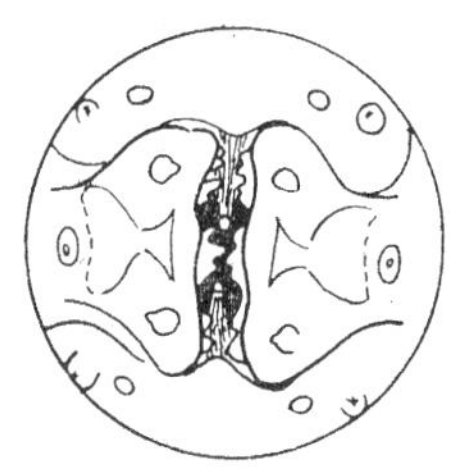
3.112

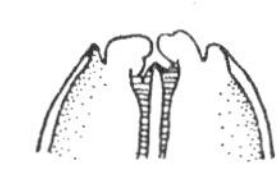
3.113

FIG. 3.109. *Histiocephalus*, head, apical view. (After Grétillat, 1967).
FIG. 3.110. *Draschia*, head, apical view. (After Chabaud, 1958).
FIG. 3.111. *Odontospirura*, head, apical view. (After Wehr, 1933).
FIG. 3.112. *Gendrespirura*, head, apical view. (After Chabaud, 1958).
FIG. 3.113. *Metacyrnea*, head, optical section, along lateral axis (After Chabaud, 1960).

5-(18) Anterior border of ventral and dorsal lips convex (Fig. 3.112) or with convex, submedian lobes, indented on median axis (Fig. 3.114).

6-(11) Lips not indented on median axis (Figs. 3.112; 3.115; 3.116).

7-(8) Median teeth absent.
Pseudolabia almost square, with numerous teeth (Fig. 3.115).
Parasites of primates.

Chitwoodspirura Chabaud & Rousselot, 1956

8-(7) Median teeth present.
Pseudolabia constricted at base (Figs. 3.112-3.116).

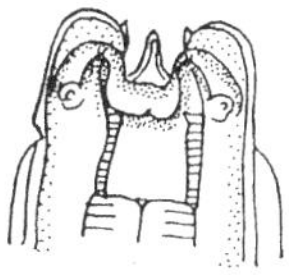
3.114

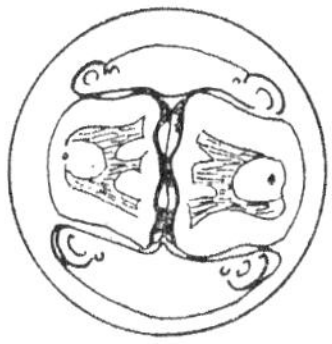
3.115

3.116

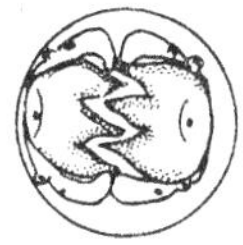
3.117

3.118

3.119

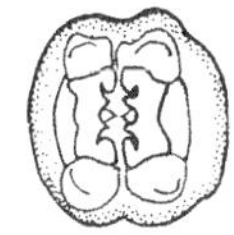
3.120

FIG. 3.114. *Procyrnea*, head, median view. (After Chabaud, 1958).
FIG. 3.115. *Chitwoodspirura*, head, apical view. (After Chabaud & Rousselot, 1956).
FIG. 3.116. *Sicarius*, head, apical view. (After Chabaud, 1958).
FIG. 3.117. *Excisa*, head, apical view. (After Chabaud, 1958).
FIG. 3.118. *Procyrnea*, head, diagram, apical view. (After Chabaud, 1958).
FIG. 3.119. *Cyrnea*, head, diagram, apical view. (After Chabaud, 1958).
FIG. 3.120. *Metacyrnea*, head, apical view. (After Chabaud, 1960).

9-(10) Buccal capsule small, limited posteriorly by the expanded bases of the two median teeth, the latter attached to the median wall by transparent cuticular blades (Fig. 3.116).
Lateral alae well developed.
Parasites of birds.

Sicarius Li, 1934

10-(9) Buccal capsule elongated, median teeth independent of bottom of buccal capsule (Fig. 3.112).
Lateral alae absent.
Parasites of Pholidotes.

Gendrespirura Chabaud, 1958
(= *Chenospirura* Kou, 1958 *nec* Hsü, 1957; = *Cheniellospirura* Kou, 1962; *Sinicaspirura* Skrjabin *et al.*, 1963)

11-(6) Lips deeply indented on median axis (Fig. 3.114).
12-(17) Projecting lobe behind pseudolabia absent.
Pseudolabia extending more anteriorly than lips.
13-(14) Median teeth absent.
Buccal cavity compressed, with little sclerotization of the wall.
Pseudolabia terminated by three large teeth which interlock with those of the opposite side (Fig. 3.117).
Parasites of birds.

Excisa Gendre, 1928

14-(13) Median teeth present (Figs. 3.112; 3.118).
Buccal capsule with strongly sclerotized wall.
Teeth on pseudolabia rather weakly developed.
15-(16) Lateral teeth inserted near anterior border of pseudolabia.
Submedian lobes of lips simple in form (Fig. 3.118).
Deirids in front of nerve ring.
Vulva in middle region of body.
Tail of male of spirurid type (Fig. 3.122) but often with torsion producing asymmetry.
Parasites of birds.

Procyrnea (Chabaud, 1958 subgen.)
(= *Aviabronema* Ali, 1961; = *Ruschiella* Freitas, 1967; = ? *Draschearia* Skrjabin, Sobolev & Ivashkin, 1965)

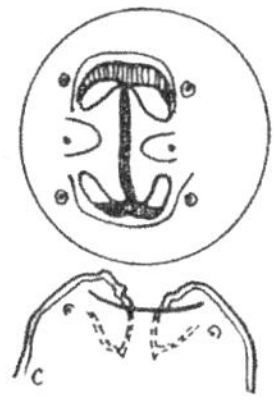

3.121

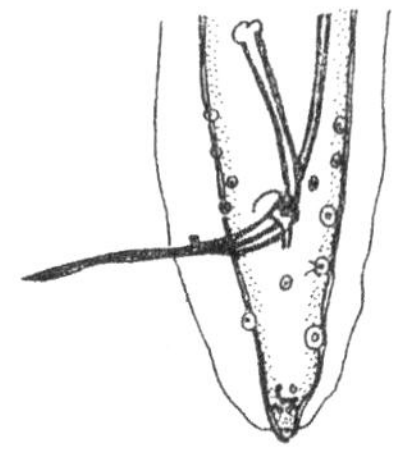

3.122

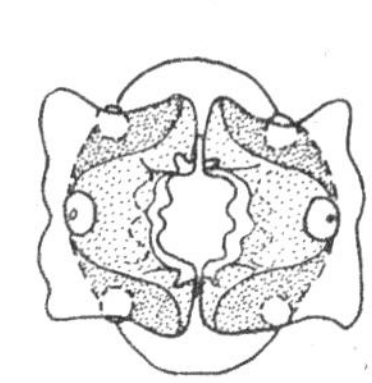

3.123

FIG. 3.121. *Habronema*, head, apical and median views. (After Chabaud, 1958).
FIG. 3.122. *Cyrnea*, male tail, ventral view. (The "spirurid type" is characterized by 9 preanal papillae, 4 postanal papillae and a terminal group of 8 small papillae and 2 phasmids). (After Chabaud, 1968).
FIG. 3.123. *Hadjelia*, head, apical view. (After Chabaud & Campana, 1950).

16-(15) Lateral teeth inserted on posterior part of pseudolabia, at the bottom of the buccal capsule (Fig. 3.119), or absent (additional teeth may be present in anterior part of buccal capsule but these are neoformations).
Lips generally provided with submedian supplementary lobe (Fig. 3.119).
Deirids habitually behind nerve ring.
Vulva markedly posterior.
Tail of male generally of acuarid type (Fig. 3.127).
Parasites of birds.

Cyrnea Seurat, 1914
(= *Seurocyrnea* Strand, 1929; = *Skrjabinochona* Guschanskaja, 1931; = *Chenospirura* Hsü, 1957 *nec* Kou, 1958)

17-(12) Projecting lobe behind pseudolabia present (Fig. 3.113). Pseudolabia less protruberant anteriorly than lips (Fig. 3.120).
Projecting median crest in buccal cavity absent.
Lateral teeth present, inserted at the bottom of the buccal cavity and replaced anteriorly by secondary teeth.
Vulva in anterior half of body.
Tail of male of acuarid type (Fig. 3.127).
Parasites of birds.

Metacyrnea (Chabaud, 1960 subgen.)

18-(5) Anterior border of ventral and dorsal lips rectilinear or slightly concave (Fig. 3.121).
Parasites of mammals.

Habronema Diesing, 1861

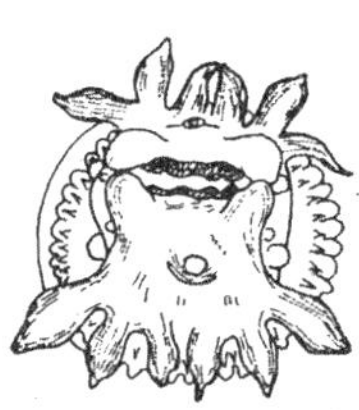

3.124

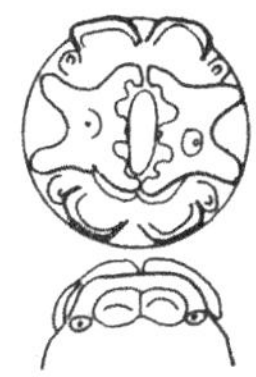

3.125

3.126

FIG. 3.124. *Histiocephalus*, head, apical view. (After Grétillat, 1967).
FIG. 3.125. *Viguiera*, head, apical and median views. (After Chabaud, 1960).
FIG. 3.126. *Torquatoides*, head, apical view. (After Chabaud, Brygoo & Durette, 1960).

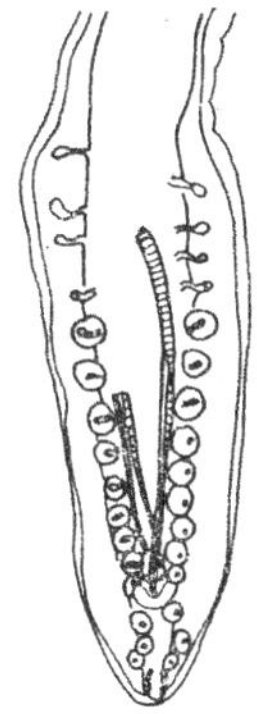

3.127

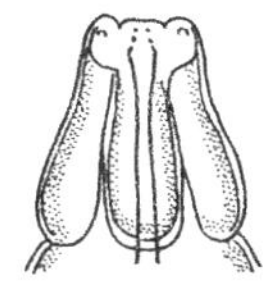

3.128

FIG. 3.127. *Torquatoides*, male tail, ventral view. (The "acuarid type" does not have the terminal group of 8 small papillae and 2 phasmids which characterizes the "spirurid type" (Fig. 3.122). The papillae tend to lie along 2 longitudinal lines. The preanal papillae are frequently very numerous. (After Chabaud, Brygoo & Durette, 1960).

FIG. 3.128. *Okapinema*, head, lateral view. (After van den Berghe & Vuylsteke, 1937).

Subfamily **Histiocephalinae** Gendre, 1922

Key to genera

1-(4) Posterior border of pseudolabia armed with simple (Fig. 3.123) or denticulate blades (Fig. 3.124) but with simple ventral and dorsal lips.
Tail of male of spirurid type (Fig. 3.122).

2-(3) Cervical collar absent.

Hadjelia Seurat, 1916
(= *Gilsonia* Gedoelst, 1919; = *Parhadjelia* Lent & Freitas, 1939; = *Sobolevicephalus* Parukhin, 1964; = *Stellobronema* Guschanskaja, 1937)

3-(2) Cervical cuticle inflated and ornamented with longitudinal striations (Fig. 3.109).
Histiocephalus Diesing, 1851

4-(1) Ornamentation on posterior border of pseudolabia.
Dorsal and ventral lips with blades or shields (Figs. 3.125; 3.126).
Tail of male of acuarid type (Fig. 3.127).

5-(8) Blades or shields in a single row (Fig. 3.125).

6-(7) Vulva posterior.
Preanal papillae numerous.

Viguiera, Seurat 1913
(= *Serticeps* Railliet, 1916; = *Spirocaudata* Sharma, 1971)

7-(6) Vulva anterior.
Three pairs of preanal papillae present.

Stellocaronema Gilbert, 1930

8-(5) Blades or shields in several rows (Figs. 3.108; 3.126).
Torquatoides (Williams, 1929, subgen.) Inglis, 1965
(= *Torquatella* Yorke & Maplestone, 1926 (preoccupied); = *Cathematella* Yamaguti, 1961; = *Leprirophus* Skrjabin, Sobolev & Ivaschkin, 1965).

Subfamily **Parabronematinae** Skrjabin, 1941

Key to genera

1-(2) Cephalic shields not divided transversely, and all in one row (Fig. 3.128).
Parasites of okapi.

Okapinema Ivaschkin, 1960

2-(1) Cephalic shields complex, and in two rows (Fig. 3.107).
Parasites of various mammals.

Parabronema Baylis, 1921
(= *Squamanema* van Thiel, 1925)

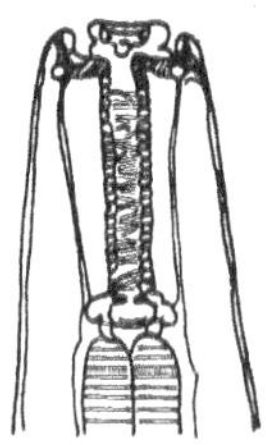

3.129

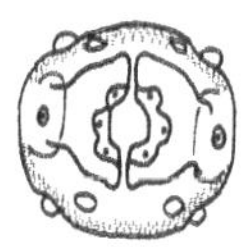

3.130

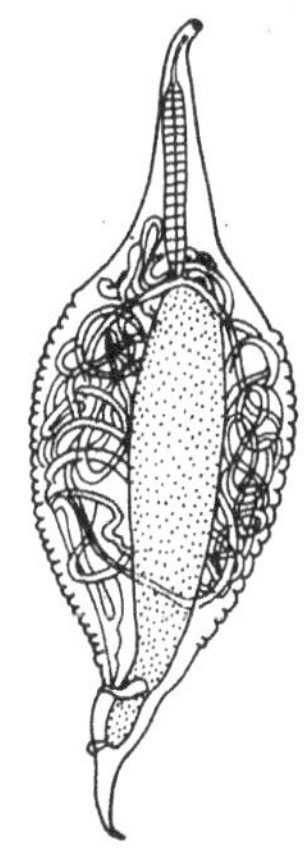

3.131

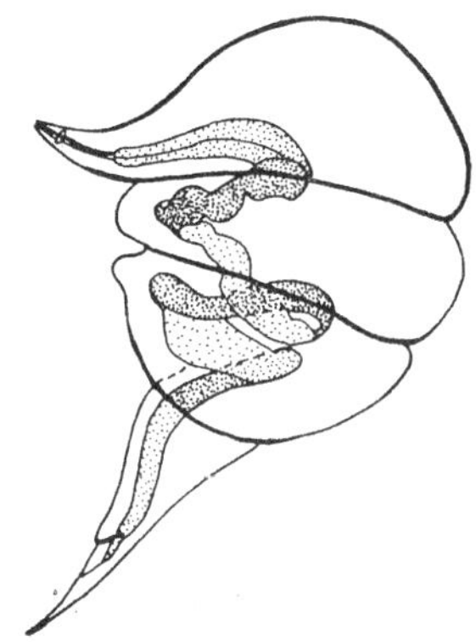

3.132

FIG. 3.129. *Microhadjelia*, head, lateral view. (After Quentin & Wertheim, 1974).
FIG. 3.130. *Microhadjelia*, head, apical view. (After Quentin & Wertheim, 1974).
FIG. 3.131. *Tetrameres (Tetrameres)*. Female body. (After Travassos, 1919).
FIG. 3.132. *Tetrameres (Microtetrameres)*, female body. (After Seurat, 1913).

Family *TETRAMERIDAE* Travassos, 1914

Key to subfamilies

1-(2) Posterior extremity of body of female without constriction followed by terminal swelling of body. Body of female spirally coiled or globular in form.
Parasites of proventriculus (rarely gizzard) of birds.

Tetramerinae

2-(1) Posterior extremity of body of female with constriction followed by terminal swelling of body (Figs. 3.103; 3.104). Body of female not spirally coiled or globular in form.

3-(4) Constricted part of posterior extremity of body of female in form of elongate peduncle (Fig. 3.103).
Sexual dimorphism marked.
Parasites of gizzard of birds, the female in tumour in wall.

Geopetitiinae

4-(3) Constricted part of posterior extremity of body of female a simple narrowing (Fig. 3.104).
Sexual dimorphism not marked.
Parasites of tissues of whales.

Crassicaudinae

Subfamily **Tetramerinae** (Travassos, 1914, fam.)

Key to genera

1-(2) Body of female spirally coiled but neverthcless cylindrical and filiform.
Pharynx rugous (Fig. 3.129).
Preanal papillae of acuarid type.
Pseudolabia large with concave anterior borders (Fig. 3.130).
Parasites of gizzard of birds.

Microhadjelia Jogis, 1965

2-(1) Body of female globular (Fig. 3.131) or in the form of an almost closed spiral (Fig. 3.132); body markedly swollen between head and tail region.
Pharynx smooth.
Anal papillae more or less atrophied.
Parasites of the proventriculus of birds, females in proventricular glands and males on mucosal surface.

Tetrameres Creplin, 1846

3-(4) Body of female globular or subglobular (Fig. 3.131).
Male usually with spines on lateral fields (Fig. 3.133).

Tetrameres (*Tetrameres*) Creplin

4-(3) Body of female twisted in tight spiral (Fig. 3.132).
Male without spines on lateral fields.

Tetrameres (*Microtetrameres*) Travassos, 1915

Subfamily **Geopetitiinae** Chabaud, 1951

One genus.
Parasites of tumours in gizzard wall of birds.

Geopetitia Chabaud, 1951

Subfamily **Crassicaudinae** Yorke & Maplestone, 1926

Key to genera

1-(2) Male without caudal alae.
Two uteri present.

Crassicauda Leiper & Atkinson, 1914

2-(1) Male with small caudal alae (Fig. 3.134).
32 uteri present.

Placentonema Gubanov, 1951

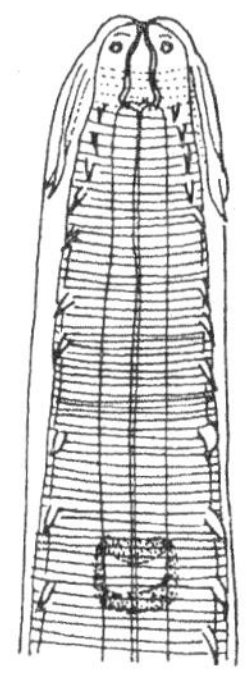

3.133

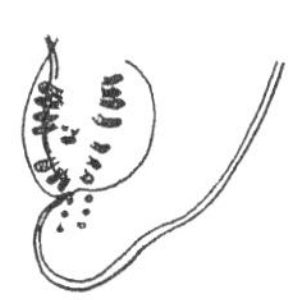

3.134

FIG. 3.133. *Tetrameres* (*Tetrameres*), male anterior extremity, median view. (After Garkavi, 1949).
FIG. 3.134. *Placentonema*, male tail. (After Gubanov, 1951).

Family *CYSTIDICOLIDAE* (Skrjabin, 1946, subfam.)

Key to genera

1-(6) Pseudolabia large with sclerotized reinforcements (Figs. 3.135; 3.137; 3.138).

2-(3) Eight cephalic papillae present.
Pseudolabia joined on median line.
Anterior extremity with short, longitudinal, cuticular crests (Fig. 3.136).
Parasites of alimentary tract of marine fish.

Cristitectus Petter, 1970

3-(2) Four cephalic papillae present.
Pseudolabia leaving ventral space and dorsal space in oral openings (Figs. 3.137; 3.138).

4-(5) Buccal cavity continuous anteriorly with sclerotized thickening which covers entire surface of pseudolabia and lips (Fig. 3.137).
Right spicule longer than left.
Parasites of tissues of marine fish.

Metabronema Yorke & Maplestone, 1926

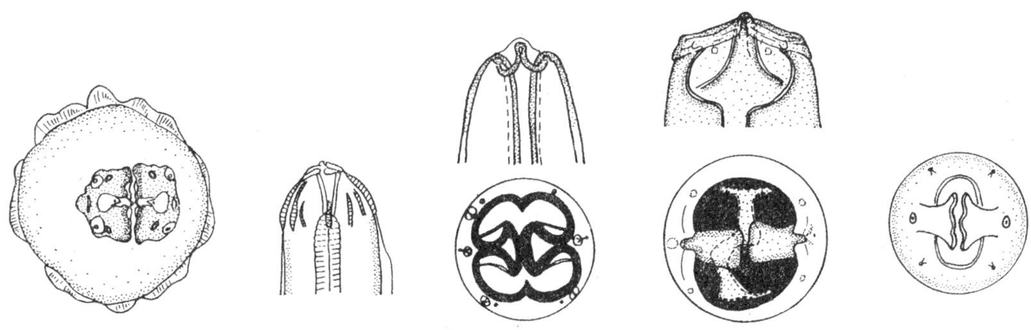

3.135 3.136 3.137 3.138 3.139

FIG. 3.135. *Cristitectus*, head, apical view. (After Petter, 1970).
FIG. 3.136. *Cristitectus*, head, lateral view. (After Petter, 1970).
FIG. 3.137. *Metabronema*, head, lateral view, apical view. (After Rasheed, 1967).
FIG. 3.138. *Salvelinema*, head, lateral view, apical view. (After Margolis, 1968).
FIG. 3.139. *Cystidicoloides*, head, apical view. (After Rasheed, 1965).

5-(4) Buccal cavity dilated anteriorly forming buccal capsule which is divided into ventral and dorsal chambers (Fig. 3.138); pseudolabia inserted on base of buccal capsule forming almost complete partition.
Spicules normal, left longer than right.
Parasites of swim bladder of salmonid fish.

Salvelinema Trofimenko, 1962
(= *Pseudometabronema* Bogdanova, 1963)

6-(1) Pseudolabia tiny, forming two poorly developed elevations on lateral axis of oral opening; sclerotized reinforcements weakly developed (Figs. 3.106; 3.139; 3.140; 3.143; 3.145; 3.147; 3.148; 3.150).

7-(20) Oral opening elongated dorsoventrally.
Pseudolabia joined on median line (Figs. 3.139; 3.140; 3.143; 3.145; 3.147; 3.148); denticulate border usually absent.
Generally 4 pairs of preanal papillae, if more are present they are not arranged in pairs.

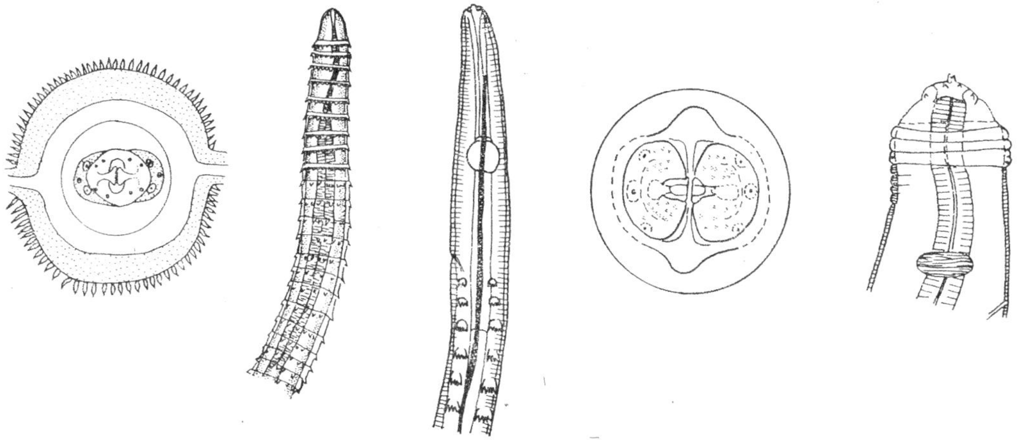

3.140 3.141 3.142 3.143 3.144

FIG. 3.140. *Spinitectus*, head, apical view. (After Rasheed, 1965).
FIG. 3.141. *Spinitectus*, anterior extremity, lateral view. (After Ali, 1956).
FIG. 3.142. *Ctenascarophis*, anterior extremity, lateral view. (After Mamaev, 1968).
FIG. 3.143. *Cyclozone*, head, apical view. (After Mokhayer, 1972).
FIG. 3.144. *Cyclozone*, head, lateral view. (After Skrjabina, 1969).

8-(13) Pseudolabia relatively large, without teeth, and with enlarged anterior border covering the greater part of the oral opening (Figs. 3.139; 3.140).

9-(10) Body without rows of spines.
Pharynx generally long.
Parasites of alimentary tract of freshwater fish (especially salmonids).

Cystidicoloides Skinker, 1931
(= *Sterliadochona* Skrjabin, 1946)

10-(9) Anterior regions of body with rows of spines or combs (Figs. 3.141; 3.142).
Pharynx generally short.

11-(12) Cuticular ornamentation in form of transverse denticulate rings (Fig. 3.141).
Parasites of alimentary tract of fish, generally freshwater.

Spinitectus Fourment, 1883

12-(11) Cuticular ornamentation in form of 4 rows of small combs (Fig. 3.142).
Parasites of alimentary tract of marine fish.

Ctenascarophis Mamaev, 1968

13-(8) Pseudolabia very small, each sometimes armed with one or two teeth, with anterior border slightly enlarged and not covering the greater part of the oral opening (Figs. 3.143; 3.145; 3.147; 3.148).

14-(15) Pharynx very short, little sclerotized.
Cuticle of cervical region inflated to form one or several rings (Fig. 3.144).
Parasites of sturgeons.

Cyclozone Dogiel, 1932

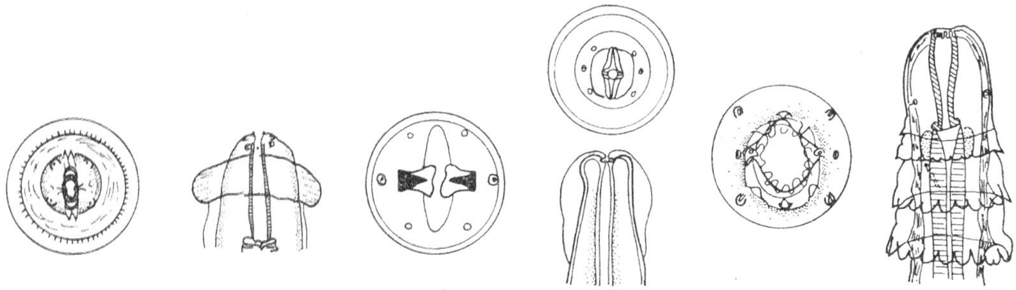

3.145 3.146 3.147 3.148 3.149 3.150

FIG. 3.145. *Pseudoproleptus*, head, apical view. (After Le Van Hoa *et al.*, 1965).
FIG. 3.146. *Pseudoproleptus*, head, median view. (After Le Van Hoa *et al.*, 1965).
FIG. 3.147. *Ascarophis*, head, apical view. (After Rasheed, 1965).
FIG. 3.148. *Parascarophis*, head, apical view, median view. (After Campana-Rouget, 1955).
FIG. 3.149. *Spinitectoides*, head, apical view. (After Petter, 1969).
FIG. 3.150. *Spinitectoides*, head, median view. (After Petter, 1969).

15-(14) Pharynx of usual shape, well sclerotized, often very long.
16-(19) Pseudolabia each with single tooth.
Vulva post-equatorial (Figs. 3.145-3.147).
17-(18) Cephalic cuticle forming collarette similar to that of physalopterids (Fig. 3.146).
Parasites of alimentary tract of marine and freshwater fish.
Pseudoproleptus Khera, 1953
(= *Notopteroides* Chakravarty & Majumdar, 1962; = *Collarinema* Sey, 1970)
18-(17) Cephalic cuticle simple (Fig. 3.147).
Parasites of alimentary tract of marine fish.
Ascarophis van Beneden, 1871
(= *Capillospirura* Skrjabin, 1924; = *Pseudocystidicola* Layman, 1933)
19-(16) Pseudolabia each armed with 2 teeth (Fig. 3.148).
Vulva anterior.
Cuticle of cephalic extremity expanded to form hood (Fig. 3.148).
Pharynx long, dilated anteriorly into cavity expanded dorsoventrally.
Parasites of selachians.
Parascarophis Campana-Rouget, 1955
20-(7) Oral opening hexagonal or oval.
Each pseudolabium reduced to denticle or tongue-like structure not joined on median line with other pseudolabium (Fig. 3.106; 3.149).
Denticulate border generally present.
Preanal papillae grouped in pairs.
21-(22) Body without cuticular ornamentations.
Preanal papillae numerous (more than 4 pairs).
Parasites of tissues of freshwater fish (especially salmonids).
Cystidicola Fischer, 1798
(= *Pseudancyracanthus* Skrjabin, 1923; = *Comephoronema* Layman, 1933)
22-(21) Cuticle of anterior part of body ornamented with festooned rings (Fig. 3.150).
Four pairs of preanal papillae present.
Parasites of alimentary tract of marine fish.
Spinitectoides Petter, 1969

REFERENCES

ALI, S. M. 1961. On some new nematodes (Habronematinae) from birds in Hyderabad, India, and the relationships of the genus *Habronema*. *J. Helminth.*, **35**, 1-48.

BAIN, O. & CHABAUD, A. G. 1965. Spirurides parasites d'oiseaux malgaches (troisième note). *Bull. Mus. natn. Hist. Nat.*, 2ème sér., **37**, 173-185.

CAMPANA-ROUGET, Y. 1955. Sur deux nouveaux genres de spirurides parasites de poissons; discussion systématique des genres voisins. *Annls Parasit. hum. comp.*, **30**, 346-362.

CHABAUD, A. G. 1959. Phénomène d'évolution régressive des structures céphaliques et classification des nématodes Spiruroidea. *Parassitologia*, **1**, 11-20.

——— 1965. In: Grassé, P. P., Traité de Zoologie, Tom. IV, Fasc. 2 et 3. *Paris: Masson et Cie*, 1497 pp.

——— 1960. Quatre spirurides parasites d'oiseaux malgaches. *Mém. Inst. scient. Madagascar*, **14**, 105-124.

———, BRYGOO, E. R. & DURETTE, M. C. 1963. Spirurides, parasites d'oiseaux malgaches. (Deuxième note). *Annls Parasit. hum. comp.*, **38**, 93-108.

CHAKRAVARTY, G. K. & MAJUMDAR, G. 1962. New nematode parasites from bird and fish. *Proc. zool. Soc., Calcutta*, **15**, 21-26.

DOLLFUS, R. P. & CAMPANA-ROUGET, Y. 1957. Une nouvelle espèce d'*Ascarophis* (Nematoda, Spirurinae) chez *Gadus luscus* L. Révision du genre. *Annls Parasit. hum. comp.*, Year 1956, **31**, 385-404.

FREITAS, J. F. TEIXEIRA DE 1967. Novo nematódeo spirurídeo parasito de beija-flor (Passeriformes, Trochilidae). *Atas Soc. Biol. Rio de J.*, **11**, 1-5.

HSÜ, G. N. [= SYUI, G. N.] 1965. [A study of the nematodes of birds in the Kwantung province.] *Acta zool. sin.*, **9**, 47-77. [In Chinese: Russian summary, pp. 71-77.]

INGLIS, W. G. 1965. The nematodes parasitic in the gizzard of birds: a study in morphological convergence. *J. Helminth.*, **39**, 207-224.

IVASHKIN, V. M. 1960. [Nematodes of the subfamily Parabronematinae Skrjabin, 1941 in the light of morphology and systematics.] *Trudy Gel'mint. Lab.*, **10**, 94-108. [In Russian.]

JOGIS, V. A. [= ÏYGIS, V. A.] 1965. [New nematodes of birds from the Kurskiĭ Spit (USSR).] *Trudy zool. Inst., Leningr.*, **35**, 208-215. [In Russian.]

JOHNSON, S. & KHERA, S. 1967. Observations on the taxonomy of the genera *Pseudoproleptus* Khera, 1953 and *Notopteroides* Chakravarty and Majumdar, 1962 (Spiruroidea; Nematoda). *Indian J. Helminth.*, **18**, 148-150.

KHERA, S. 1954. *Pseudoproleptus vestibulus* n.g., n.sp. (subfamily Physalopterinae Railliet, 1893: family Physalopteridae Leiper, 1908: Nematoda) from the fish *Mastacembelus armatus* (Lacep.). *Indian J. Helminth.*, Year 1953, **5**, 115-120.

KO, R. C. & ANDERSON, R. C. 1969. A revision of the genus *Cystidicola* Fischer, 1798 (Nematoda: Spiruroidea) of the swim bladder of fishes. *J. Fish. Res. Bd Can.*, **26**, 849-864.

KOU, C. C. 1958. [Studies on parasitic nematodes of mammals from Canton. I. Some new species from *Paradoxurus minor exitus* Schwarz, *Paguma larvata larvata* (Hamilton Smith) and *Manis pentadactyla aurita* Hodgson.] *Acta zool. sin.*, **10**, 60-72. [In Chinese: English summary pp. 68-71.]

KOU CHEN-CHUEN, 1962. [On the generic name *Chenospirura* (Nematoda Spirurida).] *Acta zool. sin.*, Year 1961, **13**, (1/4), 122. [From *Bull. signal. CNRS.*, 1963, **24**,: 16-21170.]

LEPRI, G. 1898. Elminti in rapaci della provincia di Roma. *Boll. Soc. Rom. Stud. Zool.*, **7**, 62-68; 204-219.

LE VAN HOA & BUI-THI LIEN-HUONG. 1969. Description de *Pseudoproleptus lamyi* n.sp., Nématode parasite des poissons, *Notopterus chitala* H. B. du Sud Viet-Nam. *Bull. Soc. Path. exot.*, **62**, 1106-1111.

MAGGENTI, A. R. & PAXMAN, G. A. 1971. *Sterliadochona pedispicula* sp. n. (Nematoda: Spirurinae) from *Salmo gairdnerii* Richardson, and a discussion of the genera *Sterliadochona* Skrjabin 1946 and *Cystidicoloides* Skinker, 1931. *Proc. helminth. Soc. Wash.*, **38**, 210-214.

MAMAEV, YU. L. 1968. [*Pontochona* n.g. (Rhabdochonidae, Spirurata), a new nematode genus from fish in the South China Sea.] In: Skrjabin, K. I. & Mamaev, Yu. L. [Editors], [Helminths of animals in the Pacific Ocean.] *Moscow: Izdat. "Nauka"*, pp. 30-35. [In Russian.]

MARGOLIS, L. 1968. Review of the Japanese species of *Cystidicola*, *Metabronema* and *Rhabdochona* (Nematoda) from salmonoid fishes. *Res. Bull. Meguro parasit. Mus.*, No. 2, pp. 23-44.

——— & KABATA, Z. 1967. The structure of the buccal region of *Salvelinema* Trofimenko, 1962 (Nematoda: Cystidicolinae). *Can. J. Zool.*, **45**, 1067-1072.

MORAVEC, F. 1967. The systematic status of the genus *Sterliadochona* Skrjabin, 1946 (Nematoda: Rhabdochonidae). *Folia parasit., Praha*, **14**, 371-376.

PARUKHIN, A. M. 1964. [*Sobolevicephalus chalcyonis* n.g., n.sp., a new nematode of the family Histiocephalidae Skrjabin, 1941.] *Uchen. Zap. gorgov. gos. Univ.*, No. 62, pp. 190-193. [In Russian.]

PETTER, A. J. 1969. Enquête sur les nématodes des poissons de la région nantaise. *Annls Parasit. hum. comp.*, **44**, 559-580.

——— 1970. Quelques spirurides de poissons de la région nantaise. *Annls Parasit. hum. comp.*, **45**, 31-46.

——— 1971. Redescription d'*Hedruris androphora* Nitzsch, 1821 (Nematoda, Hedruridae) et étude de son développement chez l'hôte intermédiaire. *Annls Parasit. hum. comp.*, **46**, 479-495.

QUENTIN, J. C. & WERTHEIM, G. 1974. Helminthes d'oiseaux et de mammifères d'Israel. V. Spirurides nouveaux ou peu connus. *Annls Parasit. hum. comp.* [In press.]

RASHEED, S. 1964. Observations on the spiruroid nematodes of fish with a revision of the genus *Metabronema* Yorke & Maplestone, 1926. *Z. Zool. Syst. & Evolutionsforsch.*, **3**, 359-387.

SEY, O. 1970. *Collarinema triglae* gen. et sp. nov. (Nematoda: Rhabdochonidae) from a red gurnard (*Trigla lyra* L.) in the Adriatic Sea, Yugoslavia. *Acta zool. hung.*, **16**, 209-214.

SHARMA, R. K. 1971. On a new nematode *Spirocaudata bispiculatum* gen. nov. sp. nov. (Ancyracanthidae) from the koel, *Eudynamys scolopacea*. *Parasitology*, **62**, 49-52.

SKRJABIN, K. I., SOBOLEV, A. A. & IVASHKIN, V. M. 1965. Osnov nematodologii. Vol. 14. Spirurata. Part III. Acuarioidea. *Moscow: Izdat. "Nauka"*, 572 pp. [In Russian.]

SKRJABINA, E. S. 1969. [*Cyclozone acipenserina* Dogiel, 1932 (Cyclozoninae Sobolev, 1949) and its position in the classification of parasitic nematodes.] *Mater. nauch. Konf. vses. Obshch. Gel'mint.*, Year 1969, Part 1, pp. 274-280. [In Russian.]

TROFIMENKO, V. I. 1962. [The helminth fauna of freshwater and related fish in Kamchatka.] *Trudy Gel'mint. Lab.*, **12**, 232-262. [In Russian.]

YAMAGUTI, S. 1961. Systema Helminthum. Vol. III. The Nematodes of Vertebrates. *New York: Interscience Publishers*, Pt. I, 679 pp.; Pt. II, pp. 681-1261.

KEY TO GENERA OF THE ACUARIOIDEA

The Acuarioidea is an exceptionally homogeneous group, with the result that its systematics are based almost entirely on the increasing complexity of the cephalic organs of fixation. Generic divisions are fairly arbitrary and there is the risk of creating too many genera if the definitions are too narrow. It seems important, therefore, to keep the number of genera and suprageneric categories reasonably small in order to expose the remarkable homogeneity of the group and thus avoid uselessly complicating the determination of various taxa. We retain one family divided into three subfamilies as proposed by Chitwood & Wehr (1934) and reject the families, subfamilies, and tribes proposed by Skrjabin *et al.* (1965).

Acuariinae

The new genus *Chordatortilis* was discovered by Mendonça & Rodrigues (1965). *Aliella* Ali, 1968 is regarded as a synonym of *Synhimantus* (*Synhimantus*), and *Cordonema* Schmidt & Kuntz, 1972 as a synonym of *Skrjabinoclava*. The subgenus *Syncuaria* (*Decorataria*) Sobolev, 1949 was accepted by Petter (1961) because of the thickness of cordons and because it was monodelphic, unlike the nominative subgenus. In fact, many species of the subgenus *Syncuaria* are monodelphic and it is not necessary to retain *Decorataria*.

Chordocephalus Alegret and *Skrjabinocara* Kurashvili are synonyms since they have the same type-species and both were described in 1941. *Chordocephalus* was described in August (see Index Catalogue of Medical and Veterinary Zoology, Supplement 1, 1953, p. 44) but the exact date of the distribution of Kurashvili's article is not clear (see Index Catalogue of Medical and Veterinary Zoology, Supplement 13, 1963, p. 153.) We presume that *Chordocephalus* has priority but this is not certain.

Seuratiinae

The classification of Petter (1959) and Chabaud & Petter (1959) must be amended on certain points.

The most important change derives from the work of Gibson (1964). The 4th-stage larva of *Streptocara californica* shows that the collarette around the head, characteristic of the adult, is preceded by a stage in which this structure is clearly bivalve in form. Each lateral half is attached at the labial angle on the median line as in all other Acuariidae; as Gibson indicated, *Streptocara* cannot be regarded as primitive. Contrary to that which we had earlier supposed, there is a secondary atrophy of that part of the cordons which lines the collarette at the labial angles and the genus ought to be placed among the specialized forms; this explains, moreover, the complexity of the deirids.

Ingliseria proposed by Gibson (1968), although poorly known, is particularly interesting because the cephalic collarette, interrupted on the lateral faces, is comparable to that of the 4th-stage larva of *Streptocara.*

Inglis (1965), Rijikov (1965) and Rodrigues & Mendonça (1967) have demonstrated the synonymy of *Paryseria* with *Stegophorus. Alinema* Jogis, 1968 is a homonym of the subgenus *Philometra* proposed by Rasheed (1963). We place it in the synonymy of *Rusguniella. Koriakinema* Oshmarin, 1949 is too poorly known to be classified but it may be a synonym of *Streptocara.* Similarly, *Cheilonematodum* Johnston & Mawson, 1941 is poorly known but it may be a synonym of *Rusguniella.*

Schistorophinae

The difficulties which existed in the classification of this group have disappeared as good descriptions of the cephalic structures have appeared in recent years, particularly those of *Ancyracanthopsis* by Inglis (1965) and Adams & Gibson (1969), of *Sciadiocara* by Gibson (1972) and of *Sciadiocara* and *Viktorocara* by Petter (1967). Inglis (1965) established the synonymy of *Krusadia* Sanwal, 1952 with *Schistorophus*. We regard *Smetaleksenema* Schmidt & Kuntz, 1972 as a synonym of *Ancyracanthopsis;* the "pseudointerlabia" of these authors seem to correspond to those which Inglis (1965) observed in *A. madagascariensis.*

Acuarioidea

One family.

Family *ACUARIIDAE*
(Railliet, Henry & Sisoff, 1912, subfam.)

Key to subfamilies

1-(4) Cephalic ornamentation consisting of cordons on cuticle which sometimes form cuticular collarette (Figs. 3.151; 3.153; 3.156-3.176).

2-(3) Cordons extending mainly longitudinally and expanding largely on the cervical region (Figs. 3.153; 3.156-3.169) except in *Paracuaria* (Figs. 3.151; 3.152) in which they are short).
Acuariinae

3-(2) Cordons extending mainly transversely from near point of origin, so that they expand only on the cephalic region, where they form a collarette (Figs. 3.170-3.176).
Seuratiinae

4-(1) Cephalic ornamentation consisting of cylindrical horns, blades, shields or a hood, the apex of these structures being always clearly detached from the underlying cuticle (Figs. 3.177-3.182).
Schistorophinae

Subfamily **Acuariinae** Railliet, Henry, & Sisoff, 1912

Key to genera

1-(2) Cordons very short, not recurrent or anastomosing (Figs. 3.151; 3.152).
Paracuaria Krishna Rao, 1951

2-(1) Cordons long (Figs. 3.153; 3.156-3.169).

3-(10) Cordons not markedly recurrent or anastomosing (Figs. 3.153; 3.156).

4-(5) Ventral, preanal cuticular crest present in male (Fig. 3.154).
Tuft of spines on tail of female (Fig. 3.155).
Skrjabinocerca Shikhobalova, 1930

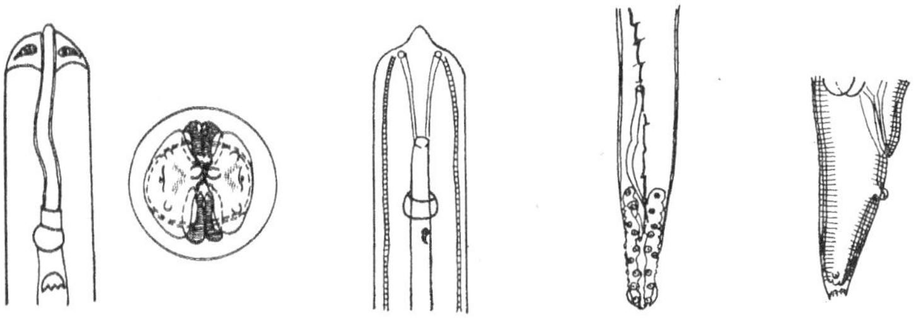

3.151 3.152 3.153 3.154 3.155

FIG. 3.151. *Paracuaria*, anterior extremity, lateral view. (After Rao, 1951).
FIG. 3.152. *Paracuaria*, head, apical view. (After Gibson, 1968).
FIG. 3.153. *Acuaria*, anterior extremity, lateral view. (After Desportes in Chabaud & Petter, 1959).
FIG. 3.154. *Skrjabinocerca*, male tail, ventral view. (After Shikhobalova, 1930).
FIG. 3.155. *Skrjabinocerca*, female tail, lateral view. (After Shikhobalova, 1930).

5-(4) Ventral, preanal cuticular crest absent in male.
No tuft of spines on tail of female.
6-(9) Cordons more or less rectilinear and extending longitudinally (Fig. 3.153).
7-(8) Spicules equal or subequal.

Acuaria Bremser, 1811

8-(7) Spicules clearly unequal.

Cheilospirura Diesing, 1861

9-(6) Cordons tortuous and in spiral, each terminating on the face opposite to the labial commissure from which each originates (Fig. 3.156).

Chordatortilis Mendonça & Rodrigues, 1965

10-(3) Cordons recurrent or anastomosing (Figs. 3.157-3.169).
11-(23) Anterior extremity of body not dilated.
Parasites of birds.
12-(24) Cordons anastomosing, not recurrent or only slightly recurrent (Figs. 3.157-3.163).
13-(22) Cordons situated symmetrically in relation to lateral line.
14-(19) No cuticular collarette in cervical region (Figs. 3.157-3.159).
15-(18) No cordons or rows of spines on lateral lines behind deirids (Figs. 3.157; 3.158).
16-(17) Deirids normal (Fig. 3.157).

Syncuaria Gilbert, 1927
(= *Decorataria* Sobolev, 1949)

17-(16) Deirids transformed into transverse row of spines behind cordons (Fig. 3.158).

Pectinospirura Wehr, 1933

18-(15) Cordons or rows of spines present on lateral lines behind deirids (Fig. 3.159).

Chordocephalus Alegret, 1941
(= *Skrjabinocara* Kurashvili, 1941)

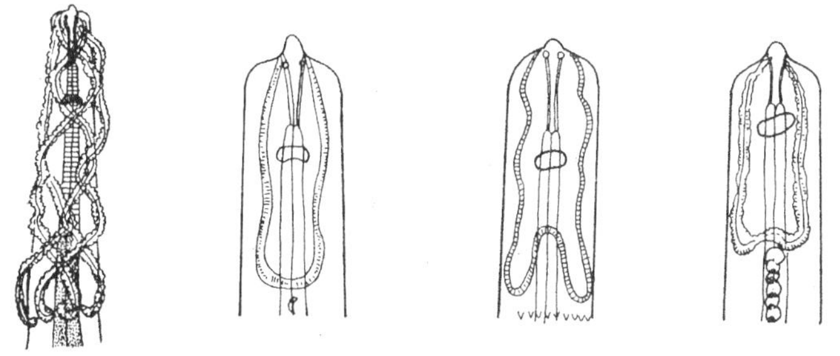

3.156 3.157 3.158 3.159

FIG. 3.156. *Chordatortilis*, anterior extremity, lateral view. (After Machado de Mendonça & Oliveira Rodrigues, 1955).
FIG. 3.157. *Syncuaria*, anterior extremity, lateral view. (After Cram, 1927).
FIG. 3.158. *Pectinospirura*, anterior extremity, lateral view. (After Wehr, 1933).
FIG. 3.159. *Chordocephalus*, anterior extremity, lateral view. (After Sobolev, 1949).

19-(14) Well developed cuticular collarette present in cervical region (Figs. 3.160-3.162).
20-(21) Body spines absent (Fig. 3.160).

Chevreuxia Seurat, 1918

21-(20) Two rows of spines present on lateral fields behind collarette (Figs. 3.161; 3.162).

Skrjabinoclava Sobolev, 1943
(= *Cordonema* Schmidt & Kuntz, 1971)

22-(13) Cordons displaced ventrally so that deirids are dorsal to cordons (Fig. 3.163).
Echinuria Soloviev, 1912
(= *Hamannia* Railliet, Henry & Sisoff, 1912)

23-(11) Anterior extremity of body markedly dilated (Fig. 3.164).
Parasites of insectivores.
Cordons anastomosing, weakly recurrent. Two rows of spines present behind cordons.
Stammerinema Osche, 1955

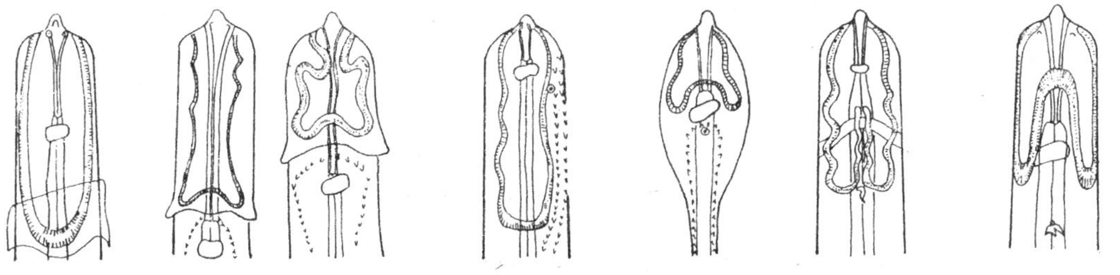

3.160 3.161 3.162 3.163 3.164 3.165 3.166

FIG. 3.160. *Chevreuxia*, anterior extremity, lateral view. (After Seurat, 1918).
FIG. 3.161. *Skrjabinoclava*, cords uncoiled, anterior extremity, lateral view. (After Schmidt & Kuntz, 1972).
FIG. 3.162. *Skrjabinoclava*, cords coiled, anterior extremity, lateral view. (After Desportes in Chabaud & Petter 1959).
FIG. 3.163. *Echinuria*, anterior extremity, lateral view. (After Desportes in Chabaud & Petter, 1959).
FIG. 3.164. *Stammerinema*, anterior extremity, lateral view. (After Tiner, 1951).
FIG. 3.165. *Synhimantus* (*Dispharynx*), anterior extremity, lateral view. (After Desportes in Chabaud & Petter, 1959).
FIG. 3.166. *Synhimantus* (*Synhimantus*), anterior extremity, lateral view. (After Desportes, 1948).

24-(12) Cordons anastomosing or not anastomosing but markedly recurrent (Figs. 3.165-3.169).
Cervical collarette absent.
Body spines on lateral fields absent.

25-(30) Cordons without anterior sinuosities (Figs. 3.165-3.167).

26-(29) Didelphic.
Cordons not enlarged posteriorly (Figs. 3.165; 3.166).
Parasites of terrestrial birds and pelicans.
Synhimantus Railliet, Henry & Sisoff, 1912

27-(28) Cordons not anastomosing (Fig. 3.165).
Synhimantus (*Dispharynx*) (Railliet, Henry & Sisoff, 1912 gen.)

28-(27) Cordons anastomosing (Fig. 3.166).
Synhimantus (*Synhimantus*) Railliet, Henry & Sisoff, 1912
(= *Aliella* Ali, 1968)

29-(26) Monodelphic.
Cordons expanded posteriorly, often with spines (Fig. 3.167).
Parasites of herons.
Desportesius (Chabaud & Campana, 1949, subgen.) Skrjabin *et al.*, 1965

30-(25) Cordons with anterior sinuosities (Figs. 3.168; 3.169).

31-(32) Cordons recurrent on lateral fields forming simple arch convex anteriorly (Fig. 3.168).
Parasites mainly of marine birds.
Cosmocephalus Molin, 1858

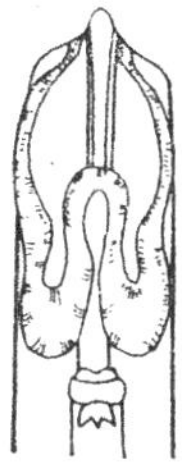

3.167

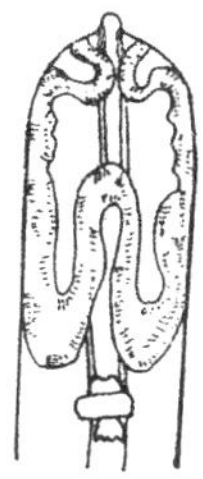

3.168

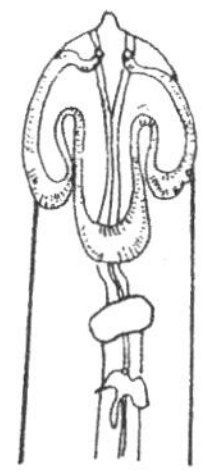

3.169

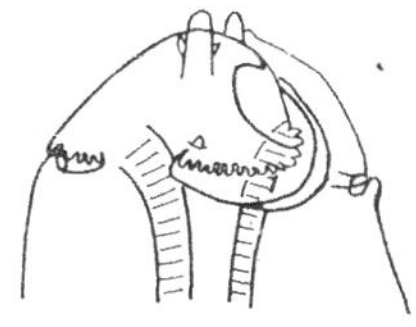

3.170

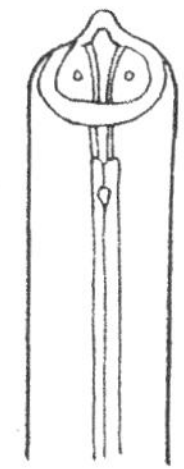

3.171

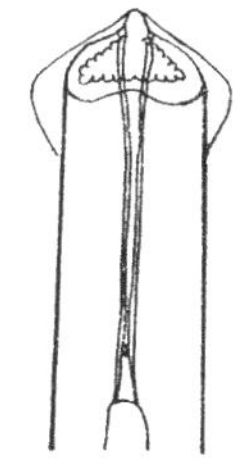

3.172

FIG. 3.167. *Desportesius*, anterior extremity, lateral view. (After Chabaud & Campana, 1949).
FIG. 3.168. *Cosmocephalus*, anterior extremity, lateral view. (After Desportes *in* Chabaud & Petter, 1959).
FIG. 3.169. *Sexansocara*, anterior extremity, lateral view. (After Sobolev & Sondarikov, 1939).
FIG. 3.170. *Ingliseria*, head, sublateral view. (After Inglis in Gibson, 1968).
FIG. 3.171. *Rusguniella*, anterior extremity, lateral view. (After Seurat, 1919).
FIG. 3.172. *Aviculariella*, anterior extremity, lateral view. (After Wehr, 1931).

32-(31) Cordons recurrent on lateral fields forming anteriorly two convex arches separated by convex posterior arch (Fig. 3.169).
Parasites of birds of prey.

Sexansocara Sobolev & Sudarikov, 1939

Subfamily **Seuratiinae** Chitwood & Wehr, 1932

Key to genera

1-(12) Cordons arising from commissures of buccal lips, the cephalic collarette being divided into two valves (Figs. 3.170; 3.175).
2-(3) Cephalic collarette interrupted on lateral lines (Fig. 3.170).
Teeth of pseudolabia with pair of small appendices at their base (Fig. 3.170).
Deirids tricuspid.

Ingliseria Gibson, 1968

3-(2) Cephalic collarette not interrupted on lateral lines (Figs. 3.171; 3. 175).
4-(7) Posterior border of cordons not detached from underlying cuticle to form denticulate collarette (Figs. 3.171; 3.172).
Deirids simple.
5-(6) Cordons expanded on lateral fields of cephalic extremity in form of crescents with smooth borders (Fig. 3.171).
Cephalic vesicle absent.

Rusguniella Seurat, 1919
(= *Rusguniella* (*Rusgunoides*) Williams, 1929; = *Alinema* Jogis, 1968 *nec* Rasheed, 1963)

6-(5) Cordons in form of triangle on lateral fields
Anterior edge of cordons generally denticulate (Fig. 3.172).
Cuticle inflated to form cephalic vesicle.

Aviculariella Wehr, 1931
(= *Alcedospirura* Oshmarin, 1959)

7-(4) Posterior border of cordons detached from underlying body cuticle to form denticulate collarette (Figs. 3.173; 3.175).
Deirids tricuspid.
8-(9) Cordons forming margins of 2 lateral leaves covering head.
Margins with numerous fine denticles.

Proyseria Petter, 1959

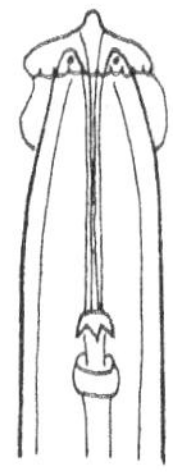
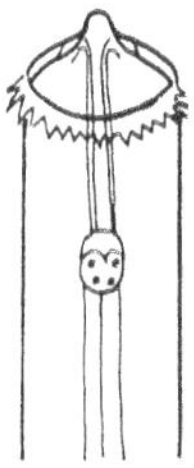
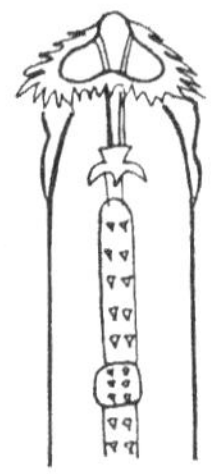
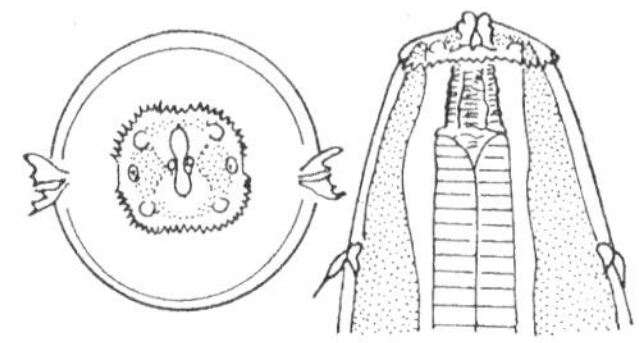

3.173 3.174 3.175 3.176

FIG. 3.173. ***Proyseria***, anterior extremity, lateral view. (After Petter, 1959).
FIG. 3.174. ***Stegophorus***, anterior extremity, lateral view. (After Petter, 1959).
FIG. 3.175. ***Seuratia***, anterior extremity, lateral view. (After Seurat, 1916).
FIG. 3.176. ***Streptocara***, head, apical and median views. (After Gibson, 1968).

9-(8) Cordons not forming margins of 2 lateral leaves covering head (Fig. 3.173) but a collarette surrounding cephalic region (Figs. 3.174; 3.175).

10-(11) No rows of spines on lateral fields.
Cordons not recurrent on lateral fields (Fig 3.174).

Stegophorus Wehr, 1934
(= *Paryseria* Johnston, 1938)

11-(10) Two rows of spines present on each lateral field.
Cordons slightly recurrent on lateral field, generally with large teeth, few in number, (Fig. 3.175).

Seuratia Skrjabin, 1916

12-(1) Cordons not arising from commissures of buccal lips but forming continuous collarette surrounding cephalic region (Fig. 3.176).
Deirids tricuspid or pectiform.

Streptocara Railliet, Henry & Sisoff, 1912
(= *Yseria* Gedoelst, 1919)

Subfamily **Schistorophinae** Travassos, 1918

Key to genera

1-(2) Anterior edge of pseudolabia free and deeply indented (Fig. 3.177).

Schistogendra Chabaud & Rousselot, 1956

2-(1) Anterior edge of pseudolabia free and not indented (=Figs. 3.178-3.182).

3-(4) Anterior extremity with 4 long (Fig. 3.178) or short (Fig. 3.179) cylindrical horns.

Schistorophus Railliet, 1916
(= *Antennocara* Wassilkowa, 1926; = *Quasithelazia* Maplestone, 1932; = *Krusadia* Sanwal, 1952)

4-(3) Head with 4 leaves (Figs. 3.180-3.182).

5-(8) Four leaves not deeply subdivided; free borders of leaves smooth or finely denticulate (Figs. 3.180; 3.181).

6-(7) Four leaves rounded at apices (Fig. 3.180).

Sciadiocara Skrjabin, 1916

7-(6) Four leaves pointed at apices (Fig. 3.181).

Viktorocara Guschanskaja, 1950

8-(5) Four leaves deeply subdivided into teeth or strips (Fig. 3.182).

Ancyracanthopsis Diesing, 1861
(= *Skrjabinobronema* Guschanskaya, 1950; = *Parahistiocephalus* Belopolskaja, 1953; *Smetaleksenema* Schmidt & Kuntz, 1972).

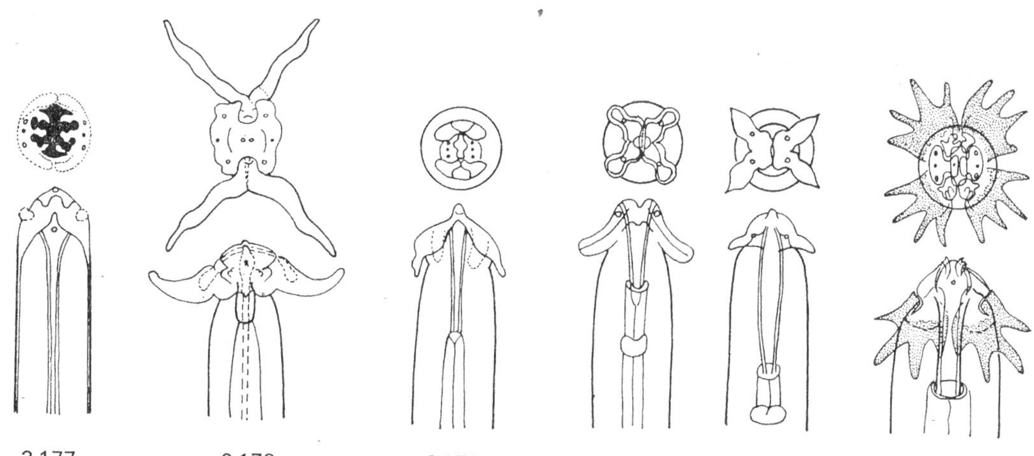

FIG. 3.177. *Schistogendra*, head, apical view. Anterior extremity, lateral view. (After Chabaud & Rousselot, 1956).
FIG. 3.178. *Schistorophus*, long horns. Head, apical view. Anterior extremity, lateral view. (After Li, 1934).
FIG. 3.179. *Schistorophus*, short horns. Head, apical view. Anteror extremity, lateral view. (After Wehr, 1934).
FIG. 3.180. *Sciadocara*, head, apical view. Anterior extremity, lateral view. (After Sobolev in Skrjabin *et al.*, 1949).
FIG. 3.181. *Viktorocara*, reconstruction, head, apical view. Anterior extremity, lateral view. (After Guschanskaja, 1950).
FIG. 3.182. *Ancyracanthopsis*, head, apical view. Anterior extremity, lateral view. (After Inglis, 1965).

REFERENCES

ADAMS, J. R. & GIBSON, G. G. 1969. *Ancyracanthopsis bendelli* n.sp. (Acuariidae: Schistorophinae) from Pacific Coast grouse, with observations on related nematode genera. *Can. J. Zool.*, **47**, 619-626.

ALI, M. M. 1968. Studies on spiruroid parasites of Indian birds. II. A new genus and five new species of Acuariidae, together with a key to the genus *Echinuria*. *J. Helminth.*, **42**, 221-242.

CHABAUD, A. G. & PETTER, A. J. 1959. Essai de classification des Nématodes Acuariidae. *Annls Parasit. hum. comp.*, **34**, 331-349.

CHITWOOD, B. G. & WEHR, E. E. 1934. The value of cephalic structures as characters in nematode classification, with special reference to the superfamily Spiruroidea. *Z. ParasitKde*, **7**, 273-335.

GIBSON, G. G. 1964. Taxonomic and biological observations on *Streptocara californica* (Gedoelst, 1919) Gedoelst and Liégeois, 1922 and the genus *Streptocara* (Nematoda: Acuariidae). *Can. J. Zool.*, **42**, 773-783.

——— 1968. Species composition of the genus *Streptocara* Railliet *et al.*, 1912 and the occurrence of these avian nematodes (Acuariidae) on the Canadian Pacific coast. *Can. J. Zool.*, **46**, 629-645.

——— 1972. *Sciadiocara denticulata* n.sp. (Acuariidae) from *Actitis macularia* (L.) and other nematodes from spotted sandpiper and black-bellied plover. *Can. J. Zool.*, **50**, 131-136.

INGLIS, W. G. 1965. The nematodes parasitic in the gizzard of birds: a study in morphological convergence. *J. Helminth.*, **39**, 207-224.

JOGIS, V. A. [= ĬYGIS, V. A.] 1968. [New and rare nematodes from migration birds of the Kurish Spit.] *Parazitologiya*, **2**, 62-70. [In Russian: English summary p. 70.]

MACHADO DE MENDONÇA, J. & OLIVEIRA DE RODRIGUES, H. 1965. Sobre uma nova subfamília e nôvo gênero de Acuariidae (Nematoda, Spiruroídea). *Mems Inst. Oswaldo Cruz*, **63**, 127-131.

OSHMARIN, P. G. 1959. [*Alcedospirura collaricephala* n.g., n.sp., a new genus and species of bird nematode from the Far East.] *Zool. Zh.*, **38**, 1310-1312. [In Russian: English summary p. 1312.]

PETTER, A. J. 1959. Redescription de *Paryseria adeliae* Johnston, 1938. Remarques sur le genre *Paryseria* et les genres voisins *Rusguniella, Aviculariella, Proyseria* (gen. nov.), *Seuratia. Annls Parasit. hum. comp.*, **34,** 322-330.

——— 1961. Un nouvel acuariidé, *Syncuaria diacantha* n.sp., parasite de la spatule: *Ajaja ajaja* (L.). *Annls Parasit. hum. comp.*, **36,** 764-769.

——— 1967. Trois espèces d'Acuariidae parasites du gésier de *Numenius phaeopus* à l'ile d'Europa. *Bull. Mus. natn. Hist. nat.*, 2è sér., Year 1966, **38,** 940-948.

RIZHIKOV, K. M. 1965. [New data about *Stegophorus stellae-polaris* (Nematoda: Acuariidae).] *Helminthologia*, **6,** 173-180. [In Russian: English and German summaries p. 180.]

OLIVEIRA DE RODRIGUES, H. & MACHADO DE MENDONÇA, J. 1967. Redescrição de *Stegophorus diomedeae* (Johnston & Mawson, 1942) Johnston & Mawson, 1945 (Nematoda, Spiruroidea). *Mems Inst. Oswaldo Cruz*, **65,** 149-152.

SCHMIDT, G. D. & KUNTZ, R. E. 1972a. Nematode parasites of Oceanica. XVI. *Cordonema venusta* gen. et sp.nov., and *Skrjabinoclava* spp. (Acuariidae: Echinuriinae) from birds. *Parasitology*, **64,** 235-244.

———, ——— 1972b. Nematode parasites of Oceanica. XVII. Schistorophidae, Spiruridae, Physalopteridae and Trichostrongylidae of birds. *Parasitology*, **64,** 269-278.

SKRJABIN, K. I., SOBOLEV, A. A. & IVASCHKIN, V. M. 1965. Osnovi nematodologii. Vol. 14. Spirurata. Part III. Acuarioidea. *Moscow: Izdat. "Nauka"*, 572 pp. [In Russian.]

14 SPIRURIDA

DIPLOTRIAENOIDEA, APROCTOIDEA AND FILARIOIDEA

by ROY C. ANDERSON* and ODILE BAIN†

The present classification of the filarioids and their allies is based on the principles followed in an earlier classification proposed by Chabaud & Anderson (1959). There are, however, a number of significant modifications made in the light of more recent work on the parasitic roundworms. It has now been proposed (Chabaud, 1974) that the oviparous superfamilies Diplotriaenoidea and Aproctoidea be recognized mainly because their basic biology relates them to certain Spirurina now divided much more extensively than previously into superfamilies (e.g. Thelazioidea). Sonin (1966, 1968) had already recognized both superfamilies although their contents in his system are fundamentally different from those proposed herein; we separate, more strictly than does Sonin, the group which produces microfilariae from those which lay eggs.

DIPLOTRIAENOIDEA

This superfamily includes those rather large, oviparous worms which inhabit the air sacs of reptiles and birds. Life-cycles which have been elucidated indicate that their thick-shelled eggs with fully differentiated first-stage larvae pass to the outside via the respiratory system and gut. The infective stage is reached in arthropods which ingest the eggs (Anderson, 1957, 1962; Bain & Vassiliades, 1969; Bain & Vaucher, 1973). Such life-cycles are typical of many spiruroids and quite distinct from those found in the Filarioidea as it is understood herein. We believe the Diplotriaenoidea is a homogeneous group representing a line of evolution fundamentally distinct from the Filarioidea.

New genera which have been proposed since 1958 and which are accepted herein are *Chabaudiella*, *Petrovifilaria* (which has marked affinities to the Diplotriaeninae although it is assigned to the Dicheilonematinae) and *Serratospiculoides*.

Other recently proposed genera which are not accepted because they do not seem to be sufficiently distinct from earlier genera are *Tytofilaria*, *Oceanifilaria* and *Heterospiculoides*. The proposed synonymies entail the following new combinations:
Tytofilaria pauloi=*Hamatospiculum pauloi* (Oliveira Rodrigues & Souza Franco, 1964) n. comb.;
Oceanifilaria verrucosa=*Hamatospiculum verrucosa* (Schmidt & Kuntz, 1970) n. comb.;
Heterospiculoides skrjabini=*Heterospiculum skrjabini* (Schigin, 1957) n. comb.

*Department of Zoology, University of Guelph, Canada.
†Museum National d'Histoire Naturelle, Paris, France.

APROCTOIDEA

The Aproctoidea includes medium-sized to small worms found in the air sacs, nasal cavities and subcutaneous tissues of the head and neck (including the orbit) of birds. Like the diplotriaenoids these nematodes produce well developed eggs containing fully developed first-stage larvae. Only one life-cycle has been elucidated: as Anderson & Chabaud (1958) thought, the eggs must pass to the outside and development occurs in an arthropod intermediate host (Quentin *et al.*, 1976).

In this classification we include the Desmidocercidae in the Aproctoidea. The former are common parasites of the air sacs of fish-eating birds and, apart from their markedly unequal spicules, they seem to have certain affinities with the Aproctidae, notably the egg and larva and the form of the tail (Anderson, 1959). It is recognized, however, that studies of the biology of both aproctids and desmidocercids will be necessary to establish the validity of these apparent affinities.

FILARIOIDEA

This superfamily is restricted to those species which produce microfilariae or seem essentially in the evolutionary line of those species which do (e.g. Filariidae). Only two families are recognized, namely, the Filariidae and the Onchocercidae. It is appreciated that some subfamilies could be raised to family rank, especially in the Onchocercidae, but it is difficult to see what would be gained by inflating the classification in this way.

FILARIIDAE

Unlike the previous classification (Chabaud & Anderson, 1959) the Stephanofilariinae is placed in the Filariidae. It is believed one of the more useful changes proposed herein is to place in the same family *Stephanofilaria*, *Pseudofilaria* (Stephanofilariinae) and *Parafilaria* (Filariinae), all from subcutaneous tissues of large mammals and associated with skin lesions in their hosts. In *Stephanofilaria* and *Pseudofilaria* the vulva is markedly anterior, in the region of the nerve ring, like that in genera of the Filariinae, and the oesophagus is short as in *Parafilaria*.

ONCHOCERCIDAE

The Onchocercidae is expanded to include the Setariinae and the Waltonellinae. Apart from the complex cephalic structures in most, but by no means all, species of the Setariinae (see Yeh, 1959) the biology of this subfamily allies it unmistakably with the microfilaria-producing Onchocercidae. We adhere to recent work which reduces the Ornithofilariinae to synonymy with the Splendidofilariinae and the Eufilariinae to synonymy with the Lemdaninae (Anderson, 1961; Anderson & Prestwood, 1969).

Oswaldofilariinae

Gonofilaria, Befilaria, Piratuboides and *Solafilaria* are added to the Oswaldofilariinae and *Conispiculum* is considered as a genus of doubtful status. The subfamily is restricted to parasites of reptiles (crocodilians and lacertilians) and is characterized by the subequatorial or posterior position of the vulva.

Waltonellinae

This subfamily, recently proposed by Bain & Prod'hon (1974), is restricted to amphibians and consists of *Waltonella, Madochotera* and *Ochoterenella*. The last-named genus was formerly (Chabaud & Anderson, 1959) placed in the Onchocercinae. The Waltonellinae is characterized mainly by a laterally flattened cephalic extremity, a pair of lateral parabuccal elevations and highly developed cephalic papillae often articulated in appearance.

Setariinae

This subfamily is characterized by its complex cephalic structures. *Skrjabinaria*, assigned earlier to this subfamily (Chabaud & Anderson, 1959), is now placed in the synonymy of *Dipetalonema* (see Anderson, 1959) and the subgenus *Acanthocheilonema* (see Chabaud & Bain, 1976). Bain & Durette-Desset (1973) and Chabaud & Bain (1976) have called attention to the affinities of *Skrjabinofilaria* with the Onchocercinae to which it is now assigned. Thus, the Setariinae now contains only *Setaria* and *Papillosetaria*.

The divisions of *Setaria* proposed by Yeh (1959) are not accepted and *Hyraconema* and *Artionema* are placed as synonyms of *Setaria*. Cephalic spines in *Setaria equina*, the generotype, are unique in the genus *Setaria*. Unfortunately, Yeh's division of remaining species into *Hyraconema* and *Artionema* is ill-conceived; the former contains one species from hyracoids and has page priority over the latter. The differences between various species of *Artionema* are often more fundamental than those between *Artionema* and *Hyraconema* and if genera or subgenera were to be recognized they would have to consist of *Setaria* for *S. equina*, and *Hyraconema* (=*Artionema*) for the remaining species. We believe this would be generally unacceptable and we prefer to ignore the proposals of Yeh.

Dirofilariinae

This subfamily, characterized by highly developed caudal alae and papillae, has been expanded since 1959 by the addition of two genera with extremely long, sacculate, glandular oesophagi, namely, *Edesonfilaria* and *Macacanema*, both from primates. *Dipetalonema spiralis* (Molin, 1880) and *Deraiophoronema freitaslenti* Yeh, 1957 have been placed in *Bostrichodera* by Chabaud & Bain (1976).

Onchocercinae

This subfamily is characterized by its long, non-alate tail and markedly dissimilar spicules. With the exception of *Macdonaldius* of reptiles, all genera occur in mammals; as mentioned earlier *Ochoterenella* has been transferred to the Waltonellinae and *Saurofilaria* is regarded by Chabaud & Frank (1961) and Sonin (1962) as a synonym of *Macdonaldius*. It was mentioned earlier that *Skrjabinofilaria* has been transferred from the Setariinae to this subfamily.

Since 1959 numerous genera have been proposed in the Onchocercinae, namely, *Filarissima*, *Cutifilaria*, *Brugia*, *Courduriella*, *Paulianfilaria*, *Chenofilaria*, *Hepaticofilaria*, *Monanema* and *Sprattia*. At the same time, however, *Skrjabinaria*, *Parlitomosa* and *Breinlia* have been regarded as synonyms of *Dipetalonema* by Anderson (1959), Esslinger & Gardiner (1974), and Spratt & Varughese (1975) respectively.

The taxonomy of the genus *Dipetalonema* has become increasingly difficult in the past few years, mainly because the genus has been defined broadly by some authors and narrowly by others. This situation has had a serious impact on the taxonomy of the Onchocercinae as a whole because it has been difficult to relate what appear to be well-defined genera to *Dipetalonema*, the synonyms of which have expanded considerably in recent years. Fortunately, however, excellent modern descriptions of many species have appeared since 1959 and Chabaud & Bain (1976) have been able to undertake an extensive revision of *Dipetalonema* and its allies. These new proposals are incorporated in the following key.

Chabaud & Bain (1976) regard *Breinlia* as valid and they recognize within it two subgenera, namely, *Breinlia* and *Johnstonema*; the latter was previously assigned (Chabaud & Anderson, 1959) as a distinct genus to the Splendidofilariinae but the careful work of Spratt & Varughese (1975) reveals clearly its affinities to *Breinlia* and the Onchocercinae even though its spicules are short, stout and subequal. *Dipetalonema* is divided into the subgenera *Dipetalonema*, *Orihelia*, *Acanthocheilonema* (=*Skrjabinaria*, =*Monnigofilaria*, =*Hepaticofilaria*), *Molinema*, *Loxodonto-*

filaria and *Chenofilaria*. *Tetrapetalonema* is regarded as a valid genus and divided into the subgenera *Tetrapetalonema* (=*Parlitomosa*), *Sandnema* and *Esslingeria*.

Elaeophora schneideri and *Cordophilus sagittus* have been redescribed by Hibler & Adcock (1968) and Bain & Haesevoets (1974); the latter also synonymized *Alcefilaria* and *Cordophilus*. These studies place in doubt distinctions made between *Elaeophora* and *Cordophilus*. One of us (R.C.A.) has compared *E. poeli*, *E. schneideri* and *C. sagittus*. Grossly *E. schneideri* and *C. sagittus* are similar. Females have markedly tapered extremities and are coiled when fixed. Both species typically occur in the lumen of the heart or arteries. Female *E. poeli* are much larger worms and the anterior end of the body is typically embedded in a nodule on the aorta. The female has tapered extremities and when fixed assumes the shape of a "C". The male tail of *E. poeli* and *C. sagittus* is rounded and spicules are rather short and robust, whereas the tail of *E. schneideri* has three large terminal protuberances and spicules are relatively long and slender. *E. schneideri* and *C. sagittus* have an *area rugosa* and a median preanal papilla, both absent in *E. poeli*. Deirids are well developed in *C. sagittus* and *E. schneideri* but absent or obscure in *E. poeli*. The oesophagus is long and divided in all species. The vulva opens near the middle of the muscular oesophagus in *C. sagittus* but well behind the junction of the muscular and glandular oesophagus in *E. poeli* and *E. schneideri*. *E. poeli* is said to be quadridelphic but this has never been confirmed. *E. schneideri*, however, is didelphic. Thus, *C. sagittus* resembles *E. poeli* in some characters and *E. schneideri* in others. We do not think, therefore, these important parasites of the vascular system of ruminants and equines should be separated generically and *Cordophilus* is placed in the synonymy of *Elaeophora*. The genus contains the following species:

E. poeli (Vryburg, 1897) Railliet & Henry, 1912
E. schneideri Wehr & Dikmans, 1935
E. sagittus (Linstow, 1907) n.comb.
E. abramovi (Oschmarin & Belous, 1951) n.comb.

The only character which distinguishes *Wehrdikmansia* from *Onchocerca* is the absence of annular thickenings on the cuticle. However, more recent studies (Bain & Schulz-Key, 1974) show that all intermediate forms exist between species with a smooth cuticle and those with annular ridges. For this reason *Wehrdikmansia* is placed in the synonymy of *Onchocerca*.

Splendidofilariinae

This subfamily is characterized by a relatively long tail and subequal spicules. As indicated earlier, the Ornithofilariinae proposed by Chabaud & Anderson (1959) is placed in the synonymy of Splendidofilariinae. The subfamily is found in reptiles, birds and mammals. The genera *Madathamugadia*, *Pseudlemdana*, *Meningonema*, and *Dunnifilaria*, described since 1959, can be added to the subfamily. Spratt & Varughese (1975) have defined *Johnstonema* much more precisely than previously and it is now regarded as a subgenus of *Breinlia* (see Chabaud & Bain, 1976).

The recently proposed genera *Brygoofilaria*, *Pseudaproctoides*, *Francolinema*, *Nicanoria*, *Francofilaria*, *Pseudaproctella*, *Gallifilaria*, and *Parornithofilaria* are not regarded as valid because they can be accommodated in genera proposed earlier. The synonymies proposed for the first time herein result in the following new combinations:

Brygoofilaria agamae = *Thamugadia agamae* (Sulahian & Schacher, 1968) n.comb.
Nicanoria ibanezi = *Paronchocerca ibanezi* (Freitas, Vincente & Pinto, 1970) n.comb.
Francolinema francolina = *Paronchocerca francolina* (Jairajpuri & Siddiqi, 1970) n.comb.
Francofilaria basiri = *Cardiofilaria basiri* (Jairajpuri & Siddiqi, 1970) n.comb.
Gallifilaria mhowensis = *Cardiofilaria mhowensis* (Jain, Alwar, Adwadhiya & Pandit, 1965) n.comb.

Vagrifilaria columbae = *Chandlerella columbae* (Sonin, 1966) n.comb.
Vagrifilaria longicaudata = *Chandlerella longicaudata* (Sonin, 1966) n.comb.
Vagrifilaria sultana = *Chandlerella sultana* (Sonin, 1966) n.comb.

Thamugadia was incompletely described by Seurat. Nevertheless, the similarity in the male tail and spicules to those in *Brygoofilaria* is striking. We suspect Seurat failed to observe caudal papillae which seem to be of general occurrence in the subfamily and the position of the vulva is not significantly different in the two genera. The host groups, site in the host, and geographical location also support the proposed synonymy of the two genera.

Two subfamilies which have been proposed for some species included in the Splendidofilariinae fall into synonymy, namely Nicanoriinae and Micipsellinae.

Some 70 to 80 species of the Splendidofilariinae from birds pose a difficult systematic problem reflected by the fact that at least 25 genera have at one time or another been proposed to accommodate them in the Filarioidea. Part of the problem stems from the existence of four poorly defined genera of rather long standing (i.e. *Paraprocta* Maplestone, 1931; *Aproctoides* Chandler, 1929; *Lerouxinema* Singh, 1949; and *Anenteronema* Oschmarin, 1949) about which some decision must be made.

(i) *Paraprocta brevicauda* was most inadequately described by Chandler (1924) and additional specimens described and assigned to the genus by Maplestone (1931) were probably not conspecific with Chandler's material. We believe this species should be regarded as a *species inquirenda* and the genus removed from the Filarioidea (see Anderson & Prestwood, 1969).

(ii) The female of *Aproctoides lissum* is unknown and the male tail inadequately described (e.g. papillae absent). We believe this species, not rediscovered and redescribed after almost half a century, should also be regarded as a *species inquirenda* and the genus removed from the Filarioidea (see Anderson & Prestwood, 1969).

(iii) The genus *Lerouxinema* is based on the stated existence of cervical alae, a character extremely rare in filarioids and requiring confirmation in the types. Apart from this character, *L. lerouxi* can be accommodated by *Chandlerella* and we still believe (see Anderson & Prestwood, 1969) that its assignation to this genus is correct since it would not then be isolated in the Splendidofilariinae.

(iv) *Anenteronema skrjabini* Oschmarin, 1949 is said to have an atrophied intestine. This is a most unusual character which requires confirmation as the species seems suspiciously close to at least one species of *Splendidofilaria* known from Corvidae. We do not think *A. skrjabini* should be isolated in the systematics of the Splendidofilariinae as at present and propose it be placed provisionally in *Splendidofilaria*, resulting in the combination of *S. skrjabini* (Oschmarin, 1949) n.comb.

Another major difficulty in the systematics of the bird filarioids of the Splendidofilariinae is the recognition of genera on the basis of the division of the oesophagus, the detailed shape of the spicules, the length of the tail, and the presence or absence of bosses on the cuticle. Unfortunately many species are inadequately described and when there is excessive splitting it becomes practically impossible to assign all species with confidence to appropriate genera. This leads inevitably to the creation of more and more genera by authors not thoroughly familiar with the group, as witnessed by the numerous obviously invalid genera proposed within the past few years.

Anderson & Prestwood (1969) pointed out that there are, in fact, only 6 genera which can clearly be recognized and characterized with confidence in the Splendidofilariinae from birds. If we assign the known species to these well defined genera we obtain the following approximate distribution: (i) *Paronchocerca*—12 species; (ii) *Striatofilaria*—one species; (iii) *Aproctella*—2 species; (iv) *Cardiofilaria*—10 species; (v) *Chandlerella*—26 species; (vi) *Splendidofilaria*—23 species. We do not feel *Chandlerella* and *Splendidofilaria* can be considered cumbersome with about two dozen species in each. At the same time, it is possible that some of the species in each genus are synonyms and this may be obscured by the type of splitting which has taken place in the group. There is considerable evidence that many species are not highly host specific. For example, *Cardiofilaria pavlovskyi* and *Aproctella stoddardi* have each been found in at least 7 families of birds, and *Splendidofilaria fallisensis* has now been found in Gaviidae as well as Anatidae. It seems to us, therefore, far more sensible to have well recognized genera capable of accommodating as many species as possible, with the hope that future work will be directed to redescribing types and new material from type hosts and localities. Perhaps then it may be possible to split larger genera if this seems desirable even after various specific synonymies are discovered.

Lemdaninae

The Lemdaninae is restricted to those onchocercids with a subterminal anus, at least in the male, regardless of the form and ratio of the spicules. As pointed out by Anderson & Prestwood (1969) *Singhfilaria* and *Ameeria* seem to unite those species with subequal spicules (e.g. *Eufilaria, Sarconema*) and those in which the left spicule is many times longer than the right (e.g. *Lemdana*). For this reason we continue to believe Eufilariinae should be retained as a synonym of Lemdaninae (cf. Chabaud & Anderson, 1959).

Since 1959, *Singhfilaria, Chiropterofilaria, Ameeria* and *Heimnema* have been added to the Lemdaninae. *Singhnema, Lemdanella, Eufilariella, Farooqifilaria, Alifilaria* and *Neofilaria* are not regarded as valid genera and the generic synonymies proposed herein result in the following new combinations:

Singhnema sonneretta = *Lemdana sonneretta* (Ali, 1969) n.comb.
Farooqifilaria pecta = *Sarconema pecta* (Jairajpuri & Siddiqi, 1970) n.comb.
Alifilaria pochardi = *Sarconema pochardi* (Ali, 1969) n.comb.
Neofilaria buckleyi = *Eufilaria utae* nom.nov. (nom.nov. for. *E. buckleyi* (Deshmukh, 1968) preoccupied).
Neofilaria alii = *Eufilaria alii* (Deshmukh, 1968) n.comb.

Eufilaria still seems somewhat heterogeneous. Sonin (1965) has proposed *Eufilariella* for species of *Eufilaria* in which the caudal end has lateral swellings and which produce unsheathed microfilariae. Unfortunately the generotype of *Eufilaria* (*E. sergenti*) is poorly described and the microfilaria of this and numerous other species is unknown. It is impossible, therefore, to divide *Eufilaria* in the way proposed by Sonin. We believe the genus contains the following species:

Eufilaria sergenti Seurat, 1921; *E. capsulata* (Annett, Dutton & Elliot, 1901) Seurat, 1921; *E. cypseli* (Annett, Dutton & Elliot, 1901) Nelson, 1965; *E. asiatica* Singh, 1949; *E. delicata* Supperer, 1958; *E. mcintoshi* Anderson & Bennett, 1960; *E. buckleyi* Rasheed, 1960; *E. longicaudata* Hibler, 1964; *E. singhi* (Sultana, 1961) Anderson & Prestwood, 1969; *E. coua* Anderson & Prestwood, 1969 nom.nov. for *E. singhi* Chabaud, Brygoo & Richard, 1964; *E. alii* (Deshmukh, 1968) n.comb. (= *Neofilaria alii*); *E. utae* nom.nov. (nom.nov. for *Neofilaria buckleyi* Deshmukh, 1968).

Heimnema has marked affinities to *Eufilaria* and we include in it those species in which the oesophagus is short and well developed. In the species regarded as belonging to *Eufilaria* the oesophagus is generally rather long and narrow, devoid of, or nearly devoid of, muscular tissue and poorly demarcated from the intestine. Unfortunately, the arrangement of the caudal papillae in some species of *Eufilaria* (including the generotype) and *Heimnema* is not known. It is impossible, therefore, to arrange the species into groups on the basis of the presence of lateral swellings or rows of discrete papillae which might result in more natural groupings. The species in *Eufilaria* and *Heimnema* should be subjected to a thorough comparative study before new genera are proposed. We include in *Heimnema* the following:

H. heimi Chabaud, Brygoo & Richard, 1964;
H. micropenis (Travassos, 1926) n.comb.;
H. lari (Yamaguti, 1935) n.comb.

A number of genera cannot presently be assigned in the classification of the Filarioidea. The following generotypes are regarded as *species inquirendae* because males are unknown: *Paralemdana clelandi* Johnston & Mawson, 1940; *Bhalfilaria badani* Bhalerao & Rao, 1944; *Indofilaria pattabiramani* Alwar, Seneviratna & Gopal, 1959: the last-named species, however, clearly belongs to the Filariidae and may be related to *Parafilaria.*

Roman (1965) has pointed out that *Encephalonema longimicrofilaria* Paruchin & Oschmarin, 1960 belongs in *Robertdollfusa* and is probably a synonym of *R. paradoxa.*

Conispiculum flavescens (Castellani & Willey, 1905), generotype of *Conispiculum* Pandit, Pandit & Iyer, 1929, is regarded as a *species inquirenda* since it is inadequately described. This species, from Agamidae, may pertain to the Oswaldofilariinae.

Dirofilarionema ulari (Gagarin, 1954), generotype of *Dirofilarionema* Sonin, 1963, is inadequately described and we are reluctant to place it in the Filarioidea at this time. From the oral mucosa of grouse (*Tetraogallus himalayensis*) in the U.S.S.R., it has markedly dissimilar spicules, caudal alae and a pointed tail, all of which are unusual in avian filarioids; it may have affinities with the Dracunculoidea, however.

A number of genera assigned to the Filarioidea belong to other nematode groups. These are as follows: (i) *Annulofilaria* Syui [Hsü], 1957 is probably a synonym of *Thelazia*; (ii) *Hamulofilaria* Chandler, 1924 (with the family Hamulofilaridae Soota & Chaturvedi, 1971) is a synonym of *Oxyspirura*; (iii) *Pesteria* Tadros, 1966 is a synonym of *Dracunculus*; (iv) *Paraicosiella* Majumdar & Chakravarty, 1963 belongs with the oxyuroids.

DIPLOTRIAENOIDEA

Family Diplotriaenidae (Skrjabin, 1916 subfam.) Anderson, 1958

Key to subfamilies

1-(2) Cephalic extremity with lateral cuticular formations in form of protrusible trident-like structures on either side of anterior end of oesophagus and opening by pores on either side of oral opening (Figs. 3.183, 3.184), or with pair of lateral blade-like elevations each with trilobed base (Fig. 3.185).
Relatively large worms.
Oral opening dorsoventrally elongated.
Oesophagus divided.
Anus subterminal.
Caudal alae absent, papillae numerous and sessile.
Vulva anterior.
Parasites of air sacs of birds.

Diplotriaeninae

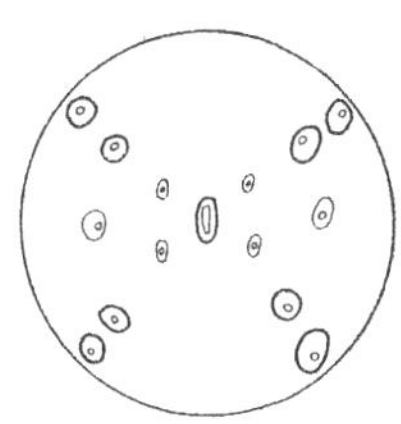

3.183

FIG. 3.183. *Quadriplotriaena hypsokysta* Crites, 1964, *en face* view. (After Anderson, 1968.)

2*

2-(1) Trident-like structures opening by pores absent. Oral opening bordered by cuticularized elevations and/or cuticular thickenings in form of epaulettes (Figs. 3.186, 3.187, 3.192).
Relatively large worms.
Oral opening dorsoventrally elongated.
Oesophagus divided.
Tail usually moderately long with lateral alae and elongate pedunculate papillae, less commonly subterminal without alae and with sessile papillae.
Vulva anterior.
Parasites of air sacs of reptiles and birds.

Dicheilonematinae

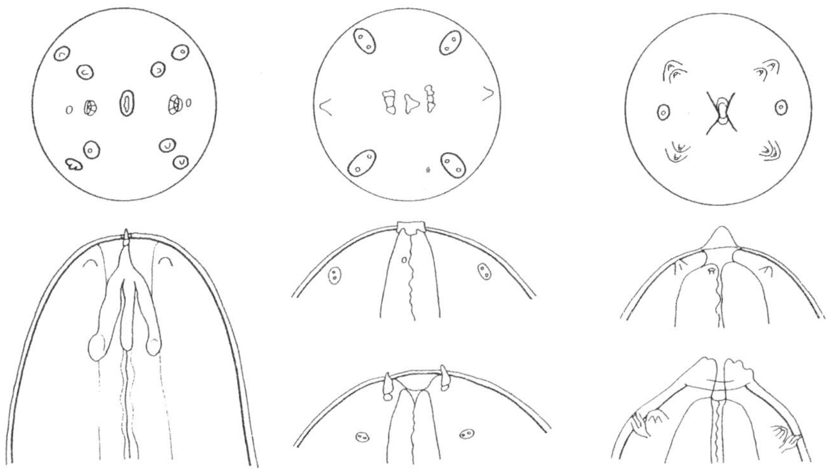

3.184 3.185 3.186

FIG. 3.184. *Diplotriaena ozouxi* Henry & Ozoux, 1909, cephalic extremity. (After Anderson, 1959.)
FIG. 3.185. *Chabaudiella cayennensis* Diaz-Ungría, 1963, cephalic extremity. (After Diaz-Ungría, 1963.)
FIG. 3.186. *Monopetalonema alcedinis* (Rudolphi, 1819), cephalic extremity. (After Anderson, 1959.)

Subfamily **Diplotriaeninae** Skrjabin, 1916

Key to genera

1-(2) Cephalic extremity with two pairs of lateral pores each opening to poorly defined digitiform structures beside anterior end of oesophagus (Fig. 3.183).
Parasites of insectivorous birds.

Quadriplotriaena Wehr, 1935

2-(1) Cephalic extremity without two pairs of lateral pores and associated structures.

3-(4) Cephalic extremity with one pair of lateral pores each opening to protrusible cuticularized structures in form of tridents beside anterior end of oesophagus (Fig. 3.184).
First-stage larva with circles of spines around cephalic extremity and on caudal extremity.
Parasites of insectivorous birds.

Diplotriaena Railliet & Henry, 1909
(= *Diplotriaenoides* Walton, 1927)

4-(3) Cephalic extremity with pair of lateral cuticularized structures, blade-like at apex and trilobed at base (Fig. 3.185).
Parasites of birds.

Chabaudiella Diaz-Ungría, 1963

Subfamily **Dicheilonematinae** Wehr, 1935

Key to genera

1-(2) Cephalic papillae (4 submedian pairs) double or papillae of each pair close to each other (Fig. 3.186).
Pair of lateral tooth-like elevations present beside oral opening.
Epaulettes absent.
Caudal alae well developed, joined anteriorly, with numerous elongate papillae and roughened ventrally.
Spicules markedly dissimilar.
Oesophagus long and divided.
First-stage larva with spines on cephalic and caudal extremities and with lateral alae.
Parasites of birds.

Monopetalonema Diesing, 1861
(= *Politospiculum* Skrjabin, 1916; = *Ornithosetaria* Sandground, 1933)

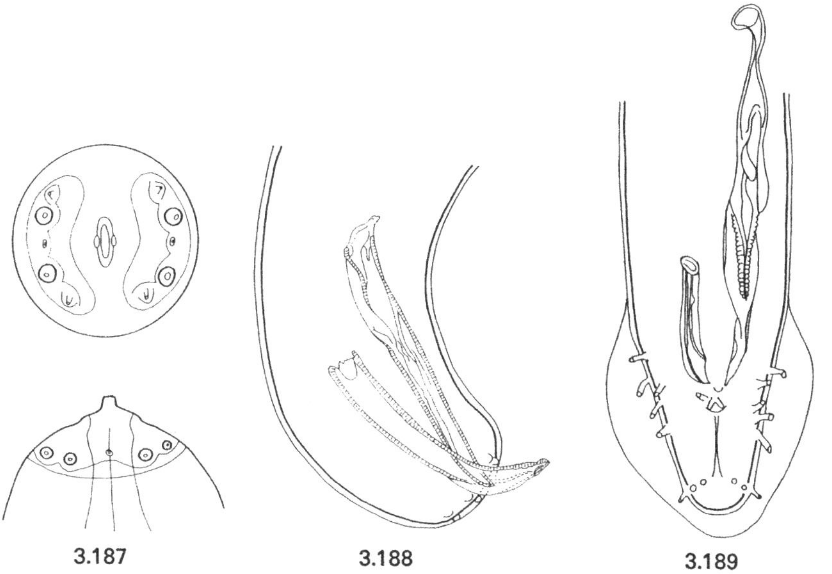

3.187 3.188 3.189

FIG. 3.187. *Dicheilonema americana* (Railliet, 1918), cephalic extremity. (After Anderson, 1968.)
FIG. 3.188. *Petrovifilaria mongolica* (Petrov & Ivaschkina, 1954), caudal extremity male, lateral view. (After Sonin, 1961.)
FIG. 3.189. *Dicheilonema rheae* (Owen, 1843), caudal extremity male, ventral view. (After Yorke & Maplestone, 1926.)

2-(1) Cephalic papillae widely separated.
3-(10) Lateral cephalic cuticularized tooth-like elevations present (Fig. 3.187).
4-(5) Anus subterminal in both sexes (Fig. 3.188).
Spicules unequal and dissimilar but rather short and stout.
Caudal alae absent.
Caudal papillae minute and sessile.
Oesophagus divided.
Parasites of birds.

Petrovifilaria Sonin, 1961

5-(4) Anus not subterminal.

6-(7) Left spicule about twice to 3 times as long as right, alate, often highly complex in form (Fig. 3.189).
Caudal alae well developed with elongate caudal papillae.
Cephalic extremity with epaulette-like formations.
Parasites of birds.

Dicheilonema Diesing, 1861
(= *Contortospiculum* Skrjabin, 1915)

7-(6) Left spicule generally much more than twice as long as right, tubular, not complex in form (Fig. 3.190).

8-(9) Caudal papillae rather large, generally extending into caudal alae (Fig. 3.190).
Cephalic extremity often with epaulette-like formations.
Parasites of reptiles.

Hastospiculum Skrjabin, 1923
(= *Setarospiculum* Mirza & Basir, 1939)

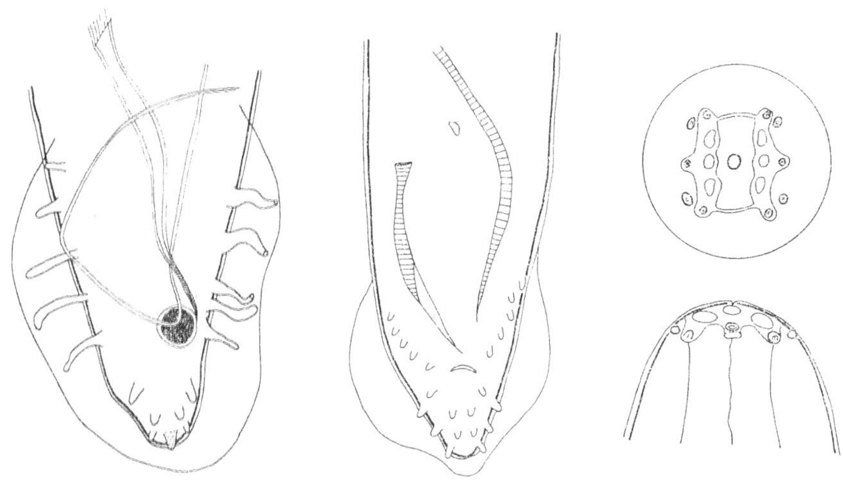

3.190 3.191 3.192

FIG. 3.190. *Hastospiculum setiferum* Chitwood, 1932, caudal extremity male, ventral view. (After Chitwood, 1932.)
FIG. 3.191. *Hamatospiculum insigne* (Schneider, 1866), caudal extremity male, ventral view. (After Skrjabin, 1916.)
FIG. 3.192. *Serratospiculum tendo* (Nitzsch, 1857) (=*S. thoracis*), cephalic extremity. (After Chitwood & Chitwood, 1950 and Anderson, 1968.)

9-(8) Caudal papillae rather small and slender, generally not extending into caudal alae (Fig. 3.191).
Cephalic extremity often with epaulette-like formations.
Parasites of birds.

Hamatospiculum Skrjabin, 1916
(= *Parhamatospiculum* Skrjabin & Petrov, 1928; = *Tytofilaria* Rodrigues & Franco, 1964; = *Oceanifilaria* Schmidt & Kuntz, 1970)

10-(3) Lateral cephalic tooth-like elevations absent (Fig. 3.192).

11-(12) Left spicule broad and complex (Fig. 3.193).
Cephalic extremity with epaulette-like formations (Fig. 3.192).
Caudal alae narrow.
Caudal papillac slcnder.
Right spicule complex, distal half serrated in appearance.
Parasites of birds.

Serratospiculum Skrjabin, 1915

12-(11) Left spicule tubular and simple in form (Figs. 3.194, 3.195).

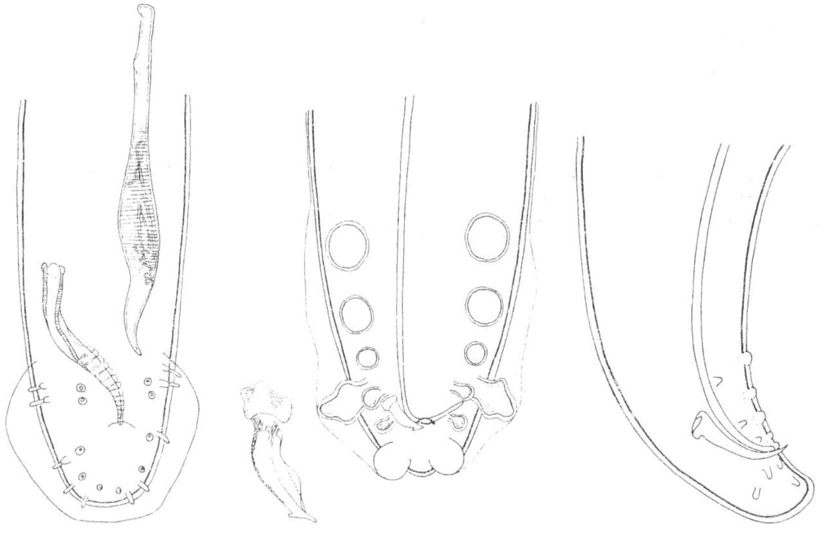

3.193 3.194 3.195

FIG. 3.193. *Serratospiculum tendo* (Nitzsch, 1857), caudal extremity male, ventral view. (After Tubangui, 1934.)
FIG. 3.194. *Heterospiculum sobolevi* Schigin, 1951, caudal extremity male, ventral view and right spicule. (After Schigin, 1951.)
FIG. 3.195. *Heterospiculum skrjabini* (Schigin, 1957) n.comb., caudal extremity male, lateral view. (After Schigin, 1957.)

13-(14) Right spicule short, tooth-like in form (Figs. 3.194, 3.195).
Caudal alae weakly developed.
Caudal papillae large and bulbous (Fig. 3.194).
Parasites of birds (herons).

Heterospiculum Schigin, 1951
(= *Heterospiculoides* Schigin, 1957)

14-(13) Right spicule rather long, not tooth-like in form (Fig. 3.196).
Caudal alae weakly developed.
Caudal papillae sessile and setiform.
Parasites of birds (hawks).

Serratospiculoides Sonin, 1968

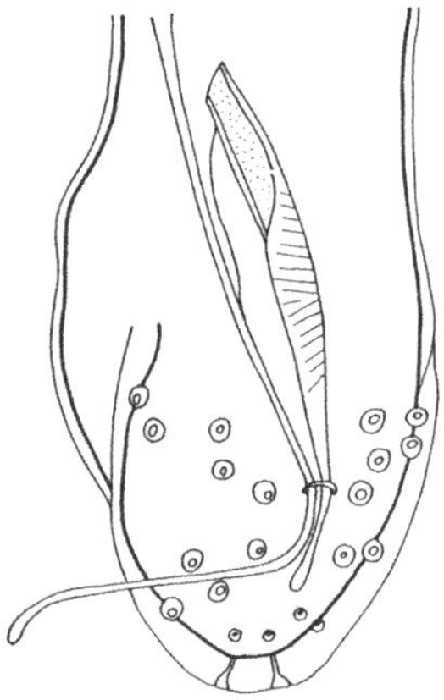

3.196

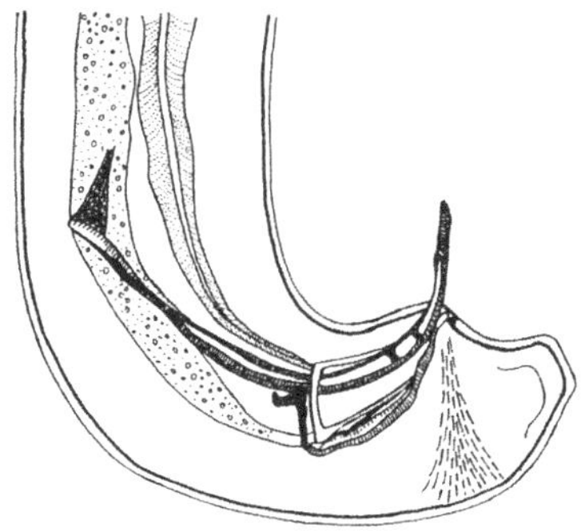

3.197

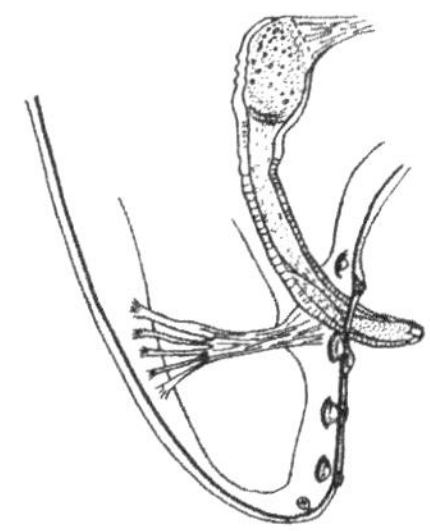

3.198

FIG. 3.196. *Serratospiculoides amaculata* (Wehr, 1938), caudal extremity male, ventral view. (After Wehr, 1938.)
FIG. 3.197. *Desmidocercella leiperi* Singh, 1948, caudal extremity male, lateral view. (After Singh, 1948.)
FIG. 3.198. *Squamofilaria sicki* (Strachan, 1957), caudal extremity male, lateral view. (After Anderson, & Chabaud 1958.)

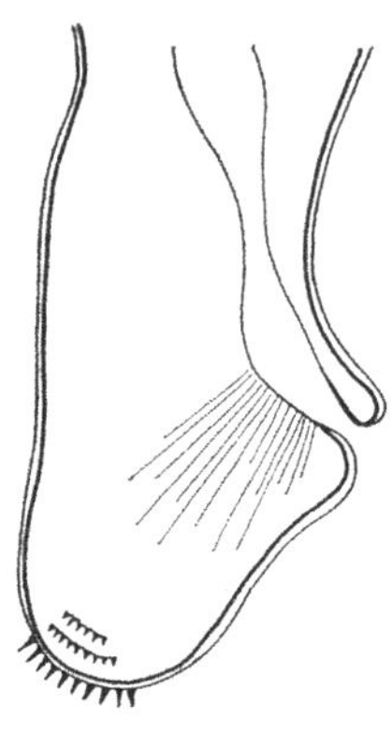

3.199

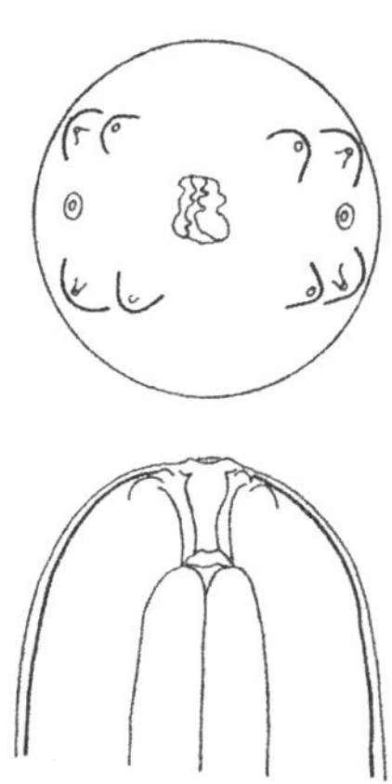

3.200

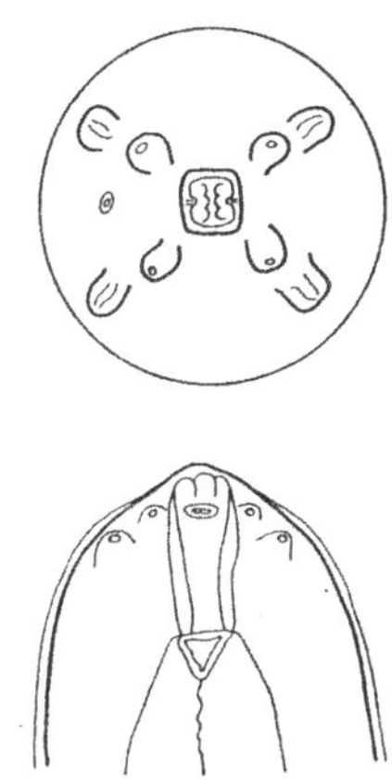

3.201

FIG. 3.199. *Desmidocerca aerophila* Skrjabin, 1916, caudal extremity female, lateral view. (After Skrjabin, 1916.)
FIG. 3.200. *Diomedenema diomedeae* Johnston & Mawson, 1952, cephalic extremity. (After Anderson, 1968.)
FIG. 3.201. *Desmidocercella numidica* (Seurat, 1920), cephalic extremity. (After Anderson, 1968.)

APROCTOIDEA

Key to families

1-(2) Spicules markedly dissimilar in size and form (Fig. 3.197).
Small worms with well developed, tubular buccal cavity, generally with two lateral teeth.
Parasites of the air sacs of fish-eating birds.

Desmidocercidae

2-(1) Spicules basically similar in size and form (Fig. 3.198).
Moderately large stout worms.
Buccal cavity highly developed or reduced, without lateral teeth.
Parasites of the air sacs, orbits and rarely subcutaneous tissue of the head and neck of birds.

Aproctidae

Family *DESMIDOCERCIDAE* Cram, 1927

Key to genera

1-(2) Oesophagus longer than half length of body.
Vulva in posterior half of body.
Caudal end of both sexes with spines or tubercles (Fig. 3.199).
Parasites of herons and cormorants.

Desmidocerca Skrjabin, 1916

2-(1) Oesophagus much shorter than half the length of the body.

3-(4) Lateral teeth of buccal cavity sharply tricuspid (Fig. 3.200).
Vulva behind oesophagus.
Caudal extremity smooth.
Cuticle with small punctations.
Parasites of albatrosses.

Diomedenema Johnston & Mawson, 1952

4-(3) Lateral teeth of buccal cavity simple or bluntly tricuspid (Fig. 3.201).
Vulva generally near posterior end of oesophagus.
Cuticle smooth or with punctations.
Caudal extremity with or without spines.
Parasites of herons and cormorants.

Desmidocercella Yorke & Maplestone, 1926
(= *Pharyngosetaria* Lubimov, 1937)

Family *APROCTIDAE*
(Yorke & Maplestone, 1926 subfam.) Skrjabin & Schikhobalova, 1945

Key to subfamilies

1-(2) Cephalic extremity with four prominent parenchymatous elevations bearing cephalic papillae (Fig. 3.202).
First-stage larva with cephalic spines and unarmed tail.
Parasites of the air sacs of birds.

Tetracheilonematinae

2-(1) Cephalic extremity rounded, or with only two poorly elevated parenchymatous reliefs (Fig. 3.203).
Oral opening more or less rounded.
Oesophagus divided or undivided.
Caudal alae present or absent.
First-stage larva with cephalic spines and tail either short with spines or long, pointed, and unarmed.
Parasites of the orbit, nasal passages, and air sacs of birds.

Aproctinae

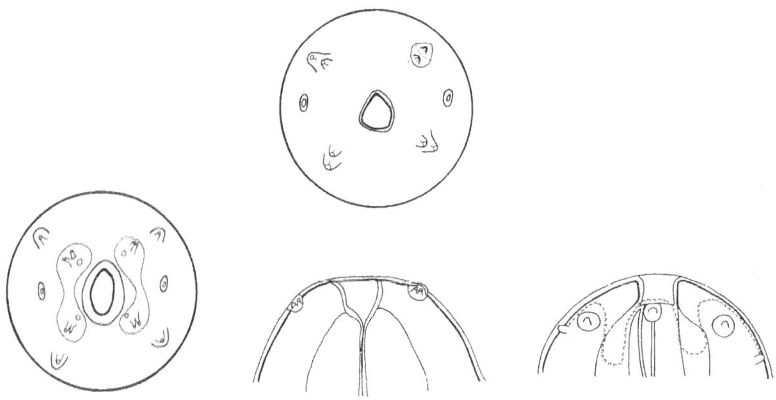

3.202 3.203 3.204

FIG. 3.202. *Tetracheilonema quadrilabiatum* (Molin, 1858), *en face* view. (After Anderson, 1968.)
FIG. 3.203. *Aprocta* (*orbitalis* according to Yeh, 1957), cephalic extremity. (After Anderson, 1968.)
FIG. 3.204. *Pseudaprocta gubernacularia* Schikhobalova, 1930, cephalic extremity, lateral view. (After Schikhobalova, 1930.)

Subfamily **Tetracheilonematinae** Wehr, 1935

One genus.
Parasites of birds.

Tetracheilonema Diesing, 1861
(= *Labiatofilaria* Adams, 1933)

Subfamily **Aproctinae** Yorke & Maplestone, 1926
(= **Squamofilariinae** Sonin, 1963)

Key to genera

1-(2) With delicate lobed cordons between cephalic papillae and amphids (Fig. 3.204).
Buccal cavity minute.
Oesophagus short and undivided.
Parasites of birds.

Pseudaprocta Schikhobalova, 1930
(= *Buckleyfilaria* Singh, 1949)

2-(1) Cordons absent.

3-(4) Buccal cavity minute or, if well developed, with thin delicate walls (Fig. 3.203).
Cephalic end rounded and smooth or with parenchymatous reliefs.
Oesophagus divided or undivided.
Parasites of birds.

Aprocta Linstow, 1883
(= *Lissonema* Linstow, 1903; = *Paramicipsella* Chow, 1939;
= *Cerebrofilaria* Kazubski, 1958)

4-(3) Buccal cavity well developed with thick, strongly cuticularized walls.

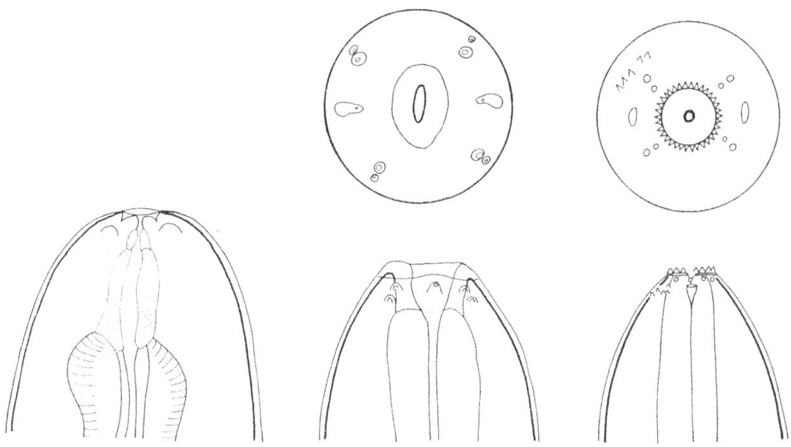

3.205 3.206 3.207

FIG. 3.205. *Mawsonfilaria rhipidurae* (Johnston & Mawson, 1952), cephalic extremity. (After Johnston & Mawson, 1952.)
FIG. 3.206. *Squamofilaria sicki* (Strachan, 1957), cephalic extremity. (After Anderson & Chabaud, 1958.)
FIG. 3.207. *Stephanofilaria stilesi* Chitwood, 1934, cephalic extremity. (After Chitwood & Chitwood, 1950.)

5-(6) Buccal cavity long and narrow, tapering anteriorly, walls divided into short anterior part and long posterior part (Fig. 3.205).
Oesophagus undivided.
Parasites of birds.

Mawsonfilaria Anderson & Chabaud, 1958

6-(5) Buccal cavity short, broad and funnel shaped (Fig. 3.206).
Oesophagus undivided.
Parasites of birds.

Squamofilaria Schmerling, 1925
(= *Coronofilaria* Yorke & Maplestone, 1926; = *Austrofilaria*
Johnston & Mawson, 1940)

3*

FILARIOIDEA

Key to families

1-(2) Vulva markedly anterior, generally anterior to nerve ring.
Spicules markedly different in size and form.
Oesophagus short and undivided or long and divided.
Eggs sometimes thick shelled.
Larva in egg differentiated or microfilarioid; larvae released into bleeding lesion or accumulating in tissues near female.
Vector attracted to skin lesions or lachrymal secretions.
Parasites of subcutaneous tissues of mammals.
Filariidae

2-(1) Vulva well behind nerve ring.
Spicules subequal or markedly different in size, rarely equal, always different in form.
Oesophagus divided or undivided.
Egg thin shelled, containing microfilaria which appears in blood, lymph or skin of host.
Vectors haematophagous arthropods.
Parasites of all tissues of amphibians, reptiles, birds and mammals.
Onchocercidae

Family *FILARIIDAE* (Weinland, 1858) Cobbold, 1879

Key to subfamilies

1-(2) Oral opening surrounded by numerous cuticularized spines (Fig. 3.207).
Small worms.
Vulva near nerve ring.
Amphids large.
Oesophagus short and undivided.
Tail short.
Gubernaculum present.
Spicules markedly different in size and structure, the left long and simple in form.
Male with numerous caudal papillae.
Microfilaria with spherical or elongate sheath.
Parasites of subcutaneous tissue of Bovidae.
Stephanofilariinae

2-(1) Oral opening not surrounded by cuticularized spines.
Large worms.
Vulva anterior to nerve ring.
Oesophagus short and undivided or long and divided.
Tail long or short.
Gubernaculum absent.
Spicules markedly different in size and structure, the left tubular and relatively simple, or broad and complex.
Male with numerous caudal papillae.
Eggs with thick or thin shells.
Larva differentiated or microfilarioid.
Parasites of subcutaneous tissue of mammals.
Filariinae

Subfamily **Stephanofilariinae** Wehr, 1935

One genus.
Parasites of Bovidae.

Stephanofilaria Ihle & Ihle-Landenberg, 1933

Subfamily **Filariinae** Weinland, 1858
(= **Pseudofilariinae** Sonin, 1971)

Key to genera

1-(4) Tail much longer than width of body at anus (Figs. 3.208, 3.209).
Oesophagus long and divided.
Left spicule with complex alate blade.

2-(3) Buccal cavity reduced (Fig. 3.210).
Amphids minute.
Vulva beside oral opening (Fig. 3.210).
Cuticular ring on anterior end of oesophagus.
Caudal end of female rounded, sometimes with delicate spines.
Left spicule complex, not unusually long (Fig. 3.208).
Eggs generally with thick shells.
Larva differentiated.
Parasites of carnivores and rodents.

Filaria Mueller, 1787
(= *Hyracofilaria* Ortlepp, 1937)

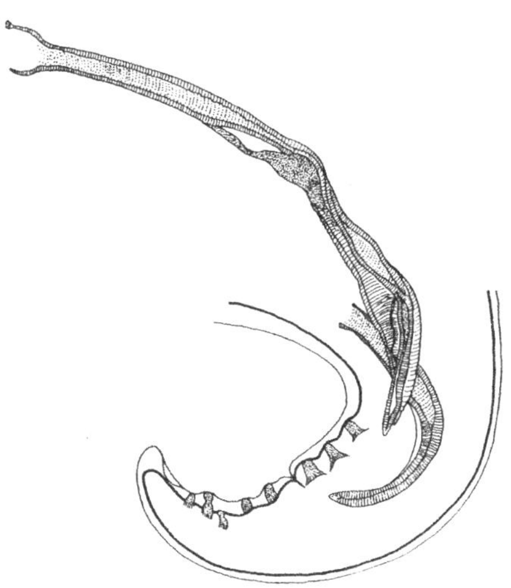

3.208

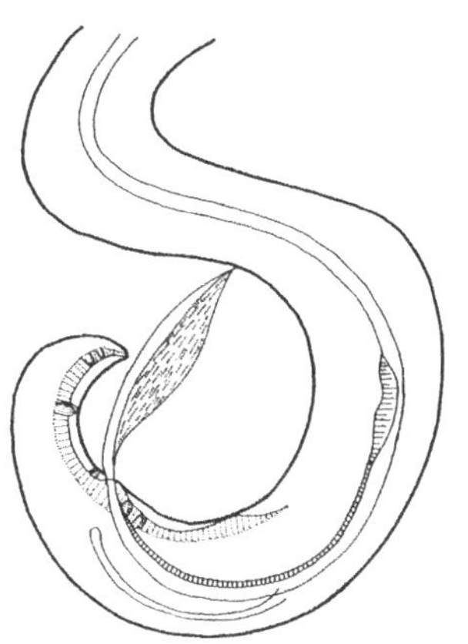

3.209

FIG. 3.208. *Filaria martis* Gmelin, 1790, caudal extremity male, lateral view. (After Anderson, 1960.)
FIG. 3.209. *Suifilaria suis* Ortlepp, 1937, caudal extremity male, lateral view. (After Ortlepp, 1937.)

3-(2) Buccal cavity elongate (Fig. 3.211).
Amphids large and annular.
Caudal end of female indented with delicate spines.
Caudal alae present, right shorter than left (Fig. 3.209).
Left spicule extremely long.
Eggs thin shelled.
Parasites of Suidae.

Suifilaria Ortlepp, 1937

4-(1) Tail shorter than width of body at anus (Figs. 3.212, 3.213).
Oesophagus short and undivided.
Left spicule tubular, with simple non-alate blade.

5-(6) Cuticle of anterior end of body with cuticular bosses, often elongated transversely (Fig. 3.214).
Eggs thin shelled.
Anus in female terminal or subterminal.
Caudal papillae numerous and often sustained by cuticular inflations (Fig. 3.212).
Parasites of ruminants and equids.

Parafilaria Yorke & Maplestone, 1926

6-(5) Cuticle without bosses.
Microfilaria unsheathed.
Anus in female not terminal or subterminal (Fig. 3.215).
Caudal papillae small and few in number (Fig. 3.213).
Cephalic extremity with inner circle of six papillae (Fig. 3.216).
Amphids large.
Vulva between nerve ring and cephalic extremity.
Parasites of antelope.

Pseudofilaria Sandground, 1935

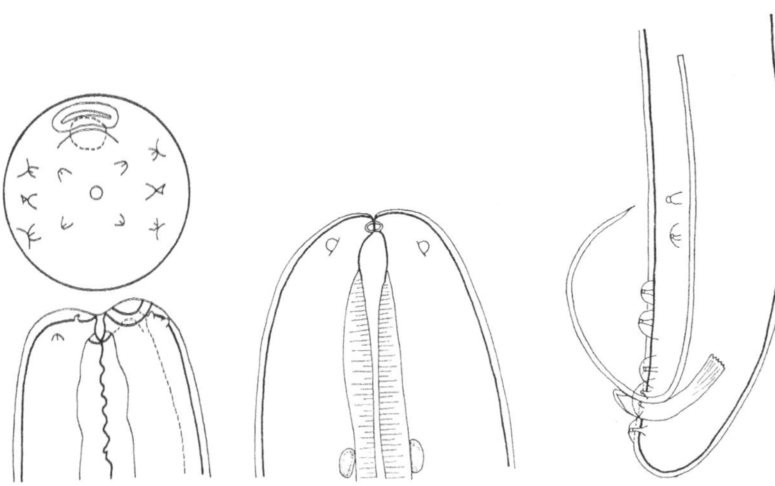

3.210 3.211 3.212

FIG. 3.210. *Filaria martis* Gmelin, 1790, cephalic extremity female. (After Anderson, 1960.)
FIG. 3.211. *Suifilaria suis* Ortlepp, 1937, cephalic extremity, lateral view. (After Ortlepp, 1937.)
FIG. 3.212. *Parafilaria multipapillosa* (Condamine & Drouilly, 1878), caudal extremity male, lateral view. (After Railliet *in* Yorke & Maplestone, 1926.)

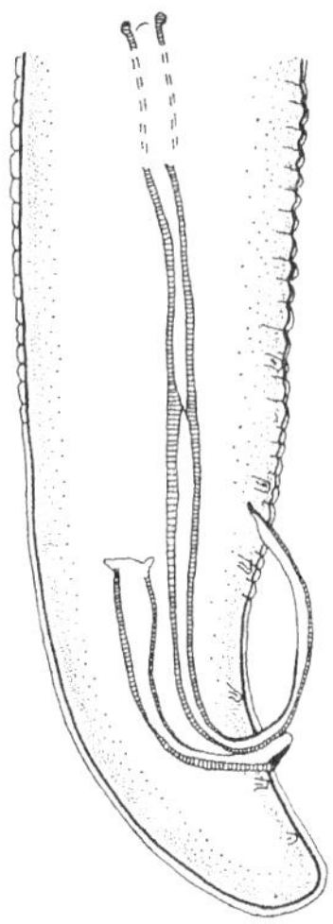

3.213

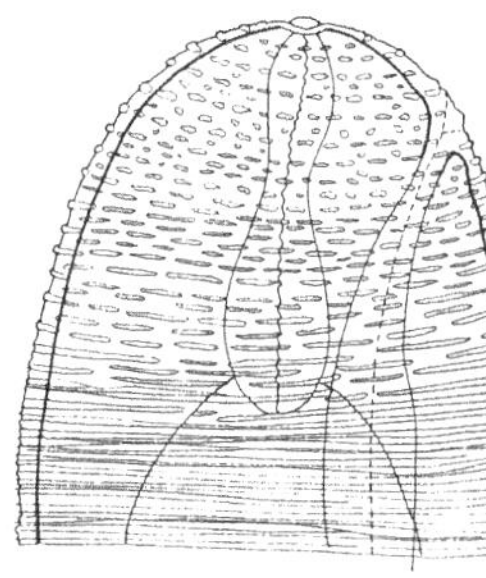

3.214

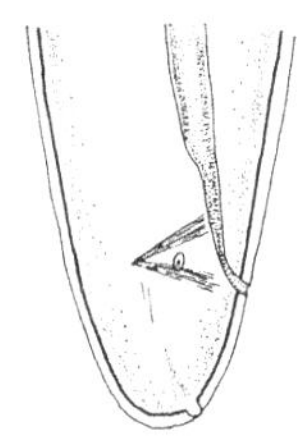

3.215

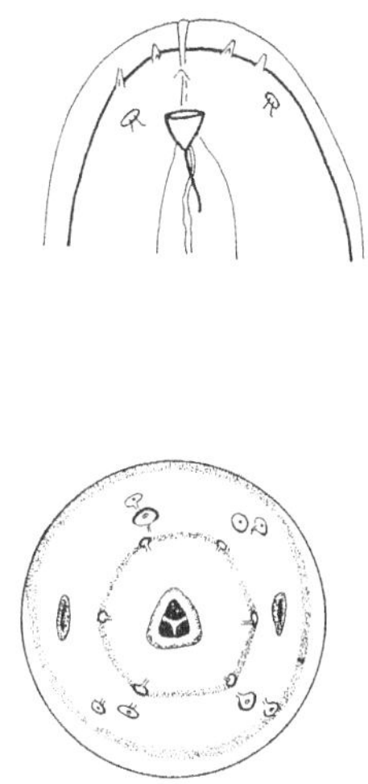

3.216

FIG. 3.213. *Pseudofilaria pertenue* (Rodhain, 1919), caudal end male, lateral view. (After Sandground, 1935.)
FIG. 3.214. *Parafilaria multipapillosa* (Condamine & Drouilly, 1878), cephalic extremity female, lateral view. (After Railliet *in* Yorke & Maplestone, 1926.)
FIG. 3.215. *Pseudofilaria pertenue* (Rodhain, 1919), caudal end female, lateral view. (After Sandground, 1935.)
FIG. 3.216. *Pseudofilaria pertenue* (Rodhain, 1919), cephalic extremity. (After Anderson, 1968.)

Family *ONCHOCERCIDAE* (Leiper, 1911)
(= *DIPETALONEMATIDAE* Wehr, 1935;
= *SETARIIDAE* Yorke & Maplestone, 1926 subfam.)

Key to subfamilies

1-(2) Vulva in middle or posterior regions of body, rarely in first third.
Cephalic extremity rounded.
Buccal capsule well cuticularized.
Oesophagus well developed and divided.
Caudal papillae numerous and large, some forming a subterminal group.
Caudal alae absent.
Spicules often stout, unequal, subequal or equal in size.
Development in adipose tissue of Diptera.
Infective stage with long tail.
Parasites of lacertilians and crocodilians.

Oswaldofilariinae

2-(1) Vulva in anterior region of body.
3-(8) Conspicuous, generally salient, cuticular formations present on anterior extremity in form of spines, flanges, or circumoral elevations.
4-(7) Cephalic spines or flanges present.
5-(6) Cephalic cuticular formations consisting of two pairs of submedian spines (Fig. 3.217).
Anus subterminal.
Four pairs submedian cephalic papillae present.
Area rugosa absent.

Spicules markedly dissimilar in size and morphology.
Caudal alae absent.
Caudal papillae small or absent.
Development in muscles of Diptera; sausage stage broad.
Infective larva with two median cephalic spines; long glandular oesophagus and short, blunt tail.
Parasites of amphibians.

Icosiellinae

6-(5) Two small, lateral parabuccal cuticular flanges generally present.
Cephalic extremity expanded laterally.
Cephalic papillae large, often salient and articulated.
Buccal capsule small.
Vulva near glandular oesophagus or behind it.
Lateral and caudal alae often present.
Male tail with two subventral rows of sessile papillae.
Spicules thin and dissimilar.
Development in adipose tissue of mosquitoes.
Infective stage without cephalic spines and with long conical tail.
Parasites of amphibians.

Waltonellinae

7-(4) Conspicuous circumoral elevation, reinforced in medial and sometimes lateral planes, present, or two lateral, flattened areas present on cephalic extremity.
Cephalic extremity sometimes with four lateromedial spines.
Cephalic extremity often expanded laterally.
Buccal capsule often well developed.
Vulva near muscular oesophagus.
Caudal alae absent.
Spicules markedly dissimilar.
Male tail usually with two long rows of subventral papillae and sometimes with isolated terminal group.
Development in Diptera.
Infective stage with large glandular oesophagus and long tail with tiny subterminal processes.
Parasites of mammals.

Setariinae

8-(3) Conspicuous and salient cuticular cephalic formations absent.
9-(10) Caudal alae of male highly developed. Caudal papillae large and pedunculate.
Tail usually short.
Spicules markedly dissimilar.
Development in various tissues of Diptera.
Infective stage with long tail and tiny subterminal processes.
Parasites of reptiles, birds and mammals.

Dirofilariinae

10-(9) Caudal alae absent or weakly developed.
Caudal papillae sessile.
11-(14) Tail long.
12-(13) Spicules markedly different in size and morphology (except *Dipetalonema* (*Johnstonema*)).
Buccal capsule clearly defined or inconspicuous.
Oesophagus generally well developed.

Caudal alae rarely present and, if so, weakly developed.
Development in various haematophagous arthropods.
Infective stage with long tail.
Parasites of reptiles, birds, and mammals.

Onchocercinae

13-(12) Spicules little different in size and morphology; rarely, left twice length of right.
Buccal cavity reduced or absent.
Oesophagus variable, often reduced.
Caudal papillae often reduced in size and number, sometimes grouped about anus.
Development in Diptera.
Sausage stage broad.
Infective laıva small.
Parasites of reptiles, birds and mammals.

Splendidofilariinae

14-(11) Anus subterminal in both sexes.
Buccal capsule usually absent.
Caudal papillae few in number.
Spicules similar in size and form or markedly different.
Posterior extremity sometimes dilated.
Development in Diptera and Mallophaga.
Sausage stage broad.
Infective larvae small.
Mainly parasites of birds, less commonly of reptiles and mammals.

Lemdaninae

Subfamily **Oswaldofilariinae** Chabaud & Choquet, 1953

Key to genera

1-(4) Glandular oesophagus long (>2 mm. in the female).
2-(3) Vulva generally pre-equatorial, rarely near end of oesophagus.
Spicules well developed, unequal in size and form (Fig. 3.218).
Ovejector long and complex (Fig. 3.219).
Caudal papillae in two rows or in circle in anal region, plus a few at end of tail.
Oral opening sometimes slightly ventral in position.
Tail of female smooth or with subterminal tubercles.
Parasites of lizards and crocodiles.

Oswaldofilaria Travassos, 1933

3-(2) Vulva a short distance anterior to anus (Fig. 3.220).
Spicules short, stout, subequal (Fig. 3.221).
Ovejector long.
Caudal papillae in circle about anus and a few near caudal extremity (Fig. 3.221).
Parasites of lizards.

Gonofilaria Mullin, 1973

4-(1) Glandular oesophagus short (1 mm. or less in female).
5-(10) Vulva usually near middle of body, rarely near end of oesophagus.
6-(7) Spicules markedly unequal in size and dissimilar in form (Fig. 3.222).
Ovejector long and complex.
Left spicule long and slender, right stout.
Anterior extremity sometimes with cuticular median shield-like elevation.

Caudal papillae in two rows or in circle about anus, plus a few near caudal extremity (Fig. 3.223).
Tail of female smooth or with cuticular subterminal elevation.
Parasites of lizards.
Befilaria Chabaud, Anderson & Brygoo, 1959

7-(6) Spicules subequal or similar in size and form.
Ovejector simple.

8-(9) Spicules subequal (Fig. 3.224).
Ovejector long but simple.
Tail of female smooth or with small terminal elevations.
Parasites of lizards.
Piratuboides Bain & Sulahian, 1974

9-(8) Spicules similar in size and form (Fig. 3.225).
Ovejector short and simple (Fig. 3.226).
Caudal papillae numerous, many grouped around anus and others at caudal extremity.
Parasites of lizards.
Piratuba Freitas & Lent, 1947

10-(5) Vulva in anterior part of posterior third of body.
Spicules delicate, markedly dissimilar in size and form, the left long and slender (Fig. 3.227).
Ovejector long and complex.
Four small lateral alae present.
Cuticle of male with numerous tiny tubercles.
Caudal papillae few in number, in circle about anus, plus two at end of tail.
Parasites of lizards.
Solafilaria Chabaud, Anderson & Brygoo, 1959

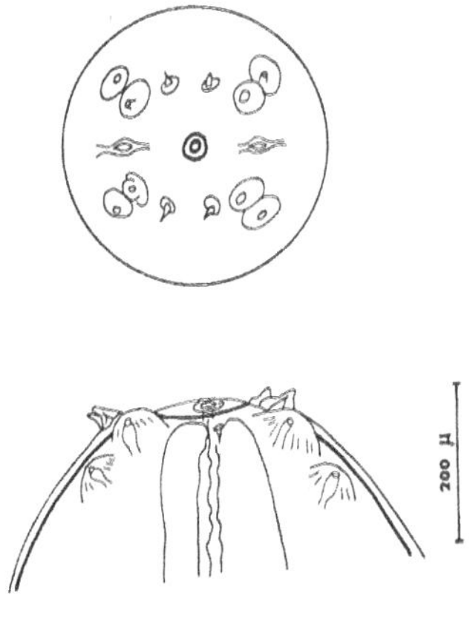

3.217

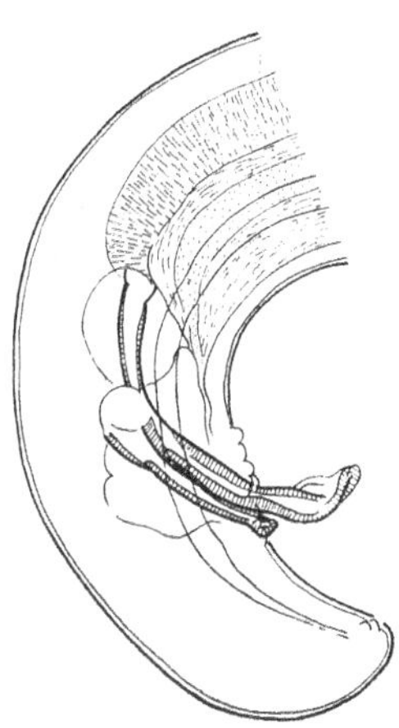

3.218

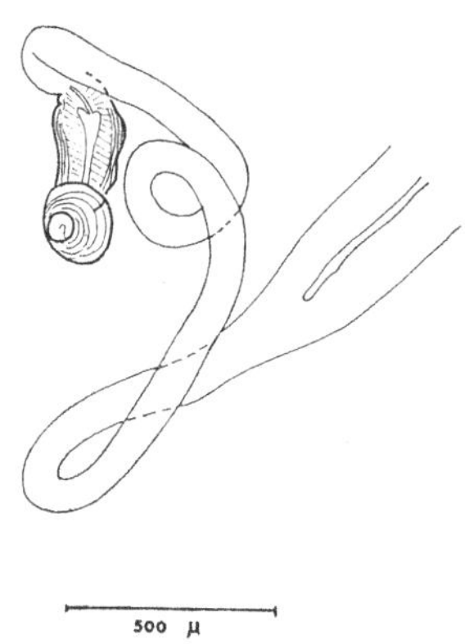

3.219

FIG. 3.217. *Icosiella neglecta* (Diesing, 1851) cephalic extremity. (After Anderson, 1968.)
FIG. 3.218. *Oswaldofilaria azevedoi* Bain, 1974, caudal extremity male, lateral view. (After Bain, 1974.)
FIG. 3.219. *Oswaldofilaria azevedoi* Bain, 1974, vagina. (After Bain, 1974.)

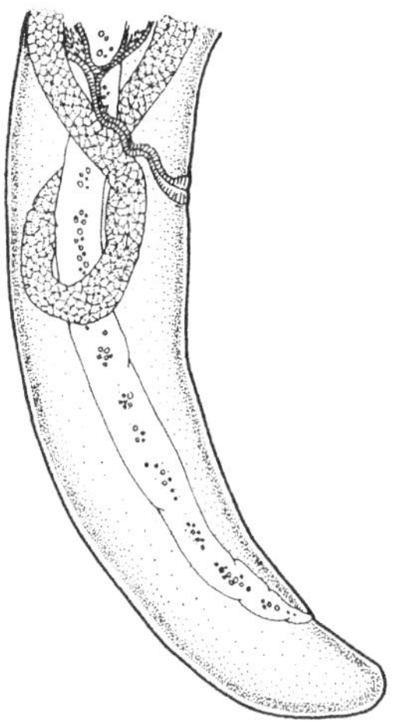

3.220

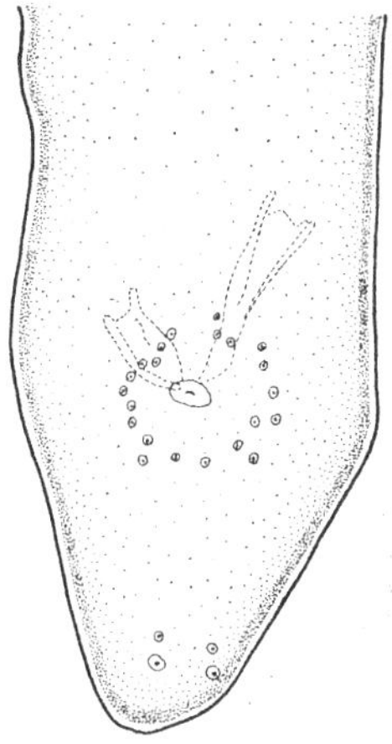

3.221

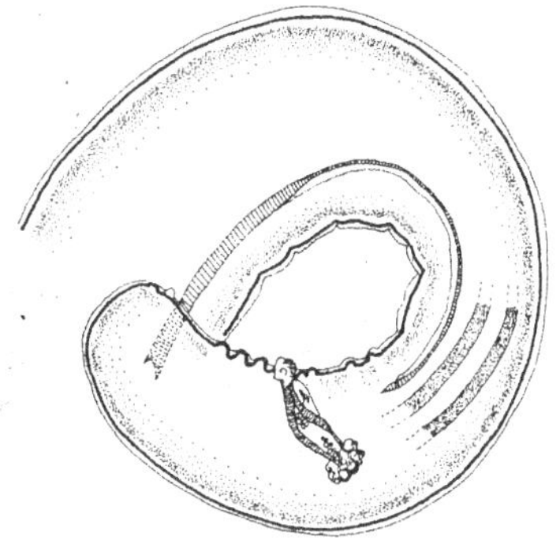

3.222

FIG. 3.220. *Gonofilaria rudnicki* Mullin, 1973, caudal extremity female, lateral view. (After Mullin, 1973.)
FIG. 3.221. *Gonofilaria rudnicki* Mullin, 1973, caudal extremity male, ventral view. (After Mullin, 1973.)
FIG. 3.222. *Befilaria africana* Bain & Ranque, 1974, caudal end male, lateral view. (After Bain & Ranque, 1974.)

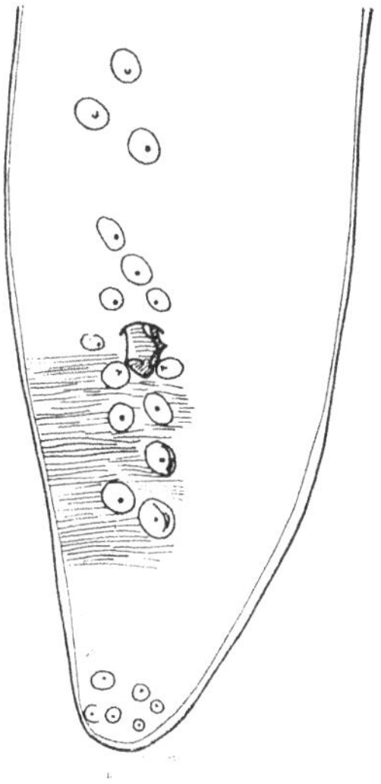

3.223

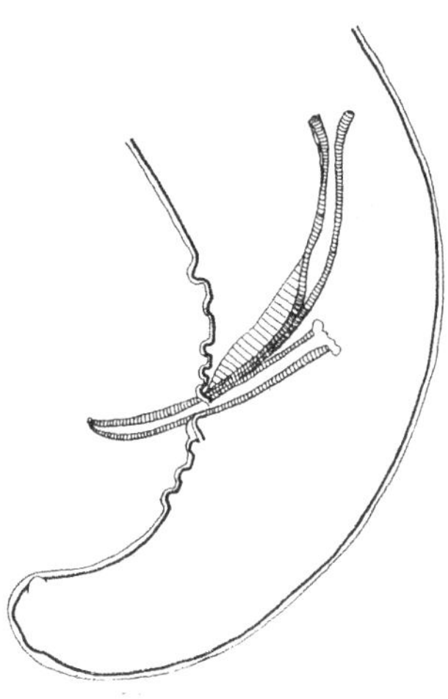

3.224

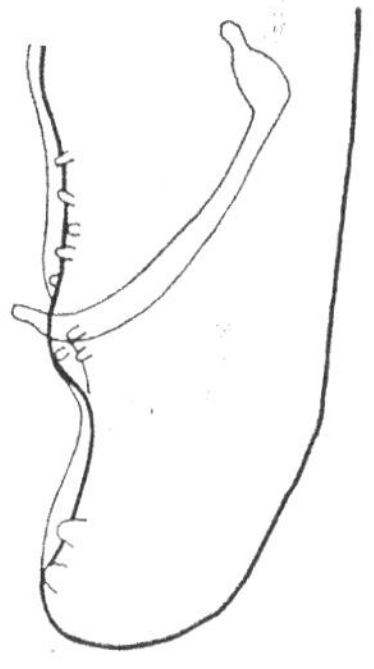

3.225

FIG. 3.223. *Befilaria africana* Bain & Ranque, 1974, caudal end male, ventral view. (After Bain & Ranque, 1974.)
FIG. 3.224. *Piratuboides zeae* (Bain, 1974), caudal end male, lateral view. (After Bain, 1974.)
FIG. 3.225. *Piratuba digiticauda* Lent & Freitas, 1941, caudal extremity male, lateral view. (After Lent & Freitas, 1941.)

4*

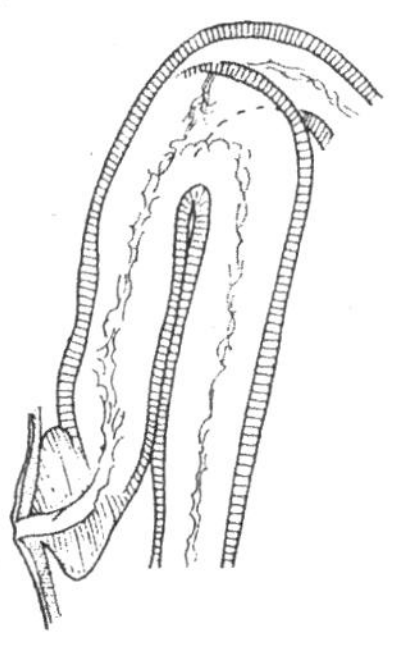

3.226

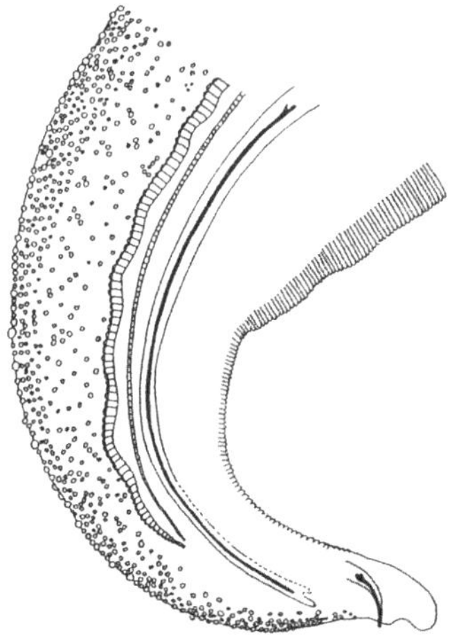

3.227

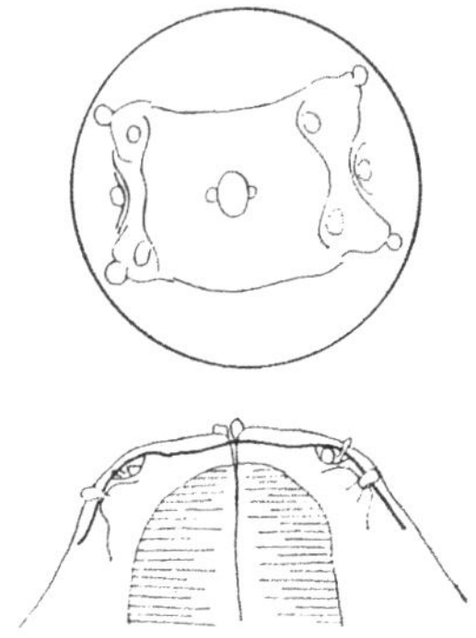

3.228

FIG. 3.226. *Piratuba scaffi* Bain, 1974, vagina. (After Bain, 1974.)

FIG. 3.227. *Solafilaria guibei* Chabaud, Anderson & Brygoo, 1959, caudal extremity male, lateral view. (After Chabaud *et al.*, 1959.)

FIG. 3.228. *Waltonella guyanensis* Bain & Prod'hon, 1974, cephalic extremity. (After Bain & Prod'hon, 1974.)

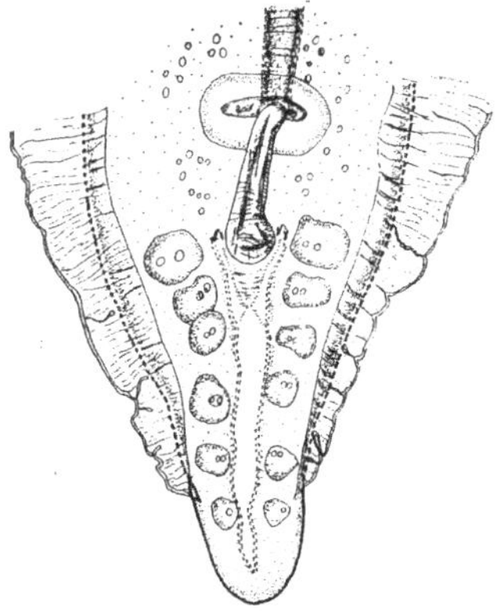

3.229

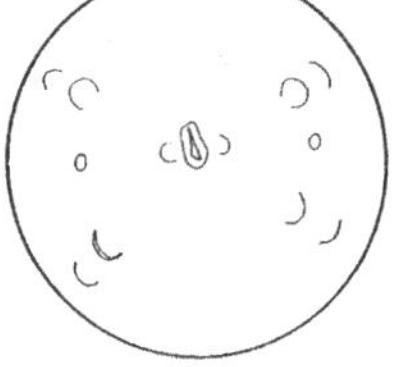

3.230

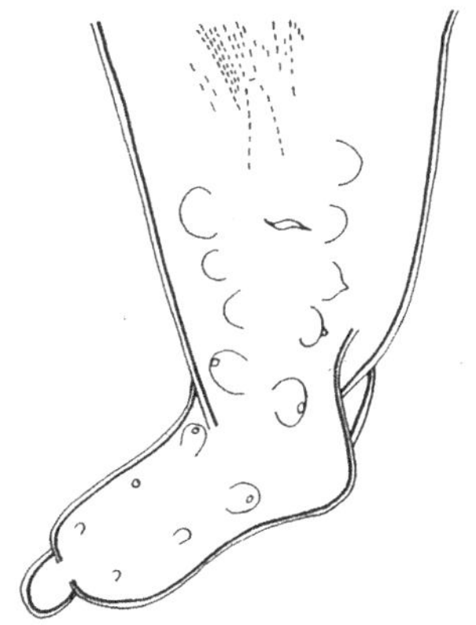

3.231

FIG. 3.229. *Waltonella flexicauda* (Schacher & Crans, 1973), caudal extremity male, ventral view. (After Schacher & Crans, 1973.)

FIG. 3.230. *Madochotera alata* Bain & Brunhes, 1968, cephalic extremity, *en face* view. (After Bain & Brunhes, 1968.)

FIG. 3.231. *Madochotera alata* Bain & Brunhes, 1968, caudal extremity male, ventral view. (After Bain & Brunhes, 1968.)

Subfamily **Icosiellinae** Anderson, 1958

One genus.
Parasites of amphibians.

Icosiella Seurat, 1917

Subfamily **Waltonellinae** Bain & Prod'hon, 1974

Key to genera

1-(4) Parabuccal formations present (Fig. 3.228).
Lateral and caudal alae generally present (Fig. 3.229).

2-(3) Cephalic papillae with wide base and articulated tip (Fig. 3.228).
Vulva in oesophageal region near cephalic extremity.
Parasites of Ranidae, Bufonidae and Leptodactylidae.
Waltonella (Schacher & Crans, 1973, subgenus)

3-(2) Cephalic papillae without articulated tip (Fig. 3.230).
Vulva behind oesophageal region (>5 mm.).
Caudal alae reduced (Fig. 3.231).
Parasites of Racophoridae (Madagascar).
Madochotera Bain & Brunhes, 1968

4-(1) Parabuccal formations absent.
Lateral and caudal alae absent.
Cephalic papillae without articulated tip.
Vulva near oesophagus.
Shaft of left spicule atrophied (Fig. 3.232).
Parasites of Ranidae, Bufonidae, and Racophoridae.
Ochoterenella Caballero, 1944

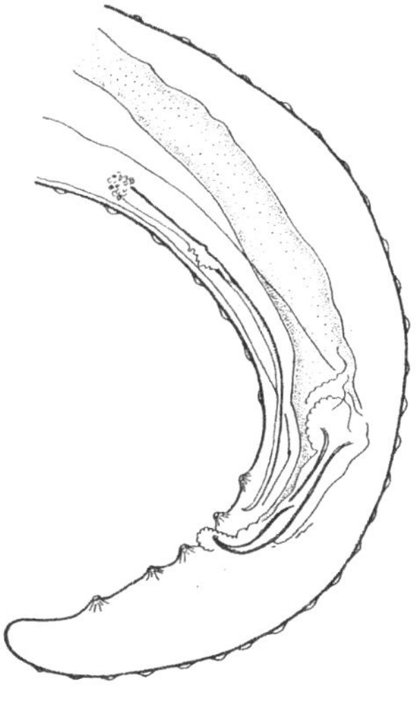
3.232

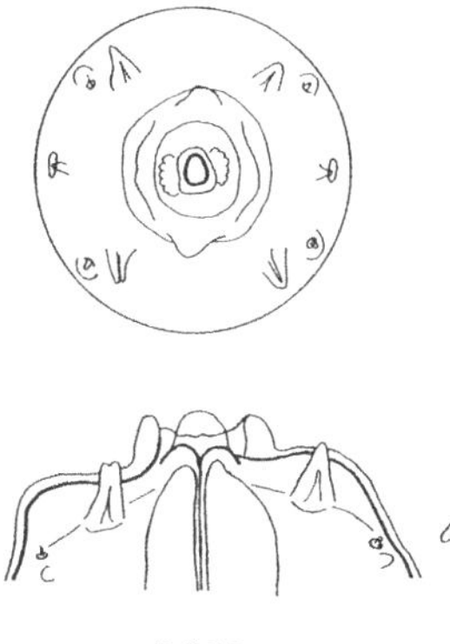
3.233

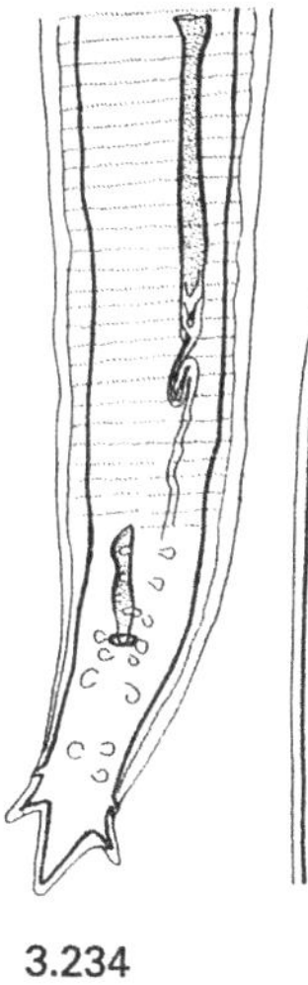
3.234

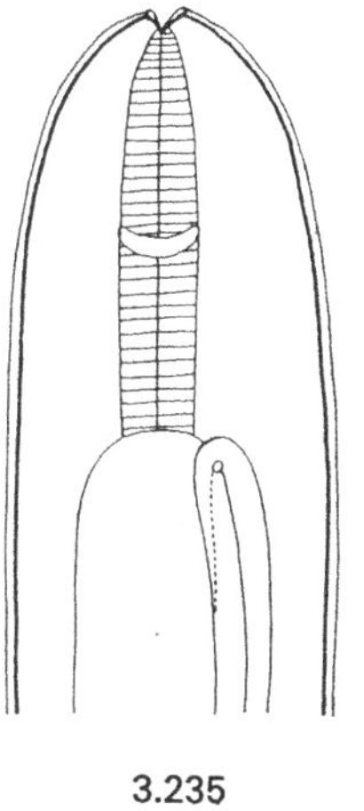
3.235

FIG. 3.232. *Ochoterenella digiticauda* Caballero, 1944, caudal extremity male lateral view. (After Caballero, 1944.)
FIG. 3.233. *Setaria equina* (Abildgaard, 1789), cephalic extremity. (After Anderson, 1968.)
FIG. 3.234. *Setaria digitata* (Linstow, 1906), caudal extremity male, ventral view. (After Yeh, 1959.)
FIG. 3.235. *Papillosetaria traguli* Vevers, 1922, cephalic extremity female, ventral view. (After Vevers, 1922.)

Subfamily **Setariinae** Yorke & Maplestone, 1926

Key to genera

1-(2) Peribuccal ring elongated dorsoventrally, thick or thin, generally provided with median or lateral cuticular elevations; cephalic cuticular spines rarely present (e.g. *S. equina*) (Fig. 3.233).
External labial papillae generally articulated.
Buccal cavity often well developed with thick walls.
Oesophagus divided.
Vulva near muscular oesophagus; ovejector highly muscular.
Area rugosa present.
Caudal papillae of male numerous, similar in size and in subventral rows (Fig. 3.234).
Tail of male tapering to rounded or pointed extremity.
Tail of female usually with terminal and subterminal tubercles or digitiform processes.
Microfilaria in blood.
Transmitted by Culicidae.
Parasites of pleural and abdominal cavity of artiodactyles, equids and hyracoids.
Setaria Viborg, 1795
(= *Hyraconema* Yeh, 1959; = *Artionema* Yeh, 1959)

2-(1) Anterior extremity with pair of lateral rounded cuticularized thickenings (Fig. 3.235).
Oesophagus divided.
Vulva near muscular oesophagus.
Tail of male short, small pre- and postanal papillae in rows present.
Female tail with digitiform appendages.
Right spicule wedge-shaped.
Cuticle with bosses except at extremities.
Parasites of peritoneal cavity of antelope (*Tragulus*).
Papillosetaria Vevers, 1922

Subfamily **Dirofilariinae** Sandground, 1921
(= **Loainae** Yorke & Maplestone, 1926)

Key to genera

1-(4) Tail of male long, at least twice as long as width of body at level of anus.
2-(3) Oesophagus divided.
Cuticle with or without longitudinal ridges or elevations (Fig. 3.236).
Spicules markedly dissimilar (Fig. 3.237).
Preanal papillae massive.
Vulva postoesophageal.
Tail of male rounded.
Parasites of edentates.
Bostrichodera Sandground, 1938

3-(2) Oesophagus undivided.
Cuticle smooth.
Spicules markedly dissimilar (Fig. 3.238).
Anal papillae large and numerous.
Vulva near posterior end of oesophagus.
Tail of male pointed.
Parasites of rodents.
Dirofilariaeformia Lubimov, 1935

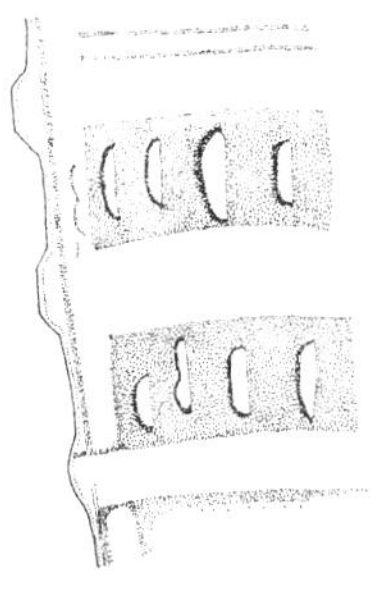

3.236

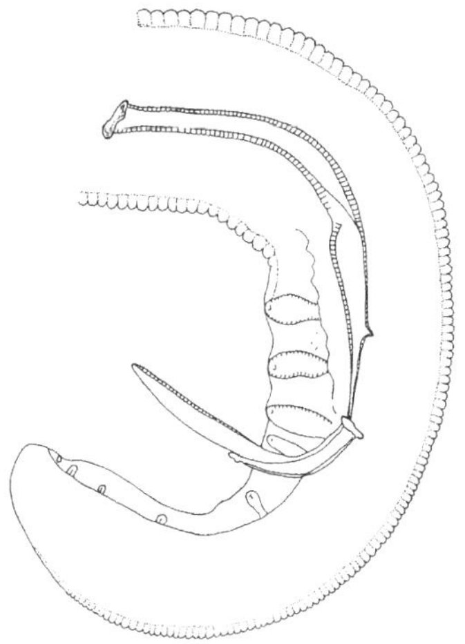

3.237

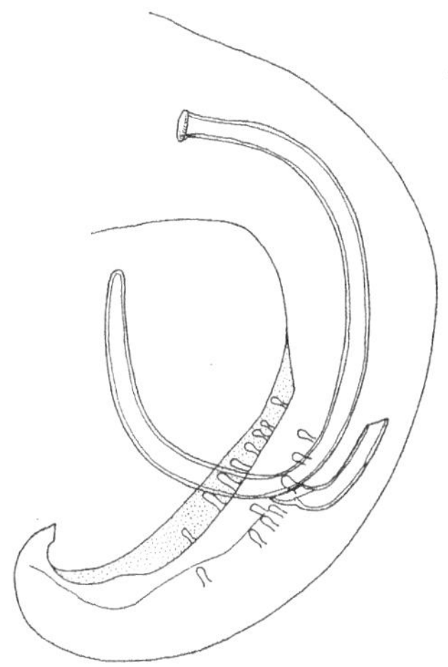

3.238

FIG. 3.236. *Bostrichodera bequaerti* Sandground, 1938, surface of cuticle with bosses and ridges. (After Sandground, 1938.)

FIG. 3.237. *Bostrichodera bequaerti* Sandground, 1938, caudal extremity male, lateral view. (After Sandground, 1938.)

FIG. 3.238. *Dirofilariaeformia sciurorum* Lubimov, 1935, caudal extremity male, lateral view. (After Lubimov, 1935.)

4-(1) Tail of male short, much less than twice width of body at anus.

5-(8) Oesophagus with long sacculate glandular part with giant bodies attached to inner wall (Fig. 3.239).

6-(7) Left spicule extremely long and filiform (Fig. 3.240).
Vagina long and convoluted.
Cuticle with fine annulations.
Vulva slightly behind muscular oesophagus.
Caudal papillae large and numerous.
Right spicule with dorsal barb.
Microfilaria unsheathed with pointed tail.
Parasites of body cavity of primates.

Edesonfilaria Yeh, 1960

7-(6) Left spicule not extremely long, with complex lamina (Fig. 3.241).
Vagina short and unconvoluted.
Cuticle with delicate transverse striations.
Vulva slightly behind muscular oesophagus.
Caudal papillae prominent.
Right spicule divided into two parts, the distal part pointed without barbs.
Microfilaria unsheathed with pointed caudal extremity.
Parasites of primates.

Macacanema Schad & Anderson, 1963

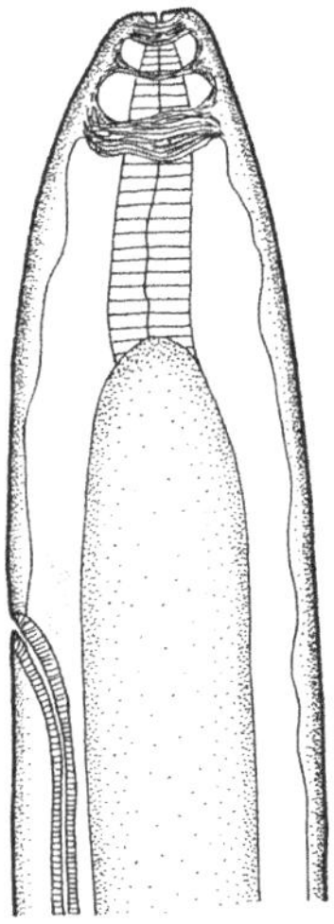

3.239

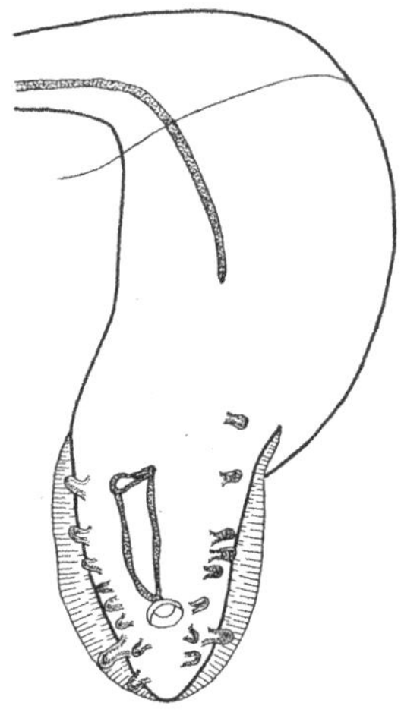

3.240

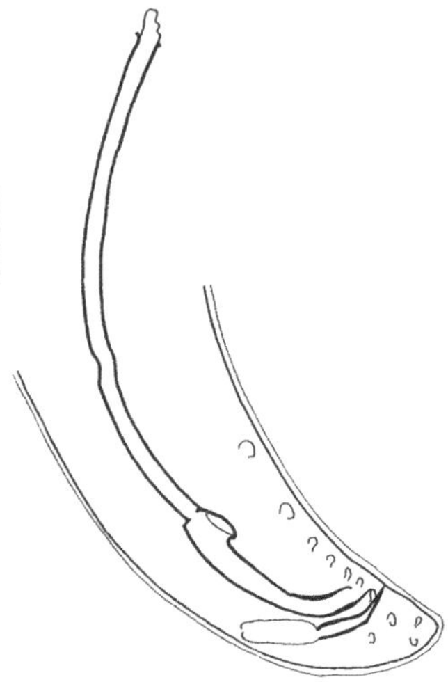

3.241

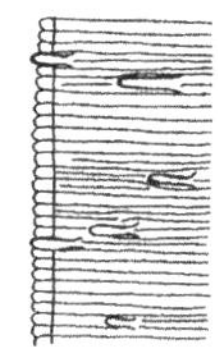

3.242

FIG. 3.239. *Edesonfilaria malayensis* Yeh, 1960, anterior extremity female, lateral view showing glandular oesophagus. (After Yeh, 1960.)

FIG. 3.240. *Edesonfilaria malayensis* Yeh, 1960, caudal extremity male, ventral view. (After Yeh, 1960.)

FIG. 3.241. *Macacanema formosana* Schad & Anderson, 1963, caudal extremity male, lateral view. (After Schad & Anderson, 1963.)

FIG. 3.242. *Skrjabinodera tanganyikae* Yeh, 1956 (=*Gazellofilaria*), cuticle with bosses. (After Yeh, 1956.)

8-(5) Oesophagus divided or undivided, but never with long sacculate glandular part with giant bodies.

9-(12) Well marked cuticularized ring present in front of oesophagus.

10-(11) Cuticle with bosses (Fig. 3.242).
Spicules markedly dissimilar (Fig. 3.243).
Caudal papillae large and numerous, of which 5 to 6 are preanal.
Oesophagus short and undivided.
Vulva postoesophageal.
Parasites of ungulates.
Skrjabinodera Gnedina & Vsevolodov, 1947 (= *Gazellofilaria* Yeh, 1956)

11-(10) Cuticle smooth.
Spicules markedly dissimilar (Fig. 3.244).
Caudal papillae large and numerous of which 6 are preanal.
Deirids well developed.
Oesophagus divided.
Vulva near posterior end of oesophagus.
Parasites of primates.
Tawila Khalil, 1932

12-(9) Without well marked cuticular ring anterior to oesophagus.
13-(14) Cuticle with bosses (Fig. 3.245).
Spicules short, dissimilar (Fig. 3.246).
Caudal papillae large.
Oesophagus short, undivided.
Vulva postoesophageal.
Parasites of primates.

Loa Stiles, 1905
(= *Paraloa* Rodhain & Van den Berghe, 1939)

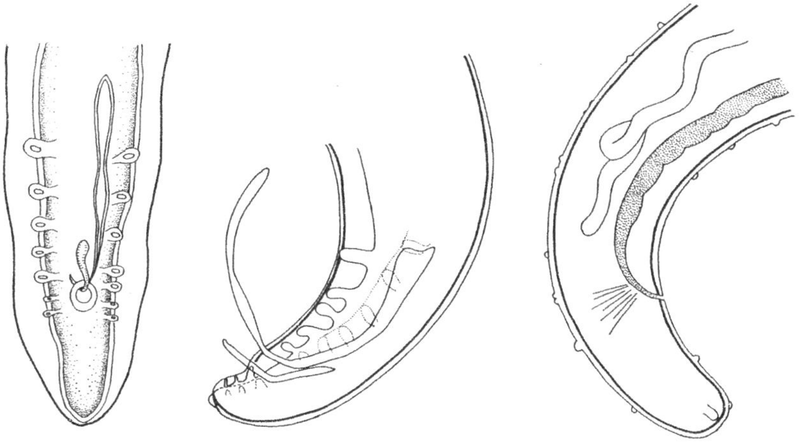

3.243 3.244 3.245

FIG. 3.243. *Skrjabinodera tanganyikae* Yeh, 1956 (=*Gazellofilaria*), caudal extremity male, ventral view. (After Yeh, 1956.)
FIG. 3.244. *Tawila tawila* Khalil, 1932, caudal extremity male, lateral view. (After Khalil, 1932.)
FIG. 3.245. *Loa loa* (Guyot, 1778), caudal extremity female with bosses. (After Yorke & Maplestone, 1926.)

14-(13) Cuticle without bosses.
15-(16) Parasites of cold-blooded vertebrates.
Oesophagus undivided and short (Fig. 3.247).
Lateral alae sometimes present.
Spicules unequal and short, or elongated and delicate (Fig. 3.248).
Vulva in oesophageal region.
Parasites of reptiles.

Foleyella Seurat, 1917
(= *Foleyellides* Caballero, 1935)

16-(15) Parasites of warm-blooded vertebrates.
17-(18) Lateral alae well developed.
Spicules short, delicate, unequal (Fig. 3.249).
Vulva preoesophageal.
Oesophagus divided or undivided.
Parasites of birds.

Pelecitus Railliet & Henry, 1910
(= *Spirofilaria* Yamaguti, 1935 = *Eulimdana* Founikoff, 1934)

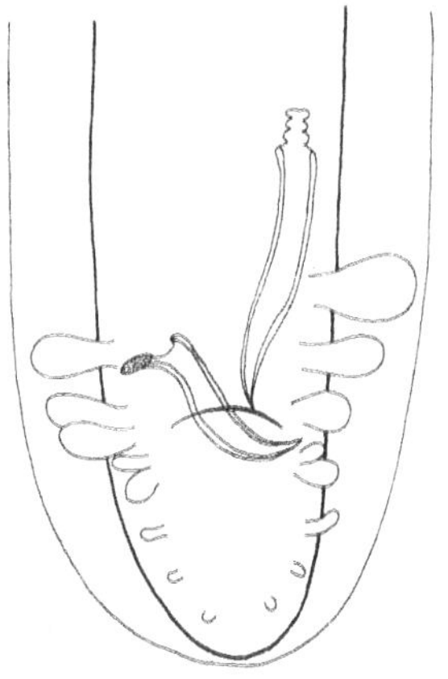

3.246

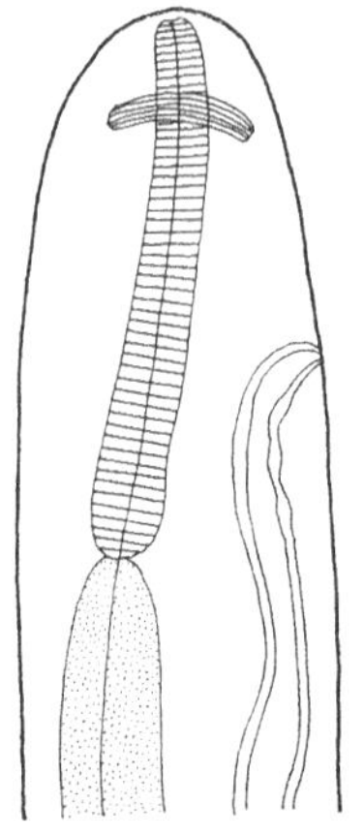

3.247

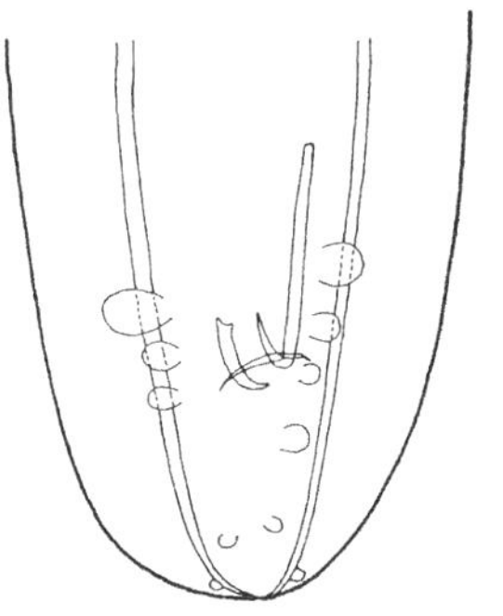

3.248

FIG. 3.246. *Loa loa* (Guyot, 1778), caudal extremity male, ventral view. (After Yorke & Maplestone, 1926.)

FIG. 3.247. *Foleyella candezei* (Fraipont, 1882), anterior extremity female, dorso-ventral view. (After Yorke & Maplestone, 1926.)

FIG. 3.248. *Foleyella candezei* (Fraipont, 1882), caudal extremity male, ventral view. (After Yorke & Maplestone, 1926.)

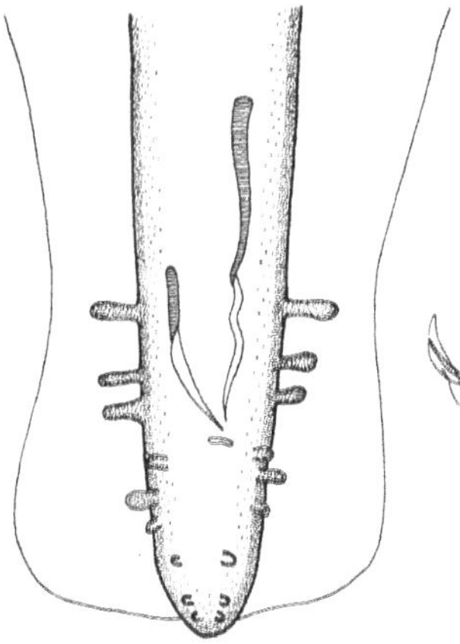

3.249

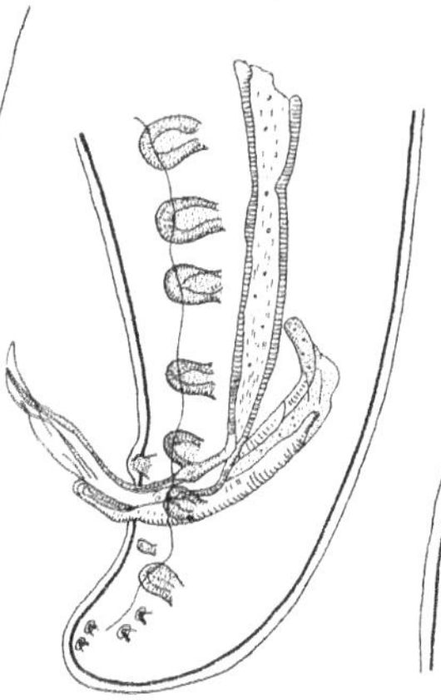

3.250

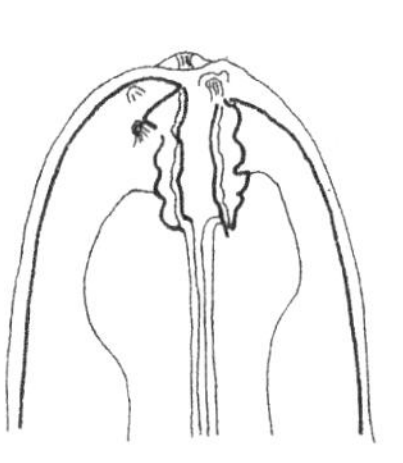

3.251

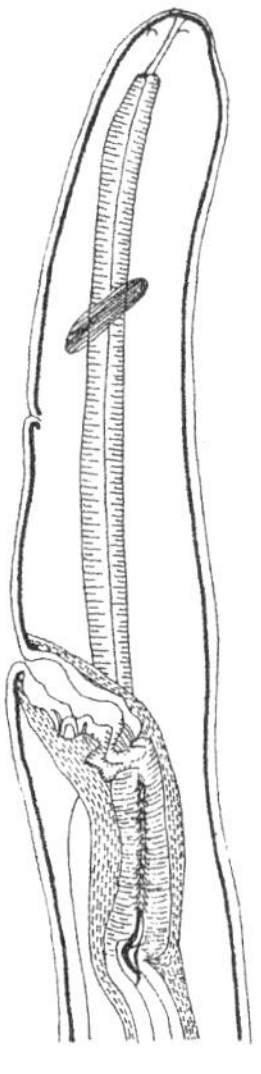

3.252

FIG. 3.249. *Pelecitus tercostatus* (Molin, 1859), caudal extremity male, ventral view. (After Skrjabin, 1916.)

FIG. 3.250. *Dirofilaria striata* (Molin, 1858), caudal end male, lateral view. (After Anderson & Diaz-Ungría, 1959.)

FIG. 3.251. *Litomosoides carinii* (Travassos, 1919), cephalic extremity. (After Anderson, 1968.)

FIG. 3.252. *Pseudolitomosa musasabi* Yamaguti, 1941, anterior extremity female, lateral view. (After Yamaguti, 1941.)

18-(17) Lateral alae absent or weakly developed.
Spicules markedly dissimilar (Fig. 3.250).
Oesophagus divided or undivided.
Vulva postoesophageal.
Caudal papillae bulky and numerous (Fig. 3.250).
Parasites of mammals.

Dirofilaria Railliet & Henry, 1910

Subfamily **Onchocercinae** Leiper, 1911

Key to genera

1-(50) Male tail long, at least twice as long as width of body at anus.
2-(7) Buccal cavity elongate and tubular.
3-(4) Buccal cavity with thick walls, complex in form (Fig. 3.251).
Oesophagus long, undivided, dilated anteriorly and enclosing part of buccal cavity.
Caudal extremity smooth.
Vulva postoesophageal or at level of oesophagus.
Preanal papillae absent.
Parasites of body cavity of rodents and bats.

Litomosoides Chandler, 1931
(= *Vestibulosetaria* Vogel & Gabaldon, 1932; = *Finlaynema* Vigueras, 1934)

4-(3) Buccal cavity with thin walls, simple in form.
5-(6) Oesophagus normal in form (Fig. 3.252).
Oesophagus undivided.
Vulva preoesophageal, near end of oesophagus.
Left spicule tubular.
Tail of female long and slender with two small protuberances (Fig. 3.253).
Caudal papillae undescribed.
Parasites of rodents.

Pseudolitomosa Yamaguti, 1941

6-(5) Oesophagus poorly developed and irregular in form (Fig. 3.254).
Vulva postoesophageal.
Spicules slender.
Caudal papillae surrounding anus.
Caudal extremity pointed.
Area rugosa present.
Microfilaria slender and unsheathed.
Parasites of Cervidae

Cutifilaria Bain & Schulz-Key, 1974

7-(2) Buccal cavity not elongate and tubular.
8-(9) Caudal extremity dilated in both sexes (Fig. 3.255).
Anterior extremity expanded dorsoventrally.
Spicules markedly dissimilar (Fig. 3.256).
Oesophagus undivided.
Vulva preoesophageal.
Parasites of bats.

Migonella Lent, Freitas & Proença, 1946

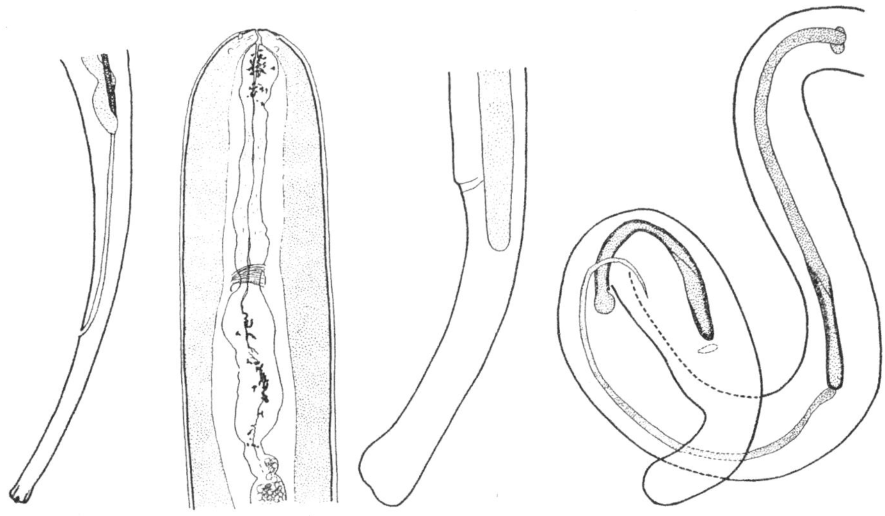

3.253 3.254 3.255 3.256

FIG. 3.253. *Pseudolitomosa musasabi* Yamaguti, 1941, caudal extremity female, lateral view. (After Yamaguti, 1941.)

FIG. 3.254. *Cutifilaria wenki* Bain & Schulz-Key, 1975, anterior extremity female, lateral view. (After Bain & Schulz-Key, 1975.)

FIG. 3.255. *Migonella fracchiai* Lent & Freitas, 1946, caudal extremity female, lateral view. (After Lent & Freitas, 1946.)

FIG. 3.256. *Migonella fracchiai* Lent & Freitas, 1946, caudal extremity male, dorsal view. (After Lent & Freitas, 1946.)

9-(8) Caudal extremity not dilated in both sexes.

10-(15) Buccal cavity well developed with thick walls, its base included in oesophagus, often constricted posteriorly and spherical in form.

11-(12) Vulva prominent and protuberant (Fig. 3.257).
Large petaloid caudal appendages present.
Anterior part of buccal cavity prominent and elevated (Fig. 3.258).
Ovejector highly muscular.
Area rugosa present.
Male tail pointed, with a single ala (Fig. 3.259).
Microfilaria sheathed.
Transmitted by Culicidae.
Parasites of American marsupials.

Skrjabinofilaria Travassos, 1925
(= *Cortiamosoides* Foster, 1939)

12-(11) Vulva not protuberant.
Large petaloid caudal appendages absent.

13-(14) Cephalic shield present, dorsoventrally elongated (Fig. 3.260).
Cephalic extremity attenuated.
Externolateral labial papillae present.
Oesophagus undivided.

Caudal extremity rounded, with tiny terminal papilliform bosses.
Microfilaria sheathed.
Parasites of Australian marsupials.

Sprattia Chabaud & Bain, 1976

14-(13) Cephalic shield absent (Fig. 3.261).
Cephalic extremity not attenuated.
Oesophagus undivided.
Vulva near end of oesophagus.
Separation of shaft and lamina of left spicule poorly defined (Fig. 3.262).
Caudal extremity of female with tiny spines.
Area rugosa present, variable in shape.
Parasites of bats.

Litomosa Yorke & Maplestone, 1926

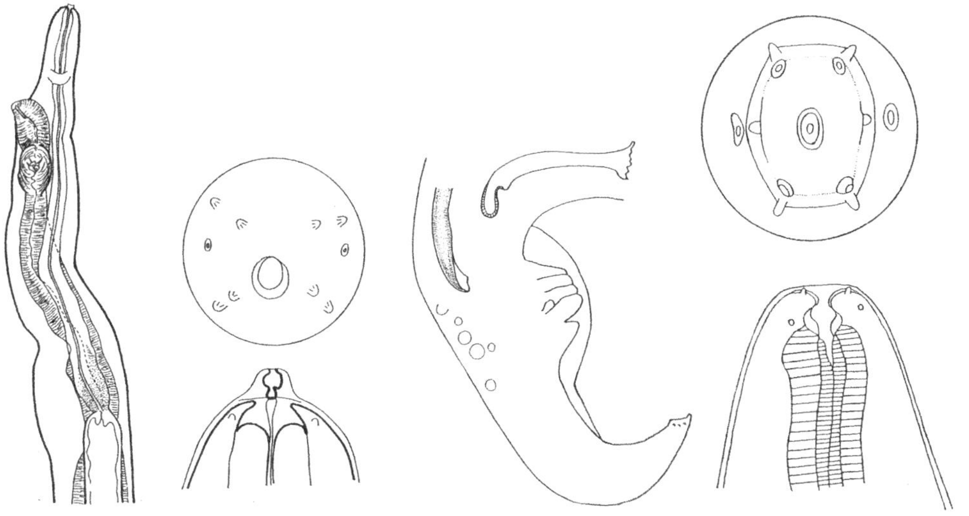

3.257 3.258 3.259 3.260

FIG. 3.257. *Skrjabinofilaria skrjabini* Travassos, 1925, anterior extremity female, lateral view. (After Bain & Durette-Desset, 1973.)
FIG. 3.258. *Skrjabinofilaria philanderi* (Foster, 1939), cephalic extremity. (After Anderson, 1968.)
FIG. 3.259. *Skrjabinofilaria skrjabini* Travassos, 1925, caudal extremity male, lateral view. (After Travassos, 1925.)
FIG. 3.260. *Sprattia venacavincola* (Spratt & Varughese, 1975), Chabaud & Bain 1976, cephalic extremity. (After Spratt & Varughese, 1975.)

15-(10) Buccal cavity absent, weakly or well developed, generally with one or two rings in front of oesophagus but never constricted posteriorly and never spherical.
16-(45) Buccal cavity generally clearly defined; if absent, oesophagus atrophied.
Caudal extremity usually with two petaloid appendages.
Area rugosa present.
17-(20) Spicules similar in length or dissimilar; if dissimilar, lamina of left spicule with initial part long and membranous and terminal thick filament. Right spicule with spatulate tip and claw-like structure.

Area rugosa generally present, composed of tiny irregularly dispersed rugosities (except in *B. dendrolagi*).
Externolateral labial papillae often present.
Buccal cavity with well developed preoesophageal ring.
Cephalic shield laterally elongated.
Oesophagus divided, at least in male.
Caudal extremity rounded, smooth or with reduced subterminal petaloid structures.
Caudal papillae often including pair in tandem in middle of tail.
Microfilariae without sheath.

Breinlia Yorke & Maplestone, 1926

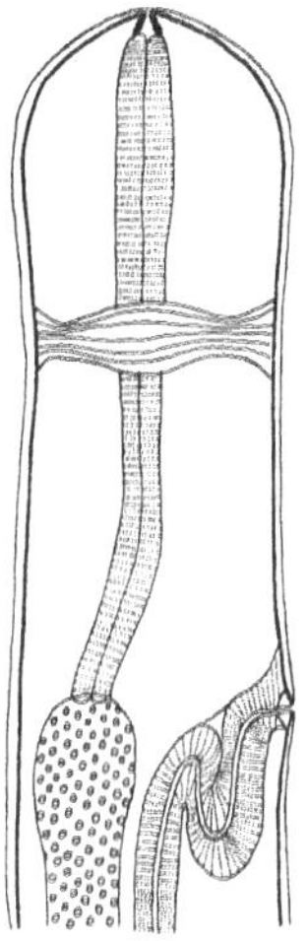

3.261

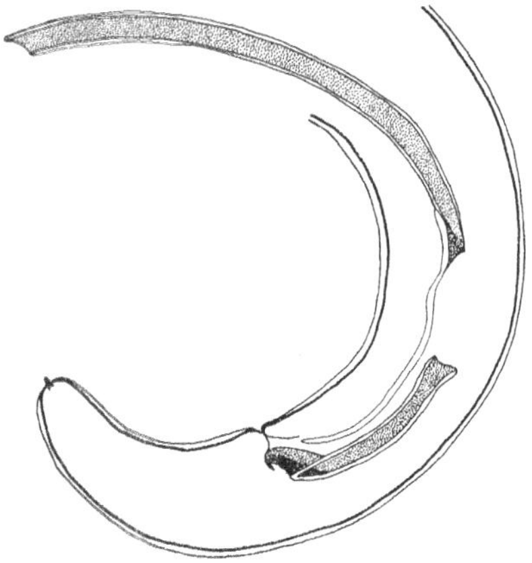

3.262

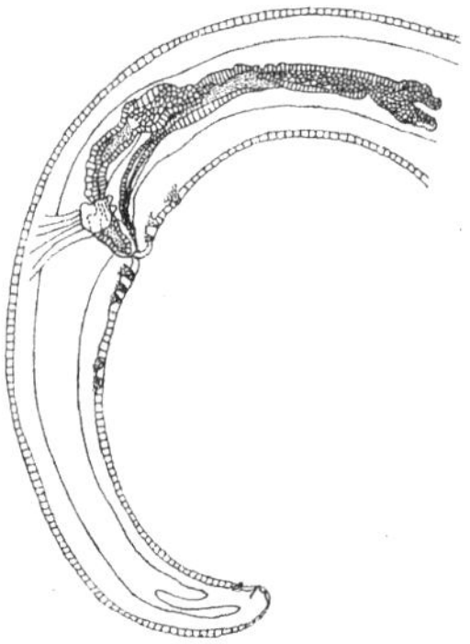

3.263

FIG. 3.261. *Litomosa filaria* (Beneden, 1873), anterior extremity female, lateral view. (After Seurat, 1921.)
FIG. 3.262. *Litomosa filaria* (Beneden, 1873), caudal extremity male, lateral view. (After Seurat, 1921.)
FIG. 3.263. *Breinlia (Breinlia) trichosuri* (Breinl, 1913) Yorke & Maplestone, 1926, caudal extremity male, lateral view. (After Spratt & Varughese, 1975.)

18-(19) Spicules unequal in size (Fig. 3.263).
Oesophagus divided.
External labial papillae arranged in form of square or laterally expanded rectangle.
Buccal cavity with thick walls.
Gubernaculum present.
Parasites of Phalangeridae, Macropodidae and Indomalaysian lemurs and rodents.

Breinlia (*Breinlia*) Chabaud & Bain, 1976

19-(18) Spicules equal or subequal (Fig. 3.264).
Oesophagus usually not clearly divided in female.
External labial papillae arranged in form of dorsoventrally expanded rectangle.
Buccal cavity with thin walls.
Gubernaculum absent.
Parasites of Phalangeridae and Macropodidae.

Breinlia (*Johnstonema*) (Yeh, 1957)

20-(17) Spicules unequal in size, the left without initial long membranous part and terminal thick filament of lamina. Right spicule without claw-like structure.
Area rugosa consisting of transverse bands of bacilliform elevations.
Externo-lateral labial papillae generally absent.

21-(22) Body extremely thin (50 μ).
Tail of male with transparent cone-like cap (Fig. 3.265).
Buccal cavity absent.
Cephalic extremity slightly dilated.
Oesophagus atrophied.
Vulva postoesophageal.
Caudal extremity without tongue-like appendages.
Spicules slender (ratio 2 : 6).
Microfilaria without sheath.
Parasites of intestinal wall of porcupines.

Filarissima Chabaud, 1974

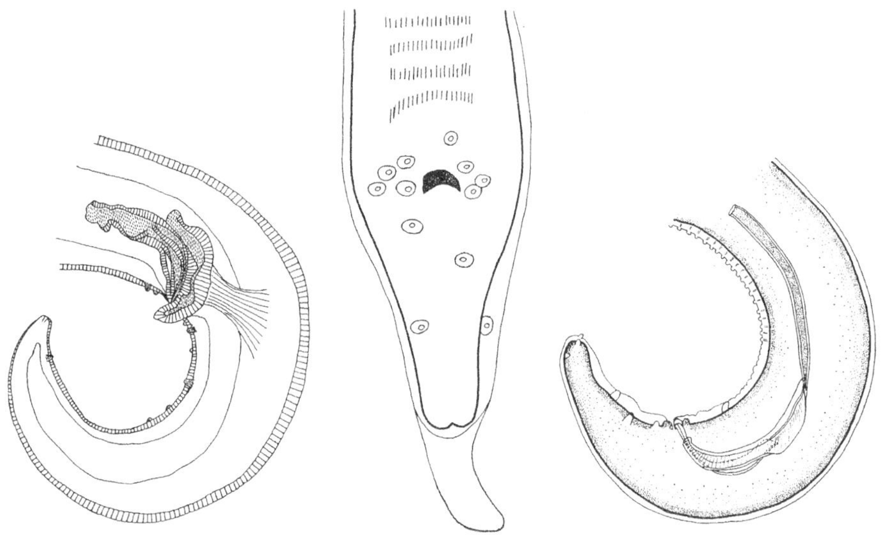

3.264 3.265 3.266

FIG. 3.264. *Breinlia* (*Johnstonema*) *woerlei* (Spratt & Varughese, 1975) Chabaud & Bain, 1976, caudal end male, lateral view. (After Spratt & Varughese, 1975.)
FIG. 3.265. *Filarissima lainsoni* Chabaud, 1974, caudal extremity male, ventral view. (After Chabaud, 1974.)
FIG. 3.266. *Macdonaldius andersoni* Chabaud & Frank, 1961, caudal extremity male, lateral view. (After Chabaud & Frank, 1961.)

22-(21) Body not extremely thin (greater than 50 μ).
Tail of male without cone-like cap.

23-(24) Terminal tongue-like caudal appendages usually absent (Fig. 3.266).
Cephalic papillae arranged in form of square.
Oesophagus short, indistinctly divided, or undivided.
Spicule ratio 1.6 to 4.1.
Parasites of reptiles.

Macdonaldius Khanna, 1933
(= *Saurofilaria* Caballero, 1954)

24-(23) Terminal tongue-like caudal appendages usually present.
Parasites of mammals.

25-(26) Preanal papillae arranged in single, irregular median row (Fig. 3.267).
Buccal cavity small, surrounded by thick ring (Fig. 3.268).
Cephalic extremity not clearly dilated.
Cephalic papillae poorly developed.
Oesophagus indistinctly divided.
Caudal tongue-like appendages absent in female.
Postanal papillae reduced or absent.
Lamina of left spicule short, filament extremely long.
Parasites of caviomorph rodents.

Ackertia Vaz, 1934

26-(25) Preanal papillae in two clearly defined ventrolateral rows (Fig. 3.269).

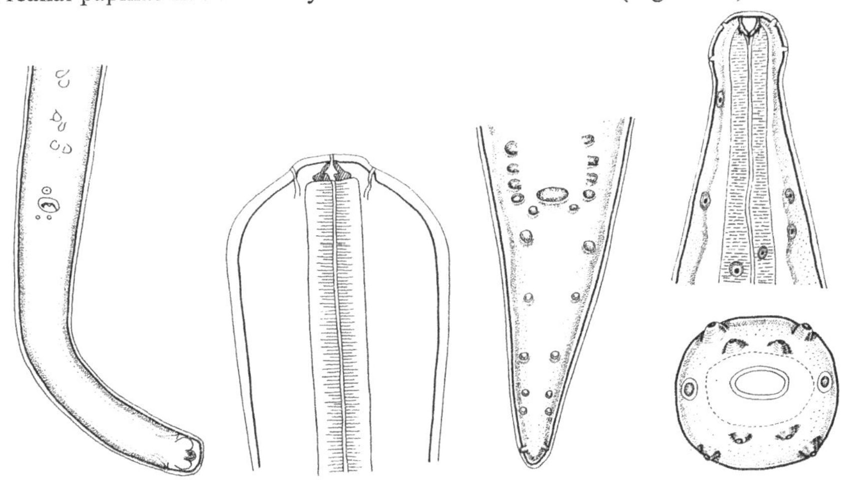

3.267 3.268 3.269 3.270

FIG. 3.267. *Ackertia dorsti* Bain & Hocquet, 1968, caudal end male, ventral view. (After Bain & Hocquet, 1968.)
FIG. 3.268. *Ackertia finlayi* (Mazza & Fiora, 1932) Chabaud & Bain, 1976, cephalic extremity of male, dorsoventral view. (After Buckley, 1933.)
FIG. 3.269. *Monanema marmotae* (Webster, 1967) Anteson, 1968, caudal extremity male, ventral view. (After Ko, 1971; Ph.D. thesis, University of Guelph, Canada.)
FIG. 3.270. *Monanema marmotae* (Webster, 1967) Anteson, 1968, cephalic extremity female, lateral and *en face* views. (After Ko, 1971; Ph.D. thesis, University of Guelph, Canada.)

27-(28) Cephalic extremity markedly dilated, bulb-like (Fig. 3.270).
Buccal cavity globular.
Cephalic papillae highly developed, arranged in form of square and well posterior to labial papillae.
Oesophagus undivided.
Caudal tongue-like appendages absent in female.
Postanal papillae well developed (Fig. 3.269).
Lamina of left spicule short, filament extremely long.
Microfilaria in skin of host.
Transmitted by ticks.
Parasites of rodents.

Monanema Anteson, 1968

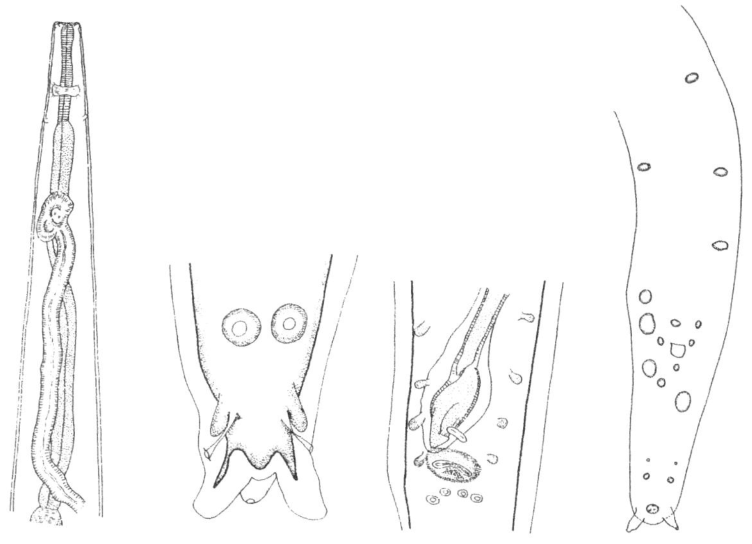

3.271 3.272 3.273 3.274

FIG. 3.271. *Dipetalonema* (*Molinema*) *sprenti* Anderson, 1953, anterior extremity female, dorsoventral view. (After Anderson, 1953.)

FIG. 3.272. *Dipetalonema* (*Molinema*) *sprenti* Anderson, 1953, caudal extremity male, ventral view. (After Anderson, 1953.)

FIG. 3.273. *Dipetalonema* (*Molinema*) *sprenti* Anderson, 1953, caudal extremity male, ventral view. (After Anderson, 1953.)

FIG. 3.274. *Dipetalonema* (*Orihelia*) *anticlava* (Molin, 1858) Lent & Freitas, 1942, caudal extremity male, ventral view. (After Lent & Freitas, 1942.)

28-(27) Cephalic extremity not markedly dilated, not bulb-like.
Cephalic papillae, when arranged in form of square only slightly posterior to external labial papillae.

29-(40) Oesophagus normally developed (Fig. 3.271).
Preoesophageal cuticular ring generally present.
Caudal extremity usually with two well developed petaloid appendages (Fig. 3.272).
Four pairs of regularly arranged preanal papillae usually present (Fig. 3.273).
Dipetalonema Diesing, 1861

30-(31) Caudal papillae including some massive asymmetrical papillae (Fig. 3.274).
Cephalic shield laterally elongated.
Buccal cavity well developed and highly cuticularized.
Oesophagus divided, with little change in width.
Caudal extremity rounded with two large petaloid appendages.
Ratio of spicules 3 : 2.
Gubernaculum present.
Microfilaria sheathed.
Parasites of Dasypodidae.
Dipetalonema (*Orihelia*) Chabaud & Bain, 1976

31-(30) Caudal papillae regular, not including massive asymmetrical papillae.

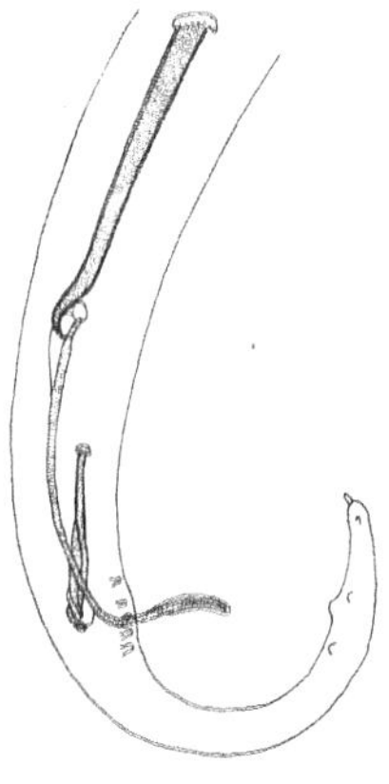

3.275

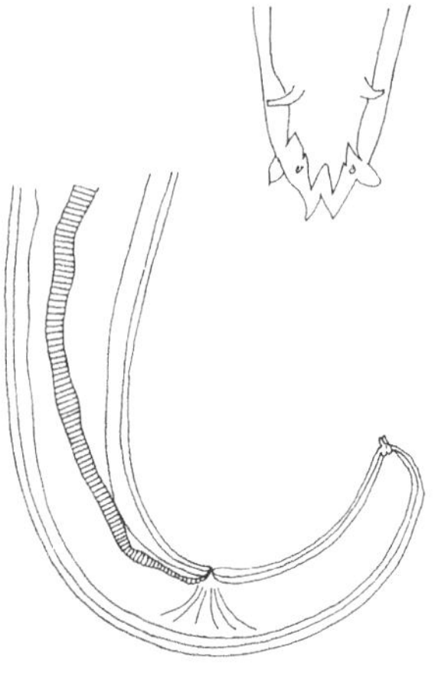

3.276

FIG. 3.275. *Dipetalonema* (*Dipetalonema*) *caudispina* (Molin, 1858) Diesing, 1861, caudal extremity male, lateral view. (After Freitas, 1943.)

FIG. 3.276. *Dipetalonema* (*Chenofilaria*) *johnstoni* Mackerras, 1954, caudal extremity female, lateral & ventral views. (After Spratt & Varughese, 1975.)

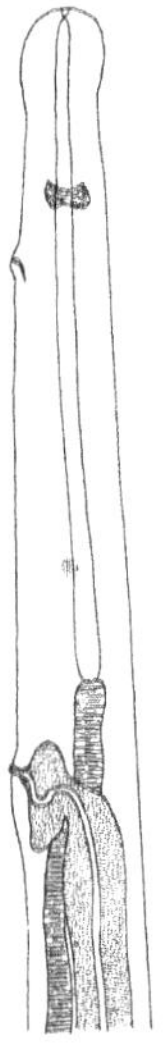

3.277

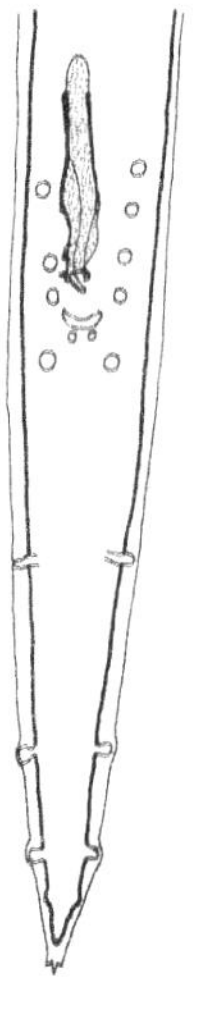

3.278

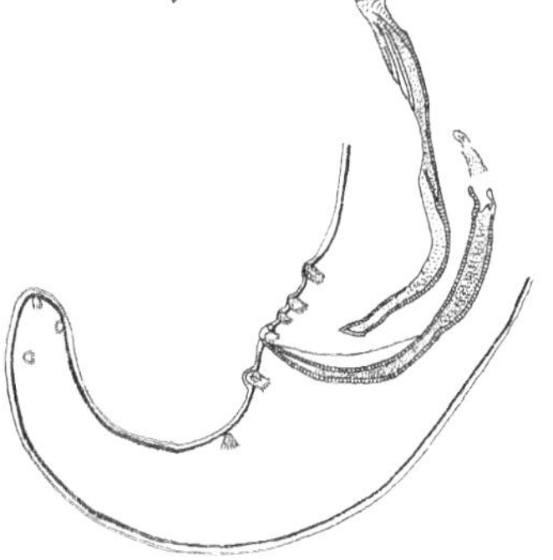

3.279

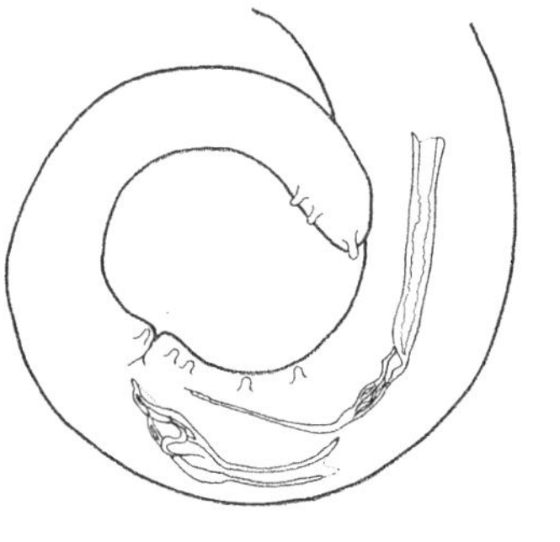

3.280

FIG. 3.277. *Dipetalonema* (*Loxodontofilaria*) *ruandae* Fain & Hérin, 1955, anterior extremity female, lateral view. (After Fain & Hérin, 1955.)

FIG. 3.278. *Dipetalonema* (*Loxodontofilaria*) *ruandae* Fain & Hérin, 1955, caudal extremity male, ventral view. (After Fain & Hérin, 1955.)

FIG. 3.279. *Dipetalonema* (*Acanthocheilonema*) *spirocauda* (Leidy, 1858) Anderson, 1959, caudal extremity male, lateral view. (After Anderson, 1959.)

FIG. 3.280. *Dipetalonema* (*Molinema*) *proechimyis* Esslinger, 1974, caudal extremity male, lateral view. (After Esslinger, 1974.)

32-(33) Gubernaculum present. Spicules slender and markedly unequal (Fig. 3.275).
Spicule ratio 3.8 to 4.8.
Cephalic extremity laterally expanded.
Oesophagus divided.
Microfilaria sheathed.
Parasites of American primates.
Dipetalonema (*Dipetalonema*) Diesing, 1861

33-(32) Gubernaculum absent. Spicules neither filiform nor markedly unequal.

34-(35) Caudal end of female bifid, ending in two conical points (Fig. 3.276).
Body slightly expanded laterally behind cephalic extremity.
Oesophagus divided (in species from Pholidota) or undivided (in species from marsupials).
Caudal papillae near anus and tail tip only.
Buccal cavity minute.
Spicule ratio 2.7 to 3.4.
Microfilaria unsheathed.
Dipetalonema (*Chenofilaria*) (Kou, 1958)

35-(34) Caudal end of female ending in simple rounded or pointed cone.

36-(37) Oesophagus undivided, rather slender (Fig. 3.277).
Caudal papillae including two or three pairs regularly distributed along length of tail (Fig. 3.278).
Parasites of African ungulates.
Dipetalonema (*Loxodontofilaria*) (Berghe & Gillain, 1939)

37-(36) Oesophagus divided, rather stout.

38-(39) Lamina of right spicule simple and spoon shaped (Fig. 3.279).
Glandular oesophagus much broader than muscular oesophagus.
Cephalic extremity flat or convex in lateral view.
Lamina of left spicule generally longer than shaft.
Ratio of spicules 2.1 to 2.9.
Microfilaria unsheathed.
Parasites of insectivores, carnivores, pinnipeds, rarely rodents.
Dipetalonema (*Acanthocheilonema*) (Cobbold, 1870)
(= *Skrjabinaria* Lubimov, 1927; = *Monnigofilaria* Skrjabin & Schikhobalova, 1948; = *Hepaticofilaria* Ortlepp, 1961)

39-(38) Lamina of right spicule in form of complex, cone-shaped structure (Fig. 3.280).
Glandular oesophagus similar in width or only slightly wider than muscular oesophagus.
Cephalic extremity flat or concave in lateral view.
Lamina of left spicule generally not longer than shaft.
Spicule ratio 1.3 to 2.4.
Microfilaria unsheathed.
Parasites of caviomorphs and beaver.
Dipetalonema (*Molinema*) (Freitas & Lent, 1939)

40-(29) Oesophagus narrow, poorly differentiated (Fig. 3.281).
Preoesophageal cuticular ring absent (Fig. 3.282).
Caudal end of female generally with four terminal lobes (Fig. 3.283).
Preanal papillae near anus and closely associated with perianal papillae (Fig. 3.284).
Tetrapetalonema Faust, 1935

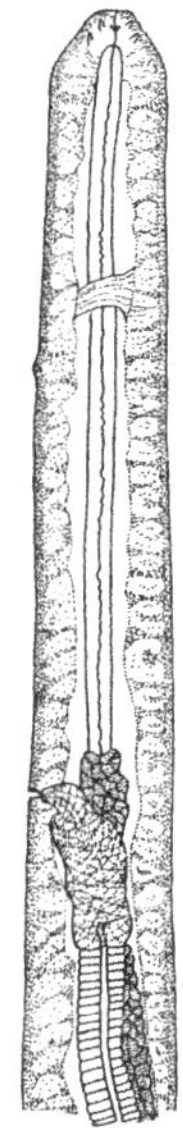
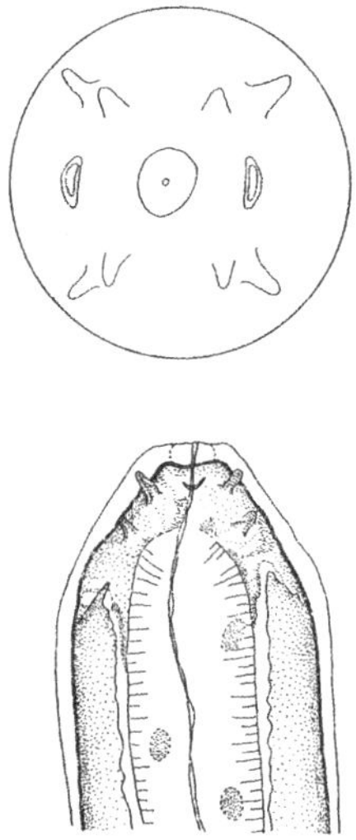

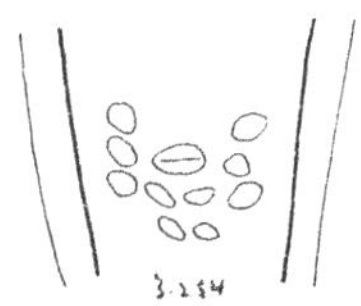
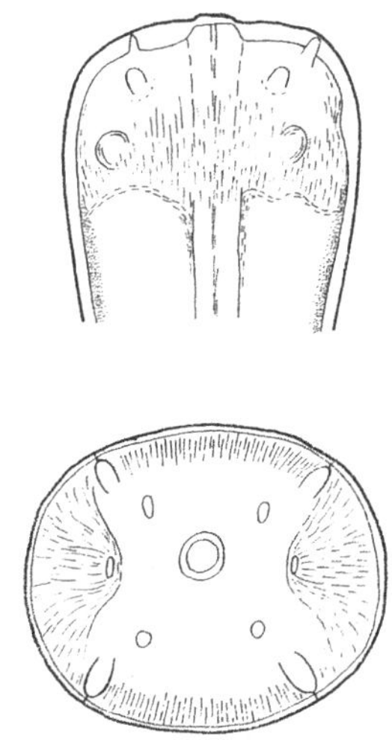

3.281 3.282 3.283 3.284 3.285

FIG. 3.281. *Tetrapetalonema* (*Tetrapetalonema*) *barbascalensis* (Esslinger & Gardiner, 1974) Chabaud & Bain, 1976, anterior extremity female, lateral view. (After Esslinger & Gardiner, 1974.)

FIG. 3.282. *Tetrapetalonema* (*Tetrapetalonema*) *barbascalensis* (Esslinger & Gardiner, 1974) Chabaud & Bain, 1976, anterior extremity female, *en face* and lateral views. (After Esslinger & Gardiner, 1974.)

FIG. 3.283. *Tetrapetalonema* (*Tetrapetalonema*) *barbascalensis* (Esslinger & Gardiner, 1974) Chabaud & Bain, 1976, caudal extremity female, ventral view. (After Esslinger & Gardiner, 1974.)

FIG. 3.284. *Tetrapetalonema* (*Tetrapetalonema*) *barbascalensis* (Esslinger & Gardiner, 1974) Chabaud & Bain, 1976, caudal extremity male, ventral view. (After Esslinger & Gardiner, 1974.)

FIG. 3.285. *Tetrapetalonema* (*Esslingeria*) *perstans* (Manson, 1891) Chabaud & Bain, 1976, cephalic extremity male, ventral and *en face* views. (After Chabaud, 1952.)

41-(42) Cephalic extremity expanded laterally (Fig. 3.285).
Body not constricted to form cephalic extremity in form of hemisphere narrower than body behind it.
Spicule ratio 2.3 to 4.7.
Parasites of anthropoid primates and man.
Tetrapetalonema (*Esslingeria*) Chabaud & Bain, 1976

42-(41) Cephalic extremity not expanded laterally.

43-(44) Cephalic extremity rounded and body not constricted to form cephalic extremity in shape of hemisphere narrower than body behind it.
Caudal papillae consisting of papillae near anus and two to three pairs regularly distributed (Fig. 3.286).
Spicule ratio 2.3 to 2.6.
Microfilaria unsheathed.
Parasites of insectivores and Asian primates.
Tetrapetalonema (*Sandnema*) Chabaud & Bain, 1976

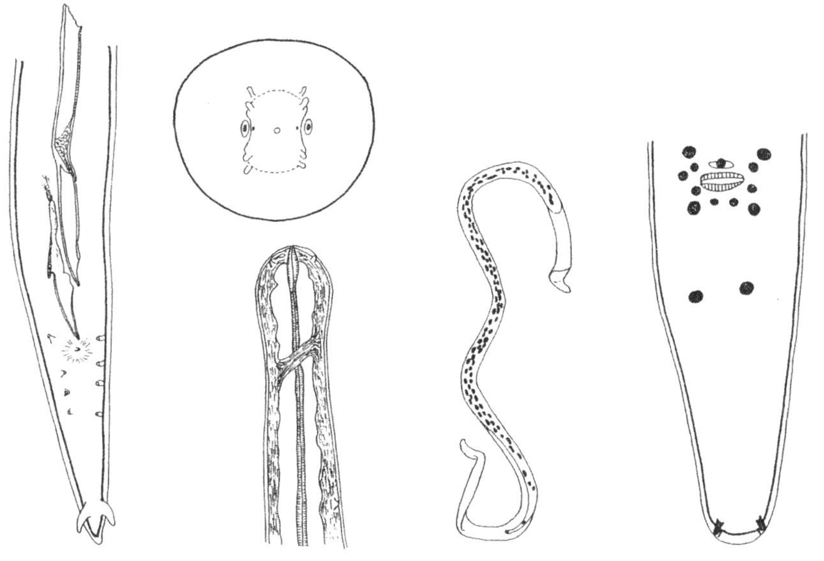

3.286 3.287 3.288 3.289

FIG. 3.286. *Tetrapetalonema* (*Sandnema*) *sunci* (Sandground, 1933) Chabaud & Bain, 1976, caudal extremity male, ventral view. (After Sandground, 1933.)

FIG. 3.287. *Tetrapetalonema* (*Tetrapetalonema*) *llewellyni* (Price, 1962) Chabaud & Bain, 1976, anterior extremity female, *en face* and lateral views. (After Price, 1962.)

FIG. 3.288. *Brugia malayi* (Brug, 1927), microfilaria. (After Buckley, 1960.)

FIG. 3.289. *Brugia malayi* (Brug, 1927), caudal end male, ventral view. (After Buckley, 1960.)

44-(43) Cephalic extremity expanded dorsoventrally (Fig. 3.287) or body constricted to form cephalic extremity in shape of hemisphere narrower than body behind it (Fig. 3.282).
Tetrapetalonema (*Tetrapetalonema*) Chabaud & Bain, 1976
(= *Parlitomosa* Nagaty, 1935)

45-(16) Buccal cavity much reduced or absent even when oesophagus not atrophied.
Caudal extremity rounded, globular in form.
Area rugosa absent.

46-(49) Oesophagus divided.
Gubernaculum present.

47-(48) Microfilaria with terminal nuclei near tip of tail (Fig. 3.288).
Caudal papillae approximately 11 (Fig. 3.289).
Cephalic extremity dilated.
Vulva preoesophageal.
Lamina of left spicule complex.
Microfilaria sheathed.
Transmitted by Culicidae.
Parasites of the lymphatics of carnivores, lagomorphs, tree shrews and man.
Brugia Buckley, 1958

48-(47) Microfilaria without terminal nuclei near tip of tail (Fig. 3.290).
Caudal papillae about 24 (Fig. 3.291).
Cephalic extremity dilated.
Vulva preoesophageal.
Lamina of left spicule not complex.
Microfilaria sheathed.
Transmitted by Culicidae.
Parasites of lymphatics of man.

Wuchereria Silva Araujo, 1877

49-(46) Oesophagus very thin (Fig. 3.292).
Gubernaculum absent.
Vulva preoesophageal.
Delicate caudal alae present near anus.
Some caudal papillae grouped near anus and extending into alae (Fig. 3.293).
Microfilaria rather short, unsheathed.
Parasites of lemurs.

Courduriella Chabaud, Brygoo & Petter, 1961

50-(1) Tail of male short, usually less than twice width of body at anus (Fig. 3.294).

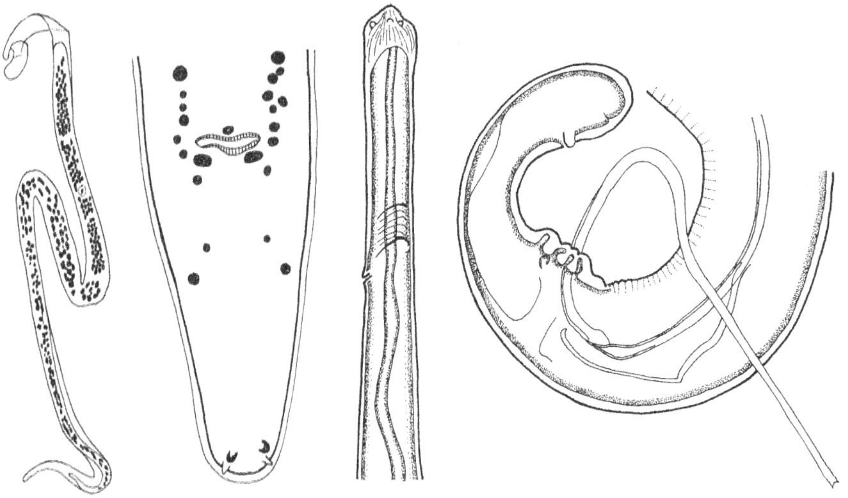

FIG. 3.290. *Wuchereria bancrofti* (Cobbold, 1877), microfilaria. (After Buckley, 1960.)
FIG. 3.291. *Wuchereria bancrofti* (Cobbold, 1877), caudal end male, ventral view. (After Buckley, 1960.)
FIG. 3.292. *Courduriella courdurieri* Chabaud, Brygoo & Petter, 1961, anterior extremity male, lateral view. (After Chabaud, Brygoo & Petter, 1961.)
FIG. 3.293. *Courduriella courdurieri* Chabaud, Brygoo & Petter, 1961, caudal extremity male, lateral view. (After Chabaud, Brygoo & Petter, 1961.)

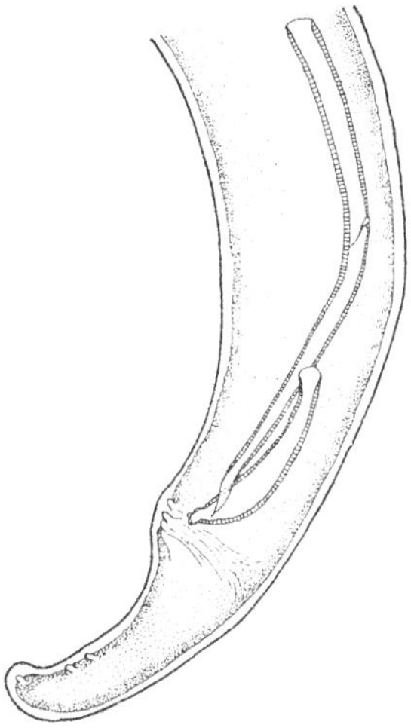

3.294

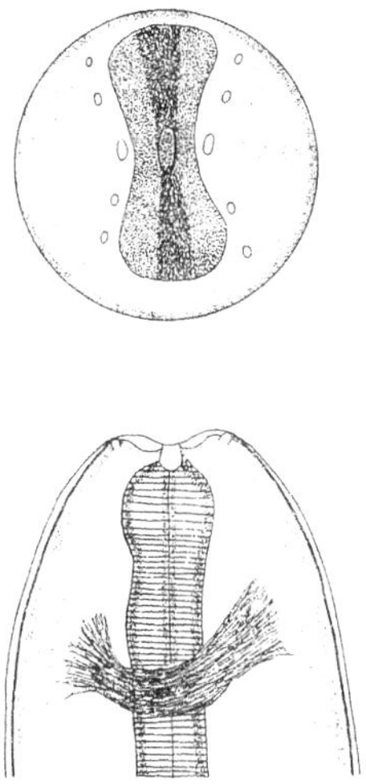

3.295

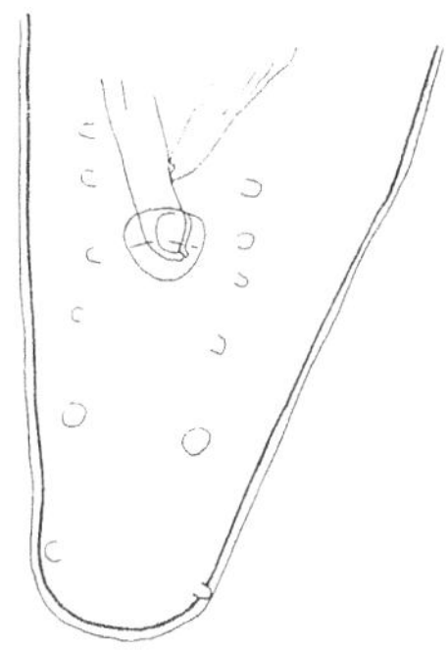

3.296

FIG. 3.294. *Paulianfilaria pauliana* Chabaud, 1961, caudal extremity male, lateral view. (After Chabaud, 1961.)
FIG. 3.295. *Deraiophoronema freitaslenti* Yeh, 1957, cephalic extremity, *en face* and lateral views. (After Yeh, 1957.)
FIG. 3.296. *Elaeophora poeli* (Vryburg, 1897), caudal extremity male, ventral view. (Original.)

51-(52) Preoesophageal ring present.
Oesophagus rather stout.
Vulva postoesophageal.
Caudal alae present.
Area rugosa present.
Caudal papillae grouped around anus and near end of tail (Fig. 3.294).
Microfilaria long, slender, unsheathed.
Parasites of lemurs.

Paulianfilaria Chabaud, Petter & Golvan, 1961

52-(51) Preoesophageal ring absent.

53-(54) Anterior extremity flattened and cuticle expanded dorsoventrally near oral opening (Fig. 3.295).
Oesophagus divided.
Vulva preoesophageal.
Caudal papillae in ventrolateral rows, well developed.
Parasites of ungulates.

Deraiophoronema Romanovitch, 1916

54-(53) Anterior extremity rounded.

55-(56) Body markedly tapered anteriorly and posteriorly in female.
Oesophagus long and divided.
Vulva preoesophageal.
Caudal papillae in subventral rows and often grouped near anus (Fig. 3.296, 3.297).

Area rugosa present or absent.
Microfilaria sheathed, found in skin.
Transmitted by Tabanidae.
Parasites of heart and arteries of ruminants.

Elaeophora Railliet & Henry, 1912
(= *Cordophilus* Mönnig, 1926; = *Alcefilaria* Oshmarin & Belous, 1951)

56-(55) Body not markedly tapered anteriorly.
Cuticle of body often with transverse ridges (Fig. 3.298).
Oesophagus divided or undivided.
Vulva pre- or postoesophageal.
Narrow caudal alae often present.
Caudal papillae prominent, in ventrolateral rows, often grouped near anus (Fig. 3.299).
Microfilaria found in skin.
Parasites of subcutaneous tissue of ruminants, equines and man.

Onchocerca Diesing, 1841
(= *Wehrdikmansia* Caballero, 1945; = *Acanthospiculum* Skrjabin & Schikhobalova, 1948)

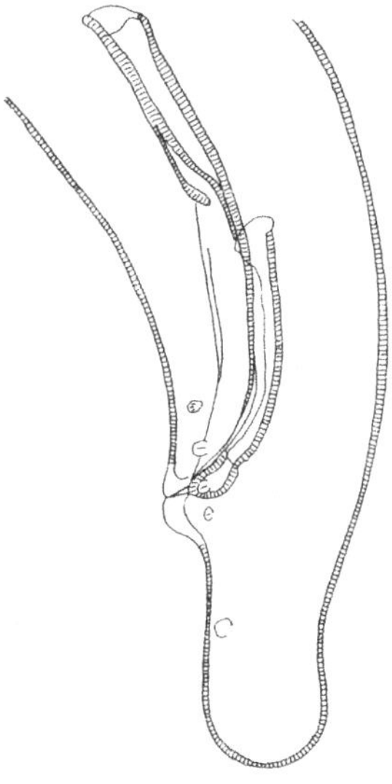

3.297

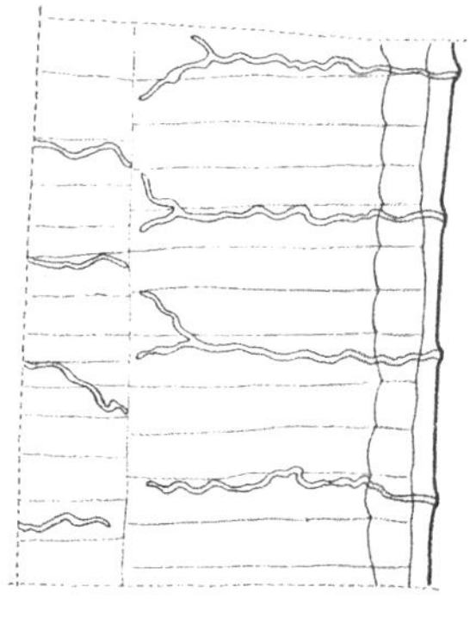

3.298

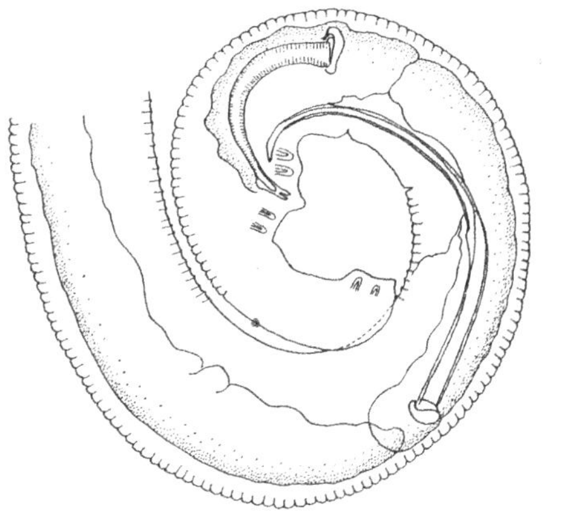

3.299

FIG. 3.297. *Elaeophora poeli* (Vryburg, 1897), caudal extremity male, lateral view. (Original.)
FIG. 3.298. *Onchocerca tubingensis* Bain & Schulz-Key, 1974, cuticle. (After Bain & Schulz-Key, 1974.)
FIG. 3.299. *Onchocerca volvulus* (Leuckart, 1893), caudal extremity male, lateral view. (After Caballero, 1944.)

Subfamily **Splendidofilariinae** Chabaud & Choquet, 1953
(= **Ornithofilariinae** Chabaud & Anderson, 1959; = **Nicanoriinae** Freitas, Vicente & Pinto, 1970; = **Micipsellinae** Sonin, 1971)

Key to genera

1-(8) Parasites of reptiles.

2-(3) Spicules rather long, the distal part complex and membranous (Fig. 3.300).
Oral opening elongated dorsoventrally.
Cuticle with fine longitudinal striations.
Tail relatively long.
Oesophagus slender and undivided.
Caudal papillae small, one pair preanal and two pairs postanal.
Vulva slightly behind end of oesophagus.
Microfilaria sheathed.
Parasites in heart of turtle.

Cardianema Alicata, 1933

3-(2) Spicules short and stout, the distal part not membranous and complex.

4-(5) Tail of male long, about twice to 3 times width of body at anus (Fig. 3.301).
Cuticle smooth.
Oesophagus short and undivided.
Vulva slightly posterior to end of oesophagus.
Microfilaria sheathed.
Parasites of subcutaneous tissues of geckos.

Thamugadia Seurat, 1917
(= *Brygoofilaria* Sulahian & Schacher, 1968)

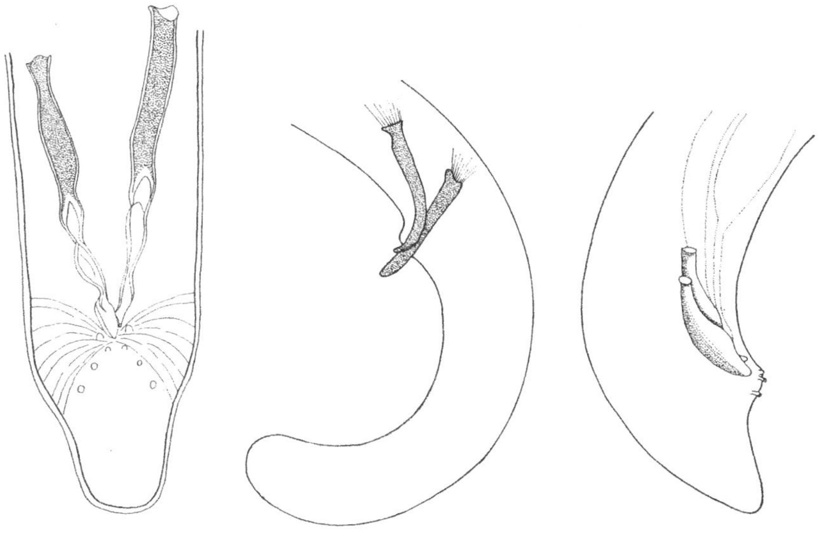

3.300 3.301 3.302

FIG. 3.300. *Cardianema cistudinis* (Leidy, 1856), caudal extremity male, ventral view. (After Alicata, 1933.)
FIG. 3.301. *Thamugadia hyalina* Seurat, 1917, caudal extremity male, lateral view. (After Seurat, 1917.)
FIG. 3.302. *Pseudothamugadia physignati* (Johnston, 1912), caudal extremity male, lateral view. (After Johnston, 1912.)

5-(4) Tail of male short, little longer than width of body at anus.

6-(7) Spicules similar in size and structure, in form of scalpel blade (Fig. 3.302).
Cuticle smooth.
Oesophagus short and undivided.
Caudal papillae consisting of indistinct preanal pair and two pairs postanal.
Vulva slightly posterior to end of oesophagus.
Microfilaria sheathed, fairly long and slender.
Parasites of lizards.

Pseudothamugadia López-Neyra, 1956

7-(6) Spicules subequal and markedly different in structure (Fig. 3.303).
Cuticle smooth.
Oesophagus rather short, indistinctly divided.
Vulva slightly anterior to end of oesophagus.
Caudal papillae consisting of three pairs preanal and one pair postanal.
Microfilaria short and stout.
Parasites of lizards (Madagascar).

Madathamugadia Chabaud, Anderson & Brygoo, 1959

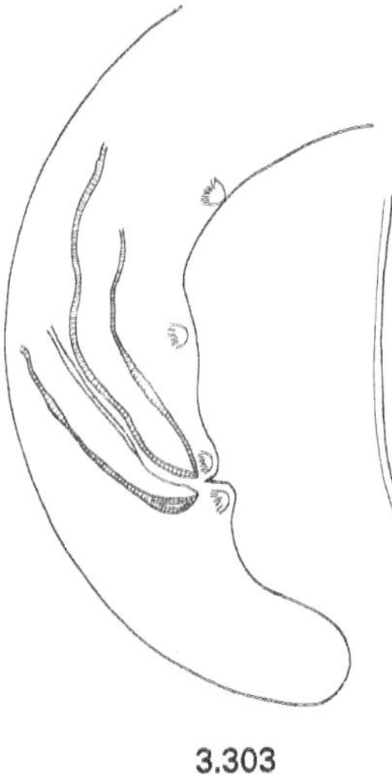

3.303

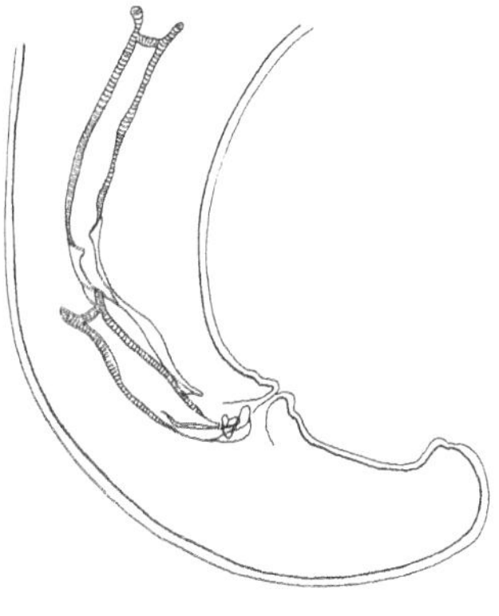

3.304

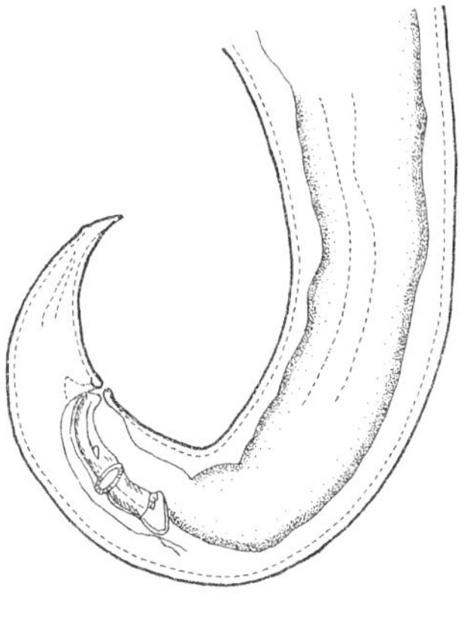

3.305

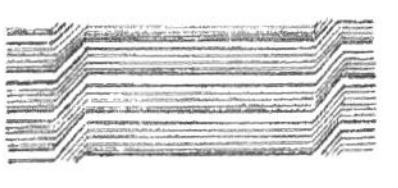

3.306

FIG. 3.303. *Madathamugadia zonosauri* Chabaud, Anderson & Brygoo, 1959, caudal extremity male, lateral view. (After Chabaud, Anderson & Brygoo, 1959.)

FIG. 3.304. *Paronchocerca bumpae* Anderson & Prestwood, 1969, caudal extremity male, lateral view. (After Anderson & Prestwood, 1969.)

FIG. 3.305. *Pseudlemdana corvicola* (Schikhobalova, 1948), caudal extremity male, lateral view. (In Sonin, 1968.)

FIG. 3.306. *Striatofilaria phoenicopteri* (Annett, Dutton & Elliot, 1901), cuticle. (After Chabaud & Campana-Rouget, 1956.)

8-(1) Parasites of birds and mammals.
9-(22) Parasites of birds.
10-(11) Spicules markedly different in size and shape, the left about twice as long as right (Fig. 3.304).
Oesophagus divided.
Vulva usually postoesophageal.
Caudal papillae usually in circle or semi-circle around anus.
Cuticular thickenings sometimes present on body.

Paronchocerca Peters, 1936
(= *Houdemerus* Chow, 1939; = *Chinesocerca* Skrjabin & Schikhobalova, 1937; = *Wymania* Wehr & Hwang, 1957; = *Pseudaproctoides* Sonin, 1961; = *Francolinema* Jairajpuri & Siddiqi, 1970; = *Nicanoria* Freitas, Vicente & Pinto, 1970)

11-(10) Spicules subequal.
12-(13) Tail pointed in both sexes.
Spicules short and stout (Fig. 3.305).
Oral opening dorsoventrally elongated, surrounded by thickened cuticle.
Oesophagus short, stout, undivided.
Vulva postoesophageal.
Parasites of subcutaneous tissues of Corvidae.

Pseudlemdana Sonin & Shumilo, 1964

13-(12) Tail rounded in both sexes.
14-(15) Cuticle with spiral striations (Fig. 3.306).
Tail of male rounded and short with numerous small papillae, many near anus.
Oesophagus divided.
Vulva preoesophageal.
Parasites of Phoenicopteridae and Pelecanidae.

Striatofilaria Lubimov, 1927

15-(14) Cuticle without spiral striations.
16-(19) Caudal papillae not in two rows along tail, generally in circle or semi-circle around anus.
Vulva well behind posterior end of oesophagus.
Oesophagus usually well developed and undivided.
17-(18) Buccal cavity with pair of lateral tooth-like structures protruding from oral opening (Fig. 3.307).
Caudal papillae in circle or semi-circle around anus.
Microfilaria moderately long.

Aproctella Cram, 1931
(= *Carinema* Pereira & Vaz, 1933)

18-(17) Buccal cavity without lateral teeth.
Caudal papillae often inconspicuous, in semi-circle around anus or irregularly distributed (Fig. 3.308).
Buccal cavity often surrounded by cuticular ring.
Microfilaria unusually long.

Cardiofilaria Strom, 1937
(= *Pseudaproctella* Anderson, 1957; = *Francofilaria* Jairajpuri & Siddiqi, 1970; = *Gallifilaria* Jain, Alwar, Awadhiya & Pandit, 1965)

19-(16) Caudal papillae in two rows along tail (Fig. 3.309).
Vulva preoesophageal.
Oesophagus well developed or poorly developed.

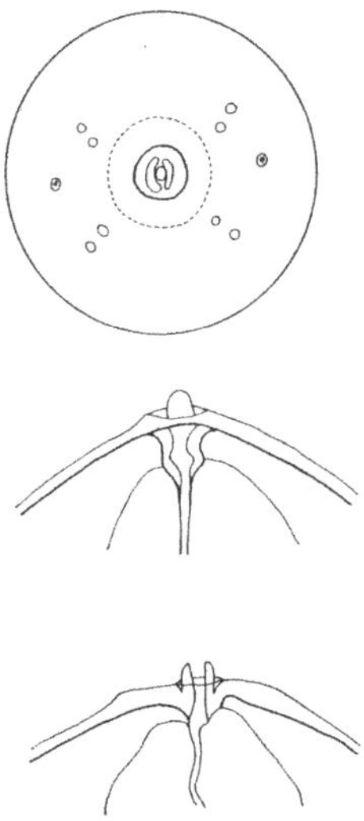

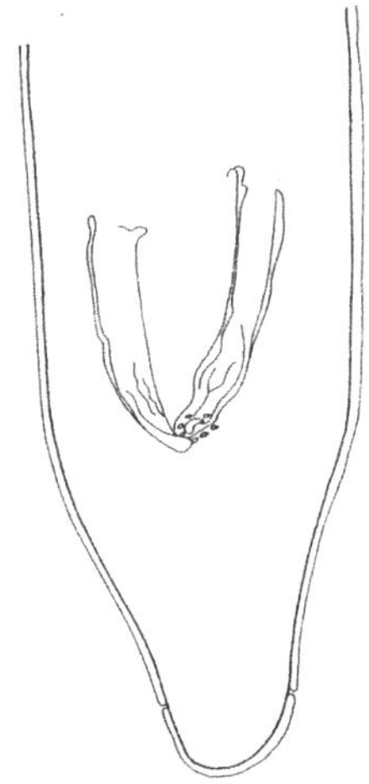

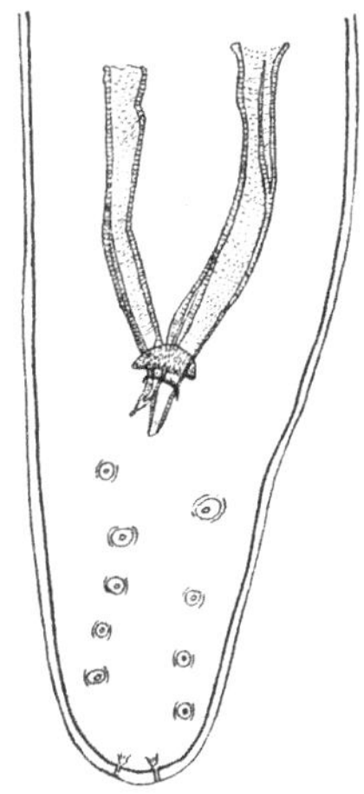

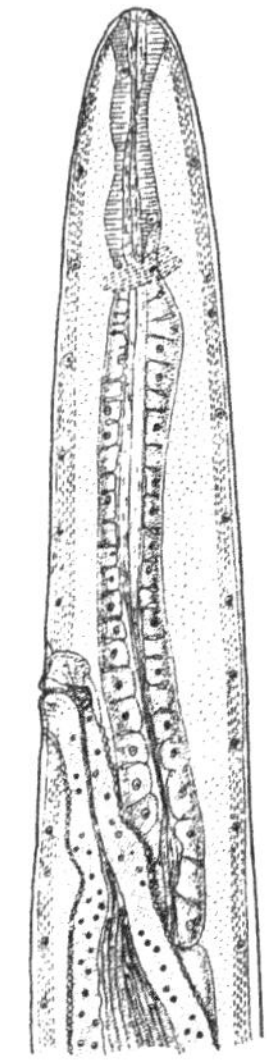

3.307 3.308 3.309 3.310

FIG. 3.307. *Aproctella stoddardi* Cram, 1931, cephalic extremity. (After Anderson, 1968.)

FIG. 3.308. *Cardiofilaria pavlovsky* Strom, 1937 (=*inornatum* Anderson, 1956), caudal extremity male, ventral view. (After Anderson, 1956.)

FIG. 3.309. *Chandlerella chitwoodae* Anderson, 1961, caudal extremity male, ventral view. (After Anderson, 1961.)

FIG. 3.310. *Chandlerella chitwoodae* Anderson, 1961, anterior extremity female, lateral view. (After Anderson, 1961.)

20-(21) Oesophagus broad and well developed, sometimes divided, clearly demarcated from intestine (Fig. 3.310).
Cuticle with or without bosses.
Four pairs of submedian cephalic papillae present.
Chandlerella Yorke & Maplestone, 1926
(= *Lerouxinema* Singh, 1949; = *Vagrifilaria* Augustine, 1937; = *Skrjabinocta* Tschertkowa, 1946; = *Parachandlerella* Caballero, 1948; = *Parornithofilaria* Sonin, 1965)

21-(20) Oesophagus reduced to narrow tube devoid of glandular tissue and imperfectly demarcated from intestine (Fig. 3.311).
Cuticle with or without bosses.
Two pairs only of cephalic papillae present (Fig. 3.312).
Splendidofilaria Skrjabin, 1923
(= *Ornithofilaria* Gönnert, 1937; = *Ularofilaria* Lubimov, 1946; = *Lophortofilaria* Wehr & Herman, 1956; = *Anenteronema* Oschmarin, 1949)

22-(9) Parasites of mammals.

23-(30) Male tail long, at least twice or 3 times as long as width of body at anus.

24-(29) Caudal papillae in regular longitudinal rows.

25-(26) Cuticle ornamented with bosses in lateral fields and on dorsal and ventral surfaces of tail (Fig. 3.313).

Spicules small.
Oesophagus slender and undivided.
Vulva near posterior end of oesophagus.
Microfilaria unsheathed.
Parasites of abdominal cavity of rodents.

Micipsella Seurat, 1921
(= *Cercofilaria* Kalantarian, 1924)

26-(25) Cuticle without bosses.
27-(28) Preanal papillae numerous (Fig. 3.314).
Quadridelphic.
Spicules small.
Oesophagus undivided.
Vulva near posterior end of oesophagus.
Microfilaria sheathed.
Parasites of central nervous system of primates.

Meningonema Orihel & Esslinger, 1973

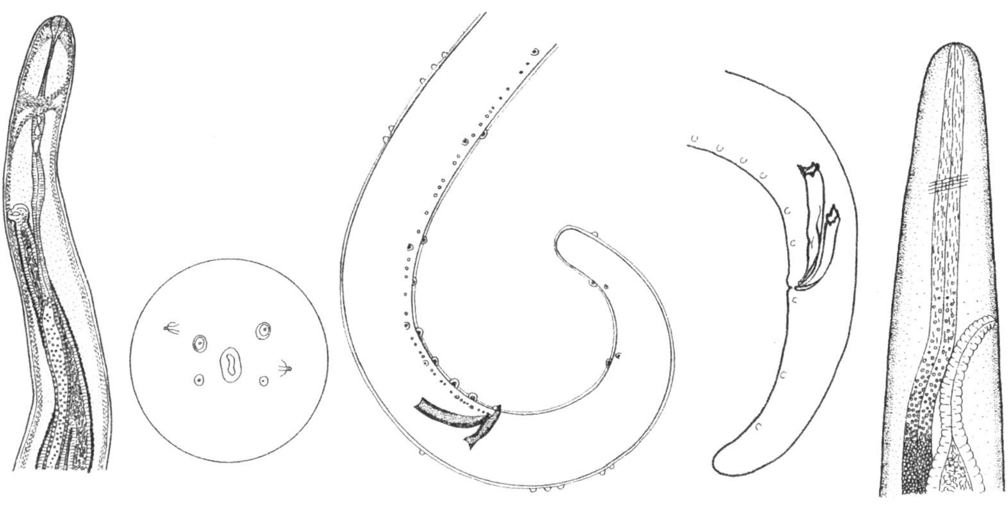

3.311 3.312 3.313 3.314 3.315

FIG. 3.311. *Splendidofilaria fallisensis* (Anderson, 1954), anterior extremity female, ventrolateral view. (After Anderson, 1954.)
FIG. 3.312. *Splendidofilaria wehri* (Anderson, 1961), *en face* view. (After Anderson, 1961.)
FIG. 3.313. *Micipsella numidica* (Seurat, 1917), caudal extremity male, lateral view. (After Seurat, 1917.)
FIG. 3.314. *Meningonema peruzzii* Orihel & Esslinger, 1973, caudal extremity male, lateral view. (After Orihel & Esslinger, 1973.)
FIG. 3.315. *Dunnifilaria ramachandrani* Mullin & Balasingham, 1973, anterior extremity female, lateral view. (After Mullin & Balasingham, 1973.)

3.316

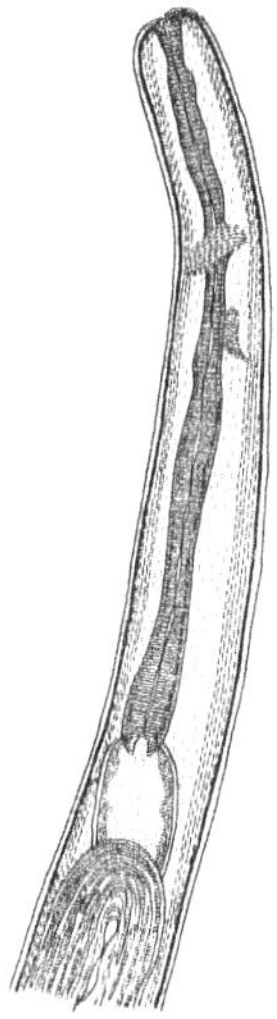

3.317

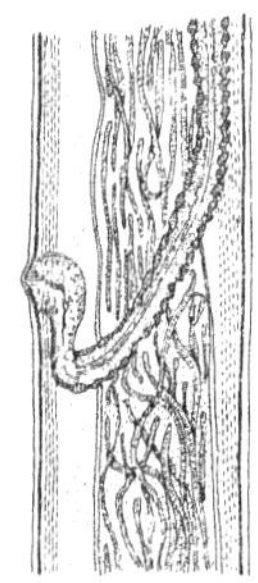

3.318

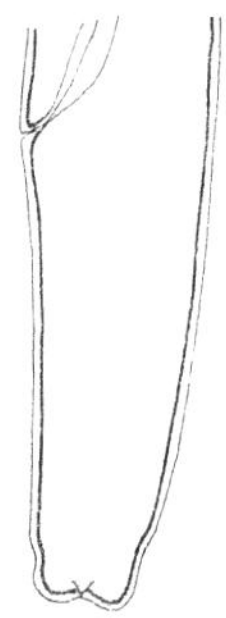

3.319

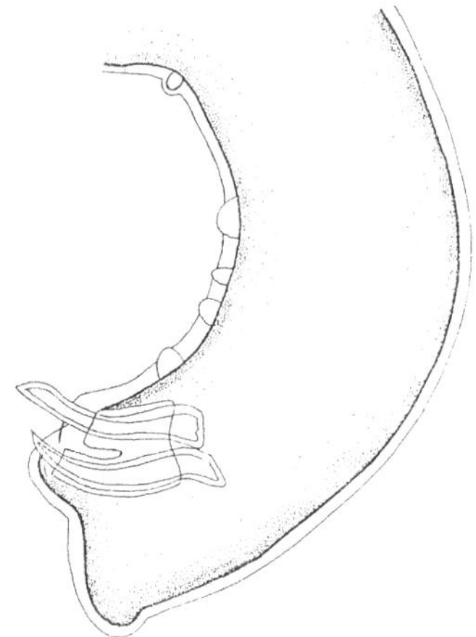

3.320

FIG. 3.316. *Onchocercella katangensis* (Gedoelst, 1916), cuticle. (After Gedoelst, 1916.)
FIG. 3.317. *Protofilaria furcata* Chandler, 1929, anterior extremity female, lateral view. (After Anderson, 1961.)
FIG. 3.318. *Protofilaria furcata* Chandler, 1929, vulva region, lateral view. (After Anderson, 1961.)
FIG. 3.319. *Protofilaria furcata* Chandler, 1929, caudal extremity female, lateral view. (After Anderson, 1961.)
FIG. 3.320. *Saurositus indicus* Deshmukh & Ali, 1965, caudal extremity male, lateral view. (After Deshmukh & Ali, 1965.)

28-(27) Preanal papillae absent.
Didelphic.
Female tail round and very short.
Spicules stout and wedge-shaped, similar in size.
Oesophagus moderately long, not distinctly divided (Fig. 3.315).
Four pairs inconspicuous postanal papillae present.
Microfilaria sheathed, short.
Parasites of heart and pulmonary arteries of rodents.
Dunnifilaria Mullin & Balasingham, 1973

29-(24) Caudal papillae not in regular longitudinal rows, most papillae generally distributed near or around anus.
Oesophagus divided.
Cuticle with irregularly arranged fusiform thickenings (Fig. 3.316).
Vulva anterior to end of oesophagus.
Spicules subequal and stout.
Microfilaria unsheathed.
Parasites of subcutaneous tissues of insectivores.
Onchocercella Yorke & Maplestone, 1931
(= *Grammophora* Gedoelst, 1916; = *Katanga* Yorke & Maplestone, 1926)

30-(23) Male tail short, much less than twice or 3 times width of body at anus.
Oesophagus undivided (Fig. 3.317).
Vulva well behind posterior end of oesophagus (Fig. 3.318).
Caudal papillae in circle around anus.
Female tail with dorsoventral cleft (Fig. 3.319).
Parasites of lemurs.

Protofilaria Chandler, 1929

Subfamily **Lemdaninae** López-Neyra, 1956
(= **Eufilariinae** López-Neyra, 1956)

Key to genera

1-(8) Oesophagus long, divided or not divided into muscular and glandular parts.
2-(7) Oesophagus clearly divided into muscular and glandular parts.
3-(4) Caudal papillae of male in two preanal rows (Fig. 3.320).
Spicules subequal, stout and partly fused.
Area rugosa punctate.
Vulva near end of oesophagus.
Oesophagus short.
Parasites of reptiles.

Saurositus Macfie, 1924

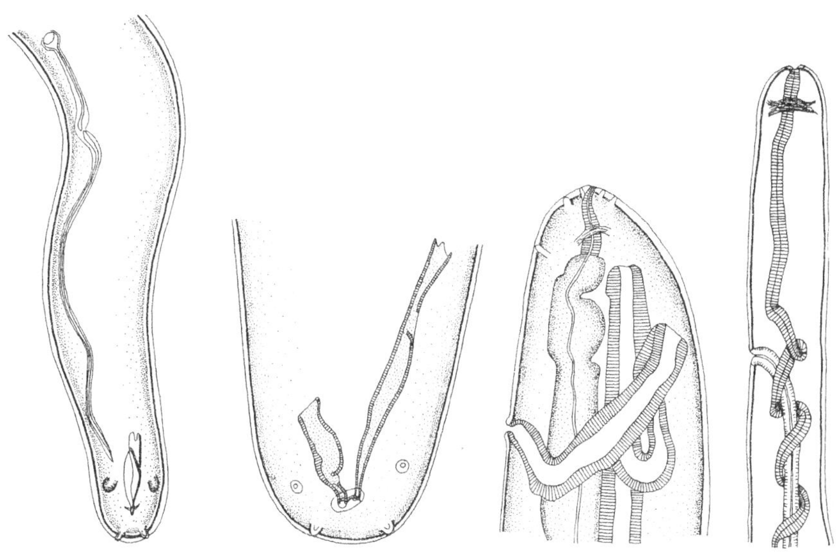

3.321 3.322 3.323 3.324

FIG. 3.321. *Lemdana sonneretta* (Ali, 1968) n.comb. (=*Singhnema sonneretta*), caudal extremity male, ventral view. (After Ali, 1968.)
FIG. 3.322. *Singhfilaria pavonica* Rasheed, 1960, posterior extremity male, ventral view. (After Rasheed, 1960.)
FIG. 3.323. *Singhfilaria pavonica* Rasheed, 1960, anterior extremity female, lateral view. (After Rasheed, 1960.)
FIG. 3.324. *Chiropterofilaria brevicaudata* Yeh, Symes & Mataika, 1958, anterior extremity female, lateral view. (After Yeh *et al.*, 1958.)

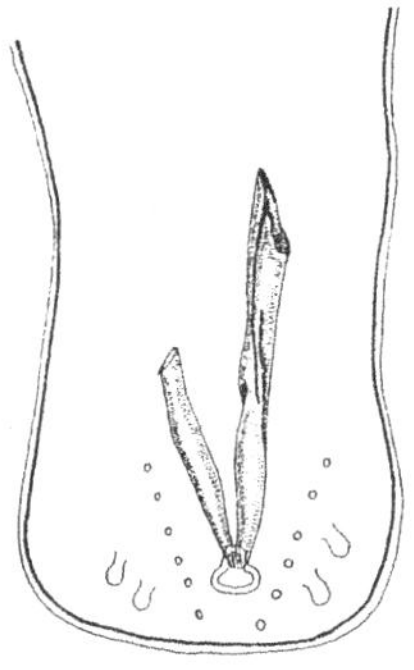

3.325

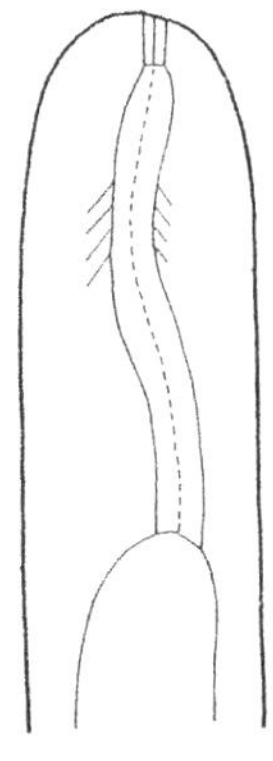

3.326

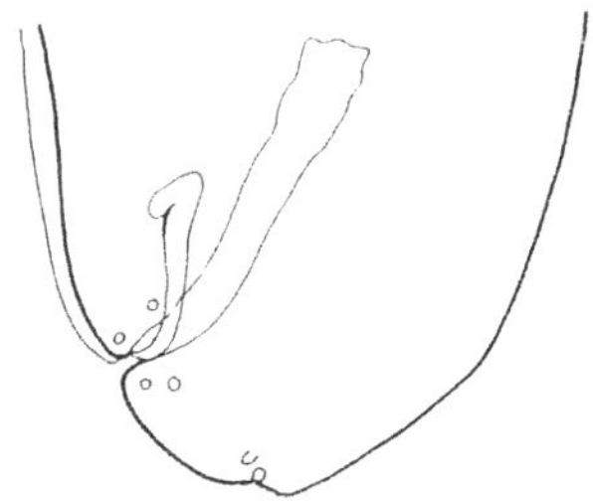

3.327

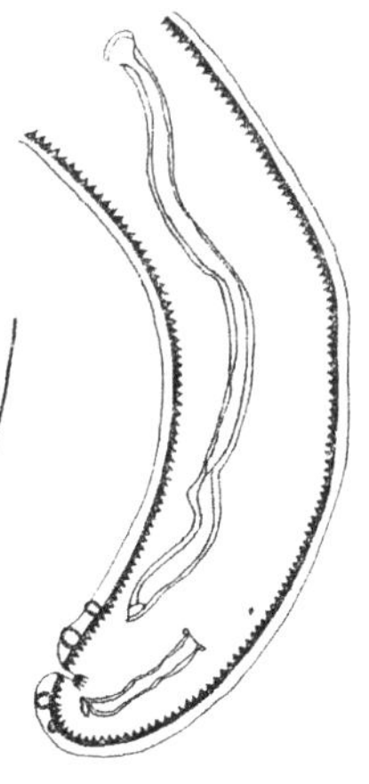

3.328

FIG. 3.325. *Chiropterofilaria brevicaudata* Yeh, Symes & Mataika, 1958, caudal extremity male, ventral view. (After Yeh *et al.*, 1958.)

FIG. 3.326. *Aproctiana meirai* (Travassos, 1930), anterior extremity male. (After Travassos, 1930.)

FIG. 3.327. *Aproctiana meirai* (Travassos, 1930), caudal extremity male, lateral view. (After Travassos, 1930.)

FIG. 3.328. *Ameeria sultanae* Ali, 1961, posterior extremity male, lateral view. (After Ali, 1961.)

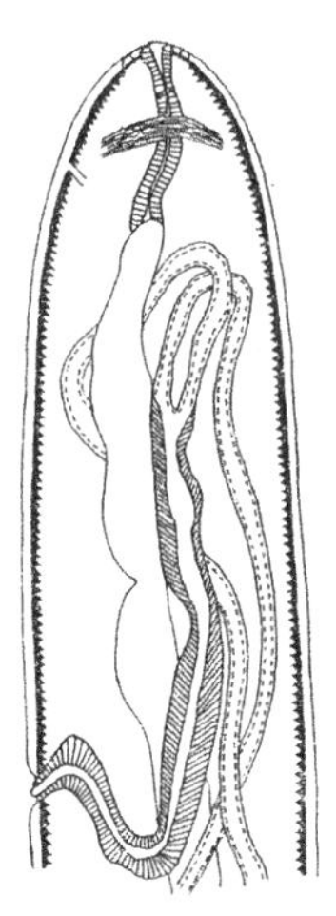

3.329

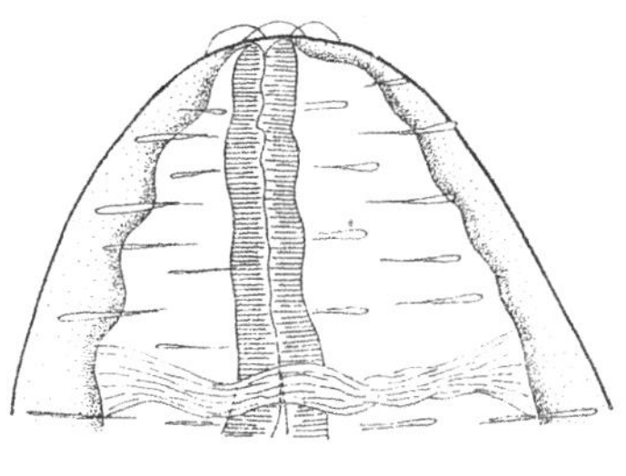

3.330

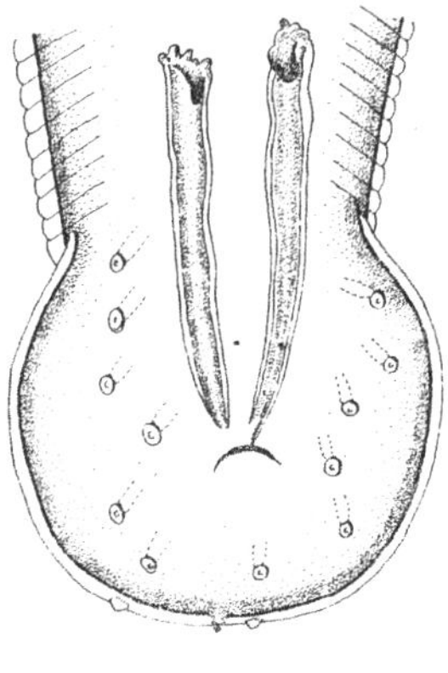

3.331

FIG. 3.329. *Ameeria sultanae* Ali, 1961, anterior extremity female, lateral view. (After Ali, 1961.)

FIG. 3.330. *Sarconema pseudolabiata* Belogurov, Daija & Sonin, 1966, anterior extremity male. (After Belogurov *et al.*, 1966.)

FIG. 3.331. *Sarconema eurycerca* Wehr, 1939, caudal extremity male, ventral view. (After Wehr, 1939.)

4-(3) Caudal papillae near anus, few in number.
Spicules different and separate.
Area rugosa absent.
Parasites of birds.

5-(6) Left spicule tubular, many times (approximately 7) as long as right (Fig. 3.321).
Vulva a short distance behind muscular oesophagus.
Lemdana Seurat, 1917
(= *Singhnema* Ali, 1969)

6-(5) Left spicule complex, about twice as long as right (Fig. 3.322).
Vulva preoesophageal (Fig. 3.323).
Cuticle often with ornamentation in form of fusiform thickenings.
Singhfilaria Rasheed, 1960
(= *Lemdanella* Sonin, 1963)

7-(2) Oesophagus not divided into muscular and glandular parts (Fig. 3.324).
Oesophagus very long (3-5 mm).
Caudal end of male slightly expanded laterally.
Two rows of small caudal papillae and two lateral pairs of large papillae present (Fig. 3.325).

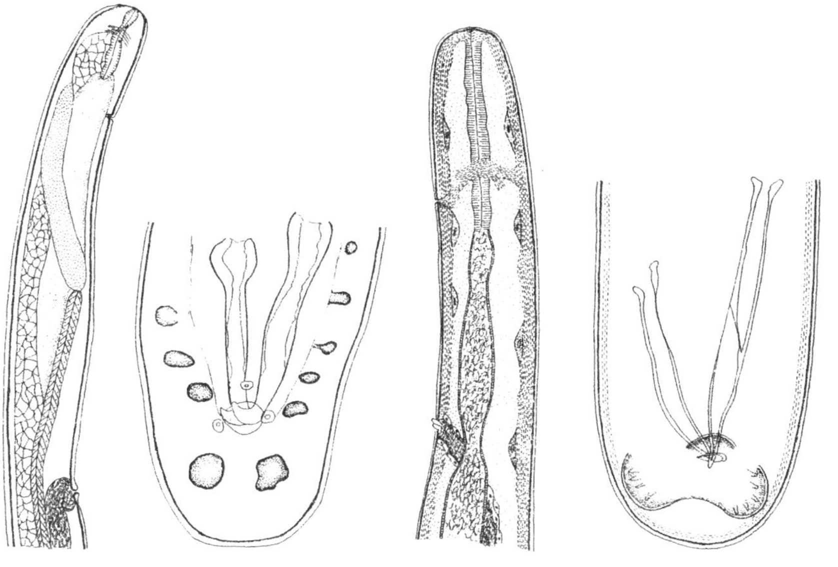

3.332 3.333 3.334 3.335

FIG. 3.332. *Heimnema heimi* Chabaud, Brygoo & Richard, 1964, anterior extremity female, lateral view. (After Chabaud *et al.*, 1964.)

FIG. 3.333. *Heimnema heimi* Chabaud, Brygoo & Richard, 1964, caudal extremity male, ventral view. (After Chabaud *et al.*, 1964.)

FIG. 3.334. *Eufilaria mcintoshi* Anderson & Bennett, 1960, anterior female, lateral view. (After Anderson & Bennett, 1960.)

FIG. 3.335. *Eufilaria mcintoshi* Anderson & Bennett, 1960, caudal extremity male, ventral view. (After Anderson & Bennett, 1960.)

Vulva near anterior third of oesophagus.
Spicules different in size and morphology.
Cuticular buccal ring present.
Parasites of Chiroptera.

Chiropterofilaria Yeh, Symes & Mataika, 1958

8-(1) Oesophagus short and undivided.
Parasites of birds.

9-(10) Buccal cavity cylindrical (Fig. 3.326).
Vulva postoesophageal.
Spicules minute (Fig. 3.327).

Aproctiana Skrjabin, 1934

10-(9) Buccal cavity absent or greatly reduced.

11-(12) Spicules markedly different in size and form (Fig. 3.328).
Vulva well behind oesophagus (Fig. 3.329).
Cuticle with rugae.

Ameeria Ali, 1961

12-(11) Spicules equal or subequal and rather similar in form.

13-(14) Cephalic extremity attenuated.
Cuticle with well-marked transverse ridges or rugae (Fig. 3.330).
Caudal extremity expanded laterally (Fig. 3.331).
Male caudal papillae minute, in two subventral rows.
Oesophagus stout and muscular.
Vulva in midregion of oesophagus.
Parasites of Anatidae.

Sarconema Wehr, 1939
(= *Farooqifilaria* Jairajpuri & Siddiqi, 1970; = *Alifilaria* Ali, 1969)

14-(13) Cephalic extremity slightly expanded.
Cuticle without ridges or rugae.

15-(16) Oesophagus stout and muscular, clearly demarcated from intestine (Fig. 3.332).
Vulva well behind posterior end of oesophagus.
Caudal papillae not on swellings, variable in number (Fig. 3.333).

Heimnema Chabaud, Brygoo & Richard, 1964

16-(15) Oesophagus usually slender, imperfectly demarcated from intestine and containing little muscular tissue (Fig. 3.334).
Vulva pre- or postoesophageal but anterior in position.
Caudal papillae variable in number and sometimes not visible; if present, some on lateral swellings near anus (Fig. 3.335).

Eufilaria Seurat, 1921
(= *Eufilariella* Sonin, 1965; = *Neofilaria* Deshmukh, 1968)

REFERENCES

ALI, M. M. 1969. Observations on some filariid worms from Indian birds with a description of two new genera. *Zool. Anz.*, **183,** 309-316.

ALI, S. M. 1961. *Ameeria sultanae* n.g., n.sp. from the common grey hornbill, *Tockus birostris* in India. *J. Parasit.*, **47,** 715-716.

ALWAR, V. S., SENEVIRATNA, P. & GOPAL, S. 1959. *Indofilaria pattabiramani* n.g., n.sp., a filarid from the Indian elephant (*Elephas maximus*), causing haemorrhagic dermatitis. *Indian vet. J.*, **36,** 408-414.

ANDERSON, R. C. 1957. Observations on the life cycles of *Diplotriaenoides translucidus* Anderson and members of the genus *Diplotriaena. Can. J. Zool.*, **35,** 15-24.

——— 1959. The egg and first-stage larva of *Desmidocercella numidica* (Seurat, 1920) with remarks on the affinities of the Desmidocercidae. *Can. J. Zool.*, **37,** 407-413.

——— 1959. The taxonomy of *Dipetalonema spirocauda* (Leidy, 1858) n.comb. (=*Skrjabinaria spirocauda*) and *Dirofilaria roemeri* (Linstow, 1905) n.comb. (=*Dipetalonema roemeri*). *Can. J. Zool.*, **37,** 481-493.

——— 1961. *Splendidofilaria wehri* n.sp. with a revision of *Splendidofilaria* and related genera. *Can. J. Zool.*, **39,** 201-207.

——— 1961. Study of two filarioid nematodes, *Chandlerella chitwoodae* n.sp. from *Padda oryzivora* (L.) and *Protofilaria furcata* Chandler, 1929. *Can. J. Zool.*, **39,** 317-323.

——— 1962. On the development, morphology and experimental transmission of *Diplotriaena bargusinica* (Filarioidea: Diplotriaenidae). *Can. J. Zool.*, **40,** 1175-1186.

——— 1968. The comparative morphology of cephalic structures in the superfamily Filarioidea (Nematoda). *Can. J. Zool.*, **46,** 181-199.

——— & CHABAUD, A. G. 1958. Taxonomie de la filaire *Squamofilaria sicki* (Strachan, 1957) n.comb. et place du genre *Squamofilaria* Schmerling, 1925 dans la sous-famille Aproctinae. *Annls Parasit. hum. comp.*, **33,** 254-266.

——— & FREEMAN, R. S. 1969. *Cardiofilaria inornata* (Anderson, 1956) from woodcock with a review of *Cardiofilaria* and related genera (Nematoda: Filarioidea). *Trans. Amer. microsc. Soc.*, **88,** 68-79.

——— & PRESTWOOD, A. K. 1969. *Singhfilaria hayesi* n.sp. from *Meleagris gallopavo* and a revision of Lemdaniinae (=Eufilariinae). *Can. J. Zool.*, **47,** 1015-1019.

——— & PRESTWOOD, A. K. 1969. *Paronchocerca bumpae* n.sp. from the brushland tinamou and the position of *Paronchocerca* within the Splendidofilariinae (Filarioidea). *Can. J. Zool.*, **47,** 1325-1331.

BAIN, O. 1974. Description de nouvelles filaires Oswaldofilariinae de lézards sud-americains; hypothèse sur l'évolution des filaires de reptiles. *Bull. Mus. natn. Hist. nat., Paris.* 3ème sér. No. 208, Zool., 138, 169-200.

——— & BRUNHES, J. 1968. Un nouveau genre de filaire, parasite de grenouilles malgaches. *Bull. Mus. natn. Hist. nat., Paris.* 2ème sér., **40,** 797-801.

——— & DURETTE-DESSET, M. C. 1973. Cycle de *Skrjabinofilaria skrjabini*, filaire de marsupial sud-americain; position systématique. *Annls Parasit. hum. comp.*, **48,** 61-79.

——— & HAESEVOETS, E. 1974. Affinités entre deux filaires de l'appareil circulatoire, l'une parasite de bovidés, l'autre de cervidés: *Cordophilus sagittus* (Linstow, 1907) et *C. abramovi* (Oshmarin et Belous, 1951) nov. comb. *Annls Parasit. hum. comp.*, **49,** 119-122.

——— & PROD'HON, J. 1974. Homogénéité des filaires de batraciens des genres *Waltonella*, *Ochoterenella* et *Madochotera;* création de Waltonellinae n.subfam. *Annls Parasit. hum. comp.*, **49,** 721-739.

——— & SCHULZ-KEY, H. 1974. Les onchocerques du cerf européen: Redescription d'*O. flexuosa* (Wedl, 1856) et description d'*O. tubingensis* n.sp. et *O. tarsicola* n.sp. *Z. Tropenmed. Parasit.*, **25,** 437-449.

——— & SCHULZ-KEY, H. 1974. Une filaire intradermique chez le cerf européen: *Cutifilaria wenki* n.gen., (Onchocercinae). *Z. Tropenmed. Parasit.* **25,** 450-453.

——— & SULAHIAN, A. 1974. Trois nouvelles filaires du genre *Oswaldofilaria* chez des lézards sud-americains; essai de classification des Oswaldofilariinae. *Bull. Mus. natn. Hist. nat., Paris.* 3ème sér. n.sp. No. 232, *Zool.*, 156, 827-841.

——— & VASSILIADES, G. 1969. Cycle évolutif d'un Dicheilonematinae, *Serratospiculum tendo* filaire parasite du faucon. *Annls Parasit. hum. comp.*, **44,** 595-604.

——— & VAUCHER, C. 1973. Developpement larvaire de *Diplotriaena tridens* (Nematoda: Filarioidea) chez *Locusta migratoria. Annls Parasit. hum. comp.*, **48,** 81-89.

BUCKLEY, J. J. C. 1960. On *Brugia* gen.nov. for *Wuchereria* spp. of the '*malayi*' group, i.e. *W. malayi* (Brug, 1927), *W. pahangi* Buckley and Edeson, 1956 and *W. patei* Buckley, Nelson and Heisch, 1958. *Ann. trop. Med. Parasit.*, **54,** 75-77.

——— 1973. A redescription of *Dipetalonema finlayi* (Mazza and Fiora, 1932) from a Viscasha, *Lagidium peruanum*, of Peru. *Annls. Parasit. hum. comp.*, **48,** 335-341.

CHABAUD, A. G. 1952. Le genre *Dipetalonema* Diesing, 1861; essai de classification. *Annls Parasit. hum. comp.*, **27,** 250-285.

——— 1974. Class Nematoda. Keys to subclasses, orders and super-families (*In* CIH Keys to Nematode Parasites of Vertebrates, edited by Anderson, R. C., Chabaud, A. G. & Willmott, S. Commonwealth Agricultural Bureaux, Farnham Royal, Bucks, England. pp. 6-17).

——— 1974. Filaire hyperévoluée, parasite de la paroi intestinale d'un coendou. *Annls Parasit. hum. comp.*, **49,** 451-455.

——— & ANDERSON, R. C. 1959. Nouvel essai de classification des filaires (superfamille des Filarioidea) II. *Annls Parasit. hum. comp.*, **34,** 64-87.

———, ——— & BRYGOO, E. R. 1959. Cinq filaires de reptiles malgaches. *Mém. Inst. scient. Madagascar.* Série A., **13,** 103-125.

———, ——— & BRYGOO, E. R. 1959. Sept filaires d'oiseaux malgaches. *Annls Parasit. hum. comp.*, **34,** 88-109.

——— & BAIN, O. 1976. La lignée *Dipetalonema*, nouvelle essai de classification. *Annls Parasit. hum. comp.* (in press).

———, BRYGOO, R. & PETTER, A. J. 1961. Les nématodes parasites de lémuriens malgaches. IV. Description de deux noveaux genres et observations sur *Protofilaria furcata* Chandler. *Bull. Mus. natn. Hist. nat., Paris*, 2ème sér. **33,** 532-544.

———, BRYGOO, E. R. & RICHARD, J. 1964. Filaires d'oiseaux malgaches (deuxième note). *Annls Parasit. hum. comp.*, **39,** 69-94.

———, PETTER, A. J. & GOLVAN, Y. 1961. Les nématodes parasites de lémuriens malgaches III. Collection récoltée par M. et Mme Francis Petter. *Annls Parasit. hum. comp.*, **36,** 113-126.

DESHMUKH, P. G. 1968. Two new species of a new genus *Neofilaria* from Indian quails. *Riv. Parassit.*, **29,** 123-128.

DIAZ-UNGRÍA, C. 1963. Description d'une nouvelle filaire, intermediaire entre les deux sous-familles Diplotriaeninae et Dicheilonematinae. *C. r. hebd. Séanc. Acad. Sci., Paris*, **256,** 4314-4316.

DIAZ-UNGRÍA, C. 1963. Nematodes parásitos colectados por la Misión Chauvancy en Guayana Francesa. *Bull. Mus. natn. Hist. nat., Paris*, 2ème sér., **35,** 441-453.

——— 1964. Estudio de una colección de nematodes de America del Sur, con descripción de un género nuevo y dos especies nuevas. *Publicaciones Ocasionales del Museo de Ciencias Naturales, Caracas.* Zoologia, **9,** 1-16.

ESSLINGER, J. H. 1974. Two new species of *Dipetalonema* (Nematoda: Filarioidea) from neotropical echimyid rodents. *J. Parasit.*, **60,** 473-479.

——— & GARDINER, C. H. 1974. *Dipetalonema barbascalensis* sp.n. (Nematoda: Filarioidea) from the owl monkey *Aotus trivirgatus* with a consideration of the status of *Parlitomosa zakii* Nagaty, 1935. *J. Parasit.*, **60,** 1001-1005.

FREITAS, J. F. TEIXEIRA DE, VICENTE, J. J., & PINTO, R. M. 1970. Sôbre uma filaria prodelfa parasita de ave (Nematoda, Filarioidea). *Atas Soc. Biol. Rio de J.*, **12,** Suppl. pp. 39-42.

HIBLER, C. P. & ADCOCK, J. L. 1968. Redescription of *Elaeophora schneideri* Wehr and Dikmans, 1935 (Nematoda: Filarioidea). *J. Parasit.*, **54,** 1095-1098.

HSÜ, G. N. *see* SYUI, G. N.

JAIN, S. K., ALWAR, V. S., AWADHIYA, R. P. & PANDIT, C. N. 1965. On a new filarid worm *Gallifilaria mhowensis* n.g., n.sp. from the heart of the fowl (*Gallus gallus domesticus* L.). *Indian vet. J.*, **42,** 895-898.

JAIRAJPURI, D. S. & SIDDIQI, A. H. 1970. On some nematodes of birds from India. Part I: Filaridae and Dipetalonematidae. *Revta bras. Biol.*, **30,** 69-74.

KAZUBSKI, S. L. 1958. *Cerebrofilaria caprimulgi* nov.gen., nov.spec. (Nematoda: Filarioidea), a parasite from the brain of the nightjar *Caprimulgus europaeus* L. *Bull. Acad. pol. Sci. Cl. II. Sér. Sci. biol.*, **6,** 73-78.

KOU, C. C. 1958. [Studies of parasitic nematodes of mammals from Canton. I. Some new species from *Paradoxurus minor* Schwarz, *Paguma larvata larvata* (Hamilton Smith) and *Manis pentadactyla aurita* Hodgson.] *Acta Zool. sin., Peking*, **10,** 60-72. (In Chinese, English summary 68-72.)

MACKERRAS, M. J. 1962. Filarial parasites (Nematoda: Filarioidea) of Australian animals. *Aust. J. Zool.*, **10,** 400-457.

MAJUMDAR, G. & CHAKRAVARTY, G. K. 1963. New nematodes from birds. Part I. *Z. ParasitKde.*, **23,** 1-10.

MULLER, R. L. & NELSON, G. S. 1975. *Ackertia globulosa* sp.n. (Nematoda: Filarioidea) from rodents in Kenya. *J. Parasit.*, **61,** 606-609.

MULLIN, S. W. & BALASINGAM, S. 1973. *Dunnifilaria ramachandrani* gen.n., sp.n. (Nematoda: Filarioidea) from the long-tailed giant rat (*Rattus sabanus*) in Malaysia. *Proc. helminth. Soc. Wash.*, **40,** 47-49.

NEVILLE, E. M. 1975. Preliminary report on the transmission of *Parafilaria bovicola* in South Africa. *Onderstepoort J. vet. Res.*, **42,** 41-48.

OLIVEIRA RODRIGUES, H. DE & SOUZA FRANCO, S. R. N. DE 1964. Sôbre um nôva gênero da subfamília Dicheilonematinae Wehr, 1935 (Nematoda, Filarioidea). *Mems Inst. Oswaldo Cruz.*, **62,** 77-80.

ORIHEL, T. & ESSLINGER, J. H. 1973. *Meningonema peruzzii* gen.et.sp.nov. (Nematoda: Filarioidea) from the central nervous system of African monkeys. *J. Parasit.*, **59,** 437-441.

ORTLEPP, R. J. 1961. On two rictularias and a filariid from South African wild carnivores. *J. Helminth., R. T. Leiper Supplement*, pp. 131-140.

——— 1962. *Parafilaria bassoni* spec.nov. from the eye of springbuck (*Antidorcas marsupialis*). *Onderstepoort J. vet. Res.*, **29,** 165-168.

PARUKHIN, A. M. & OSHMARIN, P. G. 1960. [Nematoda *Encephalonema longimicrofilaria* gen.et.sp.n. from the brain of birds.] *Zool. Zh.*, **39,** 934-936. (In Russian: English summary.)

QUENTIN, J. J., TRONCY, P. M. & BARRÉ, N. 1976. La morphogénèse larvaire d'*Aprocta cylindrica. Annls Parasit. hum. comp.*, **51,** 83-93.

RASHEED, S. 1960. The nematode parasites of the birds of Hyderabad (India). *Biologia, Lahore*, **6,** 1-116.

ROMAN, E. 1965. Super-famille des Muspiceoides (Muspiceoidea). In: Grassé, P. M., *Traité de Zoologie, Vol. IV.*, 721-726).

SCHAD, G. A. & ANDERSON, R. C. 1963. *Macacanema formosana* n.g., n.sp. (Onchocercidae: Dirofilariinae) from *Macaca cyclopsis* of Formosa. *Can. J. Zool.*, **41,** 797-800.

SCHMIDT, G. D. & KUNTZ, R. E. 1970. Nematode parasites of Oceanica VIII. Filariids of birds, with a new genus and four new species. *Parasitology*, **60,** 313-326.

SONIN, M. D. 1961. [Filariae of birds in the Turkmen SSR.] *Helminthologia*, **3,** 322-339. (In Russian: English, French and German summaries, pp. 338-339.)

——— 1961. [A revision of the subfamily Splendidofilariinae Chabaud and Choquet, 1953.] *Trudy̆ gel' mint. Lab.*, **11,** 242-250. (In Russian.)

——— 1963. [Revision of the suborder Filariata Skrjabin, 1915.] *Helminthologia*, **4,** 485-494. (In Russian: English, French and German summaries, pp. 493-494.)

——— 1963. [Filaria of birds in Far-Eastern USSR.] *Trudy̆ gel' mint. Lab.*, **13,** 227-249. (In Russian.)

——— 1963. [Revision of *Lemdana* Seurat, 1917 (Nematoda, Filariata).] [In: Helminths of man, animals and plants and their control: Papers on helminthology presented to K. I. Skrjabin on his 85th birthday.] *Moscow: Izdatelstvo Akad. Nauk SSR*, pp. 170-173. (In Russian.)

——— 1965. [*Parornithofilaria* n.g. (Filariata, Splendidofilariidae) and revision of subfamily Splendidofilariinae.] *Trudy̆ gel' mint. Lab.*, **15,** 140-144. (In Russian.)

——— 1966. Reconstruction of the systematics of the nematodes of the suborder Filariata. [Abstract.] *Proc. Int. Congr. Parasit.* (1st), Rome, Sept. 21-26, 1964, Vol. I, pp. 515-516.

——— 1966. [Principles of nematology, edited by K. I. Skrjabin. Vol. XVII. Filariata of animals and man and the diseases caused by them. Part 1. Aproctoidea.] *Moscow: Izdatelstvo "Nauka"*, 360 pp. (In Russian.)

——— 1968. [Principles of nematology, edited by K. I. Skrjabin. Vol. XXI. Filariata of animals and man and the diseases caused by them. Part 2. Diplotriaenoidea.] *Moscow: Izdatelstvo "Nauka"*, 388 pp. (In Russian.)

——— 1971. [Phylogenetic interrelationships of different groups of Filariata and a modification of the classification of nematodes of this suborder.] *Trudy̆ gel' mint. Lab.*, **22,** 162-181.

——— & SHUMILO, R. P. 1964. [*Pseudlemdana* n.g. (Filariata) from subcutaneous tissue of birds.] *Trudy̆ gel' mint. Lab.*, **14,** 194-196.

SOOTA, T. D. & CHATURVEDI, Y. 1971. On some filariid nematodes and a description of a new family Hamulofilariidae. *Zool. Anz.*, **186**, 359-367.

SPRATT, D. M. & VARUGHESE, G. 1975. A taxonomic revision of filarioid nematodes from Australian marsupials. *Aust. J. Zool. Suppl. Ser.*, No. 35, 1-99.

SULAHIAN, A. & SCHACHER, J. F. 1968. *Brygoofilaria agamae* gen.et sp.n. (Nematoda: Filarioidea) from the lizard *Agama stellio* in Lebanon. *J. Parasit.*, **54**, 831-833.

SYUI, G. N. [HSÜ, G. N.] 1957. [A study of the nematodes of birds in Kwangtung province.] *Acta. Zool. sin., Peking*, **9**, 47-77. (In Chinese: Russian summary, pp. 71-77.)

TADROS, G. 1966. On *Pesteria inglisi* gen.nov., sp.nov. (Filariidae) from the English grass snake (*Tropidonotus natrix*). *J. vet. Sci., U.A.R.*, **3**, 55-64.

YEH, L. S. 1959. A revision of the nematode genus *Setaria* Viborg, 1795, its host-parasite relationship, speciation and evolution. *J. Helminth.*, **33**, 1-98.

——— 1960. On a new filarioid worm, *Edesonfilaria malayensis* gen. et sp.nov. from the long-tailed macaque (*Macaca irus*). *J. Helminth.*, **34**, 125-128.

———, SYMES, C. B. & MATAIKA, J. U. 1958. On a new filarioid worm, *Chiropterofilaria brevicaudata* gen. et sp.nov. from the fruit bat, *Pteropus hawaiiensis* from Fiji. *J. Helminth.*, **32**, 227-232.

INDEX TO TAXA

(all taxa mentioned are included, irrespective of synonymization)

Printed and bound by CPI Group (UK) Ltd, Croydon, CR0 4YY

16/09/2026

14963056-0008